3009555854

KU-349-292

McGRAW-HILL YEARBOOK OF Science & Technology

2006

McGRAW-HILL YEARBOOK OF Science & Technology

2006

Comprehensive coverage of recent events and research as compiled by the staff of the McGraw-Hill Encyclopedia of Science & Technology

McGraw-Hill
New York Chicago San Francisco Lisbon London Madrid Mexico City Milan
New Delhi San Juan Seoul Singapore Sydney Toronto

Library of Congress Cataloging in Publication data

McGraw-Hill yearbook of science and technology.
1962- . New York, McGraw-Hill.

v. illus. 26 cm.
Vols. for 1962- compiled by the staff of the McGraw-Hill encyclopedia of science and technology.
1. Science—Yearbooks. 2. Technology—Yearbooks. 1. McGraw-Hill encyclopedia of science and technology.
Q1.M13 505.8 62-12028

ISBN 0-07-146205-8
ISSN 0076-2016

1 2 3 4 5 6 7 8 9 0 WCK/WCK 0 10 9 8 7 6 5

This book was printed on acid-free paper.

It was set in Garamond Book and Neue Helvetica Black Condensed by TechBooks, Fairfax, Virginia. The art was prepared by TechBooks. The book was printed and bound by Quebecor World/Versailles.

Contents

Editing, Design, and Production Staff

Consulting Editors

Prof. Gordon Holloway. *Department of Mechanical Engineering, University of New Brunswick, Canada.* Fluid mechanics.

Dr. S. C. Jong. *Senior Staff Scientist and Program Director, Mycology and Protistology Program, American Type Culture Collection, Manassas, Virginia.* Mycology.

Dr. Bryan A. Kibble. *National Physical Laboratory, Teddington, Middlesex, United Kingdom.* Electricity and electromagnetism.

Prof. Robert E. Knowlton. *Department of Biological Sciences, George Washington University, Washington, DC.* Invertebrate zoology.

Dr. Cynthia Larive. *Department of Chemistry, University of California, Riverside.* Inorganic chemistry.

Prof. Chao-Jun Li. *Canada Research Chair in Green Chemistry, Department of Chemistry, McGill University, Montreal, Quebec, Canada.* Organic chemistry.

Dr. Donald W. Linzey. *Wytheville Community College, Wytheville, Virginia.* Vertebrate zoology.

Dr. Philip V. Lopresti. *Retired; formerly, Engineering Research Center, AT&T Bell Laboratories, Princeton, New Jersey.* Electronic circuits.

Dr. Philip L. Marston. *Department of Physics, Washington State University, Pullman.* Acoustics.

Dr. Michelle A. Marvier. *Biology Department and Environmental Studies Institute, Santa Clara University, California.* Ecology and conservation.

Dr. Ramon A. Mata-Toledo. *Associate Professor of Computer Science, James Madison University, Harrisonburg, Virginia.* Computers.

Prof. Krzysztof Matyjaszewski. *J. C. Warner Professor of Natural Sciences, Department of Chemistry, Carnegie Mellon University, Pittsburgh, Pennsylvania.* Polymer science and engineering.

Dr. Andrew D. Miall. *Gordon Stollery Chair in Basin Analysis and Petroleum Geology, Department of Geology, University of Toronto, Ontario, Canada.* Geology.

Dr. Orlando J. Miller. *Professor Emeritus, Center for Molecular Medicine and Genetics, Wayne State University School of Medicine, Detroit, Michigan.* Genetics.

Prof. Jay M. Pasachoff. *Director, Hopkins Observatory, Williams College, Williamstown, Massachusetts.* Astronomy.

Prof. J. Jeffrey Peirce. *Department of Civil and Environmental Engineering, Edmund T. Pratt Jr. School of Engineering, Duke University, Durham, North Carolina.* Environmental engineering.

Dr. William C. Peters. *Professor Emeritus, Mining and Geological Engineering, University of Arizona, Tucson.* Mining engineering.

Dr. Kenneth P. H. Pritzker. *Pathologist-in-Chief and Director, Head, Connective Tissue Research Group, and Professor, Laboratory Medicine and Pathobiology, University of Toronto, Mount Sinai Hospital, Toronto, Ontario, Canada.* Medicine and pathology.

Prof. Justin Revenaugh. *Department of Geology and Geophysics, University of Minnesota, Minneapolis.* Geophysics.

Dr. Roger M. Rowell. *USDA-Forest Service, Forest Products Laboratory, Madison, Wisconsin.* Forestry.

Prof. Ali M. Sadegh. *Center for Advanced Engineering and Design, The City College School of Engineering, New York.* Mechanical engineering.

Dr. John L. Safko, Sr. *Distinguished Professor Emeritus, Physics and Astronomy, Associated Faculty, School of the Environment, University of South Carolina, Columbia.* Classical mechanics.

Dr. Andrew P. Sage. *Founding Dean Emeritus and First American Bank Professor, University Professor, School of Information Technology and Engineering, George Mason University, Fairfax, Virginia.* Control and information systems.

Dr. Alfred S. Schlachter. *Advanced Light Source, Lawrence Berkeley National Laboratory, Berkeley, California.* Atomic and molecular physics.

Prof. Ivan K. Schuller. *Department of Physics, University of California-San Diego, La Jolla, California.* Condensed-matter physics.

Dr. David M. Sherman. *Department of Earth Sciences, University of Bristol, United Kingdom.* Mineralogy.

Prof. Arthur A. Spector. *Department of Biochemistry, University of Iowa, Iowa City.* Biochemistry.

Prof. Anthony P. Stanton. *Carnegie Mellon University, Pittsburgh, Pennsylvania.* Graphic arts and photography.

Prof. John F. Timoney *Department of Veterinary Science, University of Kentucky, Lexington.* Veterinary medicine.

Dr. Sally E. Walker. *Associate Professor of Geology and Marine Science, University of Georgia, Athens.* Invertebrate paleontology.

Prof. Pao K. Wang. *Department of Atmospheric and Oceanic Sciences, University of Wisconsin-Madison.* Meteorology and climatology.

Dr. Nicole Y. Weekes. *Pomona College, Claremont, California.* Neuropsychology.

Prof. Mary Anne White. *Killam Research Professor in Materials Science, Department of Chemistry, Dalhousie University, Halifax, Nova Scotia, Canada.* Materials science and metallurgic engineering.

Prof. Thomas A. Wikle. *Head, Department of Geography, Oklahoma State University, Stillwater.* Physical geography.

Article Titles and Authors

Acoustooptic imaging
Charles A. DiMarzio
Alcohol production from wood
John I. Zerbe
Alzheimer's disease
Rajesh R. Tampi
Amphibious assault ships
Matthew McCarton
Animal disease surveillance and monitoring
Dominic A. Travis
Antidepressant use in minors
Lawrence Maayan; Andrés Martin
Arctic treeline change
Glen M. MacDonald
Attosecond laser pulses
Reinhard Kienberger; Ferenc Krausz
Automotive restraint systems
Ching-Yao Chan
Belowground herbivory
Sean M. Watts
Biologically inspired robotics
Roger D. Quinn; Roy E. Ritzmann
Bioluminescent fungi
Hedda J. Weitz
Biomimetic design for remanufacture
Lily H. Shu
Blackout prevention
Damir Novsel; Vahid Madani
Bose-Einstein condensation
Jozef T. Devreese; Jacques Tempere
Brain imaging of deception
Tatia M. C. Lee
Capillary electrophoresis-SELEX
Michael T. Bowser
Carbon MEMS
Chunlei Wang; Marc Madou
Cassini-Huygens mission
Trina Ray; Linda Spilker; Claudio Sollazzo
Caves and climate change
Kathleen Johnson
Chromosome painting
Martin A. Lysak
Cicada
Louie H. Yang
Click chemistry
M. G. Finn
Communications satellite failures
Mark A. Quinn
Complexity theory
David G. Green; Tania Bransden
Compound wireless services
Thaddeus J. A. Kobylarz
Computational environmental toxicology
Jerry Blancato; Daniel Vallero
Conflict analysis and resolution
Keith W. Hipel; D. Marc Kilgour; Liping Fang
Context-aware mobile systems
Ana I. Martinez-Garcia; J. Antonio Garcia-Macias; Jesus Favela
Controlled/living radical polymerization
Yves Gnanou; Daniel Taton
Crewless surface watercraft
Michael L. Bosworth
Cryptochrome
Chentao Lin
Deep-sea vertebrate species
David M. Bailey
Digital cinema
Anthony Stanton
Dino-Birds
Angela C. Milner
Dip-pen nanolithography
Clifton K.-F. Shen; Chad A. Mirkin
DNA barcoding
Diethard Tautz
DNA sequence assembly algorithms
Mihai Pop
Downscaling climate models
Edwin P. Maurer
Dry-docking and heavylift ships
Richard Hepburn
Dynamic traffic management
Yibing Wang; Moshe Ben-Akiva; Markos Papageorgiou; Jon Bottom; Apostolos Kotsialos
Earthquake sensory perception in vertebrates
Motoji Ikeya
Electric power system security
Jeffery E. Dagle
Electronic olfaction and taste systems
P. I. Gouma
Endoplasmic reticulum quality control
Jennifer L. Goeckeler; Jeffrey L. Brodsky
Energy-efficient motors
John Malinowski
Enterprise transformation
William B. Rouse
Environmental engineering informatics
Thomas I. Prudhomme
Fire safety (building design)
J. Y. Richard Liew; H. X. Yu
Fire-induced convection
J. Trentmann; M. O. Andreae
Fluorescent signaling in mantis shrimps
Thomas W. Cronin
Forensic DNA testing
Steven Lee
Forest water contamination
Roger M. Rowell
Fossil cold-seep ecosystem
Roberto Barbieri
Gene therapy (veterinary medicine)
Mark Haskins

Geochronology
Donald W. Davis
Global biogeochemical cycles
Arne Winguth
Graph-based intelligence analysis
Thayne Coffman
Herbicide resistance
Lyle F. Friesen; Rene C. Van Acker; Chris Willenborg
Heuristics in judgment and decision making
Rolf Reber
Homo floresiensis
Terry Harrison
Human papillomaviruses
John P. Harley
Hydrogen-powered flight
Timothy D. Smith
Infectious disease and human evolution
Donald J. Ortner
Infrasonic monitoring
Michael A. H. Hedlin
Insect flight
Z. Jane Wang
Intelligent vehicles and infrastructure
Christopher K. Wilson
Interaction of photons with ionized matter
Ronald A. Phaneuf
Internet communications
Daniel Heer; Michael Rauchwerk
Invasive forest species
Barbara L. Illman
Jet boring (mining)
Barry W. Schmitke
Life-cycle analysis of civil structures
Dan M. Frangopol; Min Liu
Liver X receptor (LXR)
Christopher J. Fielding
Mangrove forests and tsunami protection
F. Dahdouh-Guebas
Mars Rovers
Eric T. Baumgartner
Metagenomics
J. Handelsman
Microbial dechlorination
Claudia K. Gunsch
Microbial forensics
Randall S. Murch
Mineralogy of Mars
Jim Bell
Nanometer magnets
Denis Koltsov
Nitrogen deposition and forest degradation
Timothy H. Tear
Nobel prizes
Yearbook Editors
Noncoalescence of droplets
Pasquale Dell'Aversana
Nonconventional aircraft designs
N. Qin
Ontology (information technology)
Rubén Prieto-Díaz
Paleozoic bivalve borings
A. A. Ekdale
Permian-Triassic mass extinction
Margaret L. Fraiser
Pharmaceutical residues in the environment
Diana S. Aga
Phononic crystals
Alexey Sukhovich; John H. Page
Phylogenetic diversity and conservation
Daniel P. Faith; Kristen J. Williams
Phytoremediation
Melinda A. Klein; Ashot Papoyan; Leon V. Kochian
Plant cell walls
Breeanna Rae Urbanowicz; Sarah Nell Davidson; Jocelyn Kenneth Campbell Rose
Plume imagery
Guust Nolet
Polymer stereochemistry and properties
Yoshio Okamoto
Polymers for microelectronics
C. W. Bielawski; C. G. Willson
Post-perovskite
Thorne Lay
Primate communication
Charles T. Snowdon
Primate origins
Christophe Soligo
Printing on demand
Harold Henke
Process improvement methodologies
Donna H. Rhodes
Proteomics
Megan S. Lim; Kojo S. J. Elenitoba-Johnson
Proximal probe molecular analysis
S. S. Sheiko
Psychopathology and the Human Genome Project
Sarah Cohen; Peter McGuffin
Pterosaur
Matthew T. Wilkinson
Quantum chaos in a three-body system
Alfred S. Schlachter
Queuing theory (health care)
John Blake
Radar meteorology
Jon Eastment
Reactive oxygen species
Marcus S. Cooke; Mark D. Evans
Regional climate modeling
Wen-Yih Sun
Risk analysis (homeland security)
Vicki M. Bier
Salmon farming
James R. A. Butler
Sedna
John Stansberry
Self-recognition
Alain Morin
Size effect on strength (civil engineering)
Milan Jirasek

Space flight
Jesco von Puttkamer

Space probe communications
Charles D. Edwards

Spider silk
Nitin Kumar; Gareth H. McKinley

Steel construction
Reidar Bjorhovde

Sumatra-Andaman earthquake
Jeffrey J. Park

Sum-frequency vibrational spectroscopy
John C. Conboy

Superconductor wires
M. Parans Paranthaman

Telerobotics (mining)
Gregory R. Baiden

Tetrapod origins
Jennifer A. Clack

3D inkjet printing
Alexandra Pekarovicova; Jan Pekarovic

Tissue and organ printing
Gabor Forgacs; Vladimir Mironov

Transgressive segregation (plant breeding)
Nolan C. Kane; Briana L. Gross;
Loren H. Rieseberg

Trapped-ion optical clocks
Hugh Klein

Vehicle-highway automation systems
Steven E. Shladover

Vernalization
Richard Amasino

Very small flying machines
Robert C. Michelson

Whale-fall worms
G. W. Rouse

World Wide Web search engines
Brian D. Davison

Preface

The 2006 *McGraw-Hill Yearbook of Science & Technology* provides a broad overview of important recent developments in science, technology, and engineering as selected by our distinguished board of consulting editors. At the same time, it satisfies the nonspecialist reader's need to stay informed about important trends in research and development that will advance our knowledge in fields ranging from aerospace engineering to zoology and will lead to important new practical applications. Readers of the *McGraw-Hill Encyclopedia of Science & Technology*, 9th edition (2002), also will find the *Yearbook* to be a valuable companion publication, supplementing and updating the content of that work.

In the 2006 edition, we continue to chronicle the rapid advances in cell and molecular biology, biochemistry, and genetics with topics such as chromosome painting, DNA barcoding, reactive oxygen species, metagenomics, and proteomics. Reviews in topical areas of biomedicine, such as Alzheimer's disease, antidepressant use in minors, human papillomavirus, and infectious disease and human evolution, are presented. In chemistry we report on click chemistry, polymer stereochemistry and properties, and novel areas of spectroscopy. Advances in computing and communication are documented in articles on compound wireless services, context-aware mobile communications, Internet communications, and World Wide Web search engines, among others. Noteworthy developments in engineering and technology are reported in reviews of biologically inspired robots, carbon MEMS, digital cinema, electric power system security, energy-efficient motors, fire safety, nonconventional aircraft design, steel construction, and vehicle-highway automation systems. In the physical sciences and astronomy, we report on attosecond laser pulses, Bose-Einstein condensation, the Cassini-Huygens mission, complexity theory, the mineralogy of Mars, and the newly discovered Sedna. Reviews on regional climate models, environmental engineering informatics, geochronology, global biogeochemical cycles, mangrove forests and tsunami protection, pharmaceutical residues in the environment, and the Sumatra-Andaman earthquake are among the articles in the earth and environmental sciences.

Each contribution to the *Yearbook* is a concise article by one or more authorities in the field. We are pleased that noted researchers have been supporting the *Yearbook* since its first edition in 1962 by taking time to share their knowledge with our readers. The topics are selected by our consulting editors, in conjunction with our editorial staff, based on present significance and potential applications. McGraw-Hill strives to make each article as readily understandable as possible for the nonspecialist reader through careful editing and the extensive use of graphics, most of which are prepared specially for the *Yearbook*.

Librarians, students, teachers, the scientific community, journalists and writers, and the general reader continue to find in the *McGraw-Hill Yearbook of Science & Technology* the information they need to follow the rapid pace of advances in science and technology and to understand the developments in these fields that will shape the world of the twenty-first century.

Mark D. Licker
PUBLISHER

A–Z

Acoustooptic imaging

The need for medical diagnosis continues to demand new technologies for noninvasive imaging of the human body. While all imaging modalities continue to improve, there is an increasing emphasis on combining them in novel ways to take advantage of the best characteristics of each. Light is perhaps the oldest of medical tools. Lacking more advanced technologies, early practitioners were sensitive to the visual signatures of the patient such as skin color. Recently, technologies such as diffusive optical tomography (DOT) have led to imaging of hemoglobin concentration and oxygenation through the use of multiple wavelengths of light, and have improved resolution and depth of penetration. Sound also played an important role in early medicine, significantly advanced by R. T. H. Laennec's invention of the stethoscope in 1816. Modern uses of sound arrived with ultrasound imaging in the 1950s. The combination of light and sound is a relatively new field of medical imaging. In acoustooptic imaging, sound is used to tag light as it propagates in the human body. The goal is to combine the ability of light to characterize materials such as hemoglobin with the spatial resolution of ultrasound.

The acoustooptical effect, explained by C. V. Raman and N. S. Nath in the 1930s, has led to devices which deflect the path and shift the frequency of laser beams for a variety of applications. In the human body, this explanation is complicated by the same scattering that limits the resolution and depth of diffusive optical tomography. In recent years, significant progress has been made in understanding the propagation problem and in producing usable acoustooptic images.

Basic concepts. The interaction of light with acoustic waves is a complicated phenomenon; it involves light waves propagating in a medium with an index of refraction that varies in time and space. First, the index of refraction varies spatially because of the presence of different materials, including cells and cell components. Usually the model for this is a random array of discrete scattering particles. Second, the acoustic wave causes these scatterers to move, primarily with a sinusoidal motion at the acoustic frequency. Moreover, the scattering particles have finite dimensions which may be modified by the acoustic wave, so acoustic modulation of the scattering amplitude may also occur. Finally, the index is modulated in space and time by an amount related to the pressure field of the acoustic wave and to acoustic properties of the material.

In most biological samples of interest, the scattering is sufficiently large that wave-based computation of the scattered field is hopelessly complicated and it is necessary to resort to random ray-tracing approaches. The electric field is thus described as a sum of contributions over random paths among the scattering particles. These paths are very convoluted; the typical path length is six to ten times the path along a straight line. The contribution from a single path is the product of contributions from individual scattering and propagation events. The amplitude change produced by a single scattering is determined from the scattering cross section of that particle and the distance to the next one. More importantly, the phase ϕ is determined by the optical path length, OPL, as given by Eq. (1), where λ is the vacuum wavelength of the light, and the optical path length is given by Eq. (2), where n is the index of refraction, $d\ell$ is an

$$\phi = 2\pi \frac{\text{OPL}}{\lambda} \tag{1}$$

$$\text{OPL} = \int n \, d\ell \tag{2}$$

increment of path length, and the integral is taken from the source to the receiver. The detected field will be the sum of contributions from all the paths. Because the random optical paths are long with respect to the wavelength, the phases will be randomly distributed, resulting in a speckle pattern, familiar to those who have observed laser light scattered from a rough surface. In the presence of ultrasound, the optical path length is modulated by variations in the index and path length, resulting in a phase modulation of the light following that path, but because the paths are so convoluted, the phase of this modulation, with respect to the acoustic source, is also likely to be random.

This randomness has implications that are problematic for those who wish to observe acoustic modulation of light in tissue. It dictates that the phase

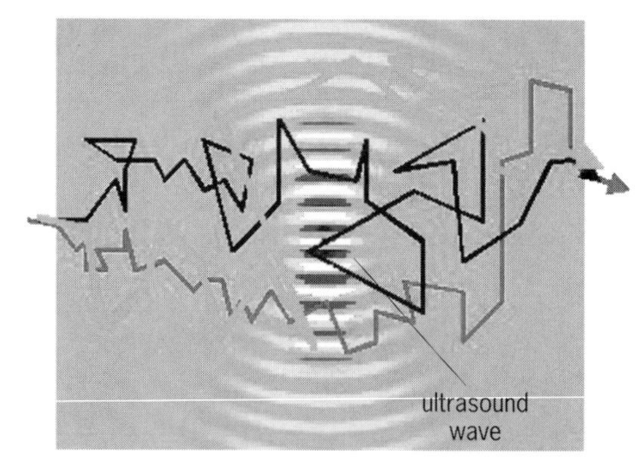

Fig. 1. Random ray traces through a turbid medium in the presence of an acoustic wave. The center region represents the ultrasound wave; light shades indicate regions of pressure above ambient, and dark shades represent such regions below.

does not provide any useful information, and it limits our ability to combine contributions to enhance the weak signals. Some representative paths are shown in **Fig. 1**. The region in the center of this figure represents the ultrasound wave, and alternating regions of pressure above and below the ambient pressure are indicated. Small changes in the optical path length, caused by variations in the pressure and by motion of the particles, will not produce appreciable changes in the amplitude of light scattered along a particular path. The changes will be mostly in the phase of the light. Because all optical detectors measure power, which is proportional to the average of the square of the field, these phase changes cannot be observed directly. However, the detected light will be a sum of contributions from many paths, and because of the random nature of these paths, each will have its own phase variation. Thus the overall signal will increase when the contributions are in phase and add constructively, and will decrease when they add destructively. The resulting modulation of the detected light will occur synchronously with the acoustic source. Thus, when we detect modulated light, we can be sure that this light must have passed through the acoustic wave.

Despite the randomness of the field contributions, several techniques have been developed for imaging. The major issues addressed have included the weak signals, their randomness, and the ability to resolve signals along the ultrasound beam direction.

Advances. Recently, coherent detection techniques involving photorefractive crystals have led to significant improvements in imaging capability. Coherent detection involves mixing the signal wave with a reference wave generally derived from the same laser source. The combined field E is thus the sum of the fields, E_{sig} and E_{ref}, associated with the signal and reference waves. The irradiance of this combined field is given by Eq. (3). Here, phasor

$$\begin{aligned}|E|^2 &= \left|E_{\text{sig}} + E_{\text{ref}}\right|^2 = \left(E_{\text{sig}} + E_{\text{ref}}\right)\left(E^*_{\text{sig}} + E^*_{\text{ref}}\right) \\ &= \left|E_{\text{sig}}\right|^2 + \left|E_{\text{ref}}\right|^2 + E_{\text{sig}}E^*_{\text{ref}} + E^*_{\text{sig}}E_{\text{ref}}\end{aligned} \tag{3}$$

notation is used, where a complex number represents the field at some angular frequency ω, and it is understood that the actual field is obtained by multiplying this number by $e^{i\omega t} = \cos\omega t + i\sin\omega t$, where $i = \sqrt{-1}$, and then taking the real part of the product. The symbol * denotes the complex conjugate of a complex number. (Thus, if a complex number A equals $a + bi$, where the real numbers a and b are, respectively, the real and imaginary parts of A, then $A^* = a - bi$.) The important terms in Eq. (3) are the last two. They are linear, rather than quadratic in E_{sig}, and this property improves the dynamic range of signal amplitudes that can be measured and provides the opportunity to measure the phase. These terms can be raised to much higher amplitudes than $|E_{\text{sig}}|^2$, which is all that would be observable with incoherent detection. The amplitude is limited only by the strength of $|E_{\text{ref}}|$.

Function of photorefractive crystal. Because of the speckle variations discussed earlier, the signal wavefront is complicated and random, so mixing it with a well-structured reference wave would not be expected to provide any significant improvement. However, the photorefractive crystal changes the situation, by making a dynamic hologram which scatters the reference wave in such a way that it has the same speckle pattern as the signal. In simple terms, the crystal's index of refraction is modulated by the local irradiance, defined in Eq. (3), to produce a hologram, E_H, which is read out by a playback wave, E_p, as given in notation (4). If the playback wave is the same as

$$E_H \propto E_p\left(E_{\text{sig}} + E_{\text{ref}}\right)\left(E^*_{\text{sig}} + E^*_{\text{ref}}\right) \tag{4}$$

the reference, then one term in this equation will be given by notation (5), and this term is matched to

$$E_H \propto \cdots + E_{\text{ref}}E^*_{\text{ref}}E_{\text{sig}} + \cdots \tag{5}$$

the signal field. Thus, the irradiance of the reference field provides gain to the signal field.

However, the response time of the crystal is on the order of 100 milliseconds, so the hologram field does not follow the temporal behavior of the acoustic modulation, or a short acoustic pulse. Thus, the crystal amplifies the stationary part of the phasor, but not the modulation part. This behavior has several advantages. First, the phase changes on individual paths now result in amplitude variations. Specifically, the signal level decreases for any modulation by the acoustic field. This behavior can be demonstrated by noting that the detected signal is $E_H + E_{\text{sig}}$, where E_H is a large field with the same phase as the mean of E_{sig}. Thus, in the absence of modulation, these fields add constructively and the amplitude of the combined field is given by Eq. (6). However, during the

$$|E_H + E_{\text{sig}}| = |E_H| + |E_{\text{sig}}| \tag{6}$$

acoustic pulse, the signal is modulated, and the amplitude of the combined field is smaller, as indicated in inequality (7). Second, because the amplitude

$$|E_H + E_{\text{sig}}| < |E_H| + |E_{\text{sig}}| \tag{7}$$

always decreases, the effects of the acoustic fields on

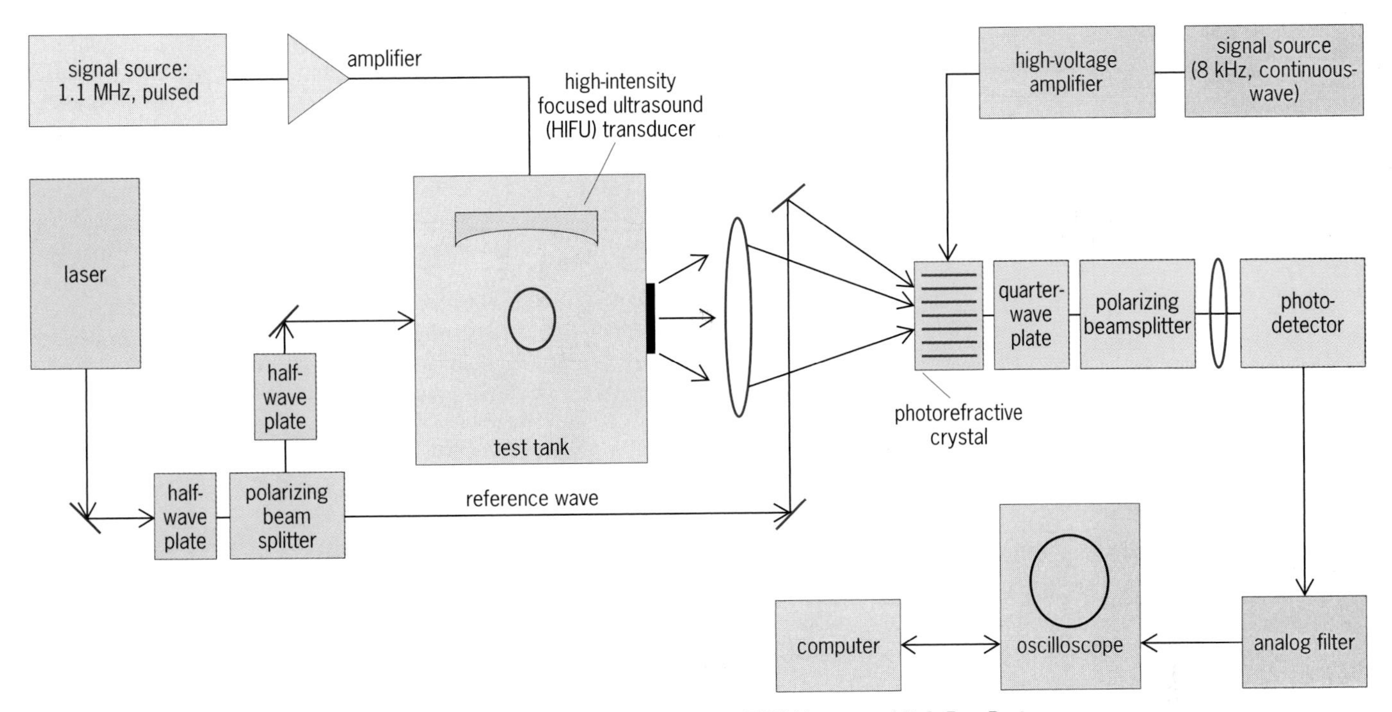

Fig. 2. Block diagram of an experiment for acoustooptic imaging. *(L. Sui, E. Bossy, T. W. Murray, and R. A. Roy, Boston University)*

multiple paths all add coherently. Third, the effect is nonlinear, occurring most strongly at the focus of the acoustic wave, so spatial resolution is possible in all three dimensions. In other words inequality (7) departs from equality by a greater amount for the large phase changes that occur when the light path passes near the acoustic focus than for the smaller random phase changes that occur elsewhere in the acoustic wave.

Implementations. The hardware for one implementation of this technique is shown in **Fig. 2**. The reference wave is derived from the same laser as the signal wave, passing to the right through the polarizing beamsplitter. The signal beam is reflected by the polarizing beamsplitter and directed into a small test tank of degassed and filtered water. An acoustic transducer which is part of an ultrasound imaging system is placed above the tank. The ultrasound transmitter provides the modulation of the light beam. The receiver (not shown in the figure) provides an ultrasound image of objects being examined. After the tank, the signal and reference beams are recombined in the photorefractive crystal, and directed to the photodetector, which converts the optical signal to an electrical one, proportional to the optical power. The resulting electrical signal is filtered appropriately to measure the decrease in signal described by inequality (7), delivered to an oscilloscope, and then to a computer.

A tissuelike object or phantom has been placed in the test tank. The phantom is fabricated of acrylamide gel with polystyrene microspheres (0.4 μm in diameter) added to provide scattering. Two smaller objects of the same material are embedded in the phantom. India ink has been added to one to provide optical contrast but little acoustic contrast. The test phantom has been cut open in **Fig. 3*a*** to show the inclusions. In Fig. 3*b*, the signals across both objects are shown, and the decrease in signal caused by the absorption is evident. Figure 3*c* shows a conventional B-mode ultrasound image, and Fig. 3*d* shows the image using the signal from the system in Fig. 2. It is evident that the absorbing object is easily detected.

An alternative implementation of this concept uses two acoustooptic modulators to produce a frequency offset between the signal and reference beams equal to the acoustic frequency. In this case it is only the signal from the ultrasound focus which produces the hologram. This provides an important advantage in improved signal-to-noise ratio, but requires that the ultrasound be on for a length of time greater than the response time of the crystal.

Both techniques show promise for surmounting the difficulties encountered in the early experiments, and the potential exists for using the spectroscopic capabilities of optics with the spatial resolution of ultrasound in biological tissues that are highly optically scattering.

Potential applications. The combination of light and sound offers new imaging opportunities not previously available. Applications are possible in a wide variety of medical imaging situations. Because the technology is new, it is not readily apparent which of these will emerge as most significant. The most promising candidates are those that are currently being explored with diffusive optical tomography, including breast, brain, and fetal imaging. Of the three, brain imaging will be the most challenging because of the need to focus ultrasound through the

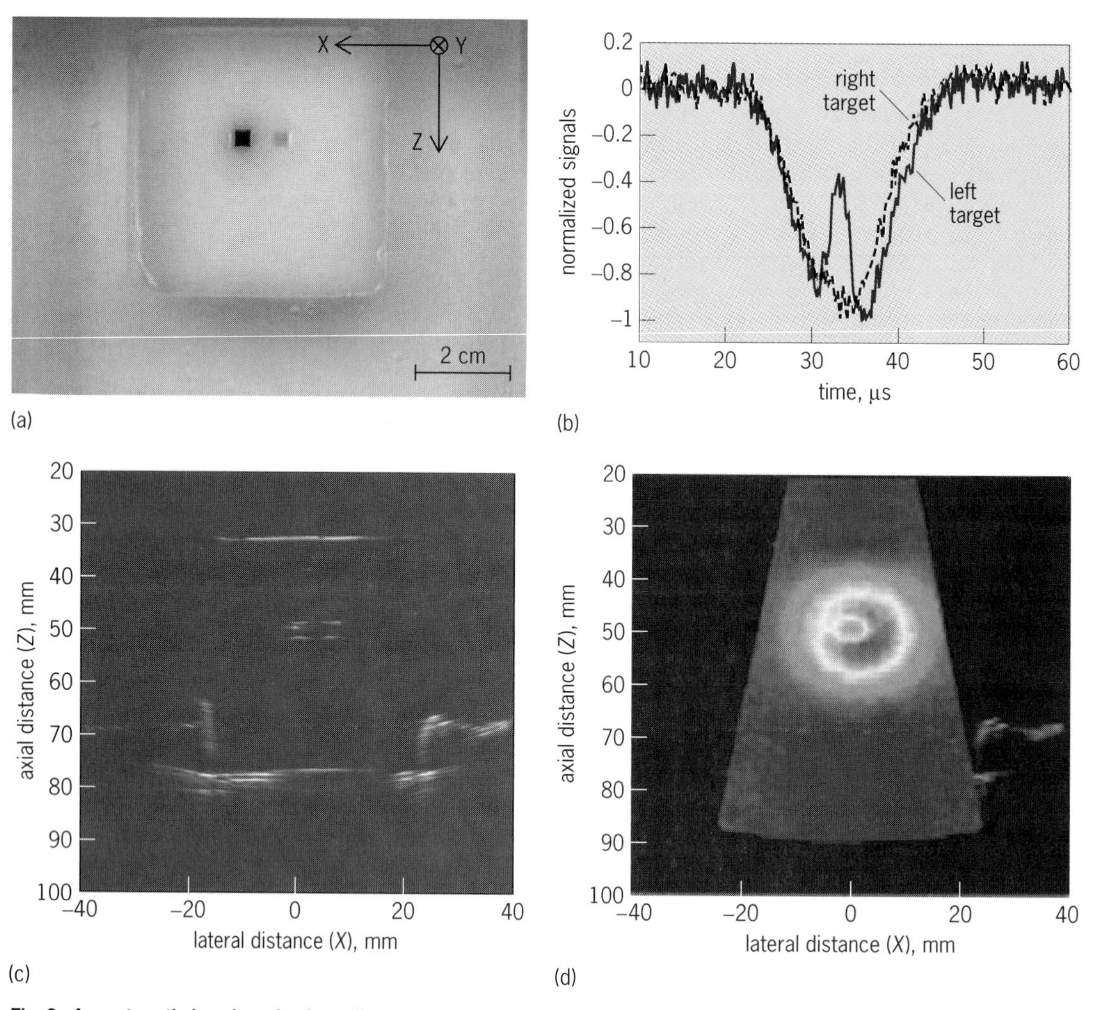

Fig. 3. Acoustooptic imaging of a tissuelike object (phantom), with two objects embedded within it. (*a*) Visible-light image of phantom cut open to show the inclusions. (*b*) Acoustooptic imaging signals across the two targets. (*c*) Conventional B-mode ultrasound image. (*d*) Acousooptic image using the system in Fig. 2. (*L. Sui, E. Bossy, T. W. Murray, and R. A. Roy, Boston University*)

skull. Thus it is reasonable to expect that the first applications will be found in breast and fetal imaging. Both applications have potential optical contrast in hemoglobin and acoustic contrast in different tissue types. The utility of this technology to different applications will only begin to emerge as tests begin on actual tissue, in living organisms.

Acknowledgments. The author wishes to thank colleagues and students at Northeastern University and Boston University, whose pioneering work has contributed acoustooptical imaging. At Northeastern University, T. J. Gaudette, D. J. Townsend, A. Nieva, and Florian Blonigen have made significant contributions to the research. This work was supported in part by the Center for Subsurface Sensing and Imaging Systems (CenSISS), under the Engineering Research Centers Program of the National Science Foundation (award number EEC 9986821).

For background information *see* ACOUSTOOPTICS; ALTERNATING-CURRENT CIRCUIT THEORY; BIOMEDICAL ULTRASONICS; COHERENCE; COMPLEX NUMBERS AND COMPLEX VARIABLES; HOLOGRAPHY; MEDICAL IMAGING; MEDICAL ULTRASONIC TOMOGRAPHY; NONLINEAR OPTICS; OPTICAL COHERENCE TOMOGRAPHY; SPECKLE; ULTRASONICS in the McGraw-Hill Encyclopedia of Science & Technology.

Charles A. DiMarzio

Bibliography. C. A. DiMarzio and T. W. Murray, Medical imaging techniques combining light and ultrasound, *Subsurface Sensing Technol. Applic.*, 4(4):289–310, October 2003; T. W. Murray et al., Detection of ultrasound-modulated photons in diffuse media using the photoreactive effect, *Opt. Lett.*, 29(21):2509–2511, 2004; G. Yao, S.-L. Jiao, and L.-H. Wang, Frequency-swept ultrasound-modulated optical tomography in biological tissue by use of parallel detection, *Opt. Lett.*, 25:734–736, 2000.

Alcohol production from wood

Wood is a source of alcohols that can be used as fuels for automobiles, among other purposes. Although these biofuels have lower fuel densities than gasoline and diesel fuel, they have higher octane ratings than regular-grade gasoline, and engines may be designed to run more efficiently with them. Methanol and ethanol are common alcohols that might be readily derived from wood if economic constraints are overcome. Propanol and butanol are other possibilities.

Environmental benefits. Production of methanol and ethanol from wood could decrease the amount of petroleum now used to manufacture motor fuels, and it would reduce some of the undesirable environmental (not to mention economic) impacts of the use of petroleum-base fuel. The burning of fossil fuels negatively impacts the environment through the atmospheric emission of oxides of nitrogen, hydrocarbons, greenhouse gases such as carbon dioxide and methane, and other pollutants, such as sulfur. Liberation of gasoline vapors contributes to formation of ozone. Excessive ozone in the troposphere during high summer heat is detrimental to human health, and has caused deaths in a number of cities, particularly among older citizens. These environmental problems could be overcome through greater use of renewable fuels with sustainable production from biomass such as wood. Carbon in greenhouse gases is recycled as growth cycles are repeated.

Use of ethanol. A growing amount of ethanol is used in the United States in an admixture of 10% ethanol with 90% gasoline, termed E10. There has also been discussion of making a fuel called E85 with 85% ethanol and 15% gasoline. E85 and neat ethanol, with only about 5% water (190 proof), are being used as motor fuels in Brazil. In the United States most ethanol is made from corn and other grains. In Brazil it is made from sugarcane bagasse.

Methanol made from wood might also be used as an alternative to petroleum-based fuel. Undiluted methanol or admixtures such as M15, M85, and M95 are possible. Methanol is also a possible fuel for fuel cells (devices that electronically convert chemical energy to electrical and thermal energy). Experimental cars with fuel cells and appliances called reformers have been built, which derive hydrogen from methanol (CH_3OH) through reaction (1). But experience using methanol as automotive fuel is limited, except in the case of racing cars.

$$CH_3OH \rightarrow 2H_2 + CO \qquad (1)$$

Ethanol production. Wood hydrolysis converts the carbohydrate polymers in wood (hemicellulose and cellulose) to simple sugars by chemical reaction with water in the presence of acid catalysts. Hydrolysis of hemicellulose produces five-carbon sugars (mainly xylose but also arabinose), reaction (2). Cellulose hydrolysis yields six-carbon sugars (mainly glucose but also mannose and galactose), reaction (3), which may then be fermented to ethanol (C_2H_5OH), reaction (4).

$$(C_5H_8O_4)_n + nH_2O \rightarrow nC_5H_{10}O_5 \qquad (2)$$

$$(C_6H_{10}O_5)_n + nH_2O \rightarrow nC_6H_{12}O_6 \qquad (3)$$

$$C_6H_{12}O_6 \rightarrow 2C_2H_5OH + 2CO_2 \qquad (4)$$

Glucose is the main sugar used for conversion to ethanol, but xylose is also fermented with enzymes. For enzyme hydrolysis, the fruits of corn (which produce glucose from hydrolysis of starch instead of cellulose) and other grains are preferred over wood because they do not contain pentoses (five-carbon sugars), which interfere with hexose (six-carbon sugar) fermentation. In addition, they do not contain lignin (the other carbohydrate polymer found in wood), from which fermentable sugars cannot be derived.

Ethanol was produced commercially from wood in the United States during World War I, in other countries during World War II, and in the Soviet Union after World War II. The technology used in these plants was based on the hydrolysis of cellulose components to 6-carbon sugars, mainly glucose, but some mannose was produced as well. The sugars were then fermented to ethanol.

Single-stage batch hydrolysis. The World War I hydrolysis technology was based on a single-stage batch process to digest wood with 80% sulfuric acid. Yields were low, since the cellulose constituent is only about 50% of wood weight. Yields in the World War I plants were only 20 gallons of ethanol per dry ton of Southern pine sawdust and chips (83.4 liters per metric ton).

Scholler process. The process used in WWII was a dilute acid percolation process developed by Scholler in Germany. This process enabled a better ratio of sugar production to sugar destruction in the hydrolysis digester and higher product yields. Yields of 95% alcohol in a well-run plant amounted to about 53 gallons per ton of wood (221 liters per metric ton).

Forest Products Laboratory Percolation Process. During World War II, the U.S. Forest Products Laboratory developed an improved version of the Scholler process known as the Forest Products Laboratory Percolation Process, FPL (or the Madison Wood-Sugar Percolation Process). Research results showed that the average ethanol yield from the sugars produced by hydrolysis corresponded to 64.5 gallons per ton of dry, bark-free wood (269 liters per ton). In 1944 construction began on a plant designed to produce 10,700 gallons (40,500 liters) of 190 proof (95%) ethanol from 220 tons (200 tons) of dry bark-free Douglas-fir wood residues per day. The plant was shut down shortly after startup, however, because the operation was not economical in the postwar period. Data from a 1975 study using the FPL process for hydrolyzing cellulose are shown in **Table 1**. [With the FPL process, the production of salable by-products (such as furfural and phenols) from hemicellulose and lignin were assumed. Pentose sugar for production of furfural and methanol were also considered as salable items, and lignin was seen as a fuel for process energy.]

Enzyme hydrolysis. In the 1980s, research began to focus on ethanol production from wood via enzyme hydrolysis. One form of this method involves acid prehydrolysis followed by enzyme hydrolysis. Another variation is known as simultaneous saccharification (hydrolysis of starch or cellulose into fermentable sugars) and fermentation (SSF), in which hydrolysis enzymes and fermentation enzymes work

TABLE 1. Estimates of costs of ethanol production from dry wood*

Study year	Technology†	Bone-dry metric ton of feedstock/day	Wood cost, $/dry metric ton	Cost of ethanol produced, $/liter
1989	SSF following xylose removal and fermentation	1745	46	0.33
1989	SSF following xylose removal and fermentation	9090	46	0.27
1975	FPL percolation	1361‡	37.48	0.41
1975	FPL percolation	2722‡	37.48	0.32

*Based on studies by different researchers using different technologies, scales of operation, and feedstock costs, and working in different years.
†FPL, Forest Products Laboratory; SSF, simultaneous saccharification and fermentation.
‡Including 25% bark.

in the same batch process. Estimates of yields and costs of ethanol derived from dry wood using simultaneous saccharification and fermentation technology are shown in Table 1.

The high cost of enzymes and difficulties associated with their recycling have impeded the use of this technology on a large scale. Currently, there are no plants that use enzyme technology in commercial operation. However, enzyme manufacturers have reported breakthroughs in cost reduction. One company proposed the building of a plant that would produce 200 million liters of ethanol per year (53 million gallons per year) from lignocellulosic material (wheat straw); however, the company is considering the use of wood as a feedstock (raw material).

Gasification. There are research projects in various stages of pilot testing that would produce ethanol via gasification of wood and other lignocellulosic materials. First, the biomass (wood and lignocellulose) is gasified to produce carbon monoxide and hydrogen. After the gas is cleared and the correct proportions of carbon monoxide and hydrogen are achieved, a synthesis gas (syngas) results. Subjecting the syngas to controlled pressure and temperature and employing a proper catalyst yields ethanol as well as other alcohols, such as methanol, propanol, or butanol. (Different alcohol products usually result from the use of different catalysts, but can also be produced by varying the temperature and pressure.) Since gasification converts lignin as well as cellulose and hemicellulose, this process results in higher yields of ethanol and other alcohols. However, there are currently no commercial plants.

Methanol production. In the 1920s and the 1930s methanol was made commercially from wood and was commonly known as wood alcohol. Methanol was used in automobiles as an antifreeze, although it was not a permanent antifreeze. (If engines overheated, it boiled away.) Later, during cold weather methanol was mixed in low concentrations with gasoline to prevent gas lines from freezing and causing engines to stall. This use was disadvantageous in a few cars that had an enamel coating on the inside of the fuel lines, because the methanol broke down the enamel and plugged the carburetor. Since the ban of lead from gasoline, methanol has been used in the manufacture of methyl tertiary-butyl ether (mtbe) as an octane enhancer or an oxygenate to improve knock resistance. However, mtbe is outlawed in many jurisdictions, and it will likely see little use in the future due to its affinity for getting into ground water and causing contamination.

Retort production. Earlier methanol was produced in retorts, vessels that enable the heating of a substance in the absence of oxygen. When wood in the retort was heated, it produced char and a mixture of chemicals termed lignosulfonic acid. Methanol could be separated from the mixture. However, yields were low. One publication reported the production of only 16 kg of methanol from 1000 kg of air-dry wood (32 pounds per ton) via the retort method.

Gasification. Today commercial manufacture of methanol from wood is nonexistent; but research to attain commercial viability is progressing. Modern technology for methanol manufacture is based on wood gasification. When wood is gasified, the main constituents of the gas are hydrogen and carbon monoxide. In the shift reaction (5), a portion

$$CO + H_2O \rightarrow H_2 + CO_2 \qquad (5)$$

of the carbon monoxide reacts with steam in the presence of an iron catalyst to form additional hydrogen. Hydrogen and carbon monoxide are reacted under controlled pressure, time, and temperature in the presence of a catalyst to produce methanol (6).

$$2H_2 + CO \rightleftharpoons CH_3OH \qquad (6)$$

Reported estimated of yields and costs of methanol production from dry wood with different gasifiers are shown in **Table 2**.

The IGT "Renugas" gasifier is a high-pressure oxygen-blown fluid-bed gasifier developed specifically for biomass. In a fluid-bed gasifier, there is commonly a combustion zone and a gasification zone. A material that may be fluidized (usually sand) is used to support the wood or wood and bark that is being gasified. Air must be supplied to support combustion and heat the sand. The hot sand is then transferred to the gasification zone, where the gasification reaction takes place. The objective is to confine nitrogen from the air to the combustion zone and transfer as little as possible with the heated sand that goes to the gasification zone.

The Koppers-Totzek gasifier is a low-pressure oxygen-blown gasifier developed originally for coal.

TABLE 2. Estimates of costs of methanol production from dry wood*

Study year	Gasifier	Bone-dry ton of feedstock/day	Wood cost, $/dry metric ton	Cost of methanol produced, $/liter
1989	IGT "Renugas"	1814	46	0.25
1989	Koppers-Totzek	1814	46	0.35
1989	Battelle Columbus	1814	46	0.20
1977	Moore-Canada	1361	37.48	0.26
1977	Moore-Canada	5443	37.48	0.22

*Based on studies by different researchers using different gasifiers, scales of operation, and feedstock costs, and working in different years.

It introduces pure oxygen to the stream of wood or of wood and bark being gasified.

The Battelle Columbus gasifier is a low-pressure, indirectly heated gasifier, in which the product char is burned to heat sand, which in turn is mixed with fresh biomass to supply heat for gasification.

The Moore-Canada gasifier is an air-blown moving-bed reactor that operates at 1200°C in the oxidation zone. In a moving-bed gasifier, wood or wood and bark particles are introduced at the top of the gasifier, along with air or oxygen to partially oxidize the wood and produce gas. Wood particles may be added to the process in a near-continuous or plug-flow mode. As the particles are added, they tend to form a bed where the gasification reaction takes place. However, the bed is not fixed in place. Depending on the mix of particle sizes, there may be difficulty in establishing a bed, or the bed may shift up or down, creating a moving bed.

Transesterification. In Europe, methanol is added to vegetable oil in the presence of a catalyst via transesterifiaction to form a methyl ester. The resulting esters, or biofuel, are compatible with conventional diesel fuel. However, it is only partly renewable, since the methanol is made from natural gas. Undoubtedly, efforts will be put forth to make methanol for this use from some form of biomass.

For background information *see* ALCOHOL; ALCOHOL FUEL; ETHYL ALCOHOL; FERMENTATION; GASOLINE; HYDROLYSIS; WOOD; WOOD CHEMICALS in the McGraw-Hill Encyclopedia of Science & Technology.

John I. Zerbe

Bibliography. Enecon Pty Ltd., Wood for alcohol fuels, RIRDC Publ. no. 02/141, a report for the RIRDC/Land & Water Australia/FWPRDC/MDBC Joint Venture Agroforestry Program, Deepdene DC Vic 3103, 2002; G. J. Hajny, *Biological Utilization of Wood for Production of Chemicals and Foodstuffs*, FPL 385, USDA Forest Service, Forest Products Laboratory, Madison, WI, 1981; A. E. Hokanson et al., Chemicals from wood waste, AN 441, Raphael Katzen Associates, Cincinnati, OH, 1975; A. E. Hokanson and R. M. Rowell, *Methanol from Wood Waste: A Technical and Economic Study*, FPL GTR-12, USDA Forest Service, Forest Products Laboratory, Madison, WI, 1977; D. Morris, Cellulose to ethanol: A progress report, *Carbohydrate Econ.*, 4(2):12–14, 2002; C. E. Wyman et al., Ethanol and methanol from cellulosic biomass, chap. 21 in T. B. Johansson and L. Burnham (eds.), *Renewable Energy: Sources for Fuels and Electricity*, pp. 865–923, Island Press, Washington, DC, 1993.

Alzheimer's disease

Dementia is a syndrome (a group of symptoms occurring together) characterized by concurrent impairment of cognition, behavior, and activities of daily living. Dementias comprise the most common type of degenerative disorders affecting the brain. Alzheimer's disease is the most common form of dementia, accounting for about 60% of all diagnosed cases. An estimated 4 million people in the United States have Alzheimer's disease, and the prevalence worldwide is close to 15 million.

Clinical features. The disease usually begins after age 60, although rare cases of earlier onset have been reported. Progression is gradual, and the subjects usually live 7–10 years after the first onset of symptoms. The most common presenting symptom of Alzheimer's disease is the impairment in short-term, or recent, memory; long-term memory is affected in the later stages of the disease. Changes in memory can lead to impairments in language and judgment. Personality or behavioral problems are seen in over 50% of patients. Behavioral problems may manifest as irritability, paranoia, physical aggression, social withdrawal, inappropriate social behaviors, or wandering. Behavioral changes usually occur in the later stages of the disease, though it is possible for them to be the presenting symptom. Patients with Alzheimer's disease also have difficulty carrying out their activities of daily living. Initially, this change may manifest as an inability to operate common household appliances, manage finances, and take care of the home. In the later stages of the disease, patients become bedridden, incontinent, and in need of total care.

Risk factors. Age is the most important risk factor for the development of Alzheimer's disease. The prevalence is about 8% among people over age 65 and doubles every 5 years to reach close to 40% among 85-year-olds. The risk of developing the disease is almost 50% in first-degree relatives of patients with Alzheimer's disease. Genetic mutations on chromosomes 1, 14, and 21 have been shown to cause early-onset, familial (inherited) Alzheimer's disease. However, these mutations are responsible for less than 1% of all known cases. In this form of the disease, the age of onset is usually earlier than 60 years, and disease progression is rapid.

Recent research has shown that there is an excess of cholesterol in plaques seen in the brains of patients with Alzheimer's disease. Apolipoprotein E (Apo-E) is a protein that helps transport blood cholesterol from the liver to various organs in the body. The gene for Apo-E is found on chromosome 9 and has three common forms (alleles): epsilon 2, 3, and 4. Having one or two copies of the epsilon 4 allele appears to increase the risk of developing the later onset form of the disease. This finding helps explain some of the variations that occur in the age of disease onset and the difference in prevalence among various racial and ethnic groups.

Female sex, diabetes, hypertension, and high cholesterol levels have also been found to be risk factors for the development of Alzheimer's disease. The use of estrogen was initially thought to protect against the development of Alzheimer's disease, but a later women's health initiative study showed that estrogen therapy actually increased the risk of developing dementia and cerebrovascular events.

Lower levels of education and early social and environmental deprivation are also thought to be risk factors for the development of Alzheimer's disease.

Thus, so far, research evidence suggests an interaction between genetic and environmental factors in the development of the disease. Controlling the environmental risk factors in a genetically predisposed individual can reduce the risk of development of Alzheimer's disease.

Neuropathology. The structural changes seen in the brains of patients with Alzheimer's disease are diffuse atrophy (that is, loss of tissue) and associated ventricular enlargement. In some subjects, there may be pronounced atrophy of a specific part of the brain, for example, the temporal lobes. Blood vessel changes may also be seen, though the presence of a major stroke is suggestive of another type of dementia known as vascular dementia.

Pathological changes in the brain include the presence of senile plaques and neurofibrillary tangles. Senile plaques are composed of a protein called beta amyloid that is produced by nerve cells in the brain and accumulates between them. Although senile plaque formation occurs in the brain due to normal aging, it is not as extensive as the plaque formation found in the brains of patients with Alzheimer's disease. Neurofibrillary tangles, on the other hand, are visualized within nerve cells and are composed of helical fragments of an intracellular protein called Tau. Although neurofibrillary tangles are less specific to Alzheimer's disease, they are important in the development of behavioral problems. Other changes include the loss of nerve cells producing the neurotransmitter (chemical messenger) acetylcholine, especially in an area called the basal forebrain nucleus. There is also an increase in the number of microglial cells, a type of connective tissue cell of the central nervous system that is activated to remove breakdown products in the brain and spinal cord during periods of stress, such as inflammation, disease, or injury.

Diagnosis. The diagnosis of Alzheimer's disease is usually made by the evaluation of the patient's clinical history; laboratory tests, brain scans, and neuropsychological tests only aid in the diagnosis. Laboratory tests help to rule out medical conditions that can cause dementia, such as hypothyroidism, human immunodeficiency virus (HIV) infection, and syphilis. Computerized tomography scans or magnetic resonance imaging of the brain indicate the presence of strokes, dilation of ventricles (hydrocephalus), tumors, and changes consistent with other types of dementia, such as vascular or frontotemporal. Neuropsychological testing is useful in diagnosis when all other evaluations are equivocal or the diagnosis is unclear. Neuropsychological tests can also help differentiate dementia from depression. Testing can also provide additional data when decisions must be made about the patient's ability to drive and work, personal safety, and decision-making capacity (competence).

Treatment. The treatment of patients with Alzheimer's disease involves the use of medication and nonpharmacologic (behavioral) methods. Usually, the two are combined for optimum management of symptoms. Unfortunately, there is no treatment or combination of treatments that can halt or reverse the progression of Alzheimer's disease. Pharmacologic agents are helpful in delaying disease onset, slowing the progression of the disease, and treatment of behavioral symptoms. Behavioral techniques are used to ensure safety, assist with self-care, prevent behavioral problems, and improve medication compliance.

Pharmacologic agents. There are two classes of medications that are approved by the Food and Drug Administration (FDA) for the treatment of Alzheimer's disease. The first class of drugs are called cholinesterase inhibitors. They prevent the breakdown of the neurotransmitter acetylcholine by inhibiting the enzyme acetylcholine esterase. The drugs in this class are donepezil (Aricept®), rivastigmine (Exelon®), and galantamine (Reminyl®). The original acetylcholine esterase, tacrine (Cognex®), has been removed from the market because of its potential to cause liver dysfunction. Though these medications are approved for the treatment of mild to moderate Alzheimer's disease, later data have shown that they might also be helpful in the treatment of more advanced disease. Research has also shown that these medications appear to slow disease progression and might aid in the treatment of behavioral symptoms. The second class of drugs used for the treatment of Alzheimer's disease are called *N*-methyl-D-aspartate (NMDA) antagonists. These drugs are thought to prevent neuronal cell death, especially in the hippocampus of the brain, by preventing overactivity of the excitatory amino acid glutamate. The only FDA-approved drug in this class is memantine (Namenda)®. Memantine alone or in combination with donepezil has been shown to be effective in the treatment of cognitive and behavioral symptoms of moderate to severe disease.

Other approaches for treating Alzheimer's disease are being explored. A trial conducted by the Alzheimer's Disease Cooperative Study (ADCS) showed that both selegiline (Carbex®, Eldepryl®), an inhibitor of the monoamine oxidase A enzyme, and vitamin E delay the progression of the disease, as measured by severity of functional decline, placement at a nursing home, or death. Both drugs are thought to inhibit nerve cell damage by preventing the formation of toxic free radicals in the brain. However, the use of vitamin E has come under scrutiny, as a recent meta-analysis found that its use was associated with a slight increase in mortality due to cardiovascular complications. (However, this was a limited study, and this finding requires replication before more definitive conclusions can be drawn.) Although some epidemiological studies have shown that the use of nonsteroidal anti-inflammatory drugs (NSAIDs) is associated with a lower risk of developing Alzheimer's disease by preventing microscopic inflammation in the brain, controlled studies of these agents have been disappointing. Researchers are investigating the efficacy of cholesterol-lowering drugs and vitamin therapy [pyridoxine (B-6), vitamin B-12, and folic acid] for the prevention of disease progression. Cholesterol-lowering drugs are theorized to reduce the progression of Alzheimer's disease by decreasing the cholesterol available for beta amyloid production and by lowering the risk of strokes. Vitamin therapy is thought to reduce the level of the amino acid homocysteine, which has been found to make the nerve cells of the brain more susceptible to the toxic effects of beta amyloid and to increase the risk of strokes. A recent trial with a vaccine that was postulated to reduce the development of beta amyloid plaques had to be halted because some subjects developed inflammation of the brain (encephalitis).

Several different psychotropic agents have been used experimentally for the treatment of behavioral symptoms of Alzheimer's disease. Antidepressants, anticonvulsant mood stabilizers, and antipsychotic medications have all been shown to be helpful in controlled studies. However, none of these medications have received FDA approval for the treatment of behavioral symptoms, as there is still no consensus on the definition of these symptoms, and the effect of these medications is modest at best.

Behavioral techniques. Behavioral treatments used in Alzheimer's disease include the management of inappropriate behaviors, assisting with activities of daily life, and the identification of resources available in the community. Supportive psychotherapy for both the patient and the caregiver help reduce frustration, guilt, and stress due to the illness. Alzheimer's disease causes immense suffering for the patients and their families. An estimated 50% of primary caregivers of subjects with Alzheimer's disease have significant psychological distress. This translates into a decline in the quality of life for both the caregiver and the patient. Recent research has shown that almost 60% of caregivers have a higher mortality rate than age-matched noncaregiver controls. Respite care gives caregivers a well-deserved break from their duties. Placement at a skilled nursing facility is needed when the care of the patient in the community becomes difficult or unsafe.

For background information *see* ACETYLCHOLINE; ALZHEIMER'S DISEASE; BRAIN; HUMAN GENETICS; MEMORY; NERVOUS SYSTEM DISORDERS; PSYCHOPHARMACOLOGY in the McGraw-Hill Encyclopedia of Science & Technology. Rajesh R. Tampi

Bibliography. J. L. Cummings, Alzheimer's disease: From molecular biology to neuropsychiatry, *Semin. Clin. Neuropsychiatry*, 8(1):31–36, 20036, 2003; J. L. Cummings, Alzheimer's disease, *New Engl. J. Med.*, 351(1):56–67, 2004; K. C. Fleming et al., Dementia: Diagnosis and evaluation, *Mayo Clinic Proc.*, 70:1093–1107, 1995; R. Mayeux and M. Sano, Treatment of Alzheimer's disease, *New Engl. J. Med.*, 341(22):1670–1678, 1999; M. Sano, Current concepts in the prevention of Alzheimer's disease, *CNS Spectrum*, 8(11):846–853, 2003; G. W. Small et al., Diagnosis and treatment of Alzheimer's disease and related disorders, *JAMA*, 278(16):1363–1371, 1997.

Amphibious assault ships

The world's navies continue to explore a variety of ship, landing craft, and amphibian vehicle designs to improve amphibious assaults. There has been an evolution over the last 60 years from ships coming close to shore to deliver troops and supplies using small craft, to ships operating far out to sea and delivering troops by aircraft, fast craft, and self-propelled amphibious vehicles. These design evolutions are discussed along with speculation on future design developments and operating concepts.

LSTs and LSDs. Amphibious warfare has existed as long as naval warfare has been practiced. In particular, the British in the Napoleonic Wars developed amphibious boat techniques and tactics. Before World War II, however, ships were not specifically designed to conduct amphibious assaults. Faced with the need to conduct amphibious assaults against Europe and Japanese-held islands, the United States and Great Britain developed new ship designs that specialized in amphibious assault. The two largest were the Landing Ship Tank (LST) and Landing Ship Dock (LSD), which worked in conjunction with a mass of existing commercial ships pressed into service as troop and cargo transports and a multitude of smaller ships and craft specially built for amphibious assault.

The LST was developed to meet the need for a ship that could beach itself and land large numbers of troops with full armor support. It had enough seaworthiness to make transits over moderate distances and was large enough to carry tanks. The 328-ft (100-m) long LST was designed with a ballast system that could be filled for ocean passage and pumped out for beaching operations. Tanks, troops, and their equipment were rapidly unloaded by means of a bow door and ramp (**Fig. 1**). The versatile LST

Fig. 1. LSTs at Leyte, D-Day, October 20, 1944.

served many other roles, including hospital ships, landing-craft repair ships, and ammunition replenishment ships.

The Landing Ship Dock was developed for amphibious landings against positions that were far from rear support bases. The LSD transported troops, landing craft, and amphibious vehicles in a large docking well that measured 44 by 396 ft (13 by 121 m) and ran almost its entire length (**Fig. 2**). To release the craft, the LSD was designed with a stern gate, which was lowered after the ship ballasted down to flood its docking well. The LSD was then used as an offshore repair dock for ships and craft. The design of the LSD type has changed little since World War II, except that new LSD classes have a main deck built over the well to accommodate helicopters.

Fig. 2. USS *Carter Hall* (LSD 3) in 1951.

Helicopter operations ships. After World War II, tests showed that atomic weapons could devastate a massed amphibious landing force. In response, the U.S. Marine Corps began using helicopters based on aircraft carriers to carry troops ashore. This mitigated the effects of an atomic blast because it allowed dispersion of ships and could bypass coastline defenses. The British first used helicopters as part of amphibious landings during the Suez Crisis in 1956. In the meantime, the Korean War showed that amphibious landings were still necessary. The conflict also set the pattern for the United States' Cold War stance of maintaining a forward-deployed rapid-response amphibious assault force. Faced with these commitments, the U.S. Navy was in the forefront of amphibious ship design in the 1950s and 1960s, while the rest of the world's navies still relied largely on World War II–era amphibious designs. However, most of the U.S. Navy's amphibious ships lacked seaworthiness, speed, range, and the ability to embark troops for extended voyages. In the 1950s the Navy began building amphibious ships with a speed of at least 20 knots (10 m/s), versus the 12 knots (6 m/s) of World War II LSTs, better sea-keeping characteristics, greater endurance, and habitability improvements to allow troops to live on board for long periods.

In the 1950s the Navy used surplus World War II–era fleet and escort carriers to embark helicopters. However, they were not designed for this mission and performed poorly. In 1961 the Navy introduced the Landing Personnel Helicopter (LPH) as the first ship built specifically for helicopter operations (**Fig. 3**). At 600 ft (183 m) long and displacing 18,000 tons (16,320 metric tons), the LPH was the largest amphibious ship built at the time. It had a deck edge elevator amidships on the port side and another aft of the island structure on the starboard side. The flight deck was 602 by 104 ft (183 by 32 m). The ships embarked 20–30 helicopters and approximately 1500 troops.

Designed solely for helicopter operations, the LPH was the last of the Navy's single-mission amphibious ships. In the 1950s and 1960s there were over

Fig. 3. USS *Iwo Jima* (LPH 2), the world's first purpose-built helicopter assault ship.

Fig. 4. USS *Saipan* (LHA 2) conducting flight tests of MV-22 tilt rotor aircraft in 1999.

100 amphibious ships, comprising 22 different ship classes providing at least 12 distinct functions during amphibious assaults. Since the LPH class, the U.S. Navy (and many other navies) has followed a strategy of designing larger, multimission ships in fewer numbers.

Multimission ships. The Landing Personnel Dock (LPD), the first multimission ship, was introduced in 1962. The LPD combined a well deck, helicopter deck, troop berthing, and cargo, vehicle, and equipment storage.

The Navy's next multipurpose ship was a new type of helicopter carrier, the LHA (Landing Helicopter Assault). The LHA differed from the LPH by combining the vehicle, cargo, and landing craft capability of the LSD with the helicopter and troop capability of the LPH (**Fig. 4**). This combination made the LHA a general-purpose amphibious assault ship versus a simple helicopter assault ship. When introduced in 1976, the LHA was the largest amphibious ship ever built, with a length of 820 ft (250 m) and 39,400 tons (35,744 metric tons) displacement. The LHA has two elevators, one amidships portside and one at the stern of the ship above the well deck. The flight deck measures 820 by 118 ft (250 by 36 m) and can accommodate simultaneous takeoff or landing of ten helicopters.

In the 1980s the LHD (Landing Helicopter Dock) was introduced. An improved LHA design, the LHD can carry three Landing Craft Air Cushion (LCAC) versus only one on the LHA, and is designed to operate and maintain improved versions of the AV-8 Harrier jumpjet and heavy helicopters. The greater weight and size of these aircraft required a larger flight deck. As a result, island superstructure width reductions from 51 to 35 ft (16 to 11 m) and other alterations created 14% more usable deck area than the LHA (**Fig. 5**). Two levels of the island structure were also removed to reduce vulnerability from missiles and to offset the growth in topside weight from the larger aircraft and flight deck. To allow embarkation of three LCACs, the LHD well deck was narrowed from 78 to 50 ft (24 to 15 m) and the LHA's centerline well deck structure was removed.

The LHA well deck measures 268 by 78 ft (82 by 24 m) and was designed to accommodate conventional (floating, displacement) landing craft (specifically the Landing Craft Utility, LCU, and Landing Craft Mechanized, LCM, classes). However, these slow-moving craft, whose basic design had not changed since World War II, had speeds of only 10 to 12 knots (5 to 6 m/s) and required assault ships to come as

Fig. 5. LHD with an LCAC entering the ship's well deck.

Fig. 6. LCAC 69 with two Marine Corps amphibious assault vehicles embarked.

Fig. 7. Russian *Zubr* class landing assault hovercraft.

close as 4000 yards (3658 m) offshore, placing the ships in extreme danger from increasingly sophisticated missiles and mines. The lethality of these weapons was highlighted (after the LHA was designed) during the Arab-Israeli War of 1973 and the Falklands War in 1982. As a result, the Navy increasingly deployed its ships over the horizon (OTH) during amphibious operations, which limited the utility of landing craft. In response, the LCAC was developed in the 1970s. A pressurized cushion of air, contained under the craft by a surrounding rubber skirt, enabled operation on both water (even marsh) and the beach. With a range of over 200 nautical miles (230 mi), a speed of over 40 knots (20.6 m/s), and the ability to carry heavy equipment, the LCAC facilitated OTH operations when it became operational in 1987 (**Fig. 6**). In addition to landing troops and equipment during assaults, the LCAC provides evacuation support, lane breaching, and mine countermeasure operations. A refurbishment program is currently in progress to add 10 years of useful life to the LCAC.

The Russian (then Soviet) Navy introduced its own landing assault hovercraft, the *Zubr* class (**Fig. 7**), in 1988. It is the largest hovercraft ever developed, with a size of 188 by 84 ft (57 by 26 m) and a displacement of 555 tons (503 metric tons). Unlike the U.S. Navy, the Russian Navy does not need to project amphibious forces over long distances. As such, the massive *Zubr*, which cannot be carried by a ship, can carry eight amphibian tanks or ten armored vehicles and 140 troops.

Fig. 8. Newly completed LPD 17 showing advanced radar cross section reduction feature.

Future developments. The world's navies continue to develop ships, landing craft, vehicles, and methods of attack to reduce the threat from missiles launched by aircraft, ships, and land-based facilities. The U.S. Navy is building a greatly improved LPD (**Fig. 8**) and developing a replacement for the LHA. Both incorporate design features to reduce the ships' radar cross section and electronic and exhaust emissions in an effort to lessen their visibility to hostile missile guidance systems. The Marine Corps is developing the Expeditionary Fighting Vehicle (EFV) to replace the Amphibious Assault Vehicle. The EFV is projected to have a range of 65 nautical miles (75 mi) at 20 knots (10 m/s), and a maximum speed of 25 knots (13 m/s) in the water and 45 mph (20 m/s) on land. The EFV's range and speed, combined with the LCAC, will enable the Navy's amphibious ships to conduct assaults from much farther out to sea. This will support new operational concepts, including Operational Maneuver from the Sea (OMFTS) and Ship to Objective Maneuver (STOM), which seek to move amphibious forces directly to inland objectives, thus bypassing coastal defenses.

For background information *see* AIR-CUSHION VEHICLE; LANDING SHIPS AND CRAFT; NAVAL SURFACE SHIP in the McGraw-Hill Encyclopedia of Science & Technology.

Matthew McCarton

Bibliography. D. E. Barbey, *MacArthur's Amphibious Navy: Seventh Amphibious Force Operations 1943–1945*, United States Naval Institute Press, Annapolis, 1969; N. Friedman, *U.S. Amphibious Ships and Craft: An Illustrated Design History*, United States Naval Institute Press, Annapolis, 2002; J. A. Isely and P. A. Crowl, *The United States Marines and Amphibious War: Its Theory, and Its Practice in the Pacific*, Princeton University Press, 1951; M. McCarton, *Amphibious Warfare and the Evolution of the Helicopter Carrier*, Naval Sea Systems Command report, March 1998; N. Polmar and P. B. Mersky, *Amphibious Warfare: An Illustrated History*, Blanford Press, London, 1988.

Animal disease surveillance and monitoring

The United States Centers for Disease Control and Prevention (CDC) defines public health surveillance as the ongoing, systematic collection, analysis, interpretation, and dissemination of data, including clinical signs and symptoms, laboratory test results, and prevalence of behavioral and attitudinal risk factors. Epidemiologists use these data to detect outbreaks, describe patterns of disease transmission, evaluate prevention and control programs, and prioritize future health-care needs. Traditionally, veterinary disease surveillance and monitoring programs have focused on control and eradication of diseases of agricultural significance, although significance has often been defined by the zoonotic (ability for spread between humans and animals) potential of the agent. Today, in an expanding global community, animal disease surveillance and monitoring is largely geared toward establishing national and regional disease status to support international trade in animals and animal products. Recently, diseases spread from wild animals have affected both public health and agricultural disease control programs, significantly draining valuable resources.

Integration of surveillance and monitoring programs. The terms surveillance and monitoring are often used synonymously, but actually have small but important differences. Surveillance systems are aimed at identifying the first case of a disease in a population (nonendemic or emerging diseases). An important part of surveillance is the ability to respond to the incursion of the identified disease. Monitoring systems are designed to detect and regularly report on the changing status of a disease that is already present in a population (endemic diseases). Usually monitoring systems help describe how a disease interacts in a population or ecosystem, but does not include outbreak response capability or eradication and control measures. The combination of surveillance and monitoring programs results in an integrated system capable of addressing both emerging and endemic diseases in a population.

Types of surveillance and monitoring systems. Several types of surveillance and monitoring systems exist.

Passive systems. Passive surveillance and monitoring systems, such as those aimed at collating results from diagnostic laboratories, are conducted by gathering data from existing medical records. Historically, passive tabulation of "reportable" diseases from hospitals and diagnostic laboratories has been the most common and affordable form of surveillance. However, since this information is gathered from only those that were obviously sick and seen by a doctor, they may not accurately portray the entire population. As technology has progressed and the ease of international travel and commerce has increased, the risk of rapid, global disease spread has grown. To combat this risk, accurate, up-to-date information must be collected, collated, and communicated to outbreak response teams in a timely manner.

Active systems. The need for real-time surveillance and monitoring has resulted in an emphasis on active surveillance and monitoring programs. Active surveillance and monitoring requires the use of statistically derived sampling to reliably assess the health status of a population. In these programs, results are not just passively tabulated from laboratory records; rather individuals that statistically represent the population are sought out for testing.

Identification of important syndromes. In today's world, in which agents of bioterrorism (such as anthrax) and deadly emerging diseases (such as ebola virus) are of great concern, surveillance and monitoring systems are being geared toward identifying specific syndromes of disease. Because many diseases, some reportable and some not, result in similar clinical signs, potentially important syndromes must be quickly identified for further testing to rule out diseases of great concern. Examples of important syndromes and diseases that are the focus of human surveillance and monitoring programs include skin lesions that may indicate anthrax or smallpox and respiratory syndromes that may indicate severe acute respiratory syndrome (SARS) or a new strain of influenza. Skin lesions are also an important potential indicator of a number of animal diseases such as foot-and-mouth disease.

Emerging zoonotic disease. Since the term "emerging disease" was popularized in the early 1990s, both well-known diseases (rabies, tuberculosis, brucellosis, tularemia, avian influenza, and plague) and emerging wildlife diseases (ebola virus in great apes, SARS in civets, monkey pox in rodents, nipah and hendra virus in bats and flying foxes, and West Nile virus in hundreds of species) have dominated the popular and scientific literature. The zoonotic nature and complex ecology of emerging diseases requires special consideration when incorporating them into existing or novel surveillance and monitoring programs.

Zoonotic disease surveillance and monitoring is complicated because of the unique, and often unknown, role of wildlife in the ecology of these diseases. With the introduction of West Nile virus into the Western Hemisphere, the public health community in North America has begun to take a more ecological approach to disease surveillance and monitoring. This includes the "canary in the coal mine" approach, in which the "canaries" are sentinels for the introduction of disease. From an ecological perspective, humans, domestic animals (pets and farm animals), and wildlife may all be sentinels for each other, depending on the goals of the program.

Captive wildlife disease. Zoonotic disease surveillance and monitoring usually benefits multiple parties. For example, surveillance and monitoring of diseases such as tuberculosis and West Nile virus in captive wildlife has begun to play an increasingly important role for both public and veterinary health.

Tuberculosis. Tuberculosis is an infectious disease caused by *Mycobacterium* spp., which infects most animals, including humans. Different species and subspecies of this bacterium commonly infect specific groups of mammals, birds, reptiles, and amphibians. For example, *M. bovis* is usually associated with disease in cattle or other ungulates (hoofed mammals), where-as *M. tuberculosis* is usually associated with disease in humans. However, both of these organisms may infect either animals or humans and are reportable to the U.S. Public Health Service (USPHS) or the U.S. Department of Agriculture (USDA). The USDA has undertaken a national *M. bovis* eradication campaign in the United States since the early twentieth century. In the 1990s, *M. bovis* was found in wild deer in Michigan, and the same strain of *M. tuberculosis* was isolated from captive elephants and animal care staff.

As a result of increased concern surrounding zoonotic tuberculosis and wildlife, zoological institutions accredited by the American Zoo and Aquarium Association (AZA), in conjunction with the USDA, began a surveillance and monitoring program for tuberculosis in elephants and ungulates. The process began with the creation of a set of standardized diagnostic testing protocols for all AZA animals. The active surveillance program for elephants includes yearly testing and bacterial culture of all zoo elephants, with guidelines for the quarantine and treatment of suspected as well as confirmed positives. The passive monitoring program for other exotic ungulates includes the collection of standardized test results through a centralized database; results are reported quarterly to both the AZA and USDA. These programs have resulted in numerous benefits. From a public health standpoint, zoos are actively protecting the health of employees that come into contact with these animals; from the agricultural standpoint, zoos are minimizing the risk of harboring or spreading a disease that the USDA has been trying to eradicate for almost 100 years; and from the zoo's standpoint, the risk of spreading tuberculosis between institutions is decreased, and the welfare of the animals in their care, many of which are endangered species, is safeguarded.

West Nile virus. West Nile virus, first isolated in 1937 in Uganda, is an arbovirus (arthropod-borne virus) that usually cycles between birds and mosquitoes, but sometimes affects accidental mammalian hosts, such as humans and horses. Clinical signs of illness do not usually appear in birds, but infected humans occasionally show clinical signs reminiscent of the flu, which sometimes progress to neurological problems and death. The first epidemic in humans was reported in Israel in the 1950s. The virus was found only in northern Africa, the Middle East, and eastern Europe until 1999, when it was isolated in the United States from wild crows and from animals at the Bronx Zoo in New York. Since the initial focus in New York, New Jersey, and Maryland, the virus has spread west, north, and south across the United States, Canada, Mexico, and the Caribbean, infecting hundreds of species of birds, mammals, and reptiles—many of which have succumbed to the disease.

In June 2001, the Lincoln Park Zoo in Chicago cohosted a meeting with the CDC that brought zoo professionals together with human and veterinary public health experts from local, state, and federal agencies in order to create a nationwide surveillance and monitoring system for West Nile virus in AZA zoos. This system provided zoos with affordable, reliable diagnostic testing, while allowing public health officials to increase the scope of West Nile virus surveillance data in the United States. Participating institutions submitted samples to the Cornell Animal Health Diagnostic Laboratory, and results were reported directly to the submitting institution and to the central Lincoln Park Zoo database. Submitting institutions were required to share results with local public health officials who recorded the data for the national system and were responsible for local outbreak response.

The initial objective of this program was to detect geographic spread of the virus. Samples were solicited from suspect animals (those experiencing systemic illness or neurologic problems after exposure to mosquitoes) in institutions along the front edge of the spread of disease. Phase II included an analysis of archived blood to determine if viral activity could be documented before the first case found in an area. The infrastructure developed for West Nile virus surveillance is being reconstructed for use with other zoonotic disease issues of importance.

This system has numerous benefits for the public, veterinary, and wildlife health communities. It has shown that a multidisciplinary approach can result in an integrated ecological surveillance and monitoring system for emerging diseases. It has greatly enhanced knowledge of the ecology of West Nile virus and identified hundreds of susceptible species, while providing a unique source of data for the public health systems in large urban and rural areas. Finally, it has provided the captive wildlife community with an affordable means of reliable diagnostic testing, while concurrently fostering relationships with local public health and agricultural officials.

For background information *see* ARBOVIRAL ENCEPHALITIDES; DISEASE ECOLOGY; EPIDEMIOLOGY; INFECTIOUS DISEASE; TUBERCULOSIS; ZOONOSES in the McGraw-Hill Encyclopedia of Science & Technology. Dominic A. Travis

Bibliography. R. G. Bengis et al., The role of wildlife in emerging and re-emerging zoonoses, *Rev. Sci. Tech. Off. Int. Epiz.*, 23(2):497–511, 2004; O. Cosivi et al., Zoonotic tuberculosis due to *Mycobacterium bovis* in developing countries, *Emerg. Infect. Dis.*, 4(1):59–70, 1998; A. Glaser, West Nile virus and North America: An unfolding story, *Rev. Sci. Tech. Off. Int. Epiz.*, 23(2):557–568, 2004; H. Kruse, A. M. Kirkemo, and K. Handeland, Wildlife as source of zoonotic infections, *Emerg. Infect. Dis.*, 10(12): 2067–2072, 2004; K. Michalak et al., *Mycobacterium tuberculosis* infection as a zoonotic disease: Transmission between humans and elephants, *Emerg. Infect. Dis.*, 4(2):283–287, 1998; T. Morner et al., Surveillance and monitoring of wildlife diseases, *Rev. Sci. Tech. Off. Int. Epiz.*, 21(1): 67–76, 2002; M. D. Salmon (ed.), *Animal Disease Surveillance and Survey Systems*, Iowa State Press, Ames, 2003; B. Toma et al., *Applied Veterinary Epidemiology and the Control of Disease in Populations*, Office International des Epizooties, Alfort, France, 1999; E. S. Williams and I. K. Barker (eds.), *Infectious Diseases of Wild Mammals*, 3d ed., Iowa State Press, Ames, 2001.

Antidepressant use in minors

The strong evidence of efficacy for the stimulants methylphenidate (Ritalin®) and amphetamine (Adderall®) in treating children and adolescents with attention deficit hyperactivity disorder (ADHD) has raised expectations that medications might be found with similar levels of efficacy to help children and adolescents with other psychiatric disorders. In the case of agents to treat depression, the results have turned out to be controversial. Confronted with equivocal data for efficacy, as well as serious questions about safety, the public has become somewhat skeptical—as have prescribing clinicians.

Current usage. The antidepressants, broadly defined, include six selective serotonin reuptake inhibitors (SSRIs): fluoxetine (Prozac®), sertraline (Zoloft®), paroxetine (Paxil®), fluvoxamine (Luvox®), citalopram (Celexa®), and escitalopram (Lexapro®); the non-SSRIs venlafaxine (Effexor®), duloxetine (Cymbalta®), and nefazodone (Serzone®), all of which work on the norepinephrine and serotonin systems; buproprion (Wellbutrin®), which works on norepinephrine and dopamine receptors; and mirtazapine (Remeron®), which stimulates alpha-2 receptors in the brain. In addition, there are the older monoamine oxidase inhibitors (MAOIs) and a range of tricyclic antidepressants (TCAs), which were the mainstay for the treatment of adult depression between the 1970s and early 1990s.

The Food and Drug Administration (FDA) holds medications to a relatively high standard, requiring (among other things) at least two large, successful, randomized, placebo-controlled trials demonstrating efficacy. Most psychotropic medications used in children do not meet these stringent criteria, often because the necessary studies have not yet been undertaken. But notable exceptions exist, including several studies on antidepressants. Indeed, four of these medications have received FDA indications for use in children and adolescents. The SSRIs sertraline and fluvoxamine and the TCA clomipramine (Anafranil®) are approved for the treatment of obsessive compulsive disorder. Fluoxetine is approved for the treatment of both major depressive disorder and obsessive-compulsive disorder. All other uses of antidepressants in children and adolescents are currently considered off-label, meaning that they are not specifically recommended for disorders in minors. Nonetheless, these medications are frequently given to youths to treat a variety of other disorders.

Data on efficacy and safety. In 2003–2005 several significant contributions were made to the evidence base for and against the use of antidepressants in children and adolescents. Solid data, based on hundreds of subjects, have supported the use of antidepressants, particularly fluoxetine, to treat adolescents with major depressive disorder. On the other hand, there has been a reevaluation of the previously compiled safety and efficacy data in children and adolescents that has led to, among other things, an FDA black box warning (a warning printed in bold type in a black box on the package insert) concerning the use of antidepressants in this age group.

Differing efficacy. Although there is extensive data supporting the efficacy of antidepressants (both TCAs and SSRIs) in adults, compelling evidence for their effectiveness in treating depression in children and adolescents has been more difficult to obtain. For example, among adults about 50% of studies show efficacy; for children and adolescents that number drops to approximately 20%. One explanation for this disparity is thought to be the high placebo response rate in children and adolescents, perhaps due to their greater trust in authority figures, relative to adults, making it harder to show that antidepressants are superior to placebo. Others have theorized that the differential neurobiology of children—particularly the fact that children's serotonergic and noradrenergic systems are still in development—may account for the disparity.

Antidepressant use and suicidality. The issue causing the most concern is information suggesting a link between antidepressant use in adolescents and suicidality. In June 2003, the Medicines and Healthcare Products Regulatory Agency (MHRA), England's counterpart to the FDA, issued a warning, based on published and unpublished data, that youngsters randomly assigned to receive paroxetine had reported suicidal ideation almost three times as often as those assigned to receive a placebo (3.4% versus 1.2%). In December 2003, the MHRA held all of the available studies of antidepressants in adolescents to the same scrutiny and determined that, in light of the limited data for efficacy, the heightened risk of suicidality associated with antidepressant use was not acceptable in adolescents, and that only fluoxetine could be recommended for use in those under age 18.

No clear answer has emerged as to why medication aimed at alleviating depression should increase suicidal ideation, which is usually a by-product of the underlying disorder. However, one hypothesis is that the medication causes a deeply depressed individual to become activated and acquire more energy before the underlying psychological suffering has abated, leading to a window of heightened risk of negative behavior or verbalization. Another possibility is that changes in the blood level of some medications, particularly those antidepressants with short half-lives, may cause unpleasant side effects and feelings of agitation. Neither of these hypotheses has been rigorously examined.

The FDA took a slightly different approach toward the concerns raised by the MHRA review. The FDA looked at the studies on all of the SSRIs—plus buproprion, venlafaxine, and mirtazapine—and, puzzled by inconsistencies in the analysis, commissioned an independent group to examine the issue of suicidality in youths taking antidepressants. The FDA findings bore out earlier concerns, indicating that SSRIs were associated with more suicidal ideation than placebo, showing that approximately 4% of those on SSRIs reported suicidal ideation versus 2% for placebo. Valid criticisms of the FDA study have noted that the agency relied exclusively on self-report data of suicidality, rather than questionnaires which, when used, did not show a similar increase. Others have complained that the predictive validity of suicidal ideation for actual suicidal behavior (let alone for completed suicide) has not been established. It is also worth noting that only 78 of the 4400 patients involved in these pooled randomized studies experienced suicidal thinking or suicidal behavior, and no completed suicides took place.

Response to findings on suicide risk. After the suicidality findings were reviewed by the FDA in September and October 2004, a black box warning was placed on all antidepressants, indicating the higher-than-placebo rate of suicidality in children and adolescents. The FDA also requires that such antidepressants be sold with a medication guide noting circumstances in which patients should be particularly cautious before continuing medication use. These include rapid changes in mood and changes in medication dose. Additionally, the guide includes a recommended schedule for physician visits to monitor side effects: one visit per week for the first month, once every 2 weeks for the second month, and then once at the end of the third month after the initiation of treatment.

The American Academy of Child and Adolescent Psychiatry (AACAP) responded to the black box warning by noting that depression is a severe illness with morbid sequelae (consequences or complications) and that, while it would be beneficial to curtail the careless prescription of antidepressants, the warning would have a chilling effect on appropriate prescriptions as well. A precipitous reduction in the rate of antidepressant use in youngsters could potentially do more harm than good, given that untreated depression has a high mortality rate. Paradoxically, whatever suicidal ideation may be prevented through lower use of SSRIs may be replaced by completed suicides and other serious sequelae of untreated depression.

Treatment of Adolescent Depression Study. In the fall of 2004 the results of the Treatment of Adolescent Depression Study (TADS) were published and supported the use of antidepressants to treat adolescent depression. TADS followed 439 children aged 12–17 with major depressive disorder. The findings gave robust support for fluoxetine as a treatment for adolescent depression. Patients who received fluoxetine and psychotherapy showed a 71% response rate, whereas those treated with psychotherapy alone had a 43% response rate. Of the groups that did not receive psychotherapy, those on placebo had a 34% rate and those on fluoxetine 61%. Findings were all statistically significant.

The AACAP took note of the TADS findings and created a website in early 2005 evaluating the risks and benefits of using antidepressants in youth. They pointed out that, based on the available data, a physician would have to treat only three adolescents to see a therapeutic response attributable to medication, but would have to treat 50 to see an increase in suicidality attributable to medication. Thus, the use of antidepressants seems to be more beneficial than harmful—especially in the context of adequate clinical monitoring and when considering the dire consequences of the alternative: untreated adolescent depression is a severe and potentially devastating disorder. Based on current data, it makes sense for clinicians to proceed, albeit cautiously, in providing treatment for depression, be it pharmacologic or psychotherapeutic in nature.

Current and future research. There are a number of recent National Institute of Mental Health initiatives to address the need for additional reliable data on the use of antidepressants in adolescents and the clinical issues confronting patients and practitioners on a regular basis. TADS is being extended by one year to provide information on long-term usage of antidepressants. In addition, the Treatment of Resistant Depression in Adolescents (TORDIA) study will consider adolescents who do not respond to their initial pharmacotherapy for depression, and examine their response to a second agent. The Treatment of Adolescent Suicide Attempters study will consider the psychopharmacological and psychotherapeutic treatment of individuals who have made suicide attempts in the prior 45 days.

For background information *see* AFFECTIVE DISORDERS; NORADRENERGIC SYSTEM; PSYCHOPHARMACOLOGY; SEROTONIN in the McGraw-Hill Encyclopedia of Science & Technology.

Lawrence Maayan; Andrés Martin

Bibliography. T. Hammad, Results of the analysis of suicidality in pediatric trials of newer antidepressants, in Department of Health and Human Services, Food and Drug Administration Center for Drug Evaluation and Research, *Joint Meeting of the CDER Psychopharmacologic Drugs Advisory Committee and the FDA Pediatric Advisory Committee, September 13, 2004*, pp. 152–200; J. March et al., Fluoxetine, cognitive-behavioral therapy, and their combination for adolescents with depression: Treatment for Adolescents with Depression Study (TADS) randomized controlled trial, *J. Amer. Med. Ass.*, 292(7):807–820, 2004; M. Olfson, Relationship between antidepressant medication treatment and suicide in adolescents, *Arch. Gen. Psych.*, 60(10):978–982, 2003; J. M. Rey and A. Martin, SSRIs and suicidality in juveniles: Review of the evidence and implications for clinical practice, *Child and Adolescent Psychiatric Clinics of North America*, 14(1), in press, 2006; V. Ruchkin and A. Martin, SSRIs and the developing brain (or, Minor problems with antidepressants?) [Commentary], *Lancet*, 365:451–453, 2005.

Arctic treeline change

The Arctic treeline is the biogeographic boundary between the boreal forest to the south, dominated by coniferous trees, and tundra vegetation to the north, dominated by small shrubs, herbaceous plants, nonvascular plants, and bare ground. The treeline stretches across North America from northern Alaska to the coast of Labrador, and across Eurasia from Scandinavia to the far east of Siberia. The geographic location of the arctic treeline zone is controlled by global climate patterns but also can influence global climate. For this reason, scientists study past changes in the Arctic treeline to understand the natural variability of climate, detect long-term changes in climate, and predict how future climate changes might affect the treeline.

Treeline zone today. The modern Arctic treeline zone lies mostly between 60 and 70°N latitude. The treeline is farthest south in northeastern Canada where it descends below 60°N latitude, and farthest north in central Siberia where larch trees are found growing at 73°N latitude. Coniferous trees of the

Arctic treeline zone of North America include white spruce (*Picea glauca*), black spruce (*P. mariana*), and larch (*Larix laricina*). In Eurasia, the dominant treeline conifers include Scots pine (*Pinus sylvestris*) and Norway and Russian spruce (*Picea abies* and *P. obovata*) in northern Europe, with larches (*L. sibirica* and *L. dahurica*) and stone pine (*Pinus pumila*) dominant eastward into Siberia. Broadleaf trees such as birch (*Betula*) and poplar (*Populus*) can also be found in the North American and Eurasian treeline zones.

The treeline is not a sharp boundary, but a zone of transition where continuous forest cover in the south gives way to increasingly scattered fragments of forest and smaller and smaller individual trees northward. At the extreme northern limits of their distributions, the coniferous trees often grow as low, ground-hugging shrubs called krummholz (German for twisted wood). Sexual reproduction is rare for such plants, and a number of treeline species propagate by vegetative layering; that is, low-lying branches develop rooting systems where they touch the ground and eventually become independently growing trees. The ability to propagate vegetatively allows tree species to persist in far northern locations even during extended periods when summer temperatures are too cold for sexual reproduction.

The ultimate reason why trees cannot grow farther north in the Arctic is because the summers are too cool and short for adequate photosynthesis to grow and reproduce. Secondary factors such as bud and needle damage by blowing snow in the harsh winters are also important in limiting tree growth. In North America and Eurasia, the treeline is found where the average July temperature declines to less than 10.0–12.5°C (50–54°F). In areas north of this, the vegetation is treeless tundra (**Fig. 1**). The treeline vegetation is very sensitive to summer temperatures. A warming of a few degrees could cause the treeline to advance northward, while a cooling could cause it to retreat southward.

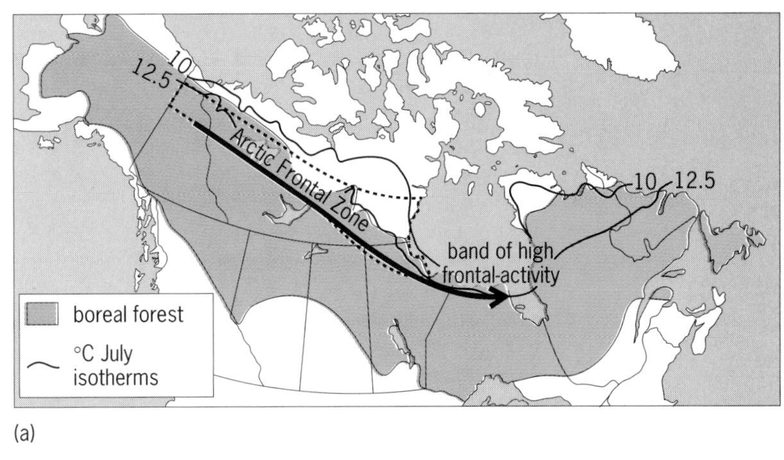

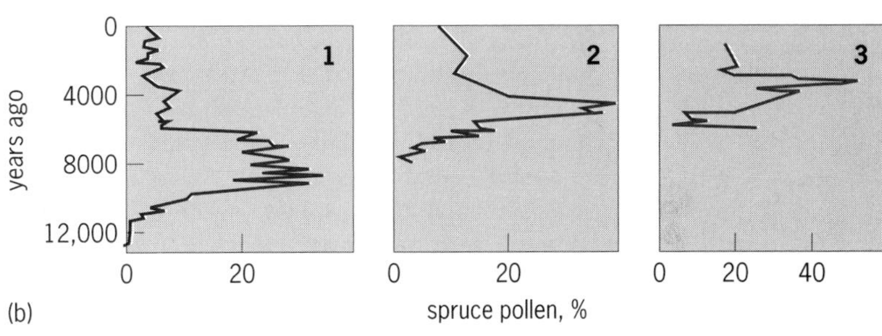

Fig. 1. **Treelines in Canada. (*a*) Relationship between the North American Arctic treeline and 10 and 12°C mean July temperatures. The July position of the arctic frontal zone is also indicated. (*b*) Fossil pollen records of the treelines advance north of its modern position in (1) western Canada, (2) central Canada, and (3) eastern Canada.**

Although the location of treeline is controlled by climate, the presence or absence of trees also affects the climate. Because of the relatively dark color of their foliage, forested areas absorb more sunlight than light-colored tundra. This leads to higher surface temperatures than are found on the treeless tundra areas to the north, where more of the incoming sunlight is reflected directly back to space. Climate models show that if the treeline were to advance northward so that all of northern North America and Eurasia were covered by forest, this could produce a significant increase in the surface temperature of the Earth due to the increased capture of solar energy by the dark forest cover.

Treeline changes in the recent past. Because the Arctic treeline is sensitive to climatic change, scientists have long been interested in studying the history of past changes in the position of treeline in order to understand the natural variability of the Earth's climate. There are some written records, maps, and pictures from explorers, scientists, and government surveyors that document the position and condition of the Arctic treeline in the recent past. In North America, most of these records come from Canada and extend back about 200 years. One example is the writings, drawings, and maps of the Englishman Samuel Hearne, who worked for the Hudsons Bay Company and explored the tundra and treeline areas of the Central Canadian Arctic between 1769 and 1772. Analysis of these and other records and comparisons of the historical documents with modern treeline conditions suggest that from the late eighteenth to mid-nineteenth century summer temperatures were colder than today and the abundance of trees in the treeline zone was lower than at present. However, it does not appear that the actual northern geographic limit of tree species changed much over this time. Cool temperatures from the seventeenth through the nineteenth century were found in many regions of the Arctic and beyond. This period is sometimes referred to as the Little Ice Age.

In order to extend the record of treeline changes into the past, paleoenvironmental researchers use techniques such as tree-ring analysis, the collection and study of preserved wood and other large remains from plants, and the analysis of preserved pollen grains and stomata from lake sediments and peat bogs. Many tree-ring studies have been done in the Arctic treeline zones of North America and Eurasia. Tree rings are formed by the annual radial growth of the cambial layer, which lies between the wood and the bark in coniferous trees. The cambial growth of trees in the far north is typically limited by the summer temperature. Years with cool summers lead to reduced growth and the production of small rings, while warm summers result in vigorous cambial growth and large rings. Most trees growing from a stand or geographic region at treeline will have a similar pattern of large and small rings. Tree-ring records are usually obtained by taking small cores from living trees. Despite their small stature, many living trees found at the Arctic treeline can be 300 to 500 years old. Specimens of wood from trees that died sometime in the past can often be dated by comparing the pattern of ring-width variations found in the dead wood sample with that observed in older living trees or specimens of dead wood that have already been cross-dated. The cross-dating of tree rings is called dendrochronology. Cross-dating of dead wood has allowed dendrochronologists to construct tree-ring records from North America that extend back more than 1000 years and from Eurasia that extend back as much as 8000 years. The dating of living and dead trees allows paleoecologists to reconstruct patterns of tree establishment and mortality back hundreds of years. Dendroclimatologists use the variations in ring widths to reconstruct temperature variations for the past 1000 years in some sites.

Tree-ring studies from many parts of the Arctic treeline show a similar pattern of tree establishment and mortality (**Fig. 2**). There appears to have been a depression in establishment success and an increase in tree mortality during the Little Ice Age. In contrast, the late nineteenth century and twentieth century experienced a general increase in establishment success. This pattern is clear in northern Canada and Siberia. Dendroclimatological studies show that summer temperatures during the Little Ice Age were depressed by 1–2°C (1.8–3.6°F). A number of treeline sites appear to have experienced an unusually prolonged warm-temperature period during the early to mid-twentieth century, a decline in the 1960s and 1970s, and renewed warming in the past 20 years. The cause of the Artic temperature variations and changes in treeline establishment and mortality patterns are likely related to natural variations in solar irradiance, the influence of volcanic eruptions, and in the twentieth century warming caused by increases in atmospheric greenhouse gases due to the burning of fossil fuels and other human activity.

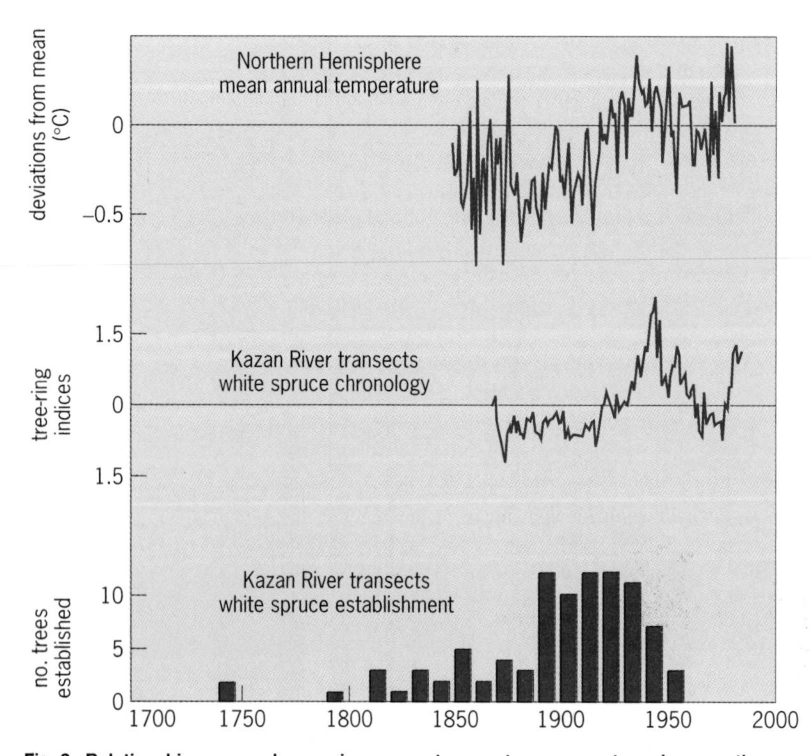

Fig. 2. Relationship among changes in summer temperature, spruce tree-ring growth, and spruce-tree establishment from a site at the treeline in central Canada.

More distant past. Most of what we know regarding long-term changes in the treeline since the close of the last glacial period of the Pleistocene Epoch and the start of the Holocene Epoch (approximately 11,500 years ago) comes from studies of preserved wood and pollen. When glacial ice retreated from northern Canada, it left in the landscape many depressions which became filled with lakes and peat bogs. These depositional environments are ideal for preserving pollen grains. Year after year, the surrounding vegetation produces pollen, some of which is incorporated and preserved in lake or bog sediments. Cores can be taken of the sediments, the pollen grains extracted and identified, and changes in the vegetation can be reconstructed. In addition to pollen, the lignified stomatal cells from coniferous tree needles are also preserved and provide a very specific record of the presence or absence of trees around a particular lake or bog. Ages are assigned to the cores based on radiocarbon dates from preserved organic materials.

Pollen analysis from sites along the treeline zone of northern Canada provides evidence that the boreal forest once extended farther north than it now does in most regions. In western Canada near the mouth of the Mackenzie River, the forest grew several tens of kilometers north of its modern position between about 11,000 years ago and retreated to its present position after about 5600 years ago. In central Canada, there was a northward advance of the forest between about 5600 and 4500 years ago. In northern Quebec, there was an advance of the forest between about 4500 and 2500 years ago. In both central Canada and Quebec, the actual northward extension of tree species range limits appears to have been very slight at best, although the treeline forest density was greater.

In northern Eurasia, fossil pollen and stomata records are augmented by a particularly rich record of well-preserved conifer tree stumps found on the tundra surface, in peat bogs, and in lakes. Hundreds of such samples have been collected from northern Fennoscandia to Siberia and have been radiocarbon-dated. These specimens provide evidence that, between about 10,000 to 3000 years ago, forest cover extended all the way to the present Arctic coastline in many regions (**Fig. 3**). In contrast to North America where the actual range extension of tree species during past treeline advances was very small, in some cases Eurasian conifer species grew tens to hundreds

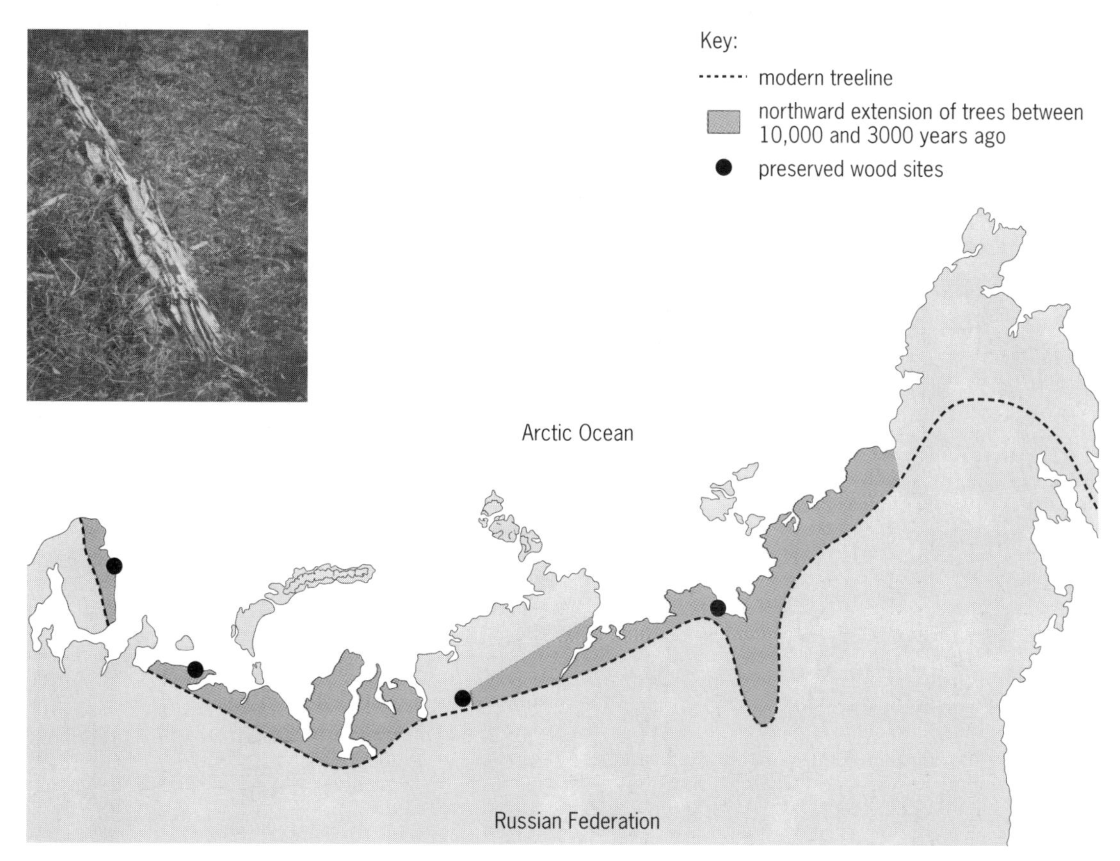

Fig. 3. Past northward extension of Arctic treeline in the Russian Federation between 10,000 and 3000 years ago, as determined from preserved trees found on the tundra. The inset shows a preserved larch stump from central Siberia.

of kilometers north of their modern range limits. It has been estimated that this northward extension of the boreal forest across Eurasia represents a warming of summer temperatures of 2.5–7.0°C (4.5–12.6°F), compared to modern conditions.

In both North America and Eurasia, the early to mid-Holocene advances of treeline likely reflect higher amounts of summer solar radiation that the Arctic received at that time due to natural variations in the orbital geometry of the Earth. Why the treeline advance in North America was more limited in scope and less synchronous than in Eurasia is still under investigation.

For background information *see* BOG; CLIMATE HISTORY; CLIMATE MODIFICATION; DENDROCHRONOLOGY; FOREST AND FORESTRY; FOREST ECOSYSTEM; HOLOCENE; PALEOECOLOGY; PALYNOLOGY; PLANT GEOGRAPHY; PLEISTOCENE; POSTGLACIAL VEGETATION AND CLIMATE; RADIOCARBON DATING; TAIGA; TREE; TUNDRA in the McGraw-Hill Encyclopedia of Science & Technology. Glen M. MacDonald

Bibliography. J. Esper and F. H. Schweingruber, Large-scale treeline changes recorded in Siberia, *Geophys. Res. Lett.*, 31:L06202, 2004; G. M. MacDonald et al., Holocene treeline history and climate change across northern Eurasia, *Quat. Res.*, 53:302–311, 2000; G. M. MacDonald et al., Response of the central Canadian treeline to recent climatic changes, *Ann. Ass. Amer. Geog.*, 88:183–208, 1998; J. Overpeck et al., Arctic environmental change of the last four centuries, *Science*, 278:1251–1256, 1997.

Attosecond laser pulses

The motion of electrons inside atoms and molecules occurs with awesome rapidity. Quantum mechanics predicts that the electron (in the language of quantum mechanics, the electron wavepacket) of a hydrogen atom takes about 400 attoseconds (1 as = 10^{-18} s) to perform an oscillation around the nucleus when it is most closely bound to it. The electron wavepacket on a molecular orbit binding two hydrogen atoms together to form a molecule takes approximately the same time to circle around the hydrogen nuclei. No one has ever been able to observe these hyperfast electron motions in real time. Attosecond pulses open up this prospect.

Efforts to access ever shorter time scales are motivated by the endeavor to explore the microcosm at ever smaller dimensions. Recently, femtosecond (1 fs = 10^{-15} s) laser techniques have allowed control and tracing of molecular dynamics and the motion of atoms on the length scale of internuclear separation. Laser light consisting of a few well-controlled field oscillations extends these capabilities to the interior of atoms, allowing control and measurement of electronic motion on an atomic time scale.

The measurement of ever shorter intervals of time and the tracing of dynamics within these intervals relies on the reproducible generation of ever briefer events and on probing techniques of corresponding resolution (**Fig. 1**). The briefest events produced until recently have been pulses of visible laser light,

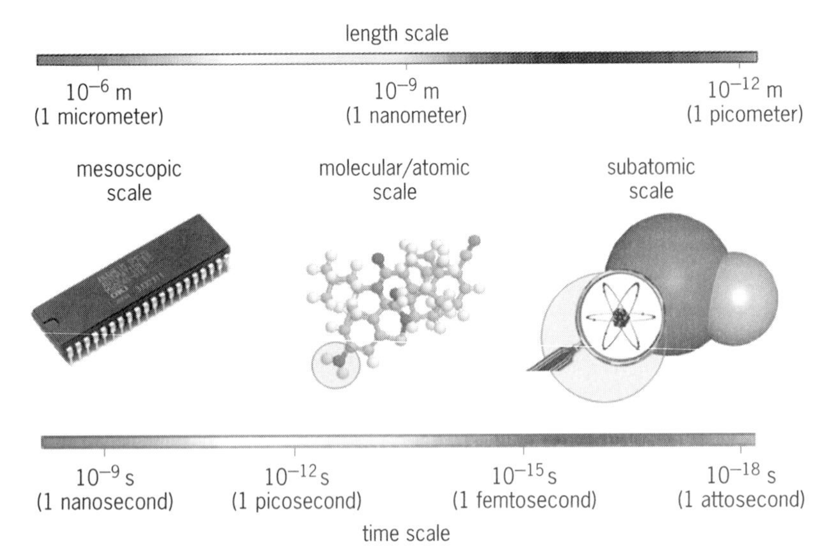

Fig. 1. Processes and intervals on different time scales.

with durations of several femtoseconds. Traditionally, the fastest measurement techniques have used the brief flashing of these pulses for sampling, with the oscillation of the electromagnetic field being irrelevant. With the advent of intense laser pulses consisting of a few, precisely controlled oscillations of the electric (and hence also magnetic) field, it becomes possible to control and trace microscopic processes that take place within a single wave cycle of visible light (that is, within less than 1000 attoseconds). This capability constitutes the key both to the generation of isolated attosecond pulses and to attosecond real-time observation of electronic motion deep in the interior of atoms and molecules.

Pump-probe technique. The motion of speedy objects can be reconstructed from a series of snapshots in slow-motion replay. Freeze-frame shots call for a camera exposure time sufficiently short to "freeze" the object's motion. In the case of a microscopic process, a short burst of infrared, visible, ultraviolet, or x-ray light (pump pulse) can be used to initiate a particular process while a second flash of light (probe pulse) is used to take snapshots of the motion at well-defined instants in time. The approach, known as the pump-probe technique, can be regarded as an extension of high-speed photography to the microcosm. There is an obvious need for synchronism between pump and probe.

If a laser pulse is used to generate—in a highly nonlinear process—an ultrashort burst of extreme ultraviolet (xuv) light, the demand for synchronism is fulfilled. Atoms exposed to a few cycles of intense visible or near-infrared laser light are able to emit a single xuv burst lasting several hundred attoseconds. Precise temporal control of this attosecond pulse can be achieved by full control of the oscillating laser field driving the xuv emission process. The laser light field can also be used to record the history of electron emission from atoms or molecules following an implusive excitation by the well-synchronized attosecond xuv pulse. Electrons are emitted by two mechanisms. Primary electrons (photoelectrons) are knocked out directly by the incident xuv burst and hence mimic the exciting xuv pulse. Secondary (Auger) electrons are ejected when the atom relaxes into a lower energy state; hence their emission history displays the subsequent relaxation processes. With these tools, attosecond photography of electronic motion deep inside atoms and molecules is now becoming feasible.

Attosecond pulse generation. If the electric field of an intense femtosecond laser pulse is polarized along a well-defined direction, it can induce a gigantic atomic dipole by pulling an electron out of an atom. Half a wave cycle later the direction of the electric field is reversed, smashing the electron back to the atomic core. The resultant giant dipole oscillation results in ultrabrief xuv emission upon the electron returning to the vicinity of the core. In a laser pulse comprising many wave cycles, this emission is repeated quasiperiodically, resulting in a series of bursts in the time domain. In the spectral domain (that is, the energy or frequency domain, time and frequency being inverse quantities as they are related by the Fourier transformation), the xuv emission is made up of narrow lines at high-order harmonics of the driving laser frequency. The xuv burst was theoretically predicted to be several hundred attoseconds in duration, and this prediction has recently been confirmed experimentally by several groups. When the laser pulse is confined to a couple of precisely controlled wave cycles, the periodicity vanishes and only a few giant atomic dipole oscillations are observed. Careful choice of the waveform and the xuv emission band results in a single isolated xuv pulse. The parameters of this pulse (its duration, energy, and timing with respect to the oscillating laser field) are well reproducible from one pulse to the next, as has been demonstrated in a series of experiments. This reproducibility is a prerequisite of key importance for the feasibility of real-time observations of electronic processes with these tools on an attosecond time scale.

The perfect synchronism of the xuv burst with the field oscillations of the generating few-cycles laser wave offers the potential for using it in combination with the laser wave for attosecond spectroscopy. This is essential because these laser-produced xuv bursts are too weak to be applicable for both triggering and probing electronic dynamics. Instead, the oscillating laser field, which changes its strength from zero to maximum within 600 attoseconds in the case of a pulse of 750-nm (red) laser light, can take over the role of the probing xuv pulse and record the temporal evolution of the electronic processes previously initiated by the sudden xuv excitation.

Observation of electron motion within atoms. Atoms can be suddenly excited by a very short pulse of energetic light (such as xuv light), resulting in the emission of primary and secondary electrons. The evolution of the emission processes mirrors exactly how the atom is excited and relaxes back into its lowest energy state. Thus, temporal images of these emission processes should provide complete information about how electrons are excited and relax

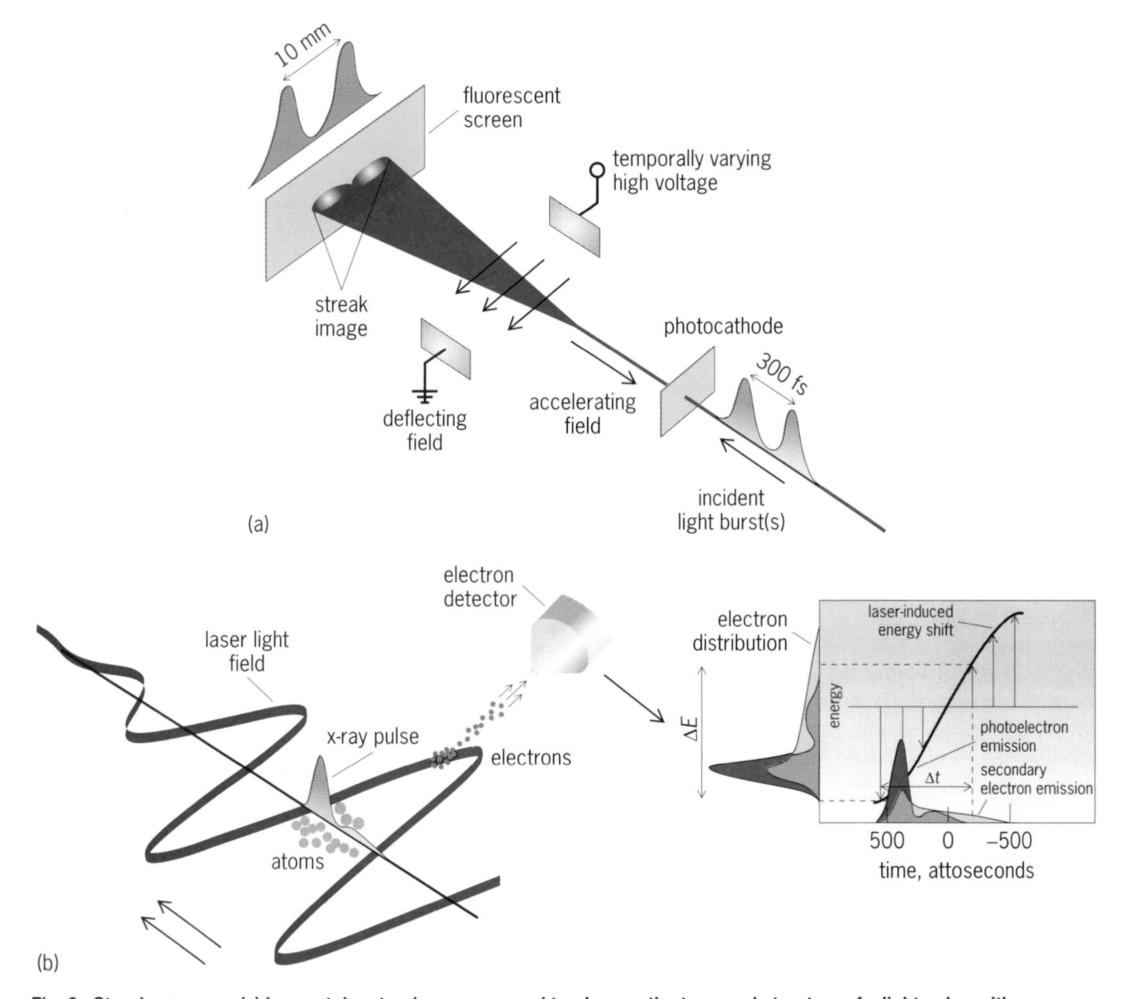

Fig. 2. Streak cameras. (*a*) Image-tube streak camera, used to observe the temporal structure of a light pulse with sub-picosecond resolution. (*b*) Light-field-controlled streak camera, used to observe electron emission from atoms with attosecond resolution.

in the interior of the atom. These hyperfast electron emissions can be recorded by a process that closely resembles the recording of a light flash in a streak camera.

The image-tube streak camera can be used to measure the duration and time structure of a light pulse. It is based on the concept of converting the temporal distribution of a bunch of rapidly moving particles into a spatial distribution by affecting their motion in a time-dependent manner. Specifically, a beam of electrons generated at a photocathode by an incident light burst (**Fig. 2*a***) is deflected by a rapidly varying voltage before being recorded at a fluorescent screen. The sudden deflection of the photoelectrons produces a streaked spot on the screen, allowing measurement of the flash duration. From the streaked image of the bunch of photoelectrons, the temporal structure of the light pulse triggering the photoemission can be inferred with subpicosecond resolution.

By drawing on the same basic concept, the temporal characterization of attosecond electron emission from atoms can be reconstructed. As previously discussed, electrons are emitted from atoms exposed to an attosecond xuv or x-ray pulse in the presence of an intense laser wave. The electric field of the laser wave is polarized along a well-defined direction (pointing toward the electron detector in Fig. 2*b*) and performs a few well-controlled oscillations. The initial speed of the freed electrons ejected toward the detector (and consequently parallel to the direction of the laser electric field) is boosted or decreased depending on the instant of release within the wave cycle. The laser-induced energy shift (shown by vertical arrows in the graph in Fig. 2*b*) varies monotonically with the instant of release of the electrons and results in a unique mapping of the temporal emission profile (curves drawn versus time, which is the horizontal coordinate in the graph) into a similar energy distribution of electrons (curves drawn versus energy, which is the vertical coordinate). The rapidly varying electric field of visible light streaks the energy of electrons emitted at slightly different instants within the half-wave cycle of the streaking laser light. The electron energy distribution can thus be regarded as a streaked image whose breadth and shape mirror the duration and temporal shape of the emission.

Probing primary or secondary electrons with this light-field-driven streak camera allows real-time observation of atomic excitation and subsequent

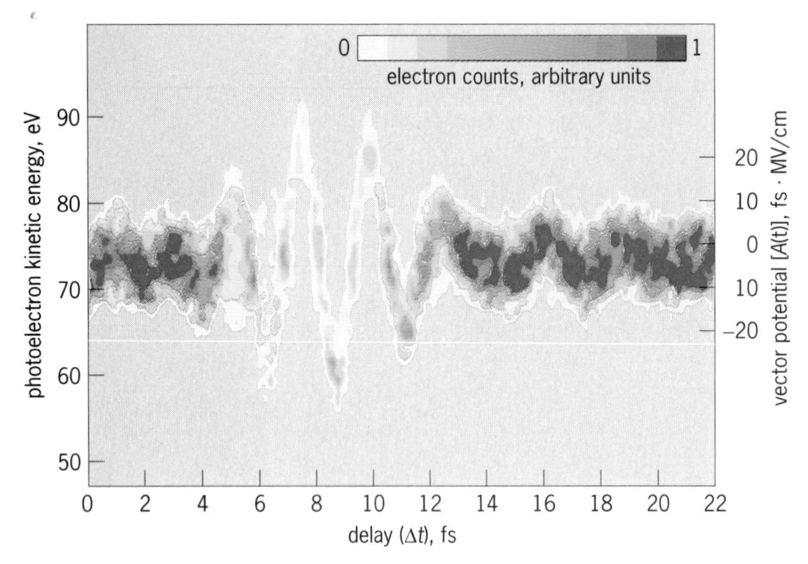

Fig. 3. Ultrafast event, the electric field of a light pulse, sampled with attosecond resolution.

relaxation processes, respectively, with attosecond temporal resolution. The duration of photoelectron emission corresponds to that of the exciting xuv pulse. As a result, the streak image of photoelectrons yields the duration of this pulse. The duration of 13-nm xuv pulses produced by the previously described technique has been determined recently by a light-field-driven streak camera measurement in this manner to be about 250 attoseconds. These pulses constitute the shortest artificial events ever produced. Applying the same apparatus to recording streak images of secondary electrons reveals how the electronic system of the atom undergoes a rearrangement that leads to relaxation into the lowest energy state of the atom. A first proof-of-principle experiment based on this concept permitted the real-time observation of how an electron from an external atomic shell underwent a transition to an inner shell within less than 8 femtoseconds.

Imaging light field oscillations. The applications of attosecond xuv pulses extend beyond triggering electronic motion in atoms and molecules. By launching an attosecond photoelectron burst, such a pulse can also be used to create a probe for tracking variations of the electric field of an electromagnetic signal or wave with a never-before-achieved attosecond resolution. The electric field of visible light has a half cycle of the order of 1 femtosecond (1000 attoseconds), and hence offers an excellent opportunity to test the capability of such an attosecond sampling apparatus. Owing to its intrinsic synchronism with the attosecond probe, the few-cycle light wave used for the generation of the attosecond xuv pulse lent itself to a proof-of-concept experiment that demonstrated this capability (**Fig. 3**).

For background information *see* ATOMIC STRUCTURE AND SPECTRA; AUGER ELECTRON SPECTROSCOPY; ELECTROMAGNETIC RADIATION; FOURIER SERIES AND TRANSFORMS; LASER; LASER PHOTOCHEMISTRY; MOLECULAR STRUCTURE AND SPECTRA; OPTICAL PULSES; PHOTOIONIZATION; STROBOSCOPIC PHOTOGRAPHY; ULTRAFAST MOLECULAR PROCESSES in the McGraw-Hill Encyclopedia of Science & Technology.

Reinhard Kienberger; Ferenc Krausz

Bibliography. P. Agostini and L. DiMauro, The physics of attosecond light pulses, *Rep. Prog. Phys.*, 67:813–855, 2004; T. Brabec and F. Krausz, Intense few-cycle laser fields: Frontiers of nonlinear optics, *Rev. Mod. Phys.*, 72:545–591, 2000; R. Kienberger and F. Krausz, Attosecond metrology comes of age, *Physica Scripta*, 110:32–38, 2004; Taking snapshots of a light wave, *New Scientist*, April 9, 2005.

Automotive restraint systems

In the past 20 years, newly produced automobiles have been equipped with occupant restraint systems, such as air bags, to reduce passenger injuries and fatalities in the event of a crash. An air bag is a passive restraint system—an automatic safety system that requires no action by the occupant. Government regulations, industrial participation, and social consciousness of safety have popularized the use of occupant restraints in vehicles worldwide. Using advanced technologies, additional restraints and enhanced functionalities are being implemented.

History. Automobiles entered the market more than 100 years ago. Since then, tremendous progress in reliability, usability, and safety of roadway vehicles has been made. However, the availability of passive safety systems was slow in coming, despite the initial conception in 1950s and experimental production in 1970s. It was not until the late 1980s that air bags became standard equipment, beginning in the United States. Over the last 20 years, the automotive safety industry has gone through considerable changes.

In recent years, passive restraint systems have been one of the fastest-growing sectors within the automotive industry. One reason for this growth is the inclusion of additional restraint devices on vehicles. For example, the average number of air-bag modules has increased from one to two per vehicle in the early 1990s to four to six per vehicle for the latest models, with more expected to come. Another reason for this growth is an increase in the complexity and sophistication of the safety systems' functional requirements. For example, to minimize out-of-position occupant injuries, a smart air-bag system needs to be equipped with occupant-sensing capabilities.

Particularly notable in the evolution of restraint systems has been the phenomenon of increasing use of electronics, which has occurred in parallel in the automotive world, as well as most technology fields. Increasingly, sensing, actuation, and control functions of air-bag systems are being implemented with electronic filters, conditioners, memory integrated circuits, microprocessors, and microcontrollers. Upcoming products implemented in electronics, such as rollover and other sensors, are being added into an integrated vehicular safety system.

Fundamental concept. In a collision, the vehicle deforms and crumbles rapidly after impact with an object, such as roadside obstacle or other vehicle. Typically, the structural deformation process takes only 0.1–0.2 second, compared to braking maneuvers, which take a few seconds. The deceleration experienced by the vehicle in a collision is at least one order of magnitude higher than normal driving and braking actions. As a result, the occupants inside a vehicle undergo a thrust in the direction of the impact. If there is no protection device in a frontal collision, the occupants will hit the interior (for example, the steering wheel or the dashboard), potentially resulting in injuries or fatalities depending on the severity of the impact. If seat belts are in place, they will exert a load on the occupants in a direction opposite to their movements. In a severe collision, the restraining force of seat belts alone will not sufficiently restrain the occupant, as the restraining forces are limited and concentrated along the narrow strips of belt fabrics. Unlike belts, air bags cushion the occupants in a manner that differ in two primary ways. First, the contact area between a bag and an occupant is larger, thus distributing the restraining load more evenly, and the total stopping force can be greater. Second, the bags have openings that allow the air to exit, thus dissipating the kinetic energy of the occupants more effectively.

The effectiveness of air bags depends on a timely deployment procedure, with all the components functioning properly. The major components in a frontal air-bag system are shown in **Fig. 1**. The sensors represent the brain and nerves of the system, which detect and decide whether and when to initiate the inflator. The inflator either generates or releases gas to deploy the bag. The bag is fully inflated before the occupant makes contact with the bag without inflicting unnecessary forces on the occupant. The vehicle interior sustains the collision with minimum penetration of structural deformation into the passenger compartment and offers supplementary support to the operation of air bags. Although the above description is for frontal systems, the fundamental concept applies to side air bags as well.

Figure 1 shows a distributed sensing system, where multiple sensors are combined to detect a collision. An electronic control unit (ECU) often monitors the triggering of multiple sensors and controls the activation of the air bags. Over the last 10 years, the increasing capabilities of electronic sensors have enabled many vehicle platforms to adopt a single-point sensing concept, with one sensing module centrally located on a vehicle. However, the success of the single-point strategy requires that a variety of crash collisions yield sufficient and distinguishable signals to be sensed at a central location. Given the diverse characteristics of vehicle structures, not all vehicles are suited for the single-point placement strategy despite its enormous success across many vehicle types in the marketplace today. One countermeasure to overcome the deficiency of the single-point sensing arrangement is to add satellite sensors in select locations, such as the frontal corners of the vehicle for frontal air bags or side doors for side air bags.

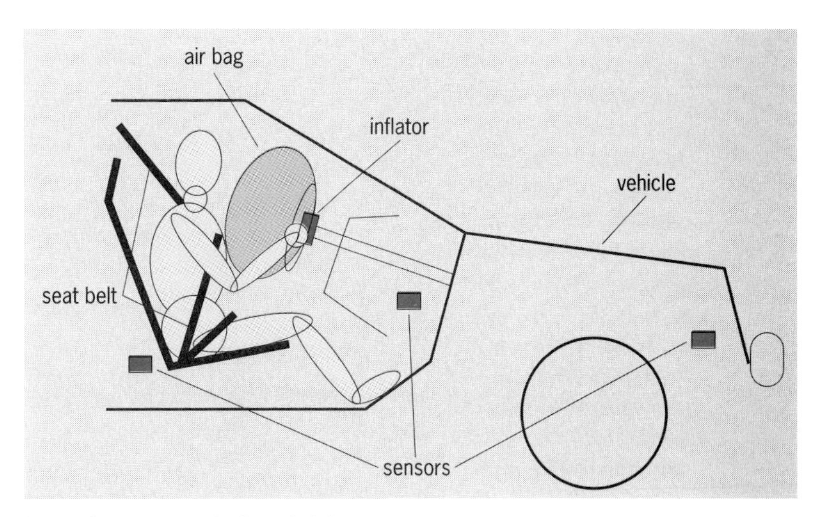

Fig. 1. Components of a frontal air-bag system.

Some concerns of air-bag safety have arisen in situations where passenger injuries or child fatalities have been caused by air-bag deployment. The countermeasures to avoid such problems include the depowering of air-bag modules or the disabling of air bags when it is warranted by vehicle interior conditions. The former can be achieved by a reduction of air-bag inflation force, if seat belts are worn by the occupants. The latter is realized by occupant sensing, which provides occupant status information to the ECU for final decisions. Smart restraints of current-generation vehicles are designed and equipped with these options.

Technology trends. Over the years, a transition in technology has occurred at several levels among the primary components of occupant restraint systems, including sensors, air-bag modules, and seat belts. In addition, product design and implementation methods have changed dramatically. The functional requirements of passive restraint systems have been elevated from a simple on-or-off deployment decision to a staged- or graded-control command, which provides a tailored choice for optimal protection in a smart system.

In the area of sensing technology, electronic sensors have become dominant and replaced first-generation mechanical and electromechanical devices. Electronic control units (ECU) have been equipped with sophisticated capabilities to allow reliable diagnosis and robust control. The ECU is coupled with a network of satellite and auxiliary sensors to acquire all the necessary information to make an appropriate crash-status decision. In addition, data from occupant sensors and seat-belt-buckle switches are fed into the ECU to provide occupant-seating conditions as a critical part of smart restraints. Through a sensor network distributed across the vehicle platform, intelligent decisions can be made for selective deployment of the proper restraints. Air-bag modules are moving quickly toward the use of dual- or multiple-stage inflators. These partitioned and staged inflators provide the benefits of minimizing loads on

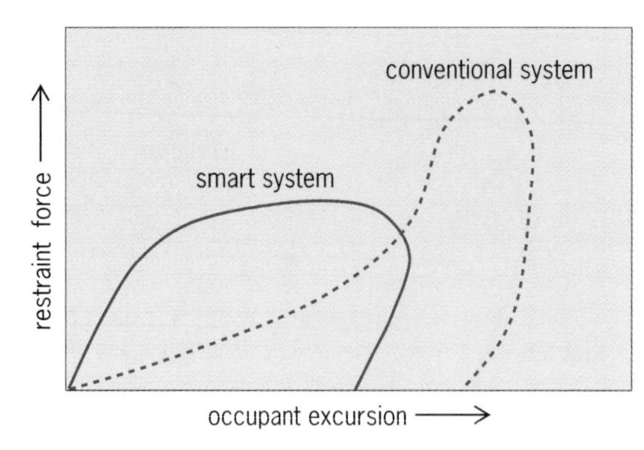

Fig. 2. Reduction of occupant excursion and restraining force by smart restraints.

the occupants and offering the flexibility of meeting the protection requirements in various crash conditions. Seat belts have been engineered with versatile options such as pretensioning and load-limiting features.

Together, sensors, actuators, and protective devices constitute the foundation of smart air bags. An optimal protection sequence begins with the pretensioning of the seat belts to engage the occupant, followed by a seat-belt load-limiting feature to minimize the restraining load, and then an effective dissipation of occupant movements with an air-bag cushion inflated with an adjusted and controlled level of deployment power. All the restraining devices in the vehicle work in a smart and coordinated manner to manage the kinetic energy carried by occupants in a collision (**Fig. 2**).

Outlook. The air-bag industry has experienced tremendous growth in the last two decades, and is already maturing in several subsectors despite its relatively short history, compared with other sectors of the automotive industry. The concept of vehicular safety, however, is being rejuvenated with a variety of new developments. For example, global positioning systems are being integrated with wireless communication and crash-sensing systems into emergency notification and response services.

Parallel to the rapid progress in passive restraint systems, significant strides have been made in active safety systems. Passive safety systems are "reactive" with the purpose of mitigating accident consequences, while active safety systems are "preventive" and target the potential to avoid accidents. In active safety systems, the actions occur prior to the accident. One example is the mild level of braking exerted automatically in some advanced cruise control system when the vehicle-following distance falls below a threshold. Another example is the use of a steering-assist function to prevent road-departure accidents.

The introduction of active safety systems will provide the foundation for integrating selective components and capabilities into automotive safety systems. The focus of safety improvements will shift from collision mitigation to collision avoidance, although the two functions are not separable. The passive restraints, originating from the concept of a simple air-bag deployment, will evolve into a part of the smart-vehicle interior that accommodates all customer-appealing features inside a protective vehicle environment.

For background information *see* AUTOMOBILE; AUTOMOTIVE BRAKE; AUTOMOTIVE STEERING; MICROSENSOR in the McGraw-Hill Encyclopedia of Science & Technology.

Ching-Yao Chan

Bibliography. C.-Y. Chan, *Fundamental of Crash Sensing for Automotive Air Bag Systems*, Society of Automotive and Aerospace Engineers (SAE), 2000; C.-Y. Chan, On the detection of vehicular crashes: System characteristics and architecture, *IEEE Trans. Vehicular Technol.*, 51(1):180–193, January 2002; C.-Y. Chan, A treatise on crash sensing for automotive air bag systems, *IEEE/ASME Trans. Mechatronics*, 7(2):220–234, June 2002; SAE Highway Vehicles Safety Database, Society of Automotive Engineers, Warrendale, PA, April 2005; 2005 SAE Occupant Protection and Crashworthiness Technology Collection on CD-ROM, Society of Automotive Engineers, Warrendale, PA, April 2005.

Belowground herbivory

Although aboveground herbivores are known to affect plant community structure and function, the role of belowground herbivores in plant ecology is not well understood. Conducting research on belowground herbivores is difficult, primarily because of the dense, opaque nature of the soil medium that impedes direct observation belowground. Root herbivory is likely to have an important influence on plant communities because 50–90% of new plant growth is allocated belowground, and the loss of belowground biomass has greater impacts on plants than an equivalent loss of leaves or stems. Belowground herbivores affect plants through both their direct consumption and their indirect effects on decomposer and plant communities.

Direct consumption. Direct consumption by belowground herbivores often causes severe damage to individual plants because roots serve the critical functions of plant anchorage, resource acquisition/storage, and synthesis of many essential hormones and defense compounds. Belowground herbivores may damage roots internally or externally by grazing, tunneling, or piercing roots with specialized mouthparts.

Nematoda. Herbivorous nematodes (roundworms), which either pierce surface cells or tunnel in roots, are particularly important because of their ubiquity and abundance—in some cases reaching densities of $0.5–6 \times 10^6/m^2$ ($10.8 ft^2$). In 1987, infestations of nematodes caused an estimated $77 billion in global agricultural losses.

Insecta. Other than nematodes, the majority of research on belowground herbivores focuses on insect species important to agriculture or biocontrol of invasive plants. The majority of economically important species are members of the order Coleoptera (beetles), such as June beetles, weevils, and wireworms. The Lepidoptera (moths and butterflies) are also important, as are the Hemiptera.

Hemipterans are characterized by highly modified piercing stylets used to access plant vascular systems. Cicadas (family Cicadidae) are large hemipterans (suborder Homoptera) with protracted subterranean nymph (juvenile) stages which last 2–17 years, depending on the genus. The little information known about the ecology of cicada nymphs suggests that they have a major impact on the deciduous forests they inhabit. Nymphs are likely to parasitize the same root system for the entire juvenile period, and infested scrub oak trees (*Quercus ilicifolia*) suffer up to 30% reductions in growth. In 1956, the synchronous emergence of periodical cicadas led to estimated belowground nymph densities of approximately 1.5 million/acre (0.4 ha). By comparison, the largest swarm of desert locusts (*Schistocerca gregaria*) ever reported reached densities of approximately 202,000/acre. *See* CICADA.

Rodentia. The vast majority of belowground vertebrate herbivores are restricted to the order Rodentia, and rodents of several families have separately evolved similar ecological functions on all continents except Antarctica and Australia. North American pocket gophers (family Geomyidae) highlight the substantial impacts of these root herbivores. For example, plant biomass has been found to be reduced by more than one-third over active burrows of the plains pocket gopher (*Geomys bursarius*). Simulations of this root herbivory on yellow salsify, or goat's-beard (*Tragopogon dubius*) showed that removal of just 25% of root mass had greater negative impacts on total biomass and flower production than the removal of 75% of total aboveground plant material.

Indirect effects. Several studies indicate that the indirect effects of herbivores may be as important as their direct consumptive effects. Indirect effects include influences of one species on another that are mediated by changes in the abundance or activity of a third species or by changes in the physical environment. Indirect effects are especially common in belowground systems in which the producer, consumer, and decomposer communities are intimately associated with one another over small spatial scales. Ultimately, rates of organic matter decomposition and the availability of plant nutrients depend on the soil microbial community—the bacterial, fungal, and protozoan decomposers. Members of this community also include pathogenic microbes with negative effects on plants and mutualistic microbes (rhizobacteria, mycorrhizal fungi), which can increase nitrogen and phosphorus availability.

Biocontrol of Centaurea. Biocontrol efforts against invasive species in the plant genus *Centaurea* (family Asteraceae) highlight both the indirect effects of belowground herbivory on plant communities and the complexity of these interactions, which sometimes leads such programs astray. Experiments with the larvae of one moth, *Agapeta zoegana*, used for biocontrol showed that root herbivory actually increased the vigor of *C. maculosa* and reduced growth and fecundity of neighboring *Festuca idahoensis*, a North American native grass. These undesirable results were attributed to three potential indirect effects: (1) herbivores may have induced compensatory growth responses in the roots of *C. maculosa* (as the plant restores root:shoot ratios after root loss), and this may have improved their ability to compete for belowground resources; (2) herbivore-induced production of herbivore deterrents may have negatively affected neighboring plants through allelopathy (detrimental interactions among plants as well as microorganisms); (3) combined root and shoot herbivory on *C. maculosa* may have induced carbohydrate release from its roots, resulting in increased beneficial mycorrhizal infection. This may have promoted the growth of *C. maculosa* directly or indirectly through the transfer of resources from neighboring *F. idahoensis* to *C. maculosa* via common networks of mycorrhizae—mycorrhizae-mediated parasitism. These negative indirect effects of a widely applied biocontrol agent emphasize the need to increase our understanding of the complexity of belowground trophic interactions.

Pocket gopher tunnel construction and mounding. Tunnels and mounds are by-products (indirect effects) of pocket gopher plant consumption, yet tunnel excavation and mounding have broader impacts on plant and soil communities (see **illustration**). These activities promote soil formation, increase water infiltration, and affect the spatial distribution of

(a)

(b)

Direct and indirect effects of pocket gopher activity. (*a*) Direct effects of root consumption by pocket gophers (*Thomomys bottae*) on a California native plant, telegraph weed (*Heterotheca grandiflora*). The tunnel mouth at the base of the wilting plant is an example of the surface access tunnels that gophers use to expel mound soil. This lateral tunnel would lead at an angle of 33–41° to a feeding tunnel 6–20 cm belowground. The damaged plant may be able to recover, but will likely suffer reduction in growth and fecundity. The apparently undamaged, nonnative ragweed (*Ambrosia artemisiifolia*) directly left of the tunnel opening will likely benefit from the damaged *H. grandiflora*. Scale bar ~5 cm. (*b*) Mound production by pocket gophers (*T. bottae*), a by-product of foraging activity with numerous indirect effects on plant communities. This photo was taken at the height of summer drought (July 2005). During the winter growing season, these mounds would present gaps for germination in otherwise dense vegetation 0.5–1 m in height, and the thatch of dead plant material smothered below these mounds will decompose, increasing soil organic matter and releasing nutrients. Scale bar ~10 cm. (*Photos by S. M. Watts*)

soil nutrients. Densities of pocket gophers can be quite high in some areas [75/ha (30.4/acre) in alfalfa fields], and soil mounds excavated from their tunnels can cover up to 30% of the soil surface annually. Importantly, the mounds and tunnel systems produced by gophers are not evenly distributed across the landscape, but result from a combination of social and physical constraints. Pocket gophers are solitary and highly territorial, maintaining approximately a 2.5–5.0-m (8.2–16.4-ft) buffer zone between neighboring burrow systems. To maintain structural integrity within burrow systems, tunnels cannot be too shallow or too close together. However, gophers must encounter enough plant material to offset the energy expenditure associated with excavation (which requires 360–3400 times more energy than surface travel). Consequently, in low-productivity habitats gophers increase overall tunnel system length, and in patches of preferred plants they increase the number of tunnel branches, resulting in locally increased rates of mound production. The resulting landscape-scale pattern is one in which tunnels of belowground herbivory tend to be uniformly distributed and mound disturbances tend to be clumped.

A general effect is that plant biomass directly above tunnels or mounds is reduced, releasing adjacent plants from competition. The increased growth of these plants then reduces the growth of neighboring plants, which in turn release their neighbors from competition. This pattern of interactions results in successive peaks and troughs of enhanced and decreased biomass that may extend more than 0.5 m (1.6 ft) from the mound or tunnel disturbance. More specifically, the clumped distributions of mounds in areas of higher plant productivity or preferred forage have several indirect effects. In areas with high availability of resources (such as sunlight and nutrients), gopher mounds promote plant diversity, because dense growth would otherwise lead to the competitive elimination of other species. They also tend to favor colonizing plant species, which are characterized by small but numerous seeds and rapid growth. Gophers prefer to eat these fast-growing plants, which are actually facilitated by increased recruitment on mounds. By the same mechanisms, mound production may also promote invasion by fast-growing exotic plants.

Pocket gophers also have broad ecosystem effects through their impacts on soil microbes and soil processes. The discrete chambers for feces maintained by gophers, plant material smothered by mounds, and belowground food caches directly incorporate organic matter into the soil and subsidize the decomposer community, potentially increasing soil fertility. Tunnel construction and mounding has the net effect of reducing vertical heterogeneity of soil profiles (mixing) and increasing horizontal patchiness of surface soils, as nutrient-poor soil is brought up from deeper soil layers.

The effects of mound soil, however, depend on the status of surface soils. An example of the contrasting effects of gopher mounds on plant populations can be found in studies of mycorrhizal fungi, which infect host plant roots, improving phosphorus acquisition. After the 1980 eruption of Mount St. Helens in southern Washington State caused the spread of ash across the region, gophers delivered soil with a higher concentration of beneficial fungi to the surface through mound production. The colonization of plant species with obligatory mycorrhizal mutualisms was promoted by these mounds. However, in a separate study of soils with established mycorrhizal networks, gophers distributed subsurface soil containing a lower concentration of beneficial fungi, and plants growing in them had reduced root length and mycorrhizal infection.

Future research. Now that belowground herbivory is receiving more attention, it is becoming clear that its effects on plants are similar to or greater than those of aboveground herbivory. Moreover, the web of indirect effects mediated by plants and soil organisms often has unpredictable results, as illustrated by the effects of the use of biocontrol agents for *Centaurea*. Incorporating indirect effects into experiments will significantly increase their complexity and difficulty, but several investigators are making headway on this issue, especially as it concerns interactions between belowground herbivores and pathogenic or mutualistic microbes. In addition, more studies are addressing plant-mediated links between above- and belowground herbivores, which often result from damage-induced changes in the nutrient or chemical content of the plant. The incorporation of such aboveground linkages in studies of the direct and indirect effects of root herbivory mark a significant step toward a whole-plant understanding of plant ecology and plant–herbivore interactions.

For background information *see* GOPHER; HERBIVORY; MYCORRHIZAE; PHYSIOLOGICAL ECOLOGY (PLANT); RHIZOSPHERE; SOIL MICROBIOLOGY in the McGraw-Hill Encyclopedia of Science & Technology.

Sean M. Watts

Bibliography. D. C. Andersen, Belowground herbivory in natural communities: A review emphasizing fossorial animals, *Quart. Rev. Biol.*, 62: 261–286, 1987; B. Blossey and T. R. Hunt-Joshi, Belowground herbivory by insects: Influence on plants and aboveground herbivores, *Annu. Rev. Ecol. Systemat.*, 48:521–547, 2003; V. K. Brown and A. C. Gange, Insect herbivory below ground, pp. 1–58 in *Advances in Ecological Research*, ed. by M. Begon, A. H. Fitter, and A. MacFadyen, Academic Press, London, 1990; P. Lavelle, Faunal activities and soil processes: Adaptive strategies that determine ecosystem function, pp. 93–132 in *Advances in Ecological Research*, vol 27, ed. by M. Begon and A. H. Fitter, Harcourt Brace, San Diego, 1997; O. J. Reichman and E. W. Seabloom, The role of pocket gophers as subterranean ecosystem engineers, *Trends Ecol. Evol.*, 17:44–49, 2002; N. L. Stanton, The underground in grasslands, *Annu. Rev. Ecol. Systemat.*, 19:573–589, 1988; D. A. Wardle, *Communities and Ecosystems: Linking the Aboveground and Belowground Components*, Princeton University Press, 2002.

Biologically inspired robotics

Natural systems offer a wealth of solutions for robot design and autonomy. However, copying an animal mechanism or control circuit is rarely possible for two reasons. First, artificial materials and components are typically very different from those found in animals, and second, necessary detailed knowledge about the animal is often lacking. Instead, roboticists work with biologists to understand a particular biological system, extract the principles that are important for its function, and then engineer a solution based on those principles. This is often referred to as biologically inspired robotics or simply biorobotics.

Biorobotic design is a multistep process. Before a robot is designed or an animal is chosen for its inspiration, the robot's mission—and the tasks needed to perform that mission—should be clearly defined. For example, would its mission tasks best be performed by a robot that can swim, run, climb, or fly, or that uses a combination of those modes of mobility? Once the robot's mission is specified, an animal is identified that performs similar tasks in a similar environment. Biological data can then be acquired and analyzed both on the real animal and on software or hardware models of the behaving animal. Only then can the biorobot be designed. However, even with a thorough understanding of the animal's behavior, the engineer faces problems of capturing those properties with very different materials and components from the natural situation. The solution is to abstract the biological principles as necessary so that they can be implemented with current technology while retaining the desired functionality.

Data gathering. To understand the necessary biological principles, biologists rely upon several techniques, including those associated with neurobiological and behavioral studies. A fundamental issue that typically must be addressed is just how the animal moves. Video images can be taken to visualize kinematics in either two or three dimensions (**Fig. 1**). Three-dimensional analysis requires multiple camera views and some form of motion analysis that combines those views to give the true joint angles. This analysis is more difficult, but may be required where a limb moves in complex three-dimensional directions. If the animal moves slowly, conventional video is sufficient. More rapid behaviors require high-speed video to image movements. Insect leg motions often require 250 to 500 frames per second (fps), and their wing motions may require 1000 fps or higher.

In addition to motion, other pieces of information can provide a better understanding of animal behavior. Neural recordings indicate the motor activity that generates appropriate muscle contractions leading to joint movements. Electromyograms (EMGs) actually record muscle potentials, but since the sources of those potentials are motor neurons, they are actually indicating neural activity. Ultimately, these data may provide insight into control issues. However, even absent this, they can be valuable. For example, electromyograms can be synchronized with kinematic data to show whether movements arose actively from muscle contractions or passively as the body moved relative to the limb. In addition, these data can provide greater detail regarding changes in activity that could lead to critical but subtle changes in movement that may not be resolved in the video records. Force data can also be synchronized with kinematic data and provide other important insights. Force plates can provide three-dimensional ground reaction forces. Gels placed between polarizing lenses can provide optical images that can be force-calibrated to provide directional force data for each leg.

(a)

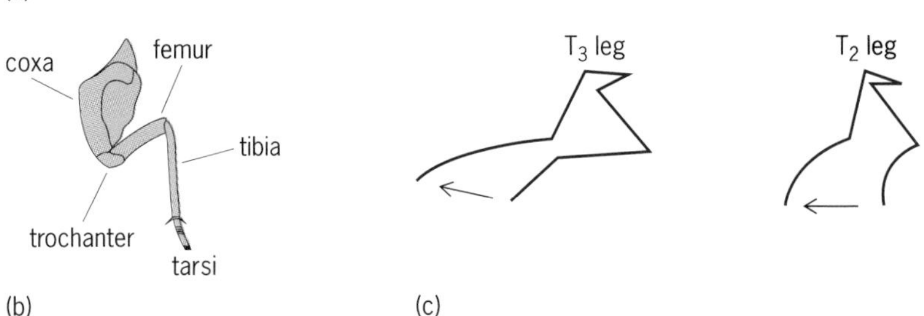

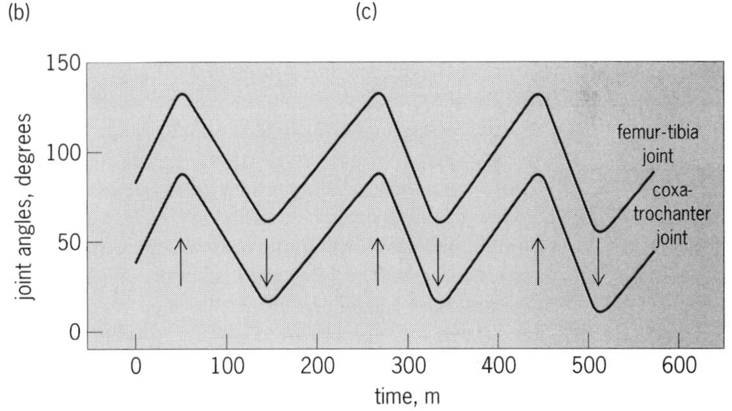

Fig. 1. Use of high-speed video to analyze the leg motions of the cockroach. (*a*) Two stills from the video. (*b*) Cockroach leg joints. (*c*) Motions of the middle (T_2) and rear (T_3) legs. (*d*) Angles between T_3 leg joints as functions of time. (*From R. E. Ritzmann, R. D. Quinn, and M. S. Fischer, Convergent evolution and locomotion through complex terrain by insects, vertebrates and robots, Arth. Struc. Dev., 33:361–379, Elsevier Press, 2004*)

Modeling. All of these data can be provided to robotics engineers, who often use modeling to guide designs. Software and hardware models are used to analyze biological data and test hypotheses about animal mechanisms that are believed to be most important for particular behaviors. Software models are also used to decide on design parameters, because they can be modified more rapidly than hardware. In a software model, mathematical equations represent

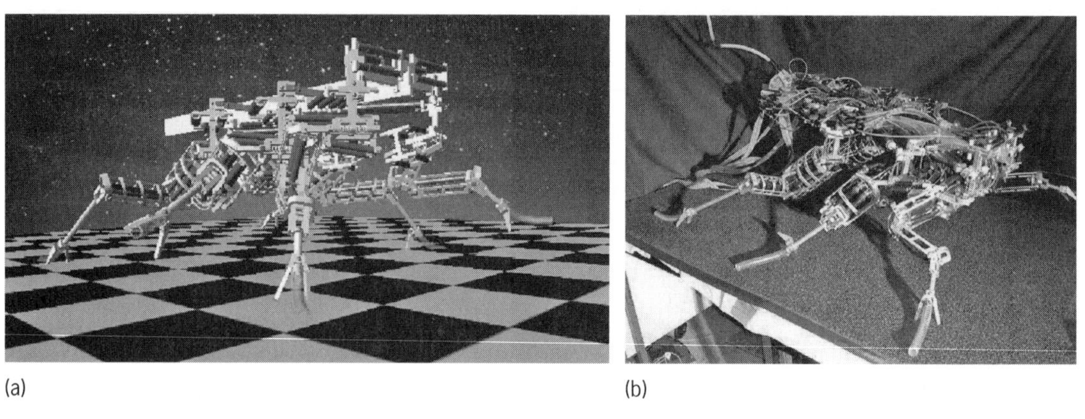

(a) (b)

Fig. 2. Cockroach-inspired robot. (*a*) Computer-aided design (CAD) model, developed in the process of designing (*b*) Robot V.

the physics of the animal or the robot and its interactions with its environment. A simulation that captures the dynamics of the system requires detailed knowledge about the forces acting on the animal and pertinent inertial and geometrical properties. As an example, geometrical and inertial properties of the various segments of a legged animal can be used to develop a dynamic model in software. It can then be used to test ideas about which joints are most necessary to perform desired functions such as running, climbing, and turning. If a joint is little used for the robot's desired tasks, that joint can then be neglected in the final design.

Hardware models are also very useful, because they are three-dimensional and can be physically manipulated by researchers. Robot V (**Fig. 2**) is essentially a hardware model of a cockroach, which is being used in sensorimotor control experiments. Hardware models are also necessarily rooted in physics. Because of their complementary nature, hardware and software models are often used in tandem. If a researcher does not know how an animal generates forces in a particular situation, a hardware model can be used to resolve the problem. For example, a scaled-up model of a fly wing was used to show that flies generate a large part of their lift force by taking advantage of unsteady air aerodynamics. Once the forces are understood, they can be represented by mathematical equations and implemented into a software model of the complete system. *See* INSECT FLIGHT.

Design. Before a robot is fabricated, all of its components must be carefully designed to ensure that they are strong enough, fit together properly, and permit the desired joint ranges of motion. Computer-aided design (CAD) tools permit engineers to design individual components, assemble them, and even animate the model robot (Fig. 2). Finite-element analysis software is used to analyze the strength of the components. Once the robot CAD model is complete, its inertial and geometrical properties can be implemented in a dynamic simulation similar to those used to study the animals. Then control systems are developed based on behavioral rules or neural organization. The robot's control circuits are tested and refined in simulation for two reasons. First, control testing in simulation can be performed rapidly without damage to hardware. Second, actuator forces necessary to drive the model can be monitored, providing data for sizing the robot's actuators and power system.

The biological accuracy implemented in a robot's design depends on its purpose and on the technology that is available. Animals are complex, and mimicking them in detail is not typically possible. The design of a robot that must run rapidly over obstacles larger than its height can gain much from observations of cockroaches that perform similar tasks with ease. Indeed, a robot with all of the neuromechanical locomotion complexities of a cockroach promises great mobility in complex, unstructured terrain. However, technical challenges, including actuators, power systems, and sensory control systems, currently limit the practicality of generating a useful, autonomous robot of this complexity.

Functionality. The primary challenge for a biorobot designer is to imbue a robot with the desired functionality observed in the animal using current technology. For example, cockroaches nominally run in a tripod gait and then modify their gait in particular ways when they climb obstacles. Two robots, RHex and Whegs^{TM}, perform these movements without the leg complexity found in the animal. They

Fig. 3. RoboLobster, designed to walk on the ocean floor using its eight multisegmented legs. (*Courtesy of Joseph Ayers, Northeastern University*)

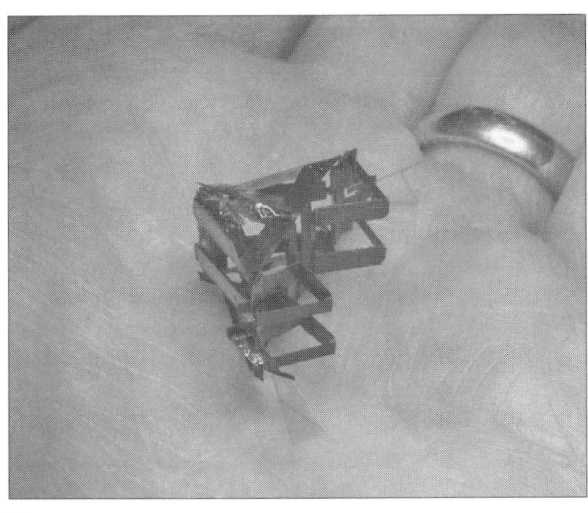

(a)

(b)

Fig. 4. Micromechanical Flying Insect (MFI), which is the size of an insect and generates lift with its fly-inspired wings. (*a*) Photograph (*courtesy of Ronald S. Fearing, University of California, Berkeley*). (*b*) CAD drawing (*courtesy of Quan Gan, University of California, Berkeley*).

generate leg movements in distinctly different ways but with similar results. On the other hand, RoboLobster (**Fig. 3**) mimics both the function and form of its namesake. It is designed to walk on the ocean floor using its eight multisegmented legs and "claws," and takes advantage of the hydrodynamic structure of the lobster.

The cockroach-inspired robots described above are much larger than their insect models because their missions require them to run over large obstacles such as standard stairs and roadside curbs. However, some missions require a robot to be small to enable movement in enclosed spaces. There the scale of an insect is appropriate. The Micromechanical Flying Insect (**Fig. 4**) is one tiny robot. It creates lift by moving its wings in a pattern and speed that mimic those of a fly, with similar wing mechanisms. *See* VERY SMALL FLYING MACHINES.

Biologically inspired robotics will continue to thrive and grow because there is a tremendous diversity of animals and because animals can be found that perform most of the activities that a roboticist might want to mimic. Engineers have much to gain from working with biologists, but the reverse is also true. Biorobotics also affords biologists a new tool to learn about animals. They are using robots as models of animals to test hypotheses about how they function.

For background information *see* COMPUTER-AIDED DESIGN AND MANUFACTURING; CONTROL SYSTEMS; ELECTROMYOGRAPHY; FINITE ELEMENT METHOD; MODEL THEORY; ROBOTICS; SIMULATION in the McGraw-Hill Encyclopedia of Science & Technology.

Roger D. Quinn; Roy E. Ritzmann

Bibliography. R. D. Beer et al., Biologically-inspired approaches to robotics, *Commun. ACM*, 40(3):30–38, March 1997; A. A. Bieweiner and R. J. Full, Force platform and kinematic analysis, in A. A. Biewener (ed.), *Biomechanics—Structures and Systems: A Practical Approach*, pp. 45–73, IRL/Oxford University Press, New York, 1992; M. H. Dickinson, F.-O. Lehmann, and S. P. Sane, Wing rotation and the aerodynamic basis of insect flight, *Science*, 284:1954–1960, 1999; R. J. Full, R. Blickhan, and L. H. Ting, Leg design in hexapedal runners, *J. Exp. Biol.*, 158:369–390, 1991; D. A. Kingsley, R. D. Quinn, and R. E. Ritzmann, A cockroach inspired robot with artificial muscles, *International Symposium on Adaptive Motion of Animals and Machines (AMAM'03)*, Kyoto, Japan, 2003; R. E. Ritzmann, R. D. Quinn, and M. S. Fischer, Convergent evolution and locomotion through complex terrain by insects, vertebrates and robots, *Arth. Struc. Dev.*, 33:361–379, 2004; U. Saranli, M. Buehler, and D. Koditschek, RHex: A simple and highly mobile hexapod robot, *Int. J. Robotics Res.*, 20(7):616–631, 2001; J. T. Watson and R. E. Ritzmann, Leg kinematics and muscle activity during treadmill running in the cockroach, *Blaberus discoidalis*: I. Slow running, *J. Comp. Physiol.*, 182:11–22, 1998; R. J. Wood, S. Avadhanula, and R. S. Fearing, Microrobotics using composite materials: The micromechanical flying insect thorax, *IEEE International Conference on Robotics and Automation 2003*, pp. 1842–1849, 2003.

Bioluminescent fungi

Bioluminescence is the emission of photons of visible light by living organisms via the chemical reaction of molecular oxygen (O_2) with a substrate (luciferin) catalyzed by an enzyme (luciferase). Organisms that exhibit bioluminescence are diverse and widely distributed in nature, for example, bacteria, dinoflagellates, fungi, insects (such as click beetles and fireflies), and fish. (Some fish, such as deep-sea angler fish, exhibit bioluminescence through bioluminescent bacterial symbionts, whereas others, such as hatchet fish, exhibit bioluminescence on their own.) The colors of the light produced by bioluminescent organisms differ; for example, bacteria emit a blue-green light (490 nm) and fireflies emit a yellow light (560 nm). The enzymes that catalyze the bioluminescence reaction show no homology to each other, and the substrates are also chemically unrelated. Molecular oxygen is the only common feature of bioluminescence reactions, indicating that the luminescent systems in most organisms may have evolved independently. The terms "luciferase" and "luciferin" are thus generic terms for the enzymes and substrates involved in the different bioluminescent reactions.

History. The natural phenomenon of bioluminescence has long been studied. Naturally bioluminescent fungi growing on rotting wood were observed in antiquity by Aristotle and Pliny the Elder. There are several accounts of fungal bioluminescence from the sixteenth, seventeenth, and eighteenth centuries. In 1853 Johann F. Heller demonstrated that the light emission from decayed wood is fungal in origin. The continuing interest in fungal bioluminescence is reflected by several accounts of bioluminescence in fungi in the twentieth century. However, relatively little research has been carried out on naturally bioluminescent fungi.

Naturally bioluminescent fungi. These fungi have been found in both temperate and tropical climates, and there are currently thought to be more than 40 species of bioluminescent fungi within nine genera, all of which are members of the Basidiomycota. Examples of luminescent fungi include *Armillaria mellea* (common name, Honey Fungus), *Mycena citricolor* (synonym, *Omphalia flavida*), and *Omphalotus olearius* (synonyms, *Pleurotus olearius* and *Clitocybe illudens*; common name, Jack-O-Lantern). They are commonly wood-degrading fungi, for example *Armillaria* species (*A. gallica, A. mellea, A. ostoyae*, and *A. tabescens*) and litter-degrading fungi, for example *Mycena* species, although *Armillaria ectypa* (common name, Marsh Honey Fungus) is found in fens and bogs.

Panellus stipticus (synonyms, *Pannus stypticus* and *P. stipticus*) is unusual in that bioluminescence is exhibited only by North American strains and not by Eurasian strains. Some apparently nonluminescent North American strains of *P. stipticus* have also been found, suggesting that bioluminescence is strain-specific. There have been reports that species of the genus *Xylaria*, for example *X. hypoxylon*, are bioluminescent; these species are the only examples from the Ascomycota that may be luminescent, but this needs to be confirmed.

Bioluminescent fungi emit a greenish light (520–530 nm), and the light emission is continuous and occurs only in living cells. No correlation of fungal bioluminescence with cell structure has been found. Bioluminescence may occur in both mycelia (vegetative bodies) and fruiting bodies, as in *P. stipticus* and *O. olearius*, or only in mycelia and young rhizomorphs (rootlike structures), as in *A. mellea*. In some *Mycena* species, such as *M. rorida* var. *lamprospora*, damp spores are bioluminescent. Some bioluminescent fungi, for example *A. mellea*, are reported to exhibit diurnal periodicity and seasonal variation of bioluminescence.

Physiology. Naturally bioluminescent fungi can be cultured both on solid medium and in liquid medium with and without agitation. The medium used to grow naturally bioluminescent fungi has been found to affect their luminescence. (A breadcrumb medium prepared using 2.5% finely ground organic white bread has been found to be particularly good for culturing bioluminescent fungi.) Conditions that affect growth, such as pH, light, and temperature, have also been found to influence fungal bioluminescence. The fact that conditions allowing growth have been found to result in bioluminescence suggests a link between metabolic activity and fungal bioluminescence.

Biochemistry. The biochemistry of the bioluminescence system of fungi is not yet fully understood. However, the preparation of bioluminescent, cell-free extracts has allowed the characterization of the in-vitro requirements of fungal bioluminescence. Cell-free luminescence was observed by combining a fungal extract containing luciferase with one containing luciferin. Fungal bioluminescence has been characterized as a luciferin–luciferase reaction requiring molecular oxygen and either reduced nicotinamide adenine dinucleotide (NADH) or reduced nicotinamide adenine dinucleotide phosphate (NADPH) in *A. mellea* and *M. citricolor*. Bioluminescence in *P. stipticus* has been shown to require molecular oxygen, and there is evidence to suggest that it may involve a luciferin-luciferase system, but an in-vitro luciferase-luciferin reaction has not been successfully demonstrated.

The chemical nature of fungal luciferin has been investigated in different bioluminescent fungi by some researchers. Several compounds have been isolated, for example from *Lampteromyces japonicus*, *M. citricolor*, and *P. stipticus* (for example, panal), which are believed to be luciferin or precursors of luciferin because (1) they are chemiluminescent, that is, they produce light as a result of a chemical reaction; (2) they are fluorescent, and their fluorescence emission spectra are similar to the fungal bioluminescence spectrum; and (3) they are active as a fungal luciferin in the in-vitro luciferin–luciferase reaction.

It is thought that naturally bioluminescent fungi emit light by the same or closely related mechanisms; however, the luciferase and luciferin in fungal bioluminescence have not yet been identified.

Genetics. Little is known about the genetics of fungal bioluminescence, although it has been shown that luminescence is a dominant characteristic. Several genes are thought to be required for bioluminescence in *Mycena* species and *P. stipticus*. The genes encoding fungal luciferase have not yet been identified.

Functional ecology. The existence of both luminescent and nonluminescent strains of the same species, species with only low-level luminescence, and species that have only a luminescent mycelium suggests that bioluminescence may not confer a selective advantage in fungi. However, numerous reasons for the occurrence of bioluminescence in fungi have been suggested.

One hypothesis suggests that the role of luminescence is to attract invertebrates to assist fungal spore dispersal. The strong presence of luminescence in the gills of some fruiting bodies or luminescent spores and the display of some level of attraction of invertebrates to luminescent fruiting bodies may support this hypothesis. It has been suggested that bioluminescence might be used to attract predators of fungivores and to repel fungivores. Although there is limited evidence to support these hypotheses, overall they remain unproven.

Another hypothesis is that bioluminescence is a by-product of a biochemical reaction and actually has no ecological value. Light production has been calculated not to be a significant energetic burden, and bioluminescent fungi may be releasing light (not heat) as an energy by-product of enzyme-mediated oxidation reactions. The superoxide radical (O_2^-), which is generated during metabolism (especially during oxidation reactions of various compounds), has been suggested to be required in the bioluminescence system of *P. stipticus*.

It has been suggested that bioluminescence may detoxify hydroxyl radicals that are formed during the breakdown of lignin. This is supported by the fact that many bioluminescent fungi, for example *A. mellea* and *P. stipticus*, are white rot fungi involved in wood decay. In addition, factors that have been shown to increase or decrease ligninolytic activity of other white rot fungi also affect the level of bioluminescence in *P. stipticus*.

Applications. The development of a range of equipment to measure bioluminescence, for example luminometers and photon-counting cameras, has enabled the application of bioluminescence in a variety of areas.

One application is the use of naturally bioluminescent organisms or genes encoding bioluminescence (for example, bacterial and firefly luminescence are encoded by *lux* and *luc* genes, respectively) for the detection and tracking of cells. Fungal bioluminescence has been used to measure the effect of fungal pathogens. Two bioluminescent fungi, *O. olearius* and *P. stipticus*, were used as a reporter system in the evaluation of biological control agents.

Another application of bioluminescence is for toxicity testing. Bioluminescence-based biological sensors, whose response to toxins corresponds to a change in light output, are rapid, reproducible, and sensitive to a wide range of toxins. Bacterial and firefly bioluminescence has been widely used in bacterial biosensors (for example *Vibrio fischeri* and *lux*-marked *Escherichia coli*) to assess potential toxicity. In order to complement the range of bioluminescence-based bioassays available, a filamentous fungal bioluminescence-based bioassay was developed using the naturally bioluminescent fungi *A. mellea* and *M. citricolor*. Bioluminescent fungi are eukaryotic, multicellular organisms, and their significance in key ecological processes makes them suitable for toxicity testing, complementing prokaryotic biosensors.

For naturally bioluminescent fungi to be used in future applications, a better understanding of the physiology and biochemistry, as well as the elucidation of the molecular biology of fungal bioluminescence, is required.

For background information *see* BASIDIOMYCOTA; BIOLUMINESCENCE; BIOSENSOR; CHEMILUMINESCENCE; FUNGI in the McGraw-Hill Encyclopedia of Science & Technology. Hedda J. Weitz

Bibliography. E. N. Harvey, Fungi, in *Bioluminescence*, pp. 96–117, Academic Press, New York, 1952; P. J. Herring, Luminous fungi, *Mycologist*, 8:181–183, 1994; E. C. Wassink, Luminescence in fungi, in P. J. Herring (ed.), *Bioluminescence in Action*, pp. 171–197, Academic Press, London, 1978; T. Wilson and J. W. Hastings, Bioluminescence, *Annu. Rev. Cell Develop. Biol.*, 14:197–230, 1998.

Biomimetic design for remanufacture

Many elegant solutions to engineering problems have been inspired by biological phenomena. Robots have been modeled after insects and other organisms, and leaves have served as models for solar membrane folding. Biological models have also been used to develop novel materials, actuators, and sensors. However, it is not always clear how the particular biological models were identified or selected. For example, an engineer open to using biological models for design might not know how to find relevant biological analogies for a given design. Therefore, a general method was developed for which analogous biological phenomena can be identified and used on any design problem.

An example of design for remanufacture will be used to illustrate the biomimetic design method. Remanufacture aims to reuse product components through processes of disassembly, cleaning, repair, and reassembly. Some obstacles to part reuse may be due to the desire to design parts for easier manufacture, assembly, and even recycling of scrap material. To overcome these problems, insight was sought from biological models.

Generalized biomimetic search. One possible approach to enable generalized biomimetic design is to build a database of biological phenomena for engineering use. However, the task of cataloging all biological phenomena for engineering purposes is an enormous and most likely subjective task. The approach chosen and described here takes advantage of the abundant biological information already available in natural language format by searching it directly for relevant phenomena.

This approach was implemented in the form of a computerized search tool that locates in biology texts instances of functional keywords describing the engineering design problem.

Source. The initial source of information selected was W. K. Purves's *Life, the Science of Biology*, which is written at a level suitable for those with little background in biology and covers a large range of organizational levels, from the molecular and cellular to the ecosystem, such that potential solutions are not limited to a particular organizational level.

Keywords. The keywords used to search for relevant text segments were verbs that described the desired effect of possible solutions. Verbs are typically preferred over nouns as keywords to initiate searches. Searching for nouns may find preconceived solutions, while searching for verbs that describe the desired action will identify biological forms that may not have occurred to the designer.

Synonyms were used in the past to increase the number of matches. Recently, it was found that troponyms produce superior alternative keywords. Troponyms are verbs that describe specific manners of another verb; for example, bake is a troponym of cook.

Remanufacture example. Several analogies for case studies in design for remanufacture and microassembly were located at various levels of biological organization by searching for keywords and synonyms related to the engineering problem.

Remanufacture is a process applied to products at the end of their life that seeks to reuse product components. An advantage over recycling for scrap material is the conservation of resources required to melt and reform components. One design guideline identified to facilitate remanufacture is that product features prone to failure should be made to be separable. In this way, the failed features can be replaced, enabling the reuse of a component without labor- and capital-intensive repairs. However, making failure-prone parts that are separable increases the part count and assembly cost.

As a specific example of the above problem, "snap fits" enable fastening without introducing additional parts or materials and are often preferred for assembly and recycling purposes. However, snap fits are problematic for remanufacture when they fail because they are difficult to repair. **Figures 1** and **2** show toner cartridge components undergoing remanufacture that contain failed snap fits. Biological phenomena analogous to remanufacture were sought to address this problem.

Analogous biological phenomena. Searching for the keyword remanufacture resulted in no matches in the biology text. Therefore, alternative keywords—repair, restore, and correct—were used.

Molecular: DNA repair. A keyword match identified using "repair" involves base excision repair, described as a repair mechanism for replacing chemically damaged abnormal bases with functional bases.

While the above description confirms the relevance of the excision repair phenomenon to remanufacture, it does not provide ideas on how to solve the problem. Further research using a more advanced text, such as E. C. Friedberg's *DNA Repair and Mutagenesis*, revealed details that could be used as stimuli for design; specifically, a conformation change associated with failure can be used to facilitate the removal and replacement of faulty components.

Organism: sacrifice and regenerate parts. The ability of plants to grow new parts to replace damaged parts was identified as an analogy at the organ/organism level comparing plant and animal repair strategies for damaged or diseased tissues. While animals generally repair tissues, plants seal off and sacrifice damaged or infected tissue to stop further harm and then replace the damaged parts by growing new parts (such as stems, leaves, or roots).

Applying this analogy to products involves incorporating a sacrificial part that can be replaced such that repairing the broken feature or replacing the entire part that contained the feature can be avoided.

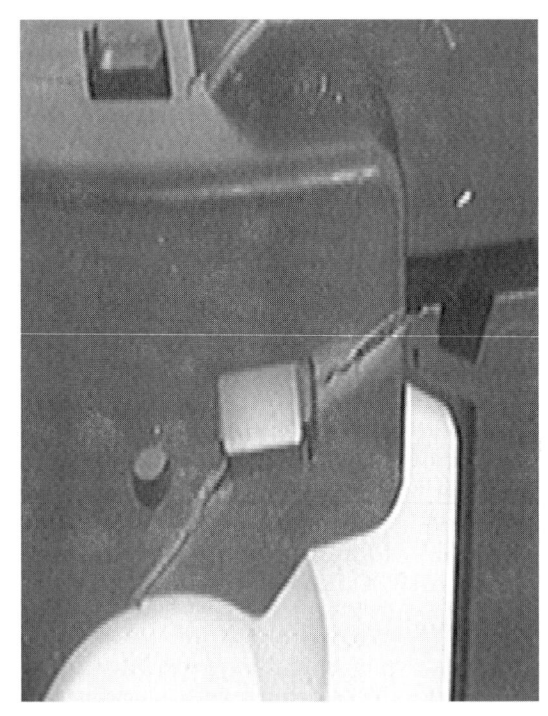

Fig. 1. Damaged part with snap fit. (*Courtesy L. H. Shu*)

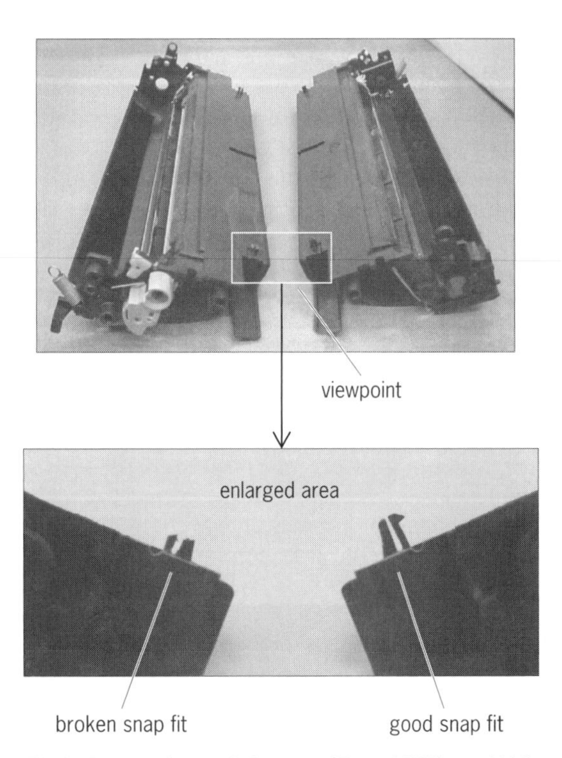

Fig. 2. Damaged snap-fit features. (*From J. Williams, 2001*)

Organ system: fainting. Another phenomenon identified using the keyword "correct" involved fainting, which occurs, for example, if on standing (vertically) the effect of gravity results in too little blood being pumped from the lower body to the brain. One consequence of fainting is falling (horizontally), which reduces the gravity effect, thereby alleviating the problem.

The strategy derived from the above example is that fainting is a form of defensive failure that prevents more serious failure. For product and part design, features could be included that induce failure modes that are easier to repair.

Ecosystem: forest restoration. The keyword "restore" identified an example of the restoration of the tropical deciduous forest in Guanacaste National Park in northwestern Costa Rica using small fragments of the tropical forest remaining in areas converted to pastures.

The strategy to be used from this analogy is that restoration builds upon a foundation or substrate that supports the restoration process. An analogous concept can be used for restoring parts in remanufacture.

Biological phenomena in design. **Figure** 3 shows a redesigned snap fit where a part containing a failed snap-fit feature can be more easily reused. This redesign incorporates the strategies identified from the above analogous biological phenomena.

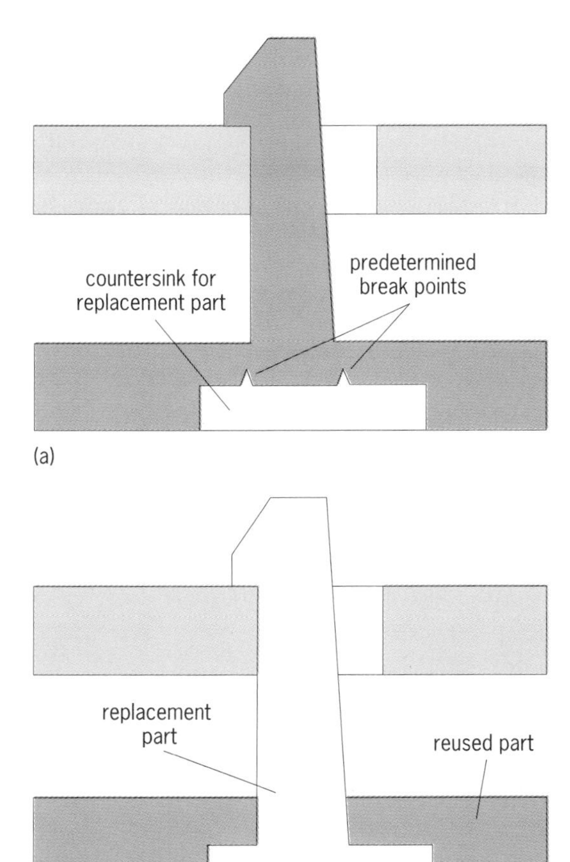

Fig. 3. Redesigned snap fit to facilitate repair. (*a*) Snap fit redesigned with countersink and break points. (*b*) Redesigned snap fit after failure and refurbishment. (*Adapted from E. Hacco and L. Shu, 2002*)

From DNA excision repair, a configuration change associated with failure enables repair processes, specifically removal and replacement of faulty components. In the redesigned configuration, failure of the snap-fit feature leads to its removal, clearing the way for replacement of the feature.

From the ability of plants to seal off and sacrifice old parts, new replacement parts are generated instead of repairing the damaged part. In the redesigned configuration, the failed snap fit constitutes a sacrificial part that is replaced rather than repaired.

From forest restoration, the rebuilding of parts may benefit from a substrate or other structure that supports and locates the replacement part. Figure 3 shows a countersink that would help support as well as locate a replacement part.

Applying the fainting strategy to the redesign, predetermined break points incorporated into snap-fit configurations may cause earlier failure than with standard snap-fit configurations, but may also induce failure along predicted locations such that the part is easier to repair. Figure 3 shows the predetermined break points that may serve such a purpose.

For background information *see* DATABASE MANAGEMENT SYSTEM; DEOXYRIBONUCLEIC ACID (DNA); ENGINEERING DESIGN; MANUFACTURING ENGINEERING; RECOMBINATION (GENETICS); REFORESTATION; REGENERATION (BIOLOGY); RELIABILITY, AVAILABILITY, AND MAINTAINABILITY in the McGraw-Hill Encyclopedia of Science and Technology. Lily H. Shu

Bibliography. I. Chiu and L. Shu, Natural language analysis for biomimetic design, *Proc. 2004 ASME Design Eng. Tech. Conf. Computers Inform. Eng. Conf., Salt Lake City, Sept. 28–Oct. 2, 2004*, DETC2004/DTM-57250, 2004; E. C. Friedberg, G. C. Walker, and W. Siede, *DNA Repair and Mutagenesis*, ASM Press, Washington, DC, 1995; E. Hacco and L. Shu, Biomimetic concept generation applied to design for remanufacture, *Proc. ASME Design Eng. Tech. Conf. Computers Inform. Eng. Conf., Montreal, Sept. 29–Oct. 2, 2002*, DETC2002/DFM-34177, 2002; T. W. Mak and L. H. Shu, Abstraction of biological analogies for design, *CIRP Ann.*, 53(1):117–120, 2004; W. K. Purves et al., *Life, The Science of Biology*, 6th ed., Sinauer Associates, Sunderland, MA, 2001; L. H. Shu, Biomimetic design for remanufacture in the context of design for assembly (Invited Paper), *Proc. Inst. Mech. Eng. Part B: J. Eng. Manufac.*, 218(3):349–352, 2004; L. H. Shu et al., Biomimetics applied to centering in microassembly, *CIRP Ann.*, 52(1):101–104, 2003; J. Vincent and D. Mann, Systematic technology transfer from biology to engineering, *Phil. Trans. Roy. Soc.: Phys. Sci.*, 360:159–173, 2002; J. Williams, *Quantification and Analysis of Remanufacturing Waste Streams for Improving Product Design*, Master of Applied Science Thesis, University of Toronto, Department of Mechanical and Industrial Engineering, 2001.

Blackout prevention

Our dependence on electrical energy and the demand for continuous and reliable power continue to grow. Recent wide-area electrical blackouts have raised many questions about the causes of such events and the weakness of interconnected power systems. The exchange of information obtained from blackouts worldwide, examination of the root causes, and application of both proven and new solutions to help prevent propagation of such large-scale events should help the electric power industry design, operate, and maintain reliable power delivery infrastructures for the future.

Although large-scale blackouts are still very low probability events, they are costly and have negative consequences for customers and society in general as well as for power companies. It is easy to misjudge the risk of such extreme cases. The high costs of extensive mitigation strategies against grid congestions, combined with inaccurate probabilistic assessments, have led to risk management not adequately focusing on cost-effective prevention and mitigation initiatives.

The power system is a complex generation, transmission, and distribution system that deals with an energy medium that cannot be stored. An understanding of the complexities of the interconnected power grid and the need to implement planning, maintenance, and operating practices, with well-balanced cost, performance, and risk, is necessary to prevent blackouts or minimize their consequence.

Increase in frequency and size of blackouts. In recent years, as power systems are pushed closer to their limits, the number and size of wide-area outages have increased, affecting many millions of customers worldwide. These wide-area outages, caused by widespread cascading failures of power system equipment, lead to many blackouts. Historically, after each widespread cascading failure in the past 40 years, the power industry has focused attention on the need to understand the complex phenomena associated with blackouts. For example, major reliability improvements have been made after major blackout events in the United States in 1965, 1977, and 1996. **Figure 1** shows some of the major blackouts and their consequences.

The North American and the European grid systems are among the most reliable systems worldwide. Transmission lines connect all the electric generation and distribution within the respective grid systems and continents. However, the same systems are subject to many challenges: aging infrastructure, knowledge attrition, need for generation sitings near the load centers, environmental and political pressures to limit conventional transmission and generation expansions to meet the growing load, and insufficient regulatory or legislative enforcements for conforming to standards.

Moreover, the bulk grids were not designed originally to transfer large amounts of power between neighboring systems. Individual power systems were interconnected to improve electrical network reliability by enabling neighboring utilities to support each other during stressed conditions. In recent years, this original intent has changed with emphasis on deregulation, imposing additional requirements of high transfers from new generation sources to the load areas. At the same time, public pressures and the "not in my backyard" sentiment make it difficult to site transmission lines or major local generation sources, especially in the more densely populated heavy-load areas, making system expansion very expensive and difficult. The challenge facing the power industry today is in finding the balance between reliability, economics, environmental, and other public-purpose objectives in order to optimize transmission and distribution resources to meet demand.

Preoutage conditions and symptoms of blackouts. The electric power grid is complex. Although there is a tendency to point at one or two significant events as the main reasons for triggering cascading outages, major blackouts are typically caused by a sequence of low-probability multiple contingencies with complex interactions. Low-probability sequential outages are not anticipated by system operators or may be too fast for human intervention, thus rendering the power system more susceptible to wide-area blackouts. As the chain of events at various locations in the interconnected grid unfolds, operators

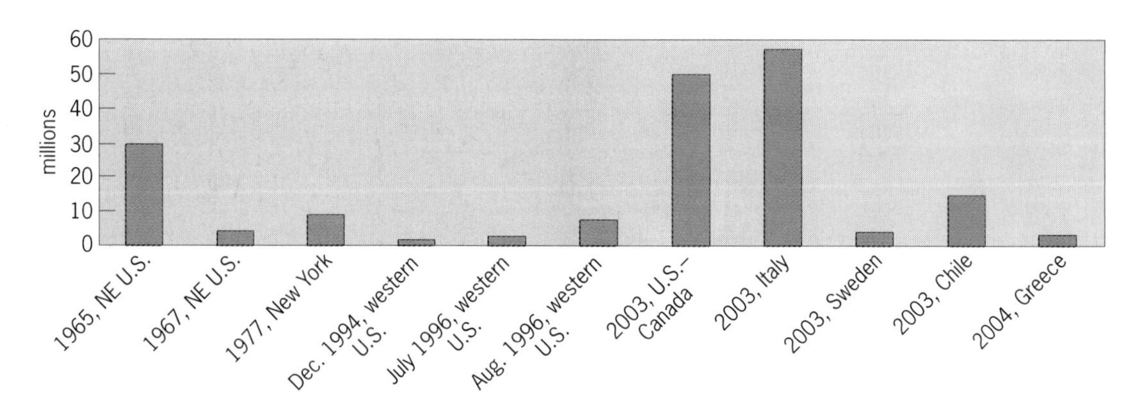

Fig. 1. History of some widespread blackouts, showing number of customers affected.

are also exposed to either a flood of alarms or incomplete information.

Power systems are designed to allow for reliable power delivery in the absence of one or more major pieces of equipment (such as lines, transformers, and generators). For example, the North American Electric Reliability Council (NERC) Planning Standard set forth the performance requirements a system must meet for various contingencies. The complexity of the grid operation, however, makes it difficult to study the permutation of contingency conditions that would lead to perfect reliability at reasonable cost. An accurate sequence of events is difficult to predict, as there is practically an infinite number of operating contingencies. Furthermore, as a system changes (for example, as a result of independent power producers selling power to remote regions, load growth, or new equipment installations changing power flows), these contingencies may significantly differ from the expectations of the original system designers.

Generally, disturbance propagation involves a combination of phenomena, such as:

1. Cascading line tripping by overloading transmission lines, leading to other equipment to disconnect, contributing further to systemwide outages.

2. Power system separation into islands, with an imbalance between generation and load causing the frequency to deviate from the nominal value and leading to more equipment tripping.

3. Loss of synchronous operation among generators and oscillatory instability, causing self-exciting interarea oscillations.

4. Voltage instability/collapse, when the power system becomes unable to allow for both power and voltage to be controllable to maintain the required voltage profile. Voltage instability leads to a very low voltage profile in a significant part of the system. Those problems usually occur when the power transfer is increased because the local resources have been displaced by remote resources without the proper installation of needed transmission lines or voltage support devices in the correct locations.

For the December 14, 1994, and the August 10, 1996, blackouts in the western part of North America, the Western System Coordinating Council investigation teams offered 38 conclusions and 28 recommendations, and 130 conclusions and 54 recommendations, respectively.

Learning from the past. The investigation reports on the blackouts of December 14, 1994, and August 10, 1996, in western North America offered 38 conclusions and 28 recommendations, and 130 conclusions and 54 recommendations, respectively. The report on the blackout of September 28, 2003, in continental Europe lists 14 observations. The investigating teams following the blackout on August 14, 2003, in eastern North America made 60 recommendations. There are striking similarities in findings among those blackouts. It is believed that if the 1994–1996 recommendations had been applied in other parts of the world, the impacts of the 2003 outages could have been reduced.

The 1996 outage in California alone cost $1 billion with 7.5 million customers having experienced power loss from just a few minutes up to 3 hours. One hour before the disturbance, three 500-kV lines tripped (as a result of short circuits). This resulted in a heavy power flow (4700 MW) from north to south. A fourth line tripped due to a fault, and a fifth line tripped due to a design flaw. As a result, the lower voltage grid experienced heavy loads. A 115-kV line tripped due to protective equipment failure, and a 230-kV line sagged and contacted a tree, again causing a line to trip. Voltages declined, and power generation units went to full excitation and tripped. Loss of synchronous operation among generators caused further tripping of the equipment and system separation. All of the above lead to a blackout.

Analysis of the August 14, 2003, incident (**Fig. 2**) revealed similar problems. A series of cascading events over the course of several hours, rather than a single instantaneous problem, initiated this major disturbance, which toppled a large part of the eastern part of the North American grid, affecting 50 million customers. The blackout was preceded by a couple of 345-kV power transmission lines tripping (as a result of short circuits). In the span of a couple of hours, another transmission line overloaded and sagged into overgrown trees, resulting in the remaining 345-kV lines disconnecting. As a consequence, as with the 1996 outage, the underlying lower voltage grid became overloaded, and the 135-kV transmission lines tripped due to protective equipment actions. Sequential outages of power system equipment due to overloads, power swings, and apparent voltage decline led to a shutdown of large proportions. These cascading events were accompanied by failures of Energy Management System (EMS) and Supervisory Control And Data Acquisition system (SCADA) alarm systems preventing operators from diagnosing problems. Other contributing factors were the apparent lack of communication and coordination between utilities and independent system operators (ISO) and the inability of operators to respond fast enough. First, although there was plenty of time to react to initial disturbances, lack of data, inadequate training and right-of-way vegetation management, poorly maintained monitoring systems, and insufficient automated systems initiated the blackout sequence. Second, although the blackout was initiated in the limited area of one utility, it should not have been acceptable that the disturbance propagated to very large and highly populated areas.

Although the western and the eastern North American regions are quite different, with the former grid consisting of long transmission lines due to long distances among load centers and the latter grid with short lines and large load centers (densely populated cities), there is a notable similarity in the chain of events leading to the North American and European blackouts described earlier. The common

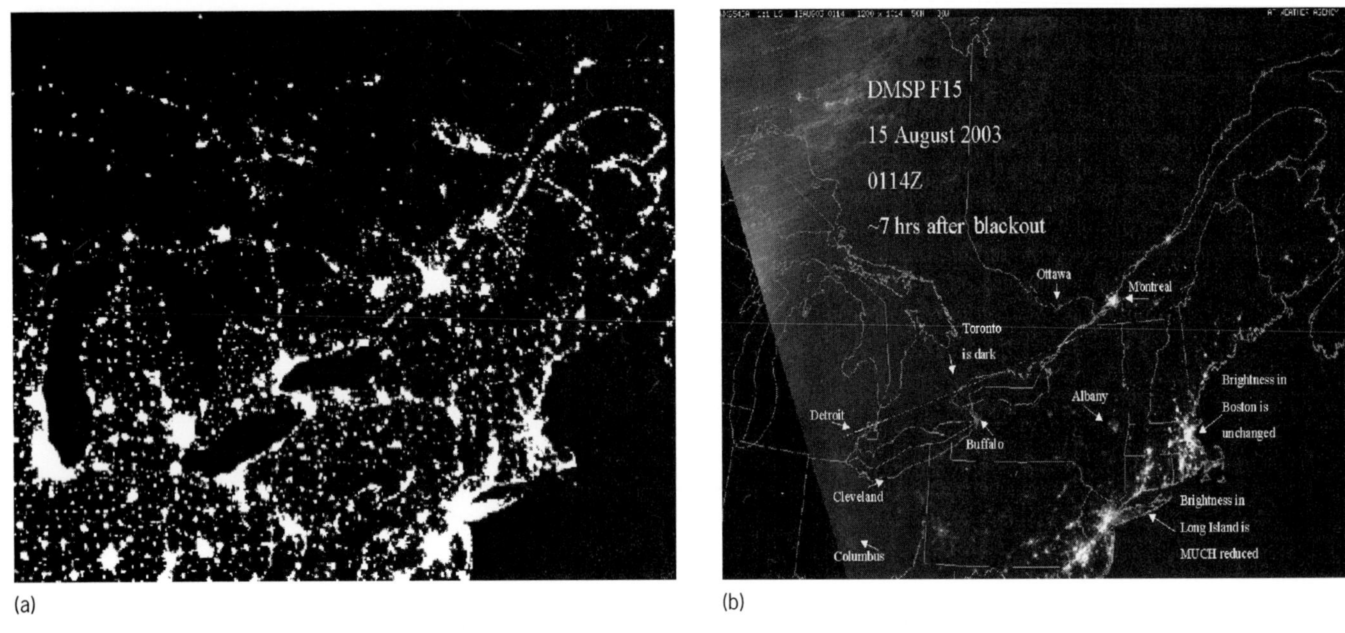

Fig. 2. Effects of blackout of August 14, 2003, in eastern North America. (*a*) Regular night. (*b*) August 14.

threads (see **table**) among the nonnatural disaster outages in western and eastern North America and Italy emphasize possibilities for preventing blackouts.

Preventive measures. The three T's—trees, tools, and training—have been identified as the leading focal points for the prevention of widespread outages. While right-of-way management can help reduce system exposures, there are other natural events, such as storms or dense fog, that have caused similar propagated disturbances in the past. Tools and training, on the other hand, are two factors required for human interactions during fast-developing cascading events. Hence, although three T's are very important, other areas also need to be addressed to arrive at the required comprehensive solutions to prevent blackouts.

Measures to prevent blackouts can be grouped into following major categories:

Regulatory and public policy. Analysis of blackouts shows that in each case reliability standards have not been met, demonstrating the need to establish more stringent compliance enforcement. Clear rules to help recover investments (such as building new transmission lines) are necessary, including how costs are to be shared based on identifying who benefits and how much.

Investing in aging infrastructure and new equipment. Power grids must be upgraded and expanded to continue meeting the growing demands. The condition of aging infrastructure must be assessed and maintenance procedures improved, including transmission line right-of-way maintenance. Timely retirement and replacement of transmission equipment is another important remedy for reducing failure rates and potential outages. Power-delivery technologies must be implemented to strengthen the grid by increasing transmission power flow control capability [such as series capacitors, single-phase operation of transmission lines, flexible ac transmission system devices (FACTS), and high voltage direct current (HVDC) links] and by other options, such as energy storage, distributed energy, superconducting materials, and microgrids. There must be increased

Common threads among blackouts

Western U.S., 1996: 7.5 M people	NE U.S.–Can., 2003: 50 M people	Italy, 2003: 57 M people
An hour before the disturbance, three 500-kV lines disconnect	Two hours before the disturbance, 500-kV lines disconnect	Heavy import to Italy (6 GW)
Heavy power flow N-S	Heavy power flow S-N	
Two lines disconnect (protection trips on fault)	One 500-kV line sags into a tree and trips	One 380-kV line sags into a tree and trips
Heavy load through 230-kV and 115-kV lines	Heavy load through 230-kV and 115-kV lines	Heavy load on parallel line that sags into a tree
230-kV/115-kV lines disconnect (overload)	230-kV/115-kV lines disconnect (overload)	220-kV/110-kV trip (overload), isolating Italy
	More 345-kV lines trip	
Voltage declines and power units trip	Voltage declines and power units trip	Voltage declines and power units trip
Power oscillations and voltage decline cause cascading separations	Power oscillations and voltage decline cause cascading separations	Power oscillations and voltage decline cause cascading separations
Blackout occurred 3 min after	Blackout occurred 3 min after	Blackout occurred 2.5 min after

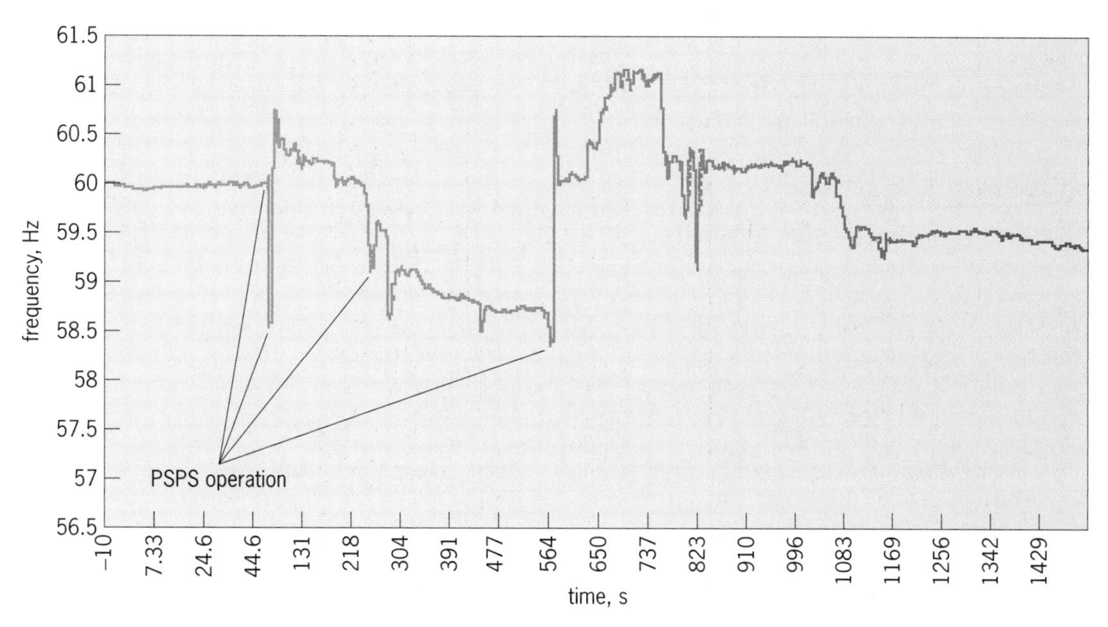

Fig. 3. System frequency in northern California. Blackout of August 10, 1996, beginning at 15:46:41 PST. Power systems protection schemes (PSPS) limited outages to 7.5 million people on one of the hottest days of the year (the western region serves more than 71 million people).

implementation of automated power system protection schemes (PSPS) designed to act during major disturbances to reduce the burden on the operators (**Fig. 3**). As the next step, expanded implementation of wide-area monitoring, protection and control systems (WAMPAC) must occur. The technology advancements today allow for an integrated, electronically controlled power system, known as a "smart grid." For example, a smart grid would utilize Phasor Measurement Unit (PMU) technology based on synchronizing power system measurement using global positioning satellites (GPS). This system can help in congestion tracking, visualization, information sharing over a wide region, and so forth. For example, poorly recognized dynamic constraints can unnecessarily narrow operating limits, endanger reliability, and prevent optimal energy transactions, resulting in lost revenues.

System planning and protection studies. Complex power system phenomena should be studied (for example, dynamic voltage and transient stability studies using appropriate models) to minimize propagation of future systemwide events. System studies should be validated against actual power system performance. Large regional geographic areas should be included in the scope of transmission planning and decision making. Protection and control systems should be properly set, maintained, and coordinated.

System operations. Many critical components, such as adequate reserve, real, and reactive power margins, must be balanced using reliable status monitoring and real-time state estimation in control centers. Improved energy management systems (EMS) should filter, display, and analyze only critical information to increase the availability of critical functions to 99.99% or better.

Assurance is required that operating capacity reserves and margins for transmission flows remain available to allow system adjustments during unintended multiple contingency conditions.

Existing operating practices and real-time data-exchange policies should be revisited and unobstructed operating visibility among control areas promoted. Well-trained operators must be allowed to take proper actions to mitigate disturbance propagation.

System restoration. Another critical step in minimizing the impact of widespread blackouts is the need for effective and fast power system restoration. Some of the key elements for responsive and intelligent restoration are (1) well-defined procedures that require overall coordination within the restoring area, as well as with neighboring electric networks; (2) reliable and efficient restoration software tools that can significantly aid operators and area coordinators to execute operating procedures and to make proper decisions; and (3) regular training sessions with practice drill scenarios to assure effectiveness of the process.

Even if advanced tools and procedures are in place to speed up restoration, there are limits on how fast the system can be restored depending on the type and distribution of generation. After the August 2003 blackout in North America, it took considerable time to restore generation. Some of the units did not have capabilities to be put in service immediately (black-start capabilities), and some units required more time to be put online with full power (for example, nuclear units due to security, and steam turbines due to allowable ramp-up rates). The types of loads served, the system configuration, and the effects of connecting the load back to the grid are of equal importance. For comparison, although most of the cities during the Italian blackout in September 2003 were restored in 9 h, it took more than a day to restore power to Detroit and New York. After the recent Swedish-Danish blackout, the last customers were reconnected in less than 6 h.

Conclusions. The complexity of power systems combined with systems being pushed closer to their limits without implementation of adequate preventive and corrective measures has resulted in an increase in the number and frequency of major blackouts in recent years. Statistically, a sequence of low-probability contingencies with complex interactions causing a blackout cannot be systematically ruled out. However, analysis reveals some common threads among wide-area disturbances, and that humans contribute to systemwide outages by suboptimal grid design and maintenance, and by operating systems too close to the edge. This leads to the conclusion that the probability of wide-area blackouts could be reduced, and, if the grid is adjusted accordingly and quickly, it is possible to arrest disturbance propagation and reduce size and impact of disturbances.

For background information *see* ELECTRIC DISTRIBUTION SYSTEMS; ELECTRIC POWER GENERATION; ELECTRIC POWER SYSTEMS; ELECTRIC POWER TRANSMISSION; ELECTRICITY; POWER PLANT; TRANSMISSION LINES in the McGraw-Hill Encyclopedia of Science & Technology. Damir Novsel; Vahid Madani

Bibliography. *Energy Infrastructure Defense Systems*, Special Issues of Proc. IEEE, vol. 93, no. 5, May 2005; S. H. Horowitz and A. G. Phadke, Boosting immunity to blackouts, *Power Energy Mag.*, September/October 2003; V. Madani and D. Novosel, Taming the power grid, IEEE Spectrum website, February 2005; NERC recommendations to August 14, 2003 blackout—Prevent and mitigate the impacts of future cascading blackouts (from www.NERC.com); D. Novosel, M. Begovic, and V. Madani, Shedding light on blackouts, *Power Energy Mag.*, January/February 2004; Western systems Coordinating Council disturbance summary reports for power system outages occurred in December 1994, July 1996, and August 1996 respectively (from www.WECC.biz).

Bose-Einstein condensation

Quantum mechanics divides particles or atoms into two classes: bosons and fermions. One of the most important characteristics of fermions is that no two fermions can be in the same quantum state; this is known as Pauli's exclusion principle. For bosons this restriction is not present, and in principle there is no limit to the number of bosonic particles that can be in the same quantum state. Electrons and protons are examples of fermionic particles. Light quanta (photons) and neutral atoms that contain an even number of neutrons are bosons.

In 1924, Albert Einstein made a theoretical investigation of the properties of a gas of noninteracting bosonic particles, based on a paper that Satyendra Nath Bose had sent him on the statistics of photons. At high temperatures, the quantum-mechanical nature of the gas is not important; it does not matter whether the atoms or particles in the gas are bosons or fermions. The atoms or particles jiggle around in the gas much like customers in a shopping mall, walking at different speeds, in different directions, and "out of step" with each other.

But as the gas is cooled down, the quantum-mechanical nature of the atoms or particles starts playing a crucial role. Based on his reasoning, Einstein found that as the gas is refrigerated below a specific temperature, the system undergoes a phase transition to a new state of matter in which the vast majority of the bosonic particles collect in the same quantum state, giving the system many unusual properties. Einstein described the process as "condensation without interactions," and the state of matter that appears became known as a Bose-Einstein condensate (BEC). In a Bose-Einstein condensed gas, the atoms or particles behave like soldiers in a military parade marching equally fast, in the same direction, and perfectly in step with each other.

The theory for a noninteracting Bose gas was later extended by Oliver Penrose and Lars Onsager to the case of interacting bosons, and the most widely used criterion for Bose-Einstein condensation at present is based on the concept of off-diagonal long-range order, as discussed by C. N. Yang.

Systems with Bose-Einstein condensation. Bose-Einstein condensation has long been studied exclusively in liquid helium-4 (^{4}He) cooled below 2.17 K (−455.76°F), where it is responsible for the frictionless flow of this liquid, known as helium-II. The superconducting state can be interpreted as a charged Bose-Einstein condensate of paired electrons. This Bose-Einstein condensate of Cooper pairs gives rise to the dissipationless flow of electric charge. Also, the fermionic isotope of helium, helium-3 (^{3}He), is known to harbor a condensate of bosonic pairs of ^{3}He atoms below approximately 2.5 mK. Finally, a laser can be considered as a Bose-Einstein condensate of photons.

It took more than 70 years before the peculiar state predicted by Einstein could be observed directly in experiments with trapped atomic gases. Although Bose-Einstein condensation is also present in liquid helium and in superconductors, the interactions present in these systems complicate the direct observation and manipulation of the condensate. Magnetically trapped atomic gases are much more dilute (typically 10^{13} atoms per cubic centimeter, as compared to 10^{22} atoms per cubic centimeter in liquid helium), and therefore the effects of interatomic interactions are weaker. The price to pay is that the temperature necessary to achieve Bose-Einstein condensation is much lower, and record cold temperatures of nanokelvins have to be reached before condensation sets in. The quest for Bose-Einstein condensation in atomic gases, sparked by experiments on stabilizing spin-polarized hydrogen gas by Isaac Silvera and Jook Walraven in 1980, culminated only after special techniques of laser cooling and trapping and evaporative cooling had been perfected.

In 1995, Eric Cornell, Wolfgang Ketterle, and Carl Wieman succeeded in the creation of Bose-Einstein condensates in dilute atomic gases. In their experiments, they first floated laser-cooled atoms in a magnetic trap. The floating cloud of atoms was then chilled further using evaporative cooling. Below a critical temperature, a peak appeared on top of the thermal density distribution, heralding the onset of Bose-Einstein condensation. This peak reflected the single-particle state, in which all the atoms condensed. The superfluid nature of the condensate expressed itself in the hydrodynamic expansion of the central peak, whereas the thermal (noncondensed) cloud expanded ballistically. Additional proof of the superfluidity of condensates was obtained later by observing matter–wave interference and by producing quantized vortices in condensates. Over 40 laboratories worldwide have achieved Bose-Einstein condensation in dilute gases of rubidium, sodium, lithium, hydrogen, helium, potassium, cesium, ytterbium, and chromium.

Macroscopic quantum-mechanical behavior. When a system is Bose-Einstein-condensed, the quantum-mechanical behavior usually associated with a single particle is transferred to the entire system. What does that mean, precisely? Quantum theory describes particles through complex wave functions whose modulus squared can be interpreted as a probability density of finding the particle at some position. The atomic orbitals are examples of wave functions for electrons in an atom. In a Bose-Einstein condensate, the quantum-mechanical description through a wave function, usually associated with a single particle, now extends to the entire system. The condensate can be described by a macroscopic wave function, known as the condensate wave function. Its modulus squared gives the density of the condensate, and the gradient of its phase is proportional to the velocity field of the condensate cloud. The condensate wave function in atomic gases obeys a nonlinear differential equation known as the Gross-Pitaevskii equation. In superconductors, the differential equation for the macroscopic wave function was derived by V. Ginzburg and L. D. Landau.

Distinguishing characteristics. Bose-Einstein condensation manifests itself most spectacularly in the phenomenon of superfluidity. The term superfluidity actually covers a range of properties that were first observed in liquid helium-II, including frictionless flow, persistent currents, and quantized vorticity. Frictionless flow, or the vanishing of the viscosity coefficient, also means that objects can be transported through condensates without dissipating any heat, provided that the velocity of these objects relative to the condensate does not exceed a critical velocity.

Quantized vorticity is related to the behavior of condensates under rotation. When a condensate is rotated below a critical stirring frequency, no flow takes place. Above a critical stirring frequency, a pattern of flow appears similar to that of a whirlpool in water or a hurricane in the atmosphere. In condensates, this flow pattern is called a vortex. It is characterized by a "hole" in the condensate: the eye of the hurricane, or vortex core. The condensate whirls around this hole, and the velocity is smaller at larger distances from the vortex core. As the condensate is rotated at even higher stirring frequencies, additional vortices appear. The observation of such quantized vortices is a telltale signature of the presence of Bose-Einstein condensation. Another unambiguous signature of the presence of a condensate is matter interference.

Interference is a typical wave phenomenon. It can occur in matter when the system obtains a macroscopic wave function, and this is precisely the case for condensates. It is one of the most striking differences between the Bose-Einstein condensed phase and other phases of matter. If two clouds of gas are combined, one (bigger) cloud emerges. Two droplets of liquid merge into one drop. Fusing two blocks of solid results in one bigger block. And yet, when two "clouds" of condensate are overlapped, they form a stack of pancakes, planes of high matter density alternating with planes of low density. Such fringe patterns have recently been observed for dilute atomic condensates, confirming the wave function description of these systems.

Research applications. Magnetic and optical trapping techniques allow exquisite control of both the geometry and the interatomic scattering properties of the atoms in the trapped cloud. Condensates can be confined to a plane or a line, and lattices of condensates can be formed. This has opened the possibility of studying and refining fundamental quantum-mechanical models of matter, by creating a controllable artificial crystal consisting of condensates on each lattice site.

The quantum-mechanical nature of the condensate makes it a natural candidate for studies in quantum information. This field of physics investigates how information can be encrypted, transmitted, and computed based on quantum-mechanical principles rather than on classical electromagnetic information technology.

The interatomic interaction strength can also be adapted experimentally, using a magnetically tunable scattering resonance known as a Feshbach resonance. This technique has allowed the creation and study of molecular condensates. The question of how atoms pair to form bosonic molecules, and how the properties of these atom pairs depend on the interaction strengths, is of fundamental importance in order to better understand superconductivity.

For background information *see* BOSE-EINSTEIN CONDENSATION; BOSE-EINSTEIN STATISTICS; INTERFERENCE OF WAVES; LASER COOLING; LIQUID HELIUM; PARTICLE TRAP; QUANTIZED VORTICES; QUANTUM MECHANICS; QUANTUM STATISTICS; RESONANCE (QUANTUM MECHANICS); SUPERCONDUCTIVITY; SUPERFLUIDITY in the McGraw-Hill Encyclopedia of Science & Technology.

Jozef T. Devreese; Jacques Tempere

Bibliography. A. Griffin, D. W. Snoke, and S. Stringari (eds.), *Bose-Einstein Condensation*, Cambridge University Press, 1995; M. Inguscio, S. Stringari, and C. E. Wieman (eds.), *Bose-Einstein Condensation in Atomic Gases*, IOS Press, Amsterdam, 1999; C. J. Pethick and H. Smith, *Bose-Einstein Condensation in Dilute Bose Gases*, Cambridge University Press, 2001; L. Pitaevskii and S. Stringari, *Bose-Einstein Condensation*, Clarendon Press, Oxford, 2003.

Brain imaging of deception

Deception, whether intentional or unintentional, adds an important dimension to social interaction. Given the theoretical as well as practical benefits of detecting deceptive behavior, much effort has been devoted to developing methods of measurement that can accurately depict the act of deception. In the past, the only possible ways to detect deception relied on indirect measurement of physiological indicators that are linked to the acts of deception—for example, involuntary arousal of the autonomic nervous system during lying. Based on this connection, the polygraph was developed; it detects deception by measuring changes in skin conductivity and variations in the heart rate and respiration rate. Other methods of physiological measurement, such as biofeedback and electroencephalography, psychological instruments (pencil and paper tests), analysis of facial expressions and other body movements, and evaluation of handwriting and voice, have been developed with the hope that they could accurately detect the act of deception. Much effort has been made to examine the psychometric properties of these methods. The validity of these indices remains a matter of constant debate. The main limitation of these instruments is that they can provide only an indirect measure of the acts of deception—that is, measurement of the changes in bodily status that result from lying. Also, many liars are skillful in applying countermeasures to avoid detection once they have learned the design and aim of such instruments. However, recent advances in imaging technology have enabled direct observation of the activities of the human brain during various cognitive operations, including lying.

Neuroimaging. Advances in functional imaging technology have made it possible to directly observe the brain activities associated with various cognitive operations. This method of measurement, thus, offers the unique opportunity to view the activity of the brain while the act of deception is being performed. The data generated are essential for unlocking the brain processes underlying deception, bringing researchers much closer to the goal of accurately detecting deception. The application of functional magnetic resonance imaging (fMRI) technology to study deception is a very new development, but it has been gaining momentum in recent years. Currently available fMRI data on the brain regions and mechanisms underlying deception have suggested that prefrontal cortex, anterior cingulate cortex, and parietal lobe (supramarginal gyrus and superior parietal lobule) activities are involved in deception (see **illustration**). The activation patterns appear to be quite robust across genders, forms of stimuli, and native language backgrounds. To understand what roles this functional network serves during deception, it is worth examining the cognitive processes involved in deception.

Cognitive processes. Deception means that the truth is known but it is manipulated to achieve certain predetermined goals. Therefore, during the act of deception, the cognitive operations underlying the act should be more sophisticated and demanding than those involved in simply making truthful responses. In other words, deception requires the accurate recall of information followed by the conscious manipulation of the recalled information. Furthermore, to lie successfully, one has to possess the ability to inhibit the impulse to tell the truth. For example, in a test of deception using a forced-choice format, a subject could lie and potentially avoid detection by using a calculated response; that is, the subject could deliberately provide some incorrect responses to make it appear as though he or she is indeed trying to recall the information but is just unable to accurately remember all the information. Therefore, to regulate the complex behavioral pattern associated with deception, the activities of the specific neural structures involved in working memory, the cognitive control of response manipulation, the selection and adoption of retrieval strategies, and calculation of the proportion of correct responses should be essential to successful deception, especially when this behavior is tested by forced-choice memory tasks.

Neuroimaging and cognition. The prefrontal cortex is involved in many neural functions and comprises many different anatomic regions. During deception, the activity of the prefrontal cortex is important for information manipulation and integration and the generation of strategies (see illustration). Indeed, the prefrontal cortex regulates motivated responses and allows the cognitive flexibility required to adjust one's behavior according to the demands of the context as well as to the anticipation of the consequences of actions. The prefrontal cortex is also the site where mapping between the goals and the corresponding cognitive strategies takes place. Furthermore, the anterior cingulate cortex, being strongly connected to the prefrontal cortex as well as to other cortical regions, plays an important role in self-monitoring as well as modulating cognitive control for the selection and execution of goal-directed responses (see illustration).

Activity of the parietal regions during deception could relate to the demands on working memory. Indeed, the supramarginal gyrus and the superior parietal lobule seem particularly relevant to deception. Previous studies have indicated that the supramarginal gyrus and the superior parietal lobule are activated by the demands on working memory (see illustration). Neuroimaging studies have also found evidence of activity in the prefrontal cortex and supramarginal gyrus regions in complex mental calculations (see illustration). Activation of the PFC-SMG network may be required for the real-time, or spontaneous, computation that is necessary to make calculated responses and to modify behavior during deception.

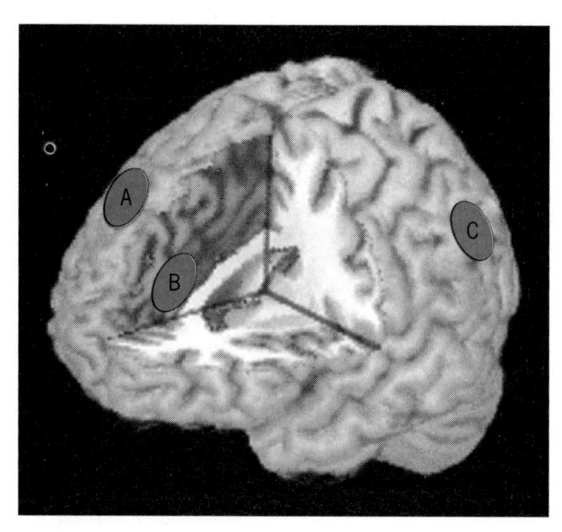

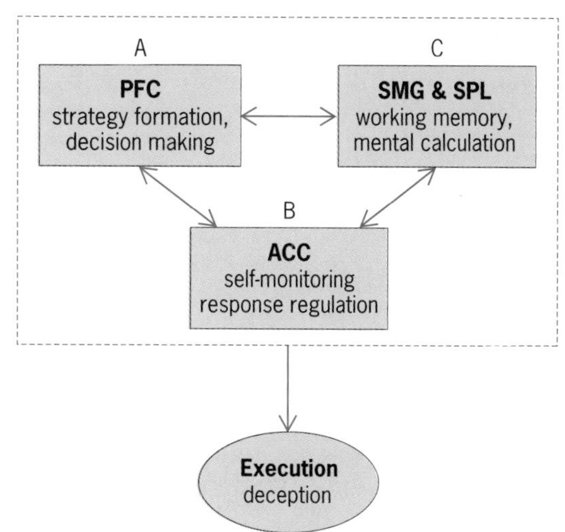

Schematic diagram of the neural areas likely involved in deception. PFC = prefrontal cortex; ACC = anterior cingulate cortex; SPL = superior parietal lobule; SMG = supramarginal gyrus.

Brain activity may vary according to the nature of the tasks involved in deception. For example, some researchers have found that the left and right prefrontal cortex regions are engaged during general deception, but that the right anterior prefrontal cortex is more involved in well-rehearsed lies that fit into a coherent story about the subject's own life history. Other investigators have suggested that the medial prefrontal cortex is involved in the carrying out of endogenous plans (previously set plans to complete a primary goal), the lateral prefrontal cortex is involved in the carrying out of exogenous plans (plans contingent upon unpredictable events), and the frontal tip of the PFC is involved in mediating the interaction between the lateral and medial prefrontal cortex regions. Activity of the amygdala has been reported in some studies of deception, which is most likely related to the emotional arousal associated with the experimental tasks of deception used in the studies.

Implications and future research. The use of functional imaging to detect deception has an advantage over behavioral methods, not only because it offers direct measurement of brain activities, but also because it may lead to more accurate detection of deception by providing knowledge of patterns of brain activities unique to specific types of deception. However, the work thus far is preliminary. While further improvement of paradigm design and image analysis methodology could increase the salience and the statistical power of the simulated deception paradigms, the question of whether the findings are generalizable to different populations in the real world awaits verification. In other words, would people suffering from damage to the parts of the brain involved in deception be unable to lie? What variation in brain activities during deception would be expected among different clinical conditions, such as antisocial personality disorder, impulse control disorders, and substance-abuse disorders? It seems that thorough experimentation and carefully designed clinical studies are needed to answer these important questions. Such knowledge will contribute to the eventual construction of a complete model describing deception.

Recently, there have been reports on the use of transcranial magnetic stimulation to study deception. Since a magnetic field emitted via transcranial magnetic stimulation could produce a transient cortical disruption, it has the potential to establish a cause-effect relationship between the roles of specific brain regions and the operation of the target cognitive task. Furthermore, the temporal brain dynamics between the different brain regions involved in deception is worth further investigation. Further studies that combine the benefits of fMRI, transcranial magnetic stimulation, and electroencephalographic technology would provide powerful paradigms that could help to unravel the mystery of the neural mechanisms underlying deception.

For background information *see* BRAIN; COGNITION; INFORMATION PROCESSING (PSYCHOLOGY); MEDICAL IMAGING; MEMORY in the McGraw-Hill Encyclopedia of Science & Technology. Tatia M. C. Lee

Bibliography. P. Ekman, *Telling Lies*, Norton, New York, 2001; G. Ganis et al., Neural correlates of different types of deception: An fMRI investigation, in *Cerebral Cortex*, vol. 13, pp. 830–836, 2003; J. Grèzes, C. Frith, and R. E. Passingham, Brain mechanisms for inferring deceit in the actions of others, *J. Neurosci.*, 24:5500–5505, 2004; F. A. Kozel, T. M. Padgett, and M. S. George, A replication study of the neural correlates of deception, *Behav. Neurosci.*, 118:852–856, 2004; A. Vrij, *Detecting Lies and Deceit: The Psychology of Lying and the Implications for Professional Practice*, Wiley, Chichester, 2000.

Capillary electrophoresis-SELEX

Every chemical reaction or biochemical pathway is influenced by how molecules interact with each other. Considering this, it is not surprising that isolating ligands with high affinity and specificity for

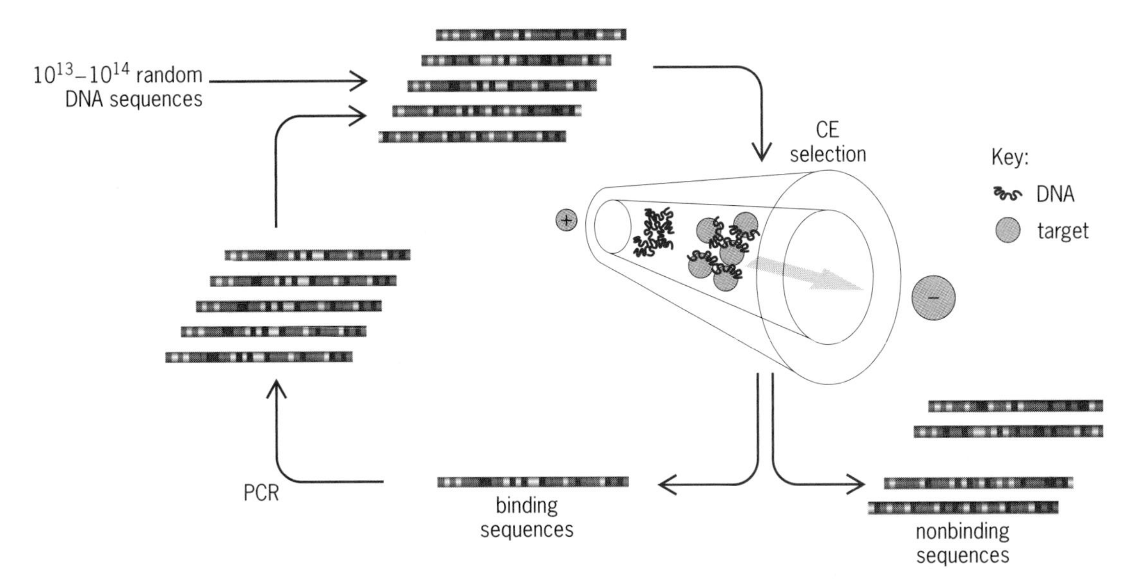

Fig. 1. CE-SELEX process used to isolate the ssDNA sequences with high affinity for a target molecule. A random-sequence ssDNA library is incubated with the target molecule. The incubation mixture is separated using CE where ssDNA that bind the target migrate through the capacity at a different velocity than nonbinding sequences, allowing them to be isolated. Binding sequences are PCR-amplified, purified, and made single-stranded, generating a new ssDNA pool ready for further rounds of enrichment. High-affinity aptamers are typically obtained after two to four rounds of CE-SELEX selection.

particular receptors is an area of intense research. Ligands for biological receptors could make excellent drug candidates. High-specificity ligands can be used to develop assays for biological, environmental, or industrial analytes of interest. Traditionally, ligands have been discovered using rational molecular design or random combinatorial methods. Both are notoriously labor-intensive processes. SELEX (systematic evolution of ligands through exponential enrichment) provides an alternative approach that uses molecular evolution to isolate high-affinity/selectivity ligands. SELEX can be simplified and shortened by incorporating capillary electrophoresis (CE) selection into the process.

SELEX. SELEX is a process for isolating high-affinity/selectivity nucleic acid ligands from large random-sequence libraries. The process starts with a large, single-stranded deoxyribonucleic acid (ssDNA) or ribonucleic acid (RNA) library (**Fig. 1**). The nucleic acids contain a 20–60-base random-sequence region, flanked by two common-sequence primer regions to facilitate amplification by the polymerase chain reaction (PCR). Depending on the nucleic acid sequence, the ssDNA or RNA can fold to form a three-dimensional structure (**Fig. 2**). If the nucleic acid folds into a shape complementary to the structure of the target, a high-affinity interaction between the two is possible. Sequences with affinity for the target molecule are selected out of the pool using one of several separation techniques, such as filtration (for proteins or other large targets) or affinity chromatography (for small molecule targets). The number of sequences in the original library that can fold into a shape that can interact with the target is typically small (on the order of thousands). These sequences are preferentially retained, while sequences with no affinity for the target are discarded, and then PCR is used to amplify those sequences with affinity for the target. The amplified sequences are purified and made single-stranded again to generate a new nucleic acid pool. Multiple rounds of selection are used to further refine the nucleic acid pool. In conventional SELEX, approximately 50% of the nucleic acids in the pool bind the target with high affinity after 8–12 rounds of selection. Then individual sequences are cloned and sequenced for further characterization. These sequences are often called aptamers.

Aptamers isolated using SELEX have great potential. Aptamers selected for large protein targets typically bind with low-nanomolar dissociation constants, similar to the affinities observed for strongly binding monoclonal antibodies. Aptamers have several advantages over antibodies. Once cloned, the exact chemical structure of an aptamer is known,

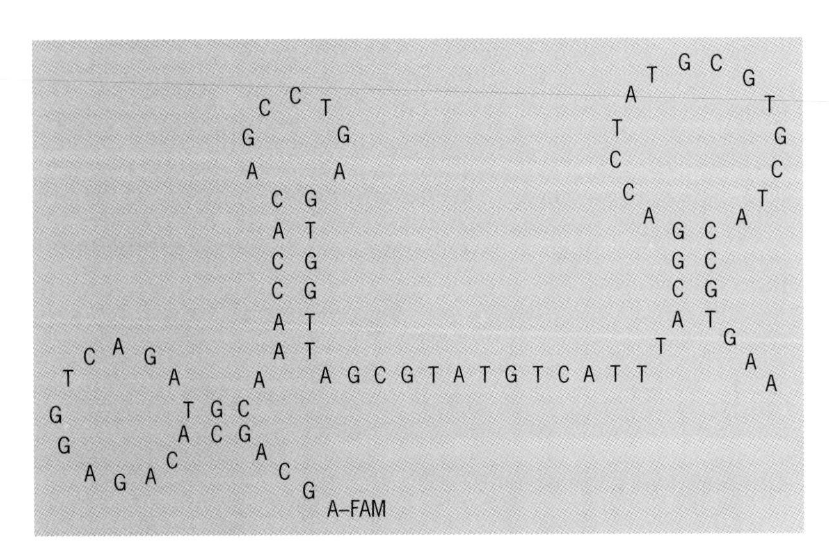

Fig. 2. Secondary structure predicted by mFOLD of a ssDNA aptamer selected using CE-SELEX with affinity for neuropeptide Y. mFOLD is a readily available algorithm for predicting nucleic acid secondary structure.

making synthesis of additional material straightforward. Aptamers are more stable than antibodies, with longer shelf lives and the ability to refold into their optimum conformation after being exposed to high temperatures or other denaturing conditions. In addition, no animals are required in the preparation of aptamers.

There are a number of drawbacks to the SELEX process, though. While less time-consuming than combinatorial approaches, SELEX is still a labor-intensive process. It will often take a well-trained researcher 4-6 weeks to complete the 8–12 selection rounds necessary to isolate high-quality aptamers. Care must be taken to avoid nonspecific background interactions that can give rise to aptamers with affinity for stationary supports or other surfaces that the nucleic acids are exposed to in the selection procedure. Affinity chromatography requires covalently linking the target to a stationary support. This linkage chemically modifies the target, potentially biasing the selection. Bias could also be introduced by elution kinetics from affinity columns. It may be difficult to elute the strongest binders from the column, biasing the selection toward the intermediate binders, which are both retained but also easily recovered from the column.

Selection of aptamers using CE-SELEX. Many of these issues can be addressed by using capillary electrophoresis for the selections (CE-SELEX). Capillary electrophoresis is an analytical technique that separates compounds based on their size and charge. A fused-silica capillary (20–50 cm long, with an inner diameter of 10–100 micrometers) is filled with a buffer solution, and then 5–50 nL of a mixture is injected onto one end of the capillary. The capillary ends are immersed in buffer solution, and 10–30 kV is applied across the capillary. Smaller and higher-charged analytes travel through the buffer faster than larger, less charged analytes. In a CE-SELEX selection, the nucleic acid library is incubated with the target molecules in free solution in a sample vial, and the incubation mixture is injected onto the capillary and separated under voltage. A key characteristic of nucleic acids is that they migrate through free solution in an electric field at the same velocity, regardless of sequence or size. This means that all sequences in the library migrate through the capillary at the same velocity. A gel must be introduced to achieve the size discrimination observed in many genetic tests. Forming a complex with the target changes the size and charge of the nucleic acid. Sequences with affinity for the target migrate through the capillary at a different velocity than the nonbinding sequences, allowing binding and nonbinding sequences to be collected in separate vials. Once collected, sequences with affinity for the target can be PCR-amplified, purified, and made single-stranded using the same procedures as in conventional SELEX. This purified pool is then ready for further rounds of enrichment.

Enrichment rate. There are a number advantages of CE-SELEX. Capillary electrophoresis is a much more powerful separation technique than filtration or affinity chromatography. The enrichment rate of the pool from round to round is therefore higher than that observed in conventional SELEX. CE-SELEX typically requires only two to four selection rounds to isolate high-quality aptamers, shortening the SELEX process from 4 to 6 weeks to several days (see **table**). Since CE-SELEX selection is performed in free solution, the target does not need to be attached to a stationary support, eliminating complicated linking chemistry and the potential for bias. Because the selection takes place in free solution, the number of surfaces where nonspecific interactions can take place is greatly reduced. In CE-SELEX, this reduction in background affinity makes negative selections, required in conventional selections, unnecessary.

Library concentration. While there are a number of characteristics that make CE ideal for SELEX, selections there are also a few concerns. The number of independent nucleic acid sequences in the initial library should be as high as is reasonably possible to increase the probability that very strongly binding sequences are present. Libraries with 10^{13}–10^{15} independent sequences are common. This is easily accommodated in techniques, such as filtration or affinity chromatography, which have limited-volume restrictions. Injection volumes in CE are limited to ~50 nL. To inject 10^{13} sequences in 50 nL, the concentration of nucleic acids in the library must be 2.3 mM. This is a very large nucleic acid concentration and degrades the peak shape in the CE separation. CE-SELEX selections can be performed with lower library concentrations if narrower peaks are necessary to separate bound sequences from nonbinding sequences. CE-SELEX has been used to successfully obtain high-affinity aptamers with as few as

Comparison of ssDNA aptamers selected for various targets using CE-SELEX and conventional SELEX selections

Target	CE-SELEX K_d, nM	Rounds of selection	SELEX K_d, nM	Rounds of selection
IgE	31 ± 18	2	6*	15
	23 ± 13	4		
HIV-RT	0.18 ± 0.07	4	1†	12
NPY	300 ± 200	4	370‡	12

*T. W. Wiegand et al., *Immunology*, 157:221–230, 1996.
†D. J. Schneider et al., *Biochemistry*, 34:9599–9610, 1995.
‡RNA aptamer. To our knowledge there is no preexisting ssDNA aptamer for NPY selected using conventional techniques: D. Proske et al., *J. Biol. Chem.*, 277:11416–11422, 2002.

10^{11} independent ssDNA sequences (10 μM) in the initial library, suggesting that this is not as significant an issue as originally thought.

Material requirements. Modifying CE-SELEX selection conditions is simplified by the fact that binding and selection both take place in free solution. Changing the stringency of the selection is as simple as changing the target concentration in the incubation buffer. Selections have been performed with target concentrations as low as 1 pM. Considering the injection volume was ~50 nL, this corresponded to ~10,000 target molecules. 10^{13} DNA molecules were injected with these 10,000 targets, suggesting that competition for binding sites on the target molecules was extremely high. The low amount of material required to perform a selection is also advantageous when performing selections against rare or expensive targets.

Target size. Another initial concern regarding CE-SELEX was the minimum size of target that could be selected against. The velocity of the nucleic acid must change significantly when it binds the target to facilitate fraction collection. Obviously, large targets will have a larger effect on velocity than small targets. Surprisingly, target size has not proved to be a significant limitation to CE-SELEX so far. Selections have been successfully performed by CE-SELEX using neuropeptide Y (NPY) as a target. NPY is approximately five times smaller than the ssDNA used in the selection. Considering mass alone, NPY would only be expected to change the velocity of a ssDNA molecule modestly, if at all. It is hypothesized that binding induces a conformational change in the ssDNA molecule, which gives rise to a change in velocity. Currently, selections against even smaller targets are being explored.

Outlook. The CE-SELEX technique greatly simplifies and shortens the procedure for isolating aptamers with high affinity and selectivity for molecular targets. Aptamers have been successfully isolated for both large and small targets. In some cases, these aptamers bind their targets even stronger than aptamers selected using conventional techniques. These aptamers have great potential for use as drugs or diagnostic agents.

For background information *see* COMBINATORIAL CHEMISTRY; DEOXYRIBONUCLEIC ACID (DNA); ELECTROKINETIC PHENOMENA; ELECTROPHORESIS; GENE AMPLIFICATION; LIGAND; LIQUID CHROMATOGRAPHY; MONOCLONAL ANTIBODIES; NUCLEIC ACID in the McGraw-Hill Encyclopedia of Science & Technology. Michael T. Bowser

Bibliography. M. T. Bowser, SELEX: Just another separation?, *Analyst*, 130:128–130, 2005; S. D. Mendonsa and M. T. Bowser, In vitro evolution of functional DNA using capillary electrophoresis, *J. Amer. Chem. Soc.*, 126:20–21, 2004; S. D. Mendonsa and M. T. Bowser, In vitro selection of aptamers with affinity for neuropeptide Y using capillary electrophoresis, *J. Amer. Chem. Soc.*, 127:9382–9383, 2005; S. D. Mendonsa and M. T. Bowser, In vitro selection of high affinity DNA ligands for human IgE using capillary electrophoresis, *Anal. Chem.*, 76:5387–5392, 2004; R. K. Mosing and M. T. Bowser, Capillary electrophoresis-SELEX selection of aptamers with affinity for HIV-1 reverse transcriptase, *Anal. Chem.*, in press, 2005; M. Rimmele, Nucleic acid aptamers as tools and drugs: Recent developments, *Chem. Biochem.*, 4:963–971, 2003.

Carbon MEMS

There has been much recent progress in the field of creating small electrical or mechanical systems known as micro-electro-mechanical systems (MEMS). Most MEMS devices are based on silicon substrates. Various silicon-based devices, such as pressure and acceleration sensors, actuators, mirrors, and gyroscopes, have been widely developed for commercial production. This is not surprising because silicon manufacturing processes are very mature, compared to such processes with any other material. However, silicon has poor mechanical (for example, brittleness) and tribological (for example, stiction) properties, which make it difficult to produce high-performance MEMS devices. Therefore, different materials are being investigated for MEMS applications, such as silicon carbide (SiC), gallium arsenide (GaAs), indium phosphide (InP), germanium (Ge), quartz, polymeric materials, and carbon.

Like silicon, carbon is an attractive and versatile engineering material and is available in more structural varieties than silicon, including diamond, graphite, amorphous carbon, fullerenes, nanotubes, nanofibers, glassy carbon, and diamondlike carbon (DLC). Carbon materials' widely differing crystalline structures and physical, chemical, mechanical, thermal, and electrical properties enable a wide variety of applications. Using carbon micro-electro-mechanical systems (C-MEMS), with sizes ranging from millimeter to micrometer (10^{-3} to 10^{-6} m), it will be possible to provide solutions—alone or in combination with silicon and other organic, inorganic, and biological materials—for microelectronic, sensor, and miniaturized power system technology.

Fabrication process. Carbon-based microstructures can be fabricated using deposition (additive) or etching (subtractive) processes. Additive processes can be used to prepare bulk, thick or thin carbon films and include deposition methods such as sputtering and evaporation. Subtractive carbon processes, such as focused ion beam and reactive ion etching, are effective in patterning various types of carbon structures, including diamond, diamondlike carbon (DLC), graphite, and glassy carbon, but are time-consuming and expensive because of the need for high-vacuum systems. Screen-printing technology is an established route for mass-producing disposable, carbon-based electrochemical sensors, which are widely used for clinical diagnostic and electroanalytic applications. However, there are severe drawbacks to screen printing since the feature resolution is very limited, properties of the printed deposit may vary depending on the batch of ink, and the resulting features have a limited aspect ratio (height-to-width ratio). These disadvantages severely limit the use of screen printing for commercial high-resolution applications.

Some interesting nontraditional carbon microfabrication methods have been explored based on the heat treatment (pyrolysis) of light-definable polymer (photoresist) patterns. These methods convert patterned polymers into carbon structures by pyrolysis at 500–1100°C (932–2012°F) under vacuum or in an inert (nonreactive gas) atmosphere. The patterns can be obtained by ultraviolet (UV) photolithography (photochemical patterning technique), laser writing, molding, or stamping. Pyrolysis of bulk polymer materials, such as phenolformaldehyde, polyimide, and polyacrylonitrile, was investigated in order to better understand their electrical properties as early as the 1970s. It was only recently that pyrolyzed photoresist for MEMS application became an active field. R. L. McCreery (Ohio State University), M. Madou (University of California, Irvine), K. Kinoshita (University of California, Berkeley) and other groups have extensively studied the characteristics of carbon produced by the pyrolysis of positive (unexposed regions remain after development) and negative (light-exposed regions remain after development) photoresists. They found that the usually smooth, resulting carbon material has electrical, chemical, and crystallographic properties that change with the pyrolysis temperature and precursor material.

The pyrolyzed photoresist structures previously described in the literature were carbon features derived from positive photoresists having very low aspect ratios. The fabrication of high-aspect-ratio and high-density C-MEMS patterns is a challenging problem because with increasing photoresist thickness, the difficulty of the photolithography process increases dramatically. Recently, the Madou group explored the microfabrication of high-aspect-ratio, three-dimensional (3D) carbon structures by pyrolyzing negative photoresist patterns in an oxygen-free atmosphere using a slow heating process. They used NANO™ SU-8, an epoxy-based photoresist. SU-8 has a very high light transparency in the UV range, which makes it ideally suited for imaging near-vertical sidewalls in very thick films.

The **illustration** shows the process for producing C-MEMS structures, as well as typical scanning electron microscope images of C-MEMS structures with aspect ratios greater than 10:1. The neat rows of carbon rods were created by exposing the photoresist through a mask peppered with holes, and then applying a chemical developer that washes away the unreacted polymer. The polymer rods were then heated at 900°C (1652°F) for 1 h in an oxygen-free atmosphere. Pyrolyzed photoresist structures shrink (depending on the initial materials) but largely maintain their shape. The resolution of the C-MEMS features can be very good and is limited only by the exposure tool's

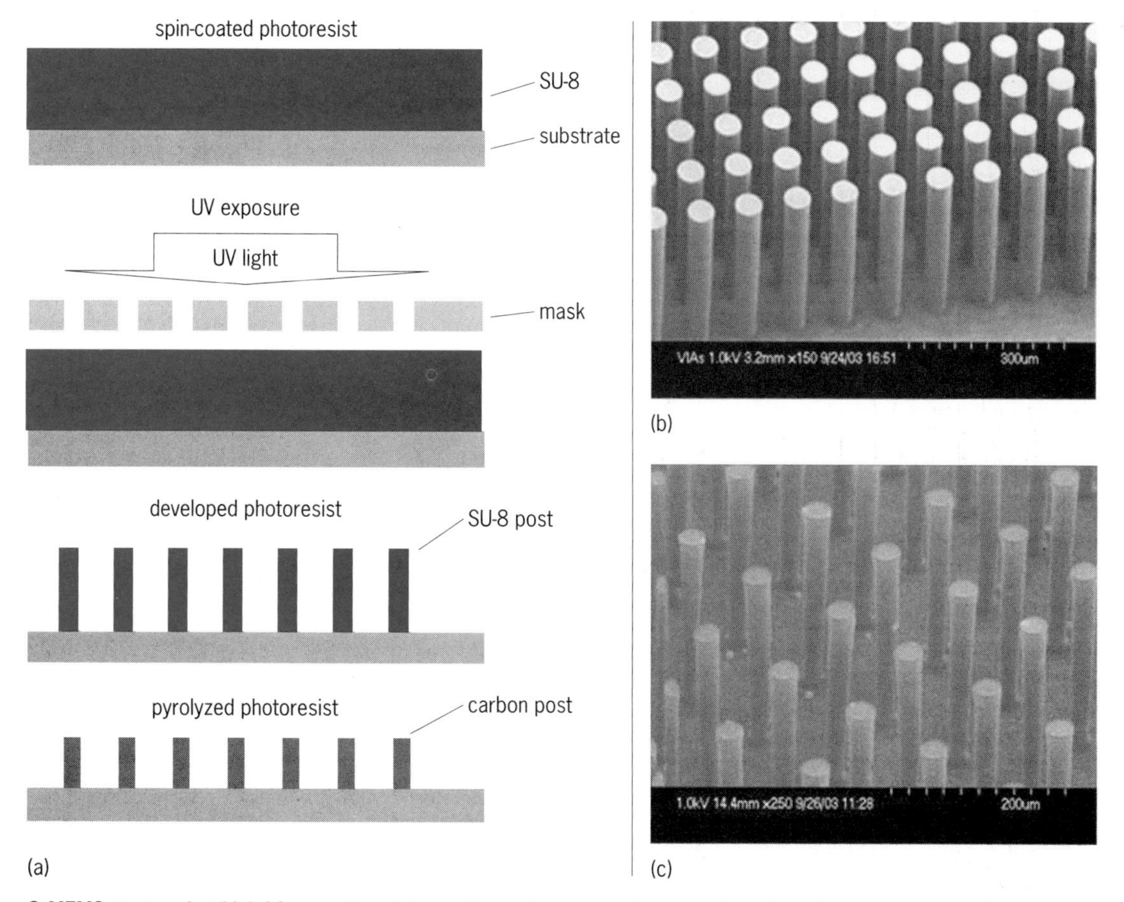

C-MEMS process, in which (*a*) a negative photoresist is patterned photochemically and pyrolyzed in an inert environment to yield carbon microstructures. Scanning electron microscope images show (*b*) photoresist arrays and (*c*) carbon arrays.

resolving power. The photoresist can be converted to different types of carbon with different specific gravity, resistivity, and crystalline structure, depending on the processing conditions. Photoresists constitute perhaps some of the most quality-controlled chemical formulations available today, so the derived C-MEMS structures are very reproducible.

Applications. C-MEMS for microbattery applications have been receiving a lot of attention recently. The trend of all advanced technologies is toward miniaturization. The future development of batteries also is toward smaller dimensions with larger storage capacity and higher energy densities. There has been a dramatic increase in demand for microbatteries for new miniature portable electronic devices, such as cardiac pacemakers, hearing aids, smart cards, MEMS devices, embedded monitors, and remote sensors with radio-frequency capability. The development of advanced microbatteries is intimately linked to the availability of new materials and the development of new miniature battery designs and manufacturing techniques. As an immediate application of C-MEMS structures, it has been demonstrated that C-MEMS posts can be charged and discharged with lithium ions (Li^+), providing a promising material and microfabrication solution to the current battery miniaturization problem. In addition, C-MEMS electrodes can be reversibly charged and discharged and have a higher capacity per unit area than unpatterned carbon films. These novel carbon-electrode arrays represent one of the critical components for building microelectrode arrays for 3D microbattery applications. Compared to traditional planar (2D) battery designs, the 3D battery concept has the advantages of (1) higher surface area on limited real estate (footprint of 25 mm^2 to 1 cm^2), (2) on-chip level compatibility with further CMOS integration, (3) switchable battery arrays, and (4) short diffusion lengths (faster charging). Ideally, such tiny battery arrays will be used in MEMS devices such as imbedded defibrillators. But the design could lead to much "smarter" batteries that could appear in other devices such as hearing aids and possibly even cell phones.

C-MEMS technology can also be applied to fuel cells, biofuel cells, and ultracapacitors, as well as biomedical applications such as DNA sensor arrays, dielectrophoresis separation devices, in vivo sensors, and drug delivery applications.

For background information *see* BATTERY; CARBON; INTERCALATION COMPOUNDS; ION SOURCES; MICRO-ELECTRO-MECHANICAL SYSTEMS (MEMS); PHENOLIC RESIN; POLYETHER RESINS; PYROLYSIS; SOLID-STATE BATTERY; SPUTTERING; VAPOR DEPOSITION in the McGraw-Hill Encyclopedia of Science & Technology. Chunlei Wang; Marc Madou

Bibliography. J. Kim et al., Electrochemical studies of carbon films from pyrolyzed photoresist, *J. Electrochem. Soc.*, 145:2314–2319, 1998; M. Madou, *Fundamentals of Microfabrication*, Boca Raton, FL, CRC Press, 1997; S. Ranganathan et al., Photoresist-derived carbon for microelectromechanical systems and electrochemical applications, *J. Electrochem. Soc.*, 147(1):277–282, 2000; C. Wang et al., C-MEMS for the manufacture of 3D microbatteries, *Electrochem. Solid State Lett.*, 7(11):A435–A438, 2004; C. Wang et al., A novel method for the fabrication of high aspect ratio C-MEMS structures, *IEEE J. Microelectromech. Sys.*, 14(2):348–358, 2005.

Cassini-Huygens mission

The international *Cassini-Huygens* spacecraft was launched on October 15, 1997, on a 7-year 3-billion-kilometer (2-billion-mile) journey to the distant planet Saturn. On July 1, 2004, the spacecraft burned its main engines for over 96 min and became the first spacecraft to go into permanent orbit around this giant planet. The next 4 months were spent on the single largest orbit (almost 120 days) of the mission, followed by the first targeted Titan flyby in October. Titan is the largest moon of Saturn, larger than the planet Mercury. The second large orbit (almost 50 days) brought the spacecraft around again to Titan in December. On the third orbit (only 32 days), the spacecraft executed a series of maneuvers that set up the release of the *Huygens* probe for an impact trajectory with Titan. Saturn orbit insertion (SOI) and the landing of the *Huygens* probe on the surface of Titan were the two critical successes of the first year in orbit, but throughout 2004 the spacecraft made 10 orbits of Saturn, traveled 63 million miles, took over 30,000 images, returned over 60 gigabytes of data, and made its first visits to several of the major satellites (see **table**). A string of remarkable discoveries have followed.

The mission is named in honor of the seventeenth-century French-Italian astronomer Jean Dominique Cassini, who discovered the prominent gap in Saturn's main rings, as well as the icy moons Iapetus, Rhea, Dione, and Tethys. The European Space Agency (ESA) Titan probe is named in honor of the Dutch scientist Christiaan Huygens, who discovered

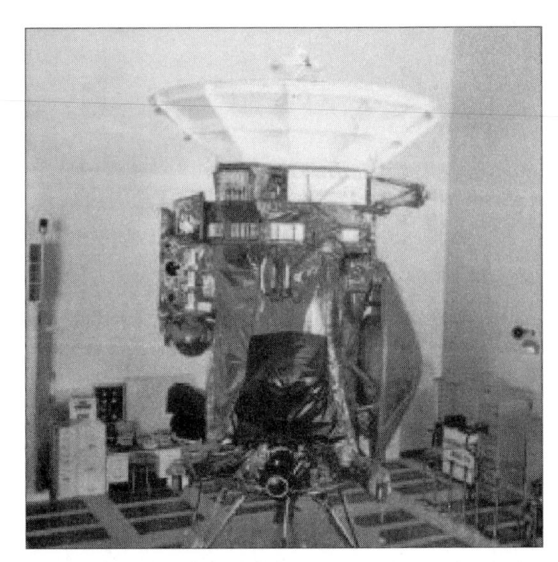

Fig. 1. *Cassini-Huygens* mission prior to launch. (NASA/JPL)

Timeline of *Cassini-Huygens* mission events for the first year in orbit

Date	Description
June 11, 2004	Close flyby (2068 km, 1285 mi) of the moon Phoebe on the way inbound to Saturn
July 01	Saturn orbit insertion (SOI) burn and closest distance (1.3 Rs) to Saturn on orbit 0, start of a 119-day orbit; Rs = radius of Saturn, roughly 60,300 km (37,400 mi)
Oct. 26	Close flyby (1174 km, 729 mi) of the moon Titan (TA)
Oct. 28	Closest distance (6.2 Rs) to Saturn on orbit A, start of a 48-day orbit
Dec. 13	Close flyby (1192 km, 741 mi) of the moon Titan (TB)
Dec. 15	Closest distance (4.8 Rs) to Saturn on orbit B, start of a 32-day orbit
Dec. 16	Probe Targeting Maneuver (puts *Cassini-Huygens* on collision course with Titan)
Dec. 25	*Huygens* release
Dec. 27	Orbit Deflection Maneuver (takes *Cassini* orbiter off collision course with Titan)
Dec. 31	Flyby (123,000 km, 76,430 mi) of the moon Iapetus
Jan. 14, 2005	*Huygens* probe descent
Jan. 14	Close flyby (60,000 km) of the moon Titan (TC)—*Huygens* landing on Titan
Jan. 16	Closest distance (4.8 Rs) to Saturn on orbit C, start of a 33-day orbit
Feb. 15	Close flyby (1579 km, 981 mi) of the moon Titan (T3)
Feb. 17	Closest distance (3.5 Rs) to Saturn on orbit number 3, start of a 21-day orbit
Feb. 17	Close flyby (1265 km, 786 mi) of the moon Enceladus
Mar. 09	Close flyby (520 km, 323 mi) of the moon Enceladus
Mar. 09	Closest distance (3.5 Rs) to Saturn on orbit number 4, start of a 21-day orbit
Mar. 29	Closest distance (3.5 Rs) to Saturn on orbit number 5, start of a 21-day orbit
Mar. 31	Close flyby (2404 km, 2404 mi) of the moon Titan (T4)
Apr. 14	Closest distance (2.6 Rs) to Saturn on orbit number 6, start of a 16-day orbit
Apr. 16	Close flyby (1027 km, 638 mi) of the moon Titan (T5)
May 03	Closest distance (3.6 Rs) to Saturn on orbit number 7, start of an 18-day orbit
May 03	*Cassini* views the Earth as it passes behind Saturn and Saturn's rings
May 21	Closest distance (3.6 Rs) to Saturn on orbit number 8, start of an 18-day orbit
May 21	*Cassini* views the Earth as it passes behind Saturn and Saturn's rings
June 08	Closest distance (3.6 Rs) to Saturn on orbit number 9, start of an 18-day orbit
June 08	*Cassini* views the Earth as it passes behind Saturn and Saturn's rings
June 26	Closest distance (3.6 Rs) to Saturn on orbit number 10, start of an 18-day orbit
June 26	*Cassini* views the Earth as it passes behind Saturn and Saturn's rings

Titan in 1655, followed in 1659 by his announcement that the strange Saturn "moons" seen by Galileo in 1610 were actually a ring system surrounding the planet. Huygens was also famous for his invention of the pendulum clock, the first accurate timekeeping device.

The *Cassini-Huygens* spacecraft is the largest interplanetary spacecraft ever flown to the outer solar system, standing 6.8 m (22 ft) high, roughly 4 m (13 ft) in diameter, and weighing over 6 tons when fully fueled (**Fig. 1**). It consists of the *Cassini* orbiter spacecraft and *Huygens* Titan probe and was specifically designed to explore the Saturn system, which includes the planet, its rings and magnetosphere, the large moon Titan, and many icy moons. The *Cassini* orbiter has 12 instruments, and the *Huygens* probe has 6, with over 17 countries and 250 scientists worldwide working on the project. During its 4-year, 75-orbit, prime mission, the orbiter is making 45 close flybys of Titan, both for gathering more information about Titan and for using Titan to make gravity-assisted orbit changes. The ability to change orbits is allowing close flybys of icy satellites, detailed studies of the rings and Saturn at various lighting geometries, and exploration of the magnetosphere at a variety of locations.

Close flyby of Phoebe. In June 2004, inbound to Saturn orbit insertion, the *Cassini* orbiter collected data on this small moon at the edge of the Saturn system during the only flyby opportunity in the tour or extended mission. The images from Phoebe show a heavily cratered world, covered mostly with water ice and a patchy clustering of silicate and organic material (**Fig. 2**). So much volatile ice tells scientists that Phoebe probably formed in the outer solar system, where water ice is abundant, and was then captured by Saturn's gravity.

Fig. 2. Mosaic of two images of Phoebe taken on June 11, 2004, at a distance of approximately 32,500 km (20,200 mi). (*NASA/JPL/Space Science Institute*)

Is Saturn's rotation slowing? Since Saturn has no visible solid surface, its rotation rate is determined by measurement of periodic radio bursts. On approach to Saturn, new measurements suggest that Saturn has slowed down by a remarkable 6 min, roughly 1%, between the *Voyager* flybys in 1980–1981 and the *Cassini-Huygens* mission 23 years later. Even crashing all of the material from all of the rings and moons into the planet could not change the rotation rate by this much. Scientists speculate that the cause might be a nonuniform rotation rate, similar to the Sun's. At this time, the length of a day is on Saturn still remains unknown.

Observations of Saturn's rings. On July 1, 2004, *Cassini* had to burn its main engines for over 96 min to be captured by Saturn's gravity and enter orbit. There are two identical main engines on *Cassini* that provide about 100 lb of thrust each. The fuel is MMH (monomethylhydrazine) and NTO (nitrogen tetroxide). If a large adult strapped both engines to his thighs, they would not lift him off the ground. The spacecraft completed the burn when it was directly over the rings of Saturn and began taking the closest-ever, highest-resolution images, along with measurements of fields and particles as the spacecraft skimmed over the rings (**Fig. 3**). Discoveries included strawlike clumps several kilometers long in the A ring, an oxygen atmosphere just above the rings, signatures of marble-sized meteoroids impacting the rings, variation of the amount of pure water ice, and evidence for slowly rotating ring particles.

Fig. 3. Closest-ever high-resolution images of Saturn's rings, obtained on July 1, 2004, immediately following Saturn orbit insertion. (*NASA/JPL/Space Science Institute*)

New radiation belt. Also, during Saturn orbit insertion, the magnetospheric imaging instrument discovered a new radiation belt between the inner edge of the D ring and the top of Saturn's atmosphere. It is much smaller and less energetic than the main belt, and is detected by the emission of fast neutral atoms created as its energetic ions interact with gas clouds in the same region.

Unusual magnetic field axis alignment. Saturn is unique in the solar system in that its spin axis is almost exactly aligned with its magnetic field, something that theorists say should not occur. The "dynamo" theory of planetary magnetic fields has the complex, constant changing of magnetic field strength driving the creation of a planet-wide field. If the rotation axis and magnetic axis are aligned, the field strength does not constantly change and the planet-wide magnetic field damps away in a matter of decades. *Cassini* has measured the alignment to an accuracy of 0.1° and still cannot see a difference.

Titan observations by Cassini orbiter. In October and December 2004, on the first and second orbits around Saturn, the *Cassini-Huygens* spacecraft made the first close flybys of Titan. Titan is the only moon in the solar system completely enshrouded in a thick atmosphere (**Fig. 4*a***). The surface of Titan is obscured by a photochemical haze. Multiple instruments on *Cassini* have special observation techniques to peer through the haze to the surface (Fig. 4*b*,*c*). The visible, infrared, and radar images of the surface taken on these first several flybys reveal a very exciting surface: complex patterns of light and dark regions, evidence of surface flows and outwash channels, perhaps wind erosion, and bright tendrils extending over dark areas. There are few craters and at least two possible cryovolcanos observed on the surface (Fig. 4*d*). A cryovolcano spews forth solid water (ice), liquid water, and vapor-phase volatiles, like ammonia and methane, instead of the familiar molten rock and lava from an Earth-based volcano.

However, the surface observations were not the highest priority of the first Titan flybys. More importantly, the *Cassini* orbiter made extensive observations of Titan's atmosphere to validate the atmospheric models the *Huygens* probe needed a few months later. Titan's atmosphere is thicker than that of Earth, roughly 1.5 times the pressure and 10 times the density of Earth's. The main constituents of Titan's atmosphere are nitrogen (~98%) and methane (~2%), very similar to the atmosphere of the Earth before life arose.

Prometheus stealing material from F ring. In October 2004, *Cassini* cameras imaged the small shepherding satellite Prometheus stealing particles from Saturn's F ring. When Prometheus was near its closest approach to the F ring, a tenuous streamer of ring particles trailed from the ring to Prometheus (**Fig. 5**). Gravitational effects from Prometheus also

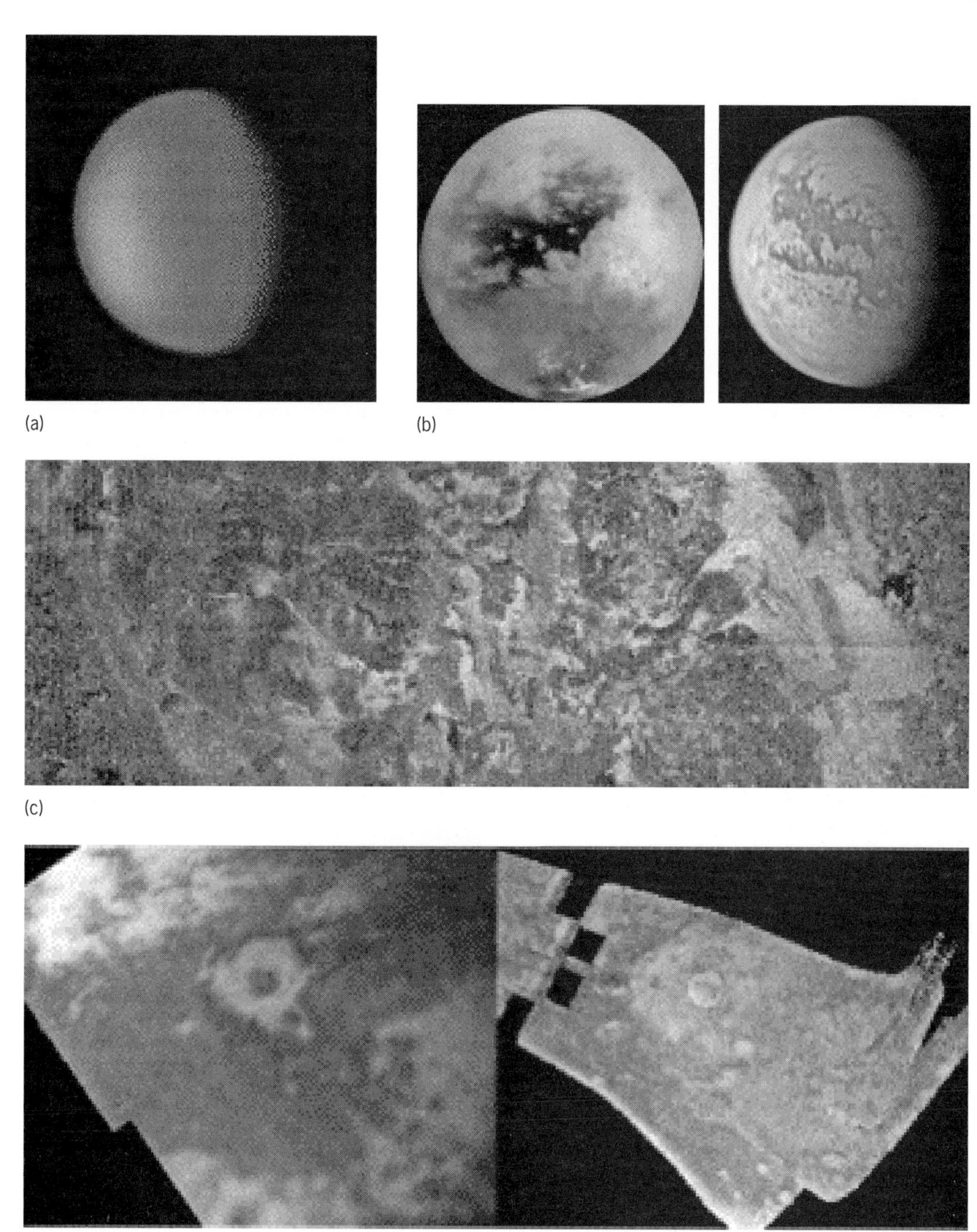

Fig. 4. Images of Titan from the *Cassini Orbiter*. (*a*) Titan as it would be seen by the human eye, completely enshrouded in a smoggy haze (*NASA/JPL/Space Science Institute*). (*b*) Two views of Titan's surface obtained with the *Cassini Orbiter* cameras, which are uniquely equipped to peer down through the haze (*NASA/JPL/Space Science Institute*). (*c*) *Cassini* radar image of Titan's surface, providing the highest resolution (*NASA/JPL*). (*d*) Same crater imaged by two different instruments (*NASA/JPL/University of Arizona*).

cause kinks and clumps in the F ring and drapelike structures in the faint material near the F ring, possibly with the help of other nearby, undiscovered moonlets.

Dione mystery resolved. In December 2004, one of the more compelling mysteries of the Saturn system was resolved. To the surprise of many *Cassini* scientists, the mysterious wispy terrain on the moon Dione was revealed to be tectonic fractures, with the clifflike walls of the fractures being particularly bright (**Fig. 6**).

Giant mountain range on Iapetus. The moon Iapetus has been a mystery almost since it was discovered in 1671. The leading hemisphere is darker than the blackest coal (reflecting 4% of the light falling on it), and the trailing hemisphere is brighter than the brightest snow (reflecting 50% of the light falling on it). One hypothesis is that the dark material comes from an external process and is deposited on the leading hemisphere as Iapetus moves around Saturn in a tidally locked rotation. A second hypothesis is that the dark material has erupted from inside

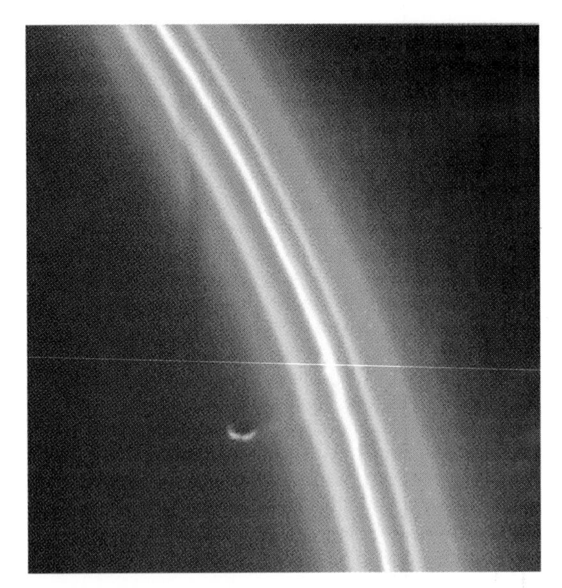

Fig. 5. Image taken on October 29, 2004, at a distance of about 782,000 km (486,000 mi), showing Prometheus, one of two small shepherding moons of the F ring (the other is Pandora), working its influence on the multistranded and kinked ring. (*NASA/JPL/Space Science Institute*)

Iapetus. The December 2004 flyby displayed a distribution of the dark material which leads scientists to favor the idea that the source is from outside Iapetus. Additionally, several large impact basins were revealed in great detail. However, most surprising was the discovery of a giant equatorial mountain range (**Fig. 7**). This "belly band" is nearly 20 km (12 mi) high, 20 km (12 mi) wide, and 1300 km (800 mi) long on Iapetus. A similarly proportioned feature on Earth would be 20 times taller than Mount Everest and 3 times longer than the Andes.

Probe release and Huygens mission. The European Space Agency's *Huygens* probe separated from the *Cassini Orbiter* on December 25, 2004. On January 14, 2005, the *Huygens* probe entered the atmosphere of Titan and successfully descended to the surface. It was the culmination of a very challenging mission that had been conceived more than 20 years earlier. During the journey to Saturn, the *Huygens* probe traveled attached to the *Cassini* orbiter. During the 7-year cruise, the probe was subjected to regular in-flight checkouts (16 in total) to monitor the health of its subsystems and scientific instruments. At these times, maintenance was performed and calibration measurements made in preparation for the mission at Titan.

Parachutes controlled the descent of the probe through Titan's atmosphere, and the aerodynamic conditions under which the main parachute should be deployed were critical. The correct instant for parachute deployment was therefore determined by the probe's onboard computers that processed the measured values of the accelerometers, which monitored the probe deceleration. Pyrotechnic devices fired a mortar, which pulled out a pilot chute and released the back cover that in turn pulled out the main parachute. The front shield was then released 30 s later when the probe under the main parachute was expected to have stabilized. At this time, roughly 160 km (100 mi) above the surface, the probe instruments began collecting and transmitting data on Titan's atmosphere. The main parachute slowed the probe down so much that it would have allowed the probe only to sample the very top of Titan's interesting atmosphere, so the main parachute was ejected in favor of a smaller chute that allowed the probe to complete its descent to the surface. The probe descended for a total of 2 h 27 min, landed in an upright position, and continued to radiate data for another 1 h 12 min from the surface of Titan.

One of the main scientific goals of the *Huygens* probe was to collect information on Titan's atmosphere. Pressure and temperature were measured all the way to the surface, and the composition was sampled at several altitudes, extracting the basic components as well as trace constituents and isotopic

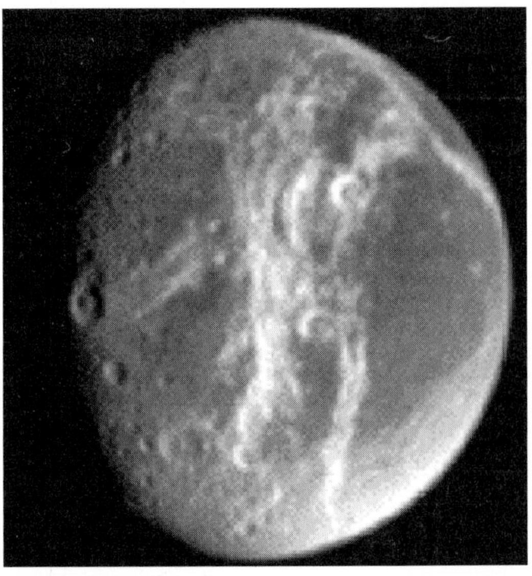

(a)

(b)

Fig. 6. *Cassini* images of Dione. (*a*) Image displaying the distinctive wispy terrain. (*b*) Very detailed image, taken on December 14, 2004, which reveals that the wispy terrain is a series of large fractures with extremely bright clifflike walls. (*NASA/JPL/Space Science Institute*)

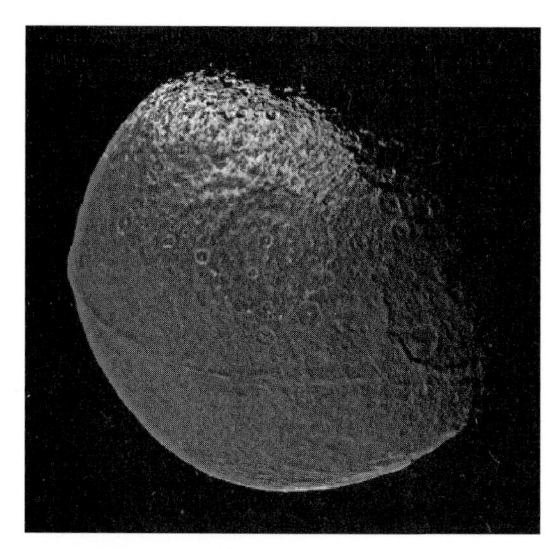

Fig. 7. Mosaic of 4 *Cassini* images of Iapetus taken on December 21, 2004, at about 172,400 km (107,000 mi), which captured the unique feature at Iapetus's equator nicknamed the "belly band." (*NASA/JPL/Space Science Institute*)

ratios. Nitrogen and methane were confirmed as the main constituents, with the methane concentration increasing steadily from 2% in the upper atmosphere to 5% at the surface. Methane clouds were detected at about 20 km (12 mi) altitude, and methane or ethane fog was seen near the surface. The probe experienced considerable turbulence above 100 km (62 mi), but near the surface the winds are weak (1–2 m/s or 2–4 mi/h).

The camera onboard the *Huygens* probe took 375 images during the entire descent and 231 on the surface. The descent images were striking, showing the same light and dark patterns that the orbiter saw, only in greater detail (**Fig. 8***a*,*b*). Short stubby drainage channels were seen, leading to what looked like a shoreline, perhaps liquid methane springs (Fig. 8*a*). Finally, the most remarkable image, that of Titan's surface (Fig. 8*c*), was returned to the Earth.

Not knowing if the probe would land on a flat surface, a mountainside, or in an ethane lake, the surface images and measurements were a shot in the dark. The first instrument to strike the surface was a penetrometer attached to the bottom of the probe. It was designed to measure the force of the impact and reveal the properties of the material on the surface. The surface of Titan resembles loose sand or clay, with either a solid thin crust or a pebble right where the penetrometer impacted. Heat from the probe warmed the surface underneath the probe, and an interesting increase in methane was detected while the probe was still relaying data to the overhead orbiter.

Enceladus observations. In February and March 2005, the *Cassini* orbiter made the first flybys over Enceladus, a small moon, only 520 km (320 mi) across, which is the shiniest object in the solar system, reflecting more than 99% of light that falls on its surface. Enceladus is embedded in the dense part of the E ring, leading some scientists to speculate that Enceladus contributes material to the E ring, a theory supported by images which clearly show evidence for endogenic geologic activity—activity generated internally to the moon. There are fractures covering much of the surface, and closer images reveal even finer-scale fracturing (**Fig. 9**). Surprisingly, Enceladus generated interesting magnetic field observations. Scientists were expecting Saturn's magnetic field to carry down to the surface, and then beyond and behind the satellite. That this did not happen implies

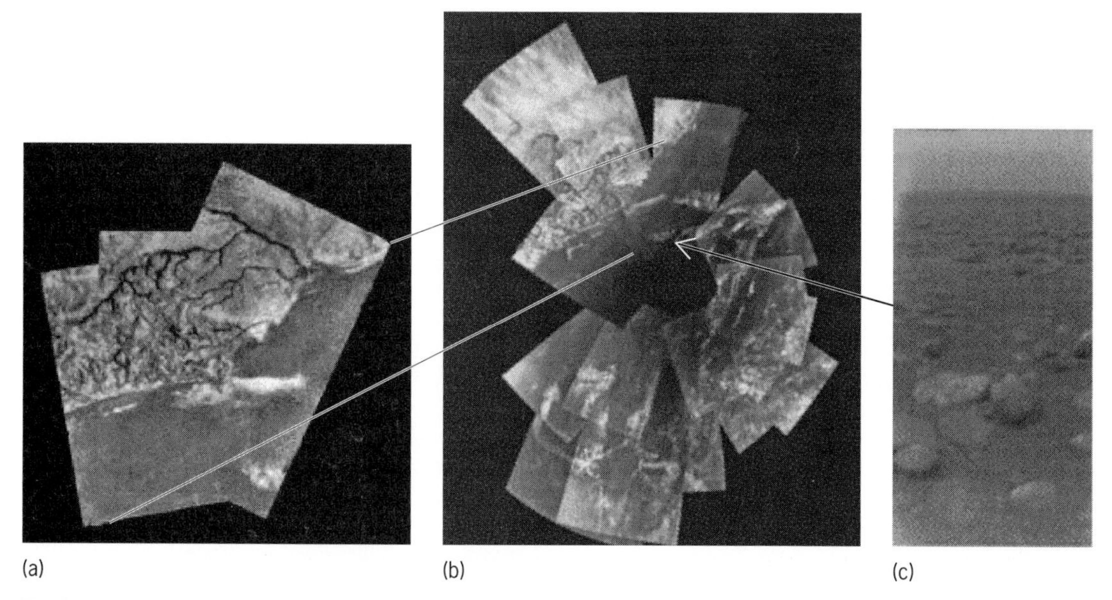

Fig. 8. Images from *Huygens* descent and landing on Titan on January 14, 2005. (*a*) Three-frame mosaic taken at roughly 17 km (11 mi) above the surface of Titan, with unprecedented detail of a high-ridge area including the flow down into a major river channel from different sources. (*b*) Composite of 30 images taken at distances of 13 km (8 mi) down to 8 km (5 mi). (*c*) Image from the surface of Titan. (*NASA/JPL/ESA/University of Arizona*)

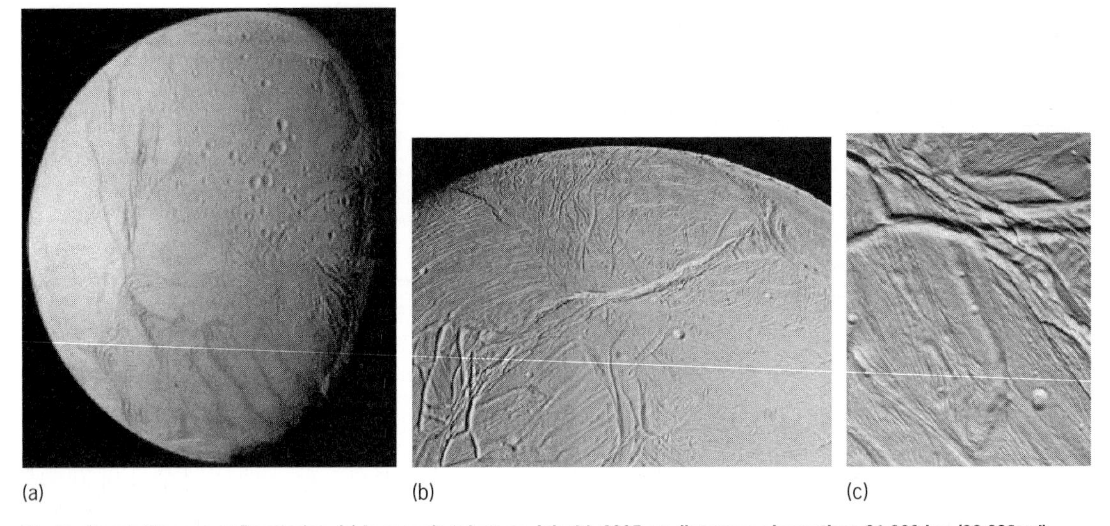
(a) (b) (c)

Fig. 9. ***Cassini*** **images of Enceladus. (*a*) In mosaic taken on July 14, 2005, at distances closer than 61,300 km (38,000 mi), Enceladus displays the myriad of faults, fractures, folds, troughs, and craters that make this satellite especially intriguing to planetary scientists. (*b*) Close-up taken on February 17, 2005, at distances closer than 26,000 km (16,000 mi), highlighting many of these features. (*c*) Close-up, also taken on February 17, 2005, at distances closer than 11,000 km (6700 mi), showing several different kinds of deformation, and a small population of craters, indicating the youthfulness of the terrain. (*NASA/JPL/Space Science Institute*)**

that something was acting as an obstacle—perhaps a thin atmosphere on this little moon.

For background information *see* SATELLITE (ASTRONOMY); SATURN; SPACE PROBE in the McGraw-Hill Encyclopedia of Science & Technology.

Trina Ray; Linda Spilker; Claudio Sollazzo

Bibliography. *Cassini* arrives at Saturn, *Science*, Special Section, 307:1222–1276, Feb. 25, 2005; *Cassini* reveals Titan, *Science*, Special Section, 308:968–995, May 13, 2005; A. Hendrix and J. Lunine, *Cassini's* cornucopia of moons: 7 satellites in 7 months at Saturn, *Planet. Rep.*, pp. 12–17, May/June 2005; D. Matson, L. Spilker, and J. P. Lebreton, The *Cassini/Huygens* mission to the Saturnian system, *Space Sci. Rev.*, 104:1–58, July 2002; J. C. Zarnecki, Destination: Titan, *Nat. Hist.*, pp. 26–32, December 2004/January 2005.

Caves and climate change

From instrumental records, we know that the Earth has warmed by 0.5°C (0.9°F) since 1860. Only through the study of the climate history over the past several centuries and millennia can we truly assess whether this warming should be expected in a naturally changing climate or if it is being caused by human activities. Paleoclimatology is the reconstruction of past climate changes. In order to accurately identify the causes and mechanisms of climatic change, it is necessary to develop high-resolution paleoclimate records which tell us about past climate changes on seasonal, annual, and decadal time scales. In addition to helping us understand the natural spatial and temporal patterns of climate variability, these high-resolution records provide an important means of testing the results of general circulation models, which are used to predict future climate changes that may result from anthropogenic carbon dioxide (CO_2) emissions. As historical and instrumental climate records cover only the very recent past, it is necessary to use indirect (proxy) records for a longer-term perspective on climatic variability. Paleoclimate records may be derived from natural archives, which incorporate climate-dependent physical or chemical proxy variables in their structure. The most complete and robust paleoclimate proxy records to date are ice-core records, marine-sediment records, and the Chinese loess record. While these records have provided key information about the Earth's climate history, they often are limited by low temporal resolution, uncertain chronologies, and poor geographic coverage.

Speleothems, or cave calcite deposits (that is, stalactites and stalagmites), are well suited for paleoclimate reconstruction, having several advantages over other established paleoclimate recorders. Speleothems, found in numerous locations around the world, tend to be very pure and well preserved. They commonly contain visible annual growth bands and record multiple types of terrestrial paleoclimatic proxy data, such as stable isotope composition ($^{18}O/^{16}O$, $^{13}C/^{12}C$) and trace-element concentration (for example, Mg, Sr, Ba, and P), on a wide range of time scales (subannual to 10^5 years). The primary advantage of speleothems is that they can be dated precisely as far back as 500,000 years ago via uranium-series radiometric dating. Speleothems typically form with uranium concentrations ranging from hundreds of parts-per-billion to several parts-per-million, but essentially no thorium due to its extreme insolubility in water. Therefore, any ^{230}Th present must have been produced by the radioactive decay of ^{234}U. By measuring the ^{230}Th to ^{234}U ratio, the age of deposition can be calculated. During the last 20 years, technological advances in mass

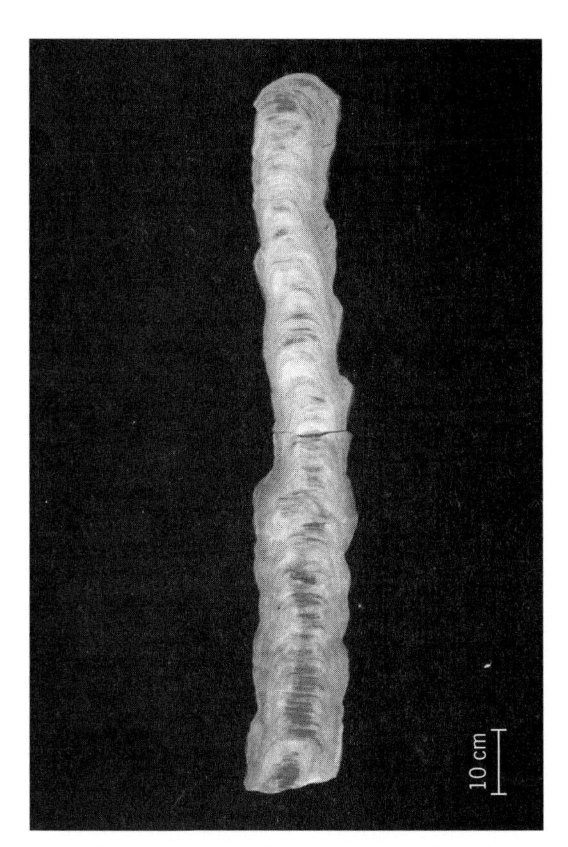

Fig. 1. Photo of stalagmite Q5, collected from Qunf Cave in Southern Oman. This stalagmite was U-series dated at 18 points along its growth axis. It grew continuously during two stages, from 10,300 to 2700 years ago, and from 1400 to 400 years ago, separated by a brief hiatus. (*Photo courtesy of Dominik Fleitmann*)

spectrometry have improved our ability to precisely date small speleothem calcite samples to within 1% or less of their age (for example, 10,000 ± 100 years).

Climatic proxies in speleothems. Stalagmites (**Fig. 1**), the type of speleothem most suitable for paleoclimate reconstruction, are calcium carbonate ($CaCO_3$) deposits precipitated from dripping water (**Fig. 2**). This water, which originally falls as precipitation, becomes slightly acidic in the soil due to increased CO_2 concentrations. The water then percolates through the carbonate bedrock (epikarst) and dissolves calcium carbonate. Once the water drips into the cave atmosphere, CO_2 degassing leads to the precipitation of calcite speleothems. The speleothem isotopic ($^{18}O/^{16}O$, $^{13}C/^{12}C$) and trace-element (Mg, Sr, P, U, and so on) geochemistry is essentially controlled by the oxygen isotopic composition of the rainfall, the carbon isotopic composition of the soil CO_2, and the trace-element and carbon isotopic composition of the soil and bedrock minerals. To derive paleoclimate records from stalagmites, the samples must be carefully selected, keeping in mind the importance of cave conservation. Generally, only a few samples are needed from a given cave, and most researchers will, if possible, use samples that have already been broken off. The speleothem samples are sectioned lengthwise along their growth axis and microsampled at approximately 0.5 mm resolution. The samples are then analyzed for stable-isotope and trace-element composition using a mass spectrometer. Interpreting variations in geochemistry in terms of past climate is not straightforward, as numerous environmental processes control how these proxies are actually incorporated in speleothem growth bands.

Oxygen isotopic composition. The oxygen isotopic composition ($^{18}O/^{16}O$) is the most understood and widely used geochemical proxy in speleothems. The $^{18}O/^{16}O$ ratio in speleothem calcite precipitated under equilibrium conditions (in the absence of evaporation or rapid CO_2 degassing) is a function only of temperature and the $^{18}O/^{16}O$ ratio of the water it precipitated from. The $^{18}O/^{16}O$ ratio in speleothems is often useful as a paleotemperature or paleorainfall proxy, since the temperature in caves deeper than 10 m (33 ft) is remarkably constant and equal to the mean annual temperature at the surface. In most cases, the $^{18}O/^{16}O$ ratio of the cave drip water is closely related to the mean annual rainfall, which itself is a function of temperature and the amount of precipitation. Obtaining quantitative paleoclimate records from oxygen isotopic composition is difficult, though, as the dependency of $^{18}O/^{16}O$ in rainfall on temperature and rainfall amount is complex and varies spatially. In general, the $^{18}O/^{16}O$ ratio

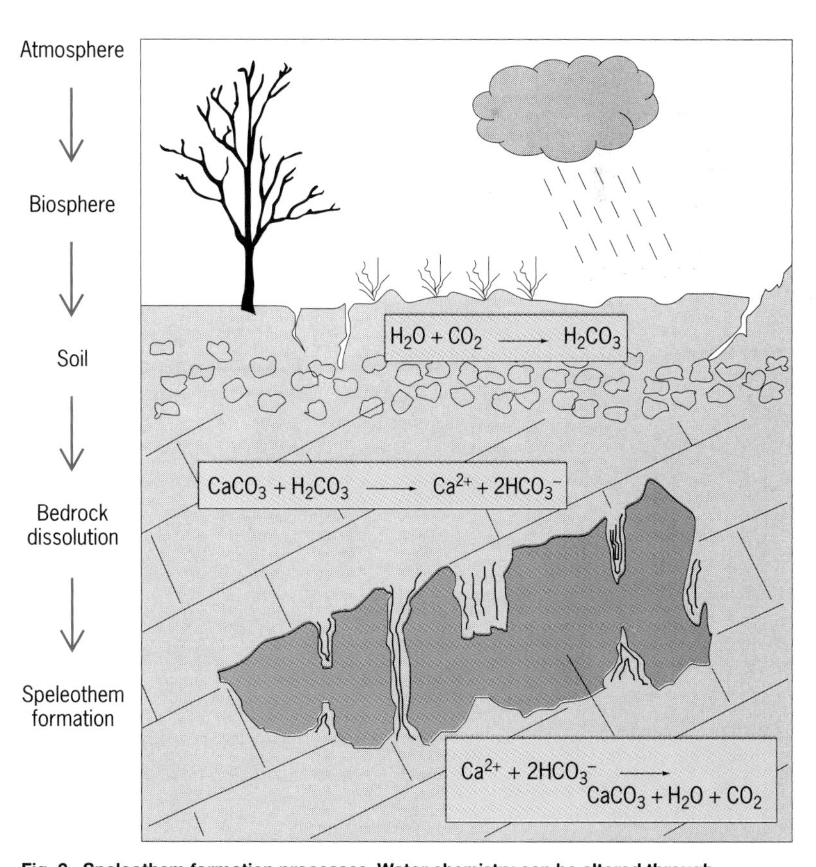

Fig. 2. Speleothem formation processes. Water chemistry can be altered through interactions with the atmosphere, biosphere, soil, and bedrock before the water enters the cave environment, where CO_2 degassing and calcite precipitation occur. The conditions at the surface (temperature, vegetation, moisture source, amount of rainfall) and in the cave (temperature, relative humidity) control how these chemical signals are incorporated in speleothems. It is this dependence that allows the use of speleothem-based geochemical proxies for paleoclimate reconstruction.

at mid-to-high latitudes is controlled more by temperature, while in low-latitude regions it is controlled more by rainfall amount. In addition, changes in atmospheric circulation, evaporation, and the composition of the source water can affect the $^{18}O/^{16}O$ ratio in speleothems.

Carbon isotopic and trace-element composition. Other speleothem geochemical proxies, which are increasingly being used, include the $^{13}C/^{12}C$ ratio and the trace-element composition of speleothem calcite. The carbon in speleothem calcite is derived from three main sources with different carbon isotopic compositions: atmospheric CO_2, soil CO_2, and the carbonate bedrock. Carbon isotope variations in speleothems are useful for reconstructing changes in the dominant vegetation. The vegetation signal may be obscured by fractionation in the cave, degassing of ground water en route to the cave, or nonequilibrium between the soil waters and soil CO_2. Trace-element variations along speleothem growth axes may reflect past changes in temperature or rainfall amount. Interpretation is complicated, as the variations may be controlled by several processes, including changes in the amount of calcite precipitated along the flow path, differential mineral dissolution (for example, calcite versus dolomite), vegetation productivity, temperature, and speleothem growth rate. Much recent work has been devoted to investigating these processes, but in many cases the trace-element controls may be highly site-specific, so calibration studies need to be done for individual caves.

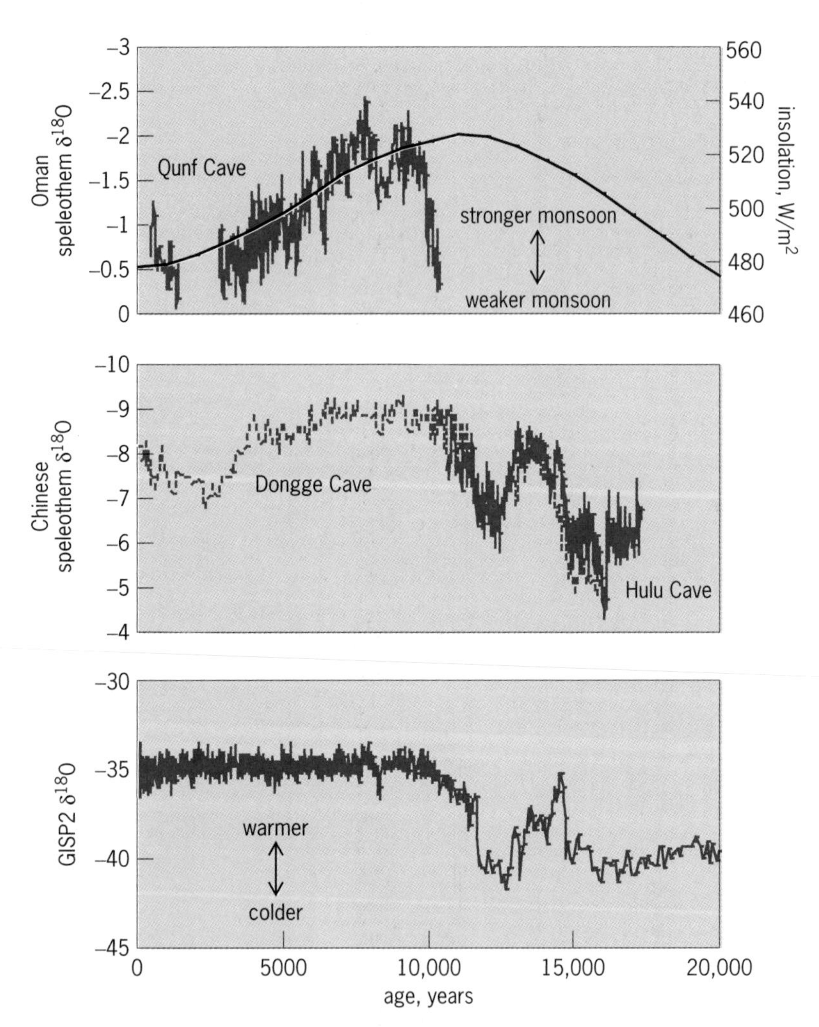

Fig. 3. Speleothem oxygen isotope records of Asian monsoon variability, compared with the Greenland ice core temperature record (GISP2) and solar insolation over the last 20,000 years. $\delta^{18}O$ is simply the $^{18}O/^{16}O$ ratio normalized to a standard. The Oman speleothem $\delta^{18}O$ (D. Fleitmann et al., 2003), and the Hulu Cave and Dongge Cave $\delta^{18}O$ (Y. J. Wang et al., 2001; D. X. Yuan et al., 2004) record Indian monsoon intensity and East Asian monsoon intensity respectively.

Speleothem records of Asian monsoon. The Asian monsoon system, consisting of the East Asian monsoon and the Indian monsoon, is characterized by seasonal changes in air pressure, wind, precipitation, and temperature gradients. During the winter, cold air over the continent and warm air over the ocean causes a pressure gradient, producing dry winds that flow over southern Asia toward the sea. The pressure gradient is reversed during the summer as the air over the continent becomes warmer than the air over the sea, producing moisture-bearing winds that sweep over the continent. The presence of the Qinghai-Tibetan Plateau intensifies the uplift and cooling of these warm air masses, leading to increased precipitation over the continent.

To date, terrestrial paleoclimate records from the region include ice-core records, loess records, lake records, tree-ring records, and speleothem records. These proxy records indicate that the Asian monsoon intensity has undergone dramatic changes throughout the past several glacial cycles, with a much weaker summer monsoon during glacial periods. A long-term record of monsoon activity will aid in predicting anomalous monsoons, which can have severe economic, environmental, and societal impacts. Even small variations in the timing and intensity of the summer monsoon can have significant consequences, such as low crop yields and famine during weak summer monsoons, or devastating floods during exceptionally strong summer monsoons.

Hulu Cave record. Y. J. Wang and others presented speleothem oxygen-isotope records from Hulu Cave, China (32°30′N, 119°10′E), which suggest a strong link between the East Asian monsoon intensity and temperature variations reconstructed from the Greenland Ice Sheet Project Two (GISP2) ice core (**Fig. 3**). In this record, oxygen-isotope variations were interpreted as reflecting the amount of summer monsoon rainfall, with lower $^{18}O/^{16}O$ ratios corresponding to increased rainfall during strong monsoons, coincident with periods of warm temperatures in the North Atlantic. The long-term trend appears very similar to the pattern of Northern Hemisphere summer insolation or the intensity of incoming solar radiation, suggesting that high-insolation periods may be characterized by an increased continent-ocean temperature gradient, and thus stronger monsoons. Several abrupt events are superimposed on this long-term record, with monsoon intensity changing rapidly over centuries or less. These events correspond to abrupt climate changes in the North Atlantic (Heinrich events) observed in GISP2, supporting the idea that these millennial-scale Heinrich events affected regions far from the North Atlantic and were most likely periods of massive reorganization of oceanic and atmospheric circulation patterns.

Dongge Cave record. D. X. Yuan and others presented a speleothem oxygen-isotope record from Dongge Cave, China (25°17′N, 108°5′E), which covered much of the past 160,000 years (Fig. 3). The major contributions of this record are that it greatly increased the time range of the Chinese speleothem record, it overlapped and agreed remarkably well with part of the Hulu Cave record, and it provided a precisely dated record of the last interglacial period. The timing of the last interglacial onset is a question of major debate, as some records suggest that warming began prior to the rise in solar insolation at this time. This calls into question the assumption that glacial cycles are predominantly controlled by insolation changes and points to some other forcing mechanism. In the Dongge Cave record, the onset occurred at 129.3 ± 0.9 thousand years ago, which is consistent with an insolation forcing mechanism.

Qunf Cave record. D. Fleitmann and others have published several records of the Indian monsoon variability based on speleothem oxygen isotopes and annual changes in layer thickness. A high-resolution oxygen-isotope record, covering much of the current interglacial period (the Holocene), was obtained from stalagmite Q5 (Fig. 1), collected from Qunf Cave (17°10′N, 54°18′E) in southern Oman. Based on this record, monsoon rainfall increased rapidly between 10,300 and 9600 years ago (Fig. 3). This was followed by a period of relatively strong summer monsoons until 8000 years ago, when monsoon intensity began to decrease in response to decreasing solar insolation. During the early Holocene, high-frequency monsoon variability was in phase with Greenland temperature variations, whereas in the mid-to-late Holocene, the high-frequency variability is linked to solar variability. This suggests that the forcing mechanisms of the Asian monsoon may be intrinsically different during glacial and interglacial periods.

For background information *see* CALCITE; CAVE; CHEMOSTRATIGRAPHY; CLIMATE HISTORY; CLIMATOLOGY; DATING METHODS; GLACIAL EPOCH; HOLOCENE; INSOLATION; ISOTOPE; LOESS; MASS SPECTROMETRY; MONSOON METEOROLOGY; PALEOCLIMATOLOGY; RADIOISOTOPE; STALACTITES AND STALAGMITES in the McGraw-Hill Encyclopedia of Science & Technology.

Kathleen Johnson

Bibliography. R. S. Bradley, *Paleoclimatology: Reconstructing Climates of the Quaternary*, Harcourt Academic Press, 1999; D. Fleitmann et al., Holocene forcing of the Indian monsoon recorded in a stalagmite from southern Oman, *Science*, 300(5626), 1737–1739, 2003; D. A. Richards and J. A. Dorale, Uranium-series chronology and environmental applications of speleothems, in B. Bourdon et al. (eds.), *Uranium-Series Geochemistry, Reviews in Mineralogy & Geochemistry*, 52:407–460, 2003; Y. J. Wang et al., A high-resolution absolute-dated late Pleistocene monsoon record from Hulu Cave, China, *Science*, 294:2345–2348, 2001; D. X. Yuan et al., Timing, duration, and transitions of the Last Interglacial Asian monsoon, I, *Science*, 304(5670):575–578, 2004.

Chromosome painting (plants)

Chromosome painting in its original sense (as applied to humans and other animals) is defined as the use of fluorescently labeled, chromosome-specific probes to visualize whole chromosomes or large-scale chromosome regions. Such probes are obtained from chromosomes or chromosome segments isolated either by flow sorting (separation of chromosomes in a suspension on the basis of their relative fluorescence using a flow cytometer/sorter) or microdissection isolation of deoxyribonucleic acid (DNA) from a particular chromosome type or its part using a fine microcapillary pipette. The isolated DNA probes are amplified by polymerase chain reaction (PCR), fluorescently labeled, and visualized on chromosome preparations on a microscopic slide. This technique is commonly known as fluorescence in situ hybridization (FISH), and it refers to the physical mapping of the recombination (hybridization) of fluorescently labeled DNA or ribonucleic acid (RNA) probes with complementary nucleic acid sequences in chromosomes or nuclei.

Since the first reports in 1988, multicolor karyotyping by chromosome painting has played a major role in clinical and tumor cytogenetics for identifying structural chromosomal aberrations and complex rearranged karyotypes. Painting probes are also extensively used in interphase cytogenetics; applications include the diagnostic identification of particular chromosome aberrations and studies of the three-dimensional organization of chromatin. Chromosome painting analyses have been hugely facilitated by improved multicolor FISH technologies allowing, for example, simultaneous analysis of all 24 human chromosomes. In comparative cytogenetics, the introduction of cross-species chromosome painting (ZOO-FISH) has enabled the identification of evolutionarily shared chromosome segments in different species and elucidation of chromosome evolution in mammals as well as reptiles and birds.

Chromosome painting in plants. In plant cytogenetics, the term chromosome painting has gained a broader meaning, being used synonymously for FISH and genomic in situ hybridization (GISH), which is the hybridization of total genomic DNA of a given species with chromosomes or nuclei of another species. Both methods use repetitive DNA sequences as "painting" probes. Chromosome painting using flow-sorted or microdissected probes (as in animal cytogenetics) is not directly applicable in plants because they have large numbers of ubiquitous repetitive sequences, which cause nonspecific hybridization. Reported painting of specific chromosomes (for example, the Y sex chromosome) in plants actually refers to the use of chromosome-specific repetitive sequences, rather than low- and single-copy sequences used to label chromosomes of animals.

Recently, pieces of genomic DNA cloned inside living bacteria in a form of bacterial artificial chromosome (BAC) have been used for FISH-based mapping in several plant species, for example, barrel medic, maize, potato, rice, and sorghum. BAC-FISH can be considered an analogous approach to chromosome painting in animals.

Chromosome painting in Arabidopsis. The tiny cruciferous weed *Arabidopsis thaliana* (thale cress) a member of the Brassicaceae (mustard) family, is one of the best characterized plant model species. Chromosome-specific BAC clones covering entire chromosome arms are available in the public domain, and these were chosen as a potential resource, enabling chromosome painting in *Arabidopsis*. *Arabidopsis* is favored for chromosome painting because of its small genome (~125–157 Mb), small amount of repetitive DNA (~10–15%), and low chromosome number ($n = 5$).

Chromosome painting in *A. thaliana* involves labeling individual BAC clones or BAC pools (mixed DNAs of several BAC clones) followed by multicolor FISH (mFISH) on *Arabidopsis* chromosomes. Typically, differently labeled BAC clones are mixed and simultaneously hybridized to chromosome preparations. Chromosomes and nuclei at different developmental stages can be painted in this way; however, largely extended meiotic chromosomes (that is, those in the pachytene stage of meiosis) provide the highest resolution of painting signals. Between 2001 and 2004, all five *Arabidopsis* chromosomes were painted using five different fluorochromes (fluorescent dyes) and chromosome-specific groups of overlapping BAC clones (BAC contigs) as painting probes, and *Arabidopsis* became the first plant with an entirely painted karyotype.

Applications of chromosome painting in *Arabidopsis* include chromosome structure analyses as well as studies of chromosome organization inside nondividing nuclei. Chromosome painting serves as a tool to study inter- and intrachromosomal rearrangements such as translocations and inversions or chromosome associations during meiosis in *Arabidopsis* wild-type and mutant plants. In interphase cytogenetics, chromosome painting enables the analysis of the spatial structure of individual chromosome territories or association of particular chromosome regions.

Comparative chromosome painting. Comparative chromosome painting (CCP) allows identification of entire chromosomes and large-scale chromosome regions evolutionarily shared between species. Brassicaceae is the first plant group in which large-scale CCP has been successfully applied, based on cross-hybridization of *Arabidopsis* chromosome-specific BAC contigs to chromosomes of other species in the mustard family. CCP within Brassicaceae takes advantage of the established multicolor chromosome painting in *Arabidopsis*, available phylogenetic information, and comparative genetic maps for *Arabidopsis* and other cruciferous species. Genetic mapping of a common set of genetic markers (genes or DNA segments whose approximate chromosomal locations are known and whose inheritance can be followed) allows alignment of evolutionarily shared chromosomes and chromosome regions between species in question. CCP was expedited by comparative genetic mapping between *A. thaliana* ($n = 5$), *A. lyrata* ($n = 8$; rock-cress) and *Capsella rubella* ($n = 8$; pink shepherd's purse), respectively. Initially, *Arabidopsis* BAC probes are differentially labeled according to the genetic maps and hybridized to extended meiotic chromosomes of other cruciferous species. If the revealed chromosomal patterns significantly differ from the expectation, the painting probes have to be modified in order to specify deviating chromosome rearrangements.

Phylogenetic applications of CCP. Since the mustard family is the only plant group in which comparative chromosome painting is currently feasible, data acquired using this technique are unique and indispensable for understanding chromosome evolution and its significance for plant speciation. CCP enables the identification of chromosome regions shared within a particular group of cruciferous species and having the same evolutionary origin. This permits specification of the chromosomal rearrangements leading to the current chromosomal structure of extant species and reconstruction of ancestral chromosomes (karyotypes). The tentative reconstruction of ancestral karyotypes is essential for the interpretation of evolutionary chromosome rearrangements.

Reconstructed chromosome patterns have to be viewed in the context of phylogenetic (evolutionary) trees that have been constructed based upon DNA sequence analysis. Molecular phylogenetic reconstructions provide invaluable information on the dating, rates, and evolutionary trends of chromosome rearrangements. However, the present phylogenetic trees show many inconsistencies. Chromosome rearrangements are considered as unique events unlikely to happen independently in different lineages; they are shared by descendant species and, therefore, should be particularly valuable for phylogenetic reconstruction. Thus, it is expected that chromosomal data will improve the understanding of phylogenetic relationships within Brassicaceae.

In *Arabidopsis* ($n = 5$) and some related cruciferous species with seven and six chromosome pairs comparative painting enabled identification of a tentative ancestral karyotype (consisting of eight chromosome pairs), as well as chromosome rearrangements (translocations, fusions, and inversions) involved in the chromosome number reduction from eight to seven, six, and five. The feasibility of comparative painting in Brassicaceae depends on the time elapsed since genome divergence between the ancestors of *Arabidopsis thaliana* and other cruciferous species. The present data indicate that chromosome homeologies (corresponding DNA sequences) arising as far back as 20 million years ago are detectable.

Beyond *Arabidopsis* and Brassicaceae, cross-species painting will become feasible only when chromosome-specific BAC libraries of other species with small genome size become available.

For background information *see* CHROMOSOME; CYTOCHEMISTRY; GENETIC MAPPING; NUCLEIC ACID; PLANT PHYLOGENY in the McGraw-Hill Encyclopedia of Science & Technology. Martin A. Lysak

Bibliography. M. A. Koch and M. Kiefer, Genome evolution among cruciferous plants: A lecture from the comparison of the genetic maps of three diploid species—*Capsella rubella, Arabidopsis lyrata* subsp. *petraea*, and *A. thaliana, Amer. J. Bot.*, 92:761–767, 2005; S. Langer et al., Multicolor chromosome painting in diagnostic and research applications, *Chromosome Res.*, 12:15–23, 2004; M. A. Lysak et al., Chromosome painting in *Arabidopsis thaliana, Plant J.*, 28:689–697, 2001; M. A. Lysak, A. Pecinka, and I. Schubert, Recent progress in chromosome painting of *Arabidopsis* and related species, *Chromosome Res.*, 11:195–204, 2003; M. A. Lysak et al., Chromosome triplication found across the tribe Brassicaceae, *Genome Res.*, 15:516–525, 2005; A. Pecinka et al., Chromosome territory arrangement and homologous pairing in nuclei of *Arabidopsis thaliana* are predominantly random except for NOR-bearing chromosomes, *Chromosoma*, 113:258–69, 2004; I. Schubert et al., Chromosome painting in plants, *Methods Cell Sci.*, 23: 57–69, 2001.

Cicada

Cicadas are a large and diverse group of insects in the family Cicadidae in the order Hemiptera. They are globally distributed, with especially high diversity in tropical regions. Most cicadas have a distinctive body form, with two pairs of wings, widely offset compound eyes on each side of the head, and three simple eyes (ocelli) on top of the head. They are generally large insects, ranging from 12 to 50 mm. At rest, the wings are held along the sides of the body, and generally extend beyond the abdomen without overlapping. Cicadas may be brightly or cryptically colored, and often have transparent wings. In many species, males produce distinctive sounds with specialized tymbal organs located on the abdomen. All cicadas are plant feeders with piercing mouthparts.

Annual cicadas. Several species of cicadas are found in North America. For example, dog-day cicadas (*Tibicen* spp.) are commonly heard in temperate forest and meadow habitats in late summer. Adult dog-day cicadas are usually cryptically colored in green, brown, or black, and many species are strong fliers. Dog-day cicadas develop belowground as root-feeding nymphs for several years prior to emergence as adults. By comparison, Apache cicadas (*Diceroprocta apache*) are abundant in the riparian woodland habitats of American southwestern deserts. These cicadas are brown or black with yellow or green markings on the thorax. They develop from egg to adulthood in 3–4 years, and exhibit adaptations for thermoregulation in a hot, arid climate. Both dog-day cicadas and apache cicadas are commonly described as "annual" cicadas. Although adults of these species are present every year, these cicadas actually experience a long, asynchronous belowground development. The term "annual cicada" is used to distinguish the vast majority of cicadas from the seven unusual species of 13- and 17-year periodical cicadas (*Magicicada* spp.).

Ecology of periodical cicadas. The periodical cicadas have the longest development of any known insect, and a synchronized life cycle that is unique among all organisms. The genus *Magicicada* includes four species of 13-year cicadas (*M. tredecim, M. tredecula, M. tredecassini*, and *M. neotridecim*) and three species of 17-year cicadas (*M. septendecim, M. septendecula*, and *M. cassini*). They share a similar pattern of black, orange, yellow, and red coloration, though different species are distinguishable by location, size, and subtle differences in markings and male courtship songs. Factors influencing the evolution and maintenance of prolonged development, 13- and 17-year periodicity, and large-scale population synchronicity include slow growth due to nutrient-poor xylem, reduced growth during past glacial events, predator or parasitoid avoidance, aboveground or belowground competition, and hybridization avoidance. (In some mathematical models, 13- and 17-year periodicities have been shown to reduce the occurrence of temporal cohort overlap, and minimize the hybridization between populations.) Although several models provide plausible explanations, many questions about the evolution of periodical cicadas remain unresolved.

Distribution. Periodical cicadas are found only in temperate forested regions of North America, east of the Great Plains. The four species of 13-year periodical cicadas have a more southern range than the three species of 17-year periodical cicadas. Periodical cicadas are likely to be the most abundant forest herbivore across this vast region. At high densities, the cumulative biomass of periodical cicadas is among the greatest of any terrestrial animal. The density of periodical cicada adults is extremely variable, ranging from 0 to more than 350 cicadas/m^2.

Belowground development. Despite these very high population densities, periodical cicadas are rarely observed due to their long belowground development. During most of this time, they feed on plant roots belowground (**Fig. 1***a*). Nymphal cicadas may feed at depths of 1 m or more, depending on the density of available plant roots. Belowground, nymphs form mud chambers or tunnels in the soil, and use piercing mouthparts to draw xylem from plant roots. Nymphal cicadas feed on the roots of a wide variety of host species, including deciduous trees, coniferous trees, grasses, and forbs. At high densities, belowground root herbivory by cicada nymphs may reduce the growth of host trees or decrease fruit yields in orchards. Root damage by periodical cicadas appears to be most common at the earliest nymphal stages.

Fig. 1. Periodical cicada life cycle. (*a*) Nymphal cicada feeding on plant roots belowground. (*b*) Ants and other detrivores feeding on dead cicada litterfall. (*c*) Teneral adult cicada recently emerged from the nymphal exuviae. (*d*) Bird predation of an adult cicada. (*Illustration by Sarina Jepsen*)

Periodical cicadas develop through five nymphal stages, or instars, before the emergence year. Recent experiments suggest that nymphal cicadas may use seasonal cycles of host plant physiology to keep track of time during this long belowground development. During the year before emergence, periodical cicada nymphs elongate the feeding cell to form sinuous tunnels. In the final months before emergence, they extend these tunnels near the soil surface. Nymphs will often form mud turrets, or "chimneys," less than 10 cm high at the soil surface in the weeks before emergence.

Synchronized emergence. The emergence of a periodical cicada cohort, or brood, is synchronized to within 5–10 days in May and June across a large area, often encompassing several 10^5 km^2. Each brood is numbered with roman numerals, beginning arbitrarily with Brood I in 1893. For example, Brood VIII emerged in part of western Pennsylvania in 2002, Brood IX emerged in southern West Virginia and western Virginia in 2003 (**Fig. 2**), and the emergence of Brood X included more than 15 U.S. states in 2004 (**Fig. 3**).

When soil temperatures approach 18°C (64°F) in the emergence year, great numbers of cicada nymphs emerge from their tunnels at dusk. These nymphs travel along the forest floor briefly before locating trees, grasses, fences, or other vertical structures. Nymphs may ascend to a height of several meters before stopping. At the dawn of the following day, the nymphal exoskeleton splits, and a teneral (pale, unhardened) adult cicada emerges (Fig. 1*c*). The cicada exoskeleton hardens and darkens within a few minutes.

Adult periodical cicadas are present in the aboveground community for 4–6 weeks. During this time, they feed, chorus, court, mate, and oviposit. Male cicadas use their tymbal organs to produce species-specific courtship songs in dense, loud aggregations called chorus centers. Mated female cicadas use their strong ovipositors to insert eggs into a series of 5–12-mm incisions in pencil-sized branches. At high densities, the ends of some tree branches may break, or flag, due to the accumulation of these incisions, and the cicada eggs in these dead branch tips will not hatch. After 6–10 weeks, a great number of small nymphs will hatch from the remaining "oviposition scars," fall to the ground, and begin digging for roots.

Mortality and predation. Adult periodical cicada mortality results from a variety of factors, including faulty emergence from the nymphal exoskeleton, stormy weather, disease, and predation. In many locations, some will die from infection with a specialized fungus, *Massospora cicadina*. This fungus infects periodical cicada nymphs in the soil, and is sexually transmitted between adult cicadas. The abdomens of

Fig. 2. Brood IX periodical cicada in Athens, West Virginia. (*Photo by Louie Yang*)

Fig. 3. Brood X periodical cicadas in Boyce, Virginia. (*Photo by Louie Yang*)

infected adults are eventually consumed by a chalky mass of fungal spores, which are redistributed to the soil to reinfect the next generation. Periodical cicadas also face a diverse community of predators, including birds (Fig. 1*d*), snakes, turtles, mammals, fish, spiders, and many types of insects.

The abundance of easily captured and nontoxic cicada prey in emergence years has been shown to increase reproductive success in several bird and small mammal species. However, the great abundance of cicadas in emergence years satiates many predators. As a result, the vast majority of periodical cicadas will die without depredation. Cicada carcasses litter the forest floor in emergence years, and contribute a substantial amount of organic detritus to forest ecosystems (Fig. 1*b*).

Enrichment of belowground forest communities. Recent experiments indicate that these pulses of cicada detrital resources may enrich belowground communities in North American forests. In controlled field experiments, the decomposition of cicada carcasses has been shown to increase microbial biomass and nitrogen availability in forest soil during the summer of emergence. In turn, these belowground changes contribute to increased nitrogen concentrations in the foliage of a common understory plant, the American bellflower (*Campanulastrum americanum*). Stable isotope analysis of bellflower foliage indicates cicada-supplemented foliage is particularly enriched with heavier nitrogen isotopes, which are generally found at higher concentrations in animal biomass, suggesting that increases in foliage nitrogen result from the decomposition of cicada carcasses. Additionally, bellflowers that were experimentally supplemented with cicada carcasses produced larger seeds than unsupplemented control plants. These results imply that cicada litterfall causes a substantial pulsed enrichment of belowground forest systems, possibly influencing forest dynamics over a large scale.

For background information *see* FOREST ECOSYSTEM; HEMIPTERA; HOMOPTERA; INSECT PHYSIOLOGY; INSECTA; POPULATION ECOLOGY in the McGraw-Hill Encyclopedia of Science & Technology.

Louie H. Yang

Bibliography. P. R. Grant, The priming of periodical cicada life cycles, *Trends Ecol. Evol.*, 20:169–174, 2005; R. Karban, C. A. Black, and S. A. Weinbaum, How 17-year cicadas keep track of time, *Ecol. Lett.*, 3:253–256, 2000; C. T. Maier, A moleseye view of 17-year periodical cicada nymphs, *Magicicada septendecim* (Hemiptera, Homoptera, Cicadidae), *Ann. Entomol. Soc. Amer.*, 73:147–152, 1980; C. L. Marlatt, The periodical cicada, *Bull. USDA Bur. Entomol.*, 71:1–181, 1907; K. S. Williams and C. Simon, The ecology, behavior, and evolution of periodical cicadas, *Annu. Rev. Entomol.*, 40:269–295, 1995; L. H. Yang, Periodical cicadas as resource pulses in North American Forests, *Science*, 306:1565–1567, 2004.

Click chemistry

Chemistry is an architectural endeavor in which function and form are intimately connected. It is the function of molecules that matters most, and the creation of a new function is the goal of all chemical research, from the discovery of pharmaceuticals to the invention of new plastics, fragrances, or dyes. The range and diversity of possible structures are limitless. It has been calculated that the number of relatively small molecules that can exist using only the nine most relevant elements is a staggering 10^{63}, approximately a million times greater than the number of atoms in the Sun.

Within this vast potential pool of molecular structures, surely there are many answers to any chemical question, that is, many different molecules that have a desired function. The problem, of course, is to find them. The creation of new functional molecules often proceeds by analogy (making a structure that resembles a known structure with the desired function) or by searching among unproven structures. Even in the latter case, the tendency among chemists is to stay with the familiar. Part of the problem is the available tools: it is difficult to make a transportation vehicle that looks a lot different than a car if all you have are the nuts, bolts, and wrenches that are used to make cars. But it is precisely the connections—the nuts and bolts—that determine what can be joined together and therefore what can be made.

In the more than 150-year history of modern chemistry, a great many techniques for joining molecular pieces to each other have been developed. Many of these are quite sophisticated, requiring the delicate handling of highly reactive reagents under tightly controlled conditions. In 2001, a group of chemists led by Nobel Laureate Prof. K. Barry Sharpless at the Scripps Research Institute gave the name click chemistry to a new way of thinking about chemical reactions. "Click" is meant to signify that joining molecular pieces is as easy as clicking together the two pieces of a buckle. The buckle works no matter what is attached to it, as long as its two pieces can reach each other and the components of the buckle can make a connection only with each other. Click reactions are easy to perform, give rise to their intended products in very high yields with little or no by-products, work well under many conditions (usually especially well in water), and are unaffected by the nature of the groups being connected to each other.

The central question posed by Sharpless and colleagues is whether desired functions can be obtained from molecules made using only click chemical reactions. In this deliberate eschewing of more "sophisticated" synthetic techniques is an implicit challenge to organic chemists that make complex structures by complex methods (can complex structures be made by simple methods?), and an explicit tip of the hat to the synthesis of functional polymers, which cannot occur without the use of reactions that meet the click chemistry standard.

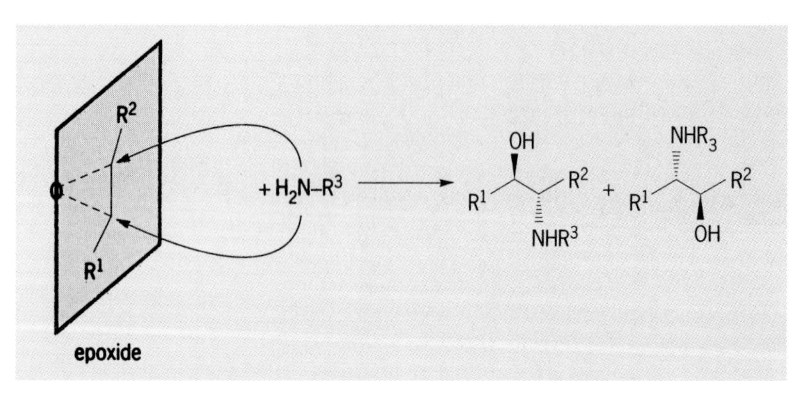

Fig. 1. Ring opening of epoxides with amines, a prototypical click reaction.

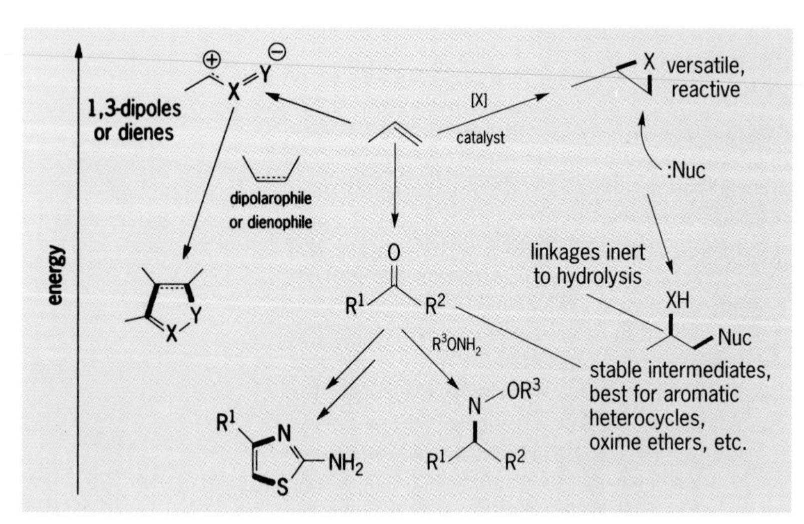

Fig. 2. Click chemistry employs highly energetic but selectively reactive groups.

Reactions. Click reactions share the following attributes. (1) Many click components are derived from alkenes and alkynes, and thus ultimately from the cracking of petroleum. Carbon-carbon multiple bonds provide both energy and mechanistic pathways to be made into reactive structures for click connections. (2) Most click reactions involve the formation of carbon-heteroatom (mostly N, O, and S) bonds. This stands in contrast to much of synthetic organic chemistry of recent years, which has emphasized the formation of carbon-carbon bonds. (3) Click reactions are strongly exothermic by virtue of highly energetic reactants or strongly stabilized products. (4) Click reactions are usually fusion processes (leaving no by-products) or condensation processes (producing water as a by-product). (5) Many click reactions are highly tolerant of and often accelerated by the presence of water.

These features are illustrated by the case of epoxide ring opening by a wide range of nucleophiles (**Fig. 1**). Since the epoxide is a strained three-membered ring, its ring opening is a strongly favored process. Yet that ring opening is required to occur in a particular way; that is, the nucleophile can attack the epoxide carbon atom only along the C-O bond axis, so by-products are avoided and yields are high. In addition, epoxides are relatively unreactive with water, while the hydrogen bonding ability and polar nature of water both serve to support the epoxide ring-opening reaction by other nucleophiles.

Most click reactions were discovered many years ago and have been widely appreciated, but have not been tapped for their full potential. They include the following classes of reactions, some of which are shown in **Fig. 2**.

Nucleophilic opening of spring-loaded electrophiles. The opening of the three-membered epoxide, aziridine, aziridinium, and episulfonium rings are all simple and versatile reactions. Included in this category are Michael addition reactions to α,β-unsaturated carbonyl compounds, which are also strongly favored processes.

Mild condensation reactions of carbonyl compounds. The formation of 1,3-dioxolane rings from aldehydes or ketones and of 1,3-diols, hydrazones, and oximes from aldehydes and hydrazines or hydroxylamine ethers, and of heterocycles from α- and β-carbonyl aldehydes, ketones, and esters are highly reliable and widely used reactions.

Cycloaddition reactions. Distinguished by the relatively nonpolar nature of their reactive groups, these fusion reactions comprise a wide range of processes, including the Diels-Alder reaction. Most useful are 1,3-dipolar cycloaddition reactions, with the cream of the crop being the reaction of organic azides with alkynes.

Azide-alkyne cycloaddition. The goals and philosophy of click chemistry have been most extensively explored with the uncatalyzed (**Fig. 3***a*) and copper-catalyzed (Fig. 3*b*) versions of the cycloaddition reaction between azides and alkynes to form triazoles. The former was first reported by A. Michael in 1893 and was later studied and popularized by the German chemist Rolf Huisgen in 1960s through the 1980s. The latter reaction was discovered independently in 2002 by the research groups of M. Meldal in Holland and K. B. Sharpless in the United States, and requires the use of a terminal alkyne. The abbreviation AAC (azide-alkyne cycloaddition) is used for the uncatalyzed process and CuAAC for the copper-mediated version.

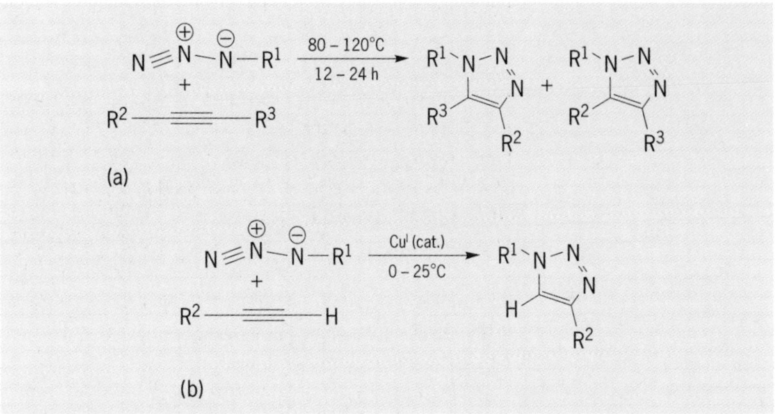

These processes are uniquely useful because of the properties of the reactive substituents. Both azides and alkynes are high in chemical potential energy, and their fusion to make triazoles is exothermic by more than 45 kcal/mol. However, the rate of this reaction is quite slow, normally requiring prolonged heating for unactivated (not strongly electron-deficient or strained) alkynes. The azide and alkyne groups are stable in the presence of the nucleophiles, electrophiles, and solvents common to standard reaction conditions, with the azide being the only 1,3-dipolar reagent to have this quality. Most importantly, azides and alkynes are almost completely unreactive toward biological molecules. They are small and incapable of significant hydrogen bonding, and thus are unlikely to significantly change the properties of structures to which they are attached. Both can be easily introduced into organic compounds.

Click chemistry in situ. Because of their special reactivity—inert toward everything else and only sluggishly reactive with each other—azides and alkynes have been used to assemble molecules that bind very tightly to enzymes by using the enzyme itself as the reaction vessel (**Fig. 4**). The technique, known as click chemistry in situ, involves the tagging of molecules that bind to adjacent locations on an enzyme with azide and alkyne groups. If the derivatized molecules can interact simultaneously with the target in the proper orientation to hold the azide and alkyne units in proximity, a triazole may form to link the enzyme-binding components to each other. Since attachments to any target by two arms is better than either arm alone, the result is invariably a molecule of much higher binding affinity. Prior knowledge of the structure of the target is not required, nor is an assay for enzyme activity necessary. Since no reaction occurs between azides and alkynes at the concentrations used in these experiments without templating by the target enzyme, the formation of a new product is easy to detect by mass spectrometry and is proof of the creation of an excellent enzyme inhibitor.

Fig. 3. Azide-alkyne cycloaddition (AAC) reaction to form triazoles: (*a*) Uncatalyzed and (*b*) copper-catalyzed version.

Click chemistry in situ has been used to create novel high-affinity binders to the important neurotransmitter enzyme acetylcholinesterase, the metabolic enzyme carbonic anhydrase, and HIV protease. It has become apparent in these and other studies that the triazole linkage has advantageous properties for drug discovery. Having a strong dipole moment, the capacity for significant hydrogen bonding, and the ability to engage in stacking interactions with aromatic groups, triazoles can bind productively to proteins. The creation of this unit from two "invisible" pieces in a pocket of an enzyme is an event that influences the selectivity of bond formation during in-situ drug discovery efforts. The click chemistry in-situ technique is used in many laboratories and pharmaceutical companies around the world as a complement to traditional methods of synthesis and screening.

While not yet used directly in living cells because of the cytotoxicity and attendant bioregulation of copper, the copper-catalyzed reaction has achieved extraordinarily wide use in organic chemistry and materials science. Applications include the synthesis of biologically active compounds, the preparation of conjugates to proteins and polynucleotides, the synthesis of dyes, the elaboration of known polymers and the synthesis of new ones, the creation of responsive materials, and the covalent attachment of desired structures to surfaces.

Conclusion. Click chemistry expands the scope of structures that can be made by expert chemists and nonchemists alike. The rationale is simply that the more tolerant is the connection reaction between chemical pieces, the more diverse the pieces that can be brought to bear on any problem. Chemists possess neither the ability to control reactions as well as living cells do nor the ability to manufacture large structures such as proteins for every purpose. Therefore, while nature can inspire us, making such

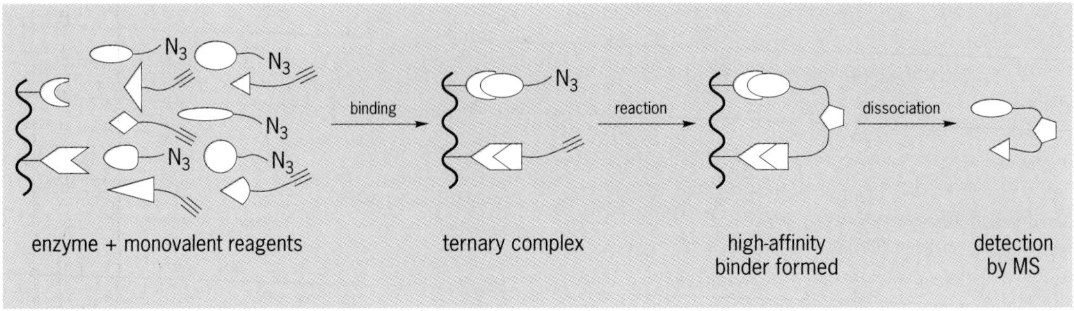

Fig. 4. Click chemistry in-situ technique.

complex molecules comes at a fearful cost of time and expense. If indeed it is chemical function that is the goal, keeping it simple is a useful rule.

For background information *see* ALDEHYDE; ALKENE; ALKYNE; AZIDE; CARBONYL; DIELS-ALDER REACTION; DIPOLE MOMENT; ELECTROPHILIC AND NUCLEOPHILIC REAGENTS; ENZYME; EPOXIDE; HETEROCYCLIC COMPOUNDS; HYDROGEN BOND; INTERMOLECULAR FORCES; KETONE; ORGANIC SYNTHESIS; OXIME in the McGraw-Hill Encyclopedia of Science & Technology. M. G. Finn

Bibliography. R. S. Bohacek, C. McMartin, and W. C. Guida, The art and practice of structure-based drug design: A molecular modeling perspective, *Med. Res. Rev.*, 16(1):3–50, 1996; Y. Bourne et al., Freeze-frame inhibitor captures acetylcholinesterase in a unique conformation, *Proc. Nat. Acad. Sci.*, 101(6):1449–1454, 2004; R. Huisgen, in A. Padwa (ed.), *1,3-Dipolar Cycloaddition Chemistry*; vol. 1, pp. 1–176, Wiley, New York, 1984; R. Huisgen, Kinetics and reaction mechanisms: Selected examples from the experience of forty years, *Pure Appl. Chem.*, 61(4):613–628, 1989; H. C. Kolb and K. B. Sharpless, The growing impact of click chemistry on drug discovery, *Drug Disc. Today*, 8(24):1128–1137, 2003; W. G. Lewis et al., Click chemistry in situ: Acetylcholinesterase as a reaction vessel for the selective assembly of a femtomolar inhibitor from an array of building blocks, *Angew. Chem. Int. Ed.*, 41(6):1053–1057, 2002; W. G. Lewis et al., Discovery and characterization of catalysts for azide-alkyne cycloaddition by fluorescence quenching, *J. Amer. Chem. Soc.*, 126(30):9152–9153, 2004; R. Manetsch et al., In situ click chemistry: Enzyme inhibitors made to their own specifications, *J. Amer. Chem. Soc.*, 126(40):12809–12818, 2004; V. P. Mocharla et al., In situ click chemistry: Enzyme-generated inhibitors of carbonic anhydrase II, *Angew. Chem. Int. Ed.*, 44(1):116–120, 2005; V. O. Rodionov, V. V. Fokin, and M. G. Finn, Mechanism of the ligand-free Cu(I)-catalyzed azide-alkyne cycloaddition reaction, *Angew. Chem. Int. Ed.*, 44(15):2210–2215, 2005; V. V. Rostovtsev et al., A stepwise Huisgen cycloaddition process: Copper(I)-catalyzed regioselective "ligation" of azides and terminal alkynes, *Angew. Chem. Int. Ed.*, 41(14):2596–2599, 2002; C. W. Tornøe, C. Christensen, and M. J. Meldal, Peptidotriazoles on solid phase: [1,2,3]-triazoles by regiospecific copper(I)-catalyzed 1,3-dipolar cycloadditions of terminal alkynes to azides, *J. Org. Chem.*, 67(9)3057–3062, 2002.

Communications satellite failures

A telecommunications satellite is a space-based amplifier that receives signals from Earth antennas and amplifies and retransmits those signals to an Earth receive antenna. The satellite operates as a communications node that allows transfer of information from one point on the Earth to another without a physical wire or cable connection. A telecommunications satellite consists of two main parts: (1) the satellite "payload," which is the hardware that actually performs the communications mission of the satellite; and (2) the satellite "platform" (or "bus"), which comprises the basic support systems of the satellite (see **illustration**).

The payload, which is unique to each satellite, consists of a chain of electronic communications equipment which receives, filters, amplifies, and transmits a signal. The platform, which is common to any satellite, consists of the following systems and subsystems:

1. *Power system:* Equipment for electrical power generation (solar cells), storage (batteries), and distribution of electrical power to operate many of the satellite's systems and subsystems.

2. *Propulsion system:* Chemical or electrical propulsion equipment needed to maneuver the satellite.

3. *Thermal system:* Mechanical and thermal equipment used to maintain the correct temperature on the satellite.

4. *Attitude and orbital control system (AOCS):* Sensors, actuators, and electronics to determine and control the satellite's orientation and position in orbit.

5. *Telemetry, tracking, and command (TT&C):* Electronic and communications equipment used to monitor and control the satellite.

The process to design, build, and launch a telecommunications satellite normally takes 2–3 years. A satellite begins its broadcast life after it has been successfully launched into space aboard a rocket, moved to its proper orbital location, and undergone in-orbit testing. In-orbit testing includes validation and configuration of the satellite platform and payload equipment as well as measurement of payload performance.

The launch, orbit raising, and initial test are usually completed within 2 months, which is generally considered to be the period of greatest risk during a satellite's 12–15-year mission life. Historically, any destruction or damage to satellites has occurred during the launch phase or during the in-orbit testing period. Once on-orbit, satellites have tended to perform reliably for their entire lives. However, in recent years satellites have exhibited poorer in-orbit performance than the historical trend line. This article tabulates these failures, explains their causes to the extent known, and describes what is being done to prevent future occurrences.

Failure descriptions. Any satellite will invariably suffer tens or hundreds of anomalous conditions during its in-orbit lifetime. Since satellite designs incorporate considerable redundancy and are built to strict standards, few of these anomalies result in degradation of the satellite payload performance. This article focuses on those in-orbit satellite anomalies that have actually reduced a satellite's performance capability.

The consequence of part or all of a satellite being damaged or failing is that it is unable to provide the communications services for which it was designed and thus cannot generate the expected amount of revenue. Most commercial communications satellites operate in a circular geostationary orbit of 35,786 km (22,237 mi) above the Earth's surface and on the Earth's equatorial plane. Since satellites in geostationary orbits cannot be retrieved or repaired, any degradation is considered permanent unless fixable through a ground-based or software uplink solution.

Most satellite in-orbit anomalies occur in satellite platform systems as opposed to satellite payload systems. Some significant platform system failures during the past few years are listed in **Table 1**, while payload failures are listed in **Table 2**.

Causes of failures. When an anomaly occurs, the satellite operator and manufacturer jointly evaluate the failure to determine its cause and initiate appropriate corrective measures. With a satellite in geostationary orbit, identifying a root cause is fraught with inherent difficulty. Most of the information available comes from satellite telemetry that is often limited in scope or indirectly related to the anomaly itself. A satellite in geostationary orbit cannot be seen from the ground, and most failure analysis is performed by extrapolation of telemetry data or recreation of failure scenarios in ground tests.

While technically specific descriptions can be provided for each of the anomalies listed in Tables 1 and 2, this information either is protected by confidentiality provisions or is not publicly available. Nevertheless, the causes of most satellite failures fall into the following categories:

Indeterminate. The cause of some failures cannot be determined. Either the data are too limited or the failure event cannot be repeated in ground tests. Frequently, a manufacturer will narrow the loss symptoms down to a few scenarios without being able to identify conclusively the root cause of the failure.

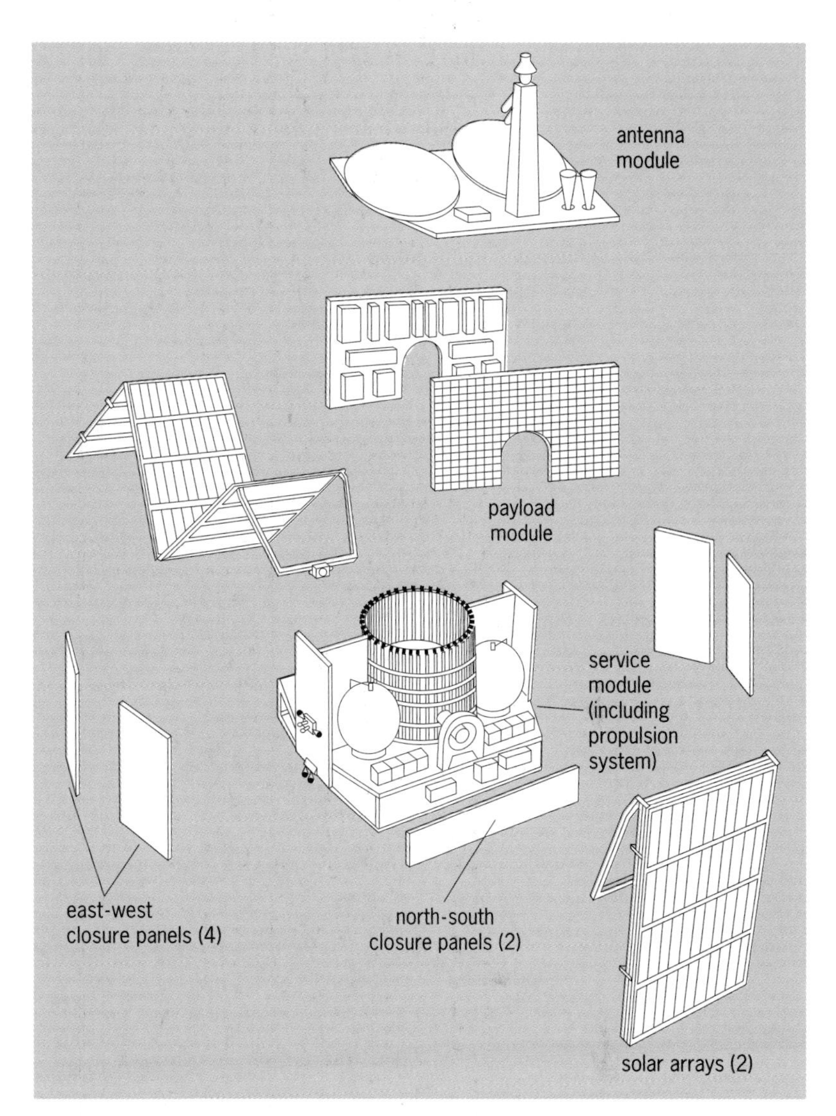

Exploded, stylized view of three-axis stabilized communications satellite, showing modular design. (*Mark Williamson, Space Technology Consultant*)

Design errors. Some failures occur because of human error, such as a component being designed or manufactured incorrectly. Other human errors have been fundamental computational mistakes that have led to failures. The occurrence of some of these errors may be related to a scarcity of engineering talent in the field.

Changes in monitoring procedures. Newer, more automated monitoring systems have recently been implemented by some manufacturers. These systems are fundamentally different from the satellite industry's traditional (and manual) in-plant inspection system. Some anomalies could have occurred only due to a breakdown in basic monitoring and oversight procedures.

Workmanship. Some failures have been caused by poor workmanship. A system or subsystem design can be correct, but incorrect construction or installation of the hardware may lead to failures.

TABLE 1. Platform system failures

System	Failure or failed unit*	Cause
Power	Solar cell concentrators	Degradation of solar concentrator, causing accelerated reduction in solar array power
	Solar array deployment, mechanics	Deployment failure, causing loss of solar array power from one solar array wing
	Solar cell strings*	Electrostatic discharges (ESD) and shorts due to insulation failures, causing loss of solar array power
	Nickel-hydrogen battery cells	Leak of hydrogen from cell pressure vessel, causing partial loss of battery capacity
	Bearing and power transfer assembly (BAPTA)	Failure in BAPTA, causing loss of power from one solar array wing
	Solar array sectional problems	Electrical connector failure, causing loss of power from one wing of solar array
	Loss of power	Short circuit of primary power system
	Loss of power	Failure to ground wiring connector hardware, causing electrical distribution anomaly
Propulsion	Premature shortening of electric propulsion thruster life	Contamination of thruster grid
	Loss of thruster capability	Injector damage due to improper operating mode
		Propulsion system contamination, degrading thruster performance
		Failure of thruster seals, causing propellant leakage
	Propellant loss	Defective valve, causing depletion of onboard fuel
Attitude and orbital control subsystem (AOCS)	Momentum wheels	Inadequate bearing lubrication or bearing roughness, reducing wheel operating life and causing occasional stiction
	Ring laser gyro	Laser intensity degradation due to contamination, resulting in premature end of life
Telemetry, tracking, and command (TT&C)	Satellite control processor	Growth of tin whiskers in electronics, causing shorts
	Distribution control units	Multiplexer failure, causing loss of parameter information to processor

*Individual solar cells are connected in series to form strings, which are then grouped into circuits of several strings in parallel, so that failures can be confined to a single string.

Random failures. Due to the limitations of analysis of anomalies aboard satellites in a geostationary orbit and in spite of extraordinary investigative efforts of satellite manufacturers, some failure causes cannot be determined and are attributed to random events.

Infant mortality. This is a general category for items that fail within a short period of time after launch. This category includes failures that have escaped the design process controls, quality inspections, and test programs.

Shorter satellite manufacturing schedules. Some satellite customers have been demanding greater technology capability and at the same time requiring manufacturing spans that have been reduced from a typical 3 years to 2 years or less. This shorter period has reduced the time for design and test verification.

Brain drain. The satellite industry is bedeviled by a lack of interest from young, promising engineers. Many satellite engineers with long experience are retiring, and satellite companies are having difficulty attracting and retaining the top engineering talent to replace them because other industries hold more attraction.

Qualification philosophy. New technology is usually subject to extensive qualification testing and is often flown in space as a noncritical demonstration before use on an operational mission. However, with shorter delivery schedules new technology may be flown that has received minimal qualification and preflight testing.

Preventive measures. To prevent future occurrences of on-orbit failures, satellite manufacturers must improve on-orbit reliability. Different manufacturers have adopted different strategies, including the following:

Quality. All manufacturers have long-standing and comprehensive quality assurance programs such as upgrading industry certifications, enhancing internal oversight and authority, and revising the process for validating new technology for new missions. Some manufacturers have shrewdly used the downturn in the number of new satellite orders of the last few years to reinforce their quality assurance programs and incorporate lessons learned in future designs.

Flight-proven components and heritage systems. Manufacturers are focused on improving satellite

TABLE 2. Payload failures

Failure	Cause
TWTAs (traveling wave tube amplifiers) damaged	Hot spot at the radiator end of the TWTA, leading to breakdown of adhesive potting compounds
Ku-band TWTA failures	Anode leakage in traveling wave tube (TWT), upsetting regulation of power converter

performance by upgrading existing systems rather than completely redesigning components. Incremental steps are more reliable and avoid some of the pitfalls of significant changes in performance.

New technologies. Previously, manufacturers introduced new technology from the supply side. Now, they will introduce new technology when a specific mission requires it and only after more extensive ground testing and internal investment and planning.

Manufacturing schedule. While there will always be competitive pressures, manufacturers realize their product's credibility is their lifeline and will strive to ensure that the manufacturing schedule is adequate to accommodate the necessary time frame for the validation of adequate testing and new technologies.

Contract price. There are some experts who believe that low profit margins on commercial satellite programs have contributed to on-orbit failures. While consolidation among satellite manufacturers has not occurred, some manufacturers have stated they will not pursue business on which they cannot earn a profit. Of course, a manufacturer cannot command a higher price unless the product that is being sold possesses an impeccable performance and reliability record.

For background information *see* COMMUNICATIONS SATELLITE; SPACECRAFT TECHNOLOGY in the McGraw-Hill Encyclopedia of Science & Technology.

Mark A. Quinn

Bibliography. B. R. Elbert, *Introduction to Satellite Communication*, 2d ed., Artech House, 1999; B. G. Evans, *Satellite Communication Systems*, 3d ed., IEE, 1999; T. Iida, J. N. Pelton, and E. Ashford (eds.), *Satellite Communications in the 21st Century: Trends and Technologies*, AIAA, 2003; J. E. Kadish and T. W. R East, *Satellite Communications Fundamentals*, Artech House, 2000; G. Maral and M. Bousquet, *Satellite Communication Systems: Systems, Techniques, and Technologies*, 4th ed., Wiley, 2002; D. H. Martin, *Communication Satellites*, 4th ed., AIAA/Aerospace Press, 2000; M. Richharia, *Satellite Communications Systems*, 2d ed., McGraw-Hill, 1999; M. Williamson, *The Cambridge Dictionary of Space Technology*, Cambridge University Press, 2001.

Complexity theory

Complexity theory arises from the need to understand the richness in structure and behavior often seen in large systems. The defining property of complexity is the emergence of global features from local interactions, as captured in the popular saying "the whole is greater than the sum of its parts." For example, a flock of birds emerges when individual birds coordinate their behavior with each other. Typically, complexity arises within systems that have strong interrelationships between components. It often appears as unpredictability, or as a tendency for a system to undergo rapid change.

Complexity theory has roots in many fields, including biology, computer science, ecology, engineering control systems, nuclear physics, and operations research. In physics, for instance, there is a need to understand emergent properties of large ensembles of atoms and other particles. General systems theory, developed by the biologist Ludwig von Bertalanffy in the 1950s, identified the role of internal interactions and processes in self-organization. In the 1960s, W. Ross Ashby and Norbert Wiener developed cybernetics, which concerned communications and feedback in the control of systems. Complexity theory came into its own during the late 1980s, as increasing computing power made large-scale simulation studies practical.

Network model. In most cases, complexity can be understood in terms of networks. The network model treats systems as sets of objects ("nodes") with connections ("edges") linking them. For instance, within a flock of birds, the nodes would be birds and their interactions would be the edges. In a nuclear chain reaction, the nodes would be atoms and the edges would be moving neutrons. The network model also applies to processes and behavior. In a game of chess, for instance, each arrangement of pieces on the board is a node in a network of board arrangements, and moves by the players provide the edges that link different arrangements.

In any network, some features arise from the nature of the objects and their relationships, and some emerge from the underlying network. Several kinds of network structure are common because they confer important properties.

Cycles or circuits. These are networks in which all the nodes lie on a single, closed path (for example, A-B-C-D-A).

Trees and hierarchies. Trees are connected networks that contain no cycles. Hierarchies are trees that have a root node, that is, a node that all edges lead away from (for example, a common ancestor in a family tree). In computing, objects can form hierarchies in two ways: whole-part hierarchies (for example, in a whole book, the parts are chapters); and gen-spec hierarchies (for example, a novel is a special kind of book, and a book is a special kind of publication).

Scale-free networks. These form when new nodes in a growing network attach preferentially to existing nodes with the highest connectivity. The distribution of links per node obeys an inverse power law. That is, some nodes have many connections, but most have few. Websites on the Internet, connected by hypertext links, provide a good example of this type of network.

Small worlds. These are connected graphs that contain both local clusters and long-range interactions. They fall between random networks at one extreme and regular networks at the other. Social networks often take this form.

Phase transitions. Complex systems have a tendency to undergo sudden change. Examples include crystallization, freezing water, and epidemics. In all cases, these phase changes occur at a fixed value (the critical point) of some order parameter. (For

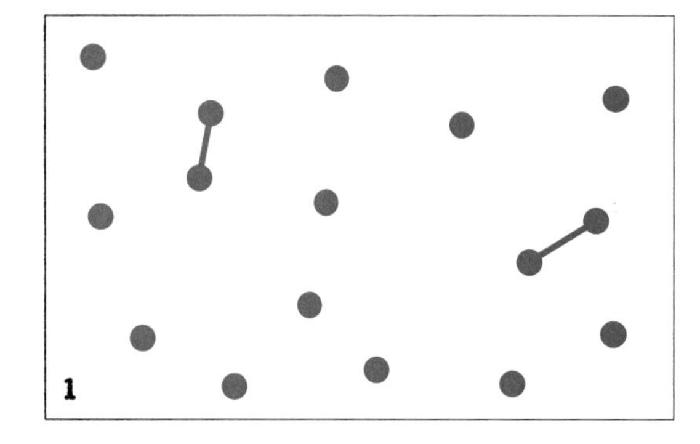

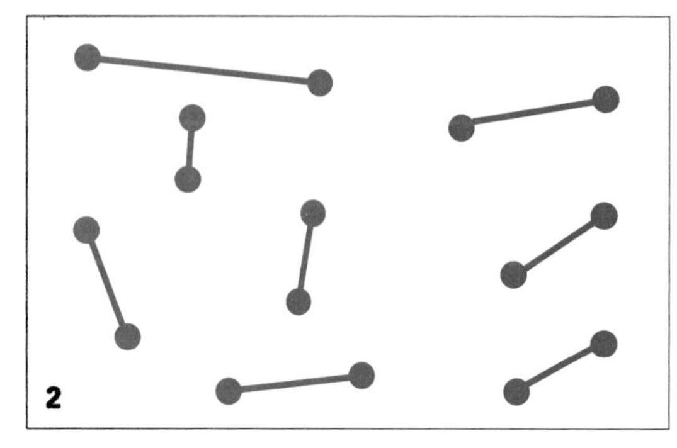

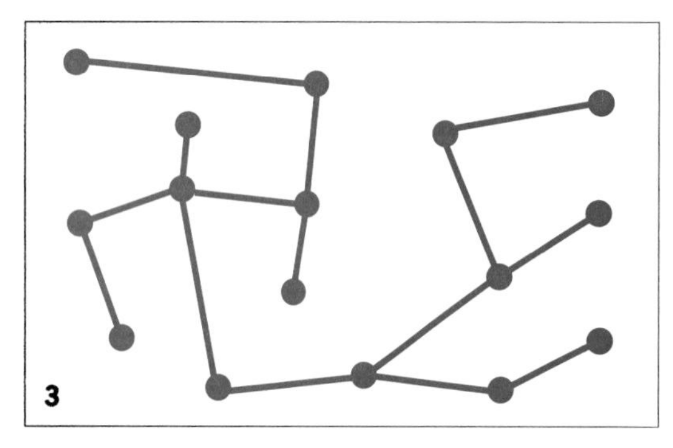

Fig. 1. Formation of a unique giant component as edges are added to a random

example, water freezes at 0°C.) Phase transitions can usually be understood as connectivity avalanches in an underlying network (**Fig. 1**). If a network is formed by adding edges at random to a set of N nodes, then a phase change (connectivity avalanche) occurs when the number of edges reaches the critical density $N/2$ (**Fig. 2**): the nodes become absorbed into a unique giant component. This connectivity avalanche translates into many physical processes. In crystallization, for instance, it involves small crystal seeds merging to form large structures. In all epidemic phenomena, the process requires a certain critical level of interaction to sustain itself. Nuclear fission, for instance, starts when the fuel reaches a "critical mass."

Automata whose state spaces lie close to the phase change in the connectivity (the so-called edge of chaos) often exhibit interesting behavior, leading to speculation that only these automata are capable of universal computation and evolvability.

Encapsulation. In both natural and artificial systems, encapsulation is the most common way of coping with complexity. It involves grouping nodes in a network into modules that are highly connected internally, but with minimal connections to the rest of the network. Development of the eye, for instance, is encoded by a set of genes whose activity is controlled by a single "eyeless gene." In human activity, large tasks are usually broken down into manageable parts: large corporations are typically divided into work units such as divisions, departments, and sections. Likewise, filing systems consist of cabinets, drawers, files, and folders.

Modular design simplifies construction. Computer software typically consists of self-contained, reusable objects and modules. Trees grow by repetition of modules, such as buds, leaves, and branches. Encapsulation also reduces complexity by eliminating long-range connections. In a human organization, for instance, specialization reduces the scope of problems that each person needs to deal with. The method can fail when long-range connections occur, for instance, by placing a file in the wrong folder.

Self-organization. Internal interactions can lead to the emergence of order within complex systems, even in the absence of external influences. Several processes are known to promote self-organization. Ilya Prigogine introduced the term "dissipative systems" to describe systems (including living things) that are far from equilibrium and share energy with their environment. He argued that in dissipative systems, local irregularities can grow and spread, allowing large-scale patterns and order to accumulate. Crystal growth is an example.

Feedback plays an important role in self-organization. Negative feedback dampens disturbances and stabilizes local patterns and behavior. Positive feedback amplifies local effects, allowing them to grow and spread. In intergalactic gas clouds, for instance, gravity leads to the formation of large centers of mass. Large centers gather material faster, and at the expense of smaller ones. Within an ant colony, piles of food grow in the same way. Firing of a laser occurs when groups of atoms synchronize their outputs by "enslaving" other atoms. Other processes that promote self-organization include

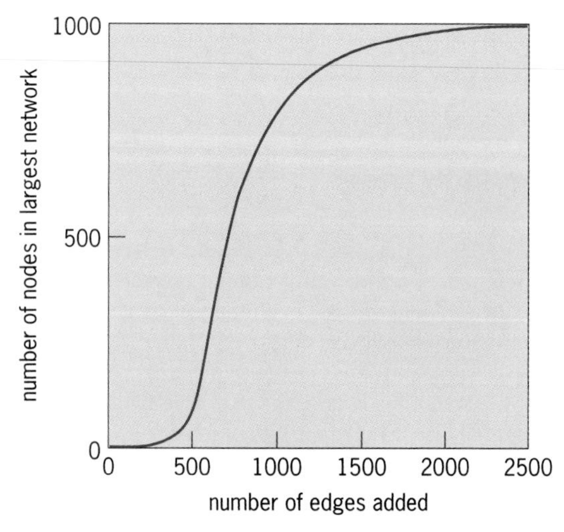

Fig. 2. Connectivity avalanche. Edges, each joining a pair of nodes, are added at random to a set of 1000 nodes. The avalanche begins when about 500 edges are added.

encapsulation, adaptation, and evolution within living systems.

Computational complexity. In computing, the complexity of a problem is measured by the time (or space) required to solve it, and the way that time increases as the size of the input increases. For instance, given a list of the locations of n towns, the time needed to find the most northerly town increases in proportion to n [denoted $O(n)$], but the time required to find the shortest distance between pairs of towns increases in proportion to n^2 [denoted $O(n^2)$]. Hard problems occur when the time complexity increases faster than any polynomial of the input. In the traveling-salesman problem, for instance, a salesman needs to find a route that minimizes the total distance traveled while visiting a number of towns (without doubling back). The solution involves checking all possible orderings of a list of the towns.

Attempts to link complexity to information and thermodynamics led to the Kolmogorov-Chaitin definition of algorithmic complexity, which is the length of the shortest program needed to compute a given pattern. For instance, in general a list of sports results contains essentially random data that cannot be compressed. In contrast, a table of (say) logarithms of the same length can be compressed by providing a formula that computes the table entries. These ideas extend into many aspects of computing. Methods such as minimum message length (MML) adapt them to provide practical methods of classification and problem solving. Data-compression algorithms typically employ a fixed program and reduce a file to the data needed to regenerate the original pattern. Fractals may be regarded as patterns that are produced by repeating a given algorithm on different scales.

Role of simulation. Complex systems usually behave in nonlinear fashion and are often mathematically intractable. Simulation is the most common way of modeling them. Certain representations are widely used to simulate specific kinds of complex systems. Notable among such representations are cellular automata and spin glasses.

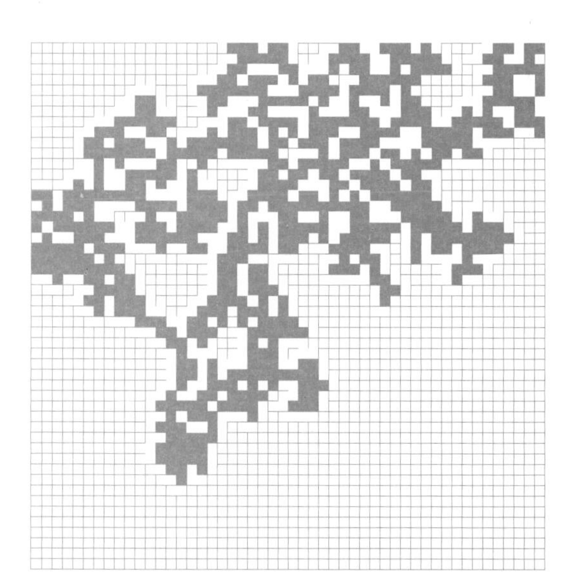

Fig. 3. Cellular automaton simulation of epidemic spread through a landscape. The cells represent areas of the land surface. Solid color cells have been infected.

Cellular automata are grids of automata that interact with their neighbors. They have been applied to many physical processes, such as flows and spread of epidemics (**Fig. 3**).

Spin glasses are regular lattices, in which each node has a "spin." They have been used to study processes, such as glass formation, involving transitions from incoherent to coherent configurations of spin states in magnetic materials.

Challenges. Complexity theory is becoming more important as the problems faced by science and society grow more complex. Is it possible to evolve designs to specification for huge networks that are both robust and efficient? The Internet, for instance, poses enormous challenges as a self-organizing system for gathering, storing, and distributing information. In biology, perhaps the greatest challenge at present is to understand how genetic regulatory networks control growth and development.

For background information *see* CHAOS; CRITICAL PHENOMENA; CRYPTOGRAPHY; CRYSTAL GROWTH; CYBERNETICS; DATA COMPRESSION; FRACTALS; GRAPH THEORY; PHASE TRANSITIONS; SIMULATION; SPIN GLASS in the McGraw-Hill Encyclopedia of Science & Technology. David G. Green; Tania Bransden

Bibliography. P. Bak, *How Nature Works—The Science of Self-Organized Criticality*, Oxford University Press, 1997; T. R. J. Bossomaier and D. G. Green (eds.), *Complex Systems*, Cambridge University Press, 2000; P. Coveney and R. Highfield, *Frontiers of Complexity*, Faber and Faber, London, 1995; S. H. Strogatz, *Sync*, Penguin, London, 2003.

Compound wireless services

Compound wireless services (patent pending) represent a future family of services for wireless devices that will evolve from mobile or wireless telephones. These services will have a greater effect on people's lives than current mobile telephones do.

Wireless mobile communication services or wireless services provide subscribers with desired activities over a wireless telecommunications network. Voice conversations are the most basic services.

A wireless terminal is an apparatus used to invoke wireless services. The common mobile telephone is an example of a wireless terminal for voice services. Voice conversation, where the two parties can speak simultaneously (full duplex), is not the only voice service; voice mail and walkie-talkie (half duplex) are other examples.

Wireless communication service providers are those companies that own the equipment that interconnects a wireless terminal to the telecommunications network. Full-duplex voice service is currently the most important subscription service. However, wireless service providers seek other applications to increase the use of their equipment, thereby increasing profits. The expectation is that the other applications will reside in another class of services called data services. Elementary examples of data services are text messages and transmission of photographs. Voice and data transmitted simultaneously to form a video is called multimedia communication. The wireless terminal used for multimedia communication is more complicated than one used for voice services only. In telecommunications, the term mobile telephone originally meant a simple terminal intended for voice services only. Today, the term wireless terminal describes a more complicated terminal for use with multimedia and complex data services. In the future, wireless terminals will rival the processing power of laptop computers.

Wireless service providers have invested heavily in data services equipment. However, applications that use data services are not commercially available and represent a financial drain for the wireless telecommunications industry. Compound wireless services (formally known as compound wireless mobile communication services) have the potential of generating widely used applications and represent an opportunity for a profitable return from these investments.

Component services. A compound wireless service (CWS) consists of a sequence of component services. It is stored in the memory of a wireless terminal, and the component services are executed from a single invocation of the CWS.

The component services include CWSs, fundamental wireless services, and utility services (see **table**). A CWS is built from all three categories of services, including the category of compound wireless services; that is, a CWS can be a component of a more complicated CWS. Users will have the means to build their own CWSs. The distinction between a CWS and a fundamental wireless service is that the latter does not contain separately identifiable component services. The component services of a CWS can be individually invoked and executed by a user with meaningful results. No such component services exist in a fundamental wireless service.

Fundamental wireless service. An example of a fundamental wireless service is the "location" service (see table). Execution of this service yields the geographic location of a wireless terminal. Note that a location service cannot be partitioned into a sequence of component services. Since locations can be provided for wireless terminals other than the one making the request, password permission is required; that is, the location request will include a wireless terminal (phone) number and a password to obtain this service.

A fundamental wireless service, such as obtaining vehicular traffic conditions for a particular region (see table), is not a service readily available through a wireless service provider alone. This service requires the assistance the state police, a state department of transportation, or a third-party service provider that monitors traffic conditions. Whether a wireless service provider provides the service itself or through another entity does not alter the service's classification as a fundamental wireless service.

Utility services. Utility services are needed for a CWS to function properly but do not constitute a wireless service. For example, a utility service is needed to invoke (start) a CWS. Ending (stopping) a CWS, determining if two parameters are equal, pausing a CWS, and so forth are all services that a CWS may need.

Serviced categories*

Fundamental wireless services	Utility services	Compound wireless services
Receive the location of a wireless terminal/telephone	Invoke a CWS	Compound wireless services built by subscribers/users
Receive a traffic report for a region	End a CWS	
Determine a best travel route according to selected criteria: fastest, shortest, etc.	Determine if two parameters are equal	Compound wireless services made available by wireless service providers
Send/receive a recorded message: image, textual, or audible, etc.	Execute a pause	
Receive weather information for a region	Assign a value	
Send/receive data to/from sensors for smoke alarms, cameras, etc. (image, textual, or audible)	Announce a value	Compound wireless services made available by manufacturers/ suppliers
Receive departure/arrival information for one or more flights: delay, gate #, terminal ID, etc.	Display a value	
Connect to a prescribed IP service provider	Determine if one parameter is greater than another	
Execute/invoke a transaction on the Web: move to a Web address, log into a Web application, etc.	Perform arithmetic operations (+, −, etc.)	Compound wireless services made available by third-party applications providers
Perform database activities in a remote computer: send/receive/edit data to/from a permitted database	Create a parameter	

*The first two columns represent example services that are not compound wireless services. Because of their recursive property, compound wireless services (third column) can be components of more elaborate compound wireless services as well.

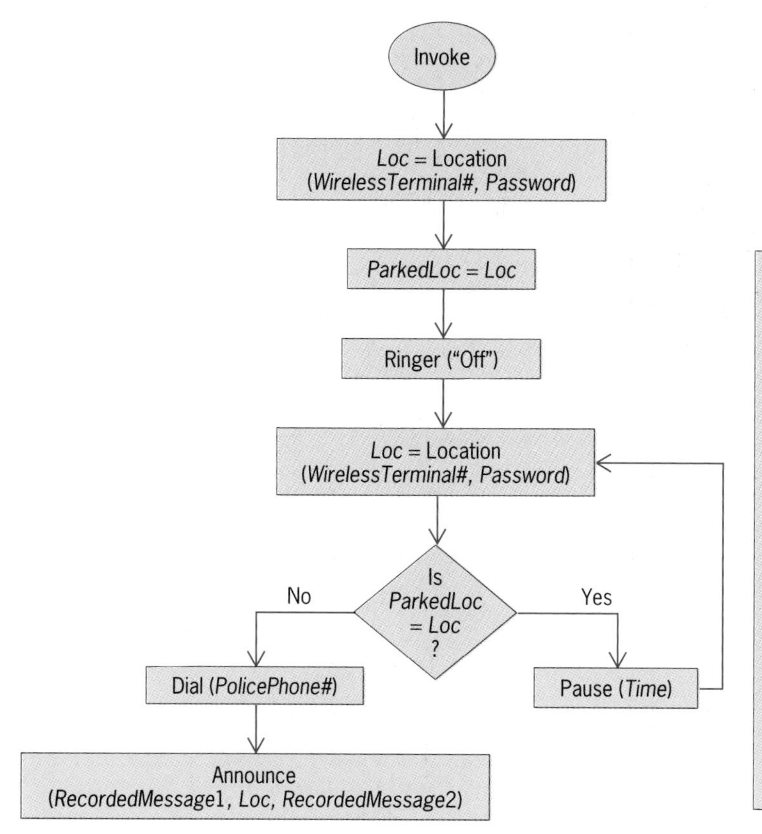

Initialized Parameters

1. *WirelessTerminal#* = ID of wireless terminal in auto
2. *Password* = Authorization for Location service
3. *Time* = Delay to next Location measurement
4. *PolicePhone#* = Local police telephone number
5. *RecordedMessage1* = Announcement that car stolen, and it is at the Location *Loc*
6. *RecordedMessage2* = Announcement that its Location will be displayed and to obtain future locations use the *WirelessTerminal#* and the *Password* that are displayed

Fig. 1. Tracking of a stolen car CWS (TrkStlnCar).

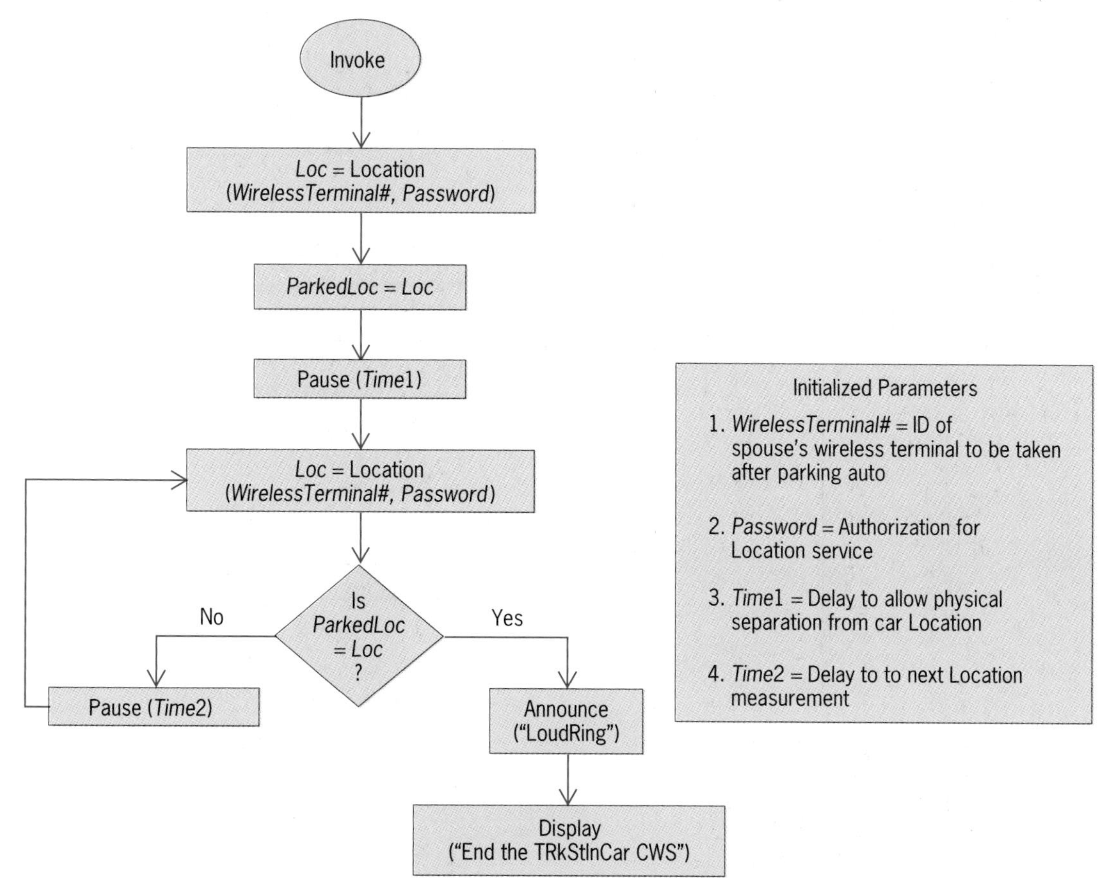

Fig. 2. Reminder to end TrkStlnCar CWS (EndTrkStlnCar).

CWS example. Vehicular traffic is a problem in urban areas. Television or radio traffic reports offer little relief because accidents and other delays occur with little advance warning, and it is difficult, if not dangerous, to search for an alternative route while driving. In a CWS, three fundamental wireless services can be used to achieve vehicular route assistance. (1) "Location" establishes the location of a user's wireless terminal in the vehicle. (2) "Best travel route" computes fastest time between the wireless terminal location and a designated destination. (3) "Traffic report" interrogates the traffic information systems available from various sources for the travel route.

The CWS continues these three services until the destination is reached. The iteration sequence consists of the following steps: (1) Determine the present location (Service 1) and provide it to the wireless terminal. If the present location is the same as the destination, inform the user and cease the iteration of services. (2) Compute the fastest route from the present location to the destination (Service 2). If the route has changed, present the user (orally/graphically/textually) with a new route and directions. (3) Retrieve traffic information for the route (Service 3) and determine if there are traffic delays (for example, due to an accident). If there are traffic delays ahead, repeat step 2 with the updated delay information. If no traffic delays exist, go to step 1.

Flowchart representation. The above verbal description of the CWS may be adequate for a human when context is applied to compensate for lack of precision. However, a microcomputer is not endowed with this human ability. Another means of representing a CWS with the attributes of precision and an intuitive appeal for humans is the software tool, the flowchart.

In a CWS flowchart, rectangles, diamonds, and ellipses represent the component services. A service name and associated parameters appear within each configuration. Arrowed lines are used to show the component services sequence (**Fig. 1**). The figure shows a CWS for tracking the location of an automobile in the event that it is stolen. Before leaving the car, the user first selects the service "Invoke" to begin the CWS. It is executed after the user selects the CWS named "TrkStlnCar" from a list on the wireless terminal screen. The next component service obtains the geographic location of the car's wireless terminal and stores it in the parameter "Loc." This geographic location is then assigned to a parameter named "ParkedLoc."

The CWS will now continually monitor the wireless terminal's location. A service named "Pause" exists within this loop to contain the cost of this

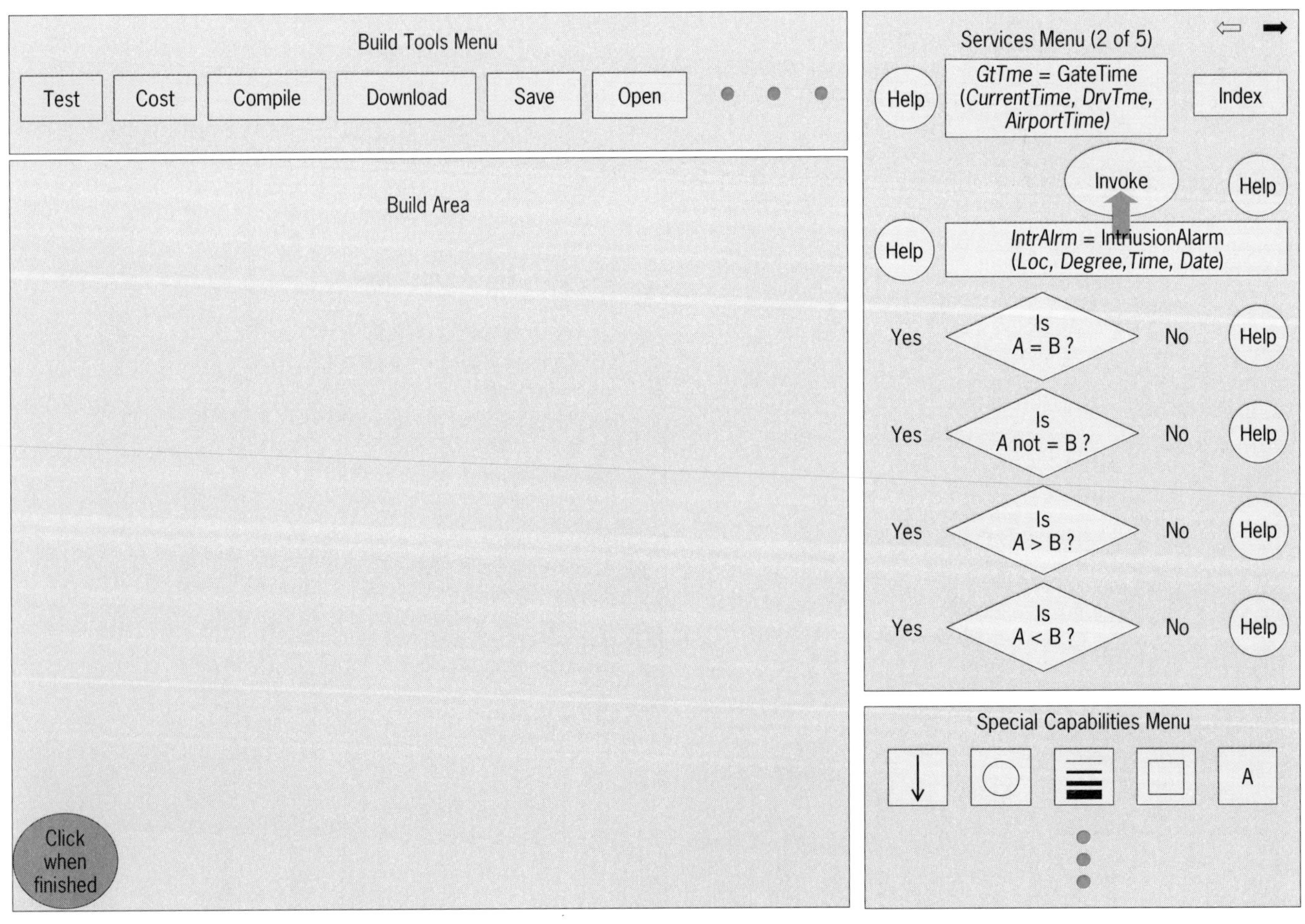

Fig. 3. CWS build layout.

CWS. Without a pause of a few seconds, thousands of location services cycles would occur within a second.

If the automobile moves several yards while this CWS is active, a service will be executed for calling the police. When the call is answered, a stored message is announced to the police, indicating the apparent theft, the automobile's current location, and the wireless terminal's number and password to obtain future locations. The police can now track the automobile with their own wireless terminal.

Note in the "Initialized Parameters" in Fig. 1, requests are made for required values in order to function. The "WirelessTerminal#" and the "Password" for permission to acquire a location represent two examples of initialized parameters.

When returning to use one's automobile, the TrkStlnCar CWS must be ended by executing "End" (**Fig. 2**); otherwise, the police will be alerted that the automobile has been stolen.

Building CWSs. The mechanisms for building CWSs will need to be user-friendly, which implies the need for extensive computational power. Most CWSs will be built on personal or other computers with the assistance of exceptionally user-friendly tools. Although it will be possible to build CWSs on a wireless terminal, these CWSs will be minimally complex. Consequently, a wireless terminal will be used principally to execute or to modify a CWS.

The mechanism for building a CWS contains a combination of textual and graphical icons, partitioned into several menus (**Fig. 3**). The "Tools Menu" (shown partially) contains the operations important for successful CWS implementation. For instance, the "Test" tool is used to determine composition errors, to warn of potential problems, and to simulate a built CWS. The "Cost" tool evaluates the cost of a CWS operation, based on input from sources of the component services. The "Cost" tool will also make recommendations to reduce the cost of executing a CWS. A flowchart is not in a language comprehensible to the wireless terminal, so the "Compile" tool translates the CWS into the wireless terminal's language. The "Download" tool will prepare a wireless terminal for the built CWS and then store it there.

Multiple pages of a "Services Menu" exist to provide the component services for a CWS. When a built CWS is "Saved," it will automatically be included in the Services Menu. Upon saving a CWS, a "Help" template will appear to assist the CWS builder in preparing the operating information.

Component services are incorporated into a CWS by the "point, click, and drag" method. Figure 3 shows pointing to the Invoke service. Clicking

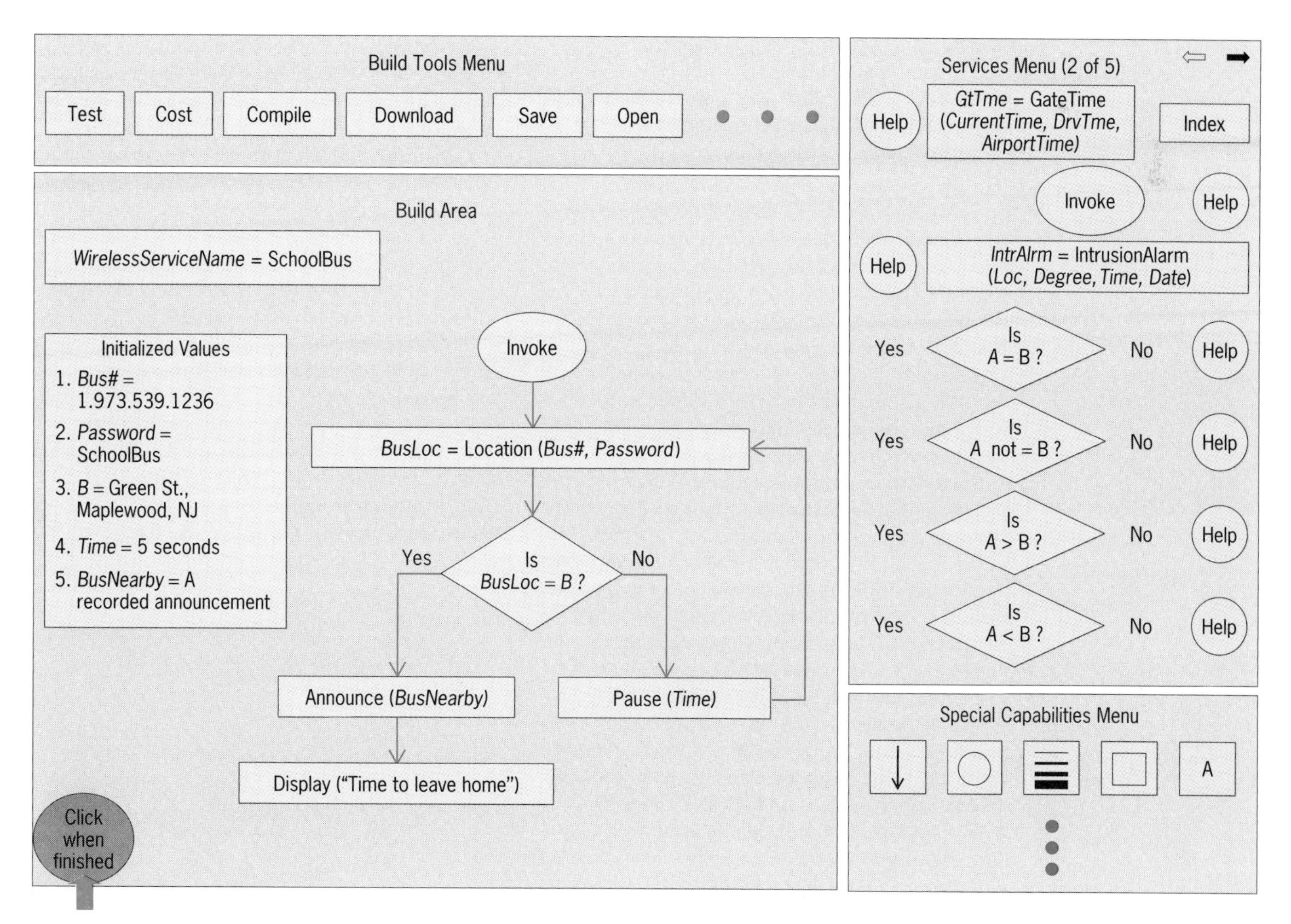

Fig. 4. Completed school bus CWS.

permits the service to be dragged to the "Build Area." Clicking on the associated "Help" icon will provide information on the use of a component service.

The "Special Capabilities Menu" contains tools for drawing and other resources that are not within the scope of the other two menus. For example, arrowed lines to interconnect the component services are "pointed to, clicked on, and dragged to" the appropriate "Build Area" location.

Figure 4 is an example of a completed CWS as it appears on a build layout for the scenario of children waiting for a school bus. This CWS will periodically determine the location of the school bus, and when the bus is approaching the child's stop, an announcement will be made on the parents' wireless terminal.

For background information *see* DATA COMMUNICATIONS; MOBILE RADIO; TELEPHONE SERVICE; WIDE-AREA NETWORKS in the McGraw-Hill Encyclopedia of Science & Technology. Thaddeus J. A. Kobylarz

Bibliography. M. Budagavi and R. Talluri, Wireless video communications, in J. D. Gibson (ed.), *The Mobile Communications Handbook*, CRC Press, Boca Raton, FL, 1999; A. H. Levesque and K. Pahlavan, Wireless data, in J. D. Gibson (ed.), *The Mobile Communications Handbook*, CRC Press, Boca Raton, FL, 1999; P. A. Stark, *Digital Computer Programming*, pp. 51–71, MacMillan, New York, 1967; B. H. Walke, *Mobile Radio Networks*, pp. 327–333, Wiley, New York, 1998.

Computational environmental toxicology

Hundreds of thousands of chemicals in current or past use are present in the environment, leaving human populations and ecosystems potentially at risk of exposure to them. The large number and various forms of chemicals preclude regulators from evaluating every chemical with the most rigorous testing strategies. Instead, standard toxicity tests have been limited to only a small number of chemicals, with the hope that the "worst" chemicals will receive specific attention. The chemicals that are tested may represent large classes of compounds, such as certain types of pesticides.

Today, advances in computational biology offer the possibility that scientists can develop a more detailed understanding of the risks posed by a larger number of chemicals. Computational toxicology is the application of computational biology, using mathematical and computer models, to the assessment of the risk chemicals pose to human health and the environment and to better understand the mechanism through which given chemicals induce harm.

Risk assessment. Early on, risk assessment was mostly a "blind" relationship between exposure levels and some observed response such as the occurrence of cancer, a neurological disorder, or a visible birth defect. The actual pathway between exposure and response, or disease, is better represented as a complex series of steps (**Fig. 1**). A chemical is absorbed (absorbed dose), distributed to internal target sites, and possibly metabolized to an active form once within the body. This results in internal toxicologically relevant doses.

Computational toxicology is a systematic approach that can model a contaminant's effect on gene expression; that is, how the contaminant exposure will affect cellular behavior and signaling, including protein synthesis (proteomics) and metabolic changes as seen in concentrations of metabolites in tissues and biofluids (metabolomics). These advances would not have been possible without the emergence of bioinformatics and computational chemistry and the opportunities they offer for transforming data into information. In particular, computational toxicology will produce risk assessments based on specific molecular changes rather than just the number of tumors, deaths, and overt clinical changes observed in test animals. Future assessments will be based on the number of DNA molecules altered at a crucial site, the change in an allosteric membrane protein that acts as a receptor, or the change in a regulating protein inside the cell. This will lead to a better understanding of how those changes cause clinical disease.

Recent advances in computational toxicology focus on breaking down the traditional dichotomy between approaches to evaluating cancer versus other disease endpoints, on addressing sensitive life stages, and on addressing aggregate and cumulative exposure to pollutants. For example, a greater understanding is needed of why certain modes of action occur more rapidly when an organism is exposed to more than one chemical (synergism), but less rapidly when other chemicals are present (antagonism).

Advancements in genomics, proteomics, and metabolomics, coupled with the advances in analytic tools, such as microarray techniques, will enable us to predict changes and evaluate which changes can initiate and promote disease. Computational techniques will help estimate the necessary quantitative information related to those changes.

Physiologic models. Probably the greatest progress in the field of computational toxicology to date has been in characterizing and quantifying relevant internal doses. Physiologically based pharmacokinetic (PBPK) models describe the time course and mass balance of chemicals entering the body (**Fig. 2**). They mathematically account for both the physiologic and biochemical processes that affect the

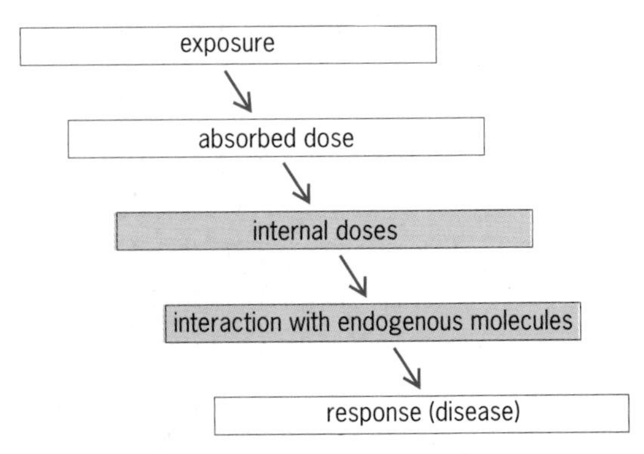

Fig. 1. Stepwise linkage of exposure to toxic response.

Example of types of parameters governing PBPK models	
Parameter	Type
Partition coefficient	Thermodynamic
Organ and body volumes	Anatomic
Blood flows, ventilation rates, absorption rates, clearance	Physiologic
Metabolic transformation rates	Biochemical

disposition of these chemicals and their products of biotransformation. As a result, these models estimate the time course of the internal doses, especially at sites relevant to toxicity.

Physiologically based pharmacokinetic models are governed by parameters such as those shown in the **table**. These parameters may be chemical- and species-specific and are from values reported in published literature, determined experimentally, or extrapolated. They can be used to give estimates of doses within the body, resulting from actual or simulated exposure conditions, at or near the location of toxic action, including subcellular sites if the proper equations are included. The estimated dose is then used in dose-response functions to predict adverse reactions. In addition, these models can estimate the dose resulting from the different routes of entry into the body and the equivalence between different exposure routes. For example, the doses at a site of toxicity in an internal organ resulting from two different sources (such as food and inhaled air) can be easily calculated and compared. It is well known that many physiologic processes are nonlinear, and that the characteristics of these nonlinear processes may differ among dose levels and species. The physiologic models account for this in a quantitative fashion.

Figure 3 shows some typical output from a physiologically based pharmacokinetic model for an inhalation exposure of 4 h, where the exposure or parent chemical (chemical 1) is metabolized in the body to a second chemical (chemical 2). The concentration profile in the blood of the parent chemical and the product of metabolism or metabolite are quite different. Assuming these modeling results are being used to design a clinical or field study, it is apparent that capturing the peak concentration of the parent would require monitoring at different times than monitoring for the metabolite.

In **Fig. 4**, if the area under the concentration (AUC) is the endpoint of interest, the time at which monitoring should cease depends upon which chemical is monitored. The AUC of "chemical 1" shows negligible increase at around 90 h, so monitoring could stop then. The AUC of "chemical 2" is still increasing at 1000 h, so monitoring would have to continue for a considerable time longer.

Computational methods. The growth in the understanding of pharmacokinetics has called for new tools to predict how contaminants will behave after exposure. The focus on improved dose calculations and understanding the basis for outcomes within the range of observation, and use of these to improve scientific judgment below the range of observation (into the range of extrapolation) will result in better environmental risk assessments. Computational toxicology information should allow the identification of hazards by providing data on measurable biochemical or cellular endpoints, which can serve as biomarkers of response for more complex adverse biological effects such as cancer or developmental disorders. Ideally, these measurable endpoints should be mechanistically linked to the biological effect, rather than simply being correlated with it. Identification of key events leading to toxicity can provide insights into the conditions necessary for response and the shape of the dose-response relationship as one goes from high to low doses. Developing the means for incorporating such "in silico" (computer-simulated) data should allow the extension of the dose-response relationship established by

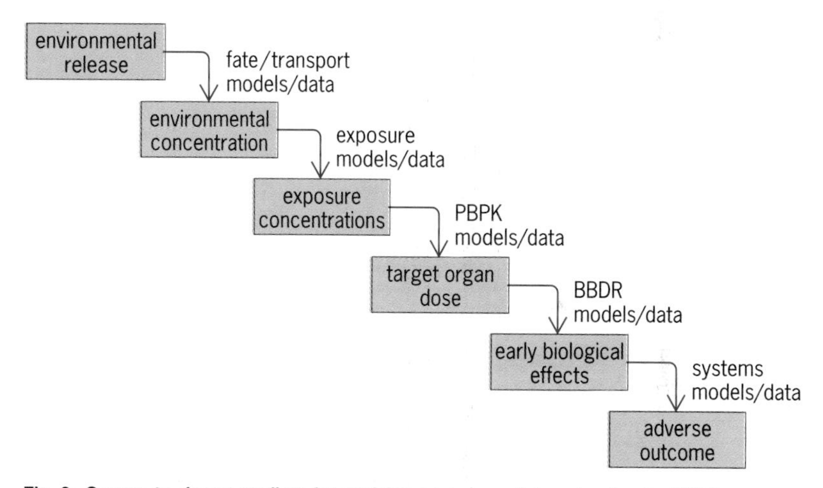

Fig. 2. Source-to-dose paradigm for studying environmental contaminants. PBPK = physiologically based pharmacokinetic; BBDR = biologically based dose response. *(U.S. Environmental Protection Agency, About Computational Toxicology, http://www.epa.gov/comptox/comptoxfactsheet.html)*

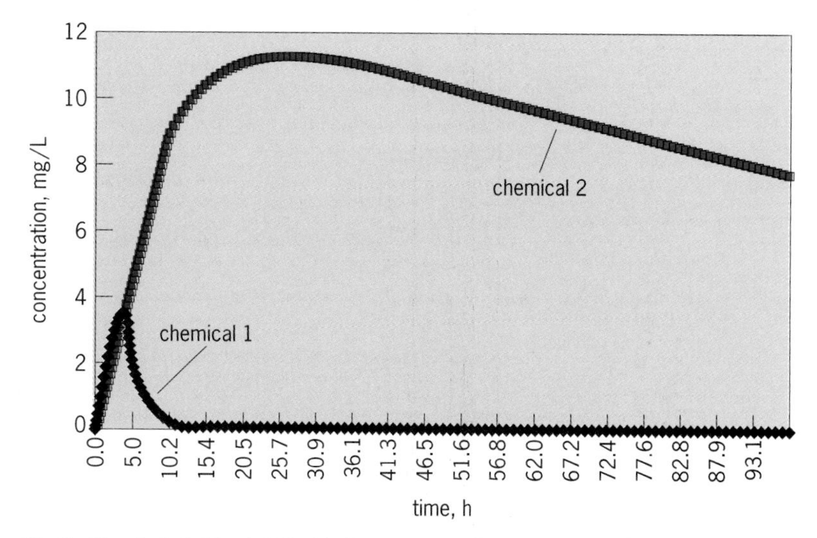

Fig. 3. Chemicals in blood. Output is from a prototypical physiologically based pharmokinetic (PBPK) model. *(PBPK simulations were performed using the U.S. EPA's Exposure Related Dose Estimating Mode, J. N. Blancato et al., 2002)*

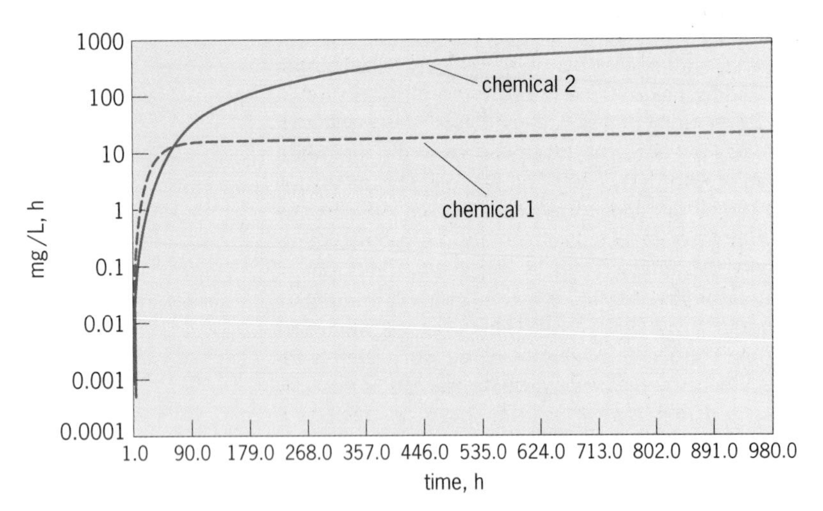

Fig. 4. Area under the concentration curve. Output is from an physiologically based pharmokinetic (PBPK) model. *(PBPK simulations were performed using the U.S. EPA's Exposure Related Dose Estimating Model, J. N. Blancato et al., 2002)*

more traditional toxicology studies to lower levels using sensitive molecular biological and computational techniques. This approach should also save steps and reduce the need for animal testing, compared to traditional toxicology.

"Omics." In the area of computational biology, recent advances have allowed for the sequencing of whole genomes, which has enhanced the understanding of the complexity of cellular biology at the molecular level. Recent technological advances in these areas have led to the development of the new discipline of toxicogenomics in which the effects of chemicals on organisms and ecosystems can be examined using genomic, proteomic, and metabolomic methods.

Omics may also be used to identify those members of a population at greater risk. Disease is considered to result from endogenous predisposition and interaction with environmental stresses, with not all individuals in a population having the same clinical outcome given the same or similar exposures. However, the exact magnitude of the role of predisposing endogenous factors remains unknown.

Omic technologies promise to help determine the molecular pathways that lead to disease after exposure to environmental stresses. The selection of the proper measure of dose within a living system is crucial and should be based on what is known about the mechanism of action, so that quantitative predictions of risk are based on the molecular interactions within the system.

Bioinformatics. Data resulting from these omic technologies are very complex and voluminous. As a result, bioinformatics has evolved for managing and analyzing the data using advanced computational techniques. Powerful software enables us to study the pattern of gene and protein expression and relate those expressions to the structure of important chemical moieties. From such analyses, connections between exposure, genetic susceptibility, and adverse effect will be made. Omics may yield specific patterns, which may be markers of potential disease and exposure. These connections may be made without necessarily understanding the details of the pathways to disease. In the future, it is hoped that both in-vitro and in-silico methods will be used. With such rapid methods, various exposure scenarios could be studied, including those where exposures to a multitude of stressors occurs. At the very least, these techniques could help prioritize which stressors need further study and which may pose the greatest risk.

Structure activity relationships. Improved quantification, such as enhanced quantitative structure activity relationships (QSAR) techniques, are helping estimate the toxicity of poorly characterized substances based on comparisons to well-studied substances having similar chemical structures. Commercially available software is used to predict toxicity endpoints based on chemical structure to predict carcinogenicity in mammals, developmental toxicity, mutagenicity, acute toxicity such as 50% lethal dose (LD_{50}), chronic thresholds, and so on.

Outlook. It is easy to imagine how schemes far more complicated than this can be used to explain the complex biochemistry within a cell, the interaction of different cells within a tissue, or the interaction of the different types of cells in neurological tissue. Such models may, for example, predict changes in brain function resulting from exposure to chemicals which are biotransformed into chemicals that in turn change membrane potentials in the brain. In the future, models may be devised to help us understand how different regions of the brain respond to changes initiated in other regions.

[Disclaimer: The U.S. Environmental Protection Agency through the Office of Research and Development funded and managed some of the research described here. The present article has been subjected to the Agency's administrative review and has been approved for publication.]

For background information *see* ENVIRONMENTAL ENGINEERING; ENVIRONMENTAL TOXICOLOGY; HAZARDOUS WASTE; HUMAN GENOME PROJECT; MATHEMATICAL BIOLOGY; MODEL THEORY; MUTAGENS AND CARCINOGENS; TOXICOLOGY in the McGraw-Hill Encyclopedia of Science & Technology.

Jerry Blancato; Daniel Vallero

Bibliography. J. N. Blancato and K.B. Bischoff, The application of pharmacokinetic models to predict target dose, in J. B. Knaak (ed.), *Health Risk Assessment Through Dermal and Inhalation Exposure and Absorption of Toxicants*, CRC Press, 1992; J. N. Blancato et al., Integrated probabilistic and deterministic modeling techniques in estimating exposure to water-borne contaminants: pt. 2: Pharmacokinetic modeling, in *Proceedings of Indoor Air 2002, 9th International Conference on Indoor Air Quality and Climate*, pp. 262–267, Monterey, CA, June 30–July 5, 2002; T. Bucheli and K. Fent, Induction of cytochrome P450 as a biomarker for environmental contamination in aquatic ecosystems, *Crit. Rev. Environ. Sci. Technol.*, 25:201–268, 1995; Z. Gregus and C. Klaasen, Mechanisms of toxicity, in C. Klaasen (ed.), *Casarett and Doull's Toxicology: The Basic Sciences of Poisons*, 5th ed., McGraw-Hill, 1996; National Research Council, *Biologic Markers in Reproductive Toxicology*, National Academy Press, Washington, DC, 1989; J. Stegeman and M. Hahn, Biochemistry and molecular biology of monooxygenases: Current perspectives on forms, functions, and regulation of cytochrome P450 in aquatic species, in D. Malins and G. Ostrander (eds.), *Aquatic Toxicology: Molecular, Biochemical, and Cellular Perspectives*, CRC Press, 1994; D. A. Vallero, *Environmental Contaminants: Assessment and Control*, Elsevier/Academic Press, 2004; S. Zakrewski, *Principles of Environmental Toxicology*, American Chemical Society, Washington, DC, 1991.

Conflict analysis and resolution

A strategic conflict is an interaction of two or more decision makers over issues such as rights or resources. Some conflicts, such as terrorist attacks, exhibit outright hostility, while others are highly cooperative situations in which disputants form coalitions or jointly act to achieve win/win solutions, that is, resolutions in which everyone gains. The key ingredients of any conflict model are the decision makers in disagreement, each decision maker's options or courses of action, how a scenario or state is determined by the decision makers' choices, and their objectives or preferences over states. Conflict analysis provides methodologies for studying these multiple participant–multiple objective decision situations systematically. By enhancing understanding and communication, it can lead to better decisions that produce resolutions that are more preferable, more stable, and more fair.

Strategic conflicts are ubiquitous; accordingly, research on conflict analysis and resolution has taken place in a wide range of disciplines including psychology, sociology, operations research, political science, and systems engineering. Many organizations—academic, governmental, or private—offer assistance with the theory and practice of conflict analysis and resolution.

Rigorous mathematical structures can provide considerable insight, which probably accounts for the success of the many game-theory-related methodologies for modeling and analyzing conflict. Techniques can be usefully classified according to the information required to calibrate a model. For example, a dinner-party host needs to know that the guest prefers red wine to white, but not how much more preferable red wine is than white. Quantitative preferences are represented on a continuous scale, and can express the extent of such differences. Usually measured in real numbers, they can also encode information about the decision maker's risk attitude: how the guest would feel if a coin toss determined red or white. Most game-theory models, including strategic form, extensive form, and characteristic function form, are quantitative techniques. Nonquantitative techniques, on the other hand, require only easier-to-obtain rankings of outcomes according to preference. Information about preference differences or preferences for randomly determined outcomes cannot be included. Nonquantitative methodologies, including metagame analysis, drama theory, conflict analysis, and the Graph Model for Conflict Resolution, are convenient for modeling societal disputes ranging from international trade to family arguments. These techniques can help resolve problems that arise in general approaches to negotiation, mediation, and arbitration. For example, nonquantitative methods are recommended for brainstorming sessions in interest-based negotiations, in part because they can be adjusted as more information becomes available or more options are recognized.

To illustrate how formal methods can be applied to actual disputes, the Graph Model for Conflict Resolution is employed here to analyze a simple sustainable development problem. This model is designed for application to both simple and complex real-world disputes, and is based on theoretical foundations formulated using the mathematics of relationships: set theory, logic, and graph theory. The decision support system GMCR II permits practitioners, researchers, teachers, and students to apply this unique decision technology conveniently to virtually any social conflict.

Decision support systems. A decision support system (DSS) is a user-friendly software package that encodes modeling and analysis capabilities for formal decision models. Decision support system technologies are an important subfield of information technology, which includes the development and application of computer software and hardware. The decision support system GMCR II allows users to readily model and analyze conflicts using the Graph Model for Conflict Resolution (**Fig. 1**). It has been applied in diverse domains including water resources, international trade, politics, and military science. GMCR II is appropriate for studying large complex disputes, but it can also be used for small models, such as the one examined below, which will provide context for a discussion of the modeling subsystem, the analysis engine, and the output interpretation subsystem.

Sustainable development conflict. Environmental conflicts can be very complex and hence require complex models. Nevertheless, in some situations it is possible to gain understanding of a conflict by studying a simple or rudimentary model. For instance, the sustainable development conflict represents a generic dispute occurring between environmentalists and developers. The developers typically wish to construct a new industrial facility, expand a residential area in a city, build a hydroelectric complex, sell genetically engineered seeds to farmers, or purchase public infrastructure. The environmentalists may include governmental agencies (often those that ensure compliance to environmental regulations), nongovernmental organizations representing specific environmental interests, and coalitions of concerned citizens. The environmentalists'

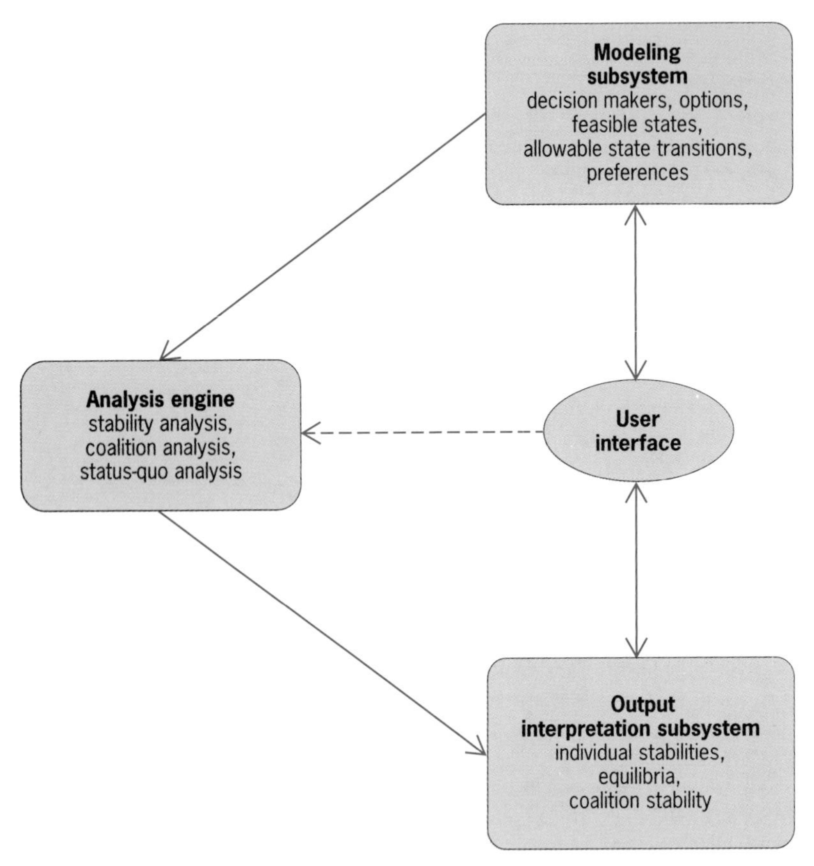

Fig. 1. Decision support system GMCR II.

objective is to ensure that the new or changed economic activities are carried out in a fashion that properly protects, or even enhances, the affected environment.

In the sustainable development conflict there are two decision makers, called environmentalists and developers. Each decision maker controls one main option or course of action, which he or she may select or not. The environmentalists can be proactive, fostering responsible environmental stewardship by the developers; or reactive, not selecting the proactive option and merely responding to environmental events after they take place. The developers can choose the sustainable option, practicing sustainable development in all of their activities and fully discharging their obligations to the environment and society. Alternatively, they can reject their sustainable option, behaving unsustainably by pursuing profit maximization as their major objective.

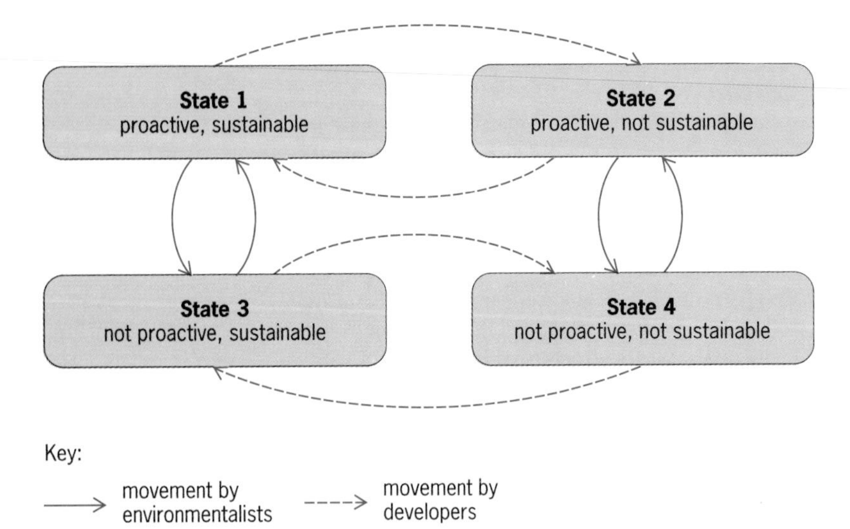

Fig. 2. Integrated graph model for the sustainable development conflict. State ranking (most preferred on left to least preferred on right) for the environmentalists: (1, 3, 2, 4); state ranking for the developers: (4, 2, 3, 1).

Figure 2 shows the integrated graph model for the sustainable development conflict. Each vertex represents one of the four possible states in the dispute. For example, state 3 is the situation in which the environmentalists reject the proactive option while the developers have chosen the option to be sustainable. The movements between states controlled by decision makers are depicted by arcs, where the direction of movement is indicated by an arrowhead. Solid and broken arrows stand for movements controlled by the environmentalists and developers, respectively.

A graph model captures the dynamic or evolving nature of a conflict, as each decision maker moves and countermoves until some resolution is reached. What motivates a decision maker to make unilateral moves is his or her value system, or preference, over the states. In fact, preference elicitation is one of the most crucial steps in the calibration of any decision model. A key advantage of the graph model for conflict resolution and other nonquantitative methods is its requirement of nonquantitative preference information only.

The ranking of states from most to least preferred for the environmentalists is [1, 3, 2, 4]. Thus, the environmentalists most prefer state 1, in which they are proactive and the developers practice sustainable development, and least prefer state 4, in which they are reactive and the developers are not environmentally responsible. A useful way to express their ranking of states is in terms of options; the highest priority of the environmentalists is that the sustainable option be chosen, and a secondary priority is that being proactive be chosen. One special capability of GMCR II is its use of such hierarchical preference statements to construct a preference ordering for a decision maker. Similarly, the preference ordering of the developers is [4, 2, 3, 1], which happens to be exactly the reverse of the environmentalists' ranking. Because state 2 is more preferred than state 1 by the developers who control movement between states 1 and 2, the movement from state 1 to 2 is called a unilateral improvement for the developers. Likewise, the movement from state 3 to state 4 is also a unilateral improvement for the developers. Similarly, the environmentalists' unilateral improvements are from state 3 to state 1 and from state 4 to state 2.

Analysis and output. A conflict model constitutes a basic framework within which the strategic interactions of the decision makers can be analyzed in detail. In GMCR II, the systematic examination of moves and countermoves during possible evolutions of the conflict, and the calculation of the most likely resolutions, are carried out at the stability analysis stage.

Stability analysis depends on a solution concept, or stability definition, which is a mathematical description of how a decision maker makes choices in a strategic conflict. Different solution concepts model different patterns of behavior, which may reflect differing levels of foresight and differing approaches to strategy. GMCR II analyzes each state for stability for each decision maker according to a rich variety of stability definitions. If a state is individually stable for a decision maker according to some solution concept, then it is not advantageous for that decision maker to move unilaterally away from that state. When a state is stable for all decision makers in a model, it is referred to as a compromise resolution or equilibrium.

For a particularly simple model such as the sustainable development conflict model, some stability calculations can be done by hand. To explain the dynamic interactions that are relevant to a stability assessment, consider state 3 in which the environmentalists are not proactive but the developers are practicing sustainable development. In a stability analysis, one attempts to determine if a decision maker would remain at a state if the conflict arrives there. Consider first whether state 3 is stable for the environmentalists. Because this decision maker can unilaterally improve from state 3 to state 1 by changing its option selection from reactive to proactive, state 3 initially appears not to be stable. However, the environmentalists must ask themselves: "If we take advantage of the unilateral improvement to state 1, will the developers change their option choice in response to our action?" According to the sequential stability solution concept, the answer is "yes." Specifically, from state 1 the developers can select their unilateral improvement at state 2 by deciding not to practice sustainable development. Since state 2 is less preferred than state 3 by the environmentalists, their possible unilateral improvement from state 3 to state 1 is credibly sanctioned. Hence, state 3 is sequentially stable for the environmentalists. As can be checked easily, however, state 3 is not sequentially stable for the developers, and is therefore not an equilibrium.

For another stability assessment, consider state 2. From this state neither the environmentalists nor the developers can make a unilateral improvement. Hence, state 2 is said to be Nash stable for each decision maker and therefore constitutes a Nash equilibrium. In fact, state 2 is the only Nash equilibrium for the sustainable development conflict model. The Nash stability definition, like most of the others included in GMCR II, predicts that the only equilibrium is state 2—a scenario in which the environment will deteriorate unless the environmentalists have sufficient political power to stop the developers. However, if the developers adopt a more positive attitude to environmental stewardship, thereby changing their preference ordering to (3, 1, 4, 2), a single Nash equilibrium occurs at state 1, predicting that a healthy environment can be maintained.

New developments. There have been several recent developments within the graph model paradigm. For instance, a formal methodology for handling "uncertain preferences" has recently been developed. Strength of preference considerations, which apply when a decision maker has one, or a few, highly preferred or strongly disliked states, have been formally incorporated into stability algorithms. Finally, flexible algorithms can now determine possible paths to "strong" or "weak" equilibria from any specified status quo state, a feature especially useful for understanding large conflict models.

Conflict is an inherent characteristic of any decision system, whether it be societal (real life), intelligent (artificial life, consisting of agents), integrated (mixed life), or environmental. Hence, research on conflict analysis and resolution is destined to continue to expand significantly in the long term with meaningful applications to an even broader spectrum of areas.

For background information *see* DECISION SUPPORT SYSTEM; DECISION THEORY; GAME THEORY; GRAPH THEORY; INFORMATION TECHNOLOGY; LOGIC; SET THEORY in the McGraw-Hill Encyclopedia of Science & Technology.

Keith W. Hipel; D. Marc Kilgour; Liping Fang

Bibliography. L. Fang, K. W. Hipel, and D. M. Kilgour, *Interactive Decision Making: The Graph Model for Conflict Resolution*, Wiley, New York, 1993; L. Fang et al., A decision support system for interactive decision making, Parts 1 and 2, *IEEE Trans. Sys. Man Cybernet.*, Part C, SMC-33(1):42-55, 56-66, 2003; K. W. Hipel et al., Strategic decision support for the services industry, *IEEE Trans. Eng. Manag.*, 48(3):358-369, 2001; K. W. Li et al., Preference uncertainty in the graph model for conflict resolution, *IEEE Trans. Sys. Man Cybernet.*, Part A, 34(4):507-520, 2004; H. Raiffa, with J. Richardson and D. Metcalfe, *Negotiation Analysis: The Science and Art of Collaborative Decision Making*, Belknap Press of Harvard University Press, Cambridge, MA, 2002.

Context-aware mobile systems

To perform everyday activities, interact, and adapt to situations, people use information from their surrounding environment such as location, time, temperature, and other people nearby. Thus, people use their context. Computers, however, are unable to understand and adapt to the environment and require users' active attention and the explicit introduction of relevant information. By providing computers with the ability to acquire and use contextual information, they will be able to provide more useful services and better human-computer communication. This has motivated the development of context-aware computing systems, which react to specific situations and anticipate users' needs.

Context-aware computing. Context-aware computing refers to an application's ability to adapt to changing situations. A system is context-aware if it uses context to provide relevant information or services to the user, where relevancy depends on the user's task. Among the main types of contextual information that can be relevant to an application are the identity and location of the user and the presence of other people or artifacts. An application could use this information, for example, to notify a user of the presence of a colleague nearby or the bus schedule and closest bus stop for the evening commute home.

The design of a context-aware application involves the following tasks: (1) identify the contextual information that is relevant to the application, (2) design the mechanisms that can be used to acquire context, and (3) define how the application will react to a change in context.

Relevant contextual information. Contextual information that can be derived directly from data acquired from equipment sensors is known as primary context. This includes location, time, and identity. Primary context can be used to derive additional, related information known as secondary context. For example, the identity (nurse) and location (in a patient's room) of a person can be used to infer the person's current task (measuring the patient's temperature).

Context can be associated with human factors or the physical environment. Context related to human factors can be structured in three categories: information about the user (such as habits and emotional state), social environment of the user (such as colocation of others and social interactions), and the task of the user (such as spontaneous activities and current task). Similarly, context related to the physical environment includes location (such as absolute position, relative position, and colocation), infrastructure (such as surrounding resources and communication), and physical conditions (such as noise, light, pressure, and temperature).

Acquiring context. Some contextual information can be acquired easily, while other requires significant processing and sophisticated computer algorithms. For example, the current time or day can be accurately and easily obtained from a computer. Similarly, the user's identity is available in most modern operating systems that require users to log on to provide access to their files and personal information. In fact, computer systems that use time to remind the user of an upcoming meeting or that give access to files based on the user's identity are context-aware.

Context-aware computing has gained attention as it has become feasible to derive more complex contextual information. In particular, information regarding the location of the user and other artifacts or services is increasingly relevant as mobile computing gains in use.

Determining the location of different objects has been possible with systems such as the Global Positioning System (GPS) for a long time now. However, these types of systems are not suitable for indoor location, as their radio signals are blocked by building walls. Other technologies have been proposed to address this problem, most of them based on triangulation with different wireless technologies such as infrared, ultrasound, radio-frequency identification (RFID), and WiFi (IEEE 802.11). In addition, pressure sensors placed in the floor or video cameras are used to locate the presence and location of individuals.

Context-aware behavior. A context-aware application might react to changes in the environment through a simple "IF-THEN" action rule. For instance, the default printer of a laptop might change to the closest printer if a user changed location. In other instances, the adaptation of the application will be more complex, as in the case of a context-aware information retrieval system that modifies the results of a search-engine query depending on the identity of the user, the device from which the query was made, and the user's previous search history.

In general, context-aware applications that need to handle several contextual parameters and react to them in nontrivial ways are difficult to build. This has motivated the development of software tools that provide useful abstractions for context acquisition, presentation, and handling.

In addition to the problem of user location and tracking, context-aware mobile systems need to deal with issues related to the limitations of the mobile device such as screen size, battery life, and frequent network disconnections, as well as the need to seamlessly adapt to the various and heterogeneous devices, such as mobile telephone, personal digital assistant (PDA), laptop, desktop computer, and public display, which might be employed or encountered by the user.

Context-aware applications. Systems that obtain information about the surrounding environment and serve as guides are the most popular context-aware applications. For example, mobile context-aware tour guides provide information to users based on their position and orientation. They could provide museum visitors with relevant information according to the object they are closest to, that is, according to the context of their location. This information could be presented on their mobile phone, PDA, or laptop computer.

Hospitals are more complex but very natural environments for using context-aware systems. The mobility, need for coordination, and information exchange required among different medical personnel makes their work highly dependent of their location, the time, and their roles. **Figure 1** shows a context-aware information system developed for a public hospital. The system provides information related to the location of hospital workers and enables them to send messages to colleagues based on contextual information such as their role or location. For instance, a physician could access the results of a patient's lab tests as he approaches the patient's bed, and a nurse could locate hospital workers in the vicinity to ask for assistance or consultation (Fig. 1*a*). In addition, a physician could send a message to the nurse who will care for a patient in the afternoon shift, even if he does not know beforehand who that nurse will be (Fig. 1*b*). This application was developed by studying the actual work practices within a hospital and identifying the contextual variables relevant to the

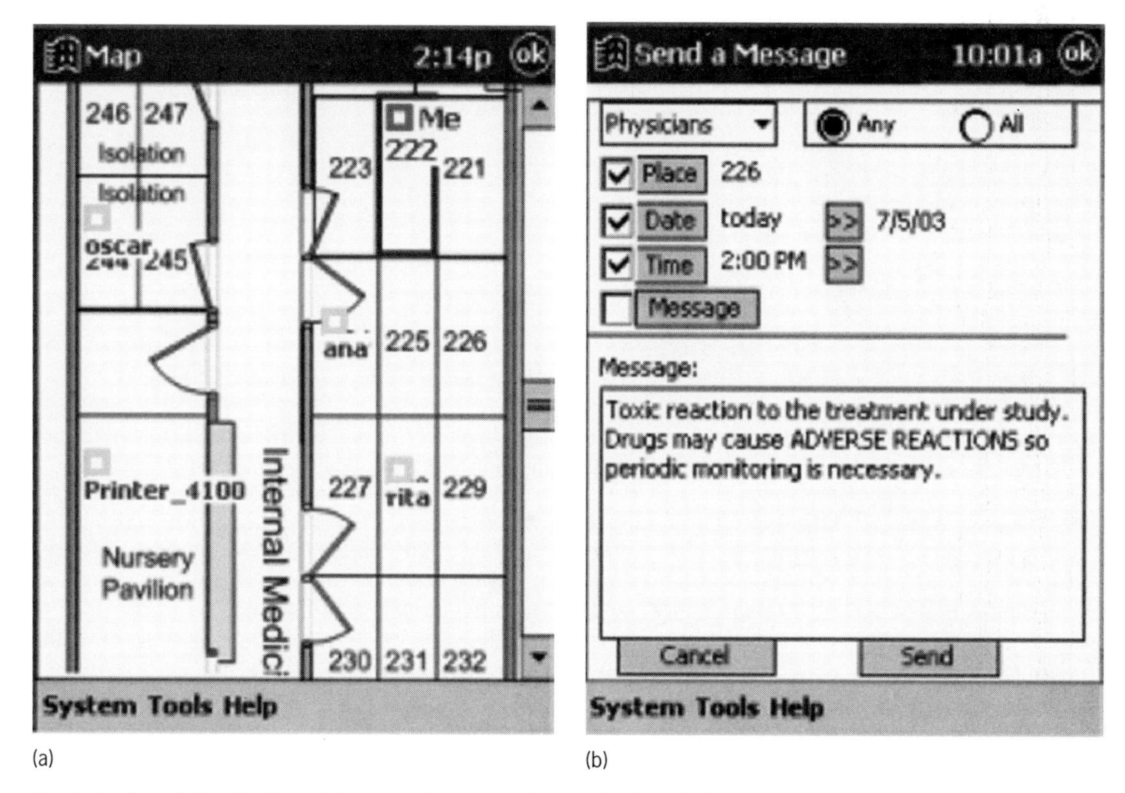

(a) (b)

Fig. 1. In a hospital application (*a*) the context-aware medical application displays, on the user's PDA, the presence and location of fellow hospital workers, and (*b*) a physician can specify the location, time, and role of the recipient, as conditions for the delivery of a context-aware message. (*From M. A. Muñoz et al., 2003*)

information exchanges. Potential users found this tool useful in supporting their daily work.

As is the case with other context-aware applications, this system integrates contextual information from different sources. The user's location is estimated on a PDA using the signal strength from at least three wireless access points using pattern recognition algorithms (**Fig. 2**). Other contextual information, such as the time of day or user's identity, is obtained directly from the device. A decision-making module receives this information, integrates it, and uses it to decide how to act. Relevant actions could include displaying certain information (such as a patient's lab results) or displaying different levels of detail according to the user's identity.

Outlook. In spite of their potential benefits, the adoption of context-aware systems raises some important issues. Chief among them is privacy. Applications that sense and share private information, such as a user's location, could be a threat or a nuisance to

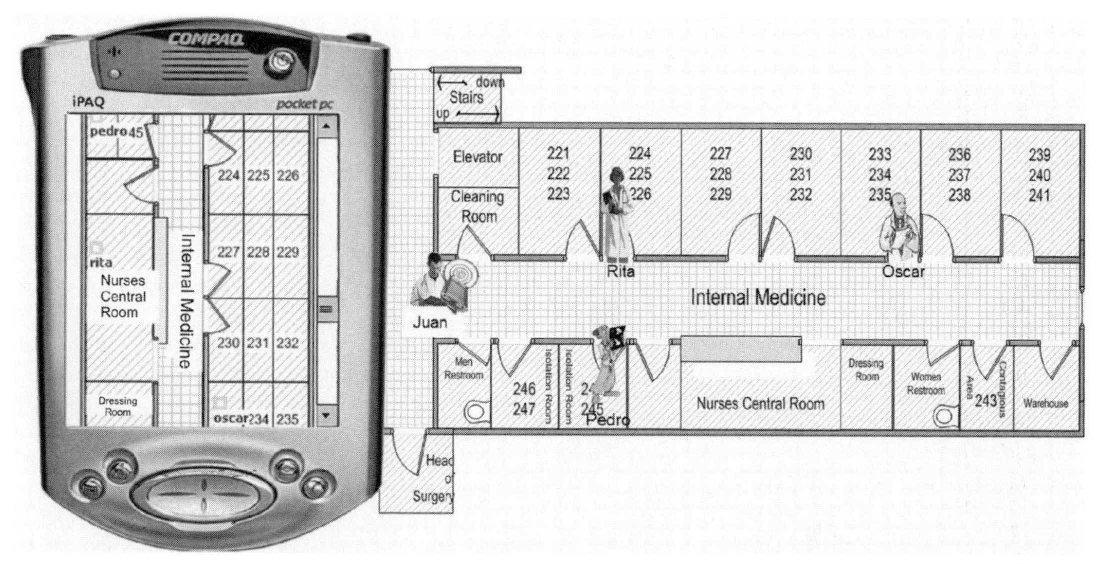

Fig. 2. Location-aware client indicating the presence and location of users in the hospital. (*From M. A. Muñoz et al., 2003*)

the user. To avoid these concerns, applications must provide mechanisms to assure users that information related to them will not shared or stored without permission.

As technologies such as sensor networks, user interfaces, and those related to the field of context-aware computing continue to evolve, we will be able to use computers without requiring our active attention. Then, computers will anticipate our needs and provide a richer environment for leisure, collaboration, and communication.

For background information *see* ALGORITHM; DATA COMMUNICATIONS; HUMAN-COMPUTER INTERACTION; HUMAN-FACTORS ENGINEERING; OPERATING SYSTEM; SOFTWARE in the McGraw-Hill Encyclopedia of Science & Technology. Ana I. Martinez-Garcia; J. Antonio Garcia-Macias; Jesus Favela

Bibliography. A. K. Dey, Understanding and using context, *Pers. Ubiquitous Comput.*, 5(1):4–7, 2001; M. Hazas, J. Scott, and J. Krumm, Location-aware computing comes of age, *IEEE Comput.*, 37(2):95–97, 2004; M. A. Muñoz et al., Context-aware mobile communication in hospitals, *Computer*, 36(8):60–67, 2003.

Controlled/living radical polymerization

Free-radical polymerization is used extensively in industry for the synthesis of a variety of polymeric materials. Its widespread use is due to its compatibility with many functional groups, its tolerance toward water and protic media, and its simplicity. Free-radical polymerization is routinely done under various conditions and processes, including bulk, solution, or aqueous environments, and is now the leading industrial method for producing vinyl polymers and copolymers.

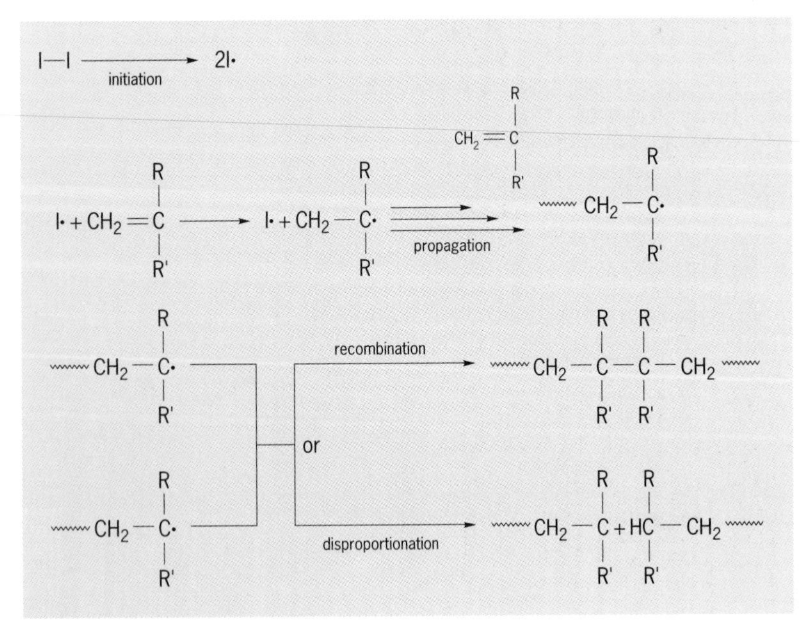

Fig. 1. Free-radical polymerization.

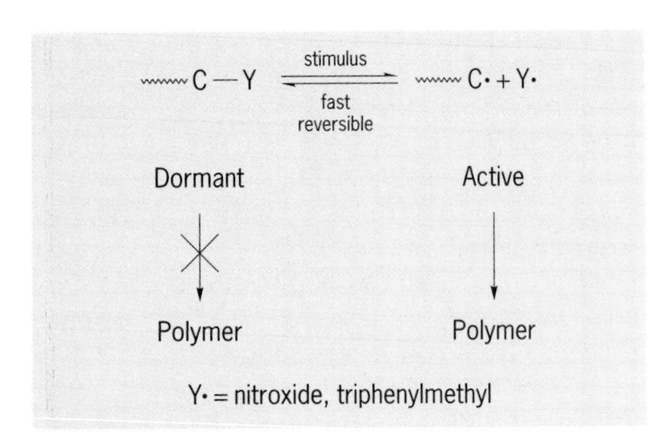

Fig. 2. Active and dormant radical equilibrium.

In spite of its utility, free-radical polymerization is considered unsuitable for the synthesis of polymers with precise compositions, architectures, and functionalities. This is because of the very limited lifetime of the propagating free-radical species (typically 0.1 to 10 s), which suffer from the propensity to undergo diffusion-controlled termination by recombination or disproportionation (**Fig. 1**), unlike growing ionic species which inherently repel each other. The polymers obtained by free-radical means are therefore generally ill defined and polydisperse (wide molecular-weight distribution), with only limited control over molecular weight or architecture.

Until recently, the only option for efficiently controlling the structure and architecture of vinyl polymers was ionic or coordination polymerizations, which do not suffer from irreversible chain-terminating (deactivating) reactions and are known as "living" polymerizations. Unfortunately, living ionic polymerizations are of limited scope since they require stringent purity conditions as well as the total absence of water and oxygen, and are generally incompatible with monomers with functional groups. The necessity to overcome these limitations fueled the development of a new polymerization procedure that would exhibit all characteristic features of regular free-radical polymerization, as well as "living" character, similar to ionic polymerizations.

When addressing this challenge, the primary difficulty was to find a way to avoid the premature termination of the growing radicals. The idea that eventually predominated was to lower the instantaneous concentration of growing radicals by trapping them in a large reservoir of covalent, dormant species and establish a fast equilibrium between the latter and a minute concentration of active, growing radical species. If such a dynamic equilibrium could be established, the probability for spontaneous bimolecular radical termination would be minimized, and all chains would be given an equal opportunity to grow through constant interconversion between their active and dormant forms (**Fig. 2**). Another interesting outcome would be a nearly uniform molecular weight, as determined by the ratio of the concentrations of converted monomer to that of the dormant species (or the initiator). First conceptualized by T. Otsu in 1982, this seminal idea has given birth to the field of controlled/living radical polymerization, which has grown explosively.

Characterized by different mechanisms and chemicals involved, three main methods (by far the most studied) of control of radical polymerization have been developed: (1) nitroxide-mediated radical polymerization (NMP), (2) atom transfer radical polymerization (ATRP), and (3) reversible addition-fragmentation chain transfer (RAFT) polymerization.

Nitroxide-mediated radical polymerization. Nitroxides are stable free radicals that are generally used in organic chemistry to trap transient radicals. They react at diffusion-controlled rates with carbon-centered radicals and are known not to initiate polymerization. Taking advantage of these features, M. K. Georges and coworkers used the commercially available nitroxide, 2,2,6,6-tetramethyl-1-piperidinyloxy (TEMPO), to trap growing polymer chains and demonstrated that the trapping process is reversible when the temperature range is appropriately chosen. The mechanism of control involves the reversible termination of growing radicals by TEMPO, which functions as a counter radical. Samples of polystyrene exhibiting narrow molecular-weight distributions could be obtained by heating styrene at 123°C (253°F) in the presence of TEMPO and a free-radical initiator. However, the reversible radical capping by TEMPO could not be used for most monomers other than styrene, since most other vinyl monomers showed a total incompatibility with this kind of nitroxide. Anticipating that the structure of the stable radical used to trap the growing radicals must play a crucial role in the success or failure of such living radical polymerization, Y. Gnanou and P. Tordo designed a more effective nitroxide to extend the scope of NMP. They developed a α-hydrogen-bearing nitroxide, namely *N-tert*-butyl-*N*-{[1-diethylphosphono-(2,2-dimethylpropyl)]nitroxide [DEPN]}, which is of slightly lower stability than TEMPO and whose main characteristics is to give alkoxyamines with much weaker C—O bonds than TEMPO. Not only could styrene be polymerized at faster rates and under living conditions in the presence of DEPN and a radical initiator, but also the polymerization of acrylates, acrylamides, and more recently methacrylates were found to be living in the presence of DEPN (**Fig. 3**).

In a similar approach, C. Hawker and coworkers developed unimolecular alkoxyamines based on 2,2,5-trimethyl-4-phenyl-3-azahexane-3-nitroxide (TIPNO), another α-hydrogen-bearing nitroxide, and used these as initiators and controlling agents for the living polymerization of the above monomers, as well as for acrylonitrile, butadiene, and others. The beauty of nitroxide-mediated radical polymerization lies in the fact that the amount of terminated chains eventually formed is exactly equal to the excess nitroxide produced by the system to minimize termination reactions. H. Fischer conceptualized this remarkable phenomenon and called it the persistent radical effect. To minimize the amount of terminated chains to a level lower than that determined by the system, the nitroxide can be used in excess with respect to the initiator.

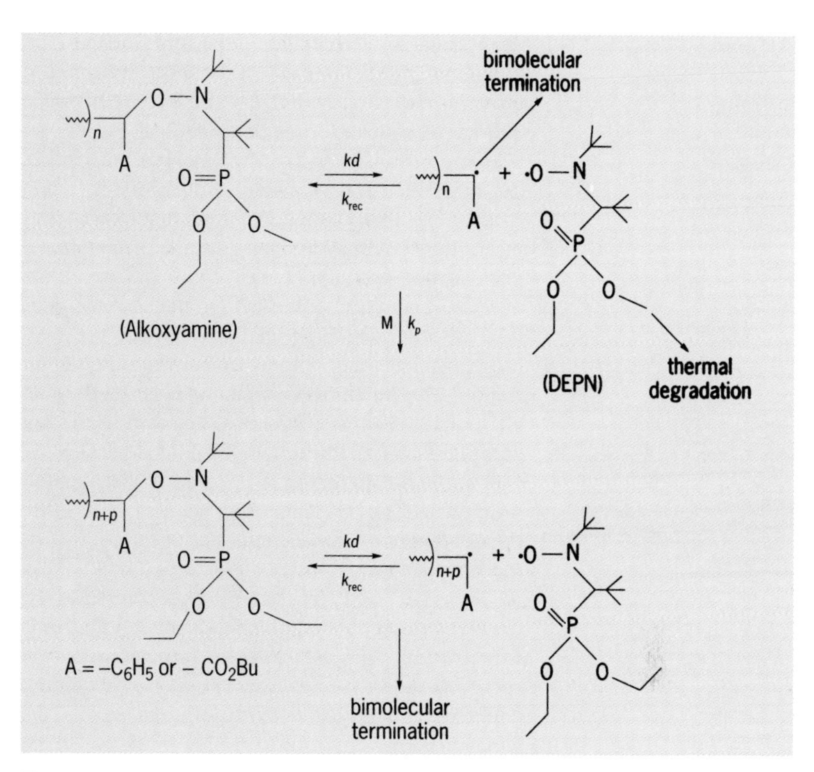

Fig. 3. Nitroxide-mediated radical polymerization of styrene or butylacrylate.

Atom transfer radical polymerization (ATRP). Discovered by K. Matyjaszewski in 1995, atom transfer radical polymerization entails a reversible one-electron oxidation of a transition metal, abstracting a halogen from an organic halide (RX), thus generating a radical species ($R^\bullet$) that can attack an unsaturated compound (CH_2=CHR′) and further propagate (R—[CH_2—CHR′]$_n$—CH_2—$C^\bullet$HR′), followed by a one-electron reduction of the now oxidized metal, transferring its halogen to the growing chain (R—[CH_2—CHR′]$_n$—CH_2—$C^\bullet$HR′), and thus producing (R—[CH_2—CHR′]$_n$—CH_2—CHR′—X) as a dormant species. Such halide abstraction (k_{act}) and subsequent transfer (k_{deact}) occur repeatedly, and polymer chains grow between each activation/deactivation cycle (**Fig. 4**).

Provided the latter process is rapid, all the chains can grow uniformly. The oxidized metal complexes X—M_t^{n+1} generated by activation of the dormant

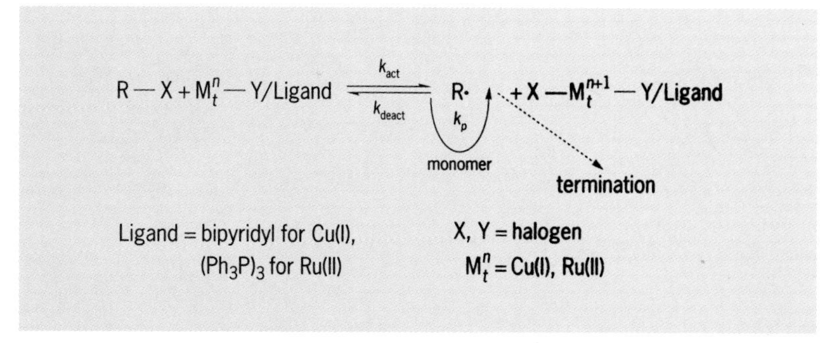

Fig. 4. Atom transfer radical polymerization.

alkyl halides behave like persistent radicals and therefore reduce the stationary concentration of growing radicals, minimizing the extent of termination. Like nitroxide-mediated radical polymerization, atom transfer radical polymerization is under the control of the persistent radical effect, which means that a successful atom transfer radical polymerization contains a small amount of terminated chains. Also in 1995, M. Sawamoto independently described metal-catalyzed living radical polymerization (MCRP), which appeared to exhibit the same mechanism as that of atom transfer radical polymerization. The initiating systems for both methodologies consist of a transition-metal/ligand complex and an initiator (typically an activated alkyl halide), the role of the former being to generate radical species from the latter by the above mechanism. In choosing the appropriate metal/ligand catalyst for controlling the polymerization, four criteria are critical: (1) the metal center must have at least two accessible oxidation states separated by one electron, (2) it must exhibit affinity toward halogen, (3) the coordination sphere around the metal must be able to accommodate the halogen upon oxidation, and (4) the ligand must form a strong complex with the metal. On these grounds, Matyjaszewski proposed copper(I)bromide associated with a variety of nitrogen-based ligands as the atom transfer radical polymerization catalysts, and Sawamoto associated Ru(II) with phosphine and other ligands as metal-catalyzed living radical polymerization catalysts. As demonstrated by its extensive use in the synthesis of polymers with precise composition and architecture, atom transfer radical polymerization is superior to metal-catalyzed living radical polymerization by its versatility and cost.

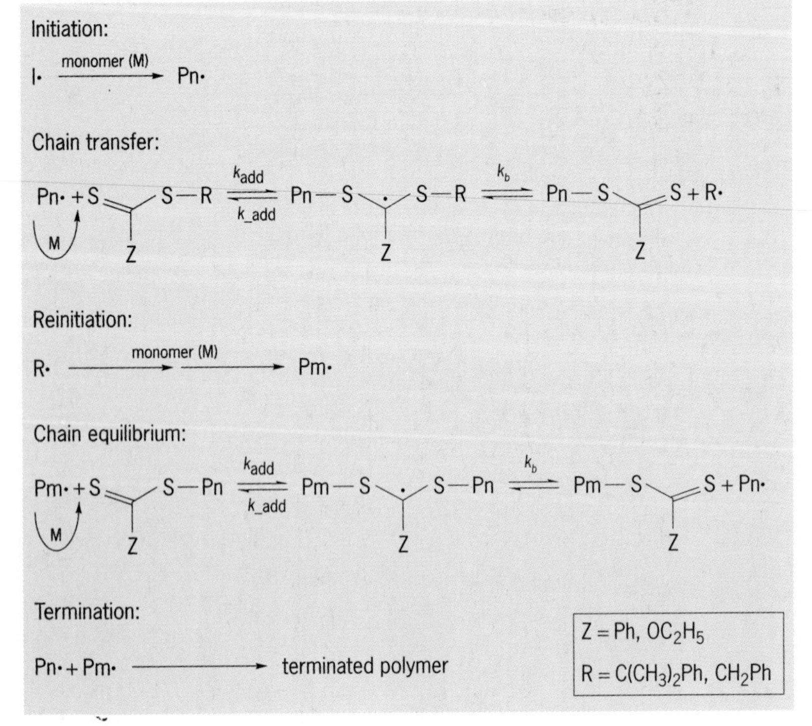

Fig. 5. RAFT polymerization.

Reversible addition-fragmentation chain transfer (RAFT) polymerization. In 1998, E. Rizzardo described a new method for controlling free-radical polymerization that involved the use of thiocarbonylthio compounds reacting by reversible addition-fragmentation transfer (RAFT) with the growing radicals. The effectiveness of these RAFT reagents, mirrored in chain-transfer constants C_{tr} larger than 1, depends on the nature of their (Z) and (R) groups (**Fig. 5**), Z determining their reactivity toward entering radicals and R its aptitude for fragmentation. The key step in a successful RAFT process is the chain transfer of growing radicals to the RAFT reagent and the subsequent formation of the intermediate radical. The latter then undergoes fragmentation, releasing a radical ($R^{\bullet}$) which can initiate polymerization. The entering polymer chain now carries a terminal [S=C(Z)S—] moiety and is transformed into a dormant species. After total consumption of the thiocarbonylthio compound, an equilibrium is established between the dormant thiocarbonylthio-carrying chains and active ones, producing a polymer with narrow polydispersity, for large C_{tr} values of the thiocarbonylthio compound.

The molecular weight is given by the ratio of monomer to thiocarbonylthio compound, and the amount of terminated chains correspond to that of the radical initiator used to trigger the polymerization. The main advantage of RAFT over nitroxide-mediated radical polymerization and atom transfer radical polymerization is its applicability to a wider range of monomers, including functional ones such as acrylic acid or those carrying an electron-donating substituent such as vinyl acetate. D. Charmot of Rhodia Recherches, France, showed that xanthates, which are also thiocarbonylthio compounds, are excellent RAFT reagents for vinyl acetate and analogues, and termed its invention MADIX for macromolecular design via interchange of xanthates.

Application. The three methodologies of controlling free-radical polymerization have their merits and limitations. ATRP is easy to carry out but it cannot be used with certain functional monomers and requires a tedious metal-removal step, whereas RAFT or NMP can be applied to functional and nonfunctional monomers but their controlling agents may necessitate complex synthesis.

With the advent of NMP, ATRP, and RAFT, novel polymeric materials—many of which are totally original and exhibit precise composition such as block and gradient copolymers and complex topology such as star-, comb-, and dendrimer-like polymers—can be obtained under the conditions of regular free-radical polymerization. This paves the way for tremendous opportunities in bio- and nanotechnologies and other high-tech areas.

For background information *see* COPOLYMER; FREE RADICAL; LIGAND; POLYACRYLATE RESIN; POLYMER; POLYMERIZATION; POLYSTYRENE RESIN; POLYVINYL RESINS in the McGraw-Hill Encyclopedia of Science & Technology. Yves Gnanou; Daniel Taton

Bibliography. D. Benoit et al., Kinetics and mechanisms of controlled free-radical polymerization of styrene and n-butyl acrylate in the presence of an acyclic β-phosphonylated nitroxide, *J. Amer. Chem. Soc.*, 122:5929–5939, 2000; J. Chefari et al., Living free-radical polymerization by reversible addition-fragmentation chain transfer: The RAFT process, *Macromolecules*, 31(16):5559–5562, 1998; C. Hawker, A. Bosman, and E. Harth, New polymer synthesis by nitroxide mediated living radical polymerization, *Chem. Rev.*, 101:3661–3688, 2001; K. Matyjaszewski and T. P. Davis (eds.), *Handbook of Radical Polymerization*, Wiley-Interscience, 2002; K. Matyjaszewski and J. Xia, Atom transfer radical polymerization, *Chem. Rev.*, 101:2921–2990, 2001.

Fig. 1. Israeli *Protector* craft. (*Rafael Ltd.*)

Crewless surface watercraft

With the development of powerful digital computers and rapidly improving information technology, especially since 1980, operating watercraft without a crew onboard has been increasingly feasible. In the development of crewless vehicles, remote autonomous aircraft have led the way in terms of numbers and technology (including autonomous control spacecraft, which have braved many frontiers), followed by ground vehicles and undersea craft; but remote and autonomously controlled surface watercraft recently have also become the subject of significant research and development. Currently there are many developmental efforts for a variety of commercial tasks and governmental missions.

Among remote and autonomously controlled aircraft (termed UAVs or unmanned air vehicles), the most widely known are the *Predator* and *Global Hawk*, but they range from minisize [wingspan the length of a pencil in the case of the Defense Advanced Research Projects Agency (DARPA) *Wasp*], to back-packable, to truck-deployable, up to full-size aircraft (such as the *Global Hawk* and the developmental *UCAS*, the size of a full combat jet). Initially relegated to surveillance roles, they have seen operational use in the past several years in weapons delivery under human remote control and ever-increasing autonomy (computer control). The current state of technology combines both aspects: computer control for the long and boring periods of the mission, and remote human control for key activities, such as combat. *See* VERY SMALL FLYING MACHINES.

The use of remote and autonomously controlled craft is especially appropriate for jobs that are dull, dirty, or dangerous. Two of the first underwater remote and autonomously controlled undersea watercraft missions were in the deployment of torpedoes and minehunting.

The factors leading designers to explore remote and autonomously controlled vehicles include weight and geometry penalties associated with human operators, danger to the operators, operator time limitations (computers can work continuously, and remote operators can work shifts, perhaps operating several vehicles augmented by computer), and the promise of eventual cost reductions in acquisition and operation. Exploiting new design freedoms, developers strive to produce smaller, simpler, and more numerous vehicles, about half the size of an equivalent, conventionally crewed vehicle. The advantages of remote and autonomously controlled craft are heightened for particularly dangerous (such as mine clearance) and high-speed, weight/geometry-constrained (such as with aircraft) applications.

The Israeli *Protector* (**Fig. 1**) is a 4-ton, 9.5-m-long (31-ft) craft which has a stabilized gun, electrooptic and radar sensors, and water jet propulsion. It was developed in 2001 and is also used in Singapore. Activities in the United States have centered on the *Spartan Scout* (**Fig. 2**), a 2-ton, 7-m-long (23-ft) remotely operated boat. It is armed with sensors (mostly cameras) and a .50 caliber machine gun. It can also operate without a remote operator [using Global Positioning System (GPS) waypoints] and is designed for modular equipment replacement for different missions, such as reconnaissance, mine detection, force protection, and antisubmarine warfare. *Spartan Scout* was developed starting in 2001 and first deployed for partial demonstration in late 2003. It is intended for extensive use on the U.S. Navy's Littoral Combat Ship (LCS).

Small marine surface craft. Small marine surface craft are used for numerous tasks. There are tens of

Fig. 2. U.S. Navy's *Spartan Scout.*

thousands of recreational vehicles as well as lighterage and tenders for larger ships, targets, and utility vehicles. However, they currently are used primarily for harbor and near-coastal duties, due to the strain on a small crew for longer-duration and longer-distance missions. If the crew can be removed (for remote control) or eliminated (for autonomous deployment), the potential exists for crewless surface vehicles (customarily termed unmanned surface vehicles or USVs) to augment larger and more expensive surface ships in certain roles, serving as a force multiplier. Moreover, the motions of small craft are less of an issue. Moreover, many systems designed for crew support (passageways, berthing, messing, some aspects of temperature control, and so forth) can be eliminated. With the removal of extensive systems, the overall size and cost of the vessel can be greatly reduced, in turn resulting in lower fuel demands. Finally, if autonomous watercraft endurance can be made long enough (through propulsion improvements and sufficient reliability), there is less need to return to base for crew relief and ship support. But some serious technology issues exist for practical, broad-based use of crewless marine surface craft: (1) agile computer algorithms are needed to handle the complexities of selected missions; (2) robust distant communications are required to enable occasional remote control, reporting, and new generalized instructions; (3) suitable speed and endurance for the particular mission are needed; smaller displacement marine surface craft have a practical speed that is limited by length, and are slower than air vehicles.

Potential for crewless surface watercraft. Surface watercraft have advantages over faster and wider-horizon aircraft and superstealthy undersea craft. Primarily, they offer affordable load-carrying capacity with long endurance. This is because floating on the surface of the water requires fewer and simpler systems than underwater or aeronautical operation (no requirement for ballasting or altitude control, for instance), and the availability of air enables less exotic, higher-power, and less costly powering solutions (such as diesel engines and gas turbines, service batteries, or fuel cells) than in undersea craft, and cheaper payload carriage than either heavier-than-air or lighter-than-air aircraft. Surface watercraft, operating at the interface between water and air, offer some advantages in being able to communicate in all three mediums of interest (undersea, air, and space) and being able to relay information from submerged assets (submarines, autonomous submersibles, undersea arrays, and so forth) to any combination of surface vessels, aircraft, or satellites.

Future trends. There are a number of experimental crewless watercraft developments under way, as well as programs for technologies that will be key to more capable crewless watercraft in the future. Some of the projects currently being designed and built are Maritime Applied Physics Corporation's USSV (Unmanned Sea Surface Vehicle), two modularly similar craft with slower heavier payload and faster lighter payload; Oregon Iron Works' (with Geneva Aerospace for one of three surface watercraft) cascading marine vehicles; and the Autonomous Maritime Navigation project, all for the U.S. Navy. The University of Plymouth (United Kingdom) is developing the Springer for shallow-water environmental research, with analogous projects ongoing at other universities. Aluminum Chambered Boats, Inc.'s commercial venture is designed to use a single remote operator for several crewless boats for harbor and nearshore monitoring.

There is a spectrum of crewing philosophies, and the USV research community is approaching the ultimate goal of total autonomy in all sizes from either end:

a. Remote controlled craft, no crew onboard but crew control offboard
b. Primarily remote control craft, augmented by some simple autonomous control
c. Primarily complex autonomous control craft, augmented by very part-time remote control for particularly cognitive tasks (for instance, unknown vessel query, casualty realignment, or weapons engagement)
d. Completely autonomous ship or craft (currently achieved for only quite simple missions and smaller craft; for instance, a homing torpedo, wide surveillance aircraft, or target boat)
e. Primarily autonomous ship with a remote "captain"
f. Primarily remote controlled ship augmented by considerable autonomous control
g. A quite small onboard crew, augmented by remote or autonomous control
h. A modestly sized onboard crew made possible by considerable machinery automation
i. A fully crewed ship

As of 2005 in surface watercraft, the state of the art is at both ends of the spectrum: *a* and *b* for smaller and shorter endurance craft, and *h* and *i* for full-sized oceangoing ships. Research and development is seriously under way for *c* and *g*, and in most cases, only concepts exist for the middle realm approaching full autonomy (*d*, except for the simplest missions, through *g*). The key technological developments for future achievements in approaching fully crewless watercraft in all sizes are control algorithms for complex autonomous missions and secure, affordable, distant communications linkages for remote control. A key incentive for fully autonomous control of more complex missions has been provided by the Defense Advanced Research Projects Agency (DARPA) through their desert terrain "Grand Challenge" autonomous control invitational field tests.

In other types of crewless vehicles (air, ground, undersea), the greater maturity and scope of the developmental efforts have led to a wide variety of sizes. This trend should extend to the remote and autonomously controlled surface watercraft development community soon. Rather than virtually all crewless surface watercraft being of the small boat size (10–40 ft, 3–12 m), efforts for smaller (mini and micro) craft and larger and ship-sized craft should gradually come under investigation. The smaller sizes might function well for some shorter-range surveillance and maintenance missions, and the larger sizes for longer-endurance logistics and ordnance delivery missions. The air delivery of smaller, mostly autonomous surface watercraft has the potential advantage of adding delivery speed. Endurance in the smallest sizes of autonomous watercraft is a key element to their eventual mission utility, and such endurance might be achievable by energy scavenging (solar, wind, or wave energy). Multiple-mode vehicles [such as the planing hydrofoil-assisted SWATH (small waterplane area twin hull) transport, the PHAST (planing hydrofoil-assisted SWATH transport) concept design, or the hybrid UAV/USV Sea Scout design being developed in 2005 by Oregon Iron Works and Geneva Aerospace] could help deal with the desire for rapid relocation combined with stationkeeping endurance.

The newly emerging field of remote and autonomously controlled surface watercraft is at the beginning of a promising future, with significant challenges yet to be mastered regarding complex control algorithms, secure remote-control linkages, and affordable long endurance. A variety of sizes will likely be developed for an ever-increasing complexity of commercial and military missions. The research will initially be led by the military services but eventually will be overtaken by competitive commercial developments as autonomous control technology, societal, and legal issues are dealt with.

For background information *see* NAVAL SURFACE SHIP; SATELLITE NAVIGATION SYSTEM; SHIP DESIGN in the McGraw-Hill Encyclopedia of Science & Technology.

Michael L. Bosworth

Bibliography. S. Cooper and M. Norton, *New Paradigms in Boat Design: An Exploration into Unmanned Surface Vehicles*, presented at Association for Unmanned Vehicles Systems International (AUVSI) Unmanned Systems North America 2002 Symposium; J. Ebken, M. Bruch, and J. Lum, *Applying UGV Technologies to Unmanned Surface Vessel's*, SPIE Proc. 5804, Unmanned Ground Vehicle Technology VII, Orlando, FL, March 29–31, 2005; L. Elkins, E. Hansen, and J. Smith, *Object Recognition for an Autonomous Maritime Navigation System*, presented at AUVSI's Unmanned Systems North America 2005 Symposium.

Cryptochrome

Cryptochrome is a blue/ultraviolet-A (UV-A) light receptor found in a wide range of organisms from bateria to humans. It was so named because the early studies of cryptochromes were largely done in cryptogamic (flowerless) plants, and because its molecular nature once remained cryptic. Cryptochromes are the only type of photoreceptor known to function in both the plant and animal kingdoms, mediating the influence of light on plant development and the biological clock in plants and animals. Cryptochromes are flavoproteins that share structural similarities with the photoactive deoxyribonucleic acid (DNA)-repairing enzyme DNA photolyase, but do not have photolyase activity.

Function in plant development. Plants possess two types of photoreceptors. The first type consists of the photosynthetic pigments that harvest light energy for carbon fixation, and the other type contains the photosensory receptors that help plants to adjust their growth and developmental processes in response to changing light conditions. There are three classes of plant photosensory receptors: the red/far-red light receptor phytochrome and the blue/UV-A light receptors cryptochrome and phototropin. Plants usually contain genes for multiple photoreceptors in each class. For example, *Arabidopsis* has at least two cryptochromes, CRY1 and CRY2, and at least one species of fern has five. In *Arabidopsis* and tomato, cryptochromes mediate light regulation of stem elongation, leaf expansion, pigment and hormone biosynthesis, chloroplast development, and flowering time. In fern, cryptochromes regulate light inhibition of spore germination as well as hormone response. Under natural light conditions, cryptochromes often act together with phytochromes to control the same developmental process.

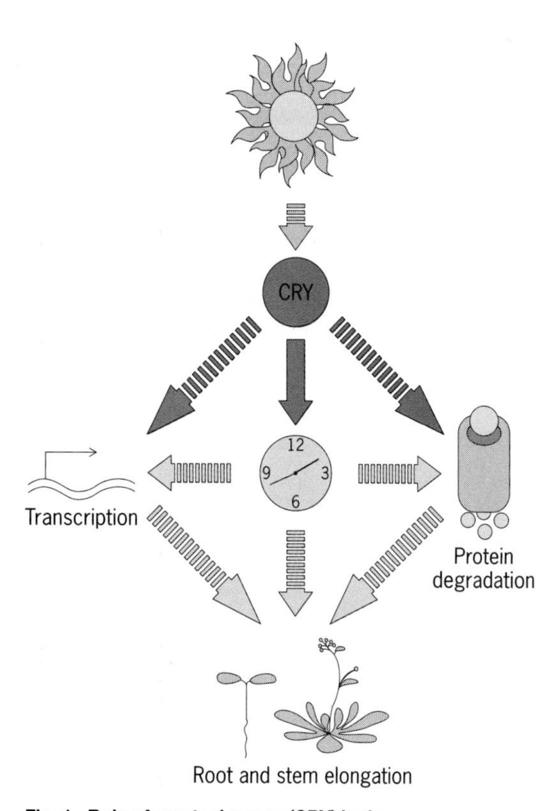

Fig. 1. Role of cryptochromes (CRY) in the sunlight-induced regulation of stem elongation and flowering time in plants. Via signal transduction through cryptochromes, sunlight directly stimulates both gene transcription in the nucleus and protein degradation in the proteosome apparatus, as well as indirectly stimulating them through the mechanism of the biological clock.

Cryptochromes are generally nuclear proteins, and the nuclear distribution of some cryptochromes is regulated by light. Cryptochromes mediate light regulation of plant development by affecting gene expression and protein stability, although the molecular mechanisms are not fully understood (**Fig. 1**). Two biochemical processes associated with the early light responses of plant cryptochromes are light-induced electron transfer and light-induced phosphorylation. Cryptochromes are phosphorylated in response to blue light but not to other wavelengths of light (such as red light). It is thought that absorption of blue light by flavin and folate bound to cryptochromes causes flavin reduction, cryptochrome phosphorylation, and changes in the protein conformation of the photoreceptor. The photoexcited cryptochrome interacts with other signaling proteins to facilitate propagation of the light signal. Although it appears that cryptochromes interact with other proteins in the nucleus to regulate gene expression, many of these proteins have yet to be identified.

Function in the biological clock. Cryptochrome has a major role in the biological clock, serving as either a photoreceptor regulating the clock or an integrated component of the clock. The biological clock, also known as the circadian clock or circadian oscillator, is the name of the self-sustaining cellular machinery that uses a negative-feedback mechanism to adjust cell activity with approximately 24-h periodicity. Regulation of the circadian clock by photoreceptors allows many biological processes to respond to the daily light/dark cycle and the seasonal day length cycle on Earth. For example, the circadian clock controls daily photosynthetic metabolism and seasonal flowering in plants, and the daily and seasonal changes of physiology and behavior in animals. As with a mechanical clock, the biological clock is set, or entrained, in this case to the solar day. Light is the most important environmental cue, or zeitgeber, for such entrainment, and cryptochromes are among the photoreceptors that mediate light input in the circadian clocks of both plants and animals.

In plants, cryptochromes function redundantly with phytochromes to regulate the level and activity of clock proteins, but the exact molecular mechanism of this regulation remains unclear. In animals, cryptochromes act redundantly with visual pigments (opsins) or related proteins (melanopsins) to entrain the clock, and also serve as integral parts of the circadian clock in mammals. At the core of the circadian clocks of animals are negative regulators of transcription: such as the PERIOD and TIMELESS proteins in *Drosophila*, or the PERIOD and cryptochrome proteins in mice. These proteins suppress the transcription of genes, including their own, to maintain appropriate levels of various proteins throughout the solar day. In *Drosophila*, cryptochrome physically interacts with TIMELESS in response to light, triggering a proteosome-dependent degradation of the TIMELESS protein. The resulting decrease in the TIMELESS protein level, before it enters the nucleus, is thought to cause a phase delay in the circadian oscillator.

In mammals, cryptochromes in different cells may have different functions. Cryptochrome is highly expressed in the inner retina of mice, as well as in the suprachiasmatic nuclei (SCN) where the body's master clock is located. Cryptochromes expressed in the inner retina cells can act as photoreceptors, whereas cryptochromes in SCN and other cells apparently play a non-photoreceptor role, by interacting with PERIOD and moving into the nucleus. Once in the nucleus, cryptochrome-PERIOD complexes interact and inhibit the activity of CLOCK and BMAL proteins, which are transcription regulators needed for the expression of many clock-regulated genes—including the ones that encode PERIOD and cryptochromes, allowing the establishment of the aforementioned negative-feedback loop (**Fig. 2**).

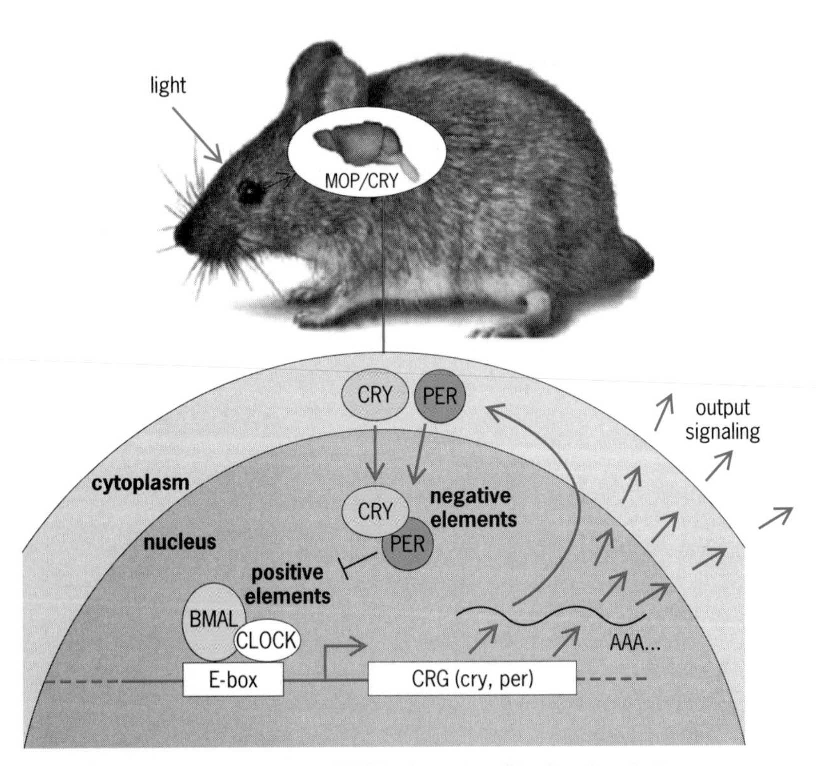

Fig. 2. Functions of cryptochromes (CRY) in the mammalian circadian clock. Cryptochromes (CRY) and melanopsins (MOP) can mediate light signals for the photo-entrainment of the circadian clock. In addition, CRY acts as a component of the negative feedback loop that suppresses the activity of the CLOCK-BMAL complex. The CLOCK-BMAL complex binds to a promoter element called E-box (CACGTG) of the clock-regulated genes (CRG), including those that encode CRY and PERIOD, to affect the expression of CRG genes.

Structure and relation to photolyase. The first cryptochrome gene was isolated in 1993 from *Arabidopsis thaliana* (mouse-ear cress), a small flowering plant commonly used as a model organism in plant biology. The *Arabidopsis* gene *CRY1* was found to be responsible for the blue-light inhibition of cell elongation. *CRY1* encodes a polypeptide of 681 amino acids that showed apparent amino acid sequence similarity to the bacterial DNA-repairing enzyme photolyase. Cryptochromes were later discovered in both eubacteria and eukaryotes, but have not yet been identified in Archaea, although photolyase has been found in all three major branches of life. Many organisms, including plants, insects, fish, and mammals, express both photolyase and cryptochrome. Different organisms have different

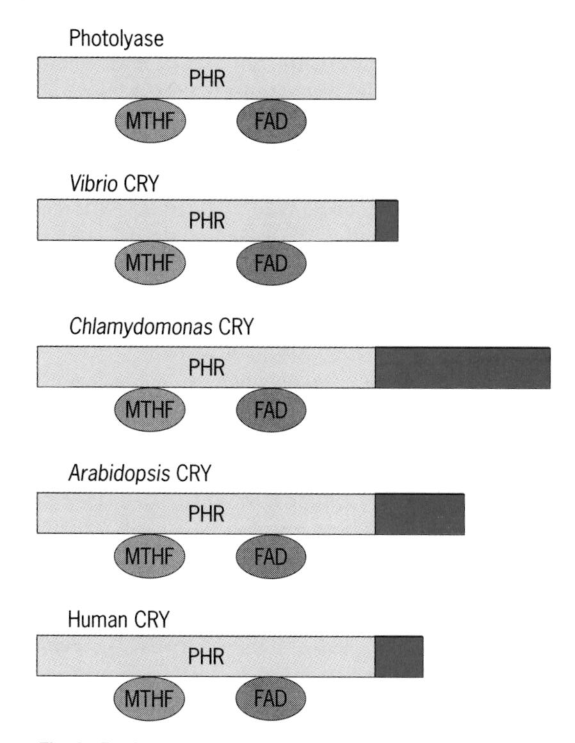

Fig. 3. Basic structure of cryptochromes, emphasizing the lack of a C-terminal extension in photolyase and the different sizes of C-terminal extensions in cryptochromes from various species. PHR indicates photolyase or photolyase-related region. MTHF (5,10-methenyltetrahydrofolate) and FAD (flavin adenine dinucleotide) are photon-absorbing cofactors of photolyase.

numbers of cryptochrome isoforms. *Drosophila*, for example, have one cryptochrome, humans and mice each have two cryptochromes, many plants have two to five cryptochromes, and zebrafish have seven cryptochromes.

The amino acid sequences of most cryptochromes can be divided into two regions, the amino-terminal photolyase-related region (PHR) and the carboxyl (C)-terminal region which is unrelated to photolyase (**Fig. 3**). The PHR domains of all cryptochromes share over 30% sequence homology with photolyases. The size and sequence of the C-terminal domain of cryptochromes, however, vary in different organisms, although it appears to be relatively conserved within many plants. Cryptochromes were most likely derived from photolyase gene duplication before the plant and animal lineages split.

DNA photolyase is a blue light-dependent DNA-repairing enzyme that repairs DNA lesions caused by ultraviolet light, including cyclobutane pyrimidine dimers and pyrimidine-pyrimidine 6-4 photoproducts. It contains two photon-absorbing cofactors: a photoactive cofactor flavin adenine dinucleotide (FAD) and a photo-antenna cofactor that can be either deazaflavin or a special form of folate [5,10-methenyltetrahydrofolate (MTHF)]. FAD is the catalytic cofactor of photolyase. It is photoreduced to catalyze cleavage of the pyrimidine dimer or the 6-4 photoproduct. All the cryptochromes studied so far contain this photoactive cofactor, FAD, and most also bind to the photo-antenna cofactor MTHF. The PHR region of cryptochromes share remarkably similar three-dimensional structures with those of photolyases—both are characterized by an α/β domain and a helical domain that are connected by an interdomain loop. The helical domain of photolyases forms a cavity to bind to FAD, and this region is also believed to play a critical role in the function of cryptochromes.

For background information *see* BIOLOGICAL CLOCKS; PHOTOMORPHOGENESIS; PHOTOPERIODISM; PHOTORECEPTION; PHYTOCHROME; SIGNAL TRANSDUCTION in the McGraw-Hill Encyclopedia of Science and Technology.

Chentao Lin

Bibliography. A. R. Cashmore, Cryptochromes: Enabling plants and animals to determine circadian time, *Cell*, 114:537–543, 2003; C. Lin and D. Shalitin, Cryptochrome structure and signal transduction, *Annu. Rev. Plant Biol.*, 54:469–496, 2003; A. Sancar, Structure and function of DNA photolyase and cryptochrome blue-light photoreceptors, *Chem. Rev.*, 103:2203–2237, 2003.

Deep-sea vertebrate species

The deep sea is defined as any marine water body deeper than 200 m, and it is divided into the bathyal (200–3000 m), abyssal (3001–6000 m), and hadal (6001–10,900 m) zones. The abyssal zone is typified by the flat abyssal plains, while the hadal zone comprises the ocean trenches, down to the Challenger Deep, the deepest point in the oceans. The deep sea is usually cold (2–4°C; 36–39°F), and the weight of the overlying water exerts massive pressure, 1 atmosphere for every 10 m of water depth. Light is strongly absorbed by seawater; thus by 1000 m, no detectable sunlight is present. Since plant photosynthesis cannot occur without light, there is no primary productivity to provide food for animals. Most of the food for deep-sea animals falls from the surface (possibly after being eaten by at least one other animal), and there is little left by the time it reaches the deep seafloor.

These environmental factors are extremely important in determining the characteristics of deep-sea animals, but the common description of the deep sea as an "extreme environment" is not justified. The deep sea is actually the most common habitat type on Earth. Although the area covered by water is very large, the shallow continental shelves make up only a very small proportion; 90% of the sea is more than 200 m deep. Until relatively recently, it was thought that no life existed in the deep sea. Little is known about deep-sea animals due to the severe technical difficulties associated with their study. The development of a wide range of technical tools has been necessary to enable the observation, capture, and study of these animals.

We now know that a diverse animal community, including many vertebrate species, does exist. Fish make up the vast majority of deep-sea vertebrates, though some species of marine mammal (notably the sperm whale, *Physeter macrocephalus*) feed in deep-sea systems. The fishes have many special adaptations to life in deep water.

Fig. 1. Grenadier fish (*Coryphaenoides* sp.) swimming across the seafloor in the North Pacific. These fish are very common, currently reaching abundances of over 20 per hectare in the North Pacific. (*Courtesy of Dr. K. Smith, Scripps Institution of Oceanography*)

Technology and deep-sea biology research. The deep-sea environment poses severe challenges to the biologist seeking to study fishes. Even collecting specimens for investigation is technically difficult, and very few deep-sea fish survive being brought to the surface. Deep-sea biologists use tools, from fishing nets to manned submersibles, to collect, observe, and conduct experiments on fishes.

Dredges and nets. The earliest studies were conducted using dredges and nets towed behind ships. After dredging in the eastern Mediterranean in 1841, Edward Forbes concluded that no life existed in the deep sea. However, the HMS *Challenger* expedition (1872–1876) circled the planet using dredges and nets to make collections of deep-sea animals, including fish, proving that life existed in all the world's deep-ocean basins. Today nets remain the most effective way of collecting samples of most deep-sea animals. Collections made in this way have shown how fish abundance and diversity decease with increasing depth. Fish species usually live within certain depth boundaries; thus the types of fish observed change with increasing depth. It is estimated that close to 1000 fish species have depth ranges that extend below 1000 m.

Using fishing nets, patterns of depth distributions have been determined for several locations around the world, and the species found differ greatly between deep-sea environments. In the North Atlantic the most abundant fish in the bathyal and abyssal depths change with increasing depth, from the cutthroat eel (*Synaphobranchus kaupii*) at 700–2300 m to the Morid cod (*Antimora rostrata*) at 1200–2600 m. At depths of 2200 m and deeper, the community is dominated by the large abyssal grenadier (*Coryphaenoides armatus*) [**Fig. 1**]. In contrast, the Mediterranean deep-sea fish community is dominated by sharks (for example, sixgill sharks, *Hexanchus griseus*) to a depth of 2500 m, after which a very small grenadier (*C. mediterraneaus*) is found in greatest abundance. The reasons for these differences in community composition are not known, but researchers believe that only a few species have been able to colonize the warm Mediterranean and enter this sea through the shallow waters of the Strait of Gibraltar.

Manned submersible vehicles. Manned submersibles were one of the earliest methods used to make direct observations in the deep sea, notably Jacques Piccard's and Don Walsh's observation of fish during their dive to the bottom of the Challenger Deep in the submersible *Trieste*. The first (and most famous) modern research submersible is the deep submergence vehicle *Alvin*, able to make dives lasting 10 h and to depths of up to 4500 m. It has made more than 4000 dives since it entered service in 1964. *Alvin* and other modern submersibles have their own propulsion systems, the ability to collect samples and operate equipment using robotic arms (called manipulators), and cameras to record video and still images. Studies of fish using *Alvin* have included research on how the whale carcasses are consumed when they reach the seafloor, and experiments to measure the metabolic rates of fishes. *See* WHALE-FALL WORMS.

Remotely operated vehicles. More recently, remotely operated vehicles (ROVs) such as JASON have been used to undertake similar studies. These vehicles are safer than crewed submersibles, as operators view the seafloor on television screens while operating the ROV from the surface. In addition, ROVs have greater underwater endurance than crewed vehicles; several current ROVs have depth ranges in excess of 4000 m and underwater endurances of more than 48 h.

Landers. One of the most effective ways of studying deep-sea fish is using equipment that works without human interaction, such as a lander. Landers are deployed from research vessels and sink to the seafloor, where they observe deep-sea animals or conduct experiments under the control of onboard computers (**Fig. 2**). An acoustic signal is sent from the research ship at the end of the experiment, and the lander drops ballast weights and floats to the surface. Many landers can be used simultaneously, and studies on fish have included photographic and video observations of scavenging behavior, biodiversity assessment and mapping, and measurement of metabolic rate and swimming performance.

Combined techniques. The various techniques developed to study the deep-sea environment and fauna are often used in combination. In 2004, a multinational team undertook the Mid-Atlantic Ridge Ecosystem study (MAR-ECO) using nets, landers, and ROVs—all backed up by detailed acoustic mapping of the ridge. The MAR is a new frontier in marine science, and the MAR-ECO scientists have only just

Fig. 2. Sprint video lander system being set up for deployment in the North Pacific. The lander consists of an aluminium tripod, at the top of which are the downward-looking video camera and the acoustic release units. Lower down the frame is the battery. The researcher is attaching bait to the lander frame to attract fish and in front of him are the syntactic foam floatation units that bring the equipment back to the surface at the end of the experiment. (*Courtesy of H. Ruhl, Scripps Institution of Oceanography*)

begun to describe the fish fauna and their habitats on this massive undersea mountain range.

Characteristics of deep-sea vertebrates. Deep-sea animals remain among the least well-known of all vertebrates, but much information has been gleaned from ecological studies focusing on distribution, diversity, life cycles, and behavior, as well as physiological studies examining metabolic rates and muscle biochemistry and performance.

Almost all deep-sea vertebrates are fishes, and these animals can be divided into two broad groups: the pelagic (or midwater) fishes, which inhabit the water column and may never contact the seafloor throughout their lives; and the benthopelagic fishes, which live near the bottom. The characteristics of midwater and bottom fish groups differ greatly, with a general trend toward lower metabolic rates and activity levels in deeper-living fishes; this pattern is best described in the pelagic fishes.

Pelagic fishes. Using a special net, researchers have collected midwater fishes from the surface to depths of more than 2000 m and measured their metabolic rates and the activities of their muscle enzymes. They concluded that as the ambient light level falls with increasing depth, the advantage of having powerful swimming muscles is reduced, as there are less frequent encounters between animals and predator chases are less prolonged. As a result, fish reduce their energy costs by having weaker muscles with lower protein contents, which allows them to decrease their overall metabolic rates. These energy savings would be highly advantageous in such a low-food environment.

Midwater fishes tend to be small and have muscles with high water content. As a result, these fish probably lack the ability for high locomotor performance and, therefore, rely on stealth and deception to escape predation and capture prey. Many species have bioluminescent organs, which are used to disguise their silhouette against the weak light from the surface and to lure prey to the area in front of the predator's mouth (for example, angler fish). The mouths of deep-sea fishes often have large teeth and the mouths themselves are large in proportion to their body. When a potential meal becomes available, the fish must take it, regardless of its size.

Benthopelagic fish. The benthopelagic fish are more like the familiar fish of the continental shelf, often including large and vigorous fish related to shallow-water cod and eels. These fish are even less well studied than the midwater species, but the little data available appear to indicate that their metabolic rates are also lower than those of related fish in shallow water.

One example are the grenadiers, relatives of the familiar Atlantic cod. Grenadiers are the dominant predators and scavengers across most of the Earth's surface (Fig. 1). They swim slowly across the deep-sea bottom, taking live prey and consuming any carrion they can find. However, they are able to survive for months without food.

Future research. Even the most common of deep-sea vertebrates remain an enigma despite all the work and advanced technology that have been applied. Further study is needed to determine how long the most common species live, where and how often they reproduce, and whether they migrate across the seafloor to breed or feed.

For background information *see* ACTINOPTERYGII; DEEP-SEA FAUNA; GADIFORMES; MARINE BIOLOGICAL SAMPLING; MARINE ECOLOGY in the McGraw-Hill Encyclopedia of Science & Technology. David M. Bailey

Bibliography. R. Kunzig, *Mapping the Deep: The Extraordinary Story of Ocean Science*, Norton, London, 2000; A. Fothergill, M. Holmes, and A. Byatt, *Blue Planet: The Natural History of the World's Ocean*, BBC, Oxford, 2001; P. Herring, *Biology of the Deep Ocean*, Oxford University Press, 2001.

Digital cinema

In recent decades, cinematography, as with other forms of media, has been undergoing a transition from photomechanical processes to digital ones. In general, digital processes offer greater flexibility, consume fewer resources, offer significant time saving, and allow electronic (rather than physical) delivery of products. In cinematography, this transition involves three distinct operations: image capture, image editing, and image display. The penetration of digital processes into each of these areas is under

Workflow comparison digital and photomechanical cinema

Digital	Photomechanical
Culture	
High-resolution video camera	35-mm film camera negative image
Immediate display	Positive film copy for display
Compress and save files	Store film strips
Editing	
Access and edit files	Cut and splice film clips
Electronically merge computer-generated effects	Output computer-generated special effects to film
Assemble electronic master	Cut and splice master negative film Intermediate positive copy
Display	
Electronic file contains meta-data for various versions	Master negative film, one for each version
Digital projector	Positive film display prints Film projector

way, although in the image editing or postproduction processes the use of digital methods is far more widespread than in digital capture or digital display (see **table**). The complete transition will result in an all-digital workflow for the production of motion pictures; however, there are several technical and economic challenges that must be overcome first.

Image capture. The image-capture process involves filming the scenes that will be edited together to make a movie. The switch to digital in this operation involves capturing the scenes with a high-definition digital camera instead of the traditional cinemagraphic 35-mm film cameras.

In the traditional process, color negative film is exposed (usually from multiple cameras simultaneously) capturing the scenes that will make up the film. At the end of each day, "dailies" are prepared for viewing the following day by first chemically processing the film, and then making the necessary color adjustments, using a color analyzer, in order to print a positive copy of the color negative. The positives are viewed by the creative team to evaluate the success of the shooting and to decide on subsequent filming.

There are several weaknesses in this traditional approach that are alleviated by digital image capture. The material cost of the film is high, and the environmental impact of chemical film processing is problematic. Since the color negative film cannot be evaluated directly (even after processing), the creative team cannot get immediate feedback on the success of a filming session. The film is easily scratched and susceptible to dust. In addition, if digital editing techniques are to be used, film images must be scanned to create digital files.

In spite of the inherent weaknesses of film, most feature movies today are captured on film. Cinematographers have been reluctant to give up film cameras because 35-mm film captures a higher-resolution image with greater dynamic range than today's high-definition digital cameras. Cinematographers also have substantial investments in film equipment and expertise in using it.

The potential advantages of digital image capture are not insignificant. *Sin City*, a 2005 release, was shot using Sony HDC-F950 digital cameras. These cameras capture images at 1920 × 1080 pixels of resolution in each of the three primary colors with 10-bit color depth at 24 frames per second. The three image-sensing devices in the camera, known as charge-coupled devices (CCDs), were linked to an optical cable to extract the data from the camera in an unprocessed form at up to 3 gigabits per second. Scenes could be viewed on a calibrated monitor immediately after shooting. The *Sin City* production team benefited from this immediate feedback. Their goal was to reproduce a graphic novel on the screen, and they were able to confirm that they were achieving this goal as they went along.

Star Wars: Episode III Revenge of the Sith was a 2005 release that used an all-digital workflow, including image capture. The production costs of $100 million are estimated to be half of what would have been required for a comparable movie shot on film. The savings came from the increased productivity by having immediate feedback during shooting and from doing much of the special-effects work before the shooting took place. This enabled director George Lucas to make decisions about which angles and perspectives were needed before (rather than after) the filming of a scene.

Today, the standard resolution for image capture is 2K (about 2000 × 1000 pixels). The Dalsa Origin, a recently developed camera, is capable of 4K image capture with 12 stops of exposure latitude and 16-bit-per-channel color depth (**Fig. 1**). Although these specifications are impressive, the Dalsa is not replacing the cameras currently in use for digital cinema because the data generated from a 4K image capture makes an impractically large data file in terms of transport and storage. Currently, 4K image capture is used only on an as-needed basis for special effects

Dalsa Origin digital cinema camera.

or background matte applications (where actors are filmed against a blue or green background which will later be replaced).

Digital cinema cameras are improving steadily, as are data transport, storage, and compression techniques. Over time, digital capture should replace film because of the creative flexibility and cost savings that it offers.

Film digitizing. Scanning rolls of film for digital editing has become a lucrative business because an increasing number of films are doing all the post-production processes digitally. Kodak, which sells 80% of the movie film used in the United States, has opened Cinesite, a subsidiary that offers film scanning as well as digital editing, special-effects creation, and imaging digital master copies for distribution.

The scanners used for rolls of film were initially developed for converting finished films to DVD format. Ideally, each frame would be scanned at 4K (4100 pixels wide by 3000 pixels high) for optimum quality, but the resulting files would be too large and, until recently, it would take days to scan a feature-length film. Today, a state-of-the-art scanner is capable of performing 2K scans in real time and 4K scans at 7.5 frames per second, or about 6 h for a feature length film.

Currently, 4K scanning resolution is reserved for critical projects and 2K (2048 pixels wide by 1556 pixels high) scanning is used for most applications. Scanners with 4K capability have an advantage for producing 2K output in that they can use over-sampling and interpolate from a 4K file to obtain 2K data.

Image editing. Nearly all special effects are achieved through digital means now, making image editing the most commonly used digital processes. By traditional methods, film editing involved cutting apart scenes from reels of movie film and taping them together to form the desired sequence. After a taped-together master was made, it was duplicated with color adjustments to form a new master on a continuous piece of film. This process was extremely time-consuming and not easily reversible.

Digital editing offers significant advantages over film editing. It is faster, more precise, and more readily reedited, and it facilitates the incorporation of digital special effects on which most modern films rely.

Digital editing offers color effects that are not obtainable with photomechanical processes. For example, the film *O Brother Where Art Though* used a restricted color palette for the background scenery that evoked the Dust Bowl era in the American South. This palette could have been achieved on film with the judicious use of lighting and filters, but it could not have been restricted to background elements only.

Digital editing allows the easy integration of digital special effects into the shot sequences and individual frames of the film. Working digitally allows this to be seamless and fully convincing on screen.

Digital editing also facilitates the restoration of classic films that have degraded over time. The 1962 film *To Kill a Mockingbird* was remastered from two partially damaged prints of the film. With digital editing, each frame could combine the undamaged portions of the two prints, the faded color could be restored, and dust and scratches could be removed.

Image distribution. Traditionally, master negative films are used to generate positive prints for distribution. Since the master film degrades with each print made from it, multiple master films are needed for a major release. The Motion Picture Association of America (MPAA) reported that in 2003 the seven major studios spent $631 million (or $3.74 million/film) on printing and distributing feature films in the United States and English-speaking Canada. This translates to about $13,000 dollars a year to service a single screen. Switching to an all-digital distribution system would cut this cost by 85%. This is a strong incentive for studios, but it is not clear that theater owners would share substantially in the savings.

Digital distribution offers the possibility of new distribution channels for studios, which could offset the declining attendances at movie houses. Studios could distribute films directly to cable and internet companies, or even to individuals willing to pay a premium to view a first-run film in their home.

Of the 37,000 movie screens in the United States today, only about 90 are equipped for digital projection. The value proposition for theater owners in switching a screen from film projection to digital is weak, since substantial investments in projection equipment, high-speed network links, servers, storage, and staff training are needed. In addition, very few movies today are being distributed in digital format.

The advantages for theater owners who invest in the digital projection equipment are that the image quality does not degrade with repeated showing and movies that are not drawing audiences can be quickly replaced. Also, the venue can be used for pay-per-view projection of sports events, concerts, and other features.

The digital supply chain will cost $4 billion in North America, with costs to each theater at about $120,000. Theater owners are not able to shoulder this cost, so Disney, Sony, and Warner have pooled $3 billion to distribute to theaters to install digital cinema. The studios are thus calculating that the investment in infrastructure will lead to substantial distribution savings in the future.

Image display. In June 1999, the first two films released for display in digital format were *Star Wars Episode I* and *An Ideal Husband*. *Star Wars* was shown in New York and Los Angeles at four venues. There were two competing projectors in use: an early version of the Texas Instruments digital light processor (DLP) technology and a Hughes/JVC ILA-12K, a liquid crystal on silicon (LCOS) projector. The DLP projectors were preferred, and all of the early digital-projection installations relied on this technology. These projectors produced a larger color gamut and greater contrast range than the competing technologies.

DLP projectors use optical semiconductors, called digital micromirror devices (DMD), to modulate the light and form images. Each DMD is an array of 1.3 million microscopic mirrors, which can be individually pivoted in and out of the projector's light path thousands of times per second. The individual mirrors are spaced 1-micrometer apart, so the device exhibits a high fill factor with good brightness and contrast. Cinema projectors are equipped with three DMD arrays (one for each primary color) to produce an image. Early projectors were limited to 1024 × 768 pixel resolution, but current equipment offer 2K pixel resolution, 15-bit color depth, and sufficient brightness for a 75-ft-wide (23-m) screen.

The recent development of the double data rate DLP (DDR DPL) promises to increase the color depth, brightness, and contrast of the projected images. In 2004, Sony released the SXRD 4K projector, which uses a variant of liquid crystal display (LCD) technology. The improved image resolution (4096 × 2160 pixels) offers a more lifelike image to the viewer.

Current issues. By 2002, there were about 40 theaters equipped for digital projection, and four different file formats were in use. Thus, *Star Wars: Episode II* was released in four different digital versions, making it more difficult to ensure color and image consistency from theater to theater. The digital cinematography industry is in need of standards for file format, compression technology, color display, and encryption to provide interoperability for the rapidly developing projector and hardware industries. Several groups are contributing to this effort, including MPAA, the Digital Cinema Initiatives (DCI) consortium (a group of seven studios formed in March 2002), the Society of Motion Picture and Television Engineering (SMPTE), and the American Society of Cinematographers (ASC).

There has been steady development of encryption and digital rights management strategies for digital cinema. Today, files are encrypted all the way into the projector, where the display files are extracted. This is of great importance to the industry in thwarting the piracy of intellectual property. It reduces the need for high security on the server since it will only receive and distribute encrypted files.

Outlook. The penetration of digital processes into filmmaking is well established but far from complete. Although it seems inevitable that digital cinema will become the normal end-to-end method for making movies, it is difficult to predict the speed with which the transition will occur. The needed hardware is available, but it is changing rapidly and has not yet reached the performance levels that would enable it to surpass film by every measure. The transition costs to implement digital cinema are high, but the potential benefits and cost savings are higher.

For background information *see* CAMERA; CINEMATOGRAPHY; CHARGE-COUPLED DEVICES; COMPUTER GRAPHICS; IMAGE PROCESSING; PHOTOGRAPHY in the McGraw-Hill Encyclopedia of Science & Technology.

Anthony Stanton

Bibliography. J. Careless, The back alleys of Sin City, *Digital Cinematography*, 1(1):22–26, 2005; C. Harrison, The evolving digital workflow in cinema, *IS&T Reporter*, 18(6): 6 pp, 2003; M. A. Hiltzik, Digital cinema take two, *Technol. Rev.*, pp. 36–44, September 2002; M. Hurwitz, Cost effective digital production—Star Wars: Episode III, Revenge of the Sith, *Digital Cinematog*, 1(2):16–19, 2005; L. Sullivan, Digital force, *Inform. Week*, pp. 39–46, May 16, 2005; C. S. Swartz (ed.), *Understanding Digital Cinema: A Professional Handbook*, Focal Press, 2005.

Dino-Birds

Dinosaurs are not completely extinct, as many might believe. It has been widely accepted for more than 30 years that birds are direct descendants of small theropods (meat-eating dinosaurs) called maniraptorans. This followed the discovery of a dromaeosaurid (of the family Dromaeosauridae) maniraptoran called *Deinonychus*, which showed a novel feature in the hand skeleton: a half-moon shaped wrist bone that allowed a wide range of hand movement and folding. This feature, which facilitated the wrist rotation and hand folding regarded as the precursor to the flight stroke, is shared uniquely with the earliest known bird, *Archaeopteryx*, discovered more than 150 years ago from 147-million-year-old latest Jurassic rocks in Germany. Many other skeletal features of *Archaeopteryx* and maniraptorans show a close resemblance. More impressive still are the impressions of wing and tail feathers, almost identical with those in modern birds, present in the fine-grained rock surrounding the skeletons of *Archaeopteryx*. The wings bear asymmetric primary and secondary feathers that confer aerodynamic properties necessary for flight.

Archaeopteryx is a demonstrable link between dinosaurs and birds, with characters of both. Thus somewhere along that evolutionary line, feathers—the hallmark of birds—must have first appeared in theropod dinosaurs, but the fossil record did not preserve such information. The recent discovery of feathered theropod fossils fills this gap, offering a whole new perspective on the origin of birds.

Liaoning fossils. During the past 10 years, a huge array of exquisitely and exceptionally preserved fossils have been discovered from a series of early Cretaceous volcanic deposits in the Sihetun region of Liaoning Province in northeastern China. The Lower Cretaceous Jehol Group, divided into the Yixian and Juifotang Formations (dated between 130 and 125 million years old), consist of a series of volcanic basalts alternating with fine-grained volcanic ash tuffs. These beds were laid down in a forested lakeside setting in a volcanically active area. The terrestrial and freshwater inhabitants were entombed rapidly in volcanic dust and ashes, sealing out oxygen and preventing decay, allowing the preservation of keratinous integumentary structures such as skin, hair, and feathers. A local fossil-collecting industry has led to the excavation of vast numbers of fossils representing a complete terrestrial and freshwater fauna. They include thousands of specimens of the most common Liaoning bird, *Confuciusornis*, which must have lived in large flocks in the trees around the margins of contemporary lakes and probably succumbed to poisonous volcanic gases and ash falls. Much rarer are a series of spectacular small, feathered dinosaurs, or Dino-Birds, that represent a snapshot of a wide span of dinosaur phylogeny.

Feathered theropods. The first Liaoning Dino-Bird, *Sinosauropteryx*, discovered in 1995, preserves a downy, featherlike covering composed of simple, hollow filaments that are considered to be the precursors of true feathers (protofeathers). True feathers (pinnate feathers) have a stiffened midrib and symmetrical vanes composed of parallel barbs linked by barbules. A recent theoretical model of feather evolution postulated that feathers began as simple hollow cylinders of keratin produced by an undifferentiated feather follicle collar. Elaboration came about by differentiation of the feather follicle collar by folding, helical spiraling, and invagination to produce barbs, barbules, and pinnate feathers. Embryological studies corroborate the developmental model and so do the fossils.

Sinosauropteryx. *Sinosauropteryx* is a conventional small coelurosaur (a type of advanced theropod) up to 1 m long, with short arms and clawed hands, much more primitive than *Deinonychus*. Mapping the feather data onto a cladogram (branching diagram showing the evolution and descent of a group of organisms) of theropods shows that protofeathers must have arisen at least as far down theropod phylogeny as the base of the major clade Coelurosauria, which includes *Tyrannosaurus rex* and maniraptorans (**Fig. 1**). This strongly suggests that feathers evolved initially not for flight but almost certainly to provide a thermoregulatory layer to help control body temperature. This may have had an important role in the development of brooding behavior. Birdlike brooding behavior is known from a dramatically preserved small maniraptoran, *Oviraptor*, from the Gobi Desert that was fossilized squatting on its nest of eggs.

Protoarchaeopteryx and Caudipteryx. Other more phylogenetically (evolutionarily) advanced Liaoning Dino-Birds illustrate that after true feathers appeared, they were soon elaborated for other functions. The turkey-sized *Protoarchaeopteryx* had long arms and hands similar to *Archaeopteryx* and long legs, and it shows traces of a tail plume and some isolated true feathers. These are not asymmetrical flight feathers, so *Protoarchaeopteryx* was certainly not a flier but a ground runner. The tail plume may have been for display. *Caudipteryx*, a 1-m-long relative of *Oviraptor* from the Gobi Desert, is known from numerous specimens. It had tufts of symmetrical pinnate feathers sprouting from very short arms and an almost peacock-like tail fan. *Caudipteryx* was certainly a cursorial (adapted for running), ground-dwelling theropod, and its feathers were likely used for display.

The suggestion that *Caudipteryx* and *Protoarchaeopteryx* represent a line of birds that had become secondarily flightless has foundered under weight of anatomical evidence and other studies that place them within the maniraptoran clade.

Juvenile dromaeosaur. One of the key finds from Liaoning was a fully feathered juvenile dromaeosaur, nicknamed "Fuzzy raptor," belonging to the family closest to *Archaeopteryx*. (Although it is an important specimen, it has no scientific name because it is a juvenile and cannot be certainly identified with any other named dromaeosaurs.) The specimen is thickly clad with pinnate feathers from the head down and

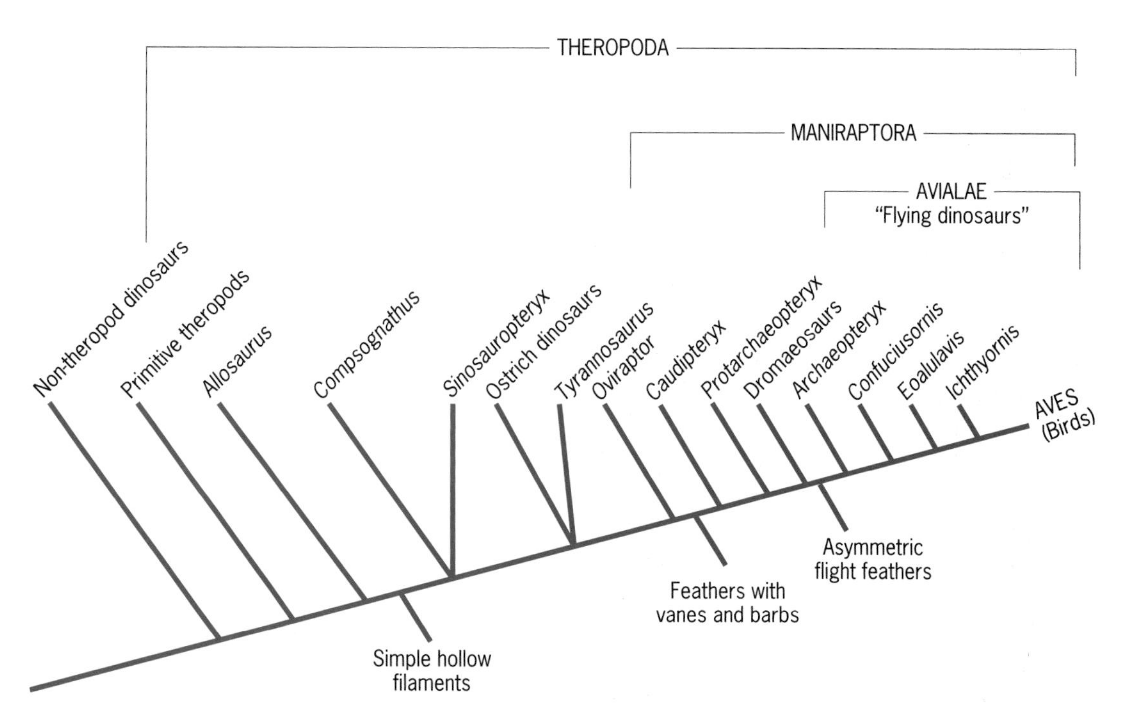

Fig. 1. Diagram of theropod dinosaur and bird evolution showing that primitive feathers, first known from *Sinosauropteryx*, go back a long way in theropod history. True feathers appeared later within the major group Maniraptora.

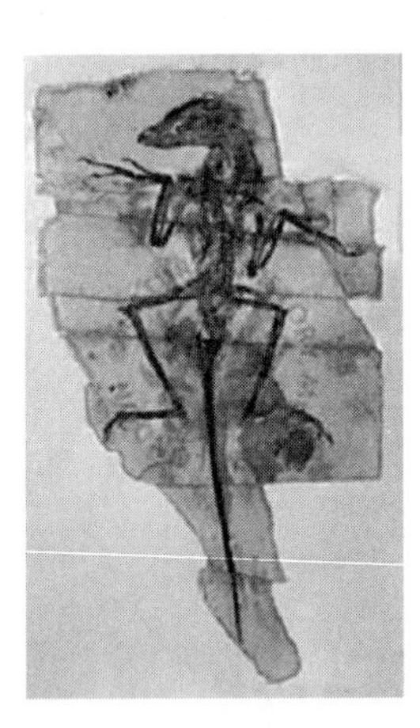

Fig. 2. Nicknamed "Fuzzy raptor," this spectacularly preserved juvenile dromaeosaur from Liaoning shows the typical long grasping hands and swiveling wrist bones together with extensive preservation of feathers around the body and limbs. (*Reprinted with permission from A. C. Milner, Dino-Birds, Natural History Museum, London, 2002*)

has long fringes of feathers on the arms and rook-like "trousers" (**Fig. 2**). This spectacularly well-preserved specimen, even by Liaoning standards, indicates that dromaosaurs, the closest relatives of birds, were fully clothed in an insulating feather coat but had no asymmetric flight feathers.

Origin of flight. The question of whether flight began from the trees down or from the ground up has been hotly debated since the discovery of agile, fast-running dromaeosaurs such as *Deinonychus*. The "trees down" hypothesis holds that flight evolved from an arboreal gliding stage. This would require the ability to climb tree trunks and also to glide from tree to tree or tree to ground. The "ground up" hypothesis suggests that flight evolved from small ground-dwelling theropods with running adaptations. It requires that wing feathers were co-opted for flight by functioning first as insect trap nets. The animals would chase small prey using sweeping motions of the forelimbs—effectively a flight stroke precursor—as seen in the range of movements in the dromaeosaur forelimb, permitted by modifications in the shoulder girdle, wrist, and hand. This motion together with running speed would eventually generate sufficient lift and thrust for the runner to become airborne.

The "ground up" hypothesis became orthodox during the 1980s and 1990s, but the discovery of a tiny Liaoning dromaeosaur, *Microraptor*, with arboreally adapted foot claws has swung the debate back into the trees. One species, *M. gui*, had some asymmetrical feathers and a very winglike feather configuration in the forelimb, and it might represent a late stage in the transition from a gliding structure to one capable of true flight. Amazingly, the single known specimen of *M. gui* also sports asymmetrical flight feathers on the hindlimbs. This has given rise to speculation about a four-winged gliding stage in flight evolution, with the hindwings acting as stabilizing canards.

Recent studies on escape behavior in modern ground-living birds, such as quails and partridges, provide an interesting addition to the debate. These birds employ "wing-assisted vertical running" to get off the ground, beating their wings rapidly to generate a down force to help them stick to the substrate while climbing a bush or tree trunk. Even the fluff on chick's wings increases the wing surface area sufficiently to allow efficient climbing. Thus it has been suggested that Dino-Bird arm feathers could have functioned in the same way—as downforce generators and "proto-wing" gliding structures. The idea that wing feathers were first used for prey catching is now giving way to the hypothesis that the selective pressures leading to flight may have arisen as a predator escape mechanism in small theropods, whereby the presence of a feathered arm would aid rapid climbing away from danger and allow gliding from perch to perch or perch to ground. *Archaeopteryx* represents a late stage in this process, with a modern wing configuration of asymmetric primary and secondary flight feathers permitting limited stable powered flight. Analyses of the claws of *Archaeopteryx* have interpreted them as multipurpose climbing and walking structures, although the claw sheaths show no signs of wear as might be expected in a ground-dweller.

True birds. True birds were also present in the Liaoning fauna, and they highlight the diversification of birds and rapid evolution of post-*Archaeopteryx* stage flight apparatus in the early Cretaceous. The majority belonged to a dominant Mesozoic-only group called enantiornithines. *Confuciusornis* had a shortened tail ending in a pygostyle, important for reducing weight (loss of teeth was another weight-saving device) and providing a controllable attachment for the tail feathers. It also had bony sternal plates for wing muscle attachment, had exceptionally large wings (an adaptation to increase surface area and lift), and is the oldest bird known to have possessed a horny beak.

Birds belonging to the Ornithurae, the clade that includes all modern birds, were also present in the fauna. *Yixianornis* and *Yanornis* were fish-eating shore waders with a keeled sternum and other skeletal modifications those suggest strong flying capabilities comparable to that of modern birds had already appeared by early Cretaceous times.

For background information *see* ANIMAL EVOLUTION; ARCHAEORNITHES; AVES; DINOSAUR; FEATHER; FLIGHT in the McGraw-Hill Encyclopedia of Science & Technology. Angela C. Milner

Bibliography. L. M. Chiappe and L. M. Witmer (eds.), *Mesozoic Birds: Above the Heads of the Dinosaurs*, University of California Press, Berkeley, 2002; A. C. Milner, *Dino-Birds*, Natural History Museum, London, 2002; K. Padian and L. Chiappe, The origin and early evolution of birds, *Biol. Rev.*, 73:1–42, 1998; G. Paul, *Dinosaurs of the Air*, Johns Hopkins Press, Baltimore, 2002; Z. Zhou, The origin and early evolution of birds: Discoveries, disputes and perspectives from fossil evidence, *Naturwissenschaften*, 91:455–471, 2004; Z. Zhou, P. M. Barrett, and J. Hilton, An exceptionally preserved Lower Cretaceous ecosystem, *Nature*, 421:807–814, 2003.

Dip-pen nanolithography

Dip-pen nanolithography™ (DPN™) is a direct-write method for nanoscale patterning in which material is transferred from a scanning probe microscope tip onto a surface. Analogous to the 4000-year-old quill pen, DPN uses an atomic force microscope (AFM) tip as a nanoscale "nib," a solid-state substrate as "paper," and molecules with a chemical affinity for the substrate as "ink." In the case of a gold substrate, the ink molecules are typically alkylthiols or arylthiols. When the thiol-coated tip comes in contact with the gold surface, molecules are delivered to the substrate via a solvent meniscus (in most cases, water), which, under ambient conditions, naturally forms between the tip and surface. The meniscus serves as a bridge over which the thiol molecules migrate from the tip to the gold surface, where they anchor chemically and form monolayer patterns (**Fig. 1**).

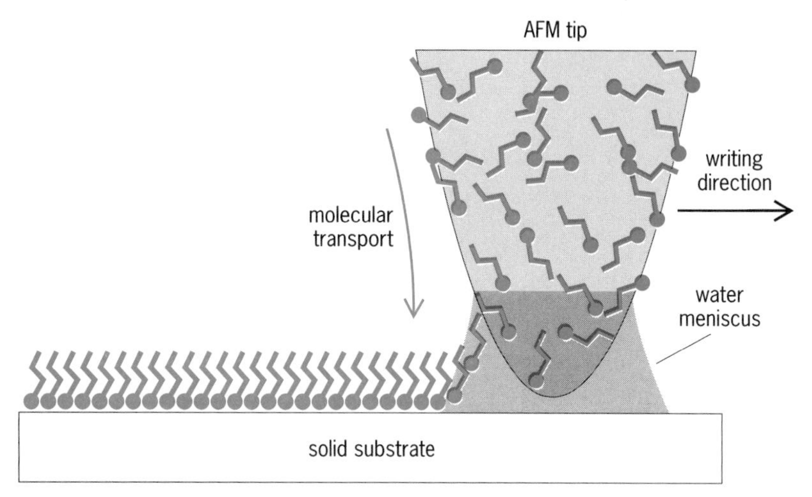

Fig. 1. DPN patterning process.

DPN offers capabilities different from conventional lithographic, robotic spotting, and printing tools. While an ink-jet printer can achieve a minimum feature size of approximately 10 micrometers (about the diameter of a human hair), DPN can print structures down to 10 nanometers, which are typically one molecule high and a few dozen molecules wide. State-of-the-art electron-beam lithography can approach this resolution, but cannot print structures made of soft matter. It is the unique direct-write capability of DPN—high-resolution and registration-alignment attributes as well as soft-matter compatibility—that makes it so powerful. With the aid of software, the DPN process is now fully automated. Through hardware efforts, DPN has been transformed into a parallel process capable of generating many structures simultaneously over the centimeter length scale.

There are many potential applications of DPN, and is quickly becoming a workhorse tool for researchers interested in fabricating and studying nanostructures on surfaces. It is also being evaluated as an inspection and repair tool in the semiconductor and flat-panel display industries, where product yield often is affected by a small number of defects, which can be repaired through direct deposition of materials. DPN has led to the development of new ways of storing information by depositing nanostructures on surfaces in a specific physical design, where the information is recorded in the physical form of the pattern and its chemical composition. In addition, DPN is opening up the opportunity to develop and study nanoarray technology. Nanoarrays are combinatorial libraries of features that consist of either inorganic or biological materials patterned on a surface on the nanometer length scale (**Fig. 2**). Such highly miniaturized arrays allow one to study processes ranging from pattern recognition between surfaces and larger biological entities (such as cells and viruses) to interactions between optically active materials.

Solid-state nanostructures. It was first demonstrated that one could use DPN to deposit alkanethiols onto gold surfaces. Since then, researchers worldwide quickly developed a variety of ink-substrate combinations for DPN, including inks composed of small organic molecules, conducting polymers, oligonucleotides, proteins and viruses, colloidal macro- and nanoparticles, metal ions and sol gels,

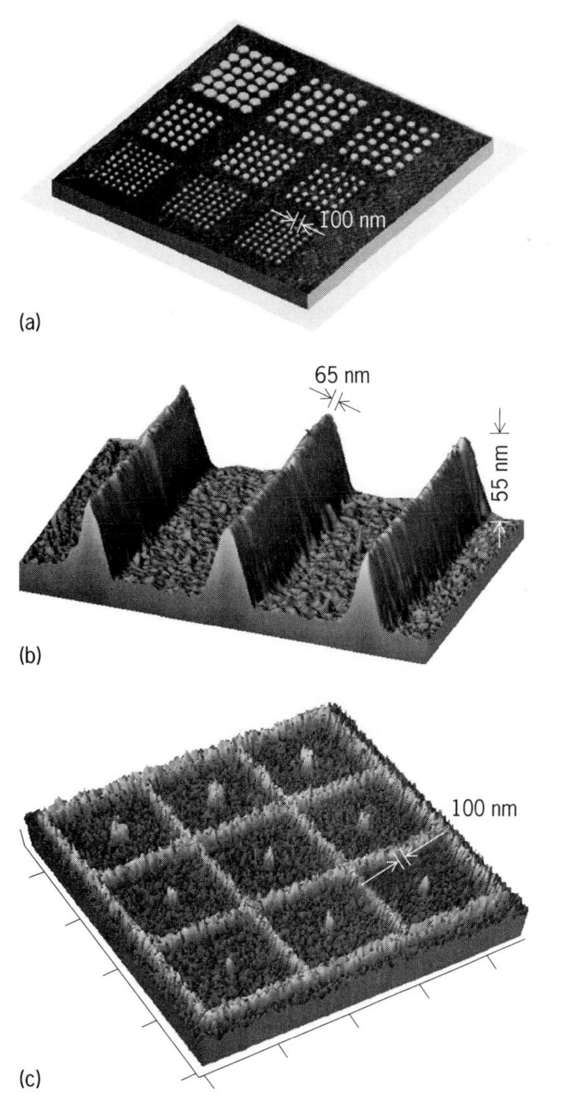

Fig. 2. Potential applications of DPN: (*a*) combinatorial DPN templates, (*b*) silicon nanostructures, and (*c*) protein nanoarrays.

and substrates composed of gold, silver, palladium, silicon, germanium, gallium arsenide, and silicon oxides. Recently, researchers have shown that DPN can be used to facilitate and control the growth of biopolymer crystals. In terms of applications, the only limitations seem to be the chemical compatibility between the environment, ink, and substrate, and the interaction between the ink and tip. For example, some macromolecules move very slowly from the tip to the substrate because of strong adhesive interactions.

One of the several promising applications of DPN is the fabrication of multilayered metal-semiconductor nanostructures. DPN-generated molecular nanopatterns on a gold-coated silicon wafer can be used as an etch resist to produce three-dimensional gold nanostructures by electrochemical wet etching, which selectively removes the unpatterned or exposed gold regions but not the underlying silicon. After etching, the remaining gold features can be used as a resist to fabricate high-quality nanoscale silicon patterns by anisotropic potassium hydroxide (KOH) etching, which selectively removes silicon. Finally, the gold remaining over the unetched silicon can be removed with aqua regia (nitric and hydrochloric acid mixture) [Fig. 2*b*]. In principle, DPN can be used to make almost any shape feature from the nanometer to the micrometer length scale. This ability to write features directly on silicon wafers opens avenues for making master stamps that can be used for printing and masks for lithography, as well as other structures that have unusual properties and depend upon metal or semiconductor nanoscale features.

Biological nanoarrays. Biological nanoarrays, just like their microarray counterparts, permit the study of gene expression, DNA sequence variation, protein-protein interactions, and other fundamental biological process in a massively parallel format, while requiring much less sample and time for production and screening (Fig. 2*c*). These arrays provide miniature biological nanoassays, which not only are on par with the length scale of single biorecognition events but also potentially allow the exploration of the genomic and proteomic profile of a single cell. So far, the technology has been used to prepare DNA, protein, and virus bioarrays on oxide and metal substrates. In the case of viruses, it allows one to site-isolate structures, opening avenues for studying virus infectivity at the single- or few-particle level.

Parallel-pen arrays. One of the limitations of "single-pen" DPN is its speed if patterning a very large area. The development of parallel-probe cantilever arrays is aimed at overcoming this shortcoming. Sophisticated parallel-pen approaches to scanning probe lithography have been developed, but the need for individual feedback on each pen has significantly hampered use of the pens. Early in the development of DPN, it was recognized that the feature size was independent of the contact force over a relatively large range. This has allowed researchers to develop massively parallel pen systems with minimum force-feedback requirements. Effectively, one can push all of the pens in contact with a surface and get clean and uniform ink transfer over a large cantilever array. The simplest implementation of parallel DPN is a passive probe array in which the pens are not actuated individually but write and scan simultaneously. Using parallel passive

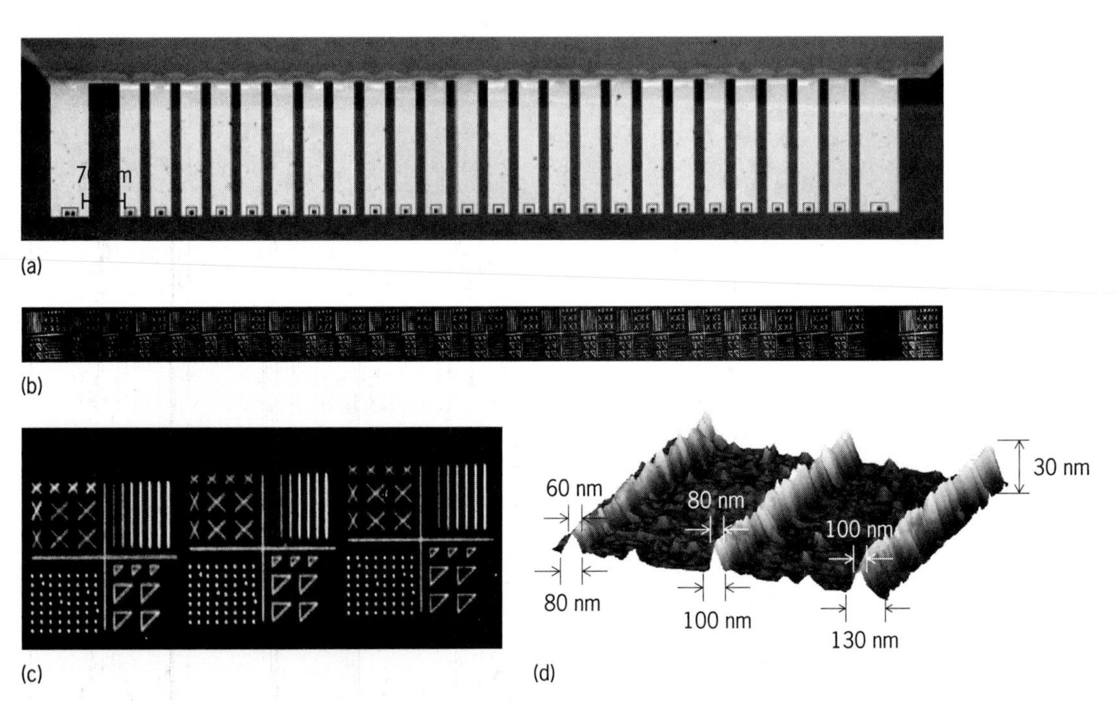

Fig. 3. Parallel pen array and resulting structures. (*a*) Optical micrograph of a 26-pen array. (*b*) Scanning electron microscope image of 26 arrays of gold structures patterned on a Si/SiO$_x$ substrate. (*c*) Dark-field optical micrograph of gold nanostructures. (*d*) Representative atomic force microscope image of one of the arrays.

pen arrays, researchers have been able to use DPN to generate libraries of nearly perfect, nanometer-sized features across square-centimeter areas within minutes (**Fig. 3**). These arrays can be quite large, with the record being 1.3 million pens. Currently, a 25,000-pen array is the largest demonstrated in a DPN experiment, with all pens simultaneously delivering ink to a substrate. Arrays with pens controlled by thermal actuation also have been developed. These arrays allow one to control each pen independently in experiments designed to make highly customized nanostructures.

Outlook. The challenges facing those working with DPN are increasing parallelization, component integration (software, pens, alignment, and inking strategies), and the identification of systems where access to the sub-100-nm length scale is critical. DPN is the ultimate tool for combinatorial nanoscience and nanotechnology, allowing one to build complex libraries of inorganic compounds, small organic molecules, and biological structures that can be used to rapidly screen for many new physical and chemical properties that relate to material size, shape, and composition. In addition to advancing the development of new tools for drug discovery, diagnostics, and fundamental research, DPN will open avenues for studying important fundamental processes, such as crystal nucleation and growth, at a scale that has never been observed. It will accelerate biomedical research by offering researchers opportunities to study important processes at the single-cell or single-particle level. Conventional technologies do not allow one to build multicomponent structures on this length scale, positioning DPN to have a major impact for many years to come.

For background information *see* COMBINATORIAL SYNTHESIS; NANOSTRUCTURE; NANOTECHNOLOGY; ORGANOSULFUR COMPOUND; SURFACE PHYSICS in the McGraw-Hill Encyclopedia of Science & Technology.

Clifton K.-F. Shen; Chad A. Mirkin

Bibliography. D. S. Ginger, H. Zhang, and C. A. Mirkin, The evolution of dip-pen nanolithography, *Angew. Chem. Int. Ed.*, 43(1):30–45, 2003; R. D. Piner et al., Dip-pen nanolithography, *Science*, 283:661–663, 1999.

DNA barcoding

Barcodes are found universally on consumer products. They are composed of unique strings of patterns that identify and classify the product. It has become clear that analogous principles can be applied to distinguish species. Each species is characterized by unique strings of nucleotides in its genome that can be used for identification and classification. These nucleotide strings can be determined easily through standardized deoxyribonucleic acid (DNA) sequencing techniques. DNA barcoding therefore has been proposed as a universal tool for studying the taxonomic diversity of organisms.

Theoretical considerations. The current technology for DNA sequencing allows determination of about 500 nucleotides in a single read. With four different nucleotides possible at each position, one can potentially code 4^{500} variants, a number that exceeds the number of atoms in the universe. In practice, however, all of the 500 nucleotides sequenced will seldom be informative. On the other hand, even a string of only 15 informative nucleotides can represent one billion variants, which exceeds even the greatest estimates of the number of species on Earth.

How do we know that species will predictably differ in their nucleotide strings? The answer is based on the theory of neutral evolution, stating that most changes in DNA sequences of living organisms will accumulate at a constant rate over time. Only beneficial changes accumulate at unpredictable rates, but these are rare. Hence the largest parts of the genomes are subject to constant divergence, a phenomenon that has been called the "molecular clock." The molecular clock is fastest in regions of low or no functional significance (the "neutral DNA") and slower in functionally important regions, because certain changes are simply not allowed in these regions and therefore cannot accumulate. Thus, after species have split from each other, their genomes start to diverge, initially mainly in neutral regions but with time also in functional regions. Their time of divergence from each other therefore will determine whether two species can be distinguished by DNA barcodes or not. The estimates of divergence rates for neutral regions range from 0.5 to 4% per million years, depending on the organismic group and type of marker analyzed. In a 500-nucleotide string, one expects therefore one new diagnostic difference every 50,000 to 400,000 years. Since most species on Earth are separated for more than one million years from each other, it is predictable that by far the majority of them also can be distinguished on the basis of DNA barcodes based on single-sequence reads. However, since speciation is a continuous process, there also always will be species pairs that are too closely related to be reliably distinguished by DNA barcodes (**Fig. 1**).

Practical considerations. In practice one does not use a pure string of neutrally evolving DNA for barcoding but a genomic region that is a patchwork of highly conserved and rapidly diverging regions. The conserved regions permit the targeted DNA region to be amplified so that the sequence of the more rapidly diverging regions can be determined. This is done by the polymerase chain reaction (PCR) which targets the region to be amplified with two short (approximately 25-nucleotide) oligonucleotide primers that flank the region. Ideally, these primers should be applicable to a broad range of species and therefore should be located in highly conserved stretches, whereas the region in between should be rapidly diverging to be as informative as possible. Fortunately, there are many regions in the genome that would fulfill this basic requirement. For the purpose of building general databases, it is necessary to standardize the region to be used in some way.

One explicit proposal for a large DNA barcoding effort proposes to use a part of a gene that occurs in the

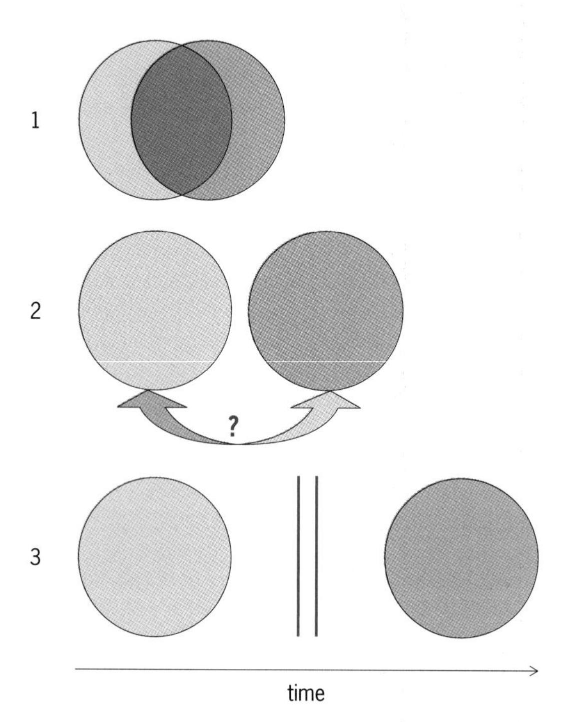

Fig. 1. Process of species splitting. The circles represent gene pools. A new species emerges when gene flow is interrupted (symbolized by different shading). However, at the earliest stages, many alleles with identical DNA sequences will still be shared between the respective gene pools (step 1) and therefore cannot be used to unequivocally assign them to only one of the two species. After some time, the gene pools will be sufficiently separated to allow a clear assignment (step 2), but there is still the risk of occasional gene flow through hybridization (symbolized by the arrow). This also can happen with mitochondria; that is, the mitochondrial DNA in one species can come from another species, which can confuse appropriate assignments. Finally, a point will be reached where the species are sufficiently different that there are no more shared sequences and gene flow can no longer occur (step 3). From this point on, DNA barcoding will always give a reliable species diagnosis. It is expected that the large majority of species on Earth are in this last stage. Note that other molecular techniques are available to reliably characterize stage 1 and stage 2 situations. However, they are too complex and costly to be used routinely.

mitochondrial genome of all eukaryotic organisms, namely the cytochrome oxidase I gene (abbreviated coxI or COI; **Fig. 2**). This is a highly conserved gene required for energy metabolism. However, because of the redundancy of the genetic code, its DNA sequence can diverge rapidly at sites where alternative codons are possible, the so-called "noncoding" or "neutral" positions (**Fig. 3**). This makes about one-third of the whole length of the sequence highly informative. Moreover, since mitochondria (and their genomes) occur in multiple copies in each cell, they are often easier to amplify than genes occurring in the cell nucleus. An international DNA barcoding approach has been started with the goal to use the COI sequences for building a comprehensive database.

Alternative sequence regions for barcoding are located in the gene clusters coding for ribosomal RNAs. These RNAs do not code for proteins but constitute the structural component of the ribosomes, which are required for protein synthesis in every cell. They are known to harbor highly conserved parts that are interspersed with more divergent, although probably not strictly neutrally evolving, parts (**Fig. 4**). Thus, they may be less suitable for distinguishing the most closely related species, although this has not been systematically explored as yet.

The use of ribosomal RNAs as signature sequences for barcoding is particularly prevalent in prokaryotes, which do not have mitochondria. Plants are also a problem, since their mitochondria show unusually low rates of change, making it difficult to distinguish closely related species. A universal alternative marker to be used in plants is therefore still not determined, but efforts are underway.

Taxonomic applications. The short sequence regions that are suggested for DNA barcoding not only are useful to distinguish species but also can help to place a new species into higher orders within a systematic taxonomic framework. This capability also relies on the principle of the molecular clock, which states that the number of nucleotide differences between sequences is a measure of evolutionary relatedness. The sequence information therefore can be used to construct phylogenetic trees (Fig. 3). These trees will reveal groups of related sequences, ideally, linked to taxonomic species names through a database. It has been proposed to call such groups of species or taxa "molecular operational taxonomic units" or MOTUs. This principle is particularly important for species that show an insufficient number of morphological characters to allow an unequivocal species determination, such as many single-celled

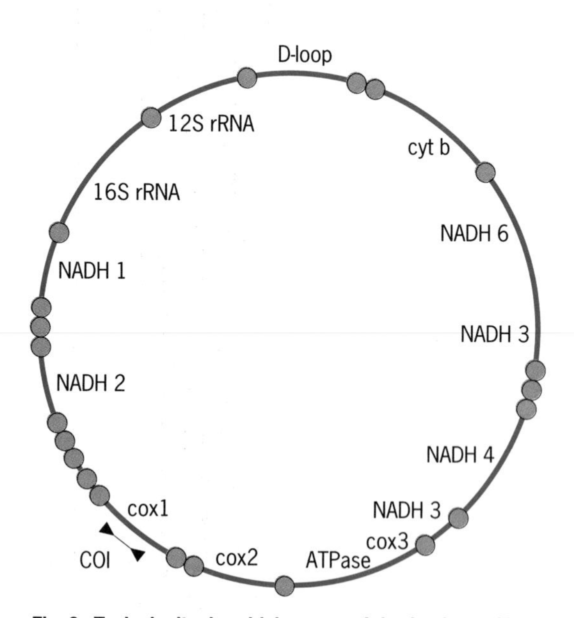

Fig. 2. Typical mitochondrial genome. It is circular and has a total length of up to 20,000 nucleotides. The approximate location of the genes is indicated. Protein coding genes are named; tRNA genes are represented by circles. The ribosomal rRNA genes (16S and 12S) code only for structural RNAs. The D-loop has no coding capacity but acts as a control region for replication. A favored region for DNA barcoding lies within the cox1 or COI gene (marked with short line segment with arrowheads at each end that represent the PCR primers). However, other regions such as 16S rRNA, cyt b, or D-loop are also often used.

prokaryotes and eukaryotes. The use of DNA sequences as reference to an inventory of organisms could affect the way taxonomic research is done in the future. Most importantly, it would seem necessary that museums establish DNA collections in parallel to the traditional collections of specimens. It would even seem possible to create a DNA taxonomy system that is built on the basis of the established taxonomic research but would rely on the new options that are provided by DNA barcoding.

An explicit goal of the DNA barcoding approach is to eventually provide a machine that can be used in the field to determine the code at low cost—basically an equivalent of a handheld barcoder used in supermarkets. However, the prospects for this seem slim at present, at least if one would rely on the current sequencing technology. This requires three crucial steps, namely PCR, followed by the sequencing reaction, followed by a separation of the resulting fragments by electrophoresis. Although the chemicals for this are relatively low-cost, the required machines are rather sophisticated and expensive. There is some hope that microfluidic technology would allow the integration of this into a small device. However, the sheer necessity of combining a PCR step with a sequencing step and electrophoresis will pose a major technological challenge. A possible way out of this would be to focus on a direct use of ribosomal RNAs, which are present in many copies in each cell, as the template for sequencing, which could eliminate the PCR step (see above). If one uses in addition a sequencing technology that does not require gel electrophoresis, such as pyrosequencing, one comes closer to the design of a machine that can be used in the field. However, as with any other development of technology, one will require a sufficiently large market to make the investment worthwhile. The success of a handheld DNA barcoder will therefore depend on the importance that society puts on the knowledge of species diversity around us.

For background information *see* DEOXYRIBONUCLEIC ACID (DNA); GENE AMPLIFICATION; GENETIC

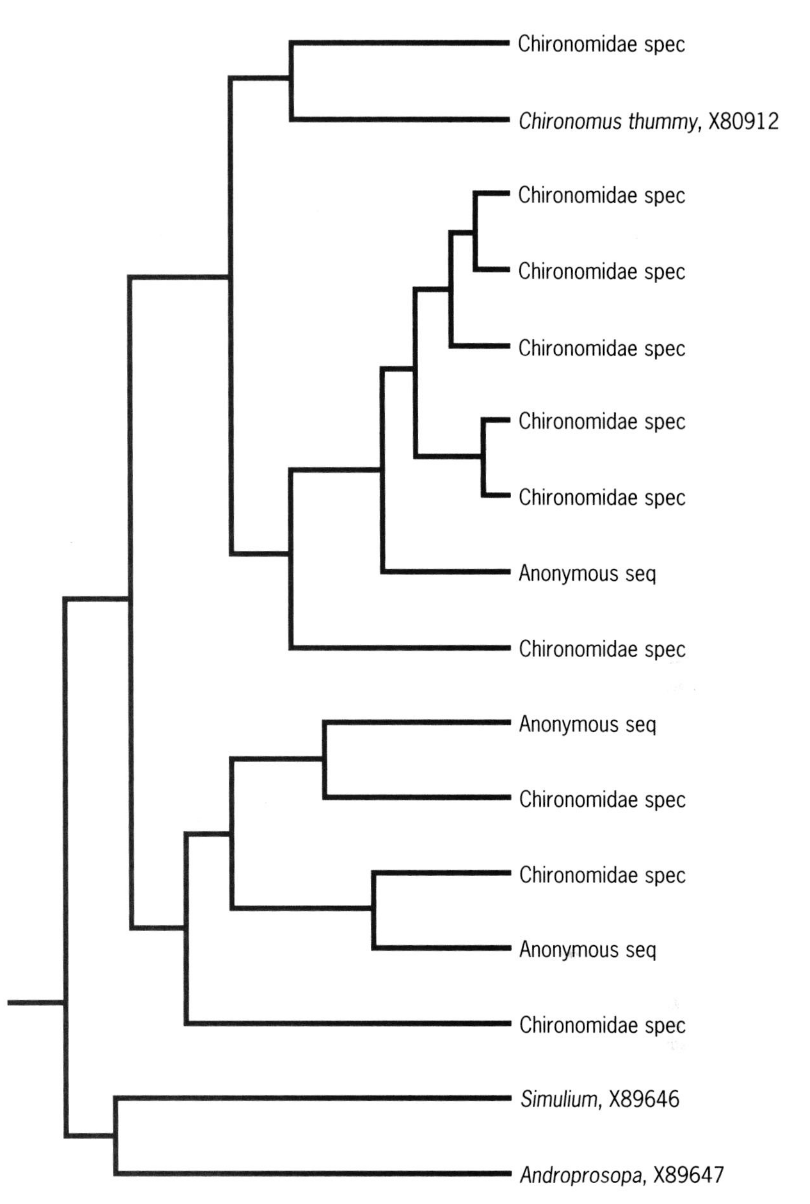

Fig. 3. Phylogenetic reconstruction with sequence fragments from unknown samples (anonymous seq), from roughly determined samples (Chironomidae spec) and database sequences (the ones with accession numbers). The group "Chironomidae" clusters well together, including the anonymous samples, which will therefore be considered to be derived from chironomids as well. Note that chironomids are a species-rich group and that the grouping here is only at the genus level.

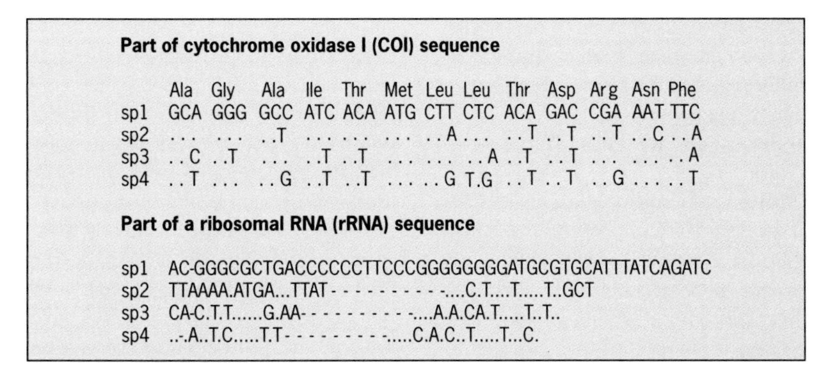

Fig. 4. Two examples of sequence regions used for DNA barcoding and their alignment between four species (sp1–sp4) are shown. Dots represent identical nucleotides, dashes deletions. The cytochrome oxidase I (COI) gene codes for a protein; the coding triplets for each amino acid (top line) are listed and aligned. It is evident that only those positions are changed which do not change the amino acid because of the redundancy of the genetic code. The ribosomal RNA (rRNA) has only a structural role and allows changes at many positions, as well as insertions and deletions, which can provide additional information.

CODE; MOLECULAR BIOLOGY; MUTATION; SPECIATION; TAXONOMY in the McGraw-Hill Encyclopedia of Science & Technology. Diethard Tautz

Bibliography. M. L. Blaxter, The promise of a DNA taxonomy, *Phil. Trans. Roy. Soc. Lond. B Biol. Sci.*, 359:669–679, 2004; P. D. Hebert et al., Biological identifications through DNA barcodes, *Proc. Roy. Soc. Lond. B Biol. Sci.*, 270:313–321, 2003; C. Moritz and C. Cicero, DNA barcoding: Promise and pitfalls, *PLoS Biol.*, 2, e354, 2004; D. Tautz et al., A plea for DNA taxonomy, *Trends Ecol. Evol.*, 18:70–74, 2003.

DNA sequence assembly algorithms

The completion of the Human Genome Project marked a significant milestone toward an understanding of the principles of life. This important step was made possible by the discovery in 1975 of a new method for determining the string of letters composing the DNA sequence of organisms. Sanger sequencing, named after its developer Fred Sanger, is an essential component of the shotgun sequencing method, the most widely used approach for DNA sequencing. Shotgun sequencing overcomes an inherent limitation of Sanger sequencing; that is, only short stretches of DNA, approximately 1000 base pairs (bp) in length, can be sequenced at one time, while the DNA of most organisms is much longer. Bacterial genomes are composed of millions of base pairs and the genomes of mammals and other vertebrates contain billions. The shotgun sequencing method involves three steps: (1) breaking up the DNA, at random, into a multitude of short fragments, most commonly between 2000 and 10,000 bp in length, (2) sequencing the ends of each fragment (the sequenced ends are commonly called reads), and (3) joining the reads together using a computer program called an assembler. The reads range 700–1000 bp in length, depending on the sequencing protocol. Sufficient reads must be generated to ensure that most reads overlap each other, thereby making assembly possible. It can be shown mathematically (Lander-Waterman model) that a genome can be assembled in relatively few pieces once enough reads are sequenced, such that each location in the genome is present in approximately 8–10 reads. The amount by which the genome is oversampled is called coverage and represents an important parameter of a sequencing project as it affects both the cost of a genome project and the ability to correctly assemble the genome.

Assembly and puzzles. Assembly programs rely on the assumption that reads that contain the same string of letters are likely to have been sequenced from the same place in the genome. The assembly problem can be compared with solving a jigsaw puzzle, where the pieces correspond to the sequencing reads and the fit between a knob in one piece and a hole in another corresponds to the similarity between the reads (**Fig. 1***a*). Mathematically, this problem was originally formulated as finding the shortest string of letters that contains a given set of strings (the original reads) [Fig. 1*b*]. The shortest common "superstring" problem was shown to belong to a class of difficult problems that cannot be efficiently solved (NP-hard problems). This class of problems is exemplified by a popular wooden log puzzle. Several wooden logs of different lengths (corresponding to reads) must be placed in such a way as to fit into a small wooden box (the shortest superstring). Readers who have attempted to solve this puzzle are familiar with its difficulty, where finding a solution requires trying all possible combinations of logs. Such a puzzle is not always difficult to solve. For example, a solution can be easily found if the logs are the same length. Similarly, in many practical cases, the assembly problem can be solved by a simple algorithm. At each step, this algorithm chooses the two reads that are most similar to each other and joins them together, repeating this procedure until the remaining reads either cannot be joined, or joining them would contradict choices made at an earlier stage. This simple greedy algorithm (so-called because of the aggressive joining of reads) formed the basis of the first assembly programs. Historically, two such programs stand out: TIGR Assembler, which was used at the Institute for Genomic Research in Maryland to obtain the first sequence of a free-living organism (*Haemophilus influenzae*) and Phrap (phragment assembly program), which was an essential element of public efforts to sequence the human genome.

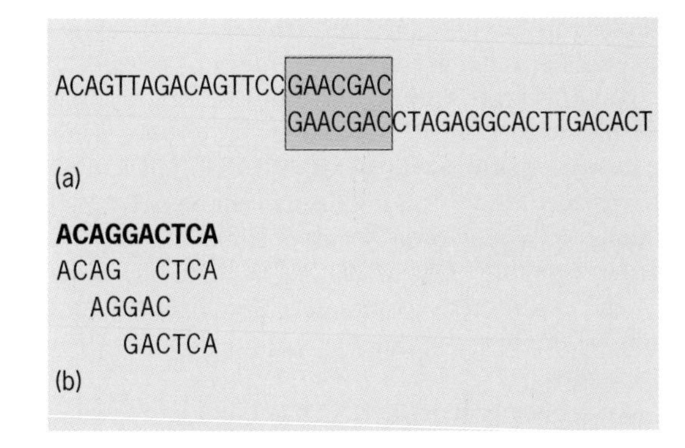

Fig. 1. Assembly of sequencing reads. (*a*) Two overlapping reads in the genome can be joined by the assembler using the sequence similarity in the shaded region. (*b*) The string in bold letters represents the shortest common superstring of the strings below it.

Repeats. While important for understanding the complexity of assembly, the shortest common superstring problem overlooks an important feature of virtually all genomes—stretches of DNA repeated in identical or nearly identical copies throughout the genome, called repeats. Reads that originate from these regions appear similar despite the fact that they represent different regions of the genome. Simple algorithms, such as greedy, can be confused by repeats and misassemble the genome (**Fig. 2***a*, *b*). The presence of complex repeats in virtually all large genomes was considered an insurmountable obstacle to sequencing an entire genome through the shotgun method (whole-genome shotgun). Until the successful assembly of the fruit fly and human genomes

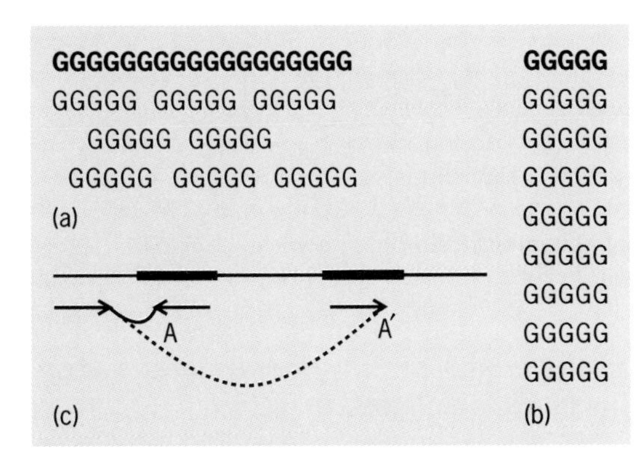

Fig. 2. Assembly programs may misassemble a genome, choosing (*b*) to collapse a repeat instead of (*a*) the correct reconstruction. Replacing G with TTAGG, we obtain a common repeat found in the telomeres at the ends of human chromosomes. (*c*) Mate-pair information allows the assembler to place-read A in the first copy of a repeat (thickened bar) instead of the second copy.

at Celera Genomics in 2001, large genome projects were performed in a two-stage process whereby the genome is first fragmented into 50–150 kb pieces (called bacterial artificial chromosomes or BACs), then each piece is sequenced through the shotgun method (BAC-by-BAC sequencing). This process reduces the complexity of the data provided to the assembler; however, it generally is substantially more expensive and labor-intensive than the whole-genome shotgun approach.

Assembly programs rely on specific characteristics of the shotgun process to correctly assemble repeats. Valuable information is provided by the pairing of reads sequenced from the ends of a same DNA fragment (mate-pairs). Randomly sheared shotgun fragments are organized into a set of libraries consisting of fragments within a narrow size range. Within a reconstruction of the genome, the distance between reads belonging to a same fragment must be consistent with the range of fragment sizes present within the specific library. Mate-pair information also provides a constraint on the relative orientation of reads. Sequencing reactions progress toward the interior of each fragment along the complementary DNA strands forming the well-known double helix, thereby providing the reads with a natural orientation (**Fig. 3**). In the assembler output, reads from opposite ends of a same fragment must be oriented toward each other (Fig. 2*c*), indicating that they were obtained from opposite strands.

Additional information is provided by the randomness of the shearing process, allowing the use of statistical methods to identify collapsed repeats due to an increased depth of coverage in these regions. Collapsed repeats are regions where the assembler incorrectly joined reads that belong to distinct copies of a repeat. If the genome looks like "A R1 B R2 C," where R1 and R2 are almost identical pieces of DNA, all the reads coming from R1 look similar enough to R2, leading the assembler to combine them into a single collapsed piece R, an incorrect reconstruction of the original DNA. In the example shown (Fig. 2*a*, *b*), an assembler expecting a genome coverage of 3 would favor the tiling on the left as the correct genome reconstruction. The genome coverage is the average number of reads that span a particular position in the genome. The coverage can be calculated as the sum of the length of all the reads divided by the expected size of the genome.

Overlap-layout-consensus. In order to overcome the challenges imposed by repeats, most modern assembly algorithms follow a strategy called overlap-layout-consensus. In the overlap stage, reads are compared to each other to determine whether they could be joined together. In the layout stage, the assembler generates a layout of the reads that satisfies a majority of the overlaps as well as the constraints imposed by mate pairs and statistical considerations. During the consensus stage, the sequence of the genome is determined by aligning the reads as specified in the layout.

Scaffolding and finishing. The output of an assembly program rarely consists of a single contiguous piece of DNA (contig), due to regions in the genome that are difficult to sequence (for example, limitations of laboratory protocols) or difficult to assemble (for example, complex repeats). Multiple contigs are usually produced, whose relative ordering along the genome is unknown. These can be ordered and oriented with respect to each other using the information provided by mate-pairs in a process called scaffolding. Mathematically, scaffolding can be formulated as finding the placement of contigs along a chromosome that satisfies most mate-pair constraints (**Fig. 4**). In this general formulation, the scaffolding problem falls within the same class of difficulty (NP-hard) as the assembly problem; however, in many practical instances it can be solved efficiently.

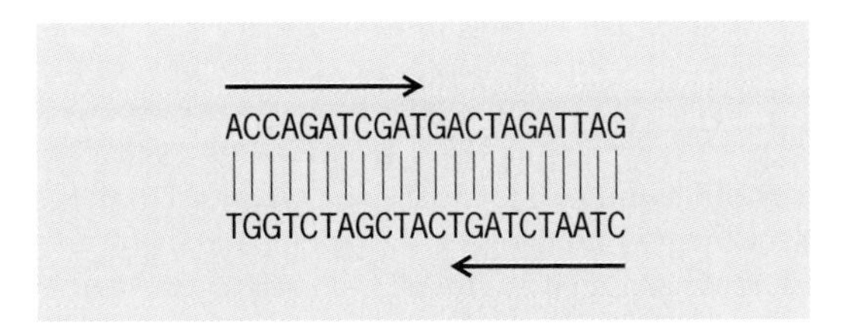

Fig. 3. Progression of sequencing reactions along double-stranded DNA. Sequencing reactions proceed in opposite directions along complementary strands, thereby providing a natural orientation to the reads being generated. The DNA double helix is held together by bonds between complementary bases (A-T and G-C).

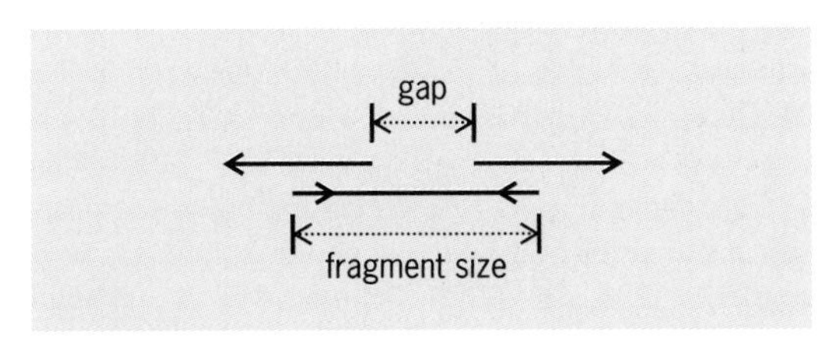

Fig. 4. Mate-pair information is used to order and orient two contigs (bold arrows). The size of the gap between the two contigs can be estimated from the known distance between the mated reads.

Scaffolding provides information about the long-range connectivity between sections of a chromosome, thereby enabling studies of genomic rearrangements even in the absence of a complete genome sequence. Scaffolding information can also be used as a guide in efforts to obtain the complete DNA sequence of an organism. As described above, the output of an assembler consists of multiple contigs, the gaps between which represent unknown regions of a chromosome. These gaps are filled in through a labor-intensive process called finishing. The contig adjacency information provided through scaffolding allows scientists to design directed sequencing experiments to bridge the gaps between contigs. Modern assembly programs use scaffolding as an integral component of the assembly algorithm, as is the case with Celera Assembler (initially developed at Celera Genomics as part of a private effort to sequence the human genome) and Arachne (developed at the Broad Institute for the assembly of mammalian-sized genomes).

Future of assembly. Before the completion of a first draft of the human genome sequence in 2001, the main goal of assembly algorithm research was the development of methods for the assembly of mammalian genomes. The algorithms and programs developed during the Human Genome Project have since been applied to many other genomes, some of which have uncovered additional assembly challenges. In particular, the basic assumption underlying all modern assembly programs that the genome being assembled represents a single piece of DNA that was uniformly sheared is being challenged. There are now efforts to sequence bacterial populations directly from an environment (for example, bacteria in the human gastrointestinal tract). Assembly programs must be modified to simultaneously assemble multiple organisms sequenced at widely varied levels of coverage. Similar problems also occur in the sequencing of single organisms; for example, in the sea squirt, the homologous chromosomes received from the parents can differ by up to 20%. Assembly algorithms must also adapt to data produced by novel sequencing technologies. Short reads and high error rates are common to virtually all such technologies.

The assembly of the human genome does not represent the end of a journey, but a milestone toward the development of assembly programs that are able to assist in our efforts to understand the diversity of nature.

For background information *see* ALGORITHM; COMPUTER PROGRAMMING; DEOXYRIBONUCLEIC ACID (DNA); GENETIC CODE; HUMAN GENETICS; HUMAN GENOME; MOLECULAR BIOLOGY; NUCLEIC ACID in the McGraw-Hill Encyclopedia of Science & Technology.

Mihai Pop

Bibliography. I. H. G. S. Consortium, Initial sequencing and analysis of the human genome, *Nature*, 409:860–921, 2001; R. D. Fleischmann et al., Whole-genome random sequencing and assembly of *Haemophilus influenzae* Rd., *Science*, 269(5223):496–512, 1995; D. B. Jaffe et al., Whole-genome sequence assembly for mammalian genomes: Arachne 2, *Genome Res.*, 13(1):91–96, 2003; E. W. Myers et al., A whole-genome assembly of *Drosophila*, *Science*, 287(5461):2196–2204, 2000; M. Pop, S. L. Salzberg, and M. Shumway, Genome sequence assembly: Algorithms and issues, *IEEE Computer*, 35(7):47–54, 2002; J. C. Venter et al., The sequence of the human genome, *Science*, 291(5507):1304–1351, 2001.

Downscaling climate models

Rising global temperatures and the related changes in climate patterns are causing impacts around the globe. Recent observations have documented unprecedented rates of glacier retreat, earlier blossoming of flowers in spring, earlier arrival of migrating birds, reduced winter snowpack, earlier arrival of snowmelt in montane river basins, and dozens of other warming-related impacts.

Anthropogenic (human-made) greenhouse gas emissions have been recognized as the leading cause of the warming experienced in recent decades, and will account for a rising proportion of warming projected over the next century. The Intergovernmental Panel on Climate Change (IPCC) has developed a suite of greenhouse gas emission scenarios that project future emissions of carbon dioxide (CO_2) and other gases that affect the radiative properties of the atmosphere, and thus influence the extent to which future warming will occur. Even under the scenario with the most rapid transition to a global economy based on sustainable energy and transportation systems, the CO_2 emissions into the atmosphere are projected to increase at least through 2050.

The current scientific questions revolve around the degree of warming to be expected, and the regional and local manifestations of the warming planet. The broad scientific consensus that temperatures have risen globally and will continue to rise is reason for concern for resource managers acting at regional to local scales. However, there is considerably greater uncertainty regarding the impacts of climate change at the smaller spatial scales at which resource managers make decisions. This uncertainty has led to a surging interest in projecting future climate change onto regional natural systems to study impacts and explore management options that might help mitigate some of the impacts.

When planning a climate change impacts study, several decisions must be made, and each decision affects to some degree the result obtained. Of particular importance are decisions regarding which global climate model to use and the assumptions about future greenhouse gas levels.

Global climate models. Global climate models (or general circulation models, as they are more formally known; GCMs) are used to project future climate. These models are physically based representations of the global climate system, and the current generation of models includes components simulating the oceans and land surface coupled to the atmosphere in a way that captures the natural feedback between these systems. The latest models are greatly improved in their ability to simulate a realistic climate without use of adjustments that were often required to nudge the simulated climates from earlier models toward more realistic states.

Global climate models have been developed by about 20 different modeling centers around the globe, and each GCM responds differently to changing atmospheric concentrations of greenhouse gases. Thus, when exploring regional or local climate change impacts, the results will be dependent on the GCM(s) chosen for the study. To characterize the response of a GCM, model sensitivity has been defined as the global change in temperature simulated by the model with a doubling of atmospheric CO_2 concentration. In their Third Assessment Report, the IPCC compiled the results of experiments determining the climate sensitivity of 30 GCMs, which can provide some guidance in model selection. While models vary in sensitivity to the levels of CO_2 and other greenhouse gases in the atmosphere, all GCMs simulate plausible climatic responses to changes in these concentrations, typically with more extreme responses at higher CO_2 concentrations.

Future greenhouse gas emissions. In addition to uncertainty about precisely how the climate will respond to changes in CO_2 concentration, it is also difficult to predict how regulation and changing technology might affect future quantities of CO_2 and other greenhouse gases released into the atmosphere. Along with modeling uncertainty, the inability to predict future anthropogenic greenhouse gas emissions is often treated as a source of uncertainty in studies of future climate. However, the two sources of uncertainty are qualitatively different in that uncertainty surrounding future emissions depends upon the choices we make as a global community, whereas modeling uncertainty arises from limits on our understanding of physics or our ability to represent the relevant processes accurately in models.

In its Special Report on Emission Scenarios, the IPCC describes a set of 40 different future scenarios of greenhouse gas emissions. These scenarios arise from four unique "narrative storylines" that describe rates and patterns of economic development, population growth, cultural and social interactions, introduction of more efficient technologies (for transportation and energy, for example), and other characteristics. While the scenarios do not include any assumptions about specific climate change policies, they can be used to examine potential futures with different emissions pathways, effectively reflecting futures with lower and higher emissions. While many past climate change impact studies focused on one "business as usual" scenario, more recent studies have found the differences in impacts under different emissions futures to be of the same magnitude as, and sometimes greater than, the differences between GCM responses. This implies that selecting different emissions scenarios for climate change impact analysis can help quantify the effect of an uncertain future for greenhouse gas emissions.

Downscaling global climate models. Once the GCMs have been selected and the emissions scenarios of interest identified, GCM model output can be obtained from a variety of sources, such as the IPCC Data Distribution Centre or directly from the modeling center performing the GCM simulations. Global climate models typically have spatial resolutions of 300–500 km. This means, for example, that a 100-km^2 area (for which a resource manager might be interested in exploring climatic impacts) would constitute less than 0.1% of a typical GCM grid box, and the GCM output is uniformly averaged over these vast areas. To illustrate the effects of aggregating data over such large spatial scales, **Fig. 1** shows how the representation of the terrain changes at different spatial resolutions, with obvious implications for local impacts analysis for which topography is important. In addition, where GCM grid boxes straddle coastlines, the climate change response can be sensitive to the proportion of land and sea in the grid box, an important consideration for impact analysis for coastal land areas.

To translate the GCM output for use in regional and local impacts analysis, there are two general approaches: dynamical and statistical downscaling.

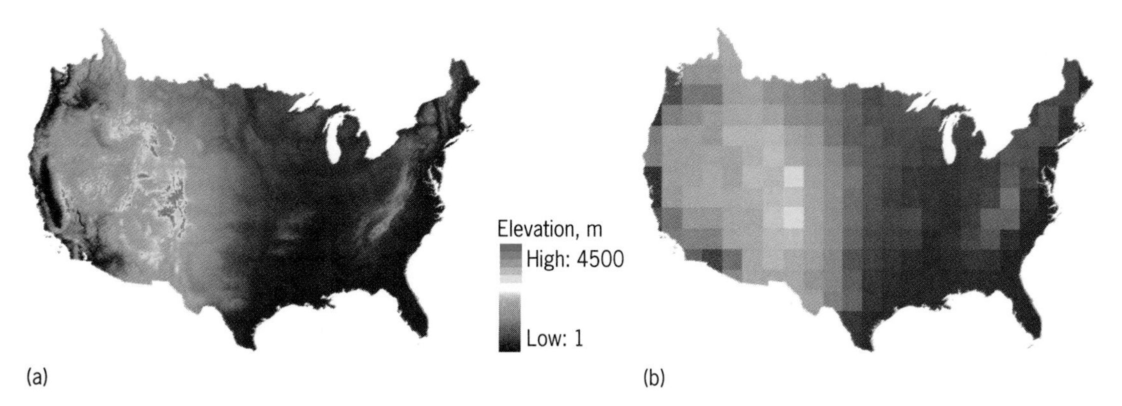

Fig. 1. Representation of topography. (*a*) Elevations of the United States (and the Columbia River Basin extending into Canada) at a spatial resolution of 1 km, with a maximum elevation over 4300 m. (*b*) Same data aggregated to a spatial resolution of 200 km (roughly that of the highest-resolution GCM), where the maximum elevation is below 2900 m. Fine-scale features are lost at the coarser resolution, as are some larger-scale features, such as the Sierra Nevada and Cascade mountain ranges, which define local spatial patterns of precipitation and temperature.

Dynamical downscaling. Dynamical downscaling uses a regional climate model (RCM), typically at a resolution of 40–50 km with a spatial domain focused on the region of interest. For each GCM and emissions scenario of interest, the GCM-simulated climate is used to define the climate at the boundary of the region, and the RCM simulates the climate within the region. The RCM, with a better representation of the terrain than the GCM, can resolve local features of climatic change. For instance, as air masses move inland from the Pacific Ocean they encounter mountain barriers (the Sierra Nevada and Cascade Mountain ranges, for example), which force them to rise and cool, producing orographic effects, that is, enhanced precipitation on the windward side of the barrier. Improved representation of the topography in the RCM, therefore, allows more accurate simulation of these effects.

While this approach is appealing, due to the use of hydroclimatic physics in the downscaling process, there are several disadvantages to dynamical downscaling that have limited its use. First, RCM simulations are computationally intensive (on the order of days of computation per year of simulation), making long continuous simulations or multiple GCM or scenario simulations impractical. Second, even at a 50-km spatial scale, many local features can be lost. This, and the approximations of many physical processes in RCMs, produce biases that must be corrected later, often by statistical methods.

Statistical downscaling. Statistical downscaling develops empirical relationships between the GCM output and the local hydroclimatic features and then applies these relationships to rescale the GCM output to the fine scale needed for local impact analysis. The scale to which the downscaling can be achieved is generally limited by the density of observations (of precipitation and temperature, for example). Statistical downscaling can typically be achieved for spatial scales of 10–50 km, with the coarser scale in locations with sparser observed records. In statistical model development, observed local variables of interest are related to concurrent GCM-simulated variables that serve as predictors of the local variables. Predictor variables can be, for example, GCM-simulated precipitation, wind speed, and temperature for the grid box overlaying the local site. Because statistical downscaling can be done quickly on an ordinary desktop computer, it has been applied over many different regions for many different GCMs and emissions scenarios, and software tools for implementing this technique are freely available.

Applications of statistical downscaling. The downscaled data can take many final forms, but most often are a predicted future shift in the mean (of precipitation, for example) or a description of predicted changes in both the mean and variability of the climatic variable. Projections of future climate developed in this way have been used in hundreds of studies to characterize future climate change impacts on many environmental features affected by a warming climate, such as water resources, habitat suitability, wildfire frequency, agriculture, and human health.

For example, in one study that focused on California, GCM data were statistically downscaled using two different methods, since the needs for different aspects of the study were different. For hydrological and agricultural analyses (for examining impacts to streamflow and snow pack, the California wine grape and dairy industries, and alpine forest ecosystem extent), monthly output from two GCMs was statistically downscaled to a higher-resolution grid (with each grid cell representing about 150 km^2). This downscaling comprised two steps. First, probability density functions were created for both the GCM and observed precipitation and temperature at the same scale as the GCM to correct for any GCM bias. (For example, if a future GCM-simulated precipitation value is equal to the median GCM value for 1961–1990, it is converted to the median observed value for 1961–1990.) Second, these "bias-corrected" future data were downscaled to the finer grid scale by taking the anomaly (the difference between the future value and the mean for the 1961–1990 period) and interpolating it onto the finer grid. **Figure 2** shows for one GCM how the coarse GCM winter precipitation is transferred onto the finer scale, recreating the precipitation patterns consistent with California topography.

In the same study, for analyses using temperature and precipitation extremes, a distinct downscaling method was employed to capture daily variability, using long-term observed records of temperature and precipitation at specific weather stations. In this second method, daily GCM-simulated temperature and precipitation from the reference period were rescaled by regression relations that were then applied to future simulations. The resulting rescaled values share the observed weather statistics at the selected stations. These downscaled GCM data were then used to analyze impacts such as changes in

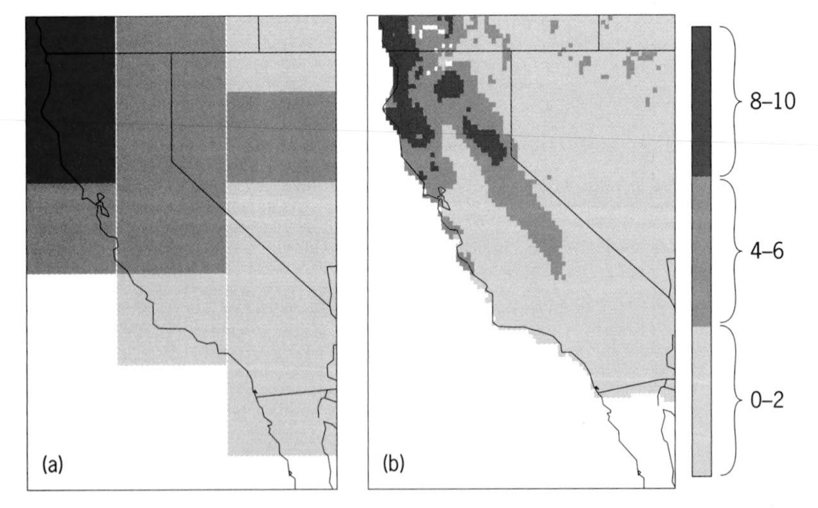

Fig. 2. Effect of statistical downscaling on GCM data to recreate physically realistic local features in precipitation over California. (*a*) Average winter precipitation projected by the HadCM3 model for 2070–2099 (after bias correction) under the A1fi emission scenario, with each grid box in this region representing approximately 90,000 km^2. (*b*) Same precipitation field statistically downscaled to a higher-resolution grid of approximately 150 km^2 per grid cell. Precipitation scale is in mm/day.

heat-wave severity and frequency and implications on human health. Due to the localized nature of these impacts, they could not be adequately studied using raw GCM output.

These downscaled models make the impacts of climate change understandable at a local level, and in turn can provide the public with a clear view of the consequences of different societal choices.

For background information *see* CLIMATE MODELING; CLIMATE MODIFICATION; CLIMATE PREDICTION; ECOLOGY; GLOBAL CLIMATE CHANGE; GREENHOUSE EFFECT in the McGraw-Hill Encyclopedia of Science & Technology. Edwin P. Maurer

Bibliography. R. E. Benestad, Empirical-statistical downscaling in climate modeling, *Eos*, 85 (42):417, 2004; K. Hayhoe et al., Emissions pathways, climate change, and impacts on California, *Proc. Nat. Acad. Sci.*, 101(34):12422-12427, 2004; IPCC, *Climate Change 2001: The Scientific Basis: Contribution of Working Group I to the Third Assessment Report of the Intergovernmental Panel on Climate Change*, ed. by J. T. Houghton et al., Cambridge University Press, 2001; L. R. Leung et al., Regional climate research: Needs and opportunities, *Bull. Amer. Meteorol. Soc.*, 84:89-95, 2003; J. Murphy, An evaluation of statistical and dynamical techniques for downscaling local climate, *J. Clim.*, 12:2256-2284, 1999; R. L. Wilby et al., Statistical downscaling of general circulation model output: A comparison of methods, *Water Resources Res.*, 34:2995-3008, 1998.

Dry-docking and heavylift ships

A dry dock is a facility that removes a ship from its water environment. A dry dock can be dug out of the ground, such as a graving dock. It can also be a floating dry dock, that is, a vessel that is initially ballasted (weighted) down and can then lift a ship brought over it out of the water after deballasting. A heavylift ship also functions this way and will be discussed later. Dry docks have been pivotal throughout the history of both merchant marine commerce and naval sea power.

Emergence of dry docks in America. Dry docks originated in America out of a desperate need for repair facilities for a new class of naval warship. This new ship class, the ship-of-the-line, at over 2000 tons was the largest ever built at that time by the United States. This ship class was the result of the War of 1812, in which the United States found the White House burned by British forces and ships at the Washington Navy Yard in flames to prevent capture. The keel of the 74-gun ship-of-the-line, USS *Delaware*, was laid at the Gosport Navy Yard (later known as the Norfolk Navy Yard) at Portsmouth, Virginia, in 1817 and was launched in 1820. However, no facility existed in North America in 1820 where a ship-of-the-line could be dry-docked. Former methods of careening (intentional grounding in a tidal area) were impractical at the larger scale. It soon became necessary to recopper the bottom of these ships; therefore, in 1827, Congress authorized the construction of two stone graving dry docks. The sites chosen were the Gosport Navy Yard at Portsmouth, and the Charlestown Navy Yard at Boston.

Fig. 1. *Gray* in a graving dry dock at Long Beach Naval Shipyard. (*Photo courtesy of U.S. Navy*)

Graving dry docks. A graving dock is generally dug out and provided with gates of various sorts, often floating/ballasting (called a caisson gate), to seal off the dock from the harbor or rivers. Water can be pumped out and the dry dock used for building or repairing a ship below its waterline (**Fig. 1**). The dock is provided with means to hold the ship upright using a blocking system when dry, and a means of dewatering using pumps. The first ship to be dry-docked in the United States was the USS *Delaware*, at Gosport Navy Yard on June 17, 1833. The graving dock at the Charlestown Navy Yard in Boston opened one week later; the first ship dry-docked there was the USS *Constitution*. Thereafter, the utility of dry docks resulted in the steady addition of more and larger dry docks. During World War II, 26 graving docks were constructed.

Floating dry docks. Another means of gaining access to the portions of a ship normally underwater is the floating dry dock, such as the wooden one first utilized for the West Coast dock at Mare Island, California (1854). A floating drydock is a large barge with hollow sidewalls that can be ballasted down to allow a ship to enter, and then deballasted to expose its underwater hull. After the Spanish-American War in 1898, the U.S. Navy needed a dry dock in the western Pacific. In 1905, USS *Dewey* (YFD 1) was towed across the Atlantic and Mediterranean, through the Suez Canal, and across the Indian Ocean, arriving at Olongapo, Philippines, $6^1/_2$ months later.

Dry docks (both graving and floating) were critical throughout World War II. The USS *Yorktown*

Fig. 2. *Yorktown* (CV 5) in dry dock at Pearl Harbor before the Battle of Midway. (*Photo courtesy of U.S. Navy*)

Fig. 3. Advanced Base Sectional Dock (ABSD) positioned in the western Pacific during World War II. (*National Archives*)

(CV 5) shown in **Fig. 2** was dry-docked at Pearl Harbor, Hawaii, following the Battle of the Coral Sea. After urgent repairs, it rushed into the Battle of Midway, where its air wing played a decisive role, although the ship was sunk one week after leaving the dry dock. The largest floating dry docks were constructed in sections that were towed, assembled, and used to dry-dock battleships in the western Pacific. Many damaged ships were repaired and made ready for battle again out in the theater. These docks were called Advanced Based Sectional Docks (**Fig. 3**). In 1955, in the wake of World War II, the U.S. Navy operated 106 floating dry docks.

Heavylift ships. Another method for lifting large floating objects out of the water utilizes heavylift ships. Enormous heavylift ships were designed to transport oil rigs from construction yards to the oil fields in distant locations. These ships are capable of being ballasted down, having another ship moved over them, and then being deballasted. The mechanism is similar to a floating dry dock; however, a heavylift ship is self-propelled and capable of speeds of over 14 knots (26 km/s). This speed enables it to get a ship on the cargo deck to its destination in half the time compared to towing. The U.S. Navy has used heavylift ships to transport damaged ships (such as the USS *Samuel B. Roberts*, *Stark*, and *Cole*) for repair. In addition, mine countermeasure ships have been transported from the United States to the Middle East for use in clearing mines. As shown in **Fig. 4**, the heavylift ship M/V *Super Servant 3* transported the USS *Impervious*, *Leader*, *Adroit*, and *Avenger* to the Persian Gulf from Norfolk in September 1990.

Current trends in dry-docking. A relatively new way to dry-dock ships is to place them into a vertical lift. **Figure 5** shows the USS *Elliott* on a vertical lift in Los Angeles harbor in 1984. This method is similar to the way an elevator works. Electric motors and cables are used to lift a platform up under a ship and then lift it out of the water.

Major U.S. shipyards have upgraded their ship construction facilities to include land-level transfer facilities. Ships are assembled on level surfaces on carriages and rolled onto floating dry docks, from which they are then undocked. This has replaced an ancient method of launching newly constructed ships on greased inclined ways.

This overview has concentrated on graving dry docks, floating dry docks, heavylift ships, and vertical lifts as means of exposing the underwater hull of the larger floating vessels for maintenance, transport, or repair. On a smaller scale, marine railways and cranes of various sorts can be used to remove vessels from the water.

The recent imperative in the construction of dry docks has been for larger sizes to accommodate larger ships, such as cruise ships and tankers carrying

Fig. 4. Heavylift ship *Super Servant 3* transported the USS *Impervious*, *Leader*, *Adroit*, and *Avenger* from Norfolk, Virginia, into the Persian Gulf in September 1990. (*Photo courtesy of U.S. Navy*)

Fig. 5. USS *Elliott* in the largest vertical lift in the world at the time (1984) in Los Angeles. (*Courtesy of SYNCROLIFT, Inc.*)

crude oil. The navies of the world have become more interested in heavylift alternatives for transporting warships to and from war zones. This has allowed small ships to get around the world ready to fight, and enabled damaged ships to return to building yards for massive reconstruction.

Even in times of reduced underwater maintenance due to advanced paint systems and corrosion-resistant materials, dry docks remain essential for the commercial and military fleets of the world. Dry docks have been a critical ingredient for cruise lines, cargo ships, and navies to keep their ships ready to work.

For background information *see* DRYDOCKING; MARINE ENGINEERING; SHIP BUILDING; SHIP DESIGN; SHIP SALVAGE in the McGraw-Hill Encyclopedia of Science & Technology. Richard Hepburn

Bibliography. N. Miller, *The U.S. Navy: An Illustrated History*, American Heritage, New York, and the United States Naval Institute Press, Annapolis, 1977; F. M. Bennet, *The Voyage of the Dewey: Proceedings of the United States Naval Institute*, vol. 32, no. 4, U.S. Naval Institute, Annapolis, December 1906; *Dry Dock Facility Characteristics*, Military Handbook 1029/3, Naval Facilities Engineering Command, Alexandria, VA, 1988.

Dynamic traffic management

Traffic congestion is a persistent problem plaguing modern road and freeway networks. It occurs when demand exceeds capacity in some network links. This may eventually block upstream network intersections (gridlock) and trap vehicles that would have exited before the bottleneck, resulting in degradation of the available infrastructure capacity. Countermeasures aim to reduce demand by using various restrictions (tolls), enhanced public transport, and high-occupancy vehicle lanes, or to enhance and protect the available infrastructure capacity by deploying dynamic traffic-management (DTM) systems.

A DTM system receives real-time measurements from sensors installed in the network and drives a number of control actuators to actively interact with network traffic. The most common sensor types used are inductive-loop detectors and video cameras, although other technologies exist such as ultrasound, radar, and magnetic devices. Loop detectors are placed below the road pavement and measure the flow (vehicles per hour), occupancy (percentage of time of occupancy of the loop by vehicles) and, in case of double loops, mean speed of vehicles. Video sensors use digital image processing algorithms to deduce similar traffic-flow measurements.

Based on the measurements and their processing, a DTM system aims to improve traffic conditions using control actuators such as traffic lights, panels advising drivers about their route or the current traffic and environmental conditions (incidents, congestion, pavement state, and so on), lane-assignment signs, and variable speed limits.

The **illustration** shows the general structure of a DTM system. Network traffic flow is viewed as a process under control, which receives two kinds of input as it evolves in time: control inputs and disturbances. The disturbances represent effects that cannot be influenced by the DTM system but may be measurable or predictable. Examples of such disturbances are the demands at the network origins, drivers' routing behavior, weather conditions, and other incidents. However, control inputs can be manipulated within certain limits by an automatic control strategy whose task is to drive the available control actuators in real time, based on measurements, estimations, and predictions, to minimize the total time spent by all vehicles in the network. The control strategy represents the "brain"

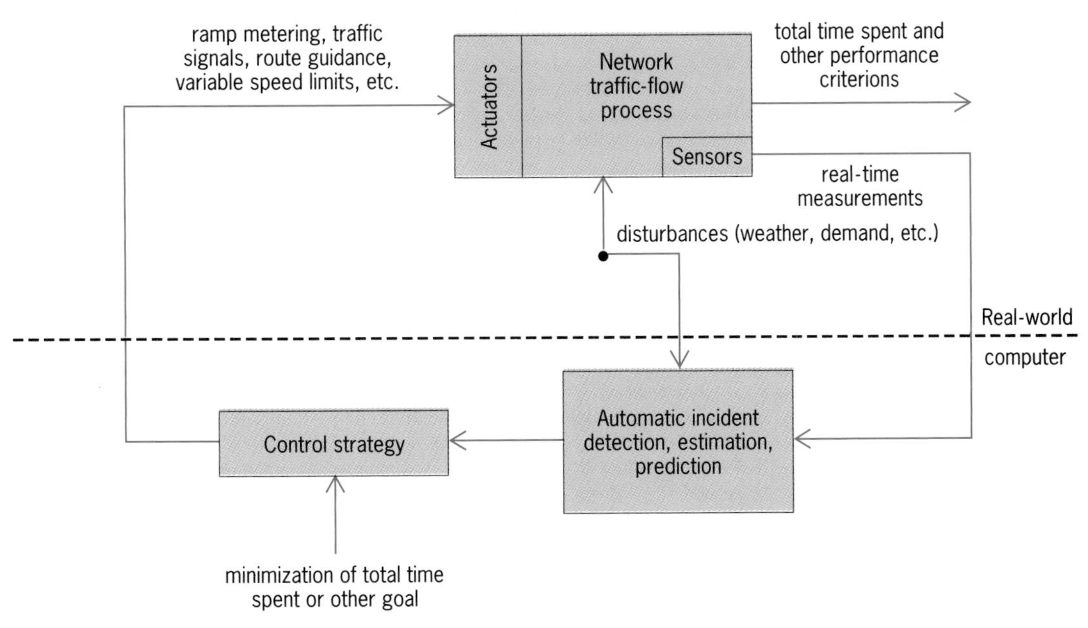

Functional structure of a dynamic traffic management system.

(decision-making) of the DTM and is based on advanced optimization and control methodologies. DTM measures, control strategies, and systems are typically developed separately for urban roads and networks, expressways and freeways, and interurban highways and motorways.

Urban traffic control. Traffic lights at intersections are the major traffic-control actuator in urban road networks. Traffic lights were originally introduced to guarantee the safe crossing of antagonistic streams of vehicles and pedestrians. With the rapid increase of traffic demand, it has been realized that traffic signals can be operated more or less efficiently (under the same safety implications) according to the control strategy used.

Traffic signal operation. In the operation of traffic lights, a signal cycle is one repetition of the basic series of signal combinations at an intersection, and its duration is called the cycle time (typically 60–120 seconds). A stage (or phase) is a part of the signal cycle, during which one set of approaching (nonantagonistic) traffic streams has the right-of-way simultaneously. Constant lost (or intergreen) times of a few seconds are necessary between stages to avoid interference between antagonistic streams of consecutive stages. The split is the relative green duration of each stage (as a portion of the cycle time) that should be optimized according to the stream demand. Longer cycle times typically increase the intersection capacity because the proportion of the constant lost times becomes accordingly smaller; however, longer cycle times may increase vehicle delays in undersaturated intersections due to longer waiting times during the red phase. The offset is the phase difference between cycles for successive intersections that may give rise to a "green wave" along an artery. The specification of the offset accounts for the possible existence of vehicle queues.

Signal strategies. The stage specification is usually decided first, whereby combinatorial-optimization-based software tools may be used with significant benefit in the case of complex intersections. The splits, cycle times, and offset values for all intersections of a traffic network are called a signal plan. Fixed signal plans may be optimized off-line for specific times, such as the morning rush hour, by using optimization codes based on historical constant demands and turning rates for each traffic stream. They are activated every day at the corresponding time (clock-based). This basic approach to traffic signal operation may not be the most efficient, as demands are usually not constant. In addition, various exceptional events (such as fairs, sport events, concerts, and roadwork) may perturb the traffic conditions in nonpredictable ways.

To address the drawbacks of fixed plans, a number of modifications have been introduced, including the installation of (usually loop) traffic sensors and communication lines to a central control room to enable real-time (traffic-responsive) operations. For example, fixed signal plans may be modified in real time by appropriate control logic at the local intersection level, based on arriving sensor data. Possible modifications include the extension of a green phase (to allow the passage of detected vehicles) and the termination or even skipping of a stage (due to detected lack of demand). A second possibility is a library of fixed plans that may be created off-line, with each fixed plan corresponding to a typical network traffic condition instead of a specific time of day. Sensor data are analyzed in real time by rule-based control logic to decide on the currently prevailing traffic pattern and activate the corresponding fixed plan from the library.

More advanced traffic-responsive signal-control strategies run optimization codes in real time to calculate optimal signal plans based on arriving sensor data, without being constrained by a limited number of precalculated fixed plans. These strategies may involve short-term demand predictions to avoid myopic actions. A problem faced by model-based, optimal signal-control strategies is the real-time feasibility of the required calculations, leading to the dilemma of either using a realistic model, in which case optimization may have to be executed for each intersection separately or a network-wide optimization model that reflects a simplified description of the network traffic phenomena.

New signal strategies. Most available signal-control strategies operate reasonably well under nonsaturated traffic conditions, that is, when the traffic demand of each stream can be served during the corresponding green phase (waiting queues occur only during the red signals and are dissipated at the end of the following green signal). In contrast, saturated traffic conditions are characterized by increasing queues in some network links, which may block upstream intersections and lead to a significant degradation of the network's capacity, particularly if gridlocks are created around topological network cycles. Saturated traffic conditions are observed more frequently during peak hours in modern metropolitan areas, but available signal control strategies are not very efficient in protecting the network from generalized congestion due to limitations in the models, optimization methods, or real-time information provided (typically one sensor per network link). Research is under way for the development of new signal-control strategies that can operate efficiently under saturated traffic conditions, for example by intentionally holding back traffic (gating) to protect saturating network links.

An important component of most operational signal-control strategies is giving priority to public transport vehicles (PTV: buses and trams) at signal intersections. A typical approach involves sensors that detect the presence of a PTV in links and subsequently activate a specific control logic that modifies the running signal settings to enable the passage of the PTV without (or with minimum) delay. Public transport priority measures may perturb traffic conditions for the rest of the traffic flow, so the adjustable degree of prioritization is decided at a policy, rather than technical, level.

Freeway network control. Freeways and motorways originally were conceived to provide virtually unlimited mobility to road users, without the annoyance of flow interruptions by traffic lights. However, the rapid increase of traffic demand soon led to congestion, both recurrent (during rush hours) and nonrecurrent (due to incidents), with long queues, degraded infrastructure use, and reduced safety. At present, the freeway network infrastructure is underutilized on a daily basis due to the lack of efficient and comprehensive traffic-control systems.

The control measures that are typically used in freeway networks include (1) ramp metering, activated via installation of traffic lights at on-ramps or freeway interchanges, (2) link control that comprises a number of possibilities including lane control, variable speed limits, congestion warning, tidal (reversible) flow, and keep-lane instructions, and (3) driver information and guidance systems that use roadside variable message signs or two-way communication with equipped vehicles.

Ramp metering. Ramp metering is the most direct and efficient way to control freeway traffic. Various positive effects are achievable if ramp metering is appropriately applied, including increased throughput due to avoidance or reduction of congestion, increased volume due to avoidance of blocked off-ramps or freeway interchanges, use of possible reserve capacity on parallel arteries, efficient incident response, and improved traffic safety due to reduced congestion and safer merging.

Some recent studies have demonstrated that efficient ramp metering strategies may deliver spectacular improvements (50% reduction in time) in large-scale freeway networks or freeways around city centers. When designing or operating ramp metering systems, two aspects should be considered: (1) Vehicle queues at on-ramps should be limited to the ramp length to avoid interference with adjacent street traffic. (2) Ramp queues should be spaced equally at different ramps to avoid driver rerouting and user dissatisfaction.

Currently, there are 3000 metered ramps worldwide, the majority located in North America, with a growing number in Europe, Australia, and elsewhere. The first ramp metering systems applied precalculated fixed rates but were soon replaced by traffic-responsive control schemes employing the same kind of traffic sensors as in the urban signal control (mostly loops). Despite research advances and the expectation of high benefits from ubiquitous application of advanced ramp metering strategies, the current state of practice is dominated by local, frequently poorly designed or maintained systems. Operating network-wide ramp metering systems are rare and have been criticized as naively designed.

Link control. Link control systems may include one action or a combination of actions, including variable speed limit, changeable message signs, lane-control measures (for example, prohibited lane use upstream of heavily used on-ramps or incident locations), incident or congestion warning, and reversible flow lanes (tidal flow).

There are many freeway stretches, particularly in Germany, the Netherlands, and more recently the United Kingdom, using a selection of these measures. It is generally thought that such control measures lead to a homogenization of traffic flow and speed, reducing the risk of congestion at high traffic densities and increasing the freeway capacity. Dedicated evaluation studies show clear benefits in traffic safety (30% reduction of traffic accidents), but the impact on traffic flow efficiency (travel time reduction) remains inconclusive. More sophisticated control strategies than the currently applied look-up table or rule-based ones are likely to lead to an efficiency increase, without affecting traffic safety.

Dynamic route guidance and driver information systems. Travelers are frequently unaware of all route options available to them, so that even experienced travelers are unable to make appropriate route decisions. The overall network capacity may be underused, with some routes heavily congested and capacity reserves on alternative routes.

Route guidance and driver information systems (RGDIS) have been implemented in some cities to improve route choices and network use efficiency. The technologies used by RGDIS include radio services, variable message signs (VMS), or special in-car equipment. RGDIS radio services and variable message signs have been in use for more than 25 years, while use of in-vehicle equipment and two-way communication with traffic-control centers are still in their infancy.

Nonpredictive systems base the guidance messages provided to drivers on measures of prevailing travel conditions in a network. For predictive systems, the key issue is whether the network predictions account for the likely reaction of drivers to guidance, and the effect of these reactions on network conditions. Some system designs extrapolate prevailing conditions and base guidance on the extrapolations. This does not account for possible driver reactions to the guidance and may lead to a situation where the predictions and guidance are invalidated by these reactions. Systems that take account of driver response when generating guidance are more complex and require dynamic traffic network models adapted for this purpose. However, such capabilities are important only when the fraction of network drivers participating in the RGDIS is enough to affect network conditions significantly.

Most current RGDIS deployments are nonpredictive, although there are a few predictive systems that use extrapolation methods. Predictive systems that can account for driver response are still prototypes, although at least two software systems have been developed and partially deployed on an experimental basis.

Outlook. As in many other engineering disciplines, only a small portion of the significant methodological advancements have really been exploited in the field as yet. The major challenge in the coming decade is the deployment of advanced and efficient traffic-control strategies in the field, with particular focus on addressing traffic saturation congestion.

For background information *see* CONTROL SYSTEMS; HIGHWAY ENGINEERING; IMAGE PROCESSING;

MULTILEVEL CONTROL THEORY; OPTIMAL CONTROL THEORY; OPTIMIZATION; TRAFFIC-CONTROL SYSTEMS; TRANSPORTATION ENGINEERING in the McGraw-Hill Encyclopedia of Science & Technology.

Markos Papageorgiou; Moshe Ben-Akiva; Yibing Wang; Jon Bottom; Apostolos Kotsialos

Bibliography. A. Kotsialos et al., *ITS and Traffic Management*, in preparation; M. Papageorgiou et al., Review of road traffic control strategies, *Proc. IEEE*, 91:2043–2067, 2003.

Earthquake sensory perception in vertebrates

There are often reports in the popular press and folk legends of unusual vertebrate animal behavior before large earthquakes and tsunamis. Retrospective reports of unusual phenomena occurring before the 1995 Kobe (Japan) and the 1999 Izmit/Turkey and Taiwan earthquakes (magnitude 7) indicated that mammals became restless, irritated, nervous, and panicky, acted "crazy," and tried to escape from their cages. A possible cause of this odd behavior could be the animals' perceptions of various physical and chemical changes that occur in the environment before earthquakes occur. For example, the fracture of underground rock structures before an earthquake causes seismo-electromagnetic signals, and there were reports of malfunctioning home electrical appliances and observations of such seismo-electromagnetic signals before the Kobe, Izmit/Turkey, and Taiwan earthquakes. It is quite possible that animals detected these signals, as laboratory experiments involving the fracture of granite rock and the generation of static electric fields and silent electromagnetic pulses have caused disturbances in animals (caused them to awake from sleep and move around). However, changes in magnetic fields were not shown to have an effect.

Animal behavior before earthquakes and tsunamis. The 2004 Southeast Asia earthquake (magnitude 9) caused a huge tsunami that killed over a quarter of a million people. News reports indicated that elephants at Khao Lak in southern Thailand, 650 km from the epicenter of the earthquake, were excited at the time of the quake and started trumpeting (as if crying) but soon calmed down. They started wailing again 1 h later but could not be comforted.

Retrospective reports indicate that approximately 20% of children (ages 8–10) at the quake's epicenter woke up more than 1 min prior to the Kobe earthquake; this percentage decreased with increasing distance from the epicenter. Catfish reportedly moved violently in turmoil a day before the Kobe earthquake as well (see **table**).

Summary of retrospective reports of unusual animal behavior before earthquakes

Animal	Behavior
Mammals	
Humans	Wake from sleep; complain of headaches, nausea, giddiness, and dizziness; vomit; heart disorders, hysteria
Dogs	Bark loudly; whine; anxious; panic and bite owners; howl
Cats	Restless; mew as if suffering; take kittens outside; climb high trees; jerk and lay ears back; hide
Rats	Hide; fuss; panic; run on wires
Horses	Stamp; snort; tremble; jump; buck
Cows	Bellow; crowd together; run in panic
Pigs	Aggressive; bite one another; attempt to run away
Deer	Leave bush and forest; do not fear humans.
Sea lions	Swim in zigzags; fuss out of the water
Dolphins	Nervous; do not obey orders; leap out of water
Bats	Fly in the daytime
Birds	Fly out of epicenter areas
Chickens	Flap wings; shriek as if in terror; crow at midnight
Hens	Lay fewer eggs or double-yolk eggs
Ducks	Avoid entering water; act aggressively
Sparrows	Flutter down while flying
Seagulls	Fly inland; mew in sky
Parakeets	High-pitched chirping; flutter wings; fly at night
Reptiles	Come out of hibernation in winter
Crocodiles	Call; leave the water for land
Snakes	Swarm in bamboo clumps in summer
Turtles	Wake from hibernation; climb on others
Fish	Float and align in one direction; leap out of water; move violently; act as if in turmoil; die; sea fish swim up rivers; fishermen have larger fish catches; deep-sea fish appear near surface

Preearthquake physical and chemical changes. The Biophysics Institute of Academica Sinica and the U.S. Geological Survey International Symposia have considered several physical and chemical mechanisms as possible causes for animal perception of an impending earthquake.

Seismic P-waves before arrival of S-waves (tremor). Seismic P-waves (primary waves) are propagated at 8 km/s, and the S-waves (secondary waves) that cause tremors are propagated at 4 km/s. Hence, animals may sense the P-waves before a tremor.

Release of odors, gases, and aerosols. The formation of microcracks in underground rocks before an earthquake may release some natural gases, such as radon, methane, nitrogen and sulfur oxides, and hydrogen sulfides. Although large quantities of gases have not been detected over wide areas before large earthquakes, there have been some reports of odors similar to ozone or odors caused by air discharges.

Serotonin syndrome (caused by effects of the neurotransmitter serotonin in the lower middle brain), a condition that results in changes in the moods of both animals and humans, may be initiated by electrical charged aerosol stimuli that can emanate from the earthquake epicenter.

Light (infrared and far-infrared light). Nighttime earthquakes may be associated with light that is generated by an atmospheric discharge from an intense electric field (similar to lightning). Some laycitizens have argued that animals might detect this light, which is invisible to humans.

Minnow fish aligned perpendicularly to the direction of electric fields to reduce their body current.

Magnetic field changes. Animals may sense magnetic field changes before large earthquakes. Although laboratory experiments have not shown animals to respond to changes in magnetic field, animal navigation for homing and migration has been attributed, in part, to their detection of the Earth's magnetic field.

Electromagnetic pulses. Researchers in submarine communication 2 weeks before the 1989 Loma Prieta (California) earthquake observed unusual noise from electromagnetic waves at ultralow frequency. The electromagnetic waves peaked a week before the earthquake and increased again 1 h prior. Animals may be able to sense electric fields of electromagnetic waves. Chirps (electromagnetic pulses) appear as noise over a wide frequency range.

Laboratory reproduction of behavior. Similar animal behavior was observed in three different experiments conducted to study vertebrate perception of electromagnetic pulses. Each study used a different method to generate the electromagnetic pulses: (1) electric discharges using Van der Graaf (100–250 kV on a metal sphere) or Wimshurst generators, which also generate sound; (2) pulsed magnetic fields using pulsed current to a loop coil; (3) a pulsed electric field applied to parallel plate electrodes by switching the dc voltage on and off. In all three studies, animals acted surprised and displayed avoidance behavior in response to the electric field; that is, to reduce the effects of the electromagnetic pulses they reoriented their body direction. (The body is an antenna, and its electric field detection efficiency changes according to the body's direction.) Flies and insects rotated simply to find a less sensitive direction and then stayed still at the less sensitive site.

In a rock fracture experiment using a 500-ton compression machine, the behavior of mice, rats, and parakeets and the brain neurotransmitter levels in rats were reported to be unusual. Electric charges and acoustic emissions were detected at the surface of granite rocks, and electromagnetic pulses were also observed by an antenna located 1 m away from the rock during compression ahead of fracture.

Electromagnetic pulses and electric fields. Experiments involving the application of a magnetic field 1 million times more intense than Earth's magnetic field, B_{Earth}, using a Neomax (Nd-B-Fe alloy) permanent magnet generated no behavioral changes. The magnetic field changes observed before earthquakes, 0.00006 millitesla (mT), are one-thousandth of B_{Earth} even at the epicenter. However, the magnetic field corresponds to an electric field of 18 V/m of electromagnetic waves ($\sim$1 W/m^2), well above the threshold detection level of vertebrates.

Catfish, which rarely move in an aquarium, have been reported to suddenly move their whiskers and move their body violently (sometimes leaping out of water) when exposed to an electric field of 3–4 V/m. Ordinary fish align perpendicularly to the direction of electric fields of 10 V/m. The alignment reduces the antenna efficiency of fish and, thus, their body current (see **illustration**). Aquatic animals with extraordinarily sensitive electrosensory organs, such as catfish and sharks, use electric fields (0.05 mV/m) for capturing prey and for communication. Humans who are standing with shoes on can feel little electric fields. Humans can also detect electric fields of 10–15 V/m in water baths, with muscle convulsion beginning at levels above 20 V/m.

The cause of elephants' agitation before the 2004 Southeast Asia tsumami is still not clear. The acoustic (gravity) waves generated by tsunamis disturb the ionosphere 100 km above the ocean and generate electromagnetic waves at ultralow frequency; the ocean floor may generate electromagnetic waves by the stress changes caused by the tsunami. These electromagnetic signals might agitate animals as preseismic electromagnetic pulses do before an earthquake.

Animal prediction of earthquakes and tsunamis. It has been argued that earthquake prediction is impossible since an earthquake is caused by rock fracture and, as such, is a chaotic phenomenon. However, earthquakes are caused by movements of geological faults following the fracture of uneven rock structures. Fluids seep into the fractured rocks several days beforehand and lubricate the fault movement, enabling the earthquake. If one can observe such underground bedrock fractures from associated electromagnetic phenomena and animal behavior, earthquake prediction may be possible. Automatic monitoring of catfish and hamsters is currently being conducted (using a PC-based charge-coupled-device camera and motion-capture software together with electromagnetic measurements) to investigate the possibility of using this information to predict earthquakes.

For background information *see* BIOELECTROMAGNETICS; BIOMAGNETISM; EARTHQUAKE; ELECTRIC ORGAN (FISH); ELECTROMAGNETIC WAVE; SENSE ORGAN in the McGraw-Hill Encyclopedia of Science & Technology.

Motoji Ikeya

Bibliography. R. E. Buskirk, J. F. Evarden, and P. D. Andriese, *Abnormal Animal Behavior Prior to Earthquakes II*, USGS Proc. Conf. XI, Menlo Park, 1980; J. F. Evarden, *Abnormal Animal Behavior Prior to Earthquakes I*, USGS Proc. Conf. I, Menlo Park, 1976; M. Ikeya, *Earthquakes and Animals—From Folk Legends to Science*, World Scientific, 2004; T. Rikitake, *Prediction and Precursors of Major Earthquakes*, Terra Science, Tokyo, 2001; H. Tributsch, *When the Snakes Awake*, MIT Press, Cambridge, 1982; R. Sheldrake, *The Sense of Being Stared At*, Random House, London, 2003.

Electric power system security

A reliable supply and delivery of electric power in North America is vital to the economic security and quality of life in modern society. The vast interconnected North American grid has been called the world's largest and most complex machine. It has achieved high levels of reliability through the nature of its interconnection by pooling of reserves and other operational efficiencies. But the grid has a significant drawback: Under the right circumstances, problems occurring in one area can cascade out of control and affect large geographical regions, as was the case on August 14, 2003, when the largest blackout in the history of the North American grid affected 50 million people and caused major economic damage. Multiple root causes of this blackout were traced to failures of computer systems that are critical to the real-time operational management of the grid. While none of the events on August 14 were found to have a malicious origin, the potential impact of a cyber attack on critical systems was nevertheless demonstrated. *See* BLACKOUT PREVENTION.

The electric power industry is capable of dealing with component failures, extreme weather, or natural disasters. But the increasing reliance of the electric power industry on information technologies is introducing a new class of cyber vulnerabilities and potential threats to the electric power infrastructure. In August 2003, the North American Electric Reliability Council (NERC) created a temporary cyber security standard which was adopted by the electric utilities. The standard identifies a set of security requirements relative to the energy industry intended to reduce risks to the reliability of the bulk electric systems from any compromise of critical cyber assets. Subsequently, NERC undertook the development of permanent cyber security standards.

In the cyber security realm, adversarial groups can be characterized as unorganized or organized. Most threats to a utility are caused by unorganized adversaries such as vandals, hackers, or malicious software, or resulting from the actions of insiders, either deliberate or accidental. On the other hand, activities of more organized adversaries are planned and methodical, involving professional organizational support, and are supported with extensive funding. Other entities, such as terrorist or criminal groups, can pose either a structured or unstructured threat depending upon their level of sophistication and resource base. Structured adversaries could pose a substantial and sophisticated cyber threat to national interests beyond what industry is normally equipped to protect against. The key challenge is to maintain reliability in a new competitive framework and under likely threats that involve multiple, distributed, and simultaneous or cascading incidents, both accidental and deliberate.

Historical perspective and trends. Electric utilities built and operated their own private communications facilities to control the electrical power grid. These systems used unique proprietary protocols, and the control center computers were isolated from other computer networks. This provided a high degree of isolation and security, but for economic reasons the trend is toward the use of shared communication with public networks, open and commonly used protocols, and general-purpose operating systems whose many weaknesses are more widely known. Market forces and technology trends are making the power system more dependent on information systems and supporting communications networks.

The Internet is used to support market operations [for example, the Open Access Same-time Information System (OASIS), where wholesale transmission capacity is allocated through an Internet-based market system]. Therefore, Internet conditions can impact the availability of power system management and trading functions. As power control responsibilities shift away from the traditional asset owners of the power systems and become concentrated in a new class of Regional Transmission Organizations, the power supply becomes more dependent on their networks and systems. Similarly, interregional power coordination relies on a nationwide network of management information systems. Reverting to the traditional method of command and control dispatch during emergencies is increasingly problematic.

Unless adequate security measures are taken, hostile individuals or organizations could exploit Internet-based points of entry. There is a growing need for organizations to ensure that this access is sufficiently protected, both for their own systems as well as for their connected partners.

Consequences. Physical sabotage of critical components has been recognized as a major threat to the electricity infrastructure. In the future, the consequences of an adversarial attack or component failure will increase because more components will be operating with less reserve margin, resulting in a smaller single element, or collection of events, being required to disrupt the system. For example, as was observed in the Northeast blackout in 1965, a single failure triggered cascading failures in other components. The two widespread disturbances that occurred in the western United States power system in the summer of 1996 were also initiated by a confluence of relatively minor events that cascaded out of control. On August 14, 2003, no fewer than four independent root causes (inadequate system understanding, inadequate situational awareness, inadequate tree trimming, and inadequate reliability coordinator diagnostic support) needed to be in place in order to provide the right conditions for the cascading failure to become possible.

It is common to plan for a single-contingency event, referred to as the N-1 planning criteria. However, in virtually all major disturbances, either faulty control settings or equipment malfunction create multiple simultaneous events that initiate a cascading failure that leads to the blackout. These conditions can overwhelm the ability of the power system to recover.

Similarly, one of the characteristics of an information attack is its ability to create simultaneous disruptions at multiple points throughout the system. Thus, a single successful attack from anywhere could result in a widespread outage.

Reducing the risk. The dependence of electric power on computing systems and computers on electric power lies at the heart of many concerns about critical infrastructures. Security is sometimes either neglected or introduced as an afterthought rather than being incorporated as a core component in the development and deployment of these new technologies and applications. The utility industry needs to develop new guidelines, policies, and standards that can aid in the selection and implementation of cost-effective security measures.

A major challenge is how to plan and execute various forms of public and private-sector collaboration and information-sharing arrangements to assist in prevention of, and aid in recovery from, events that represent unacceptable risks. All stakeholders share a common interest in deterrence, intrusion detection, security countermeasures, graceful degradation, emergency backup, and rapid recovery. To facilitate this interaction, various critical infrastructure sectors have established Information Sharing and Analysis Centers (ISACs). The role of the ISAC is to be a focal point for gathering, analyzing, and disseminating information about vulnerabilities, threats, intrusions, and other security issues within the sector and with government. The electricity sector ISAC is managed by NERC.

The electric grid must also be protected from unauthorized intrusions that could affect the confidentiality, integrity, or availability of data, services, and physical assets of utilities. Particularly as it relates to real-time control systems, integrity and availability are often more important than confidentiality. The security of real-time control systems needs to be enhanced, using sound information security practices. There remain specific challenges to effectively address the security of control systems. The U.S. government is developing a roadmap for research to enhance the security of energy-sector control systems. The roadmap will build on existing government and industry efforts to provide a vision of the requirements and a time-specific action plan to better secure both legacy and emerging control systems in the energy sector.

For background information *see* COMPUTER SECURITY; ELECTRIC POWER SYSTEMS; INTERNET in the McGraw-Hill Encyclopedia of Science & Technology.

Jeffery E. Dagle

Bibliography. *An Approach to Action for the Electricity Sector*, Working Group Forum on Critical Infrastructure Protection, North American Electric Reliability Council, June 2001; J. E. Dagle et al., *Assessment of Information Assurance for the Utility Industry*, prepared for the Electric Power Research Institute by Battelle, PNWD-2393, December 1996; *Interim National Infrastructure Protection Plan*, U.S. Department of Homeland Security, February 2005; *National Strategy to Secure Cyberspace*, U.S. Department of Homeland Security, February 2003.

Electronic olfaction and taste systems

Electronic instruments that mimic the human senses, known as electronic noses and tongues, recently have been developed and used to evaluate the quality of coffee, to assess the odor emanating from packaging materials, and to diagnose diseases such as urinary tract infections. These devices use a common feature—a set (array) of sensors, ranging from a few to 32 or more components. The sensors are nondiscriminating gas detectors of various types, including polymer-based films, electrochemical probes, and thin films. Some use changes in electrical resistance as the signal produced in the sensing process, others produce a color change, while others operate by changing their resonating frequency. In general, these devices do not recognize a specific substance; instead, they detect a signature, a "fingerprint" of an odor, or a quality of what is being measured. The availability of pattern-recognition techniques for effective signal analysis has enabled the development of complex sensor arrays for artificial olfaction.

The history of artificial olfaction systems is brief, as most prototype devices and a few commercial ones did not meet success. The primary reason for these early failures was the perception that these devices should function as analytical instruments similar to a gas chromatograph or a mass spectrometer; that is, they should recognize and discriminate analytes (substances) of all sorts. Their application was limited to situations where calibration data were available for complex gas/liquid mixtures (for example, to distinguish the smell of fragrances or detect the type of bacteria present in an assay). A single electronic olfaction system cannot address all problems. Recently, the scope of these tools has been focused to very specific applications, such as medical, forensic, or food-quality control. The trend now is to use sensor arrays with a variety of sensor types in the electronic olfaction system, where the smaller the number and the higher the selectivity of the sensors in the array, the better the response of the system. In this way, signal redundancy and noise are reduced significantly.

Sensing elements for electronic noses. There are four prominent sensor technologies used for electronic noses: (1) resistive-type semiconducting oxide sensors, (2) sorption-based polymer/composite sensors, (3) piezoelectric crystal resonator-based detectors, and (4) optoelectronic (colorimetric) detectors. Gas sensing by semiconducting oxides, such as tin dioxide (SnO_2), tungsten trioxide (WO_3), and molybdenum trioxide (MoO_3), is possible due to the change in the electrical resistance of these crystalline materials when the analyte gas is adsorbed on their surfaces or if it reacts with them (**Figs. 1** and **2**). Similarly, the adsorption of gaseous species on conductive polymer (for example, polyaniline) films and conductive polymer composites (for example, polymer with carbon black) causes them to swell, changing the material's electrical conductivity. Piezoelectric crystal detectors typically use a quartz (α-SiO_2) microbalance (often in microcantilever designs) coated with

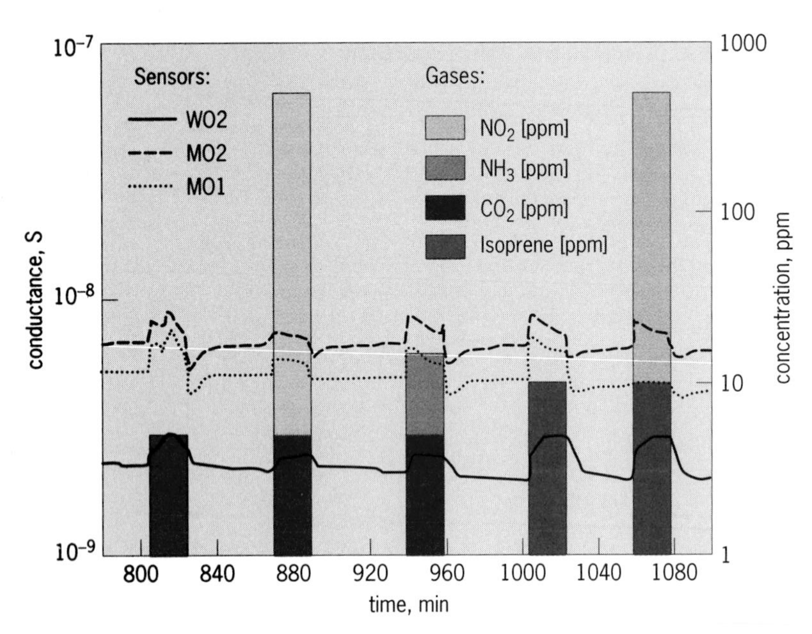

Fig. 1. **Trace gas analysis data from a three-sensor-array electronic nose. WO2 is a WO_3 sensor, and MO1 and MO2 are MoO_3 sensors.**

chemically sensitive materials. Mechanical oscillations of the miniaturized crystal slabs are excited by applying an alternating current (ac) to them. The sensing process involves measuring the change in the mass loading of the detector due to the interaction (binding or adsorption) of the gas with the chemically active coatings. This occurs indirectly by measuring the change in the oscillating frequency (resonance) of the piezoelectric crystal. Chemically responsive dyes are the active components of colorimetric sensors. Chemical interaction with these dyes causes a color change that can be observed optically.

Sensing elements for electronic tongues. Electronic tongues or taste sensors are artificial sensing systems for detecting substances in solutions. The different types of sensors are electrochemical detectors, infrared-based sensors, and mass-sensitive devices. The measurement principles for these technologies can vary; for example, there are potentiometric-, amperometric-, and conductimetric-type electrochemical sensors. Potentiometric devices measure the potential of electrodes in the solution across an ion-selective membrane (such as an ammonium electrode). Amperometric detectors are used to measure the current resulting from electron transfer reactions between the analyte and the electrode, but they are limited to measuring charged species. Conductimetric probes record changes in electrical conductivity upon the interaction of the analyte with the detector. In electronic tongue applications, these probes often use polymers (such as polypyrrole).

Optical techniques are favored when long-term stability of the sensor is desired. These can use dye/polymer membranes. Evanescent field absorption and surface plasmon resonance are among the reported optical spectroscopic techniques used in electronic taste sensing. Similarly, mass-sensitive devices, based on the microbalance principles previously described, and surface acoustic devices are frequently used to detect analytes in liquids. All of these sensors are used to measure ions or substances with low vapor pressure in solutions.

The applications of this technology are very wide, including quality monitoring in the food industry, monitoring the purity of water systems, sensing in household appliances, immunosensing, monitoring of microbial activity, and environmental control.

Pattern recognition techniques. While the sensors are the "heart" of artificial senses, the data analysis

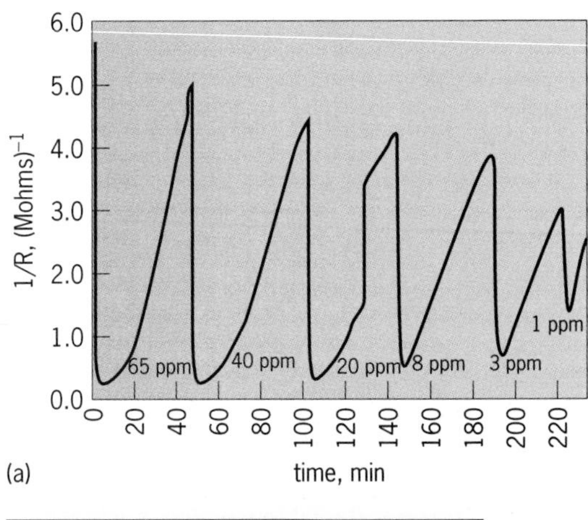

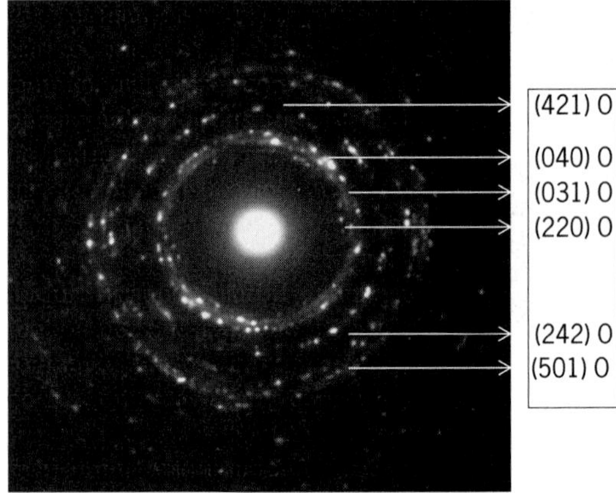

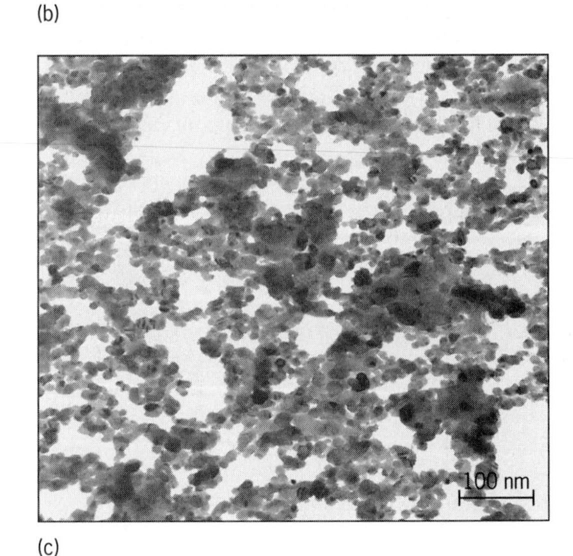

Fig. 2. **Selective sensing element based on WO_3 used to detect low concentrations of NO_2. (*a*) Sensing response, (*b*) structure (crystallography), and (*c*) morphology of nanocrystalline WO_3 films.**

and discrimination system (software) are their "brain." Multivariate analysis is most frequently used to extract useful information from the signals acquired by the sensors. Using statistical toolboxes, such as MATLAB® (mathematical software) functions, the "smell" or "taste" intensities/signatures from the samples are categorized and differentiated. Principal component analysis, artificial neural networks, and data clustering techniques are the preferred data analysis methods for separating the responses from different classes of species or qualities.

In a medical application of an electronic nose where the test subjects were individuals potentially suffering from lung cancer, multivariate data analysis was used to process the data obtained from exhaled breath (gas) samples in 4-liter disposable bags and analyzing them with an electronic nose system. According to the medical literature, alkanes and aromatic compounds (such as aniline) are volatile compounds found in the breath of patients suffering from lung cancer. In this study, complete and successful separation of those with lung cancer from the cancer-free population was achieved by measuring test subjects' breath signature and finding differences between healthy and ill subjects.

The efficiency of the chemometric methods and pattern recognition routines is rarely perfect, particularly in the absence of a precalibration process for the sensing system. Making artificial sensing systems sensitive, reliable, and reproducible for detecting odors and tastes is still a challenge. There is a lot of room for improvement in signal processing and analysis before we can substitute "sniffing" dogs and canaries with robotic counterparts using artificial senses.

Outlook. In this age of nanotechnology, the promise for new and improved materials is brighter than ever. Molecular-sized sensing probes with single-molecule sensitivities have been envisioned. Significant steps have been made toward synthesizing nanowire, nanotube, and nanocomposite systems that have high affinities for gas molecules and biomolecules, and that can respond rapidly with high specificity. In spite of the thermal and structural instability of nanomaterials, as well as environmental and health safety concerns, the future of smaller and better electronic olfaction systems is more promising than ever. Nanomanufacturing techniques have been developed to enable the "packaging" of the molecular sensors as tiny devices. Better signal processing routines and wireless data transmission capabilities have opened the way for portable sensor array systems that will monitor human health and safety and the environment. The future applications of these systems will be in implantable arrays of biosensing probes for monitoring metabolic functions, controlling drug release for therapeutic purposes, and controlling cell growth for tissue repair or cancer prevention. We are only at the beginning of a technological development that will shape the life of future generations.

For background information *see* BIOSENSOR; CHEMOMETRICS; FIBER-OPTIC SENSOR; MATHEMATICAL SOFTWARE; MICROSENSOR; NANOTECHNOLOGY; NEURAL NETWORK in the McGraw-Hill Encyclopedia of Science and Technology.

P. I. Gouma

Bibliography. J. W. Gardner et al., *Electronic Noses: Principles and Applications*, Oxford University Press, 1999; P. I. Gouma and G. Sberbeglieri (eds.), Novel materials and applications of electronic noses and tongues, special issue of *MRS Bull.*, 29(10), Materials Research Society, October 2004; E. Kress-Rogers (ed.), *Handbook of Biosensors and Electronic Noses: Medicine, Food and the Environment*, CRC Press, 1997.

Endoplasmic reticulum quality control

Endoplasmic reticulum quality control is a vital process in all eukaryotic cells that ensures that only properly folded proteins leave the endoplasmic reticulum for transport to their final destination. Under normal cellular conditions, molecular chaperones and lectins recognize aberrant or unfolded proteins and target them for destruction in a procedure known as endoplasmic reticulum–associated degradation. However, when genetic mutations, translational errors, or cell stress lead to a toxic accumulation of unfolded proteins in the endoplasmic reticulum, this degradation process may be insufficient to prevent aggregation or blockage of the endoplasmic reticulum transport machinery, and a signal transduction pathway termed the unfolded protein response is activated. This pathway leads to increased synthesis of folding and degradative enzymes while decreasing overall new protein synthesis, which allows the endoplasmic reticulum to better handle the load of unfolded protein. If endoplasmic reticulum stress cannot be abated, apoptotic pathways can be initiated, ultimately leading to cell death. Notably, much of the early research to discern the mechanisms of endoplasmic reticulum–associated degradation and the unfolded protein response was done in baker's yeast, *Saccharomyces cerevisiae*. All of the yeast components of these processes have homologs in mammals, although the mammalian systems are somewhat more complex.

Molecular chaperones. Molecular chaperones are proteins that assist in the folding and maturation of other proteins by binding to hydrophobic stretches of amino acids in unfolded proteins, which maintains their solubility and prevents aggregation. They include the heat shock proteins (Hsp's), which were first characterized by the increase in their transcription during cellular heat stress, though most have essential functions even under normal growth conditions. There are many classes of chaperones, including the well-studied Hsp70s, which often work in conjunction with other chaperones. Chaperones exist in most cellular compartments, and in addition to protein folding, some are involved in maintaining the solubility of newly synthesized proteins, transport across membranes, and in the recognition and subsequent destruction of misfolded proteins in the endoplasmic reticulum.

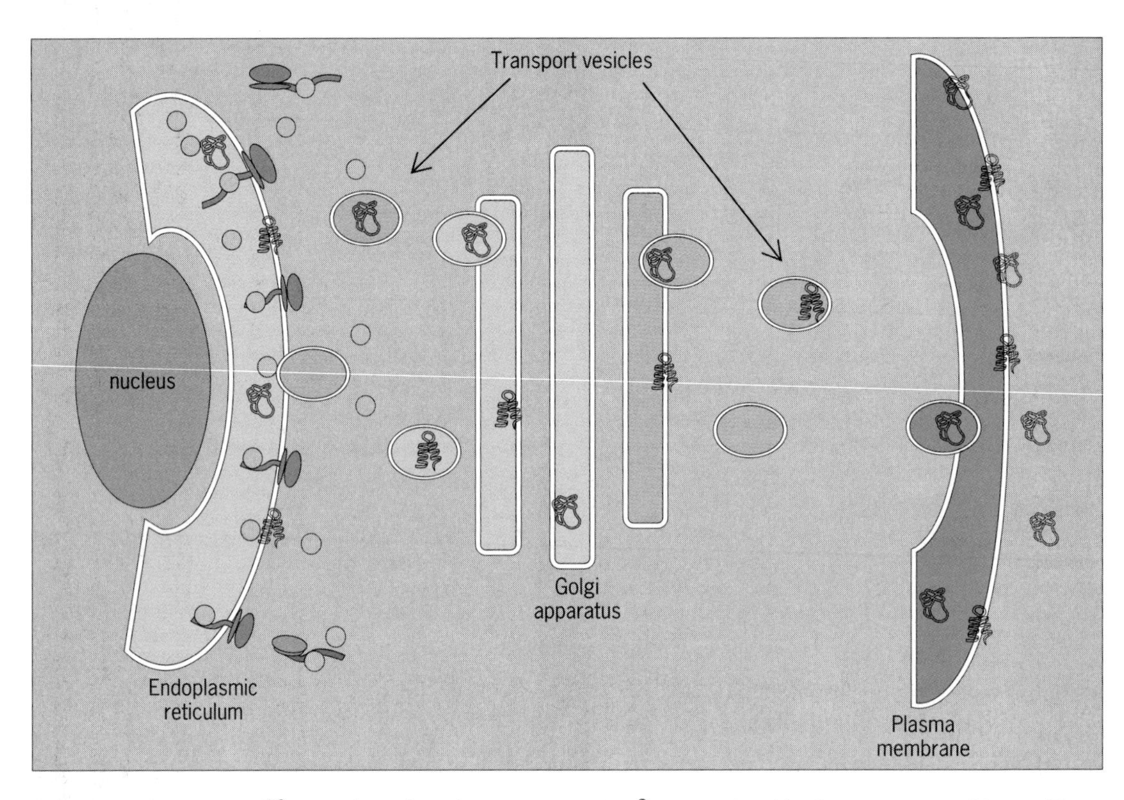

Soluble secretory proteins or membrane-bound secretory proteins are translated by ribosomes and inserted into the ER, either during or after translation. Molecular chaperones in the cytosol and ER bind to the translating proteins to keep them from aggregating and to assist other enzymes in the ER with modification and folding of the proteins. Once properly folded, the proteins can exit the ER for transport via vesicles to distal portions of the endomembrane system or for secretion from the cell.

Fig. 1. Secretory pathway of proteins.

Secretory pathway. The secretory pathway is the route by which proteins destined to be secreted from the cell, as well as those that reside on the cell surface or in organelles of the secretory pathway itself (the endoplasmic reticulum and Golgi apparatus), travel to their destinations (**Fig. 1**). Newly translated secretory proteins are recognized by the presence of an amino acid signal sequence that directs their transport into the endoplasmic reticulum, as they are being synthesized or soon after. Assisted by molecular chaperones in the cytosol and endoplasmic reticulum lumen, soluble proteins are transported into the endoplasmic reticulum via a complex of proteins known as the translocon. Integral membrane proteins move laterally through the translocon and are inserted into the endoplasmic reticulum membrane. Once in the endoplasmic reticulum, the proteins can be modified (for example, via removal of the signal sequence, disulfide bond formation, and glycosylation of asparagine residues), are folded, and (in some cases) are assembled into multiprotein complexes.

Genetic mutations, translation errors, and oxidative, heat, or other cell stresses can all contribute to protein misfolding. In addition, many wild-type proteins (that is, proteins without mutations), especially multispanning membrane proteins with many hydrophobic patches, are prone to misfolding. If allowed to accumulate in the endoplasmic reticulum, these "junk" proteins could aggregate, which would have deleterious consequences for the cell. Therefore, the endoplasmic reticulum modification and folding of secretory proteins is aided and monitored by molecular chaperones and lectins, a family of proteins that bind specific carbohydrates or carbohydrate-bound proteins (glycoproteins) and carbohydrate-bound lipids (glycolipids). This ensures that only properly folded proteins progress through the secretory pathway, while misfolded proteins are removed from the endoplasmic reticulum and destroyed.

Endoplasmic reticulum-associated degradation. When proteins enter the endoplasmic reticulum, resident lectins and chaperones bind to them.

Role of chaperones. Molecular chaperones bind both normal and aberrant polypeptides through recognition of exposed hydrophobic amino acid patches. Upon release, the substrate protein can fold; if it does not, chaperones may rebind. When a protein is properly folded, hydrophobic amino acid patches are mostly buried within the three-dimensional structure of the protein, and are therefore inaccessible to chaperones. However, if for any reason a protein does not fold properly, hydrophobic amino acids are exposed, and the chaperone cycle can repeat. This helps to hold misfolded protein in the endoplasmic reticulum and facilitates proper folding.

Role of lectins. Many secretory proteins are glycosylated, and concurrent with chaperone interaction, different endoplasmic reticulum lectins bind to proteins in a specific manner based on recognition

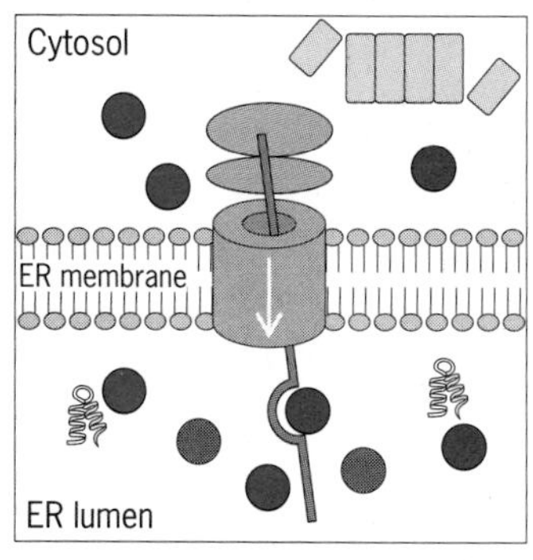

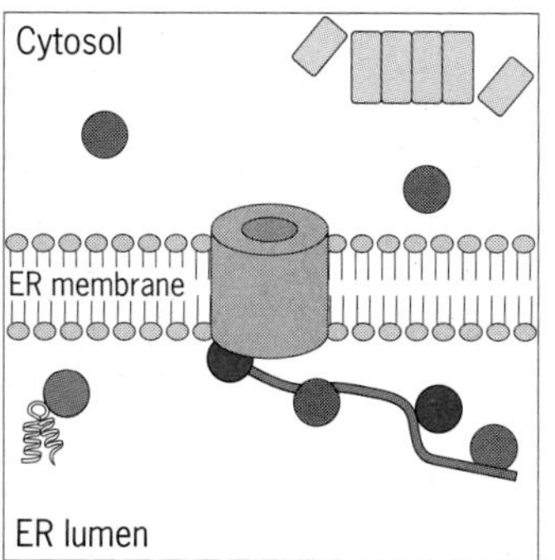

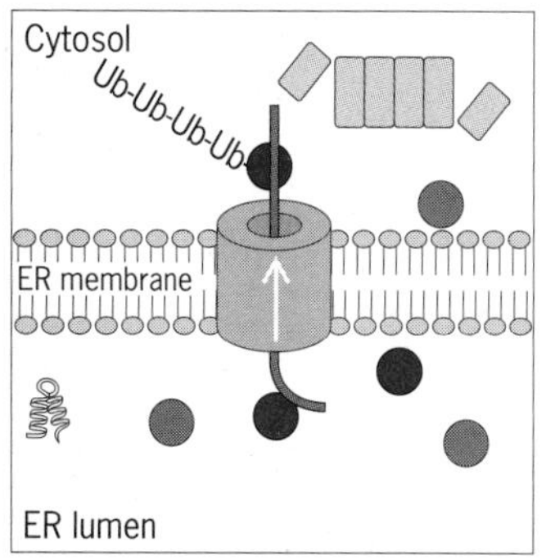

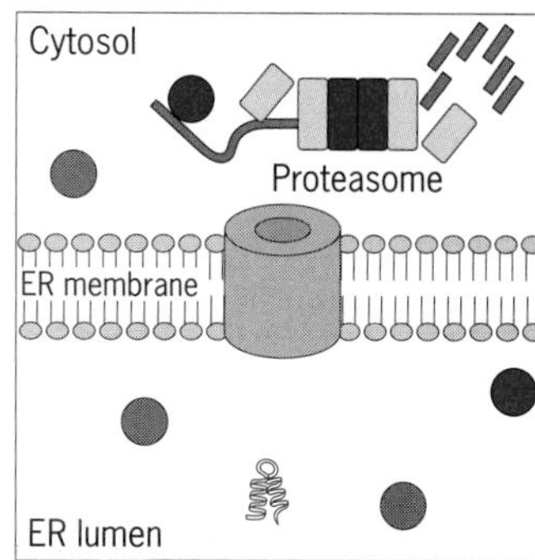

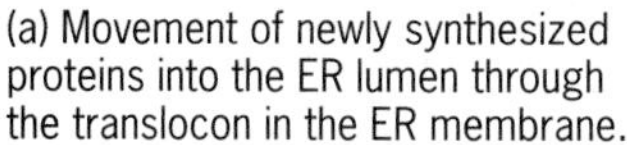

(a) Movement of newly synthesized proteins into the ER lumen through the translocon in the ER membrane.

(b) Aberrant protein is recognized as folding-incompetent by molecular chaperones and lectins.

(c) Retro-translocation of the defective protein from the ER back to the cytosol and polyubiquitination.

(d) Removal of polyubiquitin chain and degradation by the proteasome.

Fig. 2. Stages of ERAD (endoplasmic reticulum–associated degradation) for a soluble misfolded protein.

of their glycosylation state. This provides a timing mechanism for protein folding and maturation in the endoplasmic reticulum. Calnexin, a lectin that is anchored in the endoplasmic reticulum membrane, binds to immature glycoproteins to help prevent their premature exit from the endoplasmic reticulum. Over time, glucosidases trim the outer glucose residues from immature glycoproteins, which prevents recognition and rebinding by calnexin, and allows the protein to leave the endoplasmic reticulum. If held in the endoplasmic reticulum too long, due to prolonged chaperone interaction resulting from failure to fold, mannose residues can also be removed from the glycoprotein. Once the mannose residue is removed, a different lectin (called EDEM in mammals) can recognize and bind to the glycoprotein and then target the faulty protein for destruction.

Role of proteosomes. Once a protein has been recognized as hopelessly misfolded, it is transported back out of the endoplasmic reticulum through the translocon in a process called retrotranslocation (**Fig. 2**). Chaperones help to prevent aggregation of the misfolded proteins and help to deliver them to the translocon. As the protein emerges into the cytosol, ubiquitin chains are added to lysine residues via ubiquitin conjugating and ligating enzymes, some of which associate with the endoplasmic reticulum membrane. Polyubiquitin chains serve as a recognition signal for endoplasmic reticulum extraction and proteasome binding, and are removed immediately prior to degradation. Proteasomes are large, cytosolic proteases that comprise two multiprotein subunits: outer regulatory particles contain rings of protein subunits that use the energy from ATP hydrolysis to processively feed polypeptide chains into the central chamber, where proteolysis (protein degradation) occurs.

Variations. Not all endoplasmic reticulum-associated degradation substrates require the same set of factors for their degradation. For example, not all substrates are ubiquitinated, and for those that are, the ubiquitin conjugating enzymes can vary. Also, different chaperones are required for the degradation of soluble versus membrane proteins. Moreover, some substrates may exit the endoplasmic reticulum but are retrieved from the Golgi apparatus and degraded by endoplasmic reticulum-associated degradation, and some substrates may be extracted directly from the endoplasmic reticulum membrane without retrotranslocation through the endoplasmic reticulum translocon. The regulatory particle of the proteasome itself may be sufficient to extract some substrates from the endoplasmic reticulum, but other substrates require interaction with other factors. Thus, endoplasmic reticulum-associated degradation is not a "one-size-fits-all" process, but is very flexible for a wide range of substrate proteins.

Unfolded protein response. When the protein folding capacity of the endoplasmic reticulum is exceeded by an accumulation of unfolded proteins, the unfolded protein response is induced. The current model from work in yeast suggests that Ire1p, a transmembrane protein kinase/endoribonuclease, serves as the sensor of unfolded proteins in the endoplasmic reticulum. (**Fig. 3**). Normally, the endoplasmic reticulum-lumenal Hsp70 chaperone BiP binds to Ire1p and keeps it inactive. But when unfolded protein levels increase, BiP is recruited from Ire1p to the unfolded proteins, allowing molecules of Ire1p to dimerize (chemically unite to form one molecule with two identical subunits) and become activated by autophosphorylation (process by which a protein kinase adds a phosphate group to specific amino acid residues on itself). The active endonuclease removes an inhibitory intron from the messenger ribonucleic acid (mRNA) that encodes the transcription factor Hac1p. Translation of the *HAC1* mRNA then proceeds to produce active Hac1 protein. Hac1p enters the nucleus and binds to sequences of deoxyribonucleic acid (DNA) known as UPR elements (UPREs) to increase transcription of UPR target genes, many of which are involved in endoplasmic reticulum-associated degradation. UPR-induced genes include chaperones and other protein folding enzymes, components of the ubiquitin-proteasome pathway, as well as genes that increase endoplasmic reticulum volume through additional membrane synthesis.

This increase in the folding and degradative capabilities in the endoplasmic reticulum is combined with a decrease in overall new protein synthesis to provide the cell with a means to clear the endo-

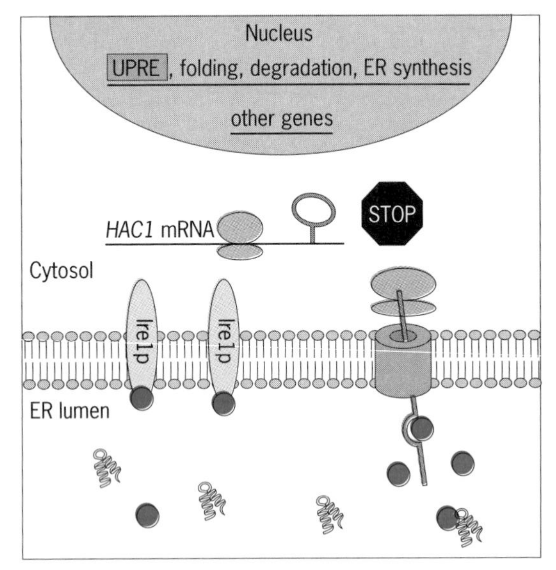

(a) No ER stress. Molecular chaperones are bound to Ire1p as well as folding polypeptides. An intron in the *HAC1* mRNA prevents translation.

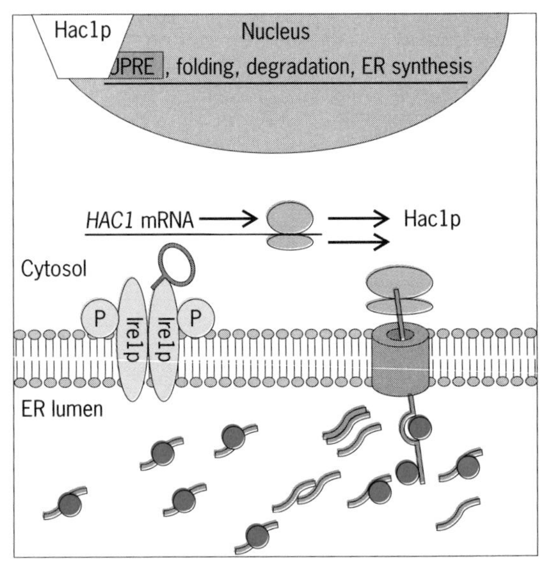

(b) ER stress caused by accumulation of unfolded proteins. Molecular chaperones are pulled away from Ire1p, allowing its dimerization and activation by autophosphorylation, (P). The active Ire1p removes the intron from the *HAC1* mRNA, and Hac1p protein is translated and enters the nucleus. There it induces transcription of folding and degradative enzymes. Transcription of other genes is decreased.

Fig. 3. Unfolded protein response in the yeast *Saccharomyces cerevisiae*.

plasmic reticulum of unfolded protein. When this is not sufficient, cell death can result in both yeast and mammals. In mammalian cells, for example, activated Ire1p recruits a kinase to the endoplasmic reticulum membrane that then initiates cell death by apoptosis. Moreover, PERK, an additional inducer of the mammalian UPR, inhibits cell cycle progression in response to prolonged endoplasmic reticulum stress.

Role in disease and normal cell physiology. Many diseases result from protein misfolding (see **table**). Often the misfolded protein is inactive, and its loss can cause the disease, as observed in α_1-antitrypsin-associated liver disease. In some of these diseases, such as cystic fibrosis, the misfolded mutant protein (cystic fibrosis transmembrane conductance regulator; CFTR) still retains partial function, but is held in the endoplasmic reticulum and degraded by endoplasmic reticulum-associated degradation so that it never reaches its target location, the plasma membrane. In other cases, such as Alzheimer's disease, the misfolded protein is not recognized as an endoplasmic reticulum-associated degradation substrate, and aggregates of the misfolded protein accumulate in the cell. Inhibition of endoplasmic reticulum quality control can also lead to ER stress. For example, in Huntington's disease, expanded polyglutamine repeat-containing protein aggregates in the cytosol of neurons, which strongly inhibits the proteasome. Consequently, misfolded protein in the endoplasmic reticulum cannot be removed and degraded, leading to endoplasmic reticulum stress and neuronal cell death. Similarly, prion diseases such as Creutzfeldt-Jakob disease result from misfolding and aggregation of a cellular protein. These aggregates are particularly resistant to degradation by the proteasome, but they also sequester chaperones and components of the ubiquitin-proteasome pathway, thereby lowering their activity in the cell. This exacerbates the toxicity of the aggregates themselves.

While destruction of potentially harmful proteins is the primary function of the endoplasmic reticulum-associated degradation machinery, some normal (correctly folded), active proteins are

Select list of disease-causing endoplasmic reticulum–associated degradation substrates

Endoplasmic reticulum–associated degradation substrate	Resulting disease
Cystic fibrosis transmembrane conductance regulator (CFTR)	Cystic fibrosis
α_1-Antitrypsin	Liver cirrhosis, emphysema
Insulin receptor	Diabetes mellitus (arising from the kidney)
Aquaporin-2	Nephrogenic diabetes insipidus
Tyrosinase	Albinism
HMG-CoA reductase	Heart disease
Factor VIII	Hemophilia A
LDL receptor	Hypercholesterolemia
Type I procollagen	Osteogenesis imperfecta
β-Amyloid	Alzheimer's disease (β-amyloid toxicity)
Apolipoprotein B	β-Lipoproteinemia
XBP-1	Bipolar disorder
β-Hexosaminidase	Tay-Sachs disease

degraded by this system as well. All three of the known examples are enzymes involved in lipid metabolism or regulation. Endoplasmic reticulum-associated degradation provides a quick way to destroy metabolic enzymes when their function is no longer needed, and more examples in which endoplasmic reticulum-associated degradation regulates a physiological pathway will likely be discovered in the future.

Research goals. Current research endeavors aim to identify all of the factors involved in the folding and degradation of a larger collection of endoplasmic reticulum-associated degradation substrates. Hopefully this will allow generalized predictions for the substrate recognition and degradation pathways of different classes of substrates, which in turn will promote the manipulation of the endoplasmic reticulum-associated degradation machinery to treat diseases caused by endoplasmic reticulum-associated degradation substrates.

For background information *see* CELL (BIOLOGY); ENDOPLASMIC RETICULUM; GENE; GOLGI APPARATUS; LECTINS; PROTEIN; RIBOSOMES; SECRETION in the McGraw-Hill Encyclopedia of Science & Technology.

Jennifer L. Goeckeler; Jeffrey L. Brodsky

Bibliography. A. Ahner and J. L. Brodsky, Checkpoints in ER-associated degradation: Excuse me, which way to the proteasome?, *Trends Cell Biol.*, 14(9):474–478, 2004; M.-J. Gething (ed.), *Guidebook to Molecular Chaperones and Protein-Folding Catalysts*, Oxford University Press (Sambrook and Tooze Publications), 1997; A. A. McCracken and J. L. Brodsky, Evolving questions and paradigm shifts in endoplasmic-reticulum-associated degradation (ERAD), *BioEssays*, 25:868–877, 2003; J. Rothblatt, P. Novick, and T. H. Stevens (eds.), *Guidebook to the Secretory Pathway*, Oxford University Press (Sambrook and Tooze Publications), 1994; B. Tsai, Y. Ye, and T. A. Rapoport, Retro-translocation of proteins from the endoplasmic reticulum into the cytosol, *Nat. Rev. Mol. Cell Biol.*, 3:246–255, 2002.

Energy-efficient motors

Electric motors are the major user of electricity, consuming 25% of all electricity in the United States. A survey by the U.S. Department of Energy (DOE) showed that electric motors consume as much as 63% of the total industrial electricity and that present technology is available to reduce this electric consumption by as much as 18%. Follow-up studies showed that actual savings were approximately 36% for companies that followed best-practices guidelines.

The importance of the efficiency of an electric motor can be understood from consideration of its life-cycle costs. The initial purchase price of an alternating-current (AC) motor is very low, around 2% of its total life-cycle cost. A small amount is added for cost of maintenance, but almost 97% of the total life-cycle cost is for electricity during operation (**Fig. 1**). The additional cost needed to make a motor more efficient will result in significant savings over its long lifetime, particularly since the costs of electricity are increasing.

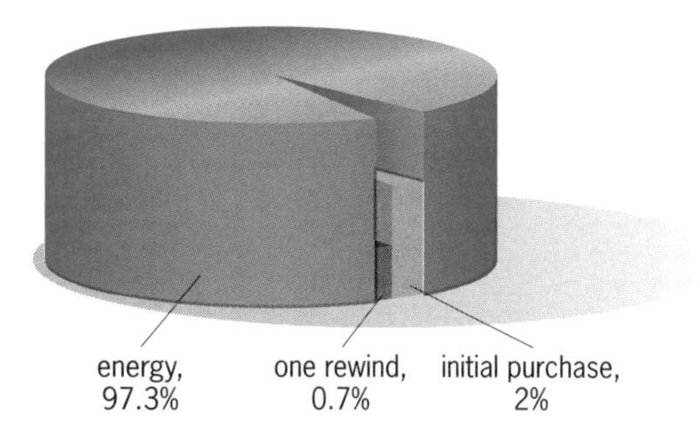

Fig. 1. Life-cycle costs of an alternating-current electric motor.

Types of electric motors. Industrial motors operate from either alternating current (AC) or direct current (DC). Three-phase AC induction motors are the most common designs used in industry (**Fig. 2**), from fractional horsepower (FHP) sizes through integral horsepower ratings of several thousand horsepower (1 hp = 0.746 kW).

Measurement of efficiency. Efficiency is a ratio of the amount of energy consumed for the amount of

Fig. 2. Alternating-current induction motors. These are typical totally enclosed, fan-cooled, premium-efficiency motors. (*a*) 10-horsepower (7.5-kW) motor. (*b*) 800-horsepower (600-kW) motor.

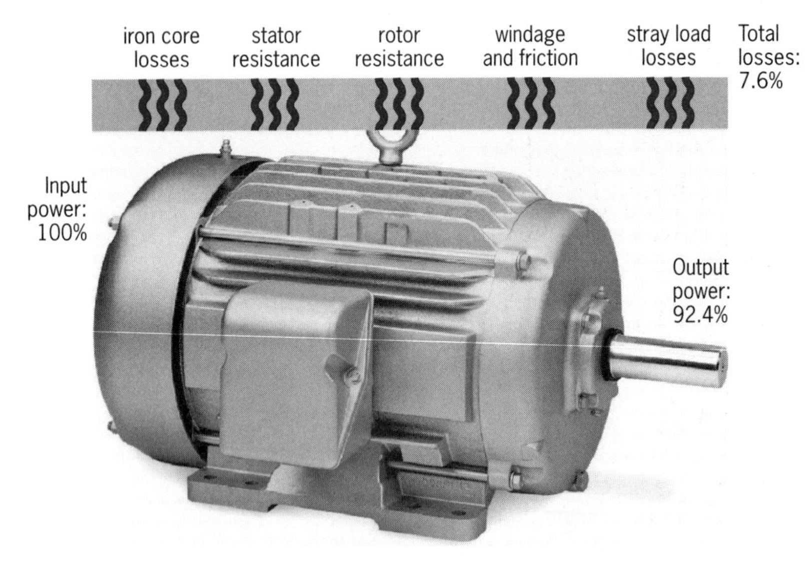

Fig. 3. Efficiency and losses in an alternating-current electric motor.

work produced (power in ÷ power out). The motor's losses are the difference between input power and output power (**Fig. 3**), and can be grouped into five categories:

1. Iron core losses: magnetic losses in the core laminations, hysteresis, and eddy current losses.

2. Stator resistance: current losses in the stator windings.

3. Rotor resistance: current losses in the rotor bars and end rings.

4. Windage and friction: mechanical drag in bearings and cooling fans.

5. Stray load losses: magnetic transfer loss in the air gap between the stator and rotor.

Improvements resulting in increased efficiency. Over the last 20 years, development and refinement of motor designs have reduced internal losses, producing premium efficient levels. The primary advance is the development of better electrical-grade steel. Lamination coatings have evolved from basic organic coatings to various inorganic-organic combination configurations and recently to oxide coatings. These coatings insulate adjacent lamination plates from one another to reduce losses. Actual losses have gone from 4–5 watts per pound (9–11 W/kg) of steel to less than 2 W/lb (4.4 W/kg).

In severe-duty and critical-service specifications for motors rated over 500 horsepower (373 kW), inorganic core plate is specified for low electrical losses. These large motors are commonly rebuilt. High-grade lamination material provides good resistance to degrading during any burnout and rewind process. This is important for motor efficiency because damage to steel laminations during an improperly performed motor rewind causes increased core losses. The **table** illustrates the effect of increased core loss on a 50-hp (37-kW), two-pole, open, drip-proof (ODP) motor. If the rewind were incorrectly performed, the operating costs of the poorly rewound motor would quickly exceed the cost of the rewind.

Efficiency standards. The high efficiency levels of today's motors have been developed over the last 25 years. Several manufacturers introduced "premium"-efficient motors in the early 1980s, but at that time there were no industry standards for "high-efficiency" motors.

In October 1997, the Energy Policy Act of 1992 (EPAct) took effect for general-purpose three-phase motors of 1–200 hp (0.75–150 kW). Any EPAct motor sold in the United States must comply with minimum nominal-efficiency, testing, and labeling standards. Similar minimum-efficiency performance standards (MEPS) are being developed and adopted throughout the world, mostly by government energy organizations.

In 1996, the Consortium for Energy Efficiency (CEE) established premium-efficiency guidelines used by many utilities for rebate programs. These guidelines were harmonized with those of the National Electrical Manufacturers Association (NEMA) in 2001, creating a NEMA Premium® efficient motor standard. NEMA sets the performance and dimensional standards for electric motors for the United States.

Benefits of premium-efficient motors. Premium-efficient motors are constructed with additional active material (laminations of higher electrical quality and more copper wire) to reduce internal losses and increase efficiency. Manufacturing tolerances and practices are held to tighter levels. Typically, vibration levels are lower, providing increased bearing life. Electric utilities in much of the United States offer rebates for premium-efficient motors and adjustable-speed drives.

Adjustable-speed operation. Most electric motors are started by direct connection to the electricity

Effect of increased core loss on motor operating cost and insulation life*

Core loss increase		Increase in annual operating cost as percentage of rewind cost	Temperature rise, °C (°F)	Approximate decrease in insulation life, percent
Percent	Watts			
50	515	28	7(13)	14
100	1030	55	14(26)	24
150	1545	83	29(39)	38
200	2060	110	29(51)	62

* For a 50-horsepower (37-kW), two-pole, open, drip-proof (ODP) motor.
SOURCE: D. Montgomery, Testing rewinds to avoid motor efficiency degradation, *Energy Eng.*, 86(3):24–40, 1989.

supply (across the line). Larger motors might be started using specialized methods such as starting by connecting only part of the winding or with special (wye-delta) connections of the stator. Solid-state starters can increase the motor voltage gradually (ramp up) to move the load and then reduce the voltage slightly. Today it is common to use an adjustable-speed drive (ASD, also known as adjustable-frequency drive or AFD) to change the motor's speed. The majority of ASD applications are for pumps and fans, presenting the motor with a variable torque load. Energy savings of over 50% are possible on these loads by eliminating pump valves and fan dampers and adjusting the flow using the ASD. If constant-torque loads (such as conveyors) are present, most premium-efficient motors are capable of at least a 10:1 constant torque speed range (CTSR). Many premium motors are supplied with a specialized insulation system, utilizing magnet wire, specifically designed to withstand voltage spikes caused by the drives.

Advantages of motor replacement. The recent doubling of electric costs has reinforced the importance of energy savings. Rewinding old motors is costly, and there is potential to cut both energy consumption and reduce the need for expensive maintenance by replacing older motors with newer, more energy-efficient models. Rewinding a motor can cost nearly as much as purchasing a new one, with the same amount of labor, and a new, energy-efficient motor continues to save money in reduced energy costs throughout its life.

Government can also provide incentives for motor replacement. For example, a California cash incentive program encourages companies to put less pressure on the energy grid through lower energy consumption. The rebate reduces the time required to realize a return on investment on new motors, typically to 6–7 months; a combination of the rebate and the ongoing energy savings of the motors can allow the investment to "pay for itself" in a short time.

Another advantage of new motors derives from the tendency of older motors with open, drip-proof frames to fill with dirt, dust, and water. This problem is especially serious in environments where there is a lot of salt and moisture in the air, as may be the case for plants located near the ocean. The maintenance costs of new motors can be cut by 50% or more since they are enclosed and do not need to be taken out of service every few years to be cleaned and refurbished. The nonenergy benefits of motor management include important factors such as reduced plant downtime due to the reduced need for motor maintenance. Many users find that a reduction in plant downtime is worth 10 times as much as the energy savings.

For background information *see* ALTERNATING-CURRENT MOTOR; EDDY CURRENT; HYSTERESIS; INDUCTION MOTOR; MAGNETIC MATERIALS; MOTOR in the McGraw-Hill Encyclopedia of Science & Technology. John Malinowski

Bibliography. P. E. Cowern, *The Cowern Papers: A Compilation of White Papers of Frequently Asked Motor Questions*, Baldor Electric Company, Fort Smith, AR, 2002; *Motor Planning Kit, Version 2.0: Strategies, Tools, and Resources for Developing a Comprehensive Motor Management Plan*, Consortium for Energy Efficiency—Motor Decisions Matter, Boston, 2005; S. Nadel et al., *Energy-Efficient Motor Systems: A Handbook on Technology, Program, and Policy Opportunities*, 2d ed., American Council for an Energy-Efficient Economy, Washington, DC, 2002; F. Parasiliti and P. Bertoldi (eds.), *Energy Efficiency in Motor Driven Systems*, Springer-Verlag, Berlin, 2003; *Recommended Practice for the Repair of Rotating Electrical Apparatus*, ANSA/EASA AR100-2001, Electrical Apparatus Service Association, St. Louis, 2001; *U.S. Industrial Motor Systems Market Opportunities Assessment*, U.S. Department of Energy, Office of Industrial Technologies, DOE/GO-10202-1661, December 2002.

Enterprise transformation

Strategic management of contemporary enterprises, whether they are in the private or public sectors, poses numerous challenges. Growing an enterprise's market or community impact perhaps reflected in increased sales and profits or services to constituencies, is a continual challenge. Evolving the enterprise's value proposition (that is, the value of what it offers consumers), the foundation for growth, is also an important challenge. Associated challenges include achieving focus, implementing change, addressing uncertainty, sharing knowledge, and managing time.

There is a wide variety of ways to address these challenges, and numerous practices and case studies to draw upon. Process improvements and other incremental changes may be sufficient for a given enterprise. However, in some cases, enterprise transformation may be necessary—that is, fundamental changes in terms of relationships to markets, product and service offerings, market perceptions, cost pressures, and the processes associated with these. Understanding and supporting transformation can be critical for enterprises to accurately assess their current situations, successfully pursue their goals, and survive the fundamental changes implied.

Transformation framework. A framework for understanding the nature of transformation is shown in the **illustration**. The goal or ends pursued tend to differentiate initiatives. The approach or means adopted for transformation pursuits relates to both the goals pursued and the nature and competencies of the enterprise. The ends and means, as well as the extent of integration of the enterprise, influence the scope of transformation.

The ends of transformation can range from greater cost efficiencies, to enhanced market perceptions, to new product and service offerings, to fundamental changes of markets. The means can range from upgrading people's skills, to redesigning business practices, to significant infusions of technology, to fundamental changes of strategy. The scope of

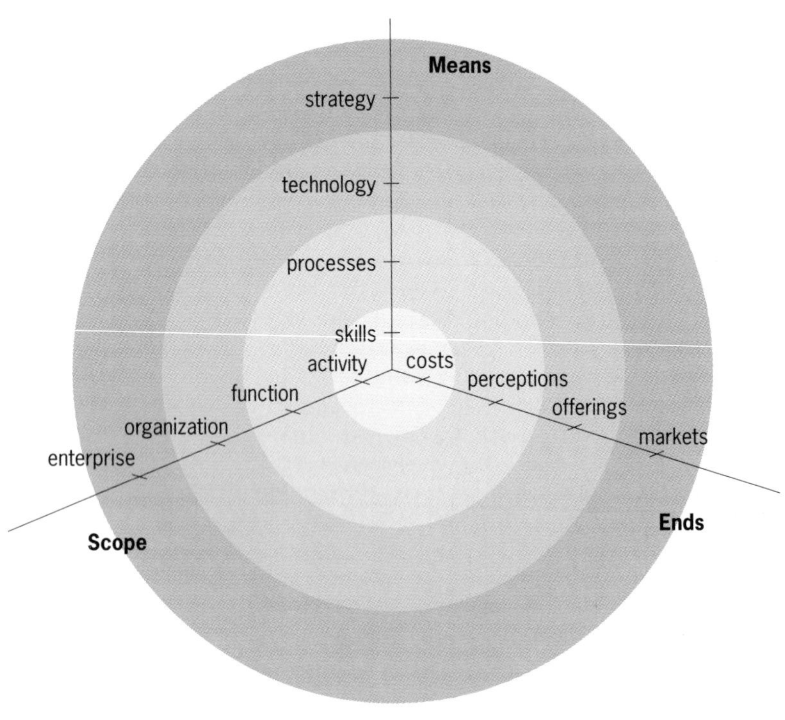

Transformation framework.

transformation can range from work activities, to business functions, to overall organizations, to the enterprise as a whole.

This framework provides a useful categorization for a broad range of enterprise transformation case studies. Considering transformation of markets, for example, a book retailer may leverage information technology to redefine book buying from a "bricks and mortar"-based activity to an online activity, while a nationwide "superstore" may leverage information technology to redefine retail supply and demand. Transformation of offerings includes 24-hour cable news broadcasters redefining news delivery, electronics producers moving from battery eliminators to radios to cell phones, package delivery companies expanding their services to provide integrated supply chain management services, and computer manufacturers moving from an emphasis on selling computers to an emphasis on selling integrated technology services. The many instances of transforming business operations include the merging of aircraft companies and numerous home products companies.

The costs and risks of transformation increase as the endeavor moves farther from the center in the illustration. Initiatives focused on the center will typically involve well-known and mature methods and tools from industrial engineering and operations management. In contrast, initiatives toward the perimeter will often require substantial changes of products, services, or channels, as well as associated large financial investments.

Successful transformations in the outer band of the illustration are likely to require significant investments in the inner bands also. In general, any level of transformation requires consideration of all subordinate levels. Thus, for example, successfully changing the market's perceptions of an enterprise's offerings is likely to also require enhanced operational excellence to underpin the new image being sought. As another illustration, significant changes of strategies often require new processes for decision making, such as for research and development investments.

Perspectives on transformation. There are basically four alternative perspectives that tend to instigate transformation:

1. Market and/or technology opportunities—the lure of greater success prompts transformation initiatives.

2. Market and/or technology threats—the danger of anticipated failure prompts transformation initiatives.

3. Competitors' initiatives—others' transformation initiatives prompt recognition that transformation is necessary to keep pace.

4. Enterprise crises—steadily declining market performance, cash flow problems, and the like prompt recognition that transformation is necessary to survive.

The perspectives driven by external opportunities and threats often encourage managers to pursue transformation long before it is forced on them, increasing the chances that resources will be available to invest in these pursuits, leveraging internal strengths, and mitigating internal weaknesses. In contrast, the perspectives driven by external competitors' initiatives and internally caused crises typically lead to very late recognition of the need for transformation, which is often forced on management by corporate parents, equity markets, or other investors. Such reactive perspectives on transformation often lead to failures.

Approaches to transformation. Transformation initiatives driven by external opportunities and threats tend to adopt strategy-oriented approaches, such as a focus on:

1. Markets targeted (for example, pursuing global markets such as emerging markets, or pursuing vertical markets such as aerospace and defense).

2. Market channels employed (for example, adding Web-based sales of products and services such as automobiles, consumer electronics, and computers).

3. Value proposition (for example, moving from selling unbundled products and services to providing integrated solutions for information technology management).

4. Offerings provided (for example, changing the products and services provided, perhaps by private labeling of outsourced products and focusing on support services).

On the other hand, transformation initiatives driven by competitors' initiatives and internal crises tend to adopt operations-oriented approaches including:

1. Supply chain restructuring (for example, simplifying supply chains, negotiating just-in-time relationships, developing collaborative information systems).

2. Outsourcing and offshoring (for example, contracting out manufacturing or information technology support; and employing low-wage, highly skilled labor from other countries).

3. Process standardization (for example, enterprise-wide standardization of processes for product and process development, research and development, finance, and personnel).

4. Process reengineering (for example, identification, design, and deployment of value-driven processes; and identification and elimination of non-value creating activities).

5. Web-enabled processes (for example, online self-support systems for customer relationship management or inventory management).

No significant transformation initiative can rely solely on either of these sets of approaches. Strategy-oriented initiatives must eventually include serious attention to operations. Similarly, operations-oriented initiatives must at least validate existing strategies, or the enterprise runs the risk of becoming very good at something it should not be doing at all.

Recognition of perspectives and adoption of appropriate stategies should be determined by a clear understanding of the current and emerging situations faced by the enterprise. Delusions about the current situation can undermine strategic thinking about opportunities, threats, competitors, and crises, resulting in the adoption of approaches that may not match the underlying needs of the enterprise.

Enterprise solutions. Many approaches to transformation, especially those that are operations-oriented, are pursued in the context of information technology solutions such as enterprise resource planning (ERP), customer relationship management (CRM), supply chain management (SCM), and sales force automation (SFA). These solutions employ information technology (IT) to automate and otherwise support work processes and workflow across the enterprise, typically requiring process mapping, database standardization and, in many cases, integration of legacy IT systems. The large investments required to deploy these types of solutions need to be understood in the context of how they help enterprises to address essential challenges and, in many cases, fundamentally transform. Implementing these solutions is only a beginning, as many enterprises have discovered after the fact.

The architecting of such enterprise information systems should reflect the enterprise as a system, or system of systems. Integration across the component systems should consider the primary value streams of the enterprise—that is, the processes whereby value is added to products and services, as perceived by the marketplace (customers or constituencies). Particular attention should be paid to how information and knowledge are shared and support creation and execution of programs of action that enhance value.

Ideally, these types of enterprise solutions are viewed as just a piece of the transformation puzzle, albeit a large one. Addressing and resolving the human and organizational issues associated with these solutions are often the thorniest part of the road to success. Understanding work processes, both as they are and should be, is usually central. Training and aiding of personnel at all levels also tends to be very important, as does alignment of incentives and rewards with new processes.

William B. Rouse

Bibliography. J. C. Collins, *Good to Great: Why Some Companies Make the Leap and Others Don't*, Harper Business, New York, 2001; ERP RIP?, *The Economist*, June 24, 1999; M. Hammer and J. Champy, *Reengineering the Corporation: A Manifesto for Business Revolution*, Harper Business, New York, 1993; W. B. Rouse, *Don't Jump to Solutions: Thirteen Delusions that Undermine Strategic Thinking*, Jossey-Bass, San Francisco, 1998; W. B. Rouse, Enterprises as systems: Essential challenges and approaches to transformation, *Syst. Eng.*, 8(2), 2005; W. B. Rouse, *Essential Challenges of Strategic Management*, Wiley, New York, 2001; A. J. Slywotsky, *Value Migration: How to Think Several Moves Ahead of the Competition*, Harvard Business School Press, Boston, 1996; J. P. Womack and D. T. Jones, *Lean Thinking: Banish Waste and Create Wealth in Your Corporation*, Simon & Schuster, New York, 1996.

Environmental engineering informatics

Environmental engineering intersects both the structure and function of environmental systems (natural and human-made) and the myriad issues that arise from human intervention in those systems. Traditionally, environmental engineers have dealt with problems such as the management of pond and stream quality, contamination from industrial or agricultural waste, fish kills, and the development of effective mitigation strategies for air, soil, and water pollution, to name some. For environmental engineers, these problems are not just technically complex, but also must take into account the social demand for the resources involved, as well as zoning, regulatory, and legal constraints.

Environmental engineering has become more complicated with the increased threat of terrorist activities. Using the environment as the vehicle for creating disasters not only could affect the availability of food and water but could trigger a public health crisis as well. Such extreme hazards might include the large-scale release of airborne toxins in urban environments or potent chemical and biotoxins in public water supplies. For chronic or extreme and immediate hazards, the ability of decision makers to consider system-wide conditions and anticipate changes in those conditions currently is limited. Critical data often are difficult to acquire or are not available. And analysis and visualization tools for presenting the available data are not well developed. In addition, the location and status of personnel in a crisis can be difficult to determine, and the organizations involved often have different requirements and protocols for communicating, sharing, and analyzing information. Yet, all of these participants need to work together and share their knowledge to ensure a positive outcome.

Informatics approach. Barriers to finding and accessing critical information, as well as difficulties in timely communication among multiple stakeholders involved in managing environmental hazards, create a wide array of challenges for environmental engineers working to remedy a polluting situation or develop strategies and plans for mitigating its effects. However, environmental engineering is beginning to embrace modern informatics approaches to address the complexity and timeliness of managing both chronic and extreme environmental hazards. A program called the Collaborative Large-Scale Engineering Analysis Network for Environmental Research (CLEANER) is being launched by the National Science Foundation to enable environmental engineers, other professionals, and decision makers managing environmental hazards to develop more rapid and effective approaches, based on a collaborative model.

Programs like CLEANER will deploy sensors and new instruments in the field and use advanced information system technologies. Sensors are transducers that convert precise conditions detected in the environment to an electrical signal (such as current, voltage, or resistance), which can be calibrated and captured as digital data. New sensors are being developed in which entire chemical analyzers or mass spectrometers are put on a chip to make sensitive measurements of complex environmental or biological variables. Sensors generate large volumes of data that need to be maintained in an accessible format and which are used in a variety of analytical tasks. These analytical tasks may be as simple as viewing data in a geographic information system (GIS) or as complex as building predictive models on the fly and testing their results against commonly used simulation models. Simulation models are computational expressions of the theoretical understanding of a system, built upon powerful mathematical expressions that describe or predict system behavior. Simulation models are commonly used to understand the complicated interactions among aspects of the physical environment (such as climate, sediment in lakes and streams, soil type, and atmospheric temperature and humidity), the chemical environment (such as naturally occurring and synthetic chemical compounds), and biological processes (such as microbial activity, photosynthesis, and respiration). These models also are used to predict the effect of changes to an environmental system, whether induced by human intervention such as the release of a large amount of fertilizer into an agricultural system or by a natural event such as a flood caused by a hurricane.

This use of data, models, and analytical tools in science is called informatics. Informatics is an approach based on the belief that new insights can result from the integration and analysis of information that is both up-to-date and appropriate to the task. The informatics approach integrates information from many sources with software tools and people. Information is critical to understanding environmental systems and to monitoring the results of mitigation efforts. The challenge in using informatics in environmental engineering is that there are no models for accurately predicting the behavior of the social systems involved. These social systems include engineers and scientists from academic institutions, engineering firms, and government offices that characterize and analyze the problem from a technical perspective. However, other key stakeholders with essential information and expertise, and with an interest in the outcome of the decision, include regulators (for example, the U.S. Environmental Protection Agency, EPA), public policy analysts (representing federal, state, and local government, academic institutions, and other advocacy organizations), and the decision makers themselves from both the public and private sectors.

Community collaboratories. As a member of a decision-making team, it is necessary to represent one's technical knowledge and to contribute to the overall decision process, with other members having potentially conflicting objectives. In the decision-making process, solutions are proposed and evaluated in an iterative fashion, and prone to compromise as a means of resolving conflicting objectives. For the environmental engineer, each proposed solution may impact the technical assessment of the problem, as well as the predicted outcome, and for each iteration the underlying assumptions need to be tested and the entire analysis may need to be recalculated. In terms of complexity, consider integrating multiple iterations of information from sensors, databases, simulation models, document repositories, and other sources, with the involvement of multiple stakeholders in several geographic locations, each having a unique and independent set of expectations or objectives.

Collaboratory projects are addressing the need for many people to work together and share critical information that is essential to gain insight into a problem, or make group decisions on how to address a problem. Collaboratories are online extensions of offices and laboratories, where collaborators from any geographical location can initiate and discuss ideas, analyze information, and solve problems interactively as a group, publish and share data/document/tools, and manage work using distributed teams. Advances in computing, information, and communication technologies are helping to meet the complex requirements of scientific community collaboratories. At the same time, there is a growing field devoted to the study of community collaboratories. By merging psychology and software engineering, the Science of Collaboratories project has made great strides since 1995 in understanding how people work together, and has created software systems to support the human collaborative process in a manner that allows technical objectives to be met.

Web services. Many new advances have enabled the development of community collaboratories or portals for environmental engineers; the most basic is the internet browser. Browser technology has enabled the exchange of data and media files seamlessly among a variety of applications, and simple Web pages that can be displayed on any computer. Standards like the Java 2 Enterprise Edition (J2EE, http://java.sun.com/j2ee) for Web programming, and the common use of the Apache Tomcat server (http://jakarta.apache.org/tomcat) to run Web-enabled applications have been adopted and are being used by development groups around the globe. These standards provide the interoperability that is supporting the rapid proliferation of useful tools, utilities, and applications distributed over the Internet. Web services (http://www.w3.org/TR/wsdl) are key to supporting advanced data analysis tools, including analysis of real-time data streams from sensors. Web services support large-scale collaborative science over the Internet by providing access through a user's Web browser to powerful databases and analytical tools, which require large-scale computational resources. For example, an application that requires a large data storage device closely coupled with a powerful computational engine can be accessed through a browser page in a community collaboratory that uses Web services.

Projects. The environmental engineering community will benefit from prior efforts to build software environments that support collaborative research and education. The space physics community was an early adopter of Web technology to support collaborative data sharing and viewing in the Space Physics and Aeronomy Research Collaboratory (SPARC) collaboratory (http://www.windows.ucar.edu/sparc). The Department of Energy has supported a number of science collaboratories (http://www.doecollaboratory.org) for scientists at federal laboratories working together over distance. The National Science Foundation supported the development of the Network for Earthquake Engineering Simulation (NEES) collaboratory, and is currently building community support for collaboratory initiatives in environmental engineering (CLEANER) and other fields in environmental sciences. There are many other collaboratory initiatives currently underway, including ecological informatics (NEON, SEEK), geosciences (GEON), and biomedical informatics (BIRN). Each of these is designed to serve the specific needs of the target community, but many of the underlying technologies are common across them.

Outlook. Environmental scientists and engineers will continue to increase their demand for powerful data discovery, analysis, and knowledge sharing tools supported by a scalable national information infrastructure. Communities are already requesting tools that are cast in a collaboration framework, that is, are focused on immediate needs but flexible and adaptable to complex problems as they evolve over time. There are a wealth of technologies now being developed that will be adapted to the specific requirements of the environmental engineering community through programs like CLEANER, providing easy-to-use software environments supporting collaborative research, education, and decision making. Ultimately, the continued development of collaboratory software and the lessons learned through in-depth community engagements will result in robust, interoperable, and easy-to-use environments for solving the most difficult problems.

For background information *see* AIR POLLUTION; DATA COMMUNICATIONS; ENVIRONMENTAL ENGINEERING; INTERNET; MODEL THEORY; SIMULATION; WATER POLLUTION; WORLD WIDE WEB in the McGraw-Hill Encyclopedia of Science & Technology.

Thomas I. Prudhomme

Bibliography. D. Estrin, W. Michener, G. Bonito, and workshop participants, *Environmental Cyberinfrastructure Needs for Distributed Sensor Networks: A Report from a National Science Foundation Sponsored Workshop*, Scripps Institution of Oceanography, La Jolla, CA, August 12–14, 2003; S. Pfirman and AC-ERE, *Complex Environmental Systems: Synthesis for Earth, Life, and Society in the 21st Century*, a report summarizing a 10-year outlook in environmental research and education for the National Science Foundation, 2003.

Fire-induced convection

Atmospheric convection can occur whenever an air parcel or air layer is heated so that it expands, becomes less dense than its surroundings, and begins to move upward by buoyancy. The large-scale (synoptic) conditions determine the buoyancy structure and stability of the atmosphere, the distribution of moisture, and overall motion patterns of the air, and thereby set the preconditions for the type and intensity of convective motion. Initiation of actual convection occurs on the local scale by small-scale processes, most commonly by local solar heating or terrain-induced lifting of air.

The heat released by natural and manmade vegetation fires is another local process that initiates atmospheric convection. The role of the local processes is often assumed to be limited to the triggering or initiating the vertical lifting of air masses from the surface, whereas the ultimate intensity and evolution of the convection depend predominantly on the atmospheric stability, humidity, and wind. In the case of fire-induced convection, the energy added by the fire can significantly modify the evolution of the convection.

In the case of large fires, the heat released by combustion can result in significant intensification of the convection process. This enhances the inflow of air into the fire, "fanning" the flames. It also causes the air to ascend to considerably greater heights than in the absence of the fire. Given the fact that large amounts of trace gases and aerosols are also released in the fires, the resulting deep convection can transport pollutants to great heights in the troposphere and even into the lower stratosphere.

History of research. Scientific research on fire-induced convection is interdisciplinary and has a long tradition. The first scientific publications of observations from fire-induced convection and the

Fig. 1. Forest fire in Siberia with a well-developed cloud formed by fire-induced convection.

formation of clouds above fires and precipitation from these clouds date back to the beginning of the twentieth century. These descriptive, observational studies were followed by mathematical approaches to determine the height of fire-induced convection, depending on the energy released by the fire, using simplified models, often constrained by observations.

Forest services have a particular interest in the interaction between the fire-induced convection and the development of the fire itself. It was early recognized that large fires create their own local wind circulation that strongly affects the spread of the fire. This is of particular importance for the containment of fires. The interest in the interaction between the local fire-induced circulation and the spread of the fire led to the development of detailed three-dimensional atmosphere–fire models that take into account these interactions and are able to reproduce some observed features of fire spreading.

Atmospheric chemistry researchers became interested in wildfires when it was discovered that the emissions from fires have a significant impact on the regional and global atmospheric budget of a number of trace gases and aerosols. With the help of satellite, airborne, and ground-based instrumentation, it was recently shown that fire-induced convection can reach well into the upper troposphere and sometimes penetrate the tropopause. These fire-induced convective events are very efficient means of vertical transport for pollutants, given that fires emit a whole suite of gaseous and particulate compounds.

Studies with research aircraft in biomass-burning regions have revealed that clouds that form as a result of fire-induced convection, called pyroclouds, have very different microphysical properties than convective clouds that are not induced by a fire (**Fig. 1**). There are indications that the precipitation-generating processes and the polarity and frequency of lightning activity are different in pyroclouds, but the causes and precise mechanisms of these effects are still under scientific debate. The high concentration of particles that act as nuclei for the formation of cloud droplets or ice particles has a pronounced impact on the microphysical evolution of pyroclouds. Furthermore, the large heat flux from the fire and the resulting high vertical velocities at the base of the cloud contribute to the different cloud and precipitation development of pyroclouds. Appropriate models that describe numerically the relevant atmospheric processes of fire-induced convection, including the formation of a pyrocloud, have only recently been developed and used for these kinds of simulations (**Fig. 2**).

Global view. The intensity of fire-induced convection varies substantially with the type of fire and the atmospheric conditions. Globally, most vegetation fires occur in tropical regions, such as Africa, South America, and Southeast Asia. In the tropical regions, fires regularly occur during the dry season. They are typically initiated and controlled by people in the context of land-use and clearing practices, rather than being ignited naturally. For this reason, they tend to be most frequent near the end of the dry season, that is, January and February in the Northern Hemisphere and September and October in the Southern Hemisphere. In boreal regions, such as

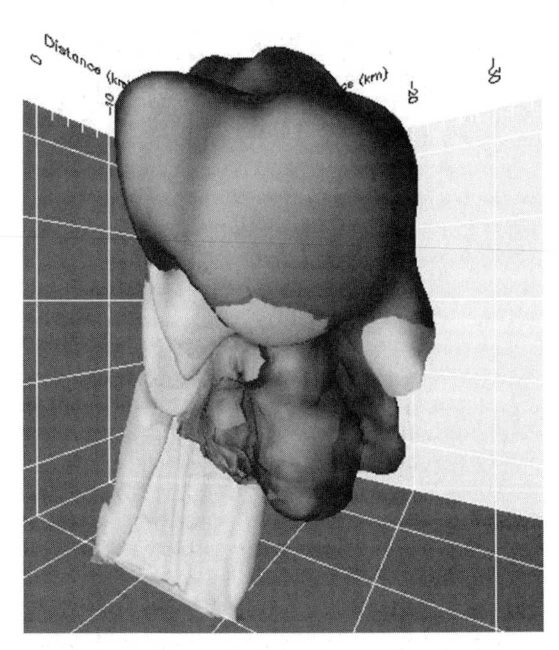

Fig. 2. Simulation of a fire-induced convective cloud that reaches into the lower stratosphere to an altitude of 13 km (43,000 ft). Smoke is indicated by light gray, cloud water by dark gray.

North America and Siberia, fires often are caused by lightning and occur during the summer season.

Tropical regions. Atmospheric conditions in tropical regions during the dry season are rather stable, especially in African savanna regions, where globally most fires occur. Combined with a limited rate of heat release due to the low loading of biomass in grasslands and savannas (about 1–10 metric tons/hectare or 0.4–4 tons/acre), the resulting fire-induced convection usually is not particularly intense. Fire plumes usually ascend only 1–2 km (3300–6600 ft) above the ground, and the smoke plumes usually do not reach more than 3–4 km (9800–13,000 ft) above sea level. In this regime, clouds are only occasionally observed to form on top of the plumes from larger fires. The fire emissions from savanna fires are consequently mostly emitted into the lower troposphere. They can reach the upper atmosphere only when they are entrained into large-scale deep convection in atmospheric convergence zones, such as the intertropical convergence zone.

Amazon region. The fire regime in South America, especially in the Amazon region, is dominated by deforestation and pasture-preservation fires. These fires are typically lit near the end of the dry season (September, October). Several factors contribute to the enhanced convective activity of South American fires, compared with African fires. The atmosphere is more humid than over Africa, especially close to the start of the wet season; that is, the potential for convection is greater. Also, the fuel load is significantly larger in the case of deforestation fires (about 40–400 metric tons/ha or 18–180 tons/acre) than in the case of savanna fires. Convection initiated by these kinds of fires under the prevailing atmospheric conditions regularly reaches the condensation level; that is, a pyrocloud forms atop these fires. The additional release of latent heat from the condensation of water leads to a significant enhancement of the convection.

Midlatitude and boreal regions. In the forests of the midlatitude and boreal regions, high fuel densities occur in periods of very favorable "fire weather" (high temperature and low humidity), that is, during the summer season in the continental climates of North America and Eurasia. For this reason, fire-induced convection can be much more severe there than in tropical regions. The most vigorous fire-induced convection has been observed when a strong atmospheric forcing (such as along a cold front) coincides with a large forest fire. This situation is common in boreal regions, where long-lasting dry and warm periods with high potential for forest fires are commonly followed by frontal passages and the potential for deep convection. Most of the convective energy in fire-induced convective events is provided through the release of latent heat present in the background atmosphere, but the initial forcing of convection by forest fires also adds a nonnegligible contribution to the overall convective energy. Also, with a large initial forcing, more latent heat can be activated from a conditionally unstable atmosphere.

Ground-based and satellite observations have shown that intense fire-induced convection events in boreal regions can penetrate up to the tropopause and into the stratosphere, thereby leading to significant pollution of the higher atmosphere. The duration of these events can be less than 2 hours, while the pollution can remain in the atmosphere for several months. The short duration of the fire-induced convective event complicates the identification of the source of the pollution.

Outlook. One may speculate that global warming may lead to enhanced potential for forest fires in mid and high latitudes. The overall larger amounts of precipitation and warmer temperatures may result in elevated biological productivity and thus higher fuel loadings in much of the region. On the other hand, the increased chance of extremely warm and dry summers provides the conditions for more frequent ignition and greater severity of vegetation fires. This can be expected to result in an increase in the frequency and intensity of fire-induced deep convection, and consequently in the transport of pollutants to the upper troposphere and lower stratosphere.

For background information *see* AEROSOL; ATMOSPHERE; ATMOSPHERIC GENERAL CIRCULATION; CLOUD PHYSICS; CONVECTIVE INSTABILITY; FOREST FIRE; HUMIDITY; LIGHTNING; PRECIPITATION (METEROLOGY); STORM; THUNDERSTORM; TROPICAL METEOROLOGY in the McGraw-Hill Encyclopedia of Science & Technology. J. Trentmann; M. O. Andreae

Bibliography. M. O. Andreae et al., Smoking rain clouds over the Amazon, *Science*, 303:1337–1342, 2004; P. L. Andrews and L. P. Queen, Fire modeling and information system technology, *Int. J. Wildland Fire*, 10:343–352, 2001; P. J. Crutzen and M. O. Andreae, Biomass burning in the tropics: Impact on atmospheric chemistry and biogeochemical cycles, *Science*, 250:1669–1678, 1990; K. A. Emanuel, *Atmospheric Convection*, Oxford University Press, 1994; M. D. Fromm and R. Servranckx, Transport of forest fire smoke above the tropopause by supercell convection, *Geophys. Res. Lett.*, 30: 1542, doi:10.1029/2002GL016820, 2003; H.-J. Jost et al., In-situ observations of midlatitude forest fire plumes deep in the stratosphere, *Geophys. Res. Lett.*, 31:L11101, doi:10.1029/2003GL019253, 2004.

Fire safety (building design)

The main objective of ensuring safety in building design is to reduce the level of damage to the building and its contents and to minimize the loss of life in the event of an accident or natural hazard. To achieve this objective, buildings should be designed to have adequate resistance against accidental loads. In the design against accidental fire, the building should be able to withstand the fire for a reasonable time, and the fire should be contained within a localized part of the building. Several safety measures must be taken to meet the expected performance, including evacuating people from the building, detecting and suppressing the fire, stopping the fire from spreading, and protecting the load-bearing members from losing their strength at elevated temperatures.

Evacuation. The governing criteria for impairment of the main safety functions are often nonstructural (for example, exposure to heat, temperature, and toxic gases). However, it is a functional requirement that the structure remain stable to allow adequate time for safe evacuation and rescue. Therefore, a quantitative assessment of a structure's fire resistance is necessary using advanced simulation tools and computational methods. A major limitation on predicting incident outcomes for building safety design is the lack of quantitative data on premovement time, evacuation time, and occupant behaviors for different accident scenarios and occupancies. There is a need to study human behavior during evacuation from a range of occupancies. Advanced computer software is available to predict the evacuation time, considering the effects of exposure to smoke and fumes on occupant evacuation behavior. However, improvements are still needed to include the effects of different warning systems, information provided to occupants, occupant characteristics, pretraining, building complexity, and level of fire safety management on premovement and evacuation time. In order to achieve effective design for buildings in emergencies, it is essential to consider occupants' characteristics, their abilities to evacuate, and the effects of exposure to fire conditions on occupant evacuation behavior.

Fire modeling. The temperature of a burning space rises according to parameters such as the combustible materials, ventilation conditions, enclosing materials, and room geometry. The standard fire curve is uniformly defined without relevance to the real burning. A more realistic parametric fire model has been proposed in the European structural design codes (Eurocodes) to predict the temperature development in a fire compartment. The main drawback of this approach is that it assumes that the fire temperature is uniform in the entire burning space. A zone model is able to give information that is more detailed by dividing the room into several zones and solving the heat balance equations for each zone to generate the fire temperature. Computational fluid dynamics (CFD) models work by dividing the burning space, as well as the boundaries, into a number of small volumes. The model's analysis systems involve fluid flow, heat transfer, and associated phenomena, following the conservation of mass for a fluid and Newton's first and second laws of thermodynamics. These equations are solved to give the temperature, velocity, pressure, and so on. The main advantage of CFD model is that it can be used to model localized fire or fire compartments with complex geometry; however, only specialists are able to run the analysis as it requires a good understanding of the input parameters and huge computing resource. A number of zone or CFD models have been developed during the last two decades. Caution needs to be taken for careful validation before application in design.

Heat-transfer analysis. Heat-transfer analysis predicts the temperature of the structural members according to the environmental temperature. Heat may be transferred from the environment to member surfaces by radiation, convection, or conduction. It is relatively easy to calculate the temperature of unprotected members. Some researchers are now working on the heat-transfer behavior of various fire-protection materials. Since concrete is often used as a fire-protection material for steel, a study on the movement of water molecules in the concrete is essential to understand the relationship between moisture content and temperature development in concrete. This will explain why spalling of concrete occurs at elevated temperature. Calculating the heat transfer in wood and the depth of charring layer also imposes a challenge to today's fire safety design.

High temperature imposes two effects on the structural members: thermal expansion and degradation of material properties (strength and stiffness). Typically, steel loses half of its strength when heated to 600°C (1100°F) and almost all its resistance at 1000°C (1800°F). Simplified methods are available in the design codes to calculate the load-bearing capacity for structural members, considering the degradation of strength at elevated temperature. However, it was later realized that structural continuity plays a big role of maintaining member stability. The effect of structural continuity on the fire resistance of structural members, such as axial and rotational end restraints, has been extensively studied. A typical example is to make use of catenary action or membrane action in the design of beams and floor slab. If high displacements are acceptable, an assessment on the likely fracture of reinforcement or connection failure due to high tensile/catenary forces should be conducted and, if necessary, more robust details should be adopted to ensure localized failure does not occur. Direct analysis of the whole building using finite element techniques provides the most comprehensive information about the structural response in a fire, with the extra cost on modeling and computer calculation. Some of the commonly used computer programs include ABAQUS, VULCAN, SAFIR, ADAPTIC (developed by B. A. Izzuddin, Imperial College, UK), and USFOS. So far, the behavior of steel frame structures in fire can be simulated using finite element software. However, the behavior of concrete and the connections at elevated temperatures is a problem yet to be addressed.

Fire protection. After the World Trade Center disaster, there has been a need to design public buildings and their workspace to allow safe evacuation in the event of fire or explosion. Certain industrial buildings have stringent fire protection requirements. For example, oil and gas production and processing, nuclear-related product storage and processing, chemical process and storage, and key infrastructure and transportation routes are facilities with greater risks. From a safety and licensing authority's point of view, structures must be capable of providing safe evacuation in the event of fire. From an owner's viewpoint, fire protection is often an expensive statutory feature, which requires initial capital investment. From an operator's point of view, fire protection must be maintained to preserve the safety margins declared in the safety documentation. The cost of fire protection is often a critical factor for design consideration. Therefore, there is a need to balance these conflicting requirements when specifying fire protection.

Fire protection can be provided as an all-encompassing scheme, or it can be functionally designed to optimize cost, weight, and maintenance. Research has shown that fire can be successfully suppressed using a properly designed and maintained active protection system without the need of passive fire protection. Active protection systems include water and chemical sprays or deluges, foam dispersion, and inert-gas dispersal. Passive systems include fire barriers, fire-resistant enclosures, fire doors, fire-retardant coatings, and fire-protective coatings. Computer models have been developed to predict the response of structures considering fire-protection materials. Such models include basic thermal transmission phenomena (such as radiation, conduction, and convection), temperature-dependent properties of materials, and the location and nature of fire-protection measures. The resulting thermal histories are then applied to the structure to predict the time to collapse or to demonstrate the degree of collapse. From this, the structure can be economically protected to meet the safety requirements.

Explosion. Assessing the response of structures to explosion is an increasingly important factor in design, particularly where the storage and processing of explosive materials are concerned. Many structures are required to be "blast-resistant" to protect personnel and adjacent facilities, and to reduce the possibility of escalation of events. These structures are designed to contain the effects of an explosion or to act as a significant barrier. There is a great difference in the structural behavior of buildings to explosion as to fire. The short duration of explosion loading implies that the material is strain-rate dependent; for example, a high strain rate will increase the yield strength of the steel. On the other hand, fire loading is associated with elevated temperature, which causes thermal strain and leads to significant deterioration of the mechanical properties of materials. A numerical approach for inelastic transient analysis of steel structures subjected to explosion loads, followed by fire, can be used effectively to solve explosion and thermal response problems, taking into account geometric and material nonlinearities. To achieve both computational accuracy and efficiency, a two-step procedure may be adopted in which fire can be treated as a separate event after an explosion. The fire resistance of the structure can be evaluated by analyzing the deformed geometry of the structure caused by the explosion. The approximate fire analysis method does not require time domain solutions, but it predicts higher fire resistance than the strict inelastic transient analysis. Nevertheless, it offers an alternative means for evaluating the performance of structures subjected to combined explosion and fire scenarios at a much-reduced computational cost.

Hand calculation procedures and computational techniques have been developed with the aim of predicting how a structure will respond under the interaction of blast and fire. A recent trend is to study the fire resistance of buildings after damage from the ballistic loading. A blast could damage the fire detection or fire alarm system, peel off the fire protection layer, and deteriorate the load resistance of structural members. Some researchers are investigating the resistance of fire protection to impacts and the effects of blast loading on the fire resistance of load-bearing members, as well as direct analysis technologies for structures subjected to combined blast and fire loading.

Integrated approach. A quantitative assessment on the safety performance of a building in fire requires the knowledge of fire science, material properties at elevated temperature, occupant behavior and evacuation procedure during an emergency, heat transient and structural response phenomena, and fire protection. All these require a multidimensional integrated approach for the performance-based design of structures. Design fire scenarios can be prescribed for standard buildings and further examined for systems that are more complex. Structural design should also consider evacuee and fire fighter interactions.

A proper fire model that interacts with the active and passive protection measures should be developed, and a relationship between emergency response and fire resistance should be established so that appropriate performance-based method can be developed to predict the building response to extreme events. Improved understanding of the real behavior of natural fire in tall buildings opens new ways of integrating fire safety and structural design. Without considering the behavior of the system's limit states, prescriptive codes are often approximate in nature. With the advance in computing technologies, there is an increasing demand for robust and efficient nonlinear analysis methods for performance-based design of structures subject to fire and explosion.

For background information *see* ARCHITECTURAL ENGINEERING; AUTOMATIC SPRINKLER SYSTEM; BUILDINGS; COMPUTATIONAL FLUID DYNAMICS; CONCRETE; FIRE DETECTOR; FIRE EXTINGUISHER; FIRE TECHNOLOGY; FLOOR CONSTRUCTION; HEAT TRANSFER; INDUSTRIAL FACILITIES; STRUCTURAL DESIGN; STRUCTURAL MATERIALS; STRUCTURE (ENGINEERING); WALL CONSTRUCTION in the McGraw-Hill of Encyclopedia & Technology.

J. Y. Richard Liew; H. X. Yu

Bibliography. European Committee for Standardization, *DD ENV 1993-1-2, Eurocode 3: Design of steel structures, Part 1.2, General rules—structural fire design*, British Standards Institution, 2001; J. Y. R. Liew, Performance based fire safety design of structures: A multi-dimensional integration, *Adv. Struc. Eng.*, 7(4):111–133, 2004; J. Y. R. Liew and H. Chen, Explosion and fire analysis of steel frames using fiber element approach, *J. Struc. Eng.*, 30(7):991–1000, 2004; L. Song et al., An integrated adaptive environment for fire and explosion analysis of steel frames, Part I: Analytical models, *J. Construc. Steel Res.*, 53: 63–85, 2000.

Fluorescent signaling in mantis shrimps

All animals face the problem of producing, detecting, and correctly interpreting signals intended for communication. In any species, effective signals must be appropriately tuned to the sensory systems of receivers, but well-designed signals must also contend with the environment within which they are transmitted. This is particularly true for visual signals that are used underwater in marine or freshwater environments. Water is a particularly difficult medium within which to use visual signals. This is because particles in the water, and the water molecules themselves, tend to scatter light, degrading images (and signals) even at relatively short distances, making long-distance transmission impossible. Furthermore, water transmits light within a restricted spectral band, primarily at blue wavelengths in clear marine systems (but varying among water types; for example, in freshwater, green or even yellow may be best transmitted). This limited spectrum, which becomes ever narrower with increasing depth, makes color signals nearly useless in waters more than a few meters deep and favors patterns that stand out well under blue illumination.

Color signaling over a range of depths. Some marine animals inhabit a range of depths, from near the surface to greater than 100 m (330 ft) below it. For them, colorful signals that are visually effective near the surface would be nearly useless—and possibly invisible—when used in deeper water. Stomatopod crustaceans, commonly known as mantis shrimps, are an example of such animals.

Stomatopods live in the ocean, most commonly in the tropics, inhabiting holes in the rocks and rubble of coral reefs or building tubelike burrows in the bottom sand or mud. Stomatopods have excellent vision, and many species rely on brightly colored marks and patterns to communicate aggression or willingness to mate. Most live at fairly shallow depths, where their color signals are highly visible and the colors are unambiguous. Since stomatopods are small animals, with most species only a few centimeters in length, they interact at short distances. Thus, the brightness of the colors is readily visible in their shallow habitats, since they need to be transmitted less than a meter through water from sender to receiver.

Stomatopods that inhabit a range of depths, however, face problems. Signals that work well near the surface become dull and colorless when used in deeper water. Such species would seem to be faced with the dilemma of using suboptimal color signals or disposing with color signals altogether. However, a few species of mantis shrimps that live both near the surface and deeper down have evolved a novel way to produce the same color regardless of the local lighting: they use fluorescent signals. Understanding how these signals function requires a brief discussion of how mantis shrimps see.

Mantis shrimp vision. Like all crustaceans, stomatopods have compound eyes. However, the visual system of mantis shrimps has several unique features not found in other crustaceans or in any other animal. Of special relevance to visual signaling, they possess the most complex color vision known, with at least eight classes of primary color receptors (compared with three in humans and no more than four in any other vertebrate species), which are activated by light in a range of wavelengths corresponding to the deep violet end to the far-red portion of the visible spectrum, and are clearly important components of a color signaling system. Other specializations enable the detection of ultraviolet light and polarized light, but probably play no role in color signaling.

One of the most unusual features of the stomatopod photoreceptors is that they are topped with strongly colored filters that intercept light traveling into the photoreceptive regions of several color classes. These filters look yellow, orange, red, purple, or blue to our eyes, and they tune each underlying receptor to just one part of the spectrum—the spectral region, or color, that they transmit. The filters probably incorporate carotenoids, a class of yellow, orange, red, and purple pigments that are widely distributed in nature, including vegetables, fruits, insects, fishes, and birds. Of special interest are the yellow filters, which produce sharply tuned yellow-sensitive receptor classes. The colors of the yellow filters are similar to the yellow markings on the bodies of many stomatopod species, suggesting that the receptors they sensitize are well-matched to the light reflected by the yellow spots and patterns. Receptors that are spectrally matched to particular signals would make the signals appear as strong stimuli seen against dim backgrounds.

Yellow fluorescent signals. The yellow patches used for signaling in some stomatopod species not only reflect yellow light, they fluoresce it. Fluorescence is the process whereby a pigment absorbs light in one spectral region and emits it at longer wavelengths. It is commonly used by humans to produce highly visible traffic marks or for visual excitement in nightclubs and similar locales. Many marine organisms, such as corals and sponges, are brightly fluorescent, but since these species have no eyes, the fluorescence is believed to be a by-product of light-active pigments that are not involved in signal production.

It has long been hypothesized that some marine animals could use fluorescence as a way to produce a bright, conspicuous signal in the blue lighting of deep water, but only recently have such fluorescent signals been seen—in mantis shrimps. Yellow markings in a number of mantis shrimp species have been demonstrated to fluoresce. In one species, *Lysiosquilla glabriuscula*, fluorescence clearly seems to be a component of a visual signal.

Signaling in *Lysiosquilla glabriuscula*. Individuals of the Caribbean species *L. glabriuscula* are large, often exceeding 20 cm (8 in.) in overall length. They live in sandy burrows and inhabit depths from the low intertidal region to at least 40 m (135 ft) down. These animals normally hunt from the entrances of their burrows, hiding just below the surface and erupting out into the water when a fish or other prey swims by. They only occasionally emerge from their burrows, sometimes to hunt on the surface of the

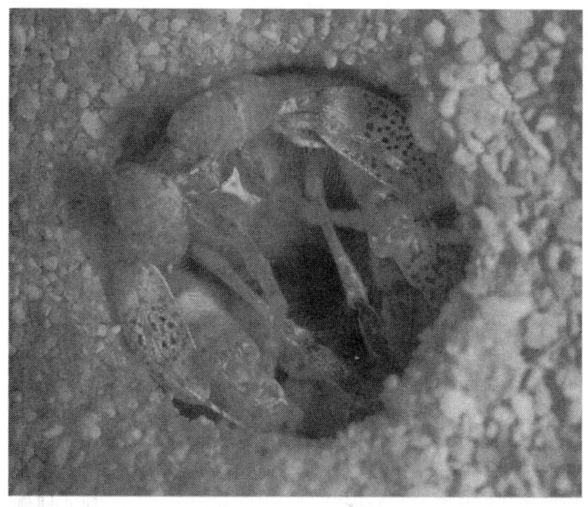

(a)

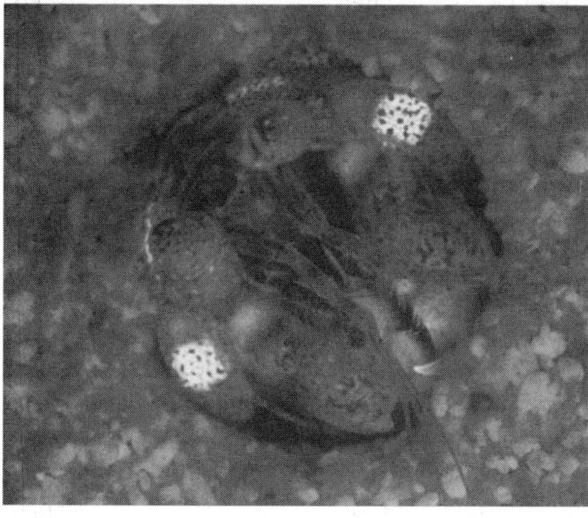

(b)

Fluorescent signals in *Lysiosquilla glabriuscula*. (*a*) A male *L. glabriuscula* in its burrow entrance illuminated by blue light, similar to the illumination occurring at depths occupied by some individuals. Note the very low contrast appearance of the animal. (*b*) The same animal, also illuminated with blue light but seen through a filter that transmits only yellow light (simulating how another mantis shrimp would perceive this image). The bright fluorescent patches on the antennal scales stand out clearly as visual signals. (*Photographs courtesy of Roy L. Caldwell*)

sand but more commonly when males travel on the surface in search of females. In regions where the animals are common, burrows are not far apart (sometimes only a few animal lengths), so males are very likely to encounter receptive females and also neighboring males who have established their own burrows. The resident male will respond by displaying to the intruder—often by emerging partway from its home—its anterior appendages (including the predatory claws) and extending large oval flaps (called antennal scales) to the sides. Another possible response is for the resident male to block the entrance of its burrow with its anterior appendages, again prominently displaying the antennal scales (see **illustration**). A similar display can be given in an emergency, for example, when a wandering male is caught out on the surface.

Large spots on the antennal scales of *L. glabriuscula*, as well as patches on its carapace in the shoulder region, look bright yellow in white light and are easily seen underwater when the animal is not far from the surface. Although for animals near the deeper parts of their vertical range there is little or no yellow light coming from the surface, the spots on the antennal scales still look yellow (see illustration). This is because the pigments in the spots absorb the illuminating blue light and reemit it as yellow, maintaining a nearly constant signal color throughout the entire range of depths *L. glabriuscula* inhabits. Computational modeling shows that below 20 m (65 ft), a significant fraction of the light in the signal color is produced by fluorescence.

Other fluorescent signals in nature. The yellow signals of *L. glabriuscula* are the first clear example of the use of fluorescence to enhance underwater communication, but it seems likely that other stomatopod species use similar signaling mechanisms. Some fishes are known to fluoresce as well. Although the use of fluorescence for communication in fishes and other marine animals has not yet been demonstrated, it is very likely that the phenomenon is used in a number of species that rely on visual signals. Fluorescence also occurs in terrestrial animals. For example, budgerigars have yellow patches of feathers on their heads that are brightly fluorescent; the signals transmitted by these patches are important for mate selection. Fluorescent signals may play wider roles in animal communication than hitherto suspected.

For background information *see* ANIMAL COMMUNICATION; CRUSTACEA; FLUORESCENCE; PIGMENT; STOMATOPODA in the McGraw-Hill Encyclopedia of Science & Technology. Thomas W. Cronin

Bibliography. R. L. Caldwell and H. Dingle, Stomatopods, *Sci. Amer.*, 234(1):80–89, 1976; T. W. Cronin, N. J. Marshall, and M. F. Land, Vision in mantis shrimps, *Amer. Scientist*, 82:356–365, 1994; J. N. Lythgoe, *The Ecology of Vision*, Clarendon Press, Oxford, 1979; C. H. Mazel et al., Fluorescent enhancement of signaling in a mantis shrimp, *Science*, 203:51, 2004.

Forensic DNA testing

Forensic biology is the scientific analysis of biological evidence to provide objective information on legal matters or those that pertain to criminal and civil law. Biological evidence such as bodily fluids or tissues that may be found at crime scenes can be analyzed through deoxyribonucleic acid (DNA) typing. Typing requires detection and screening of the biological evidence, such as blood, semen, or saliva, extracting the DNA from a specimen, amplifying specific regions of the DNA using the polymerase chain reaction (PCR), and typing the resulting PCR products to determine a DNA profile. The DNA profile from the evidence is then compared to known profiles from suspects, victims, or database samples to determine the significance of the result. Samples containing mixtures require additional interpretation to infer individual donor allele designations. Forensic biologists must also assess the statistical significance of their results, write reports, and testify in court.

The establishment of a United States national DNA database, the Combined DNA Index System or

CODIS, has facilitated the ability to compare DNA profiles from unknown biological crime scene evidence to DNA databases of known convicted criminals ("cold hits") or to DNA left at other crime scenes, resulting in the ability to link cases. As of 2005, CODIS contained over 2.5 million profiles. Many other countries, such as the United Kingdom and Germany, have DNA databases with some of the same genetic markers, permitting international database searches. In the United States alone, over 25,000 investigations have been aided through CODIS.

By comparison of the DNA profile from crime scene samples to known samples, the results can serve to link victims and suspects with the crime scene or can exclude a suspect from association with that crime. Additionally, scientific analysis of biological evidence may provide unbiased information to substantiate case circumstances, corroborate or refute an alibi, or identify a weapon used in a crime. Cases may include nonhuman samples such as botanical, fungal, entomological, or zoological specimens, which can also be used to link victims and suspects with each other or to the crime scene. Forensic biology may also be applied to homeland security, in which crimes may include weapons of mass destruction such as pathological microbes. The forensic biologist may be able not only to identify the microbe (such as anthrax) but to type the microbial DNA and link the strain of the microbe to a strain produced at a certain location or to the original progenitor strain.

Two main principles permit the use of DNA in forensics. First, no two individuals have the same DNA, with the exception of identical twins. Second, the DNA from any source (such as blood, hair, or skin) of a particular individual will be the same.

Evidence detection and screening. Detection starts with evaluating the investigative information to understand the nature of the case and samples. Visual examination of stains with alternative light sources (such as ultraviolet or infrared light and lasers) and chemical enhancement reagents may be performed. Forensic biologists can determine the nature of the biological stain (for example, blood, semen, or saliva), and whether it is human through presumptive and confirmatory tests. These tests consist of analytical procedures including microscopy, chemical tests, and immunological assays. Determination of which of the items of evidence will prove to be the most probative or informative through these tests can make or break a case.

DNA extraction. Following screening, DNA must be extracted. Among the methods used are the following:

1. *Organic extraction*. This method consists of lysis of the cells in a detergent-based buffer followed by one or more rounds of purification using an organic phase separation and concentration using column centrifugation or ethanol precipitation.

2. *Chelex resin extraction*. This method utilizes a fast, simple extraction of small amounts of sample in the presence of a chelating resin and results in a somewhat crude extract that is usually adequate for PCR amplification.

3. *Solid-phase extraction*. These methods utilize a membrane to capture the DNA. Samples are spotted onto the membranes, and the subsequent washes remove the impurities.

4. *Silica-based extraction*. In this method, nucleic acids are first adsorbed to the silica in the presence of chaotropic salts that weaken hydrophobic associations. Polysaccharides and proteins do not adsorb and are removed. Next, nucleic acids are released in a low-salt wash. This method has been automated using robot stations and is being used in several laboratories.

5. *Differential extractions*. Differential extractions are required to separate female epithelial from male spermatozoa cells in sexual assault cases. New methods are being evaluated, including laser microdissection and cell sieving.

DNA quantification. Assessing the quantity and quality of the sample is the next step. Several methods are being utilized in crime laboratories. These include agarose gel electrophoresis in the presence of quantification standards (samples with known quantities of DNA), known as yield gel electrophoresis; slot blot hybridization, using known DNA standards immobilized on a membrane followed by hybridization to a human/higher primate-specific DNA probe; homogeneous plate assays, using a DNA fluorescent dye and scanning in a plate reader; and more recently, real-time detection using quantitative PCR (QPCR). Real-time QPCR has several advantages over the other methods in that it is extremely accurate and sensitive over a broad dynamic range, and it occurs in a closed-tube system, reducing the potential for carryover contamination. Using this technique, a forensic biologist can monitor and quantify the accumulation of PCR products during log phase amplification.

DNA amplification. PCR is a fast in-vitro DNA synthesis process that can provide up to a billion (10^9) copies of a given target sequence. Specific DNA markers can be targeted for duplication by a DNA polymerase. Primers are designed to hybridize to the specific markers along the length of the DNA template during the cycling of temperatures. In the thermal cycle, DNA strands are separated, primers bind to the template, and then a special DNA polymerase that is heat-stable is used to copy and amplify the genetic markers. Through a process of 28–32 heating and cooling cycles, the DNA is increased so that it can be analyzed. The thermal cyclers contain many sample wells, permitting the amplification of multiple samples simultaneously; as many as 96 samples may be amplified in under 3 hours.

Multiplex PCR allows several different loci to be simultaneously amplified in a single tube. This enables typing from a single aliquot of the extracted genomic DNA, reducing sample consumption. Recently the ability to analyze as many as 15 autosomal short tandem repeats (STRs) simultaneously has been reported using DNA from a very small amount of degraded sample.

DNA separation and detection. Separation and detection of the amplified products is required following PCR. There are many different methods to achieve typing. These include polyacrylamide gel electrophoresis (PAGE) followed by silver staining or, if the primers are fluorescently tagged, detection by fluorescent gel scanners; and capillary electrophoresis (CE) with laser-induced fluorescence. The latter method has become the most commonly utilized method of detection, as it is highly automated (there is no gel to pour and load), samples can be easily be reinjected (robotically), and since the DNA traverses the entire length of the capillary, the resolution of the higher-molecular-weight loci is usually better than in the PAGE methods.

Forensic genetic markers. The genetic markers currently being typed in most forensic biology laboratories include autosomal short tandem repeats (STRs), mitochondrial DNA, and Y chromosome STRs.

Short tandem repeats. Short tandem repeats consist of regions of 2–7 base pairs that are tandemly repeated. Individuals may vary in the number of repeats and/or the content of the repeats. The variation in the content of the repeats occurs in either a change in the base within a repeat unit or as a deletion in the repeat unit. STRs used in forensics are either tetranucleotide or pentanucleotide repeats. STRs are highly abundant and well studied in the human genome, and their small size and the alleles' small size range facilitate typing from highly degraded, small quantities of starting material. There are 13 CODIS core loci that are being uploaded into the national DNA database.

Mitochondrial DNA. Mitochondrial DNA (mtDNA) is useful for forensic DNA in that it exists in high copy numbers in each cell, and therefore has a better chance of being detected in small samples. Moreover, mtDNA is maternally inherited, so any individual within the maternal lineage may provide an mtDNA reference sample. Finally, since the size of the amplicons (the DNA segments to be generated by PCR) is small, mtDNA can be typed from degraded DNA. Mitochondrial DNA hypervariable regions I and II are the most commonly sequenced targets in forensic DNA laboratories. The most commonly used method of mtDNA typing is fluorescent Sanger's dideoxy sequencing.

Y chromosome markers. Y chromosome markers, including Y STRs, have been recently used in many forensic DNA crime laboratories. The interest in Y chromosome markers is well supported for the following reasons:

1. The total number of male cells that are present at a rape scene may be very small in the case of rapists who are azoospermic (having no sperm) or oligospermic (having a low sperm count).

2. The total number of male cells may be low due to loss of sample or degradation.

3. Multiple semen donors may need to be identified in a multiple rape case.

4. In criminal paternity or mass disaster victim identification, determination of the haplotype of a missing individual may be conducted by typing a male relative.

5. In sexual assaults, the time-consuming and sometimes inefficient differential extraction procedure for the separation of sperm and nonsperm fractions may be bypassed.

6. Y chromosome typing may provide increased statistical discrimination in mixture or kinship analysis cases in which the discrimination obtained from autosomal markers is insufficient for identification purposes.

Interpretation. Once amplified and typed, the results need to be interpreted. Forensic biologists must have a clear understanding of the molecular methods utilized, with in-depth knowledge of the basis of typing (that is, the cell biology, technology, and genetics of the loci), the specific loci and amplification parameters (determined both externally and internally in developmental and internal validation studies), the empirically derived limitations of the system, the protocols and quality control measures implemented in their own laboratories, the instrument validations, the analytical software utilized in determining the DNA profiles, and the statistical methods and population genetics databases and software used.

In single-source samples, the interpretation requires setting a threshold of detection. That is, alleles that are at or above a certain fluorescent threshold may be designated. In highly degraded DNA samples, the interpretation requires more analysis, as higher-molecular-weight loci generally degrade more rapidly than lower-molecular-weight loci. In this case, the alleles are expected to appear in decreasing amounts as the size of the alleles increases (see **illustration**). In mixtures, the interpretation requires considering all combinations of alleles that are present, allele ratios within and among loci, and the sample type and condition.

Future technologies. Many technologies applicable to forensic biology are currently under investigation, including the following:

Hand-held, microcapillary STR typing. Development of hand-held DNA typing devices has obvious advantages for forensics in that samples might be processed quickly at the crime scene. The potential to rapidly determine the DNA type of samples left at a crime scene coupled to the growing national DNA database provides a powerful tool to law enforcement, as "cold hits" might provide leads early in the investigation. Devices have already been developed that can detect microbial pathogens and can be used out in the field. Moreover, under development are miniaturized microcapillary devices for typing human STRs.

Single nucleotide polymorphisms. Single nucleotide polymorphisms (SNPs) are single base sites that vary between individuals and as such can be used in forensic DNA typing. They have already been utilized in forensic DNA tests. The PCR-based DNA typing systems HLA-DQ Alpha and Polymarker loci were based on SNPs. The detection was by reverse dot blot hybridization, where the known samples were immobilized on the blots. Today there is interest in the detection of mtDNA and Y SNP typing using a variety of approaches, including primer extension assays, taq man assays, microarrays, liquid bead arrays (Luminex), and pyrosequencing.

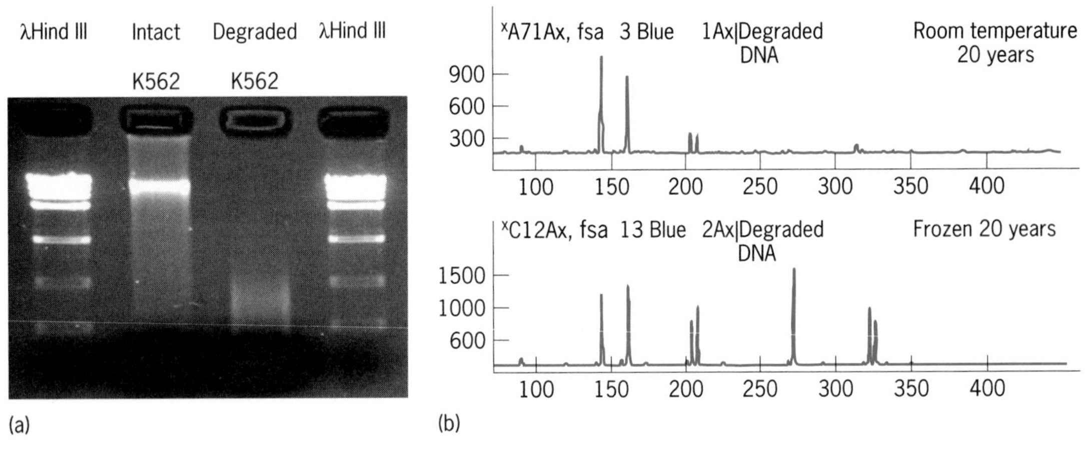

Results of degrading DNA by mechanical disruption. (*a*) Agarose gel electropherogram of intact DNA and degraded DNA, electrophoresed in lanes 2 and 3 through a 1% agarose gel, stained with ethidium bromide, and then photographed with UV light. Molecular-weight ladders of Lambda Hind III were placed on either side of the samples in lanes 1 and 4. The sample in lane 3 has been highly degraded with no detectable high-molecular-weight band as is apparent in lane 2. (*b*) Capillary electropherogram of degraded DNA stored at room temperature versus frozen. The decrease in PCR product amount as base pair size increases, as shown in the top panel, is the expected result for a degraded sample. The result demonstrates the importance of proper (frozen) sample storage.

Forensic biometrics. The ability to determine the physical characteristics of an individual by typing genetic markers has been called forensic biometrics. Inferring population of origin from DNA evidence using Y chromosome SNPs has been recently reported. The ability to determine the potential genetic origin of a perpetrator has obvious benefits to law enforcement. However, it also has obvious ethical and legal implications, as well as significant limitations, in that the genetic boundaries between races are not clear. Other applications may be in determining the age of a suspect using genetic markers.

Tissue typing using mRNA profiles. Different tissues have different genetic expression patterns. Recently the use of ribonucleic acid (RNA) has been reported for body fluid identification. The potential use of molecular technology was also reported in determining the age of a bloodstain, which could be useful in establishing the time of the crime, using analysis of messenger RNA (mRNA):ribosomal RNA (rRNA) ratios. Advantages of the mRNA-based approach, versus conventional biochemical tests, include greater specificity, simultaneous and semiautomatic analysis, rapid detection, decreased sample consumption, and compatibility with DNA extraction methodologies.

Low-copy-number amplification (LCN). Biological evidence is often found with an extremely low number of starting templates (1–15 diploid cells, <100 picograms). Analysis of LCN samples requires the development of novel strategies for its collection, analysis, and interpretation. Also known as touch sample amplification, low-copy-number amplification may require additional cycles of amplification and/or additional cleaning or sampling, but permits the analysis of extremely small amounts of sample such as might be found on the grip of a gun.

For background information *see* CRIMINALISTICS; DEOXYRIBONUCLEIC ACID (DNA); FORENSIC BIOLOGY; FORENSIC MEDICINE; GENE; GENE AMPLIFICATION; MOLECULAR BIOLOGY in the McGraw-Hill Encyclopedia of Science & Technology. Steven Lee

Bibliography. B. Budowle et al., Forensics and mitochondrial DNA: Applications, debates, and foundations, *Annu. Rev. Genomics Hum. Genet.*, 4:119–141, 2003; B. Budowle et al., Public health: Building microbial forensics as a response to bioterrorism, *Science*, 301(5641):1852–1853, 2003; J. Butler, *Forensic DNA Typing: Biology and Technology Behind STR Markers*, 2d ed., Academic Press, San Diego, 2005; A. Carracedo (ed.), *Forensic DNA Typing Protocols: Methods in Molecular Biology*, pp. 1–297, Humana Press, Totowa, NJ, 2004 (DOI: 10.1226/15925998676); M. A. Jobling and P. Gill, Encoded evidence: DNA in forensic analysis, *Nat. Rev.*, 5:739–751, 2004; E. L. Jones, Jr., The identification of semen and other body fluids, in *Forensic Science Handbook*, vol. 2, 2d ed., pp. 329–399, Pearson Education, NJ, 2005; N. Rudin and K. Inman, *Forensic DNA Analysis*, 2d ed., CRC Press, Boca Raton, FL, 2001; J. A. Siegel, G. C. Knupfer, and P. J. Saukko (eds.), *Encyclopedia of Forensic Sciences*, vols. 1–3, Academic Press, 2000.

Forest water contamination

Forests play a key role in cleaning water. Precipitation is "filtered" through the tree canopy and filtered again through the organic matter on the forest floor. The water then seeps into the subsurface to replenish the ground water. Approximately 80% of the freshwater in the United States originates in the 650 million acres (265 million hectares) of forest that cover approximately one-third of the nation. These forested watersheds are the primary source of drinking water for nearly two out of three people in the United States. The United States' national forests and grasslands, which cover nearly

192 million acres (78 million hectares), are the largest single source of freshwater in the country, and the headwaters of many large river basins originate in the national forests. Forests are thus an important source of freshwater; however, mining, forestry practices, and altered fire regimes have increasingly caused the contamination of these valuable water resources.

Sources of water contamination. Numerous low-flow-volume point and non-point sources cause significant contamination of water originating in forests. Abandoned mines are a major source of forest water contamination. Acid drainage from abandoned mines, for example, has been shown to cause serious water quality problems. There are approximately 557,000 abandoned hard-rock mine sites in the United States, more than 1 in 20 of which are located in or near national forest lands. Runoff from tailing piles containing the rejected materials from the mine also results in water contamination. Today, tailings are buried in highly engineered clay-lined or synthetic liners, but in the past when tailings were buried, it was not done with as much attention to containment. In addition, ground water can flow into an abandoned mine and come into contact with sulfide minerals (chalcopyrite, galena, pyrite, pyrrhotite, and sphalerite). The most common metals associated with sulfide minerals include lead, zinc, copper, cadmium, iron, and arsenic. In the presence of air, FeS from the minerals reacts with water and oxygen to form ferrous iron (Fe^{2+}), sulfate (SO_4^{2-}), and acid (H^+). These compounds can lower the pH of the water to 2.5–3, which causes many metal ions to be leached from the rock. When the pH is greater than 3.5, ferrous iron will oxidize to ferric iron (Fe^{3+}), which causes the water to take on a rust-stained appearance and can cause clogging of pipes and other plumbing problems. Although it affects the water's taste and smell, ferric iron is not directly hazardous to a person's health (as the body needs some iron in the diet as part of the oxygen-carrying system in the blood). The oxidation of sulfide minerals is also catalyzed by *Thiobacillus ferrooxidans* bacteria, which can increase the rate of sulfide oxidation by 10 to 100,000 times. Drainage from these mines can be a source of heavy-metal ions in watersheds that originate in forests.

Abandoned coal mine sites are also a serious problem. Pyrite, marcasite (FeS_2), and greigite (FeS_4) are the primary sulfide minerals found in coal. Oxidation of these minerals also lowers local water pH and causes mobilization and transport of metal ions to ground and surface water.

Another forest water contamination issue results from the use of cyanide in base-metal flotation and gold mining. More than 100 million pounds (45 million kilograms) of sodium cyanide were used in the United States in 1990 (the most recent year for which statistics are available). Cyanide is more toxic to aquatic organisms than to humans, although at high enough levels it can cause both short-term and long-term health effects. Cyanide dissolved in water readily complexes with metals. At a pH below 9, cyanide compounds can dissociate and hydrogen cyanide (HCN) is formed as a by-product. Hydrogen cyanide is a gas which is very toxic to humans.

There are many other sources and types of contaminants entering the water that comes from our forests, including oil and grease from vehicles, and herbicides, insecticides, and fungicides used in the treatment of lawns and other managed areas within the forest. There are also naturally occurring contaminants such as arsenic. Arsenic generally enters drinking water supplies through a process of natural leaching. Arsenic-rich rock is widely dispersed throughout the Earth's crust. The Environmental Protection Agency recently proposed lowering the maximum allowed level of arsenic from 50 parts per billion (ppb) to 5 ppb due to concerns about bladder, lung, and skin cancer. A policy compromise of 10 ppb has been scheduled to go into effect in 2006.

Forest fires and water contamination. Fires can drastically affect the role of forests in the water supply. Erosion after a fire can result in significant losses of topsoil, increases in the amount of sediments in the water (turbidity), and the release of large quantities of soluble chemicals. A major concern after a forest fire is the increase in concentration of nitrate nitrogen (NO_3–N), mostly originating from dissolved herbicides. High levels of NO_3–N, along with phosphorus, can cause eutrophication of lakes and streams (excessive nutrient richness leading to excessive plant growth).

Soil type and vegetation coverage play a critical role in the ability of the land to adsorb water. In a healthy forest, where at least 60–75% of the ground is covered with vegetation, only 2% or less of the rainfall becomes surface runoff and erosion is low (loss of less than 0.05 ton soil/acre). When the vegetation cover drops to 35–50%, the surface runoff increases to 14% and soil loss increases tenfold. Finally, when the vegetation cover is less than 10%, the surface runoff increases to 73%, and soil loss is 100 times greater than seen in the mostly highly vegetated forests. After a fire, erosion is most serious in steep terrain, where the slope exacerbates problems with runoff.

Another effect seen after a forest fire is that burned land usually adsorbs water more slowly than unburned land. A severe fire in a forest can therefore create ground conditions where surface runoff is more likely to lead to flash floods, and can also increase soil loss and contaminated water. Increases in water flow after a fire can further result in more solids and dissolved materials in the water. Inorganic compounds leached into the water increase the concentration of soluble ions, which may increase both turbidity and toxicity. Finally, both water-soluble and -insoluble nutrients can cause an increase in aquatic plant growth that may lead to decreased water flow.

Filtration. There are many types of filter systems in use, but innovations in filtration technology are needed to more effectively remove contaminants from water. The need to continually improve the efficiency of filtration has spawned an enormous industry, and its fastest-growing nonindustrial application is the generation of clean water. Global spending on all types of filtration (including dust collectors, air filtration, liquid cartridges, membranes, and liquid macrofiltration) is expected to increase from $17 billion in 1998 to $75 billion by 2020.

Currently, several types of filters are used in forests to prevent contaminants from entering the water supply. One type of filtration starts with the construction of holding ponds to contain water that has been identified as contaminated. Evaporation and sand/soil filtration are then used to clean the water, and the contaminated sediment is periodically removed. In addition, holding ponds may be treated with bacteria to degrade organic contaminants to CO_2 and water. Another type of holding pond–based filtration is used to remediate low-pH water, which can be contained in holding ponds that slowly release the water through limestone-lined pipes or stream beds to increase the pH and precipitate the metals. Biofilters are also used in many places; these filters trap organic contaminants until bacteria or fungi can degrade the compounds. Some other types of systems use bark and other waste forest biomass to filter out sediments, soluble metal ions, and toxic organics.

Forests are critical in protecting our surface water supply and sustaining it for the future. Minimizing pollution from mining and forestry processing is key, and it is also important to lower the fuel load in our forests to reduce the frequency of devastating forest fires. Otherwise, it will be increasingly necessary to rely on technological remedies to the problems of forest water contamination instead of taking advantage of the forests' natural systems.

For background information *see* FOREST ECOSYSTEM; FOREST FIRE; GROUND-WATER HYDROLOGY; SEWAGE TREATMENT; SOIL CONSERVATION; SOIL DEGRADATION; WATER CONSERVATION; WATER POLLUTION; WATER SUPPLY ENGINEERING; WATER TREATMENT in the McGraw-Hill Encyclopedia of Science & Technology. Roger M. Rowell

Bibliography. G. Dissmeyer (ed.), *Drinking Water from Forests and Grasslands: A Synthesis of the Scientific Literature*, USDA, Forest Service, Southern Research Station, Gen. Tech. Rep. GTR-SRS-39, Asheville, NC, 2000; Forest Service, *Wildland Waters*, USDA, 1400 Independence Ave SW, Washington, DC 20250-00036; U.S. Geological Survey, *The Quality of Our Nation's Waters: Nutrients and Pesticides*, USGS Circ. 1225, U.S. Department of the Interior, Box 25286, Denver, CO 80225.

Fossil cold-seep ecosystem

Cold-seep ecosystems are biological communities sustained by the seepage from the ocean subsurface of cold fluid containing chemicals (especially methane but also sulfide). Since their discovery in 1984, cold-seep ecosystems have been described from various oceans and seas. More than 200 seep species have been described so far, and many of them (approximately one-third) are symbiont-bearing species. In methane-rich seep environments, symbioses are developed between gutless bivalve mollusks, roundworms, and tubeworms, which may host in their soft tissues (gills, for example) dense communities of methane-consuming and sulfide-oxidizing bacteria. By a microbial process known as chemosynthesis, these endosymbiotic bacteria use the chemical energy derived from the oxidation of sulfide and methane to synthesize organic compounds that serve as food for their mollusk and worm hosts. In marine sediments of a methane seep, sulfide-oxidizing activity is also performed by free-living bacteria that may develop as thick mats composed of giant filaments. This microbial biomass may directly serve as nourishment for many organisms, in contrast to the symbiotic relationships in the gutless and/or mouthless animals.

Silurian cold-seep ecosystem. The remains of the oldest known fossil cold-seep ecosystem have been discovered in a high mountainous area in the Middle Atlas region of Morocco (**Fig. 1**). This seemingly anonymous assemblage of carbonate rocks has been determined to be of Upper Silurian age, dating back more than 416 million years. An understanding of the geologic history of cold-seep ecosystems may permit the evolutionary reconstruction of biological communities based on a unique combination of symbiotic strategies between invertebrate animals and bacteria associated with chemotrophic metabolism.

Recognition of fossil seeps. The rapid mineral (mostly calcium carbonate) precipitation observed in modern seep environments is induced by the increased alkalinity produced by both methane oxidation and sulfate reduction. The alkalinity increase can be attributed to the formation of bicarbonate ions (HCO_3^-) in the sulfate reduction zone. Calcium carbonate enhances the possibility that some of the biological, chemical, and sedimentary attributes of these ecosystems will be fossilized. Comparative analysis of modern cold-seep ecosystems and those from Cenozoic and Mesozoic rocks has revealed useful criteria for recognizing fossil cold seeps in the geologic (especially Paleozoic) record, even when erosive and deformation processes or lack of sufficient exposure may limit understanding of the geologic context.

^{13}C isotope levels. Since modern seep carbonates have decreased levels of the ^{13}C isotope, due to microbial metabolism of isotopically light (highly depleted in ^{13}C) methane, isotope measurements provide a good and rapid geochemical proxy for recognizing ancient seep deposits. Diagenetic processes and/or weathering of rock, however, may alter the original isotopic values, invalidating the purely geochemical approach.

Carbonate accumulations. At the field scale, an anomalous carbonate accumulation surrounded by rocks such as shales or sandstones may suggest that localized and peculiar environmental conditions were

Fig. 1. Outcrop of the Silurian carbonate deposit containing the remains of the most ancient known cold-seep ecosystem, Middle Atlas, Morocco.

established. If these carbonates host a tightly packed, monospecific or low-diversity chemosymbiotic fossil assemblage, they are the possible product of an ancient seep ecosystem.

Presence of chemosymbiotic species. The fossil cold-seep community described from the Upper Silurian of the Moroccan Middle Atlas is characterized by densely packed, monospecific brachiopod concentrations with articulated shells (**Fig. 2**) at a complete range of growth stages, as in a life assemblage. These in-situ concentrations mimic, in many respects, the assemblages of chemosymbiotic bivalve mollusks described from modern seep environments. In addition, the distribution and abundance of certain Paleozoic and Mesozoic brachiopod groups indicate that they are somewhat endemic to seep environments, suggesting a possible ancient relationship with chemosynthesis, perhaps at a trophic or metabolic level. However, little is known about living chemosymbiotic brachiopods and their past adaptations to chemosynthesis. Thus, monospecific brachiopod concentrations alone are not an indicator of cold seepage due to the lack of a modern chemosymbiotic analogue.

Microbial structures. A characteristic of the Middle Atlas Silurian ecosystem is the well-preserved and diverse microbially derived structures (**Fig. 3**). Because chemosynthetic bacteria play a significant role in the ecology of a seep site, evidence of microbially driven processes may help in the recognition of an ancient seep, especially if there is no modern analogue with which to compare it, as is the case with most Paleozoic seep ecosystems.

The number of microbial fossils and the structures produced by their metabolic processes provide evidence of the microbial activity of a fossil cold-seep ecosystem. This activity occurs at varying scales. In the field, laminations of the carbonate rock, which are interpreted as the product of bacterial mats, is a visible indicator of microbial activity. The strongest evidence, however, comes from light microscopic and scanning electron microscopic observation, which has revealed morphologies and mineral products characteristic of microbiologic origin. Mineral rods and filaments are similar to bacterial morphologies that may occur as dense concentrations, suggesting a colonial organization. Mineralized sheaths, also associated with bacterial

(a)

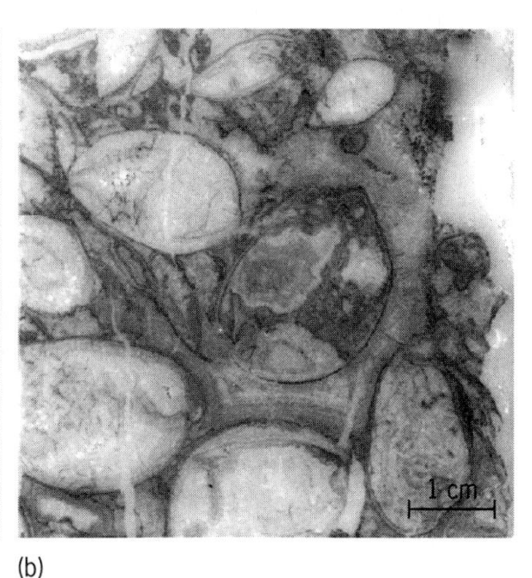

(b)

Fig. 2. Tightly packed brachiopod shells in a Silurian seep carbonate deposit, Middle Atlas, Morocco. (*a*) Brachiopod shells as they appear at the outcrop scale; the coin is 2.5 cm. (*b*) Brachiopod shells from a petrographic thin section observed through a transmitted light microscope.

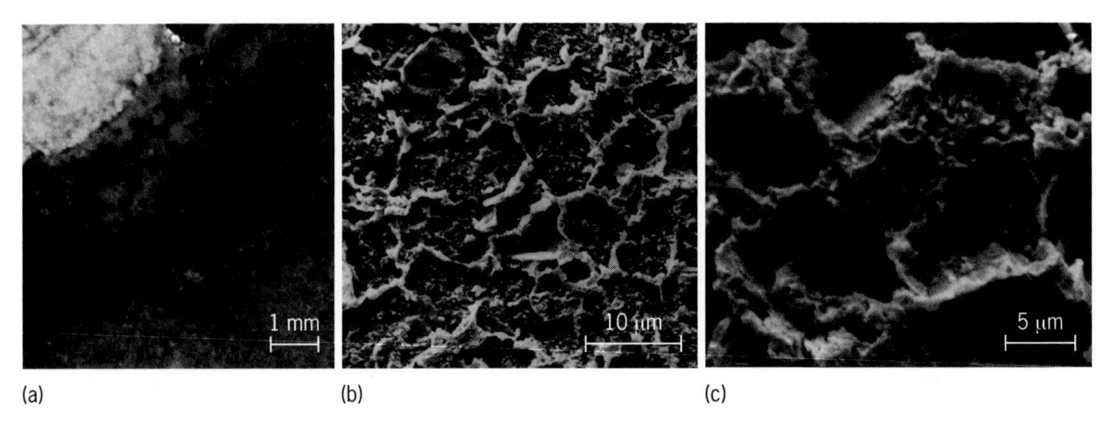

Fig. 3. Silurian seep carbonate deposit, Middle Atlas, Morocco. (*a*) Transmitted light microscope micrograph of microbially derived morphologies. Fine-grained limestone with hematite and microbial remains (dark spots) is shown. (*b, c*) Scanning electron microscope micrographs of same subject. Polygonal elements of microbial-interpreted origin are made up of hematite and filled with calcium carbonate. (*Images b and c reprinted with permission from R. Barbieri, G. G. Ori, and B. Cavalazzi, A Silurian cold-seep ecosystem from the Middle Atlas, Morocco, Palaios, 19:527–542, 2004*)

clusters, occur frequently and may represent the fossil product of the extracellular organic material of bacterial cell colonies. Other bacterial remains and products are clotted sediment concentrations, mineralized biofilms, microbial borings of shells and other hard surfaces, and mineralized crusts of uncertain origin.

A peculiarity of the Silurian ecosystem is a three-dimensional structure (Fig. 3*b, c*) recovered from fine-grained limestone beds devoid of brachiopod shells. It consists of few micrometer-sized polygonal elements made of hematite and filled with calcium carbonate. This polygonal network is unique in the fossil record and has been interpreted as a microbially derived morphology on the basis of a direct comparison with similar polygonal structures produced by the mineralization of the extracellular substances surrounding the bacterial colonies from modern coastal environments of low-latitude regions. Modern stromatolites with similar morphologies are also described from deep marine environments inhabited by sulfur-oxidizing filamentous bacteria. These textures of microbial origin in the Silurian ecosystem are probably exceptional in terms of volume, and their good preservation is a consequence of an early mineral (hematite) replacement of the original organic matter. They are believed to be the remains of the thick microbial mats that developed on the seafloor (as they do in modern environments), where fluid seepage was so highly concentrated that it prevented metazoa (especially brachiopods) from locally establishing.

Conclusion. The first appearance and geologic history of cold-seep ecosystems are still debated. The reasons are the incompleteness of the fossil record and the limited applicability of the uniformitarian principle (which assumes that processes acting in the past are the same as those acting in the present), especially for Paleozoic examples, which have only faint biotic similarities to modern counterparts. The clear connection in modern settings between methane seepage, carbonate precipitation, and chemosynthetic communities, however, is widely used for recognizing fossil seep deposits. Further understanding of these ecosystems is dependent upon the discovery of new ancient (Paleozoic) seep sites in which biotic relationships between bivalves and brachiopods, metazoans, and microbial communities can be evaluated. At the moment, however, the search for the world's oldest (older than Silurian) cold-seep community continues. Perhaps astonishing cold-seep ecosystems (such as methane seep ecosystems based solely on microbes and predating the evolutionary appearance of chemosynthetic metazoans) await discovery somewhere from ancient rocky accumulations. More information about these ecosystems over the entire Phanerozoic time interval may also allow a reconstruction of their role during critical periods of the history of life, such as mass extinctions and recovery events, when they would have acted as oases for refugee biota and actively contributed to biological evolution.

For background information *see* BIOGEOCHEMISTRY; FOSSIL; HYDROTHERMAL VENT; MARINE MICROBIOLOGY; PALEOECOLOGY; SILURIAN in the McGraw-Hill Encyclopedia of Science & Technology.

Roberto Barbieri

Bibliography. K. A. Campbell and D. J. Bottjer, Brachiopod and chemosymbiotic bivalves in Phanerozoic hydrothermal vent and cold seep environments, *Geology*, 23:321–324, 1995; K. A. Campbell, J. D. Farmer, and D. Des Marais, Ancient hydrocarbon seeps from the Mesozoic convergent margin of California: Carbonate geochemistry, fluids and paleoenvironments, *Geofluids*, 2:63–94, 2002; K. Horikoshi and K. Tsujii (eds.), *Extremophiles in Deep-Sea Environments*, Springer, Tokyo, 1999; J. Peckmann and V. Thiel, Carbon cycling at ancient methane-seeps, *Chem. Geol.*, 205:443–467, 2004; M. Sibuet and K. Olu, Biogeography, biodiversity and fluid dependence of deep-sea cold-seep communities at active and passive margins, *Deep-Sea Research II*, 45:517–567, 1998; C. L. Van Dover, *The Ecology of Deep-Sea Hydrothermal Vents*, Princeton University Press, 2000.

Gene therapy (veterinary medicine)

Gene therapy is the transfer of genetic material into specific cells of a patient to treat disease. It has been successfully used to treat several inherited metabolic diseases in pet animals, although none of the treatments is near to being routine in clinical veterinary medicine. The diseases have been studied primarily as a means to develop therapy for humans with homologous disorders.

Basic technique. Genes, the specific sequences of deoxyribonucleic acid (DNA) in chromosomes, encode the production of proteins and other molecules that provide the structure and function of cells, tissues, and organs. When an alteration (mutation) occurs in the base sequence of a gene, the result is often a disruption of the function of the protein that, in turn, can cause disease. The approach that has been used to several inherited metabolic disorders in pet animals involves the insertion of a normal copy of the relevant coding sequence (cDNA) for the protein either directly into cell chromosomes (often in a nearly random way) or nearby (episomal). This technique does not replace or correct the defective gene; rather it provides the normal coding sequence to produce the critical protein. This has been achieved by using a virus to deliver the genetic material, taking advantage of the infectious characteristics of viruses that can target cells of interest and deliver their viral genes for expression.

Viral vectors. To use viruses as gene therapy vectors (delivery vehicles), most of the viral genes are replaced with the normal genetic material to be delivered, including the cDNA of interest, and the vector is produced in large amounts, purified, and injected either locally or systemically. To promote and stabilize the transcription of the cDNA into the protein of interest, genetic elements from other viruses and mammalian species are often included in the vector.

There are various viral vectors available for gene therapy. Each has its unique advantages and disadvantages. The two viral vectors that have been used in gene therapy for the metabolic disorders described below were made from an adeno-associated virus (AAV) or a retrovirus (RV). Vectors made from AAV do not require dividing cells for transduction (the transfer of genetic material) to occur, but the genetic material may not be successfully incorporated into the patient DNA, and the amount of genetic material that can be transferred is limited. Retroviruses do integrate into the patient's chromosomal DNA and will be passed on to daughter cells, and the amount of genetic material that can be contained within the vector is relatively large; however, these vectors require dividing cells for transduction, and integration into the patient's chromosomal DNA may disrupt other genes in the cell leading to serious side effects.

Applications. Gene therapy has been successfully used to treat three inherited metabolic diseases in pet animals that also affect humans: Leber congenital amaurosis, mucopolysaccharidosis VII, and hemophilia B. The diseases are relatively rare in the pet population, but they are found in specific breeds of dogs, which increases their breed-specific prevalence.

Leber congenital amaurosis. Leber congenital amaurosis is a progressive disease of the retina that produces blindness. It is caused by a mutation in *RPE65*, a gene important to the function of the retinal pigment epithelium (RPE), which supports the photoreceptor cells of the retina. The *RPE65* gene is conserved across many species. Briard dogs with a mutation in *RPE65* have early and severe visual impairment similar to that seen in human infants with Leber congenital amaurosis. Visual impairment in affected dogs is caused by a mutation that truncates the normal protein encoded by *RPE65*, resulting in undetectable levels of the visual pigment rhodopsin, which is required for normal vision. Histopathologic examination of the retinas of dogs homozygous for the *RPE65* mutation shows prominent inclusions in the RPE and a slightly abnormal morphology of rods (a type of photoreceptor) early in life. Degeneration of the photoreceptor progresses slowly.

Gene therapy for Leber congenital amaurosis in the Briard dogs involved incorporation of the wild-type canine *RPE65* gene into a vector made from an AAV, which was then injected under the retina in one eye of the affected dogs. This approach placed the vector in direct contact with the cells of the RPE. Retinal function was measured by electroretinograms (ERGs) and showed improvement in treated eyes compared with the same eyes prior to treatment and compared with the untreated contralateral eye. As expected, retinal function did not completely return to normal, as the injection of the vector targeted only about 35% of the retina. Injection of the vector into the vitreous body, which is in contact with the retina, was also attempted but was shown not to improve retinal function.

Importantly, the results of qualitative (behavioral) assessments of visual function conducted 4 months after treatment were consistent with the ERG results. The treated dogs were scored as "normally sighted" under room lighting; that is, they consistently avoided objects placed directly in front of them and on the treated side but failed to avoid objects placed on the untreated side. The oldest treated animal has remained visually stable for 5 years.

Mucopolysaccharidosis (MPS) VII. Mucopolysaccharidosis VII is one of more than 40 lysosomal storage diseases, each of which is caused by the loss of activity of a specific enzyme necessary for the degradation and recycling of large substrates brought into cells. These large undegraded substrates accumulate (are stored) in the lysosome, resulting in disease. Fortunately for therapy: (1) a posttranslational modification (in the Golgi apparatus) adds a mannose 6-phosphate (M6P) moiety to most lysosomal enzymes, which enables their movement into the lysosome via an internal trafficking system; (2) most lysosomal enzymes are secreted from cells to some degree; and (3) many cells have M6P receptors on the cell surface, allowing uptake and trafficking of enzymes into the lysosome; and (4) only a small percentage of normal lysosomal enzyme activity is needed to produce a substantial therapeutic response in a cell. Thus, in gene therapy, providing a normal copy of the cDNA of interest to some cells allows production of the normal enzyme, secretion of the enzyme into the extracellular fluid, and distribution to and uptake by other cells in the body that have disease but have not been transduced. In essence, gene therapy creates an "enzyme factory" using the patient's own cells.

MPS VII in German shepherd dogs is caused by deficient activity of β-glucuronidase due to a single amino acid substitution in its encoding gene. The body weight of untreated MPS VII dogs is 50% of normal at 8 months of age. Affected dogs cannot stand or walk by 6 months of age due to abnormal joints, and all have corneal clouding and mitral heart valve thickening.

Affected dogs were treated at 3 days of age by the intravenous administration of a retroviral vector containing the normal canine β-glucuronidase cDNA, together with elements of human and woodchuck DNA. The age of the affected dogs was selected to allow the retrovirus, which requires dividing cells to be integrated into the patient's DNA, to transduce liver cells that are actively dividing in the rapidly growing puppies. Treated dogs have shown stable serum β-glucuronidase activity, between 40 and 6000% of normal levels, for more than 4.5 years after treatment. In addition, these dogs achieved close to 90% of their normal body weight, were capable of walking due to improvements in the bone and joint abnormalities, had little or no corneal clouding, and no mitral valve insufficiency.

Hemophilia B. Hemophilia B is a bleeding disorder caused by a lack of clotting factor IX (F.IX). The disorder has been described in more than a dozen breeds of dogs. Many experiments have been performed to evaluate gene therapy in affected dogs using a variety of vector systems and delivery approaches, usually targeting the liver (the site of normal F.IX production) or skeletal muscle. A recent method involved a novel intravascular delivery technique to transduce a large volume of skeletal muscle. The dogs in this study had disease due to a missense mutation (which converts a codon coding for one amino acid to a codon coding for another amino acid) in the portion of the gene encoding the catalytic domain of the protein, resulting in no detectable F.IX protein or activity.

The technique used to target a large volume of skeletal muscle involved vascular delivery of an AAV vector containing the canine F.IX cDNA and a cytomegalovirus promoter/enhancer to the muscle in a single limb. The approach was developed because the likelihood of forming neutralizing (inhibitory) antibodies to F.IX rises as the AAV dose per site is increased. Also, skeletal muscle has a limited capacity to execute the essential posttranslational modifications needed for F.IX activity, requiring large volumes of muscle to be transduced. Thus, safety and efficacy imposed an upper limit on the amount of vector that could be injected intramuscularly into a single site, and an excessively large number of injection sites would have been required to provide sufficient transduced skeletal muscle.

Although the protein produced from the vector was the normal canine F.IX, it was foreign to the affected animals, which had never been exposed to F.IX protein, and assays detected the production of inhibitory antibodies to F.IX. The administration of a drug that suppresses the immune system was necessary for 6 weeks following vector administration to avoid the development of the antibodies that would neutralize F.IX activity. Immunosuppression treatment was discontinued after 6 weeks, and F.IX expression persisted, with production of widespread, long-term (more than 3.5 years), stable circulating levels of canine F.IX that are 4–14% of normal human levels. This level of expression resulted in a dramatic reduction in the bleeding episodes requiring treatment in a large animal.

Conclusion. To fully understand and treat genetic diseases in human patients, the use of authentic (orthologous gene) animal models is required in studies that for ethical and practical reasons are not possible in human patients. The dogs in these gene therapy studies were selected based on several factors: (1) their phenotype—that is, they showed signs of having these genetic metabolic diseases, which also affect humans, indicating that there are similar genetic mechanisms underlying the diseases in both dogs and humans; (2) their size, which was similar to that of children; and (3) their long lifespan, which may enable the detection of limitations to therapy. Thus, they provided the translational bridge from in vitro and mouse experiments to human patients. Applications for human clinical trials are currently under consideration for all three disorders.

For background information *see* EYE (VERTEBRATE); GENE; GENE ACTION; HEMOPHILIA; HUMAN GENETICS; LYSOSOME; MUTATION; RETROVIRUS in the McGraw-Hill Encyclopedia of Science & Technology.

Mark Haskins

Bibliography. G. M. Acland et al., Gene therapy restores vision in a canine model of childhood blindness, *Nat. Genet.*, 28:92–95, 2001; V. R. Arruda et al., Regional intravascular delivery of AAV-2-F.IX to skeletal muscle achieves long-term correction of hemophilia B in a large animal model, *Blood*, 105:3458–3464, 2005; K. P. Ponder et al., Therapeutic neonatal hepatic gene therapy in mucopolysaccharidosis VII dogs, *Proc. Nat. Acad. Sci. USA*, 99:13102–13107, 2002.

Geochronology

Geochronology involves measuring absolute ages for ancient rock-forming events and applying these ages to understand the development of the Earth. The most precise methods use radioactive isotopes (unstable atoms of an element) that decay into a radiogenic daughter isotope of another element. Either the parent isotope is constantly replenished (for example, from cosmic rays, like ^{14}C), or it decays slowly enough that some still remains since the time of the Sun's formation. Such a radioactive decay system is known as a

Long-lived radioactive decay systems used for geochronometry

Parent isotope	Daughter isotope	Half-life, Ma*
	Potassium-argon	
^{40}K	^{40}Ar	12,50
	Uranium-lead	
^{238}U	^{206}Pb	4,468
^{235}U	^{207}Pb	704
	Thorium-lead	
^{232}Th	^{208}Pb	14,000
	Rubidium-strontium	
^{87}Rb	^{87}Sr	49,000
	Samarium-neodymium	
^{147}Sm	^{143}Nd	106,000
	Lutetium-hafnium	
^{176}Lu	^{176}Hf	35,000
	Rhenium-osmium	
^{187}Re	^{187}Os	46,000

*Ma = million years.

geochronometer. A list of long-lived geochronometers along with their half-lives is given in the **table**. An age is determined by measuring the ratio of radiogenic daughter over radioactive parent isotope in a mineral. This usually records the time since the atoms in the mineral became isolated as a result of initial crystallization or cooling. Ideally, the host mineral should crystallize with a very high ratio of parent element over the element to which the parent decays, so most of the latter element in the mineral will now be radiogenic. The mineral should also be inert, so the atoms of the decay system remain isolated from outside influences. Many parent and daughter elements are chemically reactive, so parent and daughter would be easily separated in rocks if they were not contained within a chemically unreactive (inert) mineral that protects them from outside influences. These conditions are most easily satisfied for the potassium-argon (K-Ar) and uranium-lead (U-Pb) systems, which are widely used for determining precise ages. The other systems are still very useful as isotopic tracers that provide clues to the origin and history of the rock reservoirs from which their magmas were derived.

Potassium-argon (K-Ar) system. Most igneous rocks contain potassium. Because argon gas is expelled from hot magma, almost all the argon in such rocks consists of radiogenic ^{40}Ar. The age of a sample can be easily measured by placing it along with a standard of known age in a nuclear reactor where neutrons convert some of the nonradioactive potassium isotope (^{39}K) to the nonradiogenic argon isotope (^{39}Ar). The sample and standard are then heated separately in an oven or with a laser, expelling the argon into a gas source mass spectrometer, which measures $^{39}Ar/^{40}Ar$. Since both sample and standard were exposed to the same neutron flux, the age of the sample can be determined by comparing its argon isotopic ratio with that of the standard. As the sample is heated, the argon is measured in batches expelled at different temperatures. The best age estimate is taken when the ages form a "plateau" at the highest temperatures. This represents the argon that was held most strongly (**Fig. 1**).

If an older mineral becomes reheated past a point called the blocking temperature, its accumulated argon will diffuse out of the crystal. The age measured today may then record the time of cooling back through the blocking temperature, rather than igneous crystallization. The K-Ar system is often used to study the thermal history of rocks, which provides important information on tectonic (mountain-building) processes. The system works particularly well on younger rocks and has largely provided the basis for the time scale of human evolution, based on dating volcanic ash beds interlayered with sediments containing early hominid fossils. The most challenging application was perhaps P. Renne's dating of the eruption of Mount Vesuvius that destroyed Pompeii. Measuring argon from the high potassium mineral sanidine, he established an age of 1925 ± 94 years, which is in agreement with historical records (AD 79).

Uranium-lead (U-Pb) system. The U-Pb system consists of two geochronometers: ^{238}U-^{206}Pb and ^{235}U-^{207}Pb. Disagreement (discordance) between these ages indicates that the system has been disturbed, a powerful test for accuracy. The success of U-Pb geochronology is largely due to zircon. This is a widespread trace mineral, which, in its pristine state, is one of the most inert minerals known. Zircon concentrates uranium but rejects lead when it crystallizes; thus its present-day lead composition is almost entirely radiogenic. Unfortunately, the crystal structure of zircon gradually degrades because of radioactive decay, and crystals with enough damage can become chemically altered, losing some of their lead. One solution is to choose only crack-free undamaged crystals for dating and abrade off exposed

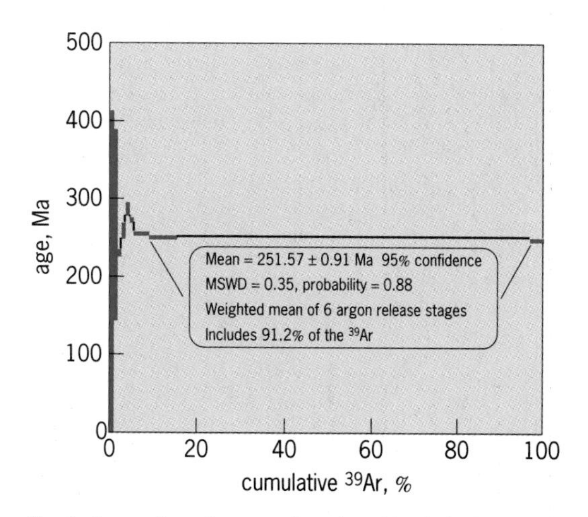

Fig. 1. Age pattern for argon from hornblende in the Noril'sk intrusion, a body related to the Siberian flood lavas. The pattern shows evidence of Ar loss (too young an age) followed by excess argon (too old an age) in the first argon fractions released at low temperatures. A more detailed analysis shows that excess argon in the plateau has biased the age slightly upward. Taking this into account, the best age estimate is 249.3 ± 1.6 Ma (million years). (*Modified from P. R. Renne, 1995*)

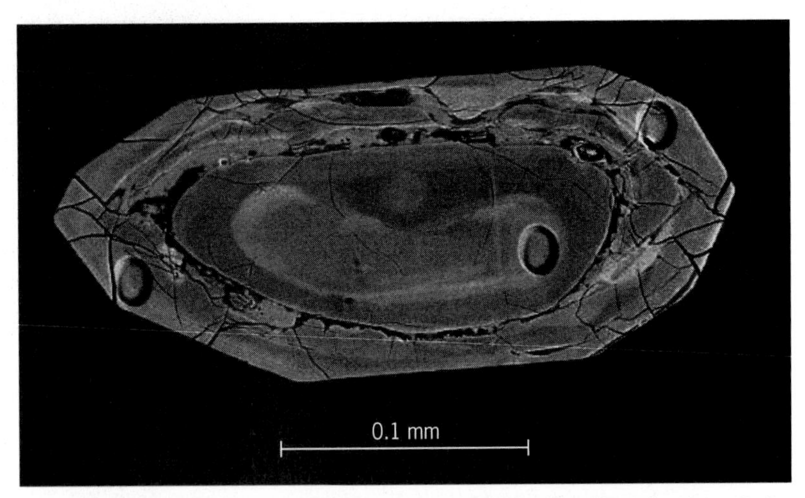

Fig. 2. Backscattered electron image of a cut and polished surface through a zircon that shows two stages of growth. A 2480-Ma-old core has been overgrown by younger zircon at 1800 Ma, the age of the granite from which this zircon was recovered. The core gives the age of at least one of the rocks that melted to form the magma from which granite crystallized. The pits were left from the ion probe beam used to date the zircon.

surfaces. This is followed by dissolution, chemical extraction of U and Pb, and isotope analysis using a thermal ionization mass spectrometer. Another approach is to examine cut and polished crystals using electron imaging methods and use an ion microbeam to sample unaltered crystal domains. The primary beam ablates lead and uranium ions from the crystal, which are focused into a secondary beam and directed into a high-resolution mass spectrometer. The first approach can produce the most precise ages (less than 0.1% error) because of the large volume of sampled material. However, it is laborious and difficult to apply to crystals with more than one age of growth. Older cores may be left from zircon in preexisting melted rocks, and younger overgrowths may have formed during metamorphism (**Fig. 2**). In such cases the ion microprobe approach (Sensitive High Resolution Ion Microprobe or SHRIMP) can be very effective. The size of the sampled spot is usually about 20 micrometers (thousandths of a millimeter) across, and the method is almost nondestructive, as the beam penetrates only a few micrometers into the crystal. However, the small volume of analyzed sample usually limits the errors to about ten times larger than with whole-grain analysis.

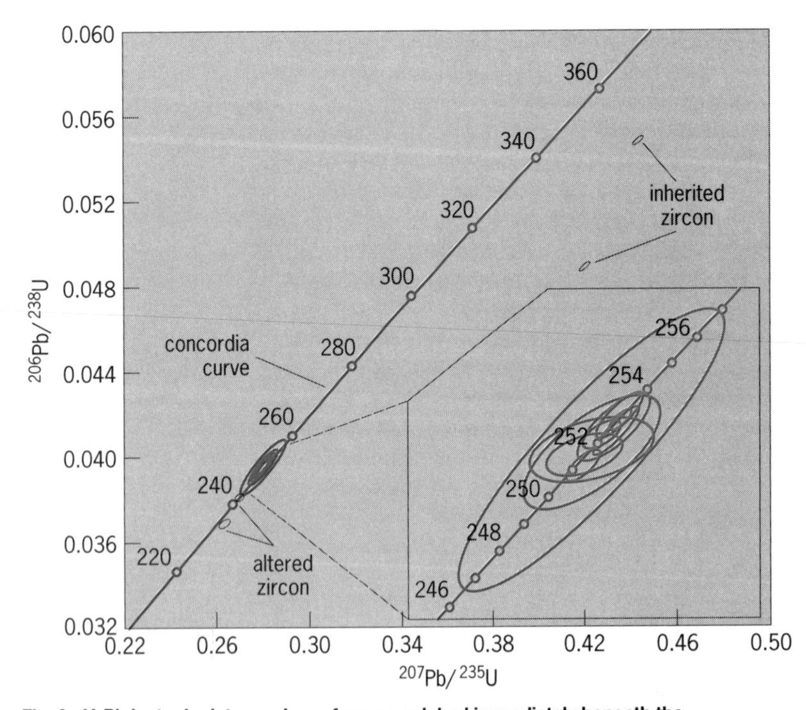

Fig. 3. U-Pb isotopic data on zircon from an ash bed immediately beneath the Permo-Triassic boundary showing the complexity of determining high-precision ages. Concordia is the curve where both U-Pb decay systems give the same age. The ellipses represent 95% confidence errors of each analysis. Two older discordant analyses are from zircon with much older cores, while two slightly younger ones show loss of lead. Even among the 10 concordant data (inset), there are two groups which give nonoverlapping ages of 252.7 ± 0.4 Ma. (contaminants?) and 251.4 ± 0.4 Ma. (*Modified from S. Bowring et al., 1998*)

Recent innovations and achievements. Y. Amelin used the U-Pb system to precisely date the first minerals to condense from the solar nebula at 4567.2 ± 0.6 Ma (million years). His age determination of a basaltic meteorite (an igneous rock) is only slightly younger at 4565.0 ± 0.9 Ma. Accretion and melting of small planetary bodies was therefore fast enough that the heat source for melting was probably radioactive now-extinct isotopes like ^{26}Al produced in earlier supernovas.

The introduction of multicollector mass spectrometers with inductively coupled plasma sources (MC-ICPMS) has opened up most of the periodic table to radiogenic and stable isotope analysis. MC-ICPMS has revealed a slight difference in isotopic composition of tungsten in meteorites compared to the Earth due to the decay of the now-extinct system ^{182}Hf to ^{182}W early in the solar system and to the fact that tungsten is strongly partitioned into the metallic cores of planets. This establishes the age of terrestrial core formation at about 30 ± 3 Ma following the earliest condensates, that is, at about 4537 Ma. This is the time when Earth reached its present size, apparently after a massive collision between a proto-Earth and a Mars-sized planet.

The oldest terrestrial rock thus far discovered is the Acasta gneiss from the Northwest Territories, Canada, dated at 4030 Ma by S. Bowring using the SHRIMP. The only other material record of the first 500 million years of Earth's history consists of detrital zircon from the Jack Hills sandstone in western Australia. SHRIMP analyses have revealed detrital zircons as old as 4400 Ma. ICPMS measurement of the ^{176}Lu-^{176}Hf system in these zircons indicates that their source magmas were derived from chemically fractionated rocks that must themselves have formed even earlier. The recent improvements in ion microprobes have also allowed precise measurement of oxygen isotope ratios from small spots within zircon grains. High $^{18}O/^{16}O$ oxygen isotope ratios measured in zircons older than 4000 Ma indicate that their magmas likely formed from melting of sedimentary rocks. This extraordinary record from one mineral suggests that weathered continents and oceans existed shortly after Earth's formation.

Understanding biological evolution is a major incentive for advances in geochronology. The largest known mass extinction occurred at the Permo-Triassic boundary. S. Bowring established an age of 251.4 ± 0.3 Ma for the extinction event by dating whole abraded zircons from volcanic ash layers (**Fig. 3**), in agreement with the age found for the largest known volcanic episode in Siberia. This suggests that intense volcanism can modify the global environment sufficiently to cause mass extinction, a controversial conclusion. To definitively establish such correlations requires accuracy on the order of ±0.1 Ma. Such errors are attainable by using mixed standards made from artificial isotopes and by assuring that all labs are calibrated to the same standard. This is one objective of the current EARTHTIME (http://earth-time.org) project. Another challenge is to improve knowledge of geochronometer decay rates. The Permo-Triassic boundary was also correlated with the Siberian flood basalts using ^{39}Ar-^{40}Ar dating but at a younger age (250.0 ± 0.3 Ma) than found from U-Pb. The difference is probably due to error in the potassium decay rate. Such uncertainties particularly limit the use of the Lu-Hf, Rb-Sr, and Re-Os systems.

Future developments in instrumentation. Rapid determination of precise ages on individual crystal domains may eventually be possible using laser ablation–inductively coupled multicollector mass spectrometers (LA-MC-ICPMS), where a finely focused ultraviolet laser ablates zircon from a polished section into plasma. These instruments can now produce ages with precision comparable to an ion microprobe but require only a few minutes (as opposed to 20 min) per analysis. ICPMS instruments are in a rapid state of development and may produce the next major advance in geochronology.

For background information *see* COSMOCHEMISTRY; DATING METHODS; GEOCHRONOMETRY; GEOLOGIC TIME SCALE; ISOTOPE; LEAD ISOTOPES (GEOCHEMISTRY); MASS SPECTROMETRY; RADIOCARBON DATING; ROCK AGE DETERMINATION in the McGraw-Hill Encyclopedia of Science & Technology.

Donald W. Davis

Bibliography. Y. Amelin et al., Pb isotopic ages of chondrules and Ca-Al-rich inclusions, *Science*, 279:1678–1683, 2002; G. B. Dalrymple, *Ancient Earth, Ancient Skies: The Age of Earth and Its Cosmic Surroundings*, Stanford University Press, 2004; A. P. Dickin, *Radiogenic Isotope Geology*, 2d ed., Cambridge University Press, 2004; G. Faure and T. A. Mensing, *Isotopes: Principles and Applications*, 3d ed., Wiley, 2004; I. MacDougall and T. M. Harrison, *Geochronology and Thermochronology by the ^{40}Ar/^{39}Ar Method*, 2d ed., 1999.

Global biogeochemical cycles

Global biogeochemical cycles can be defined as any of the natural circulation pathways through the atmosphere, hydrosphere, geosphere, and biosphere of the essential elements of living matter. These elements in various forms flow from the nonliving (abiotic) to the living (biotic) components of the biosphere and back to the nonliving again. Research on the biogeochemical cycles focuses on seven of the major elements that make up more than 95% of all living species: hydrogen, carbon, sulfur, oxygen, and the major nutrients nitrogen, phosphorus, and silicon. These elements combine in various ways. Here we will focus on carbon and sulfur due to their importance for feedback with the climate and for environmental concerns.

Earth's life is linked to climate through a variety of interacting cycles and feedback loops. Human activities, such as deforestation and fossil-fuel burning, have directly or indirectly modified the biogeochemical and physical processes involved in determining the Earth's climate over the last millennium and can disturb a variety of land and marine ecosystem interactions.

Climate system and biogeochemical cycles. The main driver for the climate system is the radiative balance at the top of the atmosphere. Incoming solar radiation is partly absorbed and partially reflected by the atmosphere. The remaining part is absorbed at the Earth's surface and reemitted as long-wave radiation into the atmosphere. Greenhouse gases such as water vapor, carbon dioxide (CO_2), methane, and nitrogen oxides absorb the long-wave radiation and serve as a blanket for the lower atmosphere.

The potential for two-way interaction and feedback between the carbon cycle and the climate system comes from the radiative properties of CO_2, methane, and other greenhouse gases in the atmosphere. Changes of atmospheric CO_2 through anthropogenic emissions may lead to significant changes in atmospheric climate, ocean circulation, and the magnitude and extent of continental ice sheets. Climate records indicate that large-scale abrupt climate changes have occurred repeatedly throughout geologic history which are related to variations in the Earth's orbital parameters, as well as to significant perturbations in the hydrological cycle. Coupled climate simulations suggest that CO_2-induced warming will lead to reduced marine and terrestrial carbon uptake. In the ocean, climate changes alter stratification, solubility, and nutrient supply to the euphotic zone, the zone in which sufficient light is available. These changes in the export of carbon into the deep sea feed back on the atmospheric carbon dioxide concentration. Several studies indicate overall positive feedback processes due to global warming, but many of these processes are quantitatively not well known.

Global carbon and nutrient cycles. The biogeochemical cycle of carbon and nutrients is of great importance to global climate change. The cycle includes four main reservoirs: the atmosphere, organic compounds in living or dead organisms, dissolved substances in the oceans and freshwater reservoirs, and storage in the geosphere (**Fig. 1**).

CO_2 in the atmosphere. The primary forces that drive the long-term drifting of climate from extremes of ice-free poles to extremes of cold with massive continental ice sheets and polar ice caps are Earth's orbital geometry and plate tectonics. The orbitally related climate oscillations vary about a climatic mean in response to changes such as continental geography and topography, seaways and bathymetry, and concentrations of greenhouse gases like CO_2 or methane. A recent review of the geologic record concluded that

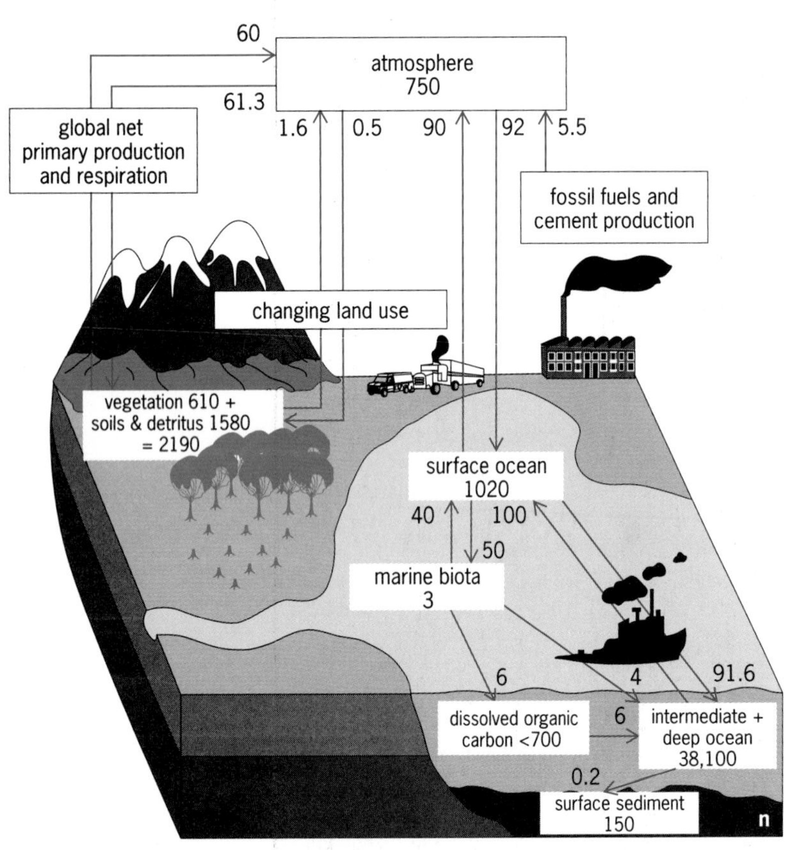

Fig. 1. Global carbon cycle storage (Pg carbon; 1 Pg C = 10^{15} g C) and fluxes (Pg C/yr) estimated from the 1980s. The numbers are similar to the recent International Panel of Climate Change Report (IPCC, 2001). Exact estimates of reservoirs and uptake of anthropogenic CO_2 are still controversial.

CO_2 and widespread continental glaciations generally are well correlated through the Phanerozoic (the last 540 million years).

The atmospheric CO_2 concentration has significantly increased from about 280 parts per million (ppm) in 1800, the beginning of the industrial age, to 380 ppm today. New evidence suggests that concentrations of CO_2 started rising about 8000 years ago, even though natural trends indicate they should have been dropping. Some 3000 years later, the same thing happened to methane. Without these changes, current temperatures in northern parts of North America and Europe would be cooler by 3–4°C, which is enough to make agriculture difficult. Recent research has tried to estimate the sources and sinks of carbon from data and with computer simulations, which is fundamental for the understanding of the natural carbon cycle and significant for formulating global CO_2 emission strategies. According to these studies, the ocean acts as a major sink for anthropogenic CO_2.

Ocean carbon cycle. The marine carbon cycle is typically broken down into two components. The first is the solubility pump, with low solubility of CO_2 in warm waters. The second is the biological pump, affecting the redistribution of biologically active elements, such as carbon, nitrogen, and silicon, within the circulating waters of the ocean. The distributions of circulation patterns (for example, eddies) and biomass are highly correlated (**Fig. 2**). Part of the biologically active elements, which is exported to deeper layers and not recycled in the water column, is buried in sediments.

Planktonic organisms use carbon during photosynthesis, reducing the dissolved inorganic carbon in seawater. This change reduces the surface-ocean partial pressure of carbon dioxide (pCO_2) and increases the uptake of CO_2 from the atmosphere. About 44% of the total gross primary production is used for the net primary production, while the remaining part is consumed by respiration. Nitrogen (N_2) fixation can substantially change the inventory of nitrogen in the ocean, thereby stimulating marine productivity and atmospheric CO_2. Micronutrients like iron delivered to the ocean by river and dust input can promote the downward export of carbon from the euphotic zone. Moreover, particle size and particle aggregation, such as mineral ballasting, help to control the flux of organic material from the euphotic zone into the deep sea.

Climate change affects the intensity of the biological pump by altering limiting factors for productivity such as nutrient availability, light, or temperature. For example, analyses of computer simulations indicate that nutrients in polar surface waters may become depleted by stratification because of global warming. These water masses are transported subsurface to the equatorial region where a reduced nutrient supply causes a decrease in productivity.

Plankton species producing calcium carbonate skeletal material (such as coccolithophores) reduce the dissolved inorganic carbon concentration. The chemistry of calcium carbonate controls the pH of the ocean and, in turn, plays a large role in regulating the CO_2 concentration of the atmosphere on time scales of thousands of years or longer. For example, recent global warming simulations predict a significant decline of the pH value by about 0.4 or more over the next hundred years. Such a dramatic change can reduce the calcification of corals necessary for the buildup of reefs and promote the dissolution of calcareous shells and exoskeletons of marine organisms.

Land biosphere carbon cycle. In contrast to the ocean, most carbon recycling through the land takes place locally within ecosystems. About half of the terrestrial gross primary production [120 Pg (1.3×10^{11} tons) carbon per year] is respired by plants. The remainder (Net Primary Production; NPP) is approximately balanced by respiration, with a smaller component of direct oxidation in fires (combustion). Most of the NPP is transformed to detritus through the degradation of plant tissue (for example, leaves). Some of the detritus decomposes and the carbon is returned quickly to the atmosphere as CO_2, while some is converted to soil carbon, which decomposes more slowly. Enhanced CO_2 levels in the atmosphere have a fertilization effect on short time scales, but the long-term effects are still not well known. Carbon dioxide uptake by the land carbon cycle may be reduced by global warming due to an increased carbon emission by soil microbes.

Carbon cycle in the geosphere. A long-term cycle on Earth involves the interaction of CO_2 with the Earth's crust. Atmospheric CO_2 dissolved in rainwater

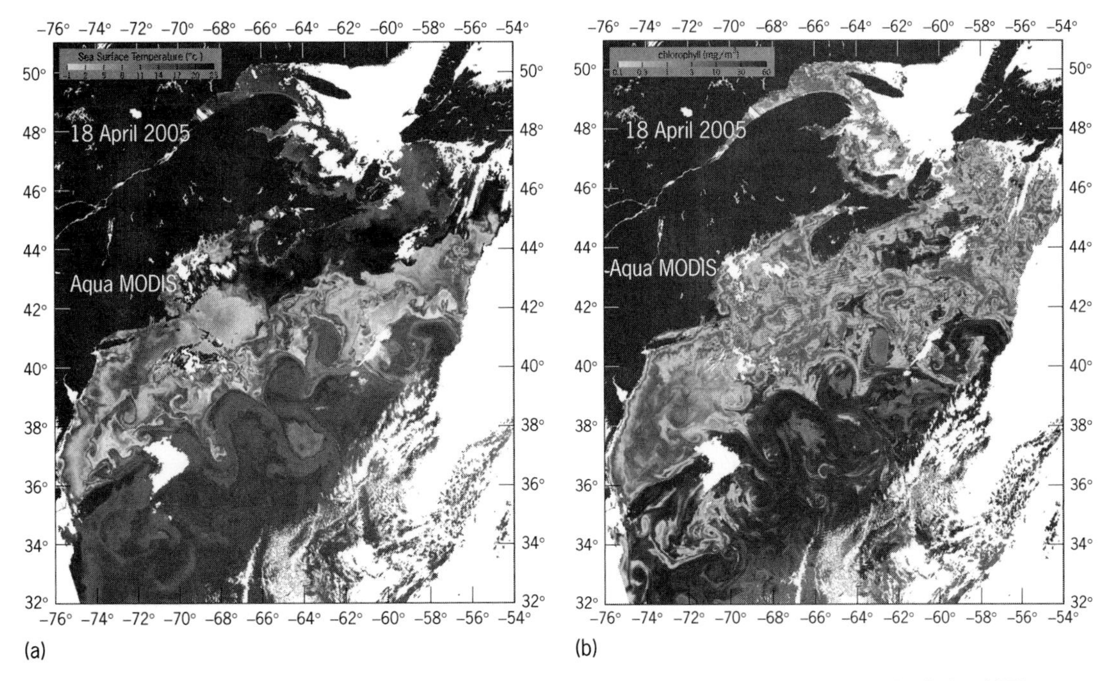

Fig. 2. Example of spatial variability of the marine biota (Fig. 1) because of fluctuations in the ocean circulation. (*a*) The warm heart of the Gulf Stream is readily apparent in the sea-surface temperature image. As the current flows toward the northeast, it begins to meander and pinch off eddies that transport warm water northward and cold water southward. (*b*) The current also divides the local ocean into a low-biomass region to the south and a higher-biomass region to the north. This is evident in the chlorophyll image. White areas are cloud coverage. The data were collected by MODIS aboard *Aqua* on April 18, 2005. (*NASA Ocean Color Research Team*, http://oceancolor.gsfc.nasa.gov/)

forms carbonic acid, which reacts with rocks, a process called weathering. On longer geologic time scales, the processes of sedimentation, chemical transformation, uplift, sea-floor spreading, and continental drift are involved in the carbon cycle.

Recently, massive methane hydrates (clathrates) which are buried in marine sediments have received attention as a possible energy source. They are not stable when the ocean water temperature rises as it did during the Paleocene/Eocene thermal maximum (about 55 million years B.P.) or if the pressure falls (for example, due to sea-level falls). These instabilities can generate undersea slumps, which can result in tsunamis and sudden methane release to the environment which might induce a positive feedback on the climate.

Global sulfur cycles. The sulfur cycle in general, and acid rain and smog issues in particular, encompasses important climatic feedbacks and socioeconomic problems. Living organisms, including plants, assimilate sulfur while at the same time sulfur is released by organisms as an end product of metabolism. Sulfur can exist both as a gas and as sulfuric acid particles. The major sulfur gases are sulfur dioxide (SO_2), dimethyl sulfide (CH_3SCH_3 or DMS), carbonyl sulfide (OCS), and hydrogen sulfide (H_2S). The lifetime of most sulfur compounds in the air is relatively short (for example, days). In its fully oxidized state, sulfur exists as sulfate and is the major cause of acidity in both natural and polluted rainwater. This link to acidity makes sulfur important to geochemical, atmospheric, and biological processes such as the natural weathering of rocks, acid precipitation, and rates of denitrification. Natural sulfur sources are SO_2 emissions from volcanoes and DMS emissions from the marine biosphere to the atmosphere. Similar to the carbon cycle, the sulfur cycle has been significantly perturbed by human influences. Today, global anthropogenic emissions constitute almost 75% of the total sulfur emissions, with 90% occurring in the Northern Hemisphere. These compounds mix with water vapor and form sulfuric acid smog. In addition to contributing to acid rain, the sulfuric acid droplets of smog form a haze layer that reflects solar radiation and can cause respiratory diseases or a cooling of the Earth's surface.

Plants take up some forms of these compounds and incorporate them into their tissues for growth. These organic sulfur compounds are recycled on land or in the water after the plants die or are consumed by animals. Bacteria are of importance since they can transform organic sulfur to hydrogen sulfide gas (H_2S). In the ocean, zones of H_2S (dead zones) can be found in areas with sluggish circulation and a high nutrient supply such as the northern Gulf of Mexico. Some of the sulfide is taken up by the lithosphere during the formation of rocks. Superimposed on the fast cycles of sulfur, the sedimentary cycle operates on long geologic time scales and includes processes such as erosion, sedimentation, and uplift of rocks containing sulfur.

Future research on global biogeochemical cycles in the interdisciplinary field of Earth system science will seek to explore interactions among the major components of the Earth system (continents, oceans, atmosphere, ice, and life) by assimilating large data sets (Fig. 2) into comprehensive climate models. These studies will be sophisticated to distinguish natural from human-induced causes of change and important to understand and predict the consequences of change.

For background information *see* ATMOSPHERE; ATMOSPHERIC CHEMISTRY; BIOGEOCHEMISTRY; BIOSPHERE; CLIMATE MODELING; GEOCHEMISTRY; GLOBAL CLIMATE CHANGE; GREENHOUSE EFFECT; HYDROSPHERE; MARINE SEDIMENTS in the McGraw-Hill Encyclopedia of Science & Technology.

Arne Winguth

Bibliography. R. B. Alley et al., Abrupt climate change, *Science*, 299:2005–2010, 2003; G. P. Brasseur, J. Orlando, and G. Tyndall (eds.), *Atmospheric Chemistry and Global Change*, Oxford University Press, New York, 1999; IPCC, *Climate Change 2001: The Scientific Basis, Contribution of Working Group I to the Third Assessment Report of the Intergovernmental Panel on Climate Change*, ed. by J. Houghton et al., Cambridge University Press, 2001; D. L. Royer et al., CO_2 as a primary driver of Phanerozoic climate, *GSA Today*, 14(3):4–10, 2004; W. F. Ruddiman, How did humans first alter global climate?, *Sci. Amer.*, 3:46–53, 2005; C. L. Sabine et al., The oceanic sink for anthropogenic CO_2, *Science*, 305:367–371, 2004.

Graph-based intelligence analysis

An intelligence analyst's job is to answer policymakers' questions on a given subject. The questions may be broad and open-ended (for example: What economic factors shape the foreign policy of another country?) or they may be very specific (for example: Is a terrorist planning to strike a particular target?). Answering these questions often requires the analyst to predict future events. There are thousands of pieces of information available on each subject, and they invariably contain incomplete and contradictory evidence on the situation. Making reliable predictions with limited and contradictory evidence is an incredibly challenging task. Many analysts are turning to new graph-based analysis approaches to help them better make predictions. These new approaches are an interdisciplinary combination of mathematical graph theory, anthropology, and sociology.

Analysts will often hand-draw diagrams when they are trying to understand a complex situation or the relationships between people in a group. The diagrams also can be created on a computer and treated formally as attributed relational graphs, an extension of the abstract directed graph used in mathematics. In a typical analyst graph, nodes represent people, organizations, objects, concepts, or events. Edges between the nodes represent communications or relationships such as collocation, ownership, or trust. Each node and edge also has attributes that record details such as a person's name or the time a conversation occurred. Graphs that represent the communication among a set of people are called sociograms.

While analysts track many pieces of evidence, most of the evidence is completely innocent when viewed in isolation. Many people rent trucks, buy fertilizer, or take pictures of public monuments, and in isolation these actions are not remarkable. In contrast, the combination of those events performed by the same person or group could signal a threat to public safety. Graphs focus analysts' attention on the relationships between pieces of information, helping them see combinations of events and the big picture of evolving situations.

Group detection. Group detection algorithms determine the cliques within a large group of people. Consider, for example, a sociogram that models the communications among students in a typical American high school. Each student is represented by a node, and an edge is drawn between every two students who talk to each other regularly. Group detection algorithms find densely connected subpopulations within the communication graph, which typically correspond to groups of friends. The algorithms would likely discover sets of people who are on the same athletic team or extracurricular activity, are mutual friends, or who have some other relationship that causes them to communicate. Group detection helps the analyst understand who knows whom, who influences whom, and the relationships between different subpopulations. It is a common starting point for further graph-based analysis.

Link analysis. The term link analysis describes a group of relatively simple graph-based search techniques. Link analysis finds indirect relationships by answering questions such as who is connected to node A by a path containing at most four edges, or are nodes A and B connected by a path containing at most five edges. These are often implemented as traditional depth-first or breadth-first searches over the edges in the graph, but more complex approaches may be required for large datasets. In other cases the analyst may need to find indirect relationships through a specific intermediary (for example: Are nodes A and B connected by a path through C?), or determine the shortest path in the graph between two nodes. Algorithms that answer these questions have been known for years in graph theory, but they have only recently been applied to intelligence analysis.

Even these very simple searches can be very useful to analysts, who are looking for hidden or indirect connections between people or events that they suspect are related. Searches through a specific intermediary can help the analyst identify intermediaries privy to covert dealings. Link-analysis searches can also be used to estimate the range of influence a given person or event may have. Shortest-path searches can identify likely paths of communication or influence. When the analyst needs to search for more complex or general patterns, they will use graph-matching algorithms.

Graph matching. Graph-matching algorithms search through a graph to find subregions that have the same connectivity or attributes as a user-defined pattern or template. Graph matching is also known as subgraph isomorphism. To use these algorithms, the analyst specifies a template graph of people, events, as well as the actions and relationships

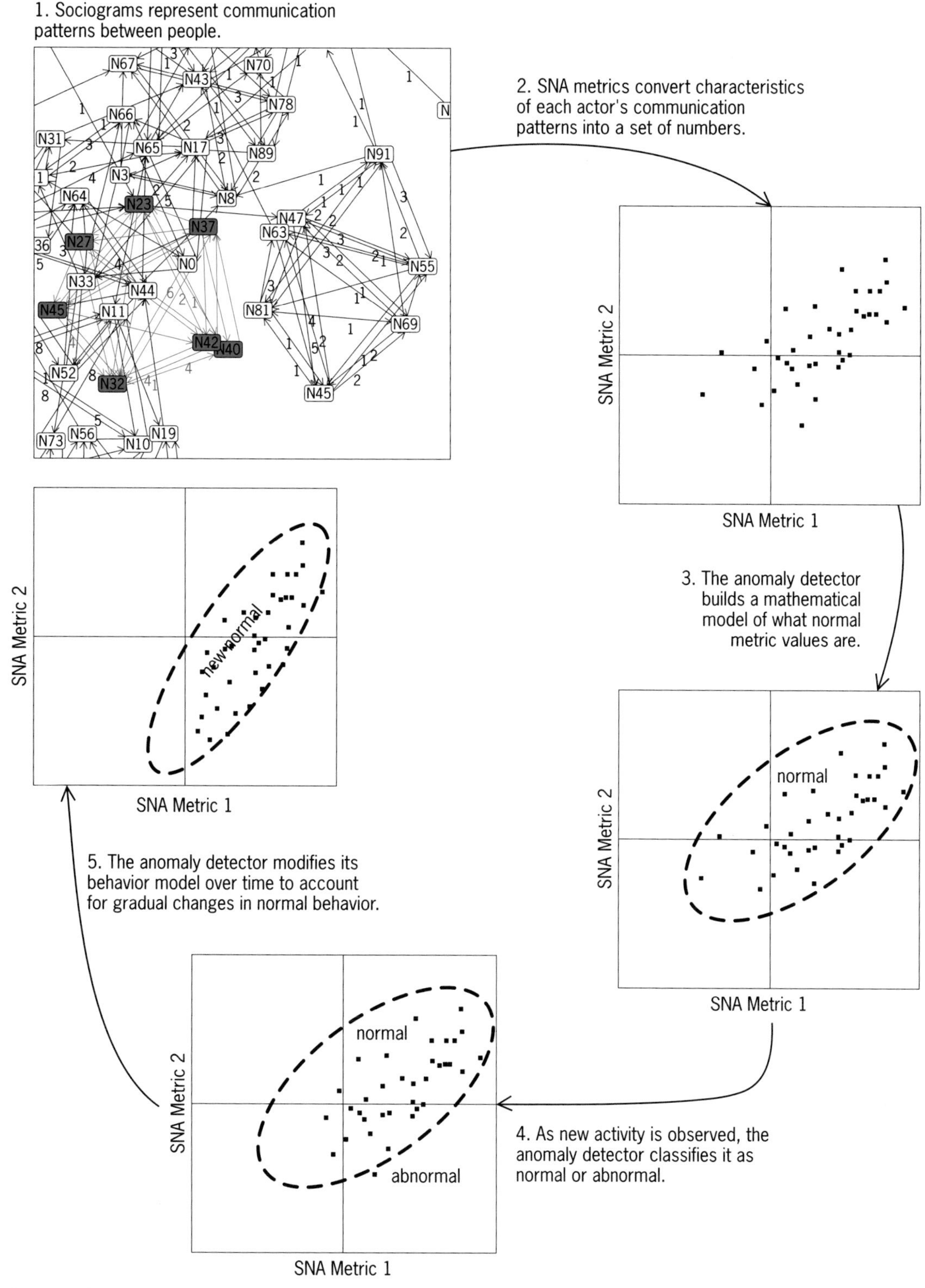
1. Sociograms represent communication patterns between people.
2. SNA metrics convert characteristics of each actor's communication patterns into a set of numbers.
SNA Metric 2
SNA Metric 1
3. The anomaly detector builds a mathematical model of what normal metric values are.
normal
SNA Metric 2
SNA Metric 1
4. As new activity is observed, the anomaly detector classifies it as normal or abnormal.
normal
abnormal
SNA Metric 2
SNA Metric 1
5. The anomaly detector modifies its behavior model over time to account for gradual changes in normal behavior.
new normal
SNA Metric 2
SNA Metric 1

Stages of SNA anomaly detection.

between them that would appear suspicious. The algorithm searches the full collection of evidence to find combinations of evidence that match the template. This is an incredible help to the analyst because performing this type of search manually is very time consuming and error prone. Using graph-matching algorithms, the analyst simply instructs the algorithm to find all instances of the threat template and then can concentrate only on matching activity. Graph-matching algorithms prevent the analyst from having to sift through all cases of events that do not occur in a threatening combination.

There are many extensions to the basic graph-matching problem that can increase its usefulness for intelligence analysis. The most important extension, inexact matching, finds evidence combinations that are very similar to the analyst's template, even if they do not match the template exactly. In any given situation, an analyst's evidence is usually incomplete; that is, the analyst may have defined the pattern slightly wrong or the attackers' behaviors might be slighty varied from the past. These factors can cause threats to show up as imperfect template matches. In addition, graph matching can be extended by using hierarchical templates, taxonomies, and logical constraints between attributes in the template.

Social network analysis. Social network analysis (SNA) is a branch of anthropology devoted to analyzing human social interaction. Over the years, researchers have designed a variety of SNA metrics to help them quantify human communication behaviors. When computed on a sociogram, SNA metrics convert abstract concepts, such as "Person X is central to this organization" or "This organization has a very redundant communication structure," into specific numbers. For example, we can compute the betweenness centrality of a specific node in a sociogram to generate a value that measures how often an organization's communications flow through a given person. Alternatively, we could compute the density or global efficiency metrics to measure how much redundancy exists in the communication structure. While the precise definitions of these metrics are beyond the scope of this article, the important concept is that each SNA metric transforms a different aspect of the communication structure into a number.

This type of analysis is useful because the SNA metric values exhibited by threat groups (groups trying to coordinate illegal activities) are usually very different from those exhibited by normal human social groups. Threat groups have fundamentally different priorities than normal social groups. Normal social groups inherently prioritize spontaneity and efficiency, so they adopt communication patterns that are irregular and unplanned, and that promote efficient information sharing. In contrast, threat groups prioritize secrecy and the ability to withstand the capture or elimination of individual group members. As a result, these groups adopt strict communication patterns that are drastically different from those of normal groups, and those differences are reflected in abnormal SNA metric values. SNA metrics transform the concepts of a communication structures' efficiency, spontaneity, secrecy, and robustness into numeric values. In doing this, they give the intelligence analyst a way to describe threatening activity that does not require the definition of specific graph templates.

Because the analyst is working with a set of numeric values instead of graphs, a wide variety of statistical pattern classification techniques can be applied. Statistical pattern classification is a broad field that we very briefly summarize. Supervised learning approaches provide algorithms in which the analyst trains the computer to differentiate between SNA metric values exhibited by threat and nonthreat groups by presenting it with examples of each. Unsupervised learning approaches provide anomaly detection, where the algorithm detects SNA metric value combinations that are unlike previously seen examples (see **illustration**). Time-series analysis provides algorithms for which the analyst uses to differentiate threat and nonthreat groups based on the way their SNA metric values change over time.

Outlook. There are a variety of graph-based approaches that can be applied to intelligence analysis, many of which are quickly gaining popularity. Graph-based approaches are very effective against the type of security threats being seen today, where illegal or terrorist organizations are attempting to coordinate complex attacks among networks of loosely connected covert agents. Almost all of the actions performed by these threat groups look innocent when viewed in isolation, but the combination of these actions, communications, and events reveals the threat. Graph-based approaches help the analyst focus on the crucial information contained in the combinations and relationships. Successfully predicting threats to national security before an attack happens is crucial and graph-based approaches will help analysts do this by providing powerful new analysis tools.

For background information *see* ALGORITHM; ANALYTIC HIERARCHY; ANTHROPOLOGY; GRAPH THEORY; GRAPHIC METHODS in the McGraw-Hill Encyclopedia of Science & Technology. Thayne Coffman

Bibliography. R. Diestel, *Graph Theory*, 2d ed., Springer-Verlag, 2000; R. O. Duda, P. E. Hart, and D. G. Stork, *Pattern Classification*, 2d ed., Wiley, 2001; R. J. Heuer, Jr., *Psychology of Intelligence Analysis*, Center for the Study of Intelligence, Central Intelligence Agency, 2001, available online at S. Wasserman and K. Faust, *Social Network Analysis: Methods and Applications*, Cambridge University Press, 1994; D. J. Watts, *Small Worlds: The Dynamics of Networks between Order and Randomness*, Princeton University Press, 1999.

Herbicide resistance

Herbicides are chemicals that kill plants, and selective herbicides are those that can kill certain plant species (weeds) while not injuring other plant species (crops). Since the introduction and commercialization of synthetic organic herbicides in the late 1940s, farmers have been increasingly reliant on herbicides for weed control. This is particularly true for agriculture in many developed countries, where there has also been a shift toward increased mechanization with larger machines and reduced tillage (less cultivation of the soil between plantings). Although herbicides have been widely used for only the past 60 years, societal changes have occurred in developed countries that would make it very difficult to return to previous weed control practices. These societal changes include a shift to fewer but larger farms with the simultaneous depopulation of rural areas, shortages of farm labor, and an increasing average wage paid to farm workers, which makes handweeding uneconomic in many instances.

It has recently been estimated that United States farmers annually apply herbicides to approximately 85% of the cropped acreage at a direct cost of $7 billion. The vast majority of herbicides are very effective in killing or suppressing weeds when applied according to label directions, and the economic cost of not using herbicides has been estimated to be at least $21 billion annually for the United States, not including the costs of increased soil erosion. Yet, decades of intensive control efforts have not resulted in complete eradication of weeds. This is a testament to the biological diversity and genetic plasticity of weeds, and these characteristics have also allowed weeds to develop resistance to many herbicides.

Evolution of resistance. Herbicide resistance is the inherited ability of a plant to survive and reproduce following exposure to a dose of herbicide that normally is lethal to its species. Evolved resistance to a formerly toxic compound is not unique to herbicides and weeds. Instances of evolved resistance to insecticides, fungicides, antibiotics, and anticancer and antiretroviral drugs are common and well documented. The operative principles that result in resistance in all these instances harken back to Darwin: the presence of genetic variability in a population and selection of the "fittest" individuals. Biocides, including herbicides, generally exert a very high selection pressure, killing almost 100% of susceptible individuals or organisms. Every biocide selects for its own failure, so it is not surprising when new individuals—and eventually entire populations—arise that are resistant. For a population of a plant species to become resistant, a population shift from predominantly susceptible to predominantly resistant individuals occurs.

Mutation. For a population to become resistant to a toxicant, there must first be heritable variation within the population. The source of this variability is mutations (random changes) in the genetic code of the organism—the deoxyribonucleic acid (DNA). When there is a mutation in the DNA that alters the order or number of amino acids in a crucial protein or enzyme, the altered protein or enzyme may become dysfunctional. If an enzyme normally catalyzes a chemical reaction that is critical to cell life and the altered enzyme cannot, then the cell will not further grow and develop and will die (a lethal mutation).

In multicellular organisms such as higher plants and animals, mutations are thought to occur primarily during meiosis in production of egg and sperm (pollen) cells. It may also be possible that the mutation occurs very early in the development of the organism in nondifferentiated cells (stem cells). If the mutation occurs later in the development of the organism in the somatic cells, the mutation would affect only one cell or a small number of cells and not the entire organism, probably would not change the response of the organism to a selecting agent (biocide), and would not be passed on to subsequent generations.

Most mutations in the genetic code are lethal or deleterious and therefore are either not passed on to the next generation (not heritable) or do not confer a selective advantage. Occasionally, though, a nonlethal mutation occurs that confers a selective advantage to an individual in its current environment. This favored individual will then give rise to more progeny than normal individuals, and over a number of generations the population will shift to comprise primarily individuals carrying the genetic mutation. Herbicide resistance is an example of such a beneficial mutation (to the weed).

The mutation rates for specific genes are currently poorly understood. In the case of genes where the vast majority of mutations are lethal and not heritable, these genes appear to be highly conserved (that is, individuals having mutations in these genes are not often encountered, because these mutations are lethal to the organism very early in its growth and development). This phenomenon is demonstrated by the rarity of resistance to certain mode-of-action herbicides. Other genes appear to be the opposite—the genetic code can be somewhat variable and the altered gene product (enzyme or protein) is still functional in the cell and organism. This is demonstrated by the high frequency of resistance to certain other mode-of-action herbicides, and has been confirmed by DNA sequencing results. A common assumption for gene mutation rates, based on some data from single-celled organisms and simple plant species, is 1 mutation in 1,000,000 per gene locus per generation. This assumption of a mutation rate of 1×10^{-6} is commonly used in herbicide resistance modeling.

The cause of random mutations in the DNA of organisms is poorly understood, and there is a common misconception that the application of pesticides may itself enhance mutation of pest DNA to produce resistance. There are some chemicals that are known mutagens, and there may be environmental factors that influence or cause mutations. However, it is generally agreed that the herbicide or pesticide does not itself cause mutations in the target pest population to which it is applied. Current standards for registering and commercializing a pesticide involve stringent tests to detect possible mutagens, carcinogens, or teratogens (primarily because of human health concerns). Pesticides that are identified as mutagens or carcinogens in these preliminary tests are not commercialized.

Target-site resistance. Herbicides have been grouped according to their mode of action—the specific life process that they interfere with which ultimately results in plant death. For example, ACCase inhibitor herbicides bind to the acetyl Co-A carboxylase enzyme in normal susceptible plants and prevent its normal functioning. Thus, the ACCase enzyme is termed the herbicide target site. This enzyme catalyzes an essential step in fatty acid synthesis in the cell, and plant death results if this enzyme does not function. Many of the weed populations that have evolved resistance to ACCase inhibitor herbicides have an altered but functional ACCase enzyme that these herbicides cannot bind to, or bind to poorly (much higher doses of herbicide are required to inhibit the activity of the altered enzyme). Therefore the altered ACCase enzyme is not affected by application of the herbicide, it is still functional in the cell, and the resistant plant continues to grow and develop despite the normally lethal herbicide dose.

There have been a number of studies of herbicide-resistant weeds where an altered DNA sequence for the appropriate gene locus has been associated with altered enzyme function which confers resistance at the whole plant level. For example, a population of green foxtail (*Setaria viridis*) was demonstrated to be resistant to ACCase inhibitor herbicides in both field and greenhouse experiments. Subsequent laboratory investigations showed that the ACCase enzyme in this resistant population had greatly reduced sensitivity to ACCase inhibitor herbicides, and that plants in this resistant population possessed an altered DNA sequence in their ACCase enzyme gene. The gene known to encode for the ACCase enzyme was then sequenced. From a total of 7710 DNA nucleotides that were sequenced, there was an A (adenine) to C (cytosine) mutation at position 5582, resulting in an amino acid substitution from isoleucine to leucine in the ACCase enzyme. It is postulated that this amino acid substitution results in a change in the conformation of the enzyme such that ACCase inhibitor herbicides no longer properly bind to the enzyme, so that even following herbicide application the synthesis of long-chain fatty acids continues and plant growth is unaffected.

Another example of a mutation in DNA resulting in an altered, herbicide-insensitive enzyme occurs with the ALS (acetolactate synthase) inhibitor herbicides. ALS inhibitor herbicides bind to and inhibit the acetolactate synthase enzyme in normal susceptible plants. The ALS enzyme catalyzes an essential step in the synthesis of three amino acids in the plant cell. When a normal, susceptible ALS enzyme is inhibited and the three amino acids are not produced, the plant dies. ALS inhibitor herbicide-resistant kochia (*Kochia scoparia*) populations have been shown to possess diverse nonlethal mutations in the ALS gene, with at least six different amino acid substitutions having been identified.

Other resistance mechanisms. In addition to target site resistance, there are several other known, but less common, mechanisms of weed resistance to herbicides. These mechanisms include (1) altered or enhanced herbicide metabolism or detoxification, (2) reduced translocation of the herbicide to the site of action within the plant, (3) target enzyme overproduction in the plant, and (4) sequestration/compartmentation of the herbicide in the cells of resistant plants. Notably, all resistance mechanisms involve some change from normal cellular function in susceptible plants, and these resistance mechanisms are heritable.

Genetically engineered herbicide-resistant crops. In the past decade, genetically engineered herbicide-resistant crops (GM crops) such as soybean, corn, cotton, and canola have been commercialized and are now widely grown in North America. In terms of the evolution of herbicide-resistant weed populations, these GM crops can worsen the situation by increasing the farmer's reliance on a certain few herbicides for weed control and increasing the selection pressure for weed resistance due to increased frequency of herbicide use. For example, glyphosate herbicide-resistant GM soybean and corn crops can be grown year after year in a crop rotation with only glyphosate herbicide used for weed control. Furthermore, glyphosate herbicide is often used in reduced tillage systems instead of tillage for general weed control prior to planting the crop. Thus, there can be an overuse and overreliance on a certain few herbicides with the simultaneous increase in selection pressure for resistant weeds. This has in fact happened in a number of states with the evolution of glyphosate-resistant horseweed (*Conyza canadensis*) populations, which are a particular problem in GM soybean crops. If additional weed species develop resistance to the few specific herbicides used in GM crops, the usefulness and value of these GM crops (and the associated herbicide) may be negated. This is particularly problematic where a herbicide such as glyphosate is a very valuable weed control tool apart from its usage in GM crops.

Outlook and tactics to delay herbicide resistance. To date, there have been approximately 180 naturally evolved herbicide-resistant weed species identified worldwide to 18 different herbicide mode-of-action groups. In fact, there are only a few herbicides for which resistant weeds have not been identified. Yet the inevitability of resistance to a wide variety of biocides, as a result of naturally occurring random mutations in DNA and subsequent selection, is still not fully acknowledged by many in the agricultural and medical communities. Herbicides are a very valuable weed control tool, and are particularly crucial to the continued expansion and sustainability of soil-conserving, reduced-tillage cropping systems. To prolong the useful lifespan of herbicides (that is, reduce selection for resistant individuals within a weed population), they should be used minimally and judiciously. Prolonging the lifespan of existing herbicides is essential because the pace of new herbicide development and commercialization—those herbicides with novel modes of action—has slowed dramatically over the past 15 years. Quite logically, herbicides can be considered a nonrenewable resource.

There are several strategies or practices that may delay the evolution of herbicide-resistant weeds. They include rotating herbicide mode-of-action groups (not using the same herbicide or herbicide group on the same field year after year), applying tank mixes of two or more herbicides with different modes of action, and the preferential use of herbicides with a short soil residue. These strategies may lower the selection pressure that a given herbicide exerts on a weed population. Most importantly, herbicides should be used as part of an integrated weed management system that includes a diverse array of nonherbicidal weed control measures. These nonherbicidal weed control measures include the planting of weed-free crop seed, growing of competitive crops and crop varieties, appropriate and diverse crop rotations, appropriate use and placement of fertilizers, appropriate tillage, mowing or other methods of mechanical weed control, and cleaning weed seeds from equipment between fields.

For background information *see* AGRICULTURAL SOIL AND CROP PRACTICES; GENETIC MAPPING; HERBICIDE; MUTATION; POPULATION GENETICS; WEEDS in the McGraw-Hill Encyclopedia of Science & Technology.

Lyle F. Friesen; Rene C. Van Acker

Bibliography. L. P. Gianessi and S. Sankula, *The Value of Herbicides in U.S. Crop Production*, National Center for Food & Agricultural Policy, Washington, DC, April 2003 [http://www.ncfap.org/whatwedo/benefits.php]; I. M. Heap, *International Survey of Herbicide Resistant Weeds* [http://www.weedscience.org]; M. Jasieniuk, A. L. Brûlé-Babel, and I. N. Morrison, The evolution and genetics of herbicide resistance in weeds, *Weed Sci.*, 44:176–193, 1996; S. B. Powles and J. A. M. Holtum (eds.), *Herbicide Resistance in Plants: Biology and Biochemistry*, Lewis Publishers, Boca Raton, FL, 1994; P. J. Tranel and T. R. Wright, Resistance of weeds to ALS-inhibiting herbicides: What have we learned?, *Weed Sci.*, 50:700–712, 2002.

Heuristics in judgments and decision making

A heuristic is a rule of thumb that serves as a mental shortcut for judgments and decisions. For example, many people are afraid of becoming victims of a crime, often more so than is justified by official crime statistics. One possible explanation for the imbalance between fear and risk is that people overestimate the prevalence of violent crime because such crimes are exhaustively covered and sensationalized by the media. This coverage enhances the availability of violent crimes in a person's memory, causing the frequency of violent crime to be overestimated. Availability is defined as the ease with which instances of a class, such as "violent crimes," come to mind, and it is often used as a heuristic in judging frequencies of that instance.

Examples of heuristics. Research in the last three decades has provided many demonstrations of the use of such heuristics. Although fast and frugal, they may result in biased judgment, which can lead to the question of whether the use of heuristics is rational.

Availability heuristic and judged frequency. As part of a study examining the effects of availability, experimental participants had to estimate the relative frequency of 41 causes of death. The participants judged unnatural causes with high media coverage, such as homicide, to be more frequent than less salient causes, such as stomach cancer, although stomach cancer deaths are actually more than five times as frequent as homicides. This finding showed that people use the high availability of vivid or sensational events as a rule of thumb for estimating event frequencies.

Representativeness heuristic and biases in judgment. When people use heuristics, they focus on some features of a problem and neglect others; this often leads to biases in judgment and decision making. In one study, respondents were asked which exact birth order of girls (**G**) and boys (**B**) is more probable in families with six children:

G-B-G-B-B-G

B-G-B-B-B-B

Seventy-five of 92 respondents judged the second sequence to be less likely than the first sequence. The participants focused on how well a given birth order represented the proportion of boys and girls in the population (the representativeness heuristic), but neglected the basic fact that in any given case the birth of a boy and the birth of a girl are equally likely statistically, and therefore both exact birth orders are equally likely.

Prior probabilities and planning fallacies. People often commit a planning fallacy where they are overly optimistic in predicting when they will finish a project. They commit this fallacy because they focus on the different steps they will take in the future to complete the task, but neglect their earlier experience (prior probability) with finishing, or not finishing, projects that early. The way to reduce such a planning fallacy is straightforward: instead of simply focusing on the future, individuals must think about the prior probabilities of having finished similar tasks within the predicted time.

Prior probabilities and base rate neglect. Another example of the neglect of prior probabilities comes from the domain of medical decision making. Researchers presented a group of medical doctors with the following situation:

> Patricia is 39 years old and feels healthy and without any signs of illness. She knows that in her age group, about one percent of all women suffer from breast cancer. She goes regularly for mammography to screen for this illness.
>
> A mammographic examination will correctly identify that a woman with breast cancer has the disease in 80% of all cases. In 10% of all women without breast cancer, the test nevertheless provides a positive result; that is, it falsely indicates that the woman suffers from breast cancer.
>
> An examination provides a positive result for Patricia.
>
> What is the probability that Patricia really suffers from breast cancer?

When this problem was introduced, an overwhelming majority of the surveyed medical doctors estimated that the probability was between 70 and 80%, although the actual probability is only around 7.5%. These doctors seem to have focused on the sensitivity of mammography when breast cancer was present and did not adjust sufficiently for the presence of false positives because they neglected the base rate, or prior probability, of breast cancer, which is 1%.

The problem can be simplified by transforming the probabilities into frequencies as follows:

> Patricia is 39 years old and feels herself to be healthy and without any signs of illness. She knows that in her age group, about 10 of 1000 women suffer from breast

cancer. She goes regularly for mammography to screen for this illness.

A mammographic examination will correctly identify that a woman with breast cancer has the disease in 8 of 10 cases. In 99 of the remaining 990 women without breast cancer, the test nevertheless provides a positive result; that is, it falsely indicates that the woman suffers from breast cancer.

An examination provides a positive result for Patricia.

What is the probability that Patricia really suffers from breast cancer?

Research has shown that calculating the probability is much easier with this frequency-based format than with the probabilistic format.

Rationality. Studies on bias have raised the question of whether or not human beings are rational. The theory of rational choice implies that identical options should lead to identical choices, regardless of how the options are framed. This principle is called description invariance. Nonetheless, many studies on human decision making have revealed that participants often violate normative rules (rules based on standards of reasoning), such as description invariance and rules of probability (such as Bayes' theorem, which takes prior probabilities into account), even if they are experts in their field.

To test whether people's decision making follows the principle of description invariance, participants were presented with two problems that had the same introduction but were then followed by different scenarios:

Imagine that the United States is preparing for the outbreak of an unusual disease that is expected to kill 600 people. Two alternative programs to combat the disease have been proposed. Assume that the exact scientific estimates of the consequences of the program are as follows:

Scenario 1. If Program A is adopted, 200 people will be saved. If Program B is adopted, there is a one-third probability that 600 people will be saved and a two-thirds probability that no people will be saved. Which of the two programs would you favor, Program A or Program B?

Scenario 2. If Program C is adopted, 400 people will die. If Program D is adopted, there is a one-third probability that nobody will die and a two-thirds probability that 600 people will die. Which of the two programs would you favor, Program C or Program D?

The two scenarios are identical: Program A corresponds to Program C—it is certain that 200 people will be saved and 400 people will die; Program B corresponds to Program D—there is a one-third probability that all people will be saved and a two-thirds probability that all people will die. Nevertheless, a majority of participants chose options A and D. This framing effect contradicts the theory of description invariance in rational choice.

There are basically four ways that participants can fail at performing a task according to normative rules: First, they may lack knowledge of the appropriate algorithms (steps to take to solve the problem) and therefore be unable to solve the task appropriately; this situation alone would count as irrationality. Second, they may know the appropriate algorithms, but commit errors due to lack of attention, memory deficits, and other temporary cognitive factors. Third, the participants might both have the accurate algorithm and apply it properly, but construe the problem differently. This can be understood by examining the above example. If participants construed "200 people will be saved" as "200 or more people will be saved" and "400 people will die" as "400 or more people will die," then the choice of options A and D is justified from a normative point of view. However, although the way a question is interpreted has been found to affect judgment, these effects do not explain the extent of the observed bias. Fourth, some scholars have claimed that when experimenters use inappropriate norms for testing, participants reject these norms to solve the problem in accordance with norms that are more appropriate. For example, the doctors who neglected the base rate in the cancer example may have assumed that Patricia is going to get a mammography because she felt a lump in her breast, which would increase the probability of breast cancer considerably. Replacing the base rate given in the description by the experienced base rate of women at risk would justify higher probability judgments (although not as high as 70–80%). If the assumption were correct that conflicting norms between experimenters and participants may result in the failure of the participants to perform the given tasks, then participants who understand the experimenter's norm would not accept it, and they would solve the task according to another, more appropriate algorithm. However, research has shown that this is not the case: in fact, participants who understand the experimenter's norm generally accept it and use it for task solution, suggesting that the experimenters generally use the same norms as knowledgeable participants do.

Adaptiveness of heuristics. Although they may cause people to commit fallacies that lead to biased judgments, heuristics are adaptive in the sense that they generally serve people acceptably well in everyday life. Frequency judgments based on the availability heuristic, for example, are often quite accurate because the ease with which people can retrieve an instance is directly proportional to the number of encounters with it. However, availability is also related to the vividness of memories; if vividness contradicts frequency, judgment becomes biased.

Even the planning fallacy can be adaptive: If people's overly optimistic time estimates lead to earlier task completion, the apparent fallacy could turn into a useful self-fulfilling prophecy. In one study, experimental participants were lured into either optimistic or pessimistic predictions by giving them an ostensibly random starting point for the prediction. As research into so-called "anchoring effects" has shown, such starting points are highly effective in guiding numerical estimates, and they were effective in guiding predictions of task completion. Optimism was indeed beneficial: Participants lured into optimistic predictions completed the tasks earlier than those lured into pessimistic predictions; however, this beneficial effect was limited to simple and controllable tasks.

Finally, the breast cancer example demonstrates that people can solve problems that involve reasoning according to Bayes' rule, but only if the problem is presented in a format that corresponds to their mental architecture. In this case, proper judgment requires the presentation of the task to be frequency-based rather than probabilistic. People do not need training in probability theory; instead they need to be taught to simplify a given problem and adapt it to use algorithms they understand and can apply.

The human mind has evolved mechanisms to deal efficiently with the allowances and constraints of the environment. The resulting heuristics often contradict normative rules, but are nonetheless a functional adaptation for increasing efficiency in everyday life.

For background information *see* DECISION ANALYSIS; DECISION THEORY; PROBABILITY; PROBLEM SOLVING (PSYCHOLOGY); RISK ASSESSMENT AND MANAGEMENT in the McGraw-Hill Encyclopedia of Science and Technology. Rolf Reber

Bibliography. G. Gigerenzer, *Adaptive Thinking: Rationality in the Real World*, Oxford University Press, 2000; T. Gilovich, *How We Know What Isn't So: The Fallibility of Human Reason in Everyday Life*, Free Press, 1991; T. Gilovich, D. Griffin, and D. Kahneman (eds.), *Heuristics and Biases: The Psychology of Intuitive Judgment*, Cambridge University Press, 2002; R. F. Pohl (ed.), *Cognitive Illusions: A Handbook on Fallacies and Biases in Thinking, Judgment, and Memory*, Psychology Press, 2004.

Homo floresiensis

In 2004 a team of Australian and Indonesian archeologists and paleoanthropologists announced the remarkable discovery of a new species of fossil hominid, *Homo floresiensis*, from the island of Flores in Indonesia. The geographical location and the anatomical specializations of the new finds were so unexpected that it has led to major debates among the scientific community and a critical rethinking about the nature of human adaptation and evolution.

Archeological excavations at Liang Bua, a large limestone cave in western Flores, yielded a partial skeleton of a fossil human, as well as an isolated tooth and a radius fragment, all associated with simple stone tools. Recent excavations have succeeded in recovering additional hominid finds, but these have not yet been described. Radiocarbon and luminescence dating indicates an age of between 18,000 and 38,000 years for the human remains. The partially complete skeleton (LB1) consists of a complete cranium, mandible, much of the limbs and pelvis, incomplete hands and feet, and fragments of the vertebral column. The isolated tooth (LB2) consists of a first premolar from the lower jaw. Study of the development of the teeth and limb bones, and the anatomy of the pelvis confirm that the skeleton belonged to an adult female.

Anatomical specializations. Anatomically the postcranial skeleton is quite similar to that of modern humans (except that the pelvis is relatively broader and the arms are somewhat longer), but the Liang Bua specimen is remarkable in being extraordinarily diminutive, having an extremely small braincase and a striking reduction of the posterior cheek teeth.

Height. Its height is estimated at only 106 cm, or 3 ft 6 in. (about the size of a 3-year-old modern human). By comparison, the female Efe pygmies of the Ituri Forest of the Congo, the modern human population with the shortest stature, have an average height of 135 cm, or 4 ft 5 in. Females of *Australopithecus*, early fossil hominids from the Pliocene of Africa, were probably similar in stature or slightly taller than the individual from Liang Bua, which makes the Flores specimen among the smallest hominids ever found.

Skull and teeth. The skull and teeth provide the most compelling evidence for the distinctiveness of the new species. In many respects the cranium is reminiscent of that of *Homo erectus* from Java, and it is certainly much more primitive than that of modern humans (**Fig. 1**). For example, the neurocranium, or braincase, is relatively long and low, with a receding forehead, a slight keel on top of the cranium in the midline, and a distinct horizontal bony bar running across the back of the braincase, and overall the bones are relatively thick. In rear view, the braincase is bun-shaped, being broadest across the base, whereas in modern human crania the greatest breadth occurs higher on the braincase, across the parietal region. The face is relatively short, but has quite strongly developed facial pillars on either side of the nasal aperture, as well as distinct brow ridges that arch over each eye socket. The lower jaw lacks a protruding chin, which is a unique hallmark of modern humans, but instead has a receding chin comparable to that found in *H. erectus* and earlier fossil hominids. The teeth are relatively small, characteristic of the genus *Homo*, although the reduced size of the second molars and the congenital absence of the third molars distinguish it from *H. erectus*. This marked reduction in the posterior cheek teeth is probably related to extreme facial shortening.

The Liang Bua hominid differs significantly from *H. erectus* in one important respect: the absolute and relative size of its brain. The volume of the braincase is tiny, being estimated at only 380 cm^3. This is among the smallest cranial capacities ever recorded for a fossil hominid and contrasts sharply with the massive braincase of modern humans, which averages about 1300 cm^3. It is also less than half the size of the smallest cranial capacity of *H. erectus* from Java, which has been estimated at just over 800 cm^3. Basically, the hominid from Flores looks like a smaller and more slender version of *H. erectus*, but with a relatively much smaller braincase. The differences are sufficient to justify the recognition of a separate

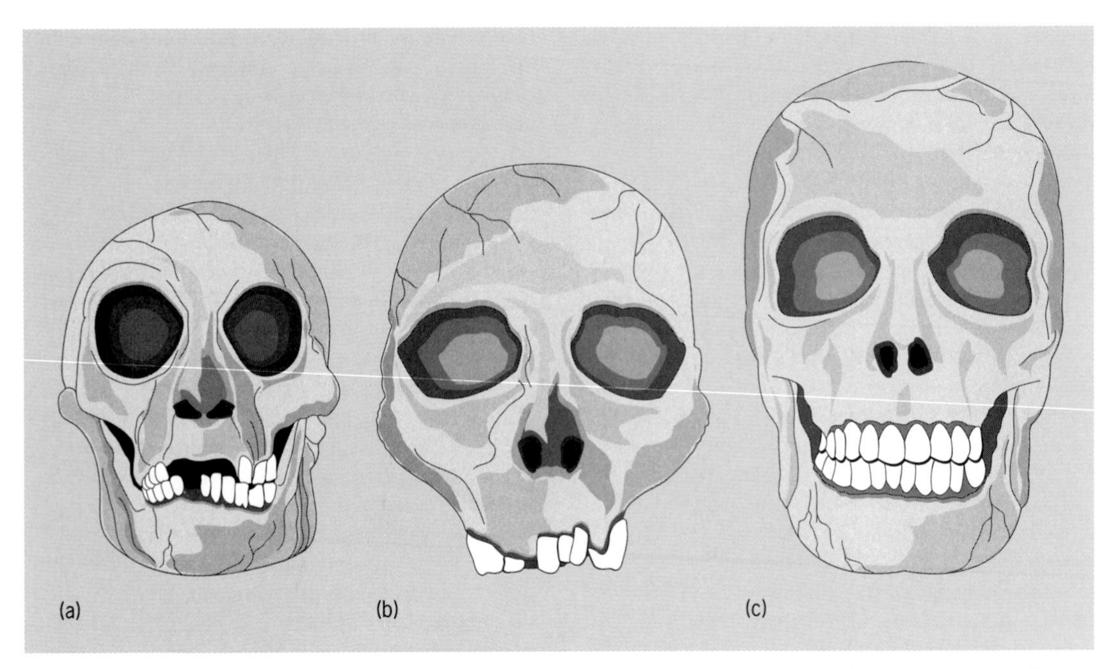

Fig. 1. Comparison of the skull of (*a*) *Homo floresiensis*, (*b*) *Homo erectus* from East Africa, and (c) modern *Homo sapiens*. Note the small overall size of the skull and the relatively very small braincase in *Homo floresiensis*.

species of hominid, but one that can be inferred to be a specialized offshoot of *H. erectus*.

Microcephaly hypothesis. Not all authorities agree with the separate species interpretation. Some researchers have argued that the Liang Bua specimen is a pygmy-sized or dwarf *H. sapiens* with a pathological condition, microcephaly, which results in an abnormally small brain size. However, the morphology of the brain case and the virtual endocast (a cast of the internal cavity of the brain case) generated from computerized tomography (CT) scans of the fossil are inconsistent with such a conclusion. The distinctive features of the skull and postcranial skeleton from Flores are unlike those found in modern humans with pathological conditions related to dwarfism and microcephaly. At present, there is no reason to suspect that the individual is pathological; and further finds of fossil hominids from Liang Bua, which have not yet been described, indicate that the same suite of specialized features were characteristic of all members of the local population, not just an isolated case.

Technological sophistication. The cultural remains associated with the Liang Bua hominids were also surprising to many prehistorians. The majority of the stone tools consist of simple flake tools, similar to those found at other sites associated with early *Homo*. However, some of the tools are much more sophisticated, and include finely worked points, blades, and awls, implements that are more characteristic of Middle and Upper Paleolithic sites associated with later species of humans. In addition, charred animal bones imply that *H. floresiensis* cooked its food. This level of technological sophistication is rather surprising for such a cerebrally challenged hominid, and it has led to a good deal of discussion and speculation among researchers about the nature of the relationship between brain size and intelligence. Of course, it is also conceivable that *H. sapiens*, which is known to have occurred throughout Southeast Asia by the end of the Pleistocene, could have coexisted on Flores (but is not represented in the paleontological record), and that it was this species, rather than *H. floresiensis*, that was responsible for the more advanced technologies.

Coexistence of three human species. The recognition of a new species of fossil hominid raises a number of problematic issues about human evolution. The fossil record from Southeast Asia provides evidence that *H. erectus* was the first hominid to colonize the Indonesian islands during the early Pleistocene, at about 1.6 million years ago. Much later, at about 50,000 years ago, *H. sapiens* or modern humans arrived, and there ensued a brief period of coexistence between the two species, until *H. erectus* became extinct about 30,000 year ago. With two species of *Homo* occurring contemporaneously in the islands of Southeast Asia during the later Pleistocene, how was it possible for a third species to have evolved and coexisted? The answer is provided by the geographical location of Flores.

Isolation and small size of flores. The island of Flores is part of a chain of smaller islands, the Lesser Sunda Islands, that extends westward from Java to within about 500 km of the northern coast of Australia (**Fig. 2**). Zoogeographically, Flores is part of Wallacea, a cluster of islands that separates the Oriental faunal province of Southeast Asia from the Australian faunal province of Australia and New Guinea. The deep-water channels separating the islands of Wallacea have provided an effective barrier between the two faunas, and this accounts for the remarkable distinctiveness of the mammals and birds in these neighboring regions, an observation that was first

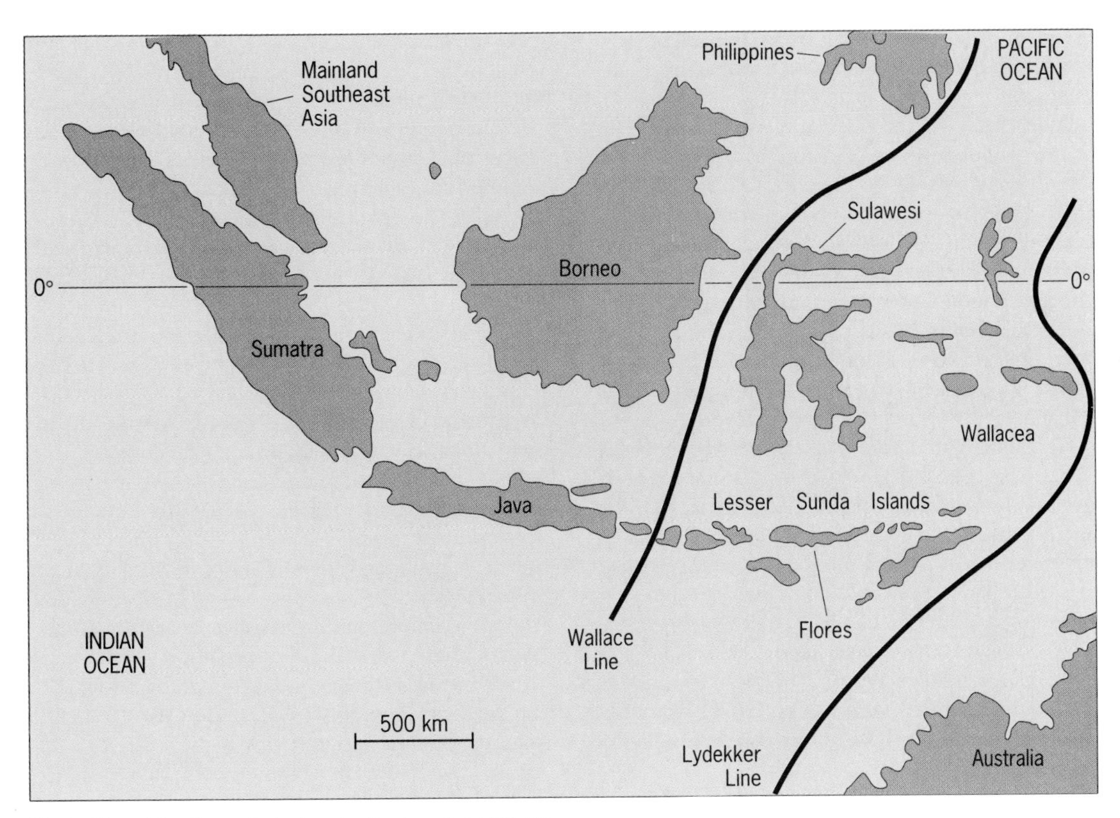

Fig. 2. Map of Southeast Asia showing the location of Flores.

made by A. R. Wallace, the nineteenth-century natural historian. As a result of this isolation, few mammal species have succeeded in colonizing Flores (at least until modern humans arrived and introduced many species common to Southeast Asia), resulting in a native fauna that was impoverished and highly endemic.

Unique ecological pressures. Paleontologists and biogeographers have long known that small, isolated islands tend to have highly distinctive faunas because only certain vertebrates can reach them and survive. Strong swimmers, such as deer, antelopes, elephants, and hippos, are common components of these faunas, while small mammals and reptiles, which can survive lengthy ocean crossings as stowaways on logs and mats of vegetation, can be washed ashore on islands by chance alone. Once colonized, remote islands offer great opportunities for vertebrates to exploit new niches in the absence of competitors, with the added advantage of being free from large mammalian carnivores. One main disadvantage, however, is that food resources are often more limited and are subject to greater fluctuations in availability through time. As a consequence, because of the combined influences of a limited food base and the reduction in predator pressure, large herbivores on islands tend toward dwarfism, while rodents become gigantic. However, the large carnivore niche may become partially occupied by large flightless birds or giant reptiles (such as the Komodo dragon on Flores), or in some cases by other mammals, such as giant hedgehogs. Researchers have suggested that *H. erectus* arriving on Flores could well have responded to the same ecological pressures as the other vertebrates, and that the small stature of *H. floresiensis* is simply another example of the well-documented phenomenon of insular dwarfism. This is certainly conceivable, but it is harder to explain why such adaptable omnivores as *H. erectus* would have suffered from the same energetic and dietary constraints as large herbivorous mammals.

The answer might lie in the fact that *H. erectus* was an extremely effective predator, which relied heavily on large vertebrates as an important source of food. The carrying capacity of mammalian carnivores is constrained by the minimum population size of its prey species, which occurs during times of unfavorable conditions. For this reason, carnivores tend to be relatively rare in comparison to their prey species. On small islands, where herbivore population sizes are likely to be extremely variable through time, the carrying capacity of carnivores is generally too low to sustain viable populations. A reduction in body size and the development of a tiny brain (an energetically expensive organ to grow and maintain) would certainly be an important strategy to reduce energy expenditure, as well as to facilitate increased reproductive output, enabling *H. floresiensis* to maintain a stable population size in the face of potentially devastating fluctuations in the numbers of their major prey species. The advantages of large body size and large brain size for *H. erectus* were significantly diminished on a small island which was devoid of large mammalian predators and competitors (Komodo dragons rarely attack humans), and where the main prey consisted of a very limited range of species, all of which would have been relatively easy to hunt.

Conclusion. The recent discovery of a new species of fossil hominid on Flores demonstrates that three human species—*H. erectus*, *H. sapiens*, and *H. floresiensis*—coexisted in Southeast Asia during the late Pleistocene. This is a level of regional diversity that was entirely unexpected in the later stages of human evolution. However, perhaps even more importantly, Flores provides a fascinating natural experiment that presents a unique new window through which to view the extraordinary cultural and biological adaptability of our own genus, *Homo*.

For background information *see* ANTHROPOLOGY; ECOLOGY; FOSSIL HUMANS; PALEOGEOGRAPHY; PHYSICAL ANTHROPOLOGY; PREHISTORIC TECHNOLOGY in the McGraw-Hill Encyclopedia of Science & Technology. Terry Harrison

Bibliography. P. Brown et al., A new small-bodied hominin from the Late Pleistocene of Flores, Indonesia, *Nature*, 431:1055–1061, 2004; M. J. Morwood, Archaeology and age of a new hominin from Flores in eastern Indonesia, *Nature*, 431:1087–1091, 2004; K. Wong, The littlest human, *Sci. Amer.*, pp. 58–65, February 2005.

Human papillomaviruses

Human papillomaviruses (HPVs) are small, nonenveloped, double-stranded deoxyribonucleic acid (DNA) viruses of the Papovaviridae family that selectively infect the epithelium of the skin and mucous membranes of a wide variety of animals as well as humans. More than 200 types of HPV have been recognized on the basis of DNA sequence data. Over 80 genotypes are well characterized and are associated with specific clinical manifestations (see **table**). For example, some cutaneous types of HPV infect epithelial cells of the skin of the hands and feet and cause papillomas (benign wartlike growths). The mucosal types infect the lining of the mouth, throat, respiratory tract, or anogenital epithelium and are a major factor in the development of cervical cancer.

Sexually transmitted HPV is one of the most common sexually transmitted viral diseases worldwide. In the United States it is estimated that over 20 million people are infected, with 1–5.5 million new cases occurring each year. Since the infection rate of HPV continues to increase, it remains an important disease.

Pathogenesis. The incubation period for HPV is usually 3–4 months, with a range of 1 month to 2 years. All types of squamous epithelium can be infected with HPV. Replication begins within the infected epithelial cells. Ultimately, viruses are assembled in the nucleus and released when the squamous cells are shed.

Epidemiology. HPV is spread primarily by skin-to-skin contact and sexual intercourse. The most important risk factor is a history of having multiple sex partners. A single sexual exposure to an infected person leads to the transfer of the virus 60% of the time. Even asymptomatic individuals infected with HPV can transmit the virus to others. Warts are often transmitted from one part of the body to another (autoinoculation). Since HPVs are fairly stable in the environment, they can also be transmitted indirectly from towels or from a shower stall, where they persist inside squamous epithelial cells that have been shed.

Age has been found to be another important determinant of risk for HPV infection in females. (There is little epidemiological information on HPV infection in males.) The greatest risk occurs at puberty and the first pregnancy and declines after menopause. In a recent study, 46% of female university students seeking routine gynecological examinations were found to be infected with HPV. HPV infection is one of the most common causes of abnormal Papanicolaou tests (Pap smears) in teenage women. HPV infection is most common among females in the 18–30 age group. There is a sharp decrease in prevalence after age 30. Some evidence indicates that people gradually develop immunity to the various types of HPV they encounter over time, as is the case with the viruses that cause the common cold. However, cervical cancer is more common after age 30. This suggests infection at a younger age and slow progression to cancer.

Clinical manifestations. HPV infection can be asymptomatic, can produce warts, or can be associated with a variety of benign and malignant neoplasms (see table).

Warts. The four major kinds are plantar warts, common warts, flat or plane warts, and anogenital (or venereal) warts. Common warts (verruca vulgaris) are found in as many as 25% of some groups and are most prevalent among young children. Plantar warts (verruca plantaris) are also widely prevalent; they

Human papillomavirus genotypes and associated diseases

Disease	HPV genotype
Common warts	1, 2, 4, 26, 27, 29
Flat warts	3, 10, 27, 29, 41
Intermediate warts*	10, 26, 28
Plantar warts	1, 2, 4, 63
Anogenital warts	6, 11, 30, 42, 70
Intraepithelial neoplasms	
Unspecified	70, 71, 74
Low-grade	45, 51, 52
High-grade	6, 11, 16
Bowen's disease	16, 31, 34
Cervical carcinoma	52, 56, 58, 66
Laryngeal papillomas	6, 11
Conjunctival papillomas	6, 11, 16
Others	6, 11, 16, 30, 33, 72, 73

*Young warts or areas on the skin that have not fully developed into visible warts.

occur most commonly among adolescents and young adults.

Anogenital warts (condyloma acuminatum) are sexually transmitted and caused by types 6, 11, 30, 42, and 70 human papillomaviruses. Once the virus enters the body, the incubation period is 1-6 months. These warts are soft, pink, cauliflowerlike growths that occur on the external genitalia, in the vagina, on the cervix, or in the rectum. They often are multiple and vary in size.

Cancer. In addition to being a common sexually transmitted disease, genital infection with HPV is of considerable importance because specific types of genital HPV play a major role in the pathogenesis of epithelial cancers of the male and female genital tracts. Over the last decade, many studies have convincingly demonstrated that specific types of HPV are the causative agents of at least 90% of cervical cancers. There is also a possible link between HPV and nonmelanoma squamous and basal cell cancers. Thirty percent of people with a rare syndrome of persistent warts (not common warts, but a particular type of wart growth) eventually develop skin cancer, and HPV viral DNA is found in malignant cells. However, the epidemiology, molecular biology, and role of HPV in the development of such cancers are largely unknown.

Development of cervical cancer. Cervical cancer is one of the best understood examples of how a viral infection can lead to a malignancy. High-risk HPV genotypes (6, 11, 16, 52, 56, 58, 66)—that is, those more likely to lead to cancer—can be distinguished from other HPV types largely by the structure and function of the proteins they encode. High-risk genotypes encode the E6 and E7 proteins, which interfere with the function of normal cell proteins and increase the expression of abnormal cellular proteins. Another distinguishing characteristic of high-risk forms of HPV is that the viral DNA is genetically integrated into the host cell genome, whereas in HPV-infected benign lesions the viral DNA is located extrachromosomally in the nucleus.

HPV replication begins with host cell factors that interact with part of the HPV genome and begin transcription of the *E6* and *E7* genes. The E6 and E7 proteins bind to and inactivate tumor suppressor proteins, cell cyclins, and cyclin-dependent protein kinases. The net result is deregulation of cell cycle and cell growth regulatory pathways and modification of the cellular environment in order to facilitate HPV replication in a cell that is terminally differentiated and has exited the cell cycle.

Cell growth is largely regulated by the tumor suppressor protein (p53) and the retinoblastoma protein (pRB). The E6 protein binds to p53 and causes it to be inactivated via enzymatic degradation; thus, tumor suppression is blocked and abnormal cell proliferation (growth, mutation) is stimulated. The E7 protein binds to the pRB protein. This disrupts the complex between pRB and the cellular transcription factor E2F1, liberating E2F1, resulting in stimulation of cellular DNA synthesis and abnormal cell proliferation (mutations). Eventually the mutations accumulate and lead to fully transformed cancerous cells. Progression to cervical cancer generally takes place over a period of 10–20 years. However, some lesions become cancerous more rapidly, sometimes within a year or two.

Diagnosis. Most warts that are visible to the naked eye can be diagnosed correctly by a physical examination and review of a patient's medical history. The use of a colposcope is invaluable in assessing vaginal and cervical lesions and is helpful in the diagnosis of oral and cutaneous HPV disease as well. Pap smears prepared from cervical scrapings often show cytological evidence of HPV infection. The most sensitive and specific methods of HPV diagnosis involve the use of techniques such as the polymerase chain reaction and the hybrid capture method to detect HPV nucleic acids and identify specific virus types.

Treatment and prevention. Decisions regarding the initiation of therapy should be made with the knowledge that currently available modes of treatment are not completely effective and some have significant deleterious side effects. In addition, treatment may be expensive, and some HPV lesions resolve spontaneously. The most common therapies include cryosurgery, application of caustic agents, electrodesiccation, surgical excision, and ablation with a laser. Topical medications such as 5-fluoracil (which interferes with abnormal cell reproduction in the skin's top layer) have also been used with some success. Both failure and recurrence have been well documented with all of these treatment methods.

No effective methods for the prevention of HPV infections are presently available other than avoidance of contact with infectious lesions. Barrier methods of contraception may be helpful in preventing transmission of condyloma acuminatum and other HPV diseases of the genital tract. Vaccine preparations have shown some promise in the prevention of HPV in some animal models.

For background information *see* ANIMAL VIRUS; CANCER (MEDICINE); ONCOGENES; REPRODUCTIVE SYSTEM DISORDERS; SEXUALLY TRANSMITTED DISEASES; VIROLOGY in the McGraw-Hill Encyclopedia of Science & Technology. John P. Harley

Bibliography. E. M. Burd, Human papillomavirus and cervical cancer, *Clin. Microbiol. Rev.*, 16(1): 1-17, 2003; E. L. Franco, Cancer causes revisited: Human papillomaviruses and cervical neoplasia, *J. Nat. Cancer Inst.*, 87:779-780, 1995; D. Richmond, *Clinical Virology*, ASM Press, Washington, DC, 2002; H. Zur Hausen, Papillomaviruses in human cancers, *Proc. Ass. Amer. Physicians*, 111:581-587, 1999.

Hydrogen-powered flight

In upcoming years aviation technology could begin to change dramatically to achieve the benefits of reduced pollutants emission and decreased dependence on hydrocarbon fuels. To accommodate a switch to hydrogen power, for example, future aircraft designs, propulsion, and power systems will look much different from the systems of today. Hydrogen fuel will enable a number of new aircraft capabilities, from high-altitude long-endurance remotely operated aircraft (HALE ROA) that will fly weeks to months without refueling, to clean, zero-emissions transport aircraft. However, the transition will require decades of research and new technology breakthroughs to fulfill the vision. Additionally, there are a number of design and development challenges that must be overcome before such systems can become operational.

Reducing NOx. The major benefit of hydrogen aircraft is that they will not produce harmful greenhouse gases such as carbon monoxide (CO), carbon dioxide (CO_2), sulfur oxides (SOx), unburned hydrocarbons, and smoke. While these aircraft emissions are a small percentage of the amount produced daily by other sources, their release into the upper atmosphere makes them particularly harmful. Another troublesome gaseous emission from aircraft is nitrogen oxides (NOx), which contribute to ozone depletion in the upper atmosphere. Impacts from NOx reduction will be seen not only in the upper atmosphere but also at airports, where NOx has led to significant production of smog from ground ozone. Depending upon the hydrogen propulsion system selected, emission of NOx can be reduced dramatically or even eliminated.

Nitrogen oxide emissions are produced during the combustion process and are primarily a function of combustion temperature and residence time. The introduction of hydrogen to a conventional gas turbine propulsion system will not eliminate NOx emissions. However, because of hydrogen's wide temperature flammability range it is feasible to create systems that are lean-burning (require less fuel) and low in NOx production. A revolutionary approach to completely eliminating NOx would be to fly all-electric aircraft powered by hydrogen–air fuel cells. The fuel cells systems would only produce water, which could be captured on board or released in the lower altitudes. Currently fuel cell systems do not have sufficient energy densities for use in large aircraft, but the long-term potential of eliminating greenhouse gas emissions makes it an intriguing and important field of research.

Fuel storage. Using hydrogen fuel will require significant changes to airframe design. This is due to the fundamental differences in properties between hydrocarbon-based liquid fuels, such as Jet-A, and hydrogen. Hydrogen has nearly 2.8 times the energy density (heat of combustion) per pound as Jet-A. However, liquid hydrogen has 1/11 the volumetric density of Jet-A. To produce equivalent energy, liquid hydrogen must be four times the volume of Jet-A.

Unlike current aircraft fuels, hydrogen fuel cannot be conveniently stored in wing tank bladders. The most efficient state to store hydrogen for flight application remains as a liquid. However, in the liquid state hydrogen is a cryogenic fluid and must be maintained at −423°F (−252.8°C or 20.4 K). Although practical tank designs that maintain fluid in the cryogenic state are available, fitting them into an aircraft structure is not as simple as for conventional jet fuel tanks. Tank geometries needed to maintain a fluid in the cryogenic state are much more limited than those to store fluids at ambient conditions. To minimize structural weight, spherical and cylindrical tank shapes are the most efficient.

A number of airframe design concepts have been developed or proposed to solve the problem of storing liquid hydrogen in an aircraft. Most of the design ideas address the problem by incorporating various tank configurations in the fuselage, such as separate tanks fore and aft, a tank along the top running the length of the fuselage, or running tandem to the passenger compartment. In all cases, the result is an increase in fuselage diameter and surface area with a corresponding increase in aircraft drag. An example is shown in a NASA artist's concept of a fore-and-aft tank configuration (**Fig. 1**).

Gas turbine power. The gas turbine engine is the most likely candidate for practical hydrogen-powered aircraft. However, any propulsion system based on the combustion of fuel with air will not eliminate the production of NOx. In general, NOx is formed progressively at high combustion temperatures and comparatively long burn times.

Fig. 1. Conceptual 300-passenger hydrogen transport aircraft. (*Terry Condrich, NASA GRC/Indyne*)

Conventional hydrocarbon-fueled jets produce a NOx-rich hot zone in the combustion chamber, before dilution, because the fuel/air ratios must be mixed close to an equal molecular ratio of oxygen and fuel—one mole of fuel to an equivalent mole of oxygen. This is called burning near the ideal stoichiometric point, which helps maintain stable combustion. The problem is that reducing the fuel-to-oxygen ratio, called lean burning, can quench combustion and cause a loss of all power. This is a phenomenon called lean-burn blowout.

The problem of lean-burn blowout can be solved for hydrogen systems by taking advantage of the wide flammability range of hydrogen, which permits combustion systems to run with excess oxygen (lean-burn). Equivalent turbine inlet temperatures are reached without creating a hot core, while maintaining combustion and avoiding lean-burn blowout. Lean direct injection (LDI) concepts take advantage of the rapid diffusivity of hydrogen to achieve quick mixing. Combining quick mixing with hydrogen's inherently faster burn kinetics (compared to jet fuel) will shorten combustion residence time, which sharply limits NOx production. Another advantage is that combustion chamber lengths can be reduced.

An additional key to an efficient hydrogen system will be the effective use of hydrogen throughout the engine. Several possibilities include using cold hydrogen for inlet air cooling through the compressor, turbine blade cooling for higher combustion temperatures, and combustor liner cooling; and using nozzle heat exchangers to preheat the hydrogen. Considerable design work is under way in this area. Integration of any or all of these concepts will require changes to existing engine designs.

Fuel cells. Fuel-cell-powered aircraft are attractive as a long-term solution to completely eliminate aircraft emissions. However, current technology requires a significant increase in fuel-cell-power density to permit design of electrically powered regional or commuter-size aircraft and an even larger increase for large commercial passenger aircraft. In conjunction with the increases to fuel cell power density, power management technologies would need improvement to supply megawatts of power throughout the aircraft.

The HALE ROA missions represent near-term applications for hydrogen-powered electric aircraft. Two missions of greatest interest are a 14-day mission with hydrogen-fed consumable fuel-cell systems and a 6-month mission with a solar-regenerative hydrogen-oxygen fuel-cell system. In a regenerative system, the fuel cell powers the aircraft during nighttime operations and the water produced is converted back to hydrogen and oxygen during the day when the vehicle is on solar power.

The two most promising fuel-cell types for aircraft applications are the proton exchange membrane (PEM) fuel cell and the solid-oxide fuel cell (SOFC). Each type has advantages and disadvantages depending upon the application. For either system, future aircraft implementation requires increasing the specific power (kW/kg) of the cells.

PEM fuel cells. PEM fuel cells are low-temperature devices, ~175°F (80°C), offering quick startup times, but require pure gaseous hydrogen fuel. Increasing the PEM operating temperature will improve tolerance to impurities (permit some hydrocarbons in the fuel) and may also improve specific power for the system. PEM fuel-cell stacks produce a significant amount of heat that is difficult to dissipate,

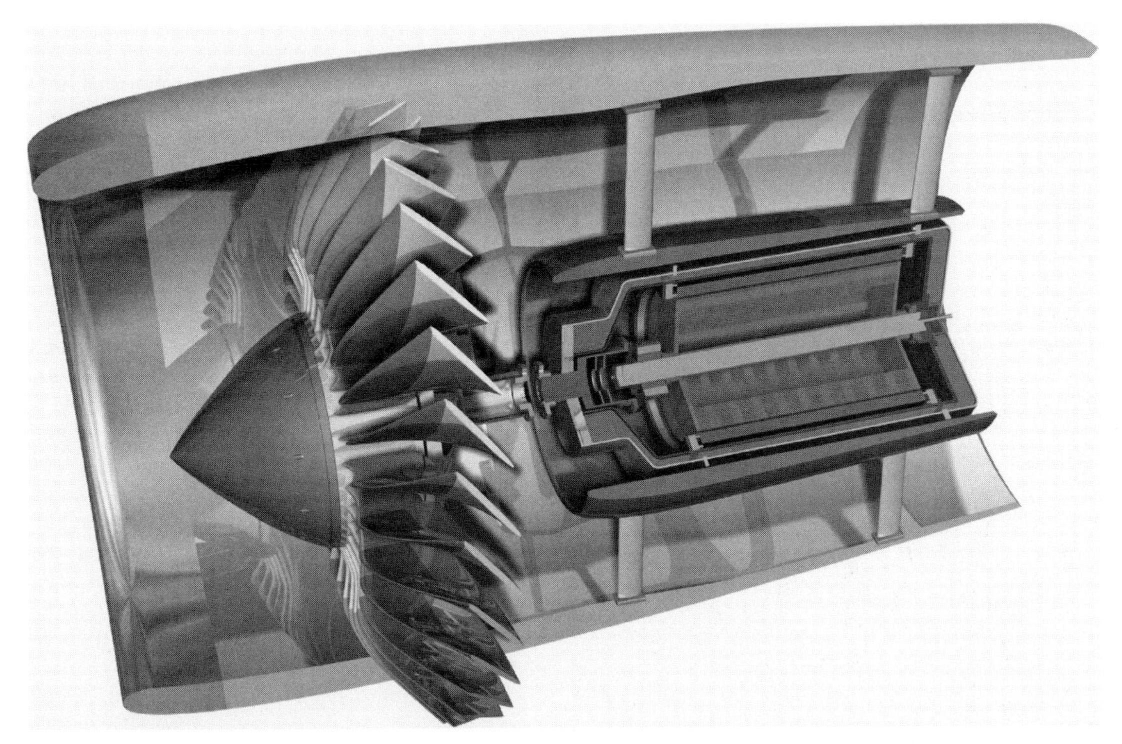

Fig. 2. Conceptual cyrogenically cooled electric motor ducted fan. (*Terry Condrich, NASA GRC/Indyne*)

and overheating stops energy output. This results in the need for liquid cooling systems. Achievement of higher PEM operating temperatures would increase the heat-transfer temperature differential for improved heat dissipation, resulting in system size and weight reductions.

SOFCs. SOFCs operate at high temperatures, in the range 1292–1832°F (700–1000°C), and tolerate higher levels of impurities. Current research on SOFCs is focused on planar (stacked plate) designs due to the higher potential specific power. The solid-oxide system could be used as a stand-alone power source or combined with a turbine in a hybrid system to achieve higher efficiencies. In contrast to PEM, SOFCs operate with significantly more airflow through the stack, which provides heat removal, eliminating the need for and corresponding weight of a liquid cooling system.

Although continued research and development has yielded promising results, both fuel-cell types will require significant design investments for incorporation into aircraft. Along with increases to the specific power, both operability and durability improvements will be required for flight applications.

Electric propulsion. A long-range goal is development of fuel cells that produce sufficient electric power to propel full-size aircraft without a turbine engine. This objective is theoretically possible, but significant technical advances are required and the configuration of the propulsion system will change. A turbine engine will be replaced by fuel-cell electricity powering an electric motor that drives an advanced propeller or ducted fan (**Fig. 2**). The major technological challenge is to achieve high electric motor power density. High-current-density motor windings will be needed.

The most promising candidate for large aircraft applications is a cryogenic synchronous motor with either high-purity aluminum conductors or superconducting windings, such as yttrium barium copper oxide (YBCO) or magnesium diboride (MgB_2). The new materials will reduce motor weight partly by eliminating the heavy iron components commonly found in electric motors. In addition, improved cooling from the cryogenic hydrogen will lower the resistance (or eliminate it for superconductors) and increase the amount of current in the motor. For smaller aircraft, either switched reluctance or permanent magnet motors that do not require cryogenic cooling may be acceptable choices. The keys will be designing high-current-density windings for switched reluctance machines or more advanced axial-gap permanent-magnet machines with high-field magnet arrays. High motor-shaft speeds will help improve motor power density, but advanced gear boxes may be required to match the lower propulsor shaft speeds. This is due to aerodynamic or structural limits for the propeller or fan. Though the electric motor is a smaller package, the equivalent fan diameter will grow in size to compensate for the loss of thrust from the hot gas core. Another option would be to power aircraft from a number of smaller, distributed propulsors along the wings or fuselage.

For background information *see* AIR POLLUTION; COMBUSTION; CRYOGENICS; FUEL CELL; HYDROGEN; JET PROPULSION; NITROGEN OXIDES in the McGraw-Hill Encyclopedia of Science & Technology.

Timothy D. Smith

Bibliography. D. S. Alexander, *Advanced Energetics for Aeronautical Applications*, MSE Technology Applications Inc., NASA CR-2003-212169, February 2003; J. J. Berton, J. E. Freeh, and T. J. Wickenheiser, *An Analytical Performance Assessment of a Fuel Cell-powered, Small Electric Airplane*, NASA TM-2003-212393, June 2003; J. Brand et al., *Potential Use of Hydrogen in Air Propulsion*, AIAA/ICAS International Air & Space Symposium and Exposition, Dayton, OH, AIAA-2003-2879, July 2003; G. D. Brewer, *Hydrogen Aircraft Technology*, CRC Press, 1991; M. D. Guynn and E. D. Olson, *Evaluation of an Aircraft Concept with Over-wing, Hydrogen-fueled Engines for Reduced Noise and Emissions*, NASA TM-2002-211926, September 2002; A. K. Sehra and W. Whitlow, Jr., Propulsion and power for 21st century aviation, in *Progress in Aerospace Sciences 40*, pp. 199–235, 2004.

Infectious disease and human evolution

Infectious pathogens have undoubtedly been a significant factor throughout the evolutionary history of humans, causing illness and death. It is equally certain that exposure to infectious agents and illness from infection became a more serious problem for human groups during the past 10,000 years. During this period animals and plants were domesticated, giving rise to an agricultural economy that provided the resources that stimulated sedentism and urbanism to become important environmental realities in human life. The risk of infectious disease was enhanced by exposure to pathogens carried by domestic animals (zoonoses) as well as increased human-to-human transmission of pathogens in larger concentrations of people.

Although exposure to infectious pathogens is the necessary factor for transmission of infectious disease, other factors influence its expression, including the number of pathogens that infect the host and the diet and general health of the individual. However, the most important factor for most people is their immune response to an infectious pathogen. An individual's immune response is affected by his or her genetic heritage and, thus, will be influenced by evolutionary mechanisms. People with an immune response better able to destroy, or at least limit, the adverse effect of infectious pathogens will tend to survive and reproduce, passing on to the next generation the genes that provide this increased resistance to infectious disease.

Host/pathogen coevolution. The coevolution between host and pathogen is seen in the relationship between all infectious pathogens and the human host population. A tension exists in this evolutionary process because infectious pathogens tend to evolve better mechanisms for maintaining access to resources in the host as they reproduce. From the host's perspective, the elimination of all harmful infectious pathogens is an obvious strategy. However, the options for pathogenic infectious agents are more complex. At one end of the spectrum are pathogens that live in the host without causing significant illness, or at least not rapid death. At the other end are highly virulent pathogenic organisms that can cause acute host illness and rapid death.

Archeological evidence and genetic studies. Unfortunately, the physical evidence for changes in the human immune response to infectious pathogens is not directly linked to anatomical features that can be studied in the fossil and archeological record of humans. However, some inferences can be made from analysis of human remains excavated from archeological sites and our knowledge about human responses to infectious pathogens in living human groups. Knowledge of the evolutionary changes that occurred in human infectious pathogens depends on inferences based on genetic, mostly deoxyribonucleic acid (DNA), studies of modern pathogens.

Research on both archeological and modern genetic evidence have revealed some general principles that help in making these inferences. In most cases, evidence of infectious disease in archeological human skeletal remains occurs only if the individual with the disease survives for an extended period of time. This means that most direct evidence of infectious disease in past human groups is a relatively chronic manifestation of disease (**Fig. 1**), in contrast with acute diseases that kill the individual quickly (viral infectious diseases such as Spanish influenza, smallpox, and plague as well as some bacterial and protozoan diseases such as cholera, dysentery, and malaria). Most, if not all, of the more chronic diseases, such as leprosy, tuberculosis, and treponematosis (one syndrome of which is syphilis), probably represent coevolutionary adaptation in which an improved immune response in the host population is combined with decreased virulence of the infectious pathogen. It is important to emphasize that relatively chronic infectious diseases may have an acute phase in an individual, particularly in early stages of the disease process or when first introduced into a new population. In either circumstance, significant illness and death can occur.

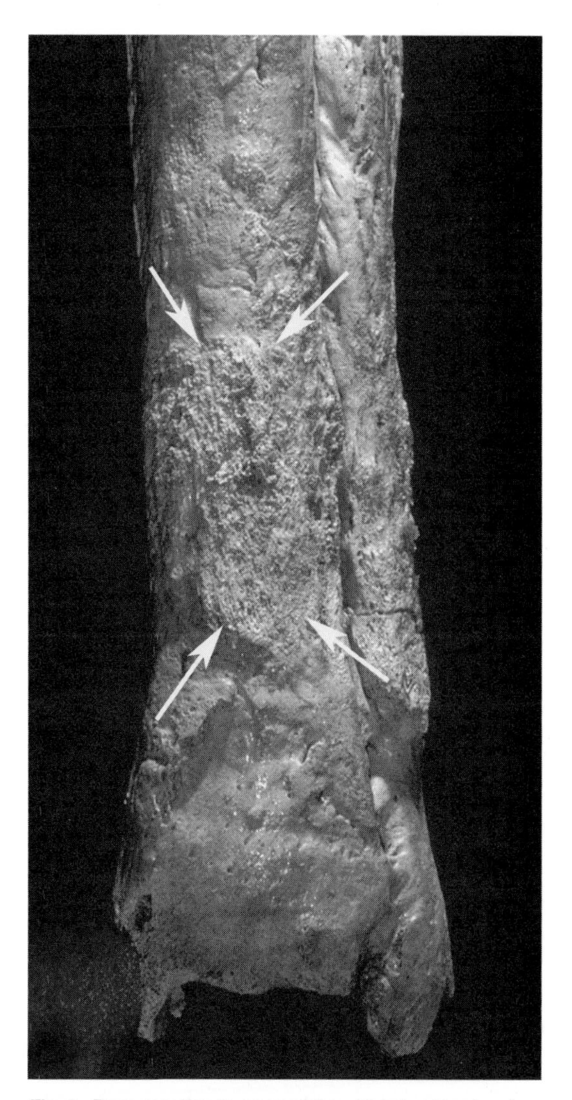

Fig. 1. Bone reaction to an overlying, chronic skin ulcer in the lower, anterior diaphysis (shaft) of the left tibia and fibula in the skeleton of a woman age 50 or older at the time of death. Note the oval boundary (arrows) of the reactive bone formation associated with the ulcer. The most common cause of ulcers of this type is *Staphylococcus aureus* bacteria. From Tomb chamber A100E, burial 2 at the site of Bab edh-Dhra' in Jordan dated to about 3250 BCE. (*Photo by Donald J. Ortner*)

Two infectious agents that illustrate important aspects of this coevolutionary process are *Mycobacterium tuberculosis*, the species of bacteria that most commonly causes tuberculosis, and human immunodeficiency virus (HIV), the virus associated with acquired immune deficiency syndrome (AIDS).

Tuberculosis. The recent emergence of an antibiotic-resistant strain of *M. tuberculosis* is a troublesome reminder of the reality of pathogen evolution. High mortality is associated with this pathogen. However, clinical experience over many decades demonstrates that most people exposed to tuberculosis will never have any symptoms of disease. Of those who do become ill, many will have very mild disease and low risk of death from tuberculosis. This relatively benign relationship between an infectious pathogen and its host also occurs in other chronic infectious diseases, including treponematosis, brucellosis, and leprosy. The pathogens causing these diseases may have had mild effects on people exposed to them from their earliest involvement with human groups. However, it is more likely that the recent relationship between these infectious pathogens and the host population is the result of coevolution in which the organism evolves a less virulent strain while the host population evolves a more effective immune response.

This coevolutionary process involves several variables and biological strategies on the part of the pathogen and host. In cases of direct pathogen transmission between human hosts (such as in sexually transmitted diseases and many types of airborne diseases), it is clearly beneficial for the pathogen to not kill the host quickly, since the pathogen tends to die with the death of the host. In situations in which the pathogen is not dependent on human-to-human transmission (as occurs with many of the water-borne diseases), the long-term survival of the host is much less important and its death does not adversely affect the survival of a species of pathogenic organisms. In the latter situation, it may be beneficial for the pathogen to be highly virulent (as occurs, for example, in cholera) even though the host typically dies within a few days unless effective treatment is available.

The history of host/pathogen coevolution in tuberculosis is of interest in understanding this process. Plausible skeletal evidence of tuberculosis is reported in the Near East at least as early as 3300 BCE (**Fig. 2**), but it probably occurred at least by the end of the Neolithic Period (approximately 4000 BCE) in that area of the world. It is unknown for how long *M. tuberculosis* has been a human pathogen. One hypothesis is that the pathogen that causes the human type of tuberculosis (*M. tuberculosis*) evolved from the pathogen that causes the cattle-borne form of the disease, *M. bovis*, after the domestication of cattle, about 8000 years ago. However, the results of current DNA research suggest that *M. bovis* is a more recent pathogenic variant of tuberculosis than *M. tuberculosis*.

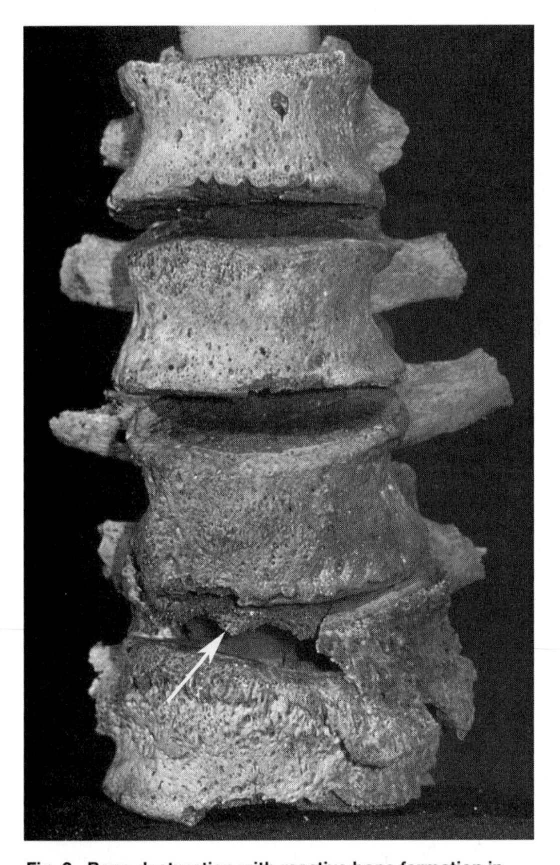

Fig. 2. Bone destruction with reactive bone formation in the lower vertebrae of a young male about age 18 at the time of death. The inferior two-thirds of the fourth lumbar vertebral body (arrow) was destroyed by the infectious disease process. The most common cause of destructive lesions of the spine is tuberculosis. Both the location and characteristics of the lesion argue for this diagnosis. The reactive bone formation at the margins of the destructive focus and on the adjacent vertebral body is indicative of survival with the disease for considerable time. From Tomb chamber A100E, burial 73 excavated from the archeological site of Bab edh-Dhra' in Jordan dated to about 3250 BCE. (*Photo by Donald J. Ortner*)

Human immunodeficiency virus. HIV is a much more recent human pathogen, commonly thought to have become a human disease sometime in the past 100 years. Current research indicates that the virus was initially transmitted from chimpanzees to humans via ingestion of contaminated meat. In AIDS, the time between pathogen exposure and host death tends to be lengthy, providing ample time for sexual transmission between hosts.

An interesting development has occurred in the host population that illustrates how genetic variability can lead to differences in disease resistance. Research on humans with a natural immunity to HIV indicates that they lack a gene encoding a protein (chemokine receptor 5, or CCR-5) that is a normal part of white blood cells. The protein appears to provide the entryway for the HIV to infect white blood cells. Thus, the absence of the CCR-5 gene and the protein it synthesizes ensures immunity to HIV. In a situation in which HIV is endemic, individuals lacking the CCR-5 protein in their white blood cells will have a major selective advantage over those who do have it, and the frequency of individuals in the population who lack the gene encoding CCR-5 will increase.

Conclusion. Human host/pathogen coevolution is an important factor in defining an effective response to infectious diseases. The emergence of antibiotic-resistant tuberculosis bacteria resulted, in part, from the failure of patients to take the prescribed antibiotic for the recommended length of time. This allowed bacteria somewhat more resistant to the antibiotic to survive, stimulating the emergence of bacterial strains better able to withstand antibiotics. Similar evolutionary processes take place in the changes that occur in infectious pathogens, as they react to improvements in the host population's biological response to pathogens.

The emergence of AIDS as a human viral disease illustrates the vulnerability of the human species to infectious pathogens. After infection, HIV causes slow development of illness, with an extended period of time before death. This means that there is minimal evolutionary benefit for the virus to develop less virulent strains, since there is ample time for transmission of the pathogen to other hosts before the death of the host. Thus, it seems likely that an effective human response to HIV will require more than a reliance on host/pathogen coevolution.

For background information *see* ACQUIRED IMMUNE DEFICIENCY SYNDROME (AIDS); BIOARCHEOLOGY; INFECTIOUS DISEASE; MEDICAL BACTERIOLOGY; ORGANIC EVOLUTION; PALEOPATHOLOGY; TUBERCULOSIS in the McGraw-Hill Encyclopedia of Science & Technology.

Donald J. Ortner

Bibliography. P. Ewald, *Plague Time*, Free Press, 2000; J. Lederberg, Infectious disease as an evolutionary paradigm, *Emerging Infectious Diseases* (serial on the Internet), October-December 1997; A. J. McMichael, Environmental and social influences on emerging infectious diseases: Past, present and future, *Phil. Trans. Roy. Soc. Lond. B*, 359:1049–1058, 2004; D. J. Ortner, Human palaeobiology: Disease ecology, in D. R. Brothwell and M. A. Pollard (eds.), *Handbook of Archaeological Sciences*, pp. 225–235, Wiley, 2001; D. J. Ortner, *Identification of Pathological Conditions in Human Skeletal Remains*, Academic Press, 2003.

Infrasonic monitoring

Sound waves in the atmosphere at frequencies below the hearing threshold of humans are termed infrasound. These subaudible waves are generated by a wide variety of natural phenomena, such as volcanic eruptions, and some human-made sources, such as large chemical or nuclear explosions. Infrasound has become an important component of the global effort to monitor nuclear testing activity. A global network of 60 infrasound arrays is currently being constructed for this purpose. Infrasound is a particularly valuable remote-sensing tool since it decays slowly across thousands of kilometers.

Infrasound. Infrasound relates to the audible part of the acoustic spectrum as infrared relates to the visible part of the electromagnetic spectrum. Human hearing is sensitive to sound at frequencies from approximately 20 to 20,000 Hz. Sound waves at frequencies below 20 Hz are subaudible and are termed infrasound. Infrasound waves above and below 1 Hz are termed near- and far-infrasound respectively. It is possible to feel intense near-infrasound. Studies indicate infrasound can also induce feelings of uneasiness.

Sources. Infrasound was discovered following the eruption of the volcano Krakatoa in Indonesia on August 27, 1883, which generated the loudest sound in recorded history. The eruption caused short-lived changes in air pressure worldwide that were detected on barometers (used by meteorologists to identify changes in atmospheric pressure) and pointed to the presence of infrasound waves. On June 30, 1908, an explosion (generally believed to be due to the explosion of a large meteor; the details are still being debated) over the Tunguska region of Siberia registered on barometers in Britain, several thousand kilometers away.

A broad suite of sources produce sound waves in Earth's atmosphere in the subaudible range. Natural sources of infrasound include volcanic eruptions, tornadoes, avalanches, earthquakes, meteors, aurora, significant storms, and atmospheric turbulence. Human-generated sources include large chemical or nuclear explosions, rockets, and supersonic aircraft. These sources must be energetically significant to

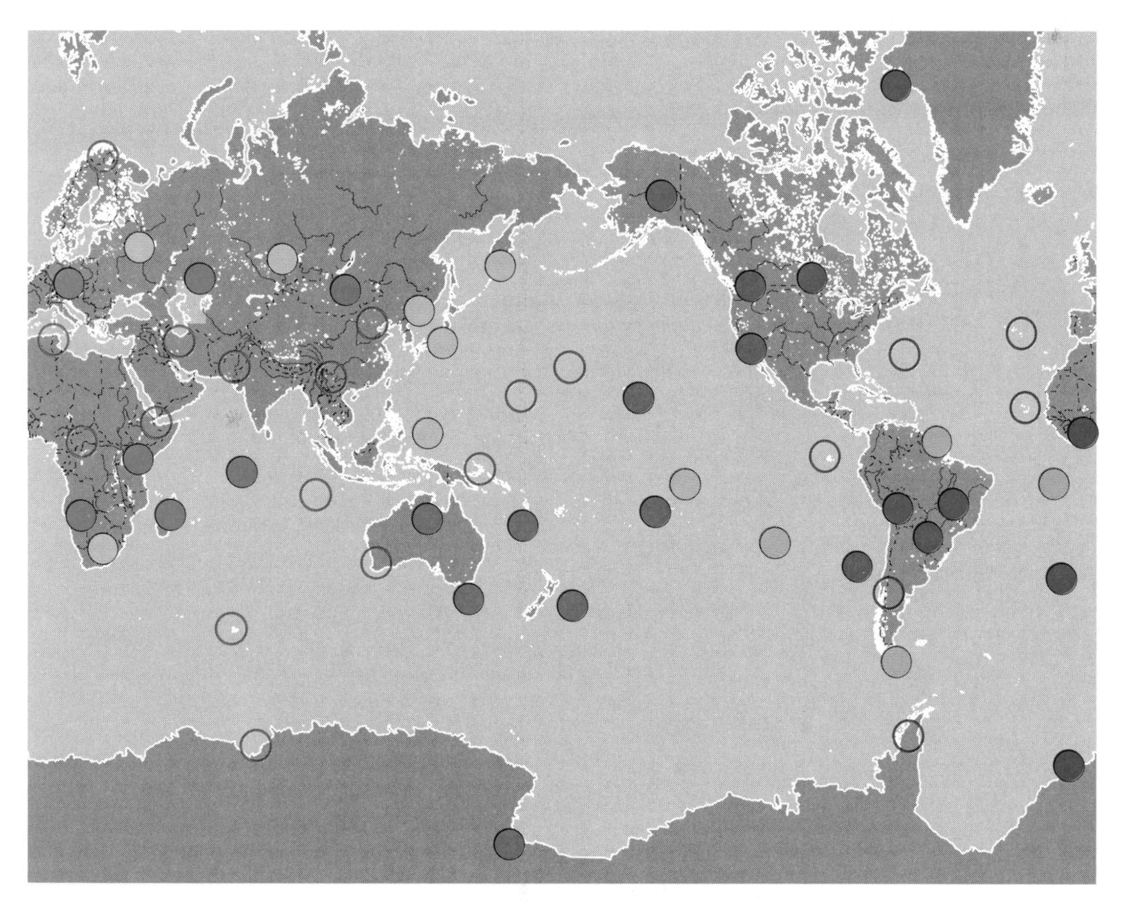

Fig. 1. Global infrasound network. Of the 60 planned stations, 27 are built and recording data (dark color), 12 are under construction (light color), and 21 are not yet complete (open circles).

move the required air volume to effectively generate observed signals. Some large animals, including elephants, rhinoceroses, and whales, are known to use infrasound to communicate over distances of many kilometers.

With the advent of the nuclear age, and testing of nuclear weapons in the atmosphere, the modern era of infrasound-based technologies began. Atmospheric explosions generate large infrasound waves, and so infrasound was adopted as one means to monitor for atmospheric nuclear tests. Recently, interest in infrasound was increased by the Comprehensive Nuclear Test Ban Treaty (CTBT), which bans all nuclear tests. The treaty establishes a network of sensors that continuously monitors the Earth's atmosphere for infrasound signals from atmospheric explosions (**Fig. 1**), as well as other networks that are used to monitor the Earth's solid interior and oceans.

Propagation. Infrasound is a mechanical wave phenomenon and is similar to seismic waves that travel through the solid Earth. The propagation of infrasound waves, however, differs significantly from seismic in that the velocity structure of the atmosphere changes rapidly. As with any other wave phenomenon, infrasound refracts, or changes direction, due to changes in the velocity of the medium it passes through. Infrasound refracts away from regions of high velocity. The velocity of infrasound is directly proportional to the square root of the absolute temperature of air and increases in the direction of wind due to advection. Both temperature and wind depend on altitude, geographic location, time of day, and time of year. Infrasound is efficiently refracted back to the Earth's surface from within the stratosphere and the thermosphere due to the increase in temperature with increasing altitude in these layers. Temperature decreases with increasing altitude in the troposphere, the lowest atmospheric layer; however, infrasound can be refracted back to the ground from within this layer if the infrasound propagates in the direction of strong winds.

Infrasound is valuable for monitoring for nuclear testing activity and other atmospheric phenomena because it propagates through the atmosphere with relatively little dissipation. The use of infrasound for the remote study of atmospheric phenomena, however, requires that the dynamic structure of the atmosphere be taken into account. Atmospheric models based on climatologies, numerical weather prediction models, and pinpoint measurements such as from radiosondes are used to simulate the propagation of infrasound.

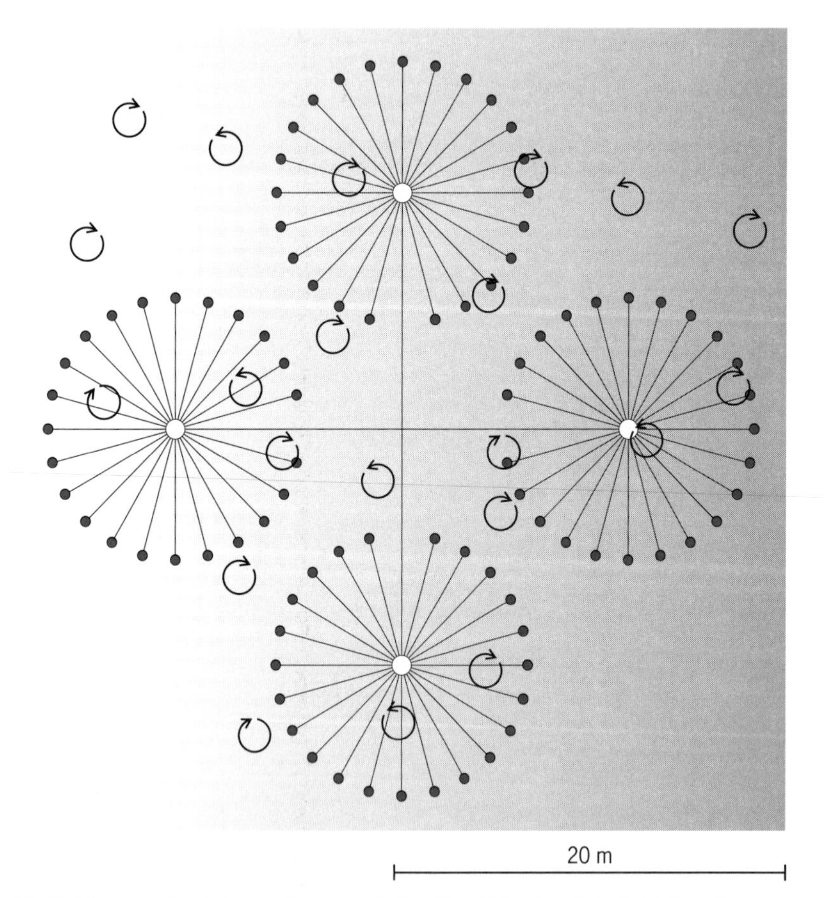

Fig. 2. Diagram of an infrasound noise-reducing filter. Noise (small circles) is largely incoherent between inlets and is averaged out in the sum. The signal (shading) varies gradually across the filter and is largely retained in the sum.

Infrasound detection. Our ability to detect infrasound waves is limited by noise due to atmospheric turbulence. Most infrasonic noise in the frequency band of interest to the nuclear monitoring community changes significantly (is coherent) across distances of less than several meters, while infrasonic signals can be largely unvarying (coherent) over distances in excess of 100 m (328 ft). The ratio of coherent signal to incoherent noise can be increased by averaging the infrasonic pressure field over an area smaller than the coherence limit of signals but greater than the coherence limit of noise. The most common method used to remove incoherent noise while preserving coherent signal is to sample atmospheric pressure at many points distributed spatially (across an area ranging up to 100 m across), transmitting the sound through an array of pipes to a microbarometer where the samples are summed acoustically. An infrasound noise-reducing filter (**Fig. 2**) comprises an array of pipes connecting inlets (small circles) to a microbarometer at the center. Sound, including noise and signals, enters the pipe system via each inlet. This design is used at many of the infrasound arrays in the global network.

There are several drawbacks to reducing noise mechanically. For example, signals are modified slightly while propagating through narrow pipes. There are small time delays between many samples of the signal, leading to the loss of high frequencies in the sum. These issues can be avoided by integrating sound waves at the speed of light without using pipes. One system under development uses fiber-optic cables for this purpose. Another approach is to record infrasound using a spatial grid of instruments and sum data from the instruments optimally to reject noise while preserving coherent signals.

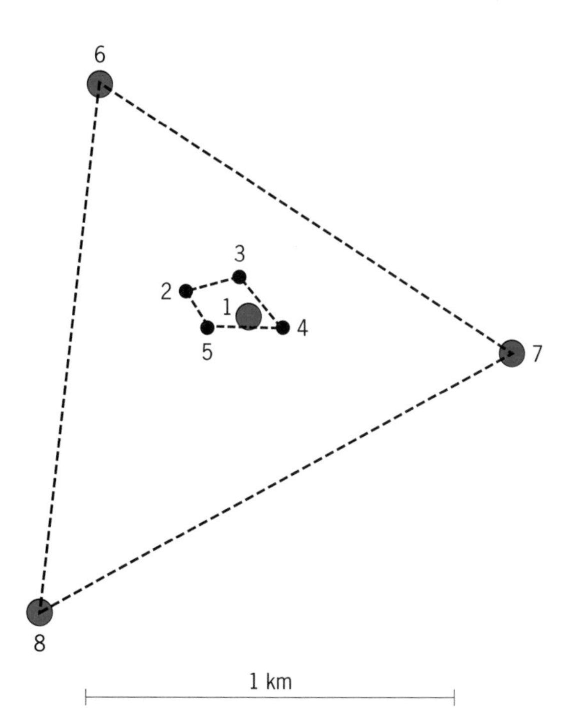

Fig. 3. Diagram of the infrasound array in southern California. The array is 1.5 km (1 mi) across and includes a large low-frequency optimized array (color circles) and a smaller high-frequency optimized subarray (black circles).

Global monitoring network. The global infrasound network will eventually comprise 60 continuously operating arrays. The network is designed to provide uniform coverage of the Earth's atmosphere, given the constraint that all arrays must be located on land. Each array in the network will comprise 4 to 8 microbarometers with an aperture of 1 to 3 km (3281 to 9840 ft; **Fig. 3**). Each microbarometer in an array is equipped with a noise-reducing filter. The array includes a large low-frequency optimized array and a smaller high-frequency optimized subarray. The high-frequency elements use noise filters of the type shown in Fig. 2. The low-frequency elements use larger filters that suppress noise at lower frequencies. The use of arrays of sensors, rather than single instruments, allows the detection of weaker infrasound signals and estimation of the speed and direction of the incident waves. The network is designed for optimal detection of signals between 0.1 to 1.0 Hz.

Array processing. Array processing methods take advantage of the coherence, or similarity, of a signal recorded at different elements in an array to provide information about the signal, such as the velocity at which the signal crossed the array. The velocity can be used to infer the direction to the source and the path the signal took through the atmosphere. The most commonly used method in the infrasound field is based on the progressive multi-channel correlation (PMCC) algorithm. The PMCC method is based on the cross-correlation of data from different elements in an array. Cross-correlation is a mathematical measure of similarity of two segments of data. If two elements in an array record the same signal, the cross-correlation procedure will provide an estimate of how similar the recordings are, and how much in time the signal is shifted between the two recordings due to the propagation delay. The PMCC method extends this idea to subarrays of several elements and defines a closure relation as a more demanding test of whether a signal is present in the data. Consider a subarray of three elements (i,j,k). The PMCC method defines signal consistency (r_{ijk}) for the subarray as

$$r_{ijk} = \Delta t_{ij} + \Delta t_{jk} + \Delta t_{ki}$$

where Δt_{ij} is the time delay between the arrival of a signal at sensors i and j. A signal is considered consistent if the sum of the time delays is below a predefined threshold. The PMCC method applies this test to several small subarrays within an array. If a highly correlated and consistent signal is found, the test is repeated for progressively larger subarrays to refine the estimate of the velocity of the signal across the array. This procedure is repeated in several frequency bands and in numerous overlapping windows spanning the time of interest. The use of small subarrays avoids ambiguity problems inherent in correlating distant signals. For example, the PMCC method has been applied to a recording of a bolide signal. A bolide (bright meteor) exploded over the Pacific at a distance of 1800 km (1100 mi) from the array in southern California diagramed in Fig. 3. The array recordings of this signal (**Fig. 4**) were highly consistent and provided accurate estimates of the signal's azimuth and speed as a function of frequency and time.

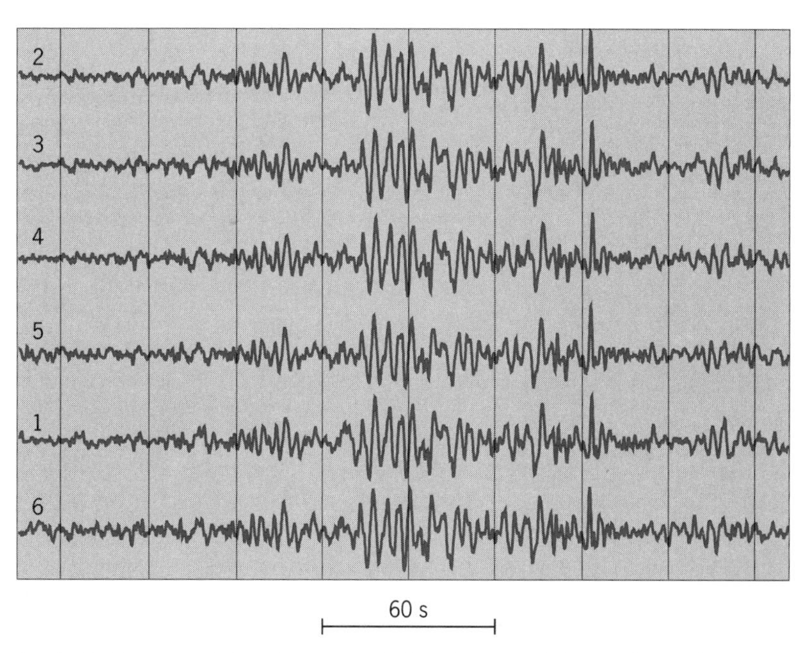

Fig. 4. Highly coherent signal from the explosion of a large bolide recorded by elements 1–6 at the array in California diagramed in Fig. 3. The other two elements, 7 and 8, were not operating at the time of the explosion.

A single-array recording can be a very accurate estimate of the direction to the source of the signal. The velocity of the signal across the array can be used to infer the path through the atmosphere taken by the infrasound; however, pinpointing the location of the source requires recordings made by more than one array.

Continuing basic research. In monitoring the atmosphere continuously, the infrasound network offers an unprecedented opportunity to better understand human-made and natural atmospheric phenomena on a global scale. Research continues on how sound propagates through our unsteady atmosphere and how clear recordings of distant events can be made despite noise due to atmospheric turbulence. Researchers are improving models of the atmosphere and are collecting information about significant atmospheric sources to provide a basis for this research.

For background information *see* ACOUSTIC NOISE; ATMOSPHERIC ACOUSTICS; INFRASOUND; SHOCK WAVE; SONIC BOOM; SOUND in the McGraw-Hill Encyclopedia of Science & Technology.

Michael A. H. Hedlin

Bibliography. A. J. Bedard and T. M. Georges, Atmospheric infrasound, *Phys. Today*, pp. 32–37, March 2000; M. A. H. Hedlin et al., Listening to the secret sounds of Earth's atmosphere, *EOS Transactions of the American Geophysical Union*, vol. 83, pp. 557, 564–565, 2002; Y. Cansi, An automated seismic event processing for detection and location: The P.M.C.C. method, *Geophys. Res. Lett.*, 22:1021–1024, 1995; D. P. Drob, J. M. Picone, and M. Garces, Global morphology of infrasound propagation, *J. Geophys. Res.*, 108 (D21), ACL 131-1 to ACL 13-12, 2003; M. A. H. Hedlin, B. Alcoverro, and G. D'Spain, Evaluation of rosette infrasonic noise-reducing spatial filters, *J. Acous. Soc. Amer.*, 114:1807–1820, 2003; M. A. Zumberge et al., An optical fiber infrasound sensor: A new lower limit on atmospheric pressure noise between 1 Hz and 10 Hz, *J. Acous. Soc. Amer.*, 113:2474–2479, 2002.

Insect flight

Insects flap and twist their wings to stay aloft, to dart forward, to turn, and to hover. It is the swirls of air stirred by a flapping wing that generate an insect's lift and thrust. To understand how insects fly, it is necessary to understand the dynamics of unsteady airflows created by a flapping wing, which is significantly different from the steady airflows around a classical airfoil. Although the aerodynamics of a steady airfoil has been well understood for almost a century, it is only recently, with the advance of new experiments, computations, and analyses, that scientists have began to unravel some of the mysteries of unsteady aerodynamics.

Reynolds number. How are insects different from birds and airplanes? Insects are obviously smaller and they flap their wings faster, but what differences do these characteristics make? The typical wing length ranges from 1 mm (0.04 in.) to 5 cm (2 in.), and the wing flapping frequency from 1 to 1000 Hz. For example, a dragonfly wing (**Fig. 1**) has a length (span) of about 4 cm (1.6 in.) and a width (chord, c) of about 1 cm (0.4 in.), and its wing frequency is about 40 Hz; thus the wing tip speed (u) is about 1 m/s (40 in./s). Based on these parameters, the dimensionless Reynolds number (Re) is about 1000, where $\mathrm{Re} = uc/\nu$, with ν being the air viscosity. A much smaller insect, the chalcid wasp, has a wing length of about 0.5–0.7 mm and beats its wing at about 400 Hz, which gives a Reynolds number of about 25. From the chalcid wasp to the hawk moth, the Reynolds number of insect flight ranges from 10 to 10,000. This range lies between the two classical limits for which the fluid dynamics is better understood: inviscid theory for steady laminar flow around an airfoil at high Reynolds numbers, and Stokes flow experienced by a bacterium swimming at very low Reynolds numbers. In other words, insect flight is neither like gliding in air nor like swimming in honey.

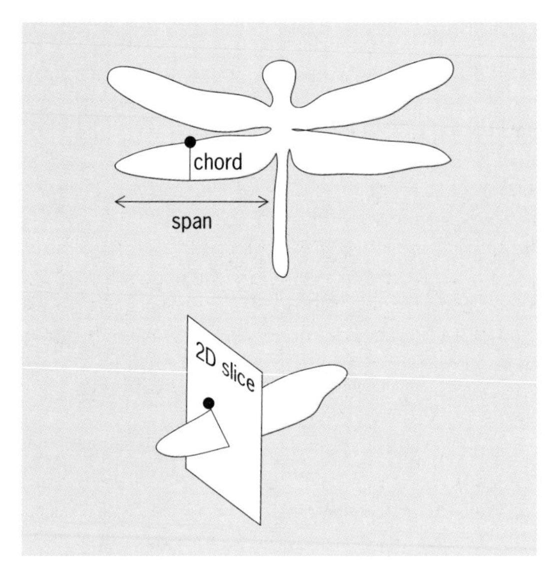

Fig. 1. Terms used to describe insect wing motion. (*After Z. J. Wang, Dissecting insect flight, Annu. Rev. Fluid Mech., 37:183–210, 2005*)

Wing motion and aerodynamics. When the wing motion is slowed down using a high-speed camera (**Fig. 2**), it is found that the typical wing motion consists of two basic modes: the wing translates back and forth along a curved stroke plane, and it rotates about the longitudinal axis, reversing its pitch near the end of each half-stroke. This ability to reverse wing pitch makes insects different from most birds (except hummingbirds), and this enables them to hover.

The combination of the wing translation and rotation produces jets of air that propel the insect. A flying insect leaves a "footprint" in air in the form of a stack of vortex rings. A two-dimensional slice of this stack reveals a succession of the clockwise and counterclockwise rotating vortices paired up in such a way (**Figs. 3** and *4b*) as to shoot jets of air away from the wing and thus induce a thrust forward and upward. This is in contrast to the more familiar vortex configuration seen in a normal Kármán wake shed behind a moving cylinder, where the vortices are paired to suck the flow in toward the cylinder to induce a drag (Fig. *4a*). The reversed Kármán wake is a signature of thrust generation in a fluid, and it is also seen in bird flight and fish swimming. With the advance of visualization techniques, it is now possible to view the intricate details of the three-dimensional structure of the flow on various insects, birds, and fish.

Force on an insect wing. How much force does an insect wing generate? The myth that the bumblebee cannot fly according to conventional aerodynamics has lived surprisingly long. Crude calculations, some dating back to 1919, have denied insects the right to fly. It is difficult to calculate the force generated by an insect wing from first principles. Depending on the approximations made, the calculated lift may or may not be sufficient to support the weight of an insect. Worse, the fact that a theory predicts enough averaged lift does not necessarily prove that the assumptions made in the theory are correct. Direct measurements of the unsteady force on a single wing of an insect are also difficult. In recent years, biologists have developed experiments using dynamically scaled-up wings, and theorists have devised computational tools to quantify the time-dependent forces and unsteady flows in flapping flight.

Utilization of dynamic stall. A distinctive feature of insect wing motion is that it employs an angle of attack (about 30°) much higher than that of an airplane wing (about 10°). At such a high angle of attack, a strong leading-edge vortex forms on the upper surface of the wing. If the wing is in steady translation, as is the case with an airfoil, the leading-edge vortex eventually separates from the wing. When this happens, the lift drops, the drag increases, and the airplane stalls. However, the separation occurs over a time proportional to the wing chord, and during this transient the airfoil experiences enhanced lift and drag. This phenomenon is known as the dynamic stall. An advantage of a flapping motion over a fixed wing is its ability to take advantage of the high transient force during dynamic stall by reversing its course before the leading-edge vortex detaches. This suggests that the flapping period should be roughly proportional to the time scale of the separation of the leading-edge vortex or the wing size, which is indeed the case. At each half-stroke, a new leading-edge vortex forms and remains attached before the wing reversal. Taking into account the dynamic stall, an insect wing can generate at least 50% more force in a transient state than in the later steady state. In addition, a flapping wing can take advantage of the coupling between wing rotation and translation, wing acceleration, and wing wake interactions, among other phenomena, to boost its force.

(a)

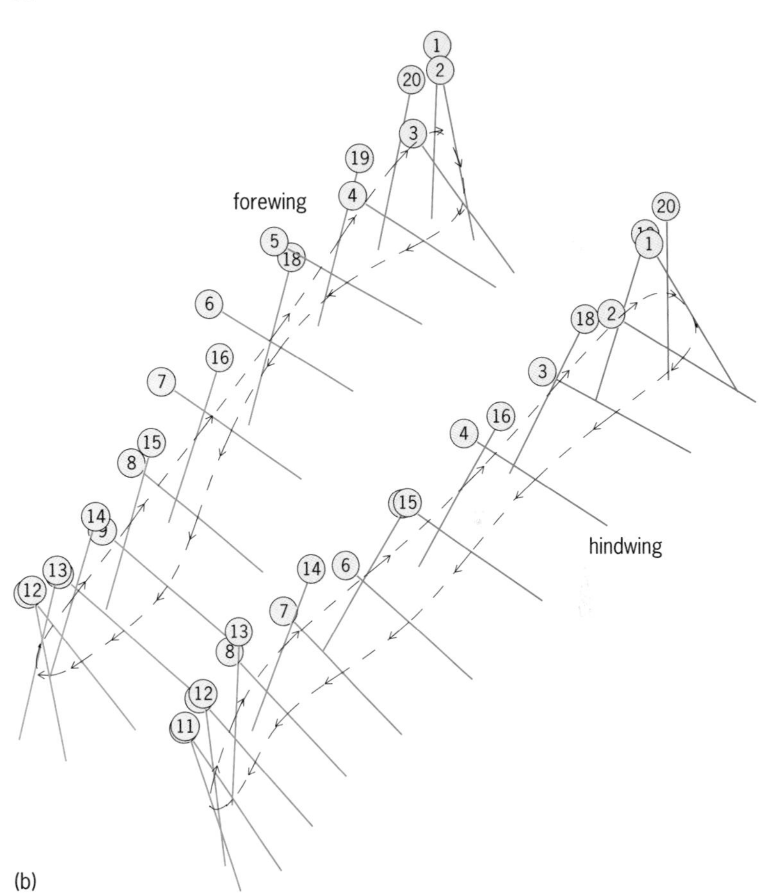

(b)

Fig. 2. Wing motion of a tethered dragonfly. (*a*) Motion during one wing beat, filmed at 1600 frames per second. (*b*) The time sequence of the chord positions of fore- and hindwings projected onto the two-dimensional slice shown on the lower part of Fig. 1. (*From Z. J. Wang, Dissecting insect flight, Annu. Rev. Fluid Mech., 37:183–210, 2005*)

Efficiency. Generating enough force is a necessity, but to do so efficiently is an art. The efficiency of an airfoil is characterized by its lift-to-drag ratio, and the design of an efficient airfoil centers on minimizing drag. On the other hand, insect wings are much smaller, the magnitudes of lift and drag are more comparable, and few insects glide. While airplanes and helicopters use only aerodynamic lift to fly, it is not *a priori* clear that insects should do the same. One of the most maneuverable species of insects, the dragonflies, use primarily drag to fly by pushing

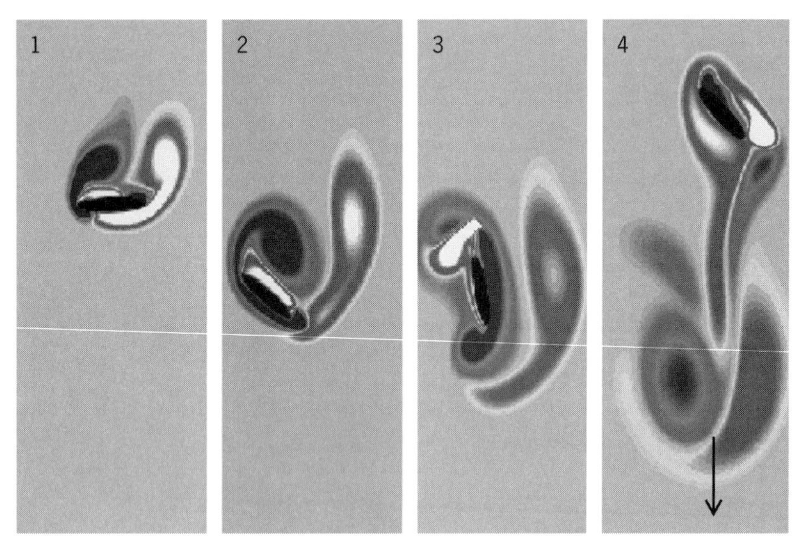

Fig. 3. Vorticity field created by dragonflylike wing motion. (*From Z. J. Wang, Two dimensional mechanism of hovering, Phys. Rev. Lett., 85:2216–2219, 2000*)

down the wing and then feathering on upstroke as if they row in air.

Understanding the unsteady aerodynamics of insect flight, though a difficult feat, is only the beginning of understanding flapping flight in nature. The lessons learned in recent studies will help to tackle a new set of questions including "Why do insects or birds flap their wings the way they do?", "How does flapping flight come about in the course of evolution?", and "When is flapping flight more efficient and stable than fixed-wing flight?"

For background information *see* AERODYNAMIC FORCE; AIRFOIL; CREEPING FLOW; FLIGHT; FLUID FLOW; KÁRMÁN VORTEX STREET; REYNOLDS NUMBER; VORTEX; WAKE FLOW in the McGraw-Hill Encyclopedia of Science & Technology. Z. Jane Wang

Bibliography. M. H. Dickinson, Unsteady mechanisms of force generation in aquatic and aerial locomotion, *Amer. Zool.*, 36:537–554, 1996; C. P. Ellington, The aerodynamics of hovering insect flight, I.–V., *Phil. Trans. Roy. Soc. Lond.*, B305:1–181, 1984; T. Maxworthy, The fluid dynamics of insect flight, *Annu. Rev. Fluid Mech.*, 13:329–350, 1981; Z. J. Wang, Dissecting insect flight, *Annu. Rev. Fluid Mech.*, 37:183–210, 2005; T. Weis-Fogh and M. Jensen, Biology and physics of locust flight, I. Basic principles in insect flight: A critical review, *Proc. Roy. Soc. B.*, 239:415–458, 1956.

Intelligent vehicles and infrastructure

Intelligence is the capacity to acquire and apply knowledge. In the context of intelligent vehicles, it refers to the ability of a vehicle to proactively support the driver or the driver's intent. This is an explicit move from the traditional vehicle manufacturer's goal of realizing the driver's intent in an accurate, linear, and reliable manner to a new goal of actively modifying the driver's intent, with the eventual possibility of completely automatic vehicle control.

The immediate intent of the driver is largely a function of the surroundings, including the location of other vehicles and the location and condition of the roadway. Therefore, the definition of an intelligent vehicle implies that it has some knowledge of its surroundings. Sensors that can derive the structure of the vehicle environment are the tools for gathering this "knowledge." They are what make vehicles or infrastructure "intelligent." Sensors may

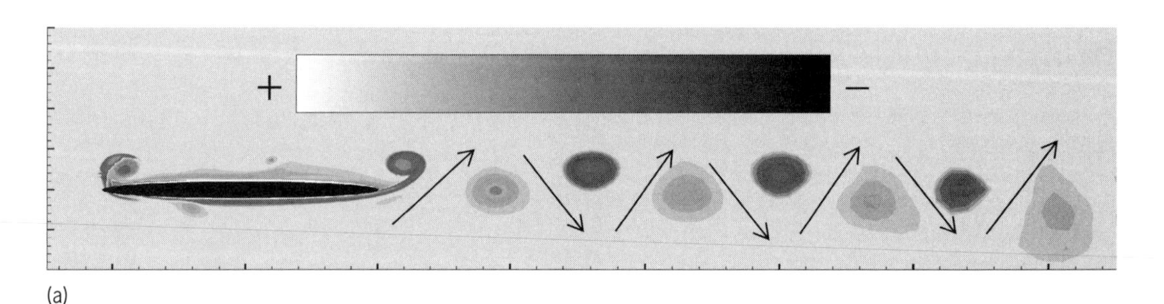

(a)

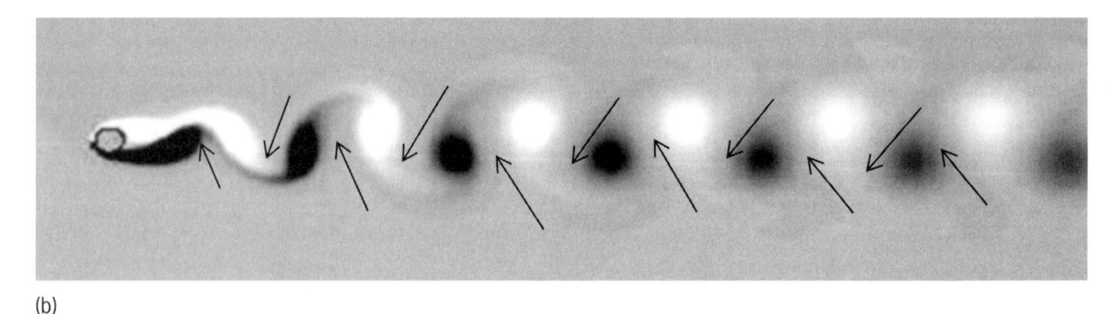

(b)

Fig. 4. Wake flows. Light gray indicates positive (counterclockwise) vorticity, dark gray indicates negative (clockwise) vorticity. Arrows indicate the direction of flow jet. (*a*) Reserved Kármán wake behind a flapping wing in forward flight. (*from Z. J. Wang, Vortex shedding and frequency selection in flapping flight, J. Fluid Mech., 410:323–341, 2000*). (*b*) Kármán wake shed behind a moving cylinder.

be autonomous (such as vehicular radars and vision systems) or cooperative, enabling communication between vehicles or between vehicles and the infrastructure. Sensors detect constraints on the vehicle's surroundings in critical dimensions, so that the near future can be reliably inferred by an onboard computer. Once dangerous situations are detected, drivers can be warned and possibly modify their intent (as exemplified by their control actions). With sufficient high-quality information, it may be possible for a vehicle to determine the proper intent in a well-constrained environment. This leads to autonomous driving.

Today's vehicles have nowhere near the capability required to enable fully autonomous driving. Processors and algorithms do not work in the unconstrained environment of the average roadway. However, there are circumstances where existing systems appear to be adequate for some intelligent actions on the part of a vehicle, such as the following examples in which the driver's intent is clear from the given situation. When following a vehicle, the driver's intent generally is to stay some distance behind. In stop-and-go traffic, the driver's actions are quite constrained. And when approaching a red light, one can assume the driver's intent is to stop. In each case, the actions of the driver are so constrained by the situation that the intent is fairly well defined. The above examples are leading to the development of applications such as adaptive cruise control (ACC), which has now been in production for several years; stop-and-go ACC, which is expected to be available soon; and signal violation warning, a current topic of research. As sensors and algorithms become available, the scope of these applications and the situations in which they are enabled will increase.

The primary goals for intelligent vehicles are twofold. The first goal is to improve the safety of the transportation system, largely by supporting drivers at the rare times when they do not behave correctly and thus cause accidents. Approximately 95% of all accidents are due to driver error. The second goal is to improve the mobility of the transportation system. For most drivers, mobility enhancement is equivalent to traffic reduction or the ability to make better traffic predictions and thus reduce the variability of the system.

Safety. Traditional vehicle safety enhancements have focused on increasing the driver's ability to control the vehicle in a variety of situations, as well as the ability of the vehicle to protect the occupants in the event of a crash. Intelligence is in the driver, and intent is communicated to the vehicle through a simple set of controls (steering wheel and pedals). Safety systems have focused on collision mitigation, but this approach is generally acknowledged to have reached the point of diminishing returns. In the last 10 years, several systems have been deployed to change this paradigm. There are now intelligent vehicles with forward-looking radars for automatic cruise control (which maintain a constant distance from the leading vehicle) and vision systems for lane-departure warning. There are also efforts to develop deployment models for communication between vehicles and car-to-infrastructure safety applications. While still very early in their development, these and related technologies are providing intelligence to vehicles, allowing the vehicles to detect hazards that drivers may not have seen. In the near term, sensors will be used to prepare a vehicle for an impending impact by pretensioning seatbelts and deploying airbags and bolsters. Longer term, vehicles will apply brakes and other nonreversible interventions, but these require a high degree of certainty in the situational assessment (**Fig. 1**).

A characteristic of intelligent vehicle technologies is that they provide the vehicle with some ability to interpret the surroundings and provide the driver with critical information or warnings about situations that the driver may not be aware of.

The driving environment is quite unconstrained, and the machine intelligence to operate properly in unconstrained situations is not yet available. A key objective of intelligent systems is to constrain the environment in such a way that the vehicle can act in a reliable and helpful manner, and above all do no harm. Constraints may come from sensor systems onboard the vehicle, or they may be communicated to the vehicle from the infrastructure via a communications link or possibly through a "static" map database. Information useful to a vehicle includes road geometry, status of traffic controls, and locations of the surrounding vehicles. If some of this information is constrained, it becomes much easier to identify a finite set of options for the intelligent system to choose the best option.

Mobility. One persistent problem in the current transportation system is congestion, which has been increasing dramatically over the past 20 years because of a 72% increase in vehicle miles traveled and only 1% increase in total road miles. It is no longer a problem isolated in large metropolitan regions; it is everywhere. One partial solution is to better manage vehicle flow across the highway network and to avoid breakdown or gridlock in the system (such as through active timing of traffic signals and ramp metering onto freeways). To be effective, this approach requires adequate knowledge of the current situation on the roadway and empirical knowledge of the effects the activating controls will have. Currently, this information is not available in most places. Intelligent vehicles seek to ameliorate this problem through vehicles communicating current traffic patterns to the infrastructure and by the infrastructure advising vehicles on routes to take.

Movement patterns in vehicles may also be used to infer accidents and other safety problems on the roadways, as well as for upgrading and planning the transportation network.

Technology. Hardware developments in radar vision systems and communications systems are proceeding at a very rapid pace, and cost is becoming less of a factor in the deployment of these technologies.

Radar. For ranges of about 100 m (330 ft), radar at 77 GHz has been deployed in vehicles for adaptive

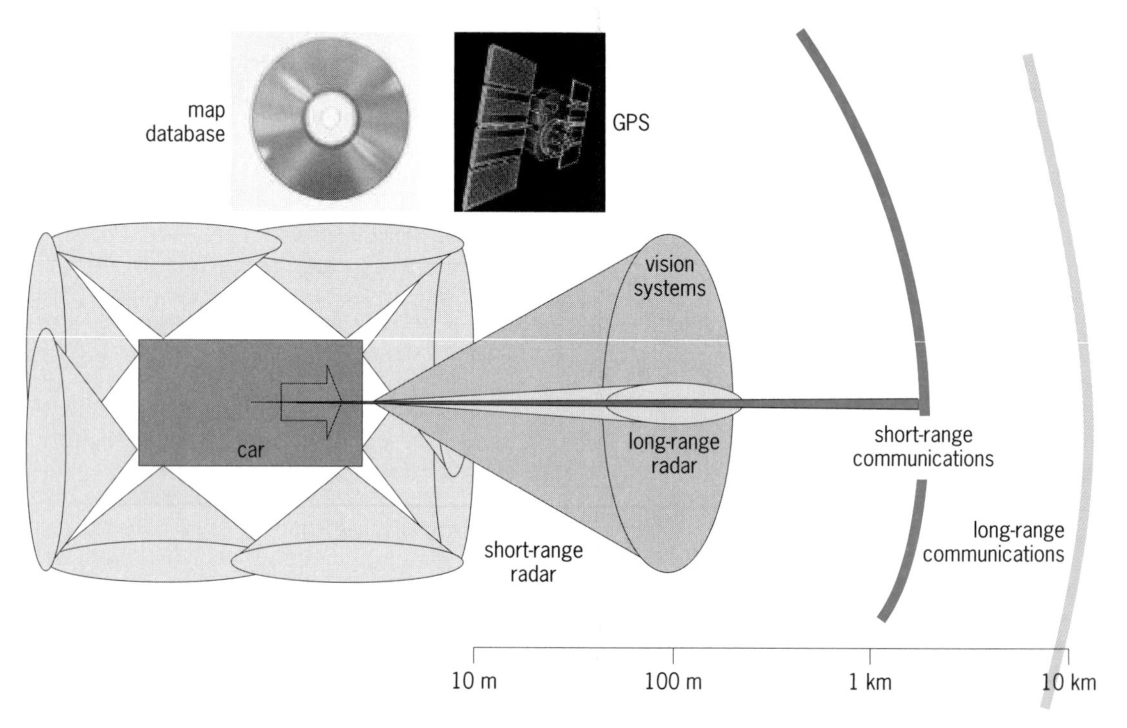

Fig. 1. Sensors on a future intelligent vehicle. Additional long-range systems may be introduced to cover rear views.

cruise control and will likely be extended to support forward-collision warning applications. This radar is also expected to be deployed rearward from vehicles to identify overtaking vehicles in applications such as lane-change warnings. Short-range radar at 24 GHz is nearing deployment for monitoring the environment adjacent to the vehicle at a range of about 10 m (30 ft). This technology will support blind-spot warnings, stop-and-go cruise control, and imminent collision mitigation.

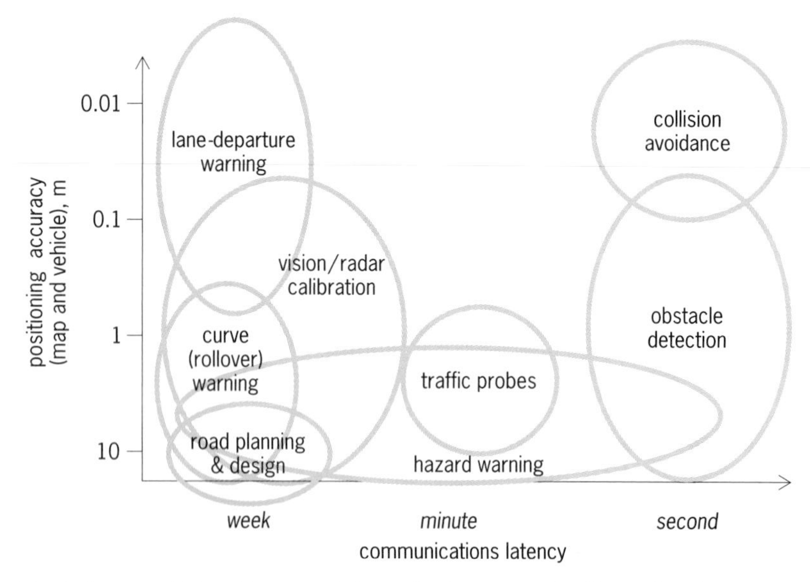

Fig. 2. Various vehicle safety and mobility applications are enabled by combinations of communications latency and positioning accuracy. Positioning accuracy may result from static objects (such as road edges captured in a map database) or dynamic objects (such as other vehicles). Communications latency for given information must be less than the characteristic time for that information.

Vision systems. Vision systems or cameras are currently being used for lane tracking in many commercial trucks and some passenger cars. Currently these systems look at the road and track the paint markings on the pavement to warn of lane departure or possibly road departure. Eventually these cameras may be used to recognize pedestrians or animals on the road ahead, especially if they use infrared wavelengths.

Communication. Communication systems that support intelligent vehicles are not available with the required latency, reliability, and short-range and high-bandwidth characteristics required. In the United States, the 5.9-GHz band has been allocated for transportation safety applications, and may be used for mobility and commercial applications. Programs are under way within the industry to evaluate the suitability of the 802.11p standard for using the 5.9-GHz band to support safety applications. In conjunction with global positioning systems (GPS) or other positioning systems, this technology could support most vehicle-to-vehicle and vehicle-to-infrastructure safety and mobility applications.

Positioning/mapping. This technology supports intelligent vehicles by allowing the vehicle's current position to be compared with a map of the region. Current positioning capability provides an accurate assessment of the road, which is sufficient to support applications such as curve warning and signal-violation warning for simple intersections. Possible future capability to support "which lane" or "where in lane" positioning would enable applications such as electronic emergency brake lights, lane-departure warning, and true collision avoidance when used in conjunction with a communications link.

Some of these applications are currently limited by the availability of highly accurate and reliable digital maps. In the future, these maps may be built by collecting data on where vehicles have traveled historically and by statistically combining this information to derive highly accurate, statistically validated maps of the road network, including traffic controls (**Fig. 2**).

Deployment issues. Radar and vision technologies are being deployed through a standard automotive deployment model. As these technologies gain consumer acceptance and prices decrease due to volume and general technology advances, they will become available on lower-priced vehicles. Due to their interactions with infrastructure [reflections from roadside furniture (lamp poles, signs, and so on) or paint], the use of the sensors for safety critical applications will be limited until a feedback loop can be put in place to ensure compatibility with the infrastructure everywhere in which they operate. The feedback loop will enable the sensors' algorithms to be updated (to reflect real conditions) or to direct infrastructure authorities to change or fix the roadside furniture or markings. This leads to the need for communications capability.

The communications requirements for applications, such as a feedback loop or traffic and mapping probes, can be met using existing cellular networks; however, the cost of the data is prohibitively high. The short lifetime of cellular communications technology and standards makes them relatively difficult to integrate with systems in vehicles, which have a lifetime of perhaps 20 years from design to junkyard.

For vehicle-to-vehicle safety communication, there is a consensus that the only communications technology suitable is the 802.11p standard at 5.9 GHz. The range of the system (up to 1000 m or 3300 ft) makes it difficult to deploy, since there is little value to the purchaser unless there are many others in operation. One current approach under investigation is to initiate a massive deployment of the technology in the infrastructure, along with a commitment to deploy it in the vehicle fleet once some infrastructure threshold is reached. This has the advantage of providing the first customers with a valuable service, so that vehicle manufacturers can recover some of their costs. This also provides stability of the communications network since both transportation departments and vehicle manufacturers have design lifetimes of about 20 years.

Outlook. Traditionally ground vehicles and the roadway infrastructure have been designed and operated independently. This is largely because there was no great need, and the technology did not exist to implement cooperation. In the last few years, this situation has changed. It is no longer feasible to build our way out of congestion, and traditional approaches to passive safety are producing diminishing returns. Recent advances in technology—radar, vision systems, vehicle positioning, communications, and information management—have provided the technological means for exchanging information and coordinating vehicles with the infrastructure. The basic technology is being deployed today, with plans for ever-increasing levels of integration. However, the institutional mechanisms are not in place to support cooperation on a wide scale. Meanwhile, many social issues must be addressed, such as privacy and the responsibilities of drivers.

For background information *see* ALARM SYSTEM; ALGORITHM; AUTOMOBILE; COMPUTER-BASED SYSTEMS; CONTROL SYSTEMS; GUIDANCE SYSTEMS; HIGHWAY ENGINEERING; MICROPROCESSOR; MICROSENSOR; RADAR; RADIO SPECTRUM ALLOCATIONS; SATELLITE NAVIGATION SYSTEMS; TRAFFIC-CONTROL SYSTEMS; TRANSPORTATION ENGINEERING in the McGraw-Hill Encyclopedia of Science & Technology.

Christopher K. Wilson

Bibliography. R. Bishop, *Intelligent Vehicle Technology and Trends*, Artech House, 2005; *VSC (Voluntary Sector Commission) Final Report*, IEEE Paper.

Interaction of photons with ionized matter

Most of the known matter in the universe exists in ionized (charged) form, and nearly all knowledge about it is carried to us by light (photons). Important to astrophysicists is an understanding of the origin of those photons and their encounters during journeys taking light-years to reach our telescopes. Interactions of photons with ionized matter are also critical in diagnosing and understanding the properties of hot ionized gases (plasmas) that are encountered in research to harness nuclear fusion, the energy source of the Sun. As an atom successively loses electrons and becomes highly ionized, it is

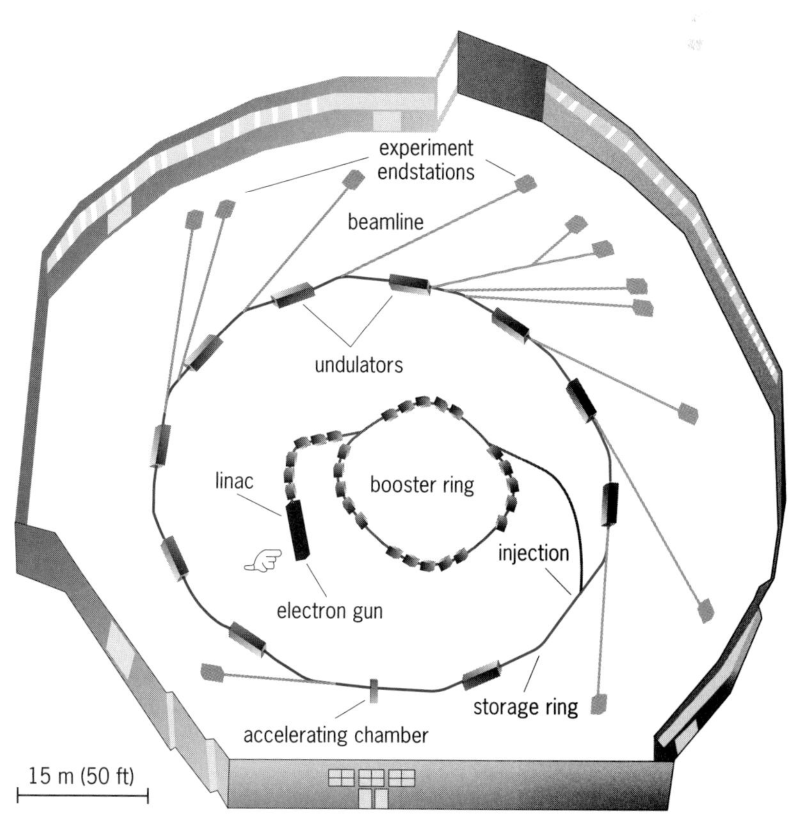

Fig. 1. Schematic layout of Advanced Light Source synchrotron radiation facility. (*Advanced Light Source*)

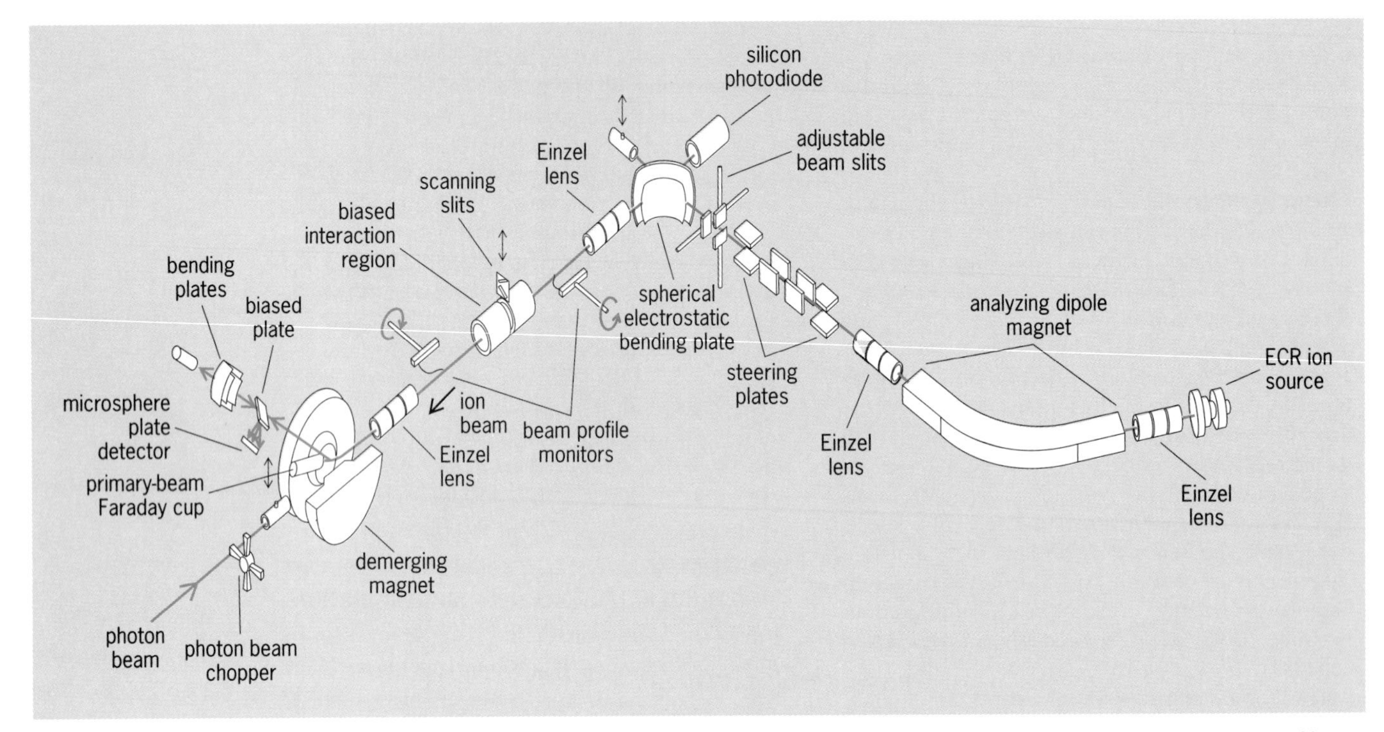

Fig. 2. Ion-photon merged-beams end station at the Advanced Light Source. (*After A. Aguilar et al., Photoionization of ions of the nitrogen isoelectronic sequence: Experiment and theory for F^{2+} and Ne^{3+}, J. Phys. B: At. Mol. Opt. Phys., 38:343–361, 2005*)

left with electrons that are more strongly attracted to the nucleus than those that have been stripped away, and transitions between electronic energy levels shift to shorter wavelengths. Correspondingly, the spectral fingerprints of multiply charged ions shift from the visible to the extreme ultraviolet and x-ray regions of the electromagnetic spectrum.

Synchrotron light sources. Not until the end of the twentieth century did light sources become available with sufficient brightness in those spectral regions to probe the internal electronic structure of ions in detail. The light, called synchrotron radiation, is produced by an accelerator the size of a football field that circulates an intense electron beam moving close to the speed of light. The photon intensity is further increased and concentrated in space by passing the electron beam through insertion devices consisting of periodic arrays of permanent magnets that cause the beam to wiggle or undulate from side to side. The result is a beam of light with properties similar to a laser, but at much shorter wavelengths than today's lasers can produce. Dispersive devices separate the light according to wavelength, so that the photon beam may be made nearly monochromatic. Photon beams produced by synchrotron light sources like the Advanced Light Source in Berkeley are ideal for examining the electronic fingerprints of highly charged ions (**Fig. 1**).

Orbiting x-ray and EUV observatories. During this same period of light-source development, orbiting x-ray and extreme-ultraviolet (EUV) observatories such as *Chandra*, *XMM-Newton*, and the *Extreme Ultraviolet Explorer* (*EUVE*) satellite have opened exciting windows into the distant universe, yielding electronic fingerprints of the ions that make up the stars and the interstellar medium. The infrastructure and motivation for laboratory experiments to probe the internal electronic structure of ions are therefore strong.

Merged beams. To study interactions between photons and ions, merged-beams experiments have been performed at several synchrotron radiation facilities around the world. Because the particle densities in ion beams are limited by the mutual repulsion of like charges (space charge) to values that are even less than those achieved in ultrahigh vacuum systems, beams of photons and ions are merged over an extended path length. A schematic of the experimental setup at the Advanced Light Source is shown in **Fig. 2**. It is located at the end of one of the beamlines shown in Fig. 1. Ions formed in the electrical discharge of an electron-cyclotron-resonance (ECR) ion source are accelerated and focused to form an ion beam that is subsequently analyzed by a dipole magnet to select ions of a specific mass and charge. A spherical electrostatic deflector places the ion beam onto the axis of a beam of synchrotron radiation. The counterpropagating ion and photon beams interact over a common path of approximately 1 m (3 ft). A second magnet demerges the beams and disperses the ion beam according to ion charge state. Ions whose charge has increased due to a process called photoionization are counted by a single-particle detector. The yield of photo-ions is measured as the wavelength or energy of the photon beam is stepped, revealing an ion fingerprint that reflects its internal electronic structure. The product of such an experiment is called a photoionization cross section, which

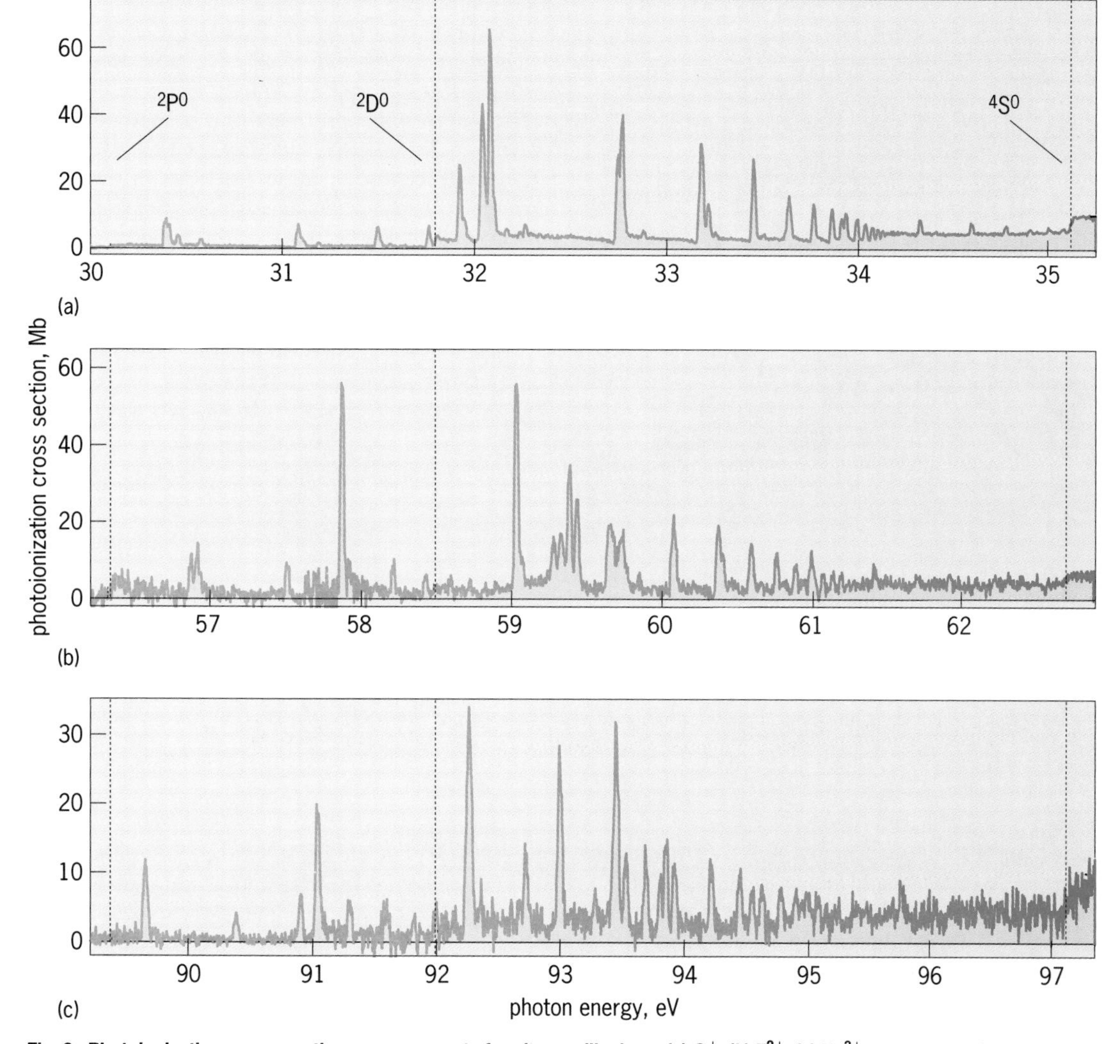

Fig. 3. **Photoionization cross-section measurements for nitrogenlike ions. (*a*) O^+. (*b*) F^{2+}. (*c*) Ne^{3+}. 1 megabarn (Mb) = 10^{-22} m^2. (*From A. Aguilar et al., Photoionization of ions of the nitrogen isoelectronic sequence: Experiment and theory for F^{2+} and Ne^{3+}, J. Phys. B: At. Mol. Opt. Phys., 38:343–361, 2005*)**

is an absolute measure of the probability that one or more electrons will be ejected from the ion when it absorbs a photon.

Photoionization of metastable ions. Typical photoionization cross-section measurements are presented in **Fig. 3** for an isoelectronic sequence of nitrogenlike ions. These ions each have seven bound electrons, but differ in their nuclear charges of +8 for O^+, +9 for F^{2+}, and +10 for Ne^{3+}. In this comparison, the ion beams consist of a mixture of ions initially in their lowest or ground electronic state ($^4S^0$) or in long-lived metastable excited states ($^2P^0$ or $^2D^0$). The vertical broken lines represent the ionization threshold energies for each of these states, as they are labeled in Fig. 3. For this comparison, the respective photon energy scales have been shifted and expanded so that these threshold energies line up with one another.

The spikes are resonances corresponding to the excitation of electrons to discrete unbound levels, called autoionizing states, which subsequently decay by ejecting an electron. The continuous background level beneath the resonances corresponds to direct photoionization to the continuum (bound-free transitions). In the O^+ case, some of the resonances have an asymmetric shape caused by quantum interference between the resonant and direct photoionization pathways. The O^+ ion in both its ground and metastable states is an important constituent of the Earth's ionosphere, where it is irradiated by extreme ultraviolet photons from the Sun.

Time-reversal symmetry. The processes of photoionization and electron–ion recombination are time-reversed analogs. Recombination has been widely studied in merged-beams experiments at heavy-ion storage-ring facilities in Europe. Due to time-reversal symmetry in nature, the cross sections for photoionization and electron–ion recombination are fundamentally related on a state-to-state basis by the principle of detailed balance. One case of special interest is the recombination of electrons with Ti^{4+} ions to form Ti^{3+} ions, which was studied at the Test Storage Ring accelerator facility in Heidelberg, Germany. The electron–ion merged-beams experiment noted an unusually large recombination cross section at relative collision energies approaching

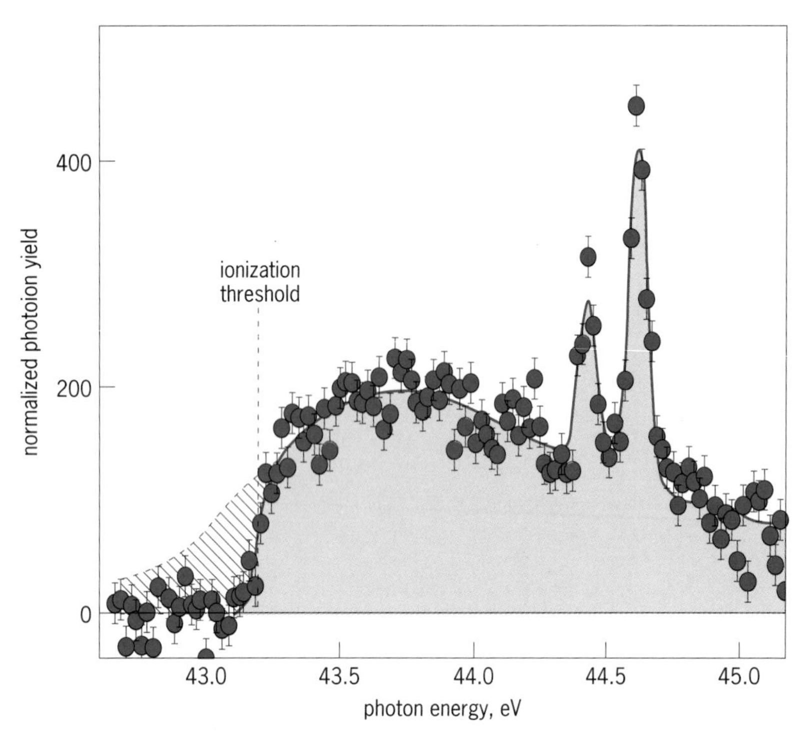

Fig. 4. Truncation of a broad resonance by the ionization threshold in photoionization of Ti^{2+}. (*From S. Schippers et al., Threshold truncation of a 'giant' dipole resonance in photoionization of Ti^{3+}, J. Phys. B: At. Mol. Opt. Phys., 37:L208–L216, 2005*)

zero, suggesting that a broad resonance might exist just at the threshold for photoionization of Ti^{3+}. The time-reversed experiment was subsequently performed and indeed showed such a broad resonance that was in fact being truncated by the Ti^{3+} ionization threshold (**Fig. 4**). The resonance is so broad because all the active electrons occupy the same shell with principal quantum number 3 and interact strongly, resulting in a superfast autoionization decay process. By the uncertainty principle, the energy width of the associated resonance is correspondingly large. Two "normal" resonances that are much narrower also occur in this energy range at photon energies near 44.5 eV. Comparison of the recombination and photoionization experiments made possible absolute photoexcitation measurements on a state-selective basis.

Collective excitations in buckeyball ions. Electrons in atoms and molecules move more or less independently, while those in solids and plasmas may move as a group. Merged-beams measurements of cross sections for photoionization of C_{60} ions provided evidence for a previously undiscovered collective motion of electrons in fullerenes (buckeyballs), large macromolecules containing 60 carbon atoms arranged in a soccer-ball-like structure. Fullerenes are of special scientific interest because their properties are intermediate between those of free molecules and of solids. It is known that the 240 valence electrons that bond C_{60} molecules may be set into collective motion by the absorption of extreme ultraviolet light, producing what is termed a giant plasmon resonance. When a beam of fullerene ions was accelerated and merged with an intense light beam, evidence of two broad resonances was revealed, corresponding to two distinct modes of collective electron motion (**Fig. 5**). These were identified by theory as surface and volume plasmon excitations, similar to those that are known to occur in solids. The surface plasmon refers to a periodic displacement of the spherical electron shell relative to the positively charged carbon ion core, whereas the volume plasmon refers to a compression of the electron shell analogous to a breathing mode. The volume plasmon excitation occurs only because of the hollow-shell electron structure of the C_{60} cage and is observed in photoionization of singly, doubly, and triply ionized C_{60} ions. The volume plasmon excitation of the valence shell of the C_{60} ion survives for approximately one oscillation period before ejecting

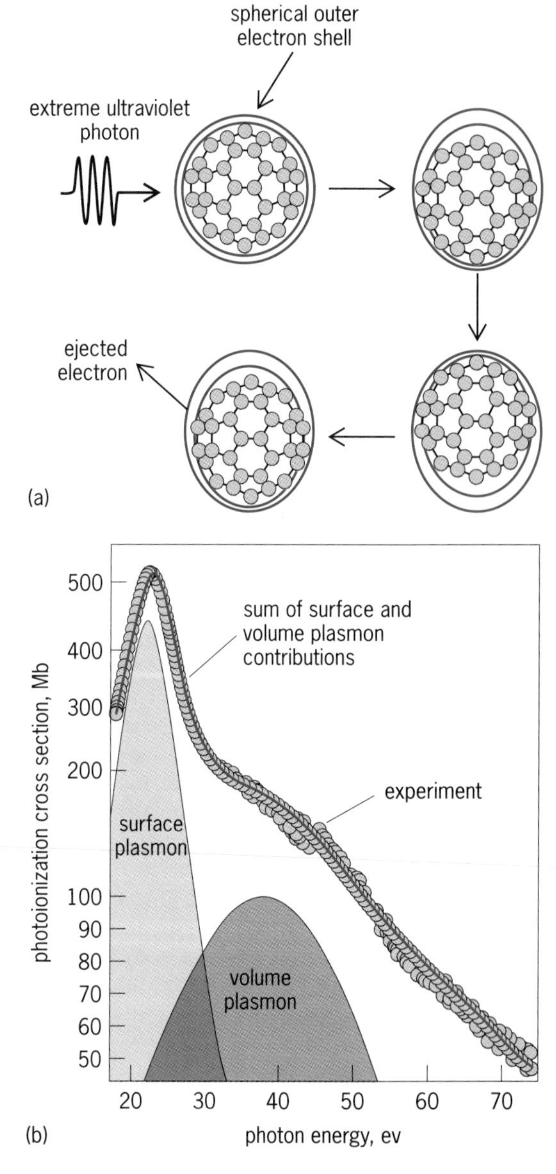

Fig. 5. Excitation of surface and volume plasmons in photoionization of the C_{60}^{+} molecular ion: $C_{60}^{+} + h\nu \rightarrow C_{60}^{2+} + e^{-}$. (*a*) Diagram of volume plasmon excitation. After approximately one period of oscillation of the outer electron shell, one electron may be ejected, resulting in photoionization. (*b*) Energy spectrum of photoionization. 1 megabarn (Mb) = 10^{-22} m^2. (*From S. W. J. Scully et al., Photoexcitation of a volume plasmon in C_{60} ions, Phys. Rev. Lett., 94:065503, 2005*)

an electron, resulting in ionization. The experiments indicate that the C_{60} molecular structure is quite stable, holding together even if as many as four electrons are removed.

For background information *see* ATOMIC STRUCTURE AND SPECTRA; FULLERENE; ION SOURCES; IONOSPHERE; PHOTOIONIZATION; PLASMA (PHYSICS); PLASMA DIAGNOSTICS; PLASMON; SATELLITE ASTRONOMY; SYCHROTRON RADIATION; TIME REVERSAL INVARIANCE; ULTRAVIOLET ASTRONOMY; UNCERTAINTY PRINCIPLE; X-RAY ASTRONOMY in the McGraw-Hill Encyclopedia of Science & Technology.

Ronald A. Phaneuf

Bibliography. A. Aguilar et al., Photoionization of ions of the nitrogen isoelectronic sequence: Experiment and theory for F^{2+} and Ne^{3+}, *J. Phys. B: At. Mol. Opt. Phys.*, 38:343–361, 2005; D. Attwood, *Soft X-Rays and Extreme Ultraviolet Radiation*, Cambridge University Press, 1999; S. Schippers et al., Threshold truncation of a "giant" dipole resonance in photoionization of Ti^{3+}, *J. Phys. B: At. Mol. Opt. Phys.*, 37:L208–L216, 2005; S. W. J. Scully et al., Photoexcitation of a volume plasmon in C_{60} ions, *Phys. Rev. Lett.*, 94:065503, 2005; J. B. West, Photoionization of atomic ions, *J. Phys. B: At. Mol. Opt. Phys.*, 34:R45–R91, 2001.

Internet communications

The Internet originated with the planning for ARPANET in 1967 at the United States government's Advanced Research Projects Agency (ARPA), inspired by prior advances in packet switching, computing, and data communications. ARPANET was developed in response to the need for a distributed communications network that could survive a nuclear attack as well as to the need for research institutions to be able to share their processing power (at a time when powerful computers were very expensive and not affordable by many institutions). The Internet has evolved into a much different and more ubiquitous communications medium than its original designers could have envisioned. Some important developments that have shaped the Internet are summarized in the **table**.

The ARPANET backbone (the primary interconnections of the network) initially used 50 kilobits-per-second (kbps) leased telephone data line services. It evolved to higher-speed wired and optical communications in the megabit range, and more recently the gigabit range.

Internet Protocol (IP) traffic was traditionally carried over time-division multiplexed (TDM) systems designed for voice traffic by adapting the variable packets into fixed frames. Later, backbones made use of asynchronous transfer mode (ATM), developed to better handle mixed voice and data traffic. Today, as all traffic migrates to IP, new protocols have been designed to map IP traffic directly onto the physical medium.

Business and consumer users can connect through either dial-up modems or broadband connections. Dial-up speed has increased from 300 bits per second to more than 50 kbps due to advances in modem technology and improved telephone line quality. Affordable broadband services offering rates of hundreds of kilobits to megabits per second have become available, including cable modems, digital subscriber line (DSL), and satellite modems.

The Internet has evolved from a text-dominated medium primarily focused on email and Web sites to one in which image, video, and audio use are enhancing the traditional applications. As the Internet evolves, offering lower-cost access, higher speeds, and wider availability, new applications such as IP TV and Voice over IP are emerging.

Backbone. Most of the Internet backbone (long-distance link) is now composed of SONET (synchronous optical network) and DWDM (dense wavelength division multiplexing).

SONET provides a number of advantages over older technologies. It provides a means for multiplexing lower-level digital signals into much faster synchronous signals, greatly simplifying the interfaces to switches, cross connects, and add/drop multiplexers. Most of the systems in place today operate primarily at10 gigabits per second (Gbps), with some at 2.5 Gbps and 40 Gbps.

DWDM provides a means for transmitting multiple wavelengths of light over the same fiber. When it is used with SONET, systems are able to transmit up to 128 wavelengths at 10 Gbps for a cumulative rate of 1.28 terabits per second. Optical add/drop multiplexers and optical switches allow termination or switching of individual wavelengths rather than processing of all the data on the fiber.

Access (user connection to Internet). While dial-up modems are still used, home and business access is gradually migrating to broadband methods that provide lower cost per bit. These methods include cable modems, DSL, optical (Ethernet, SONET, and ATM), and wireless (cellular and wireless fidelity or WiFi).

Cable modems provide high-speed Internet access via the cable television system. The modems operate downstream (from the Internet to the user) at up to 40 megabits per second (Mbps) over a 6-MHz television channel. The upstream (user to Internet)

Some important developments in Internet history

Year	Development
1972	E-mail
1972	Telnet (remote log-in) protocol
1973	File Transfer Protocol (FTP)
1973	Transmission Control Protocol/Internet Protocol (TCP/IP), a more sophisticated replacement for Network Control Protocol (NCP)
1974	Ethernet Local Area Network protocol
1978	Unix to Unix Copy Protocol (UUCP), the basis of USENET (newsgroups)
1983	DNS (Domain Name System), the Internet address naming system
1989	Archie, the precursor to Internet search engines
1989	World Wide Web and HTML (Hypertext Markup Language) for information distribution using hypertext
1993	Graphical browser Mosaic

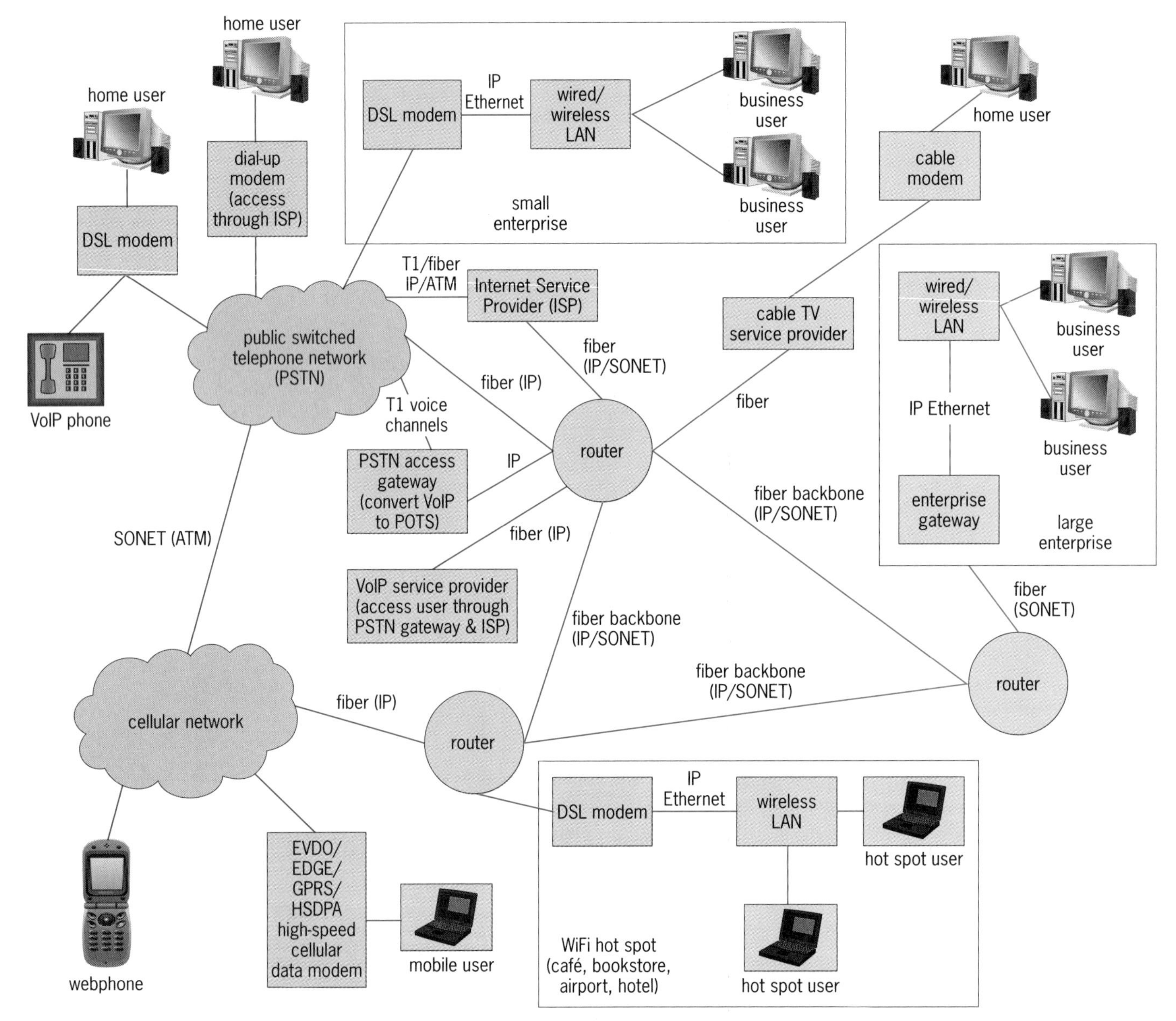

Internet access. IP = Internet Protocol; DSL = digital subscriber line; ISP = Internet service provider; VoIP = Voice over Internet Protocol; LAN = local-area network; PSTN = public switched telephone network; POTS = plain old telephone service; SONET = synchronous optical network; WiFi = wireless fidelity; EVDO = Evolution Data Only; EDGE = Enhanced Data Rates for Global Evolution; GPRS = General Packet Radio Service; HSDPA = High-Speed Downlink Packet Access.

direction provides up to 10 Mbps. Both directions are shared among multiple users.

Digital Subscriber Line (DSL) uses standard phone lines, dedicated or shared with a voice telephone. DSL comes in numerous flavors providing data rates from 384 kbps to 10 Mbps in the downstream direction and 128 kbps to 2 Mbps in the upstream direction.

Mobile wireless Internet access is currently undergoing an evolution from a relatively low-rate access of 9.6–64 kbps for second-generation cellular networks to 2 Mbps for third-generation networks. Data rates vary depending on the location and traveling speed of the terminal. Other wireless networks include various forms of 802.11(WiFi), notorious for its lack of security, which is used at homes, offices, coffee shops, bookstores, airports, and hotels.

Due to cost, optical Internet access is primarily limited to businesses. The technology used is dependent on the data rates required by customers. These systems use Ethernet, ATM, and SONET at rates from less than 2 Mbps to over 155 Mbps. The **illustration** shows different access methods and the networks they utilize to communicate over the Internet.

Applications. Voice telephony is migrating to Voice over Internet Protocol (VoIP), a major new application of the Internet, from traditional circuit switched landline and cellular networks.

Spurred by the competition of cable companies in their markets, telecommunications service providers are compelled to provide competing video services using IP networks. DSL provides data rates that can barely support video services. Advances in video coding technology such as H.264 can reduce required bandwidth, but it is generally accepted that VDSL (very high speed DSL) or fiber rates are necessary to support broadcast video services. Unlike broadcast service, IP service can support video on demand (VOD) services, which allow a user to select from a large movie database and view the selection instantly. Due to the flexibility of IP, video content delivery is moving to IP.

Improved video coding techniques, personal computer (PC) processor speed, higher bandwidth, and graphics capability have dramatically improved video quality over the Internet. Many sites provide streaming video for business and entertainment use. Videoconferencing is becoming better and cheaper, and can save business travel costs. Usage is expanding to cellphones and personal digital assistants (PDAs).

Businesses use the Internet extensively to communicate with other businesses and support new and existing customers. Payment transactions, customer support via email, chat rooms, and software upgrades have become commonplace. Audio, video, and image content in Web sites is increasingly used to advertise new products. Media of many types can be exchanged, lowering the cost of business communication: design information in electronic form, images, high-quality audio, and high-quality video. In many cases the product itself can be distributed via the Internet (software, publishing, graphic arts, music, movies, and so forth). Use of instant messaging, computer gaming, and music/video downloading have greatly increased.

Prospects. The outlook for Internet access is focused on increased data rates, ubiquitous access, and customized applications.

Higher bandwidth and expanded access will enable development of new applications. New features include interactive voice recognition, access anywhere, and instant messaging containing video. Internet Multimedia Subsystem (IMS) is an industry standard network architecture that will enable features to be available over whatever access method and terminal are available to users at the time.

Lower costs will stimulate the use of e-commerce for low-price transactions. User-specific advertisements in media will enable businesses to direct their commercials to those that would most likely be interested in hearing about their products.

Future users will access the Internet from virtually anywhere with increased data rates, using smaller, cheaper, easier-to-use devices. To compete with cable modems and DSL, fiber to the home (FTTH) and WiMAX will become more prevalent. FTTH is just one form of fiber access that provides for a tenfold improvement in data rates over cable modems, providing voice, data, and video (often called the triple play). WiMAX is a fixed wireless technology that provides data rates of greater than 130 Mbps with greatly simplified installation. MIMO (multiple in multiple out) wireless technology has the potential to increase wireless data rates by orders of magnitude.

Backbone data rates will increase with the G.709 protocol replacing SONET and 40 Gbps becoming standard. G.709 extends the capabilities of SONET, providing service transparency for the various types of access methods and enhanced operations, administration, and maintenance (OAM) features. Ethernet is expected to become more common for backbones with data rates up to 100 Gbps as more real-time applications use IP as a network protocol. Use of General Multi Protocol Label Switching (GMPLS), a protocol that allows better control over quality of service (QOS) on IP networks, will spread as networks move away from circuit switching and ATM to IP. These technologies will be transmitted over fiber, carrying an ever-increasing density of wavelengths. Optical switches will provide switching on a packet level, as contrasted to wavelength (circuit) switching used today.

For background information *see* DATA COMMUNICATIONS; INTERNET; MOBILE RADIO; MULTIPLEXING AND MULTIPLE ACCESS; OPTICAL COMMUNICATIONS; PACKET SWITCHING; TELECONFERENCING; WORLD WIDE WEB in the McGraw-Hill Encyclopedia of Science & Technology. Daniel Heer; Michael Rauchwerk

Bibliography. S. V. Ahamed and V. B. Lawrence, *Intelligent Broadband Multimedia Networks*, Kluwer Academic, 1997; W. Stallings, *Data and Computer Communications*, Prentice Hall, 7th ed., 2004.

Invasive forest species

Nonnative organisms that cause a major change to native ecosystems—once called foreign species, biological invasions, alien invasives, exotics, or biohazards—are now generally referred to as invasive species or invasives. Invasive species of insects, fungi, plants, fish, and other organisms present a rising threat to natural forest ecosystems worldwide. Invasive animals are transferred to ecosystems in which there are no natural predators to keep them from spreading. Many species escape into the new environment where they become established, upsetting the ecological balance of the native forest ecosystem. Some invasives are competitors or predators of native species, whereas others cause disease.

Transfer into new ecosystems. The transfer of invasive species into new ecosystems can be either unintentional or intentional.

Unintentional introduction. Unintentional introductions of invasive species are generally a result of international trade and global human travel for immigration, business, and tourism. The invasive species problem is becoming a greater threat to native forests worldwide because international trade and human mobility are increasing, leading to an increase in entry pathways for invasives. Pathways for unintentional introduction of invasive species to new ecosystems include world trade of products (unprocessed wood logs, timber, lumber and chips; seeds; soil and plants for nursery stock; and biological organisms) and product packaging (wood pallets, crates, spools, and dunnage). For example, the Asian longhorned

Perennial vine legume kudzu (*Pueraria lobata*) growing in California. (*Reprinted with permission, © 2001 CDFA*)

beetle was introduced into the United States on wood packing material, and invasive earthworms were unintentionally introduced into many countries in ballast water dumped from ships or from rooted soil of imported plants. Another pathway for unintentional introduction is the natural movement of species from one ecosystem to another within the same geographic region.

Intentional introduction. Some invasive plant species have been intentionally transferred to new areas as ornamentals, food, fiber, and erosion control agents. Several plants introduced into the United States are now causing problems on a monumental scale. Though not all nonnative species become invasive threats to forests, examples that do include the perennial vine legume kudzu (see **illustration**), which was introduced for erosion control; Johnson grass, introduced as an agricultural crop; purple loosestrife, originally introduced as an ornamental in the nineteenth century; and the gypsy moth, introduced for silk production.

Impact on forests. Invasive species are a force of global change. They negatively affect natural ecosystems throughout the world by outcompeting native species for habitat, resources, and growing space. The problem is wide-ranging, impacting almost all terrestrial and aquatic forest habitats. Some invasives are predators to native species; others cause or spread disease. Invasives spread quickly because there are no natural predators or consumers to prevent their population growth. They disrupt native ecosystem function by altering nutrient cycling and habitat quality of native species. Alteration and loss of habitat can lead to displacement and reduction of native species, with major impacts on ecosystem diversity. Displacement of natural species can be permanent. Invasive species negatively impact already threatened and endangered species, further reducing the biodiversity of native organisms.

Invasive species also cause major spatial and temporal landscape level changes to native forest ecosystems. They are known to remove important ecosystem elements, altering forest composition at all levels. Trees, microorganisms, and animals (including fish, birds, amphibians, and reptiles) are affected as food webs are disrupted. Invasives can eliminate tree species as a functioning component of forests. Removal of bird habitats threatens extinction of rare species in isolated ecosystems. For example, earthworms from Europe, Asia, and South America became invasive species in the Philippines, Russia, and the United States. Brazilian earthworms are harming Philippine rice terraces. Nonnative earthworms degrade the leaf litter on the forest floor, disturbing the biodiverse habitat of insects, isopods, bacteria, fungi, and other litter-inhabiting organisms. The ecosystem food chain is removed for ground-nesting birds and for small mammals such as shrews and voles. Populations of native microorganisms are at risk of becoming extinct in some areas, especially fungi that are needed for nutrient uptake by roots of trees and other forest plants. The decrease in nutrient cycling is harmful to trees and plants in the forest understory.

Disease. Invasive insects and fungi degrade forest health by causing or contributing to disease outbreaks. The majority of the most devastating and destructive forest diseases are caused by invasive species. Tree diseases have practically eliminated the dominant native species from some of the world's major forest ecosystems. For example, chestnut blight and Dutch elm disease have killed the American chestnut and elm trees, dramatically eliminating forest ecosystems in the eastern United States. At least 26 invasive insects (such as the Asian longhorned beetle, hemlock woolly adelgid, emerald ash borer, and gypsy moth) and fungi (such as *Discula destructiva*, which causes the disease dogwood anthracnose) threaten to wipe out 70 million acres of forests in the United States.

Forest fires. Tree diseases contribute to increased fire risk by causing extensive mortality rates that result in dead and dry trees that can easily burn. This "fuel load" contributes to a high risk of fire and produces extremely hot flames during forest fires. The invasive plant species European cheatgrass contributes to fuel load by competing with natural plants in piñon-juniper forests. Cheatgrass grows faster and dries before native plants, forming highly flammable ground cover that burns completely. This combination of an invasive plant and fire hinders reestablishment of native plants and prevents forest sustainability.

Watershed and riparian areas. Invasive species have invaded land around streams, lakes, and other waterways. Plants and animals that depend on these riparian areas are threatened with extinction. The fast-growing purple loosestrife plant has displaced riparian area species, causing soil erosion and depleted soil nutrients. An invasive reed canary grass is spreading in riparian areas, blocking water flow in streams. Invasive fish in riparian areas and watersheds have disrupted the ecological balance of aquatic ecosystems.

Multiple invasions. Native forests can be impacted by a symbiotic relationship between two invasive species originating from the same ecosystem. An example is the impact on North American maple forests by the two European invasives, earthworms and buckthorn. Earthworms deplete biomass on the forest floor, adding extra nitrogen to the soil. The buckthorn plant spreads rapidly in high-nitrogen soils, providing more native food for the earthworm. The symbiosis between the two species prevents essential carbon cycling and nutrient availability. These impacts cause shifts in biodiversity and threaten extinction of native understory vegetation, maple trees, and species that depend on them.

Human health. In addition to impacting human livelihood, some invasive species can have a direct deleterious effect on human health. An example is the "itchy grub" caterpillar, native to Australia, which has invaded New Zealand. The caterpillars or gum leaf skeletonizers have poisonous hairs that cause a persistent itchy rash. The reaction to the poison varies with each person. The caterpillar causes extensive damage to gum (*Eucalyptus*) trees in New Zealand and Australia.

Island ecosystems. Island ecosystems are especially vulnerable to invasives due to their unique, fragile ecosystems and evolutionary isolation. Ancient palm forests of the Seychelles are threatened by the spread of the tree *Cinnamonum verum*, introduced for commercial and economic reasons. Forests on the islands of Tahiti and Hawaii are threatened by overgrowth of the invasive plant miconia (*Miconia calvescens*). Introduced as an ornamental, miconia grows rapidly with 3-ft-long (90-cm) leaves that shade native plants, blocking needed sunlight. The plant's small seeds contribute to its rapid spread.

Societal effects. Invasive species degrade recreational areas, decreasing access to and reducing the quality of public and private forests. Cultural and ecological losses are compounded by the social and economic loss of jobs, loss of long-term supply of forest products, and monetary costs associated with possible solutions to the problem.

Prevention and control. A global strategy is needed to address the invasive species issue. International organizations are recognizing the ecological and economic problems caused by invasive forest species. The recently established Global Invasive Species Program (GISP) will compile and publish international information about invasive species prevention and control methods, including forest species. The Food and Agriculture Organization (FAO) of the United Nations has an increased focus on destabilization of forest ecosystems by invasive species, promoting appropriate measures for their control. Management of invasive forest species includes several approaches at the international, national, and local community levels.

Risk assessment. Risk assessment involves determination of the potential introduction of invasives; identification of the priority invasives and ecosystems that are at risk; identification of the pathways by which invasives may be transported from one ecosystem to another; assessment of the risks associated with each pathway and invasive species; and recommendation of short-term and long-term control methods.

Prevention. The first step in controlling invasive species is to prevent the transfer of invasives from one ecosystem to another. One way to achieve this is via the establishment of international quarantine standards that require control measures for a given invasive species pathway. One successful example is the recent international quarantine standard for heat or methyl bromide treatment to kill organisms on wood packing material, thereby slowing global transport of potential invasives.

Public awareness. An important step toward protecting global forest ecosystems is through awareness and education. The public can help identify and eradicate invasives (especially in urban areas), refrain from transporting invasives during travel, and support international, national, and local efforts to address the invasives problem. New Zealand has a major public awareness campaign that uses "Max the Biosecurity Beagle" as a mascot to provide information about how the public can help with early detection of all invasives that pose a threat to the environment. Although New Zealand has no current forest plantation pests, the forests are at risk from three species (gypsy moth, fall webworm, and gum leaf skeletonizer).

Detection. Detection methods include survey, sampling, and identification protocols. Early detection systems are used to identify newly introduced species that were not intercepted by quarantine methods. Detection strategies are used to survey forest ecosystems and regularly monitor forests for invasives. Deoxyribonucleic acid (DNA) analysis is the identification protocol currently used for *Phytophthora ramoru*, the causal agent of "sudden oak death" disease in California.

Eradication. Removing invasive species from forests includes the use of chemical (insecticides, fungicides, herbicides, and attractants), mechanical (such as harvesting and hand pulling), and biological control methods. Biological control (also known as biocontrol) uses natural enemies to manage and control pests. The commercially available bacterium *Bacillus papillae* was the first insect pathogen to be registered in the United States as a biological control agent against the Japanese beetle. The bacterium *B. thuringiensis* var. *kurstaki* is used as a biological control agent against the gypsy moth. New introductions can usually be eradicated. Containment and control of some established species is necessary when total eradication is not possible.

Restoration. Long-term measures are required to restore degraded ecosystems that have been negatively altered by invasive species. Although many approaches are used around the world to reclaim ecosystems, all require removal of invasives and replanting native trees. Additional approaches have included reconstruction of stream banks in riparian areas and watersheds; addition of soil amendments (fertilizer, erosion control mat, leaf compost, wood mulch, riparian plants) and root symbionts (mycorrhizae fungi); and reintroduction of native animals, microbes, and other organisms.

For background information *see* ECOLOGICAL COMPETITION; ECOLOGICAL SUCCESSION; FOREST ECOSYSTEM; INVASION ECOLOGY; POPULATION ECOLOGY in the McGraw-Hill Encyclopedia of Science & Technology. Barbara L. Illman

Bibliography. J. F. Franklin, Challenges to temperate forest stewardship: Focusing on the future, in *Toward Forest Sustainability*, ed. by D. B. Lindenmayer and J. F. Franklin, CSIRO Publishing, Collingwood, Australia, 2003 (also due for publication in U.S. by Island Press); D. B. Lindenmayer and J. F. Franklin, *Conserving Forest Biodiversity: A Comprehensive Multiscaled Approach*, Island Press, Washington, DC, 2002; H. A. Mooney and R. J. Hobbs (eds.), *Invasive Species in a Changing World*, Island Press, Washington, DC, 2002; D. Pimentel (ed.), *Biological Invasions: Economic and Environmental Costs of Alien Plant, Animal, and Microbe Species*, CRC Press, Boca Raton, FL, 2002; D. Pimentel et al., Environmental and economic costs associated with non-indigenous species in the United States, *BioScience*, 50:53–65, 1999; G. M. Ruiz and J. Carlton, *Invasive Species: Vectors and Management Strategies*, Island Press, Washington, DC, 2003; L. J. Sauer, *Once and Future Forest: A Guide to Forest Restoration Strategies*, Island Press, Washington, DC, 1998 (available online from Google Print); D. S. Wilcove, Quantifying threats to imperiled species in the United States, *BioScience*, 48:607–615, 1998.

Jet boring (mining)

The Cigar Lake deposit, in northern Saskatchewan, Canada, 660 km (410 mi) north of Saskatoon, contains approximately 350 million pounds (160 million kilograms) of triuranium octoxide (U_3O_8) reserves, with ore grades of about 20%. It is the second largest and richest uranium orebody in the world. A project was recently undertaken at Cigar Lake to develop an innovative nonentry method for mining and handling high-grade uranium ore. The process commenced at the proof-of-concept stage and ended with a full-scale industrial test. As a result, the jet-boring mining method will be used to extract the ore for the future Cigar Lake operation. The Cigar Lake project is now in the construction phase and could be operational in 2007.

The crescent-shaped Cigar Lake deposit averages 6 m (20 ft) in thickness and is located 450 m (1500 ft) below the surface (**Fig. 1**). The ore and surrounding rock is weak, highly fractured, and water-saturated. The basement rock below the ore zone is much tighter, with less ground water and fracturing. The major technical factors influencing the mining method selection were ground stability, control of ground water, radiation exposure, and ore handling and storage. Various studies on ground conditioning and nonentry mining methods were done and field-testing was recommended. In order to demonstrate that the ore could be mined safely and economically, a test-mining program was launched. As a result, the jet-boring mining method was selected.

The jet-boring mining method involves four major steps (**Fig. 2**). In step 1, the ore zone is frozen in bulk from below and an intermediate production level is developed. In step 2, a pilot hole is drilled from the production level to the top of the ore and lined with a casing. Step 3, the actual ore extraction, involves inserting a drill string with a nozzle inside the casing and, while rotating the nozzle, excavating a cavity using a high-pressure waterjet. The ore slurry exits the cavity by gravity and is pumped away from the mining area. In step 4, the cavity is backfilled with concrete.

Test-mining highlights. Construction began in 1984 on the various surface facilities and a 500-m-deep (1640-ft) shaft was constructed. Two levels of development underground were started, one above and one below the deposit. Extremely poor ground encountered on the upper level was frozen to advance development. After an expensive advance, a decision was made to concentrate on extraction methods from below the orebody.

Ground freezing was done to support the weak rock associated with the orebody and to minimize the potential for a large inrush of water while mining the ore. It also prevented the majority of the free water from exiting the orebody. Ground freezing was achieved by circulating calcium chloride brine, at about −40°C (−40°F), through the freeze pipes. This enabled a 6-m-wide (20-ft) orebody section to be frozen to a temperature of about −10°C (14°F) after about 4 months. Future production panels of 12–18 m (40–60 ft) width will require 1.5–2.5 years of freezing.

Proof-of-concept testing of the jet-boring mining method commenced underground in 1991. It involved installing a rotary drill equipped with a high-pressure swivel that rotated a single, side-firing nozzle at the top end of the drill string, positioned above a casing within the ore. A 520-kW pump jetted about 300 L/min (80 gal/min) of water at 80 MPa (12,000 lb/in.^{2}). A "preventer," consisting of seals

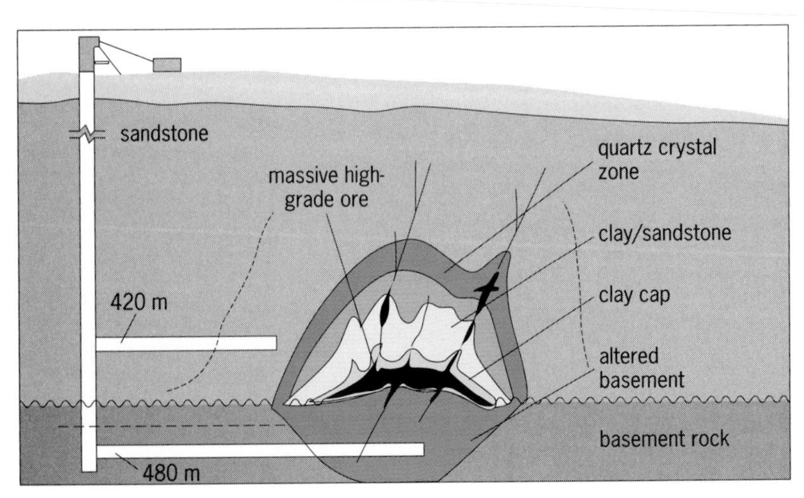

Fig. 1. Cross section through the orebody.

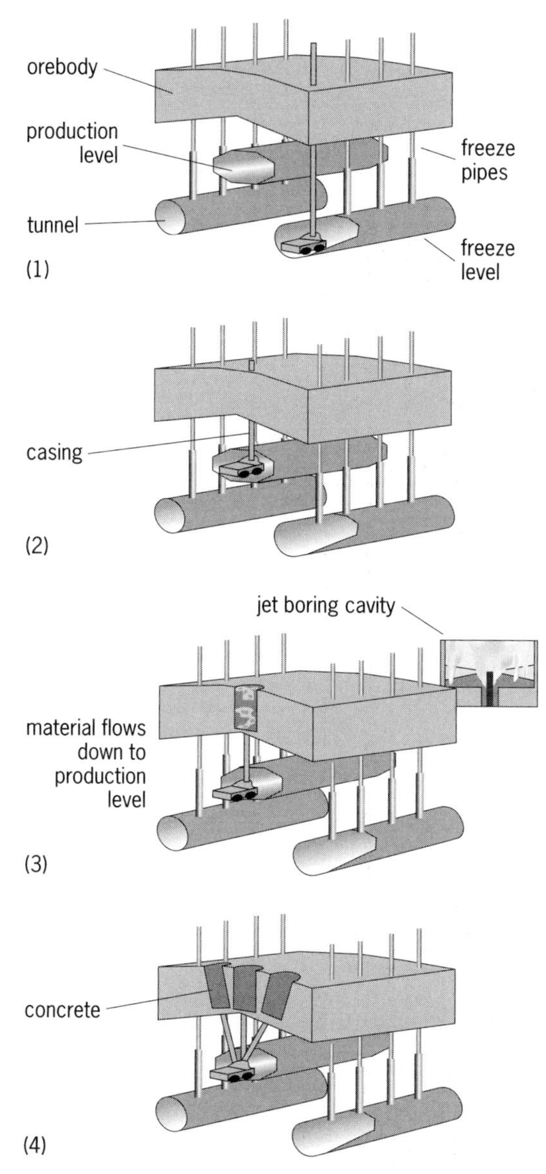

Fig. 2. Sequence of the jet-boring mining method. (1) Freeze pipes are installed in rows up through the orebody from tunnels below, and refrigerated brine is circulated through the pipes to freeze a block of ground. (2) A pilot hole is drilled to the top of the orebody, destructible fiberglass pipe is installed in the upper part of the pilot hole (that is, in the ore), and a steel pipe string, called a casing, is installed below the ore. (3) A steel pipe is inserted into the casing to the top of the orebody. High-pressure water is pumped into this pipe and mines out a cavity in the orebody. The mined material flows out of the cavity by gravity into the casing and then is pumped away from the mining area. (4) After a cavity is mined out, it is refilled with concrete. Adjacent areas can be mined when the concrete is set.

around the jet rod and a shutoff valve on the outlet, was fixed to the pilot-hole casing to contain the slurry. The preventer, rated at one-and-a-half times the formation water pressure, also provided "collar" security in the event of an ice wall failure above the orebody. The initial tests demonstrated the method's potential, as well as its slurry-handling challenges.

Proof-of-concept testing identified several significant advantages. (1) Collar security could be provided, preventing water inrushes if the ice cap was breached. (2) Slurry containment, which is inherent within the pipes, provided radiation control, dust elimination, and cost-effective ore transport. (3) Jet boring is selective, allowing ore extraction next to backfilled cavities without significant dilution. (4) The jet-boring method has the ability to mine in a fan pattern, reducing the access requirements below the orebody. (5) The small-diameter pilot hole below the orebody minimized the waste rock generated.

In order to enhance the productivity of the jet-boring mining method, it was necessary to optimize the jetting tool (**Fig. 3**). The main jetting tool has an instrument to survey the size of the cavity during the mining process. A nozzle subassembly allows the high-pressure water to turn a corner and exit the nozzle without losing a significant amount of energy. A blade screen restricts the size of the particles falling down the casing and prevents plugging. And a high-pressure swivel allows the passage of the jet water to the jetting rods.

Upon manufacturing the tools, lab-scale testing commenced and several iterations of design improvement were undertaken. During this period, the preventer was optimized to maintain collar security during the jetting process.

The lab testing results provided the information for a jetting-tool test, which would provide data on how the prototype tools would operate in a mining environment. As part of the underground mining development, the prototype jetting tools and the high-pressure (100 MPa or 14,500 lb/in.^{2}) pumping system were tested between two levels underground.

A layer cake representing the deposit's variable geology was constructed in a 3-m-diameter (10-ft) culvert (**Fig. 4**). The materials used included clay, sand, cobbles, steel, and concrete in various combinations. The test cavity was readily accessible to personnel via a removable lid, allowing for regular checks on the progress and status of the tools.

Minor improvement were made to the tools, and all materials, including granite boulders with a compressive strength of 250 MPa (36,000 lb/in.^{2}), were mined by the jet to the full 3-m diameter. The success of the tool test provided the confidence for a full-scale jet-boring test.

In 2000, a complete mining test of all the equipment and procedures was done in frozen ground. A prototype jet-boring system was manufactured, extensively shop-tested, and installed underground. Four cavities in frozen waste rock and four cavities in frozen ore were excavated. Mining began at the top of the cavities. A typical jetting sequence per increment involved rotating the jet at 20 rpm, while moving up and down 30 cm (12 in.) for a period of 30 minutes. The inclined jet created a sloped cavity bottom, enabling direct jetting of loose, harder boulders. Interim ultrasonic surveys and final laser surveys provided detailed cavity profiles.

The four cavities in the ore were larger than anticipated and approached 6 m (20 ft) in diameter in several locations. After each cavity was mined, it was filled with rapid-setting concrete, having approximately 40-MPa (5800-lb/in.^{2}) compressive strength.

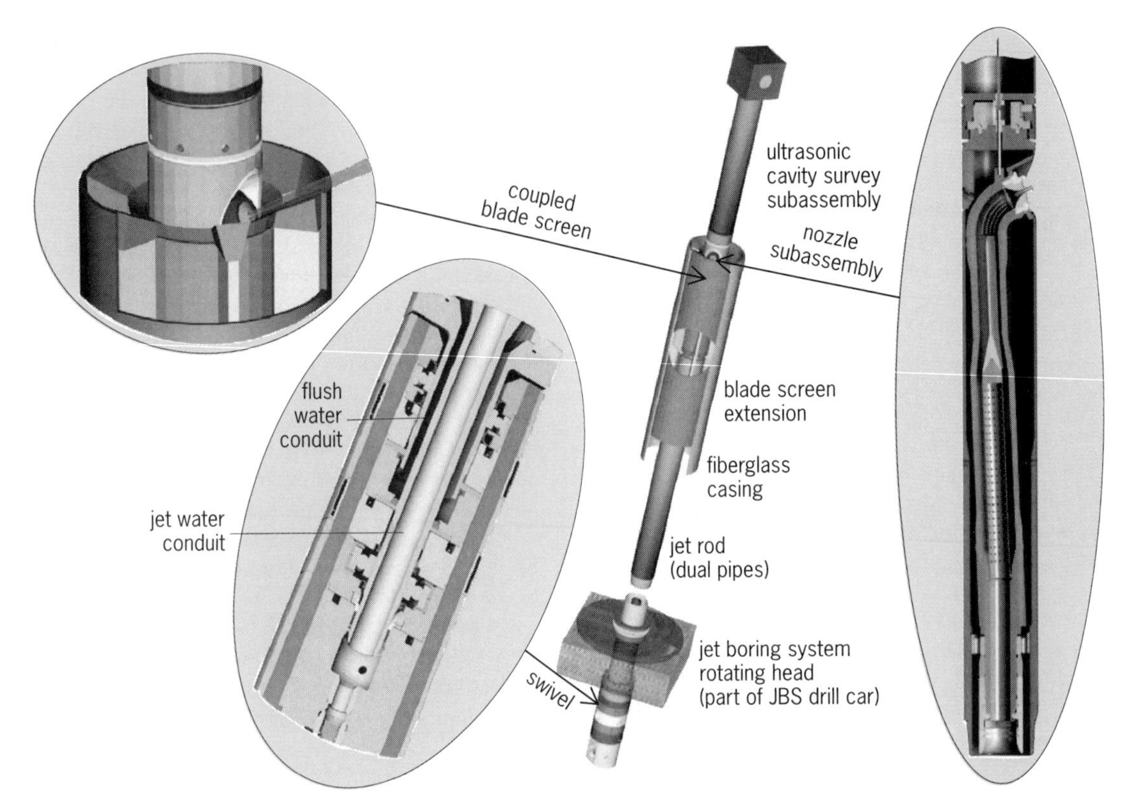

Fig. 3. Jet-string tools.

Adjacent cavity surveys revealed the concrete backfill was resistant to the jet.

The tests were highly successful. The ore cavities were nearly circular in shape and averaged 4.5 m (15 ft) in diameter. Approximately 600 metric tons (660 tons) of ore containing 90 metric tons (99 tons) of uranium was mined. The ore slurry was pumped to an underground storage area without any significant issues.

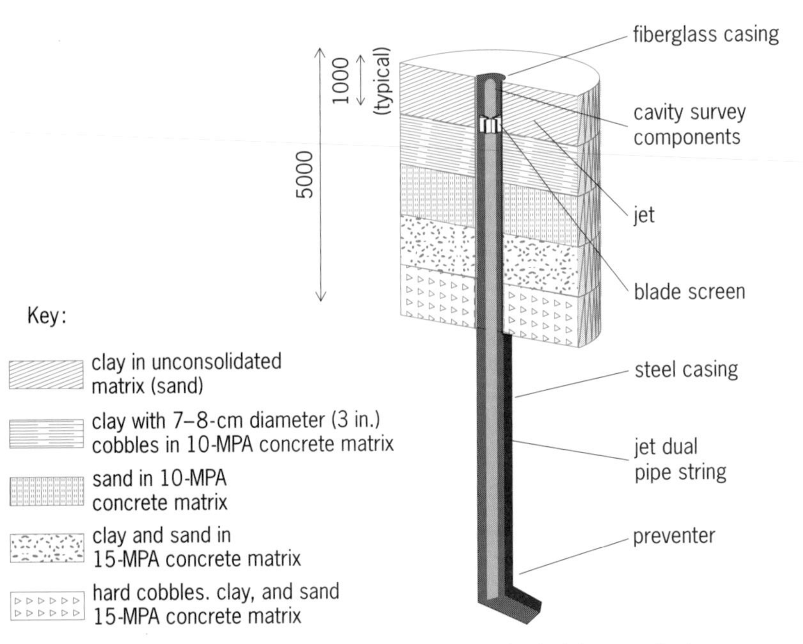

Fig. 4. Cavity matrix materials and prototype jetting tools for the laboratory test.

Productivity. The estimated productivity rate of 10 metric tons (11 tons) per hour was confirmed, with two jet-boring systems in operation, which equates to an average production rate of 100–150 metric tons (110–165 tons) per day.

During testing, numerous cavity ultrasonic video and laser surveys were completed. These observations identified the main functions of the jet: the jet removes material from the wall of the cavity, reduces the size of the broken material, cleans the top of the blade screen, and flushes the ore from the cavity.

Production system. The production jet-boring system is complex and made up of a number of subsystems, including a high-pressure pumping system capable of delivering approximately 1000 L/min (260 gal/min) of water at a pressure of 100 MPa (14,500 lb/in.^{2}); in-hole jetting tools with the surveying instrument; the drilling rig which incorporates a preventer; the slurry-handling circuit; and the backfilling system. The production jet-boring system also has five cars on rails, including the slurry car which collects and pumps the mined material to a storage area; the drill car which drills the holes into the orebody, installs casings, and drives the jet-boring rods; storage cars which store the drill rods and supporting materials; and the shuttle car which transports supplies and materials to the storage cars. The pumps that supply the high-pressure water to the jet-boring system are located in a room near the entrance to the tunnel where the jet-boring system is located. Backfilling will be conducted from a separate car and is not connected to the jet-boring system.

The jet-boring mining method produces a slurry, which requires further underground processing. Following the underground processing, the slurry will be pumped to the surface. During the first phase of production, the Cigar Lake ore slurry will be trucked to another location for processing.

For background information *see* DRILLING, GEOTECHNICAL; MINING; ORE AND MINERAL DEPOSITS; RADIOACTIVE MINERALS; UNDERGROUND MINING; URANIUM; URANIUM METALLURGY in the McGraw-Hill Encyclopedia of Science & Technology.

Barry W. Schmitke

Bibliography. C. R. Edwards, The Cigar Lake project: Mining, ore handling and milling, *CIM Bull.*, Canadian Institute of Mining, Metallurgy, and Petroleum, 97(1078):103–114, 2004; E. Ozberk and A. J. Oliver (eds.), Uranium 2000: Process metallurgy of uranium, *Proceedings of the 30th Annual Hydrometallurgical Meeting*, Canadian Institute of Mining, 2000; B. W. Schmitke, Cigar Lake's jet boring mining method, *Proceedings of the World Nuclear Association*, 29th Annual Symposium, London, 2004.

Life-cycle analysis of civil structures

In developed countries, sustainable economic growth and social development are intimately linked to the reliability and durability of civil structures such as buildings, bridges, and dams. The intended service lives of civil structures are typically decades or even centuries. During this time, civil structures are possibly exposed to abnormal loads of different types, ranging from natural hazards (such as earthquakes, floods, and hurricanes) to human-made disasters (such as terrorist attacks, fires, or vehicular collisions). At the same time, structural safety and condition undergo gradual deterioration because of material aging, harsh environmental conditions, and increasing loads. This long-term deterioration may not only impair the functionality of civil structures but also undermine their capacity against future extreme loads. The associated social, economic, and political consequences can be enormous and may exert serious, widespread, and prolonged adverse impacts on various societal sectors.

In response to these concerns, maintenance and risk mitigation are necessary to ensure satisfactory performance over a structure's life cycle (**Fig. 1**). Since maintenance needs often outpace available funds, advanced inspection/monitoring techniques, innovative maintenance strategies, and improved asset management practice become very important. Computationally, one needs to allocate these resources in the most cost-effective manner such that lifetime structural performance can be optimally improved under budget constraints. This requires (1) reliable modeling of loadings, including extreme loads and continuous deterioration processes and their effects on structural capacity, (2) accurate prediction of structural safety and performance evolution, (3) good estimation of costs of interventions such as maintenance, repair, and replacement over the specified time, and (4) generation of solutions that balance life-cycle costs and lifetime structural performance in an optimum way.

Life-cycle analysis. Life-cycle analysis is a systematic method for evaluating the impact of various time-dependent loads, deterioration, and relevant interventions (such as operation, inspection, maintenance, and replacement) on the performance and consequences over the entire or remaining service life of civil structures. Using life-cycle analysis, one can obtain long-term cost-competent design or preservation solutions. The highly uncertain structural capacity deterioration process becomes even more complex under probable extreme loads and maintenance interventions. Therefore, probabilistic treatment becomes indispensable to accounting for various sources of uncertainty in modeling time-dependent variation of structural capacity and demand and in predicting lifetime structural safety and performance. Two general types of uncertainty exist: inherent (aleatory) and knowledge-based (epistemic) randomness. The aleatory uncertainty cannot be reduced, while the epistemic uncertainty can be reduced when new information or more

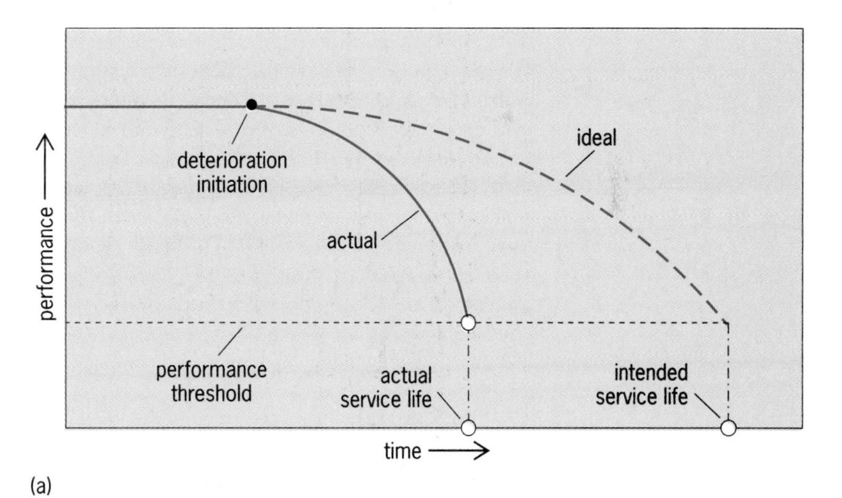

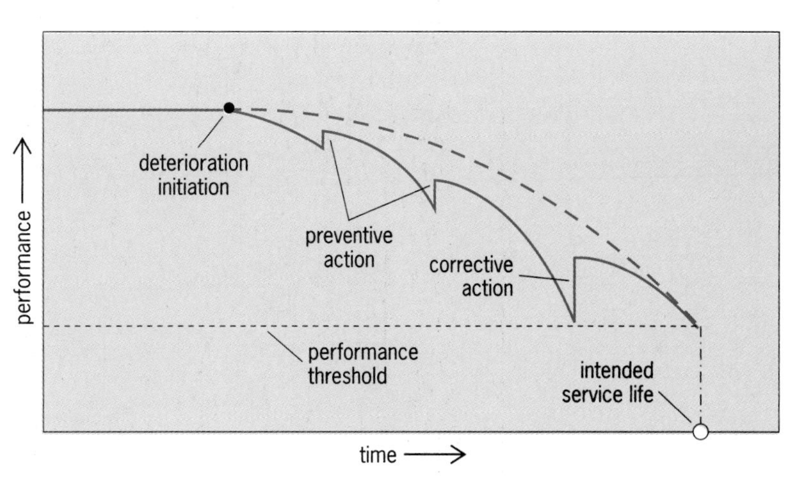

Fig. 1. Lifetime structural performance (*a*) without maintenance and (*b*) with maintenance.

refined models for performance and cost prediction are available.

Time-dependent capacity and loads. Concrete and steel are two of the most common materials used for civil construction. Structural components deteriorate progressively under harsh environmental stressors. For steel corrosion in concrete, chloride and carbonation contamination are most frequently observed. Deterioration of the concrete is in the form of cracking and spalling-induced debonding of the steel reinforcement (rebar) from internal pressures; it occurs because of chemical and physical processes such as alkali-aggregate reaction, sulfate attack, and freeze-thaw cycling. These lead to the formation of rust and loss in rebar cross section. Different diffusion processes are often used to simulate chloride-induced concrete deterioration for which the corrosion initiation time can be estimated. After that, different deterioration models exist to predict the percentage of corrosion in steel at any time. Relevant parameters may be treated as random variables to account for various sources of uncertainty. Monte Carlo simulation is then carried out to obtain probabilistic prediction of structural capacity deterioration profiles.

The probabilistically time-varying loads over the life cycle may be modeled with appropriate stochastic processes. Alternatively, simplified representations are available for practical design purposes. For example, seismic load intensities for earthquake-resistant design are customarily described in terms of exceedance probabilities during a specified period. As another example, the time-dependent bridge live load due to trucks depends on a variety of parameters (such as truck weight, axle loads, span length, position of vehicles, number of vehicles, speed of vehicles, stiffness of superstructure, and bridge geometry). The load model in current bridge design specifications of the United States is based on the mean value and standard deviation of a particular distribution for moments and shears in bridge components at a future time, corresponding to a sample size of trucks.

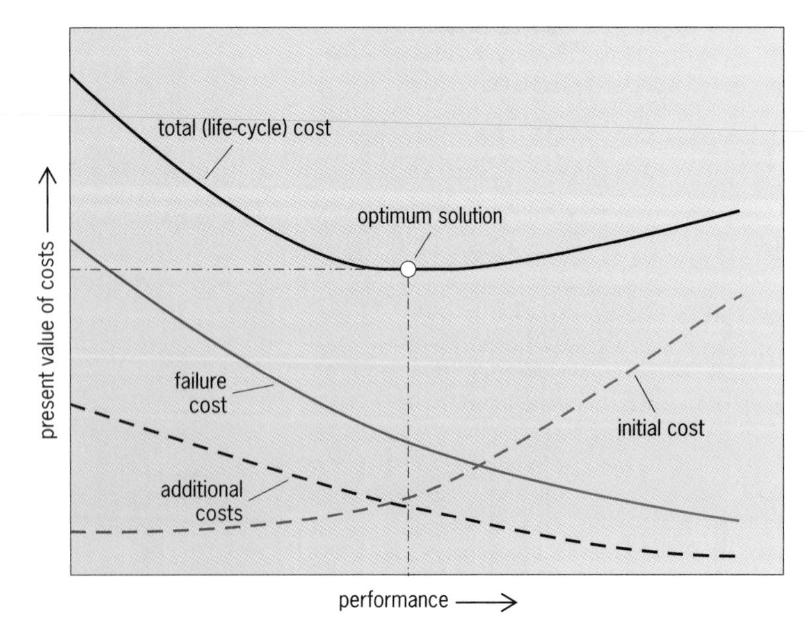

Fig. 2. Optimum solution based on life-cycle cost minimization.

Prediction of lifetime performance. The deterioration model (Markov model) currently used in civil structure management systems assumes that the structural component condition can be described by a limited number of condition states. A transition probability relates the current state with a maintenance action to a future state. The Markov property indicates that the probability of deteriorating to another state depends only on the last condition and action but not on the history of the process. The time of transition from one state to another may follow specified probability distributions. Therefore, the condition-based Markov model is flexible so that it can adapt to visual inspection data. This model, however, cannot capture the propagation of uncertainties during the entire service life, nor can it accurately predict the relation with the applied stress because of the limited number of condition states.

In contrast, the advanced structural reliability method offers a more rigorous and systematic approach to quantifying the evolution of structural safety, provided probabilistic information on structural resistance, mechanical and environmental loadings, and other contributing factors is available. Given the prescribed limit-state functions, structural reliability describes the probability of structural capacity exceeding load demand. Various numerical techniques are now available to evaluate this probability. In practice, this probability is usually converted to a so-called reliability index through Gaussian (normal) probability distribution. Consequently, the time-dependent reliability index needs to be considered for lifetime structural performance deterioration.

Prediction of deterioration using simulation provides a baseline for performance evaluation. Owing to uncertainty propagation over time, deterioration prediction alone cannot produce sufficiently accurate results. Structural condition and performance must be monitored on a regular basis and the structural defects, as symptoms of deterioration, can be detected and corrected as necessary. When new information on structural performance becomes available, for example, by inspection and monitoring, one can update the performance assessment and prediction using the Bayesian technique. This technique provides a rational method for incorporating the prior information or judgment into predicting future outcomes. Probabilistic models for the structural components and load demands can be updated, and the structural system reliability can then be recalculated accordingly.

Bridge management systems. In the United States and other countries, bridge management systems have been developed to optimally maintain an acceptable performance of deteriorating bridges. In most existing bridge management systems, bridge performance is depicted by discrete condition states using visual inspection. This subjective indicator is unable to faithfully reflect the actual structural deterioration and remaining structural capacity. Failure to reliably assess safety levels of deteriorating bridges can result in erroneous bridge-management decisions, with enormous safety and economic consequences. In addition, current bridge management systems have been mostly developed at the project level (that is, for individual bridges or a group of similar bridges). Because of the integral role of individual bridges in a highway transportation network, it is important and rational to integrate network effects into bridge management systems by directly relating the performance of different geographically located bridges to the overall functionality of the network. This can be realized by innovatively combining the time-dependent structural reliability analysis and the highway transportation network theory. The resulting methodology enables one to evaluate the overall performance of highway networks with and without different scenarios of probabilistic bridge failure.

Life-cycle cost and optimum solution. The traditional design approach relies only on the initial cost of the structure. In contrast, the life-cycle cost approach relies on all costs, including the initial cost, maintenance cost, inspection and repair costs, and cost of failure, over the useful life of the structures. A. C. Estes and D. M. Frangopol indicate that the failure cost typically includes those costs associated with structural failure multiplied by their probability of occurrence. Therefore, probabilistic assessment of the structure is necessary. Once the costs have been computed and converted to present values using an appropriate discount rate (interest rate), an optimum solution, as shown in **Fig. 2**, produces the lowest total life-cycle cost.

Optimization with conflicting objectives. Life-cycle cost minimization is the most widely used criterion for generating cost-effective design and maintenance planning solutions for civil structures. In reality, multiple and conflicting objectives usually need to be considered simultaneously to obtain a well-balanced solution. In seismic design, for example, the initial investment and the lifetime seismic-damage cost need to be appropriately balanced based on designers' risk-taking preference. Also, bridge management decisions should be made by improving the overall bridge system/network performance and by reducing various long-term costs (for example, agency cost and user cost), while ensuring satisfactory safety and condition levels of individual bridges in the highway network. Therefore, design and maintenance planning of civil structures can be best formulated as multiobjective optimization problems in order to optimize life-cycle structural performance, based on simulated performance profiles and life-cycle costs of different origins. This leads to a group of alternative solutions that exhibit the optimized tradeoff among conflicting objectives (**Fig. 3**). Decision makers can then actively compare different solution options and choose the one that preferably balances structural performance enhancement and cost reduction. It has been demonstrated that evolutionary computation techniques (for example, genetic algorithms) are effective in finding solutions to this type of problems.

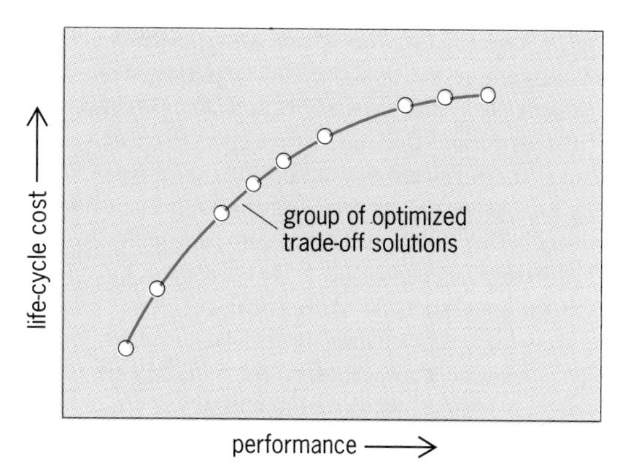

Fig. 3. Trade-off solutions between conflicting objectives.

For background information *see* BAYESIAN STATISTICS; BRIDGE; BUILDINGS; CIVIL ENGINEERING; CONCRETE; DAM; ESTIMATION THEORY; GENETIC ALGORITHMS; MONTE CARLO METHOD; PROBABILITY; REINFORCED CONCRETE; STEEL; STOCHASTIC CONTROL THEORY; STOCHASTIC PROCESS; STRUCTURAL MECHANICS; STRUCTURAL STEEL in the McGraw-Hill Encyclopedia of Science & Technology.

Dan M. Frangopol; Min Liu

Bibliography. S. E. Chang and M. Shinozuka, Life-cycle cost analysis with natural hazard risk, *J. Infrastruc. Sys.*, 2(3):118–126, 1996; A. C. Estes and D. M. Frangopol, Life-cycle evaluation and condition assessment of structures, chap. 36 in W.-F. Chen and E. M. Liu (eds.), *Structural Engineering Handbook*, 2d ed., pp. 36-1 to 36-51, CRC Press, 2005; D. M. Frangopol, J. S. Kong, and E. S. Gharaibeh, Reliability-based life-cycle management of highway bridges, *J. Comput. Civil Eng.*, 15(1):27–34, 2001; D. M. Frangopol, K.-Y. Lin, and A. C. Estes, Life-cycle cost design of deteriorating structures, *J. Struc. Eng.*, 123(10):1390–1401, 1997; M. Liu and D. M. Frangopol, Multiobjective maintenance planning optimization for deteriorating bridges considering condition, safety and life-cycle cost, *J. Struc. Eng.*, 131(7):833–842, July 2005.

Liver X receptor (LXR)

The liver X receptor (LXR) is a nuclear hormone receptor that is involved in the regulation of fatty acid and cholesterol metabolism. Nuclear hormone receptors constitute a family of specialized proteins in the nucleus that can switch on individual genes in response to the changing needs of the organism. These proteins contain two functional units: (1) a deoxyribonucleic acid (DNA)-binding domain that binds to target DNA sequences within the promoter region of genes that regulate many aspects of vertebrate metabolism and (2) a ligand-binding domain, to which hormones or other small molecules (ligands) attach. Binding of the ligand activates the receptor complex, displaces corepressor proteins from the promoters, and helps recruit coactivator proteins that link the receptor complex to the basic transcription machinery of the gene. This activates the production of multiple copies of its messenger RNA (mRNA) that, in turn, stimulate protein synthesis.

LXR proteins bind to the promoter DNA of multiple genes, particularly those whose products regulate fatty acid and cholesterol metabolism. There are two related LXR proteins, LXRα and LXRβ; they are products of different genes (*Nr1h3* and *Nr1h2*, respectively). Though both are widely distributed, LXRα is the major species present in liver cells. Oxidation products of cholesterol (oxysterols) are the natural ligands for both LXRα and LXRβ. The structure of the ligand-binding domain of LXRα was recently identified, and based on this many sterol and nonsterol synthetic ligands for this receptor have also been developed.

Transcriptional regulation. Nuclear receptors bind to a six-base consensus DNA sequence (-TGACCT-). Many of these promoter sites consist of a sequence of two of these motifs, to which the nuclear receptor binds as a homo- or heterodimer, that is, as a complex of two of the same receptor (homodimer) or two different receptors (heterodimer). The repeat may be direct or inverted, and its motifs separated by one to five bases. This pattern defines the specificity with which nuclear proteins bind to receptors. LXR binds to a DR-4 element (that is, a direct repeat of the six-base motif separated by four bases). Almost invariably, LXR binds as a heterodimer with a second nuclear receptor, retinoid X receptor (RXR) [see **illustration**]. RXR is usually present in excess. As a result, activation of LXR-dependent genes depends on the induction of LXR and the presence of a competent ligand. Oxysterols effective in activating the transcription of LXR-dependent genes include 22-OH, 24-OH, and 27-OH sterols. Because these ligands are oxidation products of cholesterol, their activation of LXR permits this nuclear receptor to act as a "cholesterostat," sensing sterol levels in tissues and promoting catabolism.

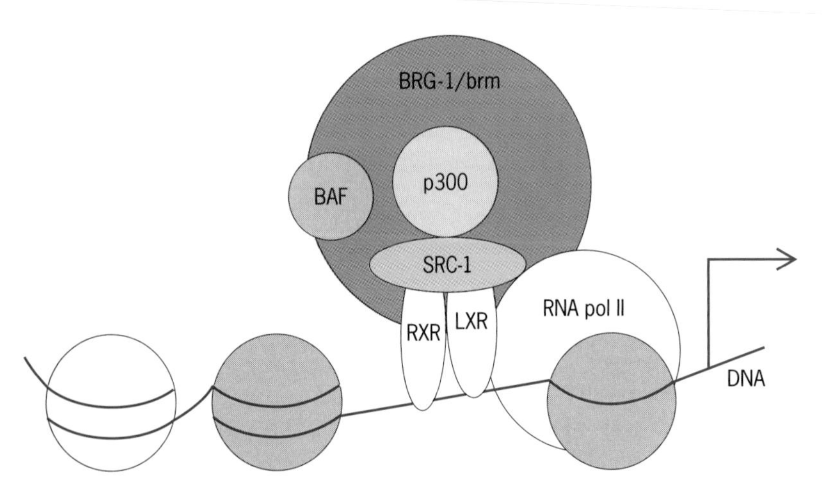

Schematic representation of an LXR-dependent transcription complex. BRG-1, BAF, SRC-1, and p300 are coactivator proteins. RNA pol II is part of the basic transcription machinery.

Transcription complexes contain multiple coactivator proteins in addition to the one or more proteins that bind directly to DNA. These coactivators play roles in remodeling DNA and in linking distant promoter sites to proteins of the basic transcription machinery at the start site for mRNA synthesis. The illustration shows a model of a typical LXR-dependent transcription complex, the one which regulates the transcription of the gene encoding the adenosine triphosphate (ATP) binding cassette A1 (ABCA1) transporter protein.

More than 50 different nuclear receptor proteins, regulating a wide range of metabolic processes, have now been identified. An additional layer of regulation is created by competition between these proteins for the same DNA-binding site as LXR. Increasing the level of one of these proteins in the nucleus can displace LXR from its binding site. The effect of displacement depends on whether the new complex activates or inhibits mRNA synthesis. For example, the LXR–RXR heterodimer activates the gene encoding the transporter protein ABCA1. Another nuclear receptor, thyroid receptor (TR), displaces LXR from its complex and inhibits activation of the *ABCA1* gene. In contrast, a second receptor, the retinoic acid receptor (RAR), displaces LXR but forms an active complex that stimulates the formation of the *ABCA1* gene's mRNA. Thus, the transcriptional regulation of a nuclear receptor–dependent gene is based on a complex network of such factors.

LXR-dependent genes. Most genes shown to be LXR-dependent promote reactions in lipid and carbohydrate metabolism. Liver and adipose tissues express many LXR-dependent genes. Some of these genes have LXR-binding sites as part of their promoter DNA sequences. In other cases, the effect of LXR is indirect; the target gene codes not for an enzyme or structural protein but for another transcription protein, with its own set of targets. The genes that encode sterol regulatory element binding protein 1a (SREBP-1a), cholesterol 7-alpha-hydroxylase, and glucose transporter-4 (GLUT-4) are examples of LXR-dependent genes.

SREBP-1c is a transcription factor that regulates the mRNA levels of a variety of genes involved in lipid and carbohydrate metabolism. In the presence of LXR, oxysterols stimulate the transcription of the *SREBP-1c* gene at a DR-4 site, thereby activating both cholesterol and fatty acid synthesis and leading to the accumulation of fat in the liver.

The enzyme cholesterol 7-alpha hydroxylase catalyzes the rate-limiting step of cholesterol catabolism. The promoter of its encoding gene in rodents (but not in humans) contains a classical LXR recognition site, and transcription is activated by oxysterols.

GLUT-4 is a transporter protein that is responsible for glucose uptake by adipocytes (fat cells) and muscle cells. In both the human and rodent genomes, *GLUT-4* gene transcription is strongly LXR-dependent, and LXR ligands stimulate glucose clearance, though the GLUT-4 protein is not completely absent in LXR −/− mice, that is, in mice in which both copies of the LXR gene have been inactivated.

Pharmacologic applications. Nuclear receptor proteins have been an attractive target for drug research because of the ability of their ligands to stimulate the synthesis of mRNA from specific genes or groups of genes and the ability of many of these small molecules to enter cells. Synthetic LXR ligands have been developed for the treatment of cardiovascular and other diseases. However, several practical difficulties have hindered their development. Although each ligand activates only a subset of genes, these ligands promote diverse and sometimes competing pathways. Synthetic LXR ligands have been developed that stimulate the regression of atherosclerotic lesions and increase high-density lipoprotein (HDL) levels (the "good" fraction of blood cholesterol) by activating the lipid transporter ABCA1, which moves lipids from peripheral tissues into the blood. Unfortunately, the same ligands usually stimulate fatty degeneration (steatosis) of the liver by activating the gene encoding fatty acid synthase.

There are differences in the DNA base sequence surrounding the LXR binding site of different genes that probably affect the competition between LXR and other receptor proteins. The identity of coactivator proteins in the transcription complex (which can number 30–50) also influences the response of individual genes to ligands. So far, the composition of these complexes for individual genes is not well defined, and it has not been exploited for drug development. However, a recent account of an experimental LXR ligand in mice with differential effects on steatosis and atherosclerosis offers hope that LXR-directed drugs can be developed.

For background information *see* CHOLESTEROL; DEOXYRIBONUCLEIC ACID (DNA); GENE ACTION; LIPID METABOLISM; NUCLEAR HORMONE RECEPTORS; NUCLEOPROTEIN; NUCLEOSOME in the McGraw-Hill Encyclopedia of Science & Technology.

Christopher J. Fielding

Bibliography. J. Huuskonen et al., Activation of ATP-binding cassette transporter A1 transcription by chromatin remodeling complex, *Arter. Thromb. Vasc. Biol.*, 25:1180–1185, 2005; B. A. Janowski et al., An oxysterol signalling pathway mediated by the nuclear receptor LXR alpha, *Nature*, 383:728–731, 1996; A. C. Li and C. K. Glass, PPAR- and LXR-dependent pathways controlling lipid metabolism and the development of atherosclerosis, *J. Lipid Res.*, 45:2161–2173, 2004; B. Miao et al., Raising HDL cholesterol without inducing hepatic steatosis and hypertriglyceridemia by a selective LXR modulator, *J. Lipid Res.*, 45:1410–1417, 2004.

Mangrove forests and tsunami protection

Mangrove forests thrive in the intertidal zones of tropical and subtropical coasts. They have several ecological, socioeconomical, and physical functions that are essential in maintaining biodiversity and protecting human populations. Their complex architecture, combined with their location on the edge of land and sea, makes mangrove forests strategic greenbelts that have a doubly protective function. They protect seaward habitats against influences from land, and they protect the landward coastal zone against influences from the ocean. The tsunami that occurred on December 26, 2004, revealed the valuable buffering functions of mangroves.

Ecological and socioeconomic benefits. The intertidal mangrove forest exhibits a unique biodiversity with uncommon adaptations such as vivipary in trees (young plants develop while still attached to the parental tree) and the amphibious lifestyle of certain fish. Mangroves are adapted to intertidal environmental conditions such as high-energy tidal action, high salt concentrations, and low levels of oxygen (hypoxia). The large aboveground aerial root systems not only offer improved breathing for the plant but also protect more seaward and more landward areas. On the one hand, mangroves protect seagrasses and coral reefs by trapping sediments and nutrients from overland fresh-water sources that would otherwise be deposited more seaward and cause turbidity and/or eutrophication (excessive nutrient concentration with periods of oxygen deficiency). On the other hand, mangrove trees can protect the landward area against the fury with which meteorologic or oceanologic processes, such as cyclones or tsunamis, may strike. Mangrove tree species that inhabit lower tidal zones, such as *Avicennia* spp. (Grey or Black mangrove) or *Sonneratia* spp. (Mangrove apple), can block or buffer wave action with their stems which can measure 30 m (100 ft) high and are several meters in circumference (**Fig. 1**). Although probably rare, mangrove trees reaching heights between 30 and 50 m (100 and 165 ft) have been reported from Latin America, Africa, and Asia, including Mexico, Panama, Venezuela, Brazil, Colombia, Ecuador, Sierra Leone, Nigeria, Gabon, Democratic Republic of Congo, Angola, Mozambique, Indonesia, and Papua New Guinea (Fig. 1*b*, *d*). Mangrove forests are feeding grounds and a refuge for many fish, crustaceans, and other lagoon and marine species at some stage in their life cycle. Mangroves are also known to enhance the biomass of coral reef fish communities and thus help maintain their biodiversity. Important biogeochemical services of mangroves include the entrapment of sediments and pollutants, filtering of nutrients, remineralization of organic and inorganic matter, and export of organic matter. Mangroves also function as carbon dioxide sinks by removing and storing carbon dioxide from the atmosphere, which is a major contributor to global warming.

In addition to these ecological benefits, mangroves provide natural resources and perform a number of functions that are of socioeconomic importance to humans. For example, mangrove degradation has been associated with a decline in the function of lagoon and offshore fisheries, which most fishing communities rely on to provide their main supply of dietary protein. Such degradation also impacts local inhabitants, who derive important resources—such as fuelwood, timber, food items, and ethnomedicinal products—from mangroves. Mangroves maintain

Fig. 1. Density and architecture of mangrove trees and their aboveground root complex in a healthy state. (*a*) Waterfront edge of 30-m-tall mixed *Camptostemon schultzii–Avicennia* spp. mangroves in the Tipoeka and Kamora estuaries, Irian Jaya, Indonesia. These pristine forests range 10–25 km in depth (dominated by *Rhizophora* spp. and *Bruguiera* spp. in the interior) and extend for hundreds of kilometers east and west. (*b*) Seafront mangroves composed of 45–50-m-high *Rhizophora mangle* and *R. racemosa* in Darien, Panama. (*c*) Inside view of the physiognomy (physical appearance) and density of seaward mixed *A. marina–R. mucronata* formation in Gazi Bay, Kenya. Seaward *A. marina* can reach 20–30 m in height here, and the stems can measure several meters in circumference. (*d*) Density of frontal *R. mucronata* mangroves in Rekawalagoon, Sri Lanka. (*e*) >30-m-tall *Rhizophora* spp. from Indonesia with a human reference point of 1.5 m, and (*f*) bottom-up perspective of *Rhizophora* spp. and (*g*) *Bruguiera* spp. in the same assemblage, all showing the mightiness of these mangrove trees. (*h*) Frontal *Sonneratia alba* mangrove fringe in Gazi Bay (Kenya), which are rather low (5–10 m) and sparse (29 trees 0.1 ha^{-1}) but very thick (D_{130} up to 100 cm). Imagine such a healthy mangrove greenbelt of just 500 m or even 250 m width separating a lagoon or bay from human settlements, and then picture a tsunami discharging its energy on this living dyke. This virtual image of how it could be and how it should be is in strong contrast with the December 26, 2004, media images showing huge waves discharging on "naked" beachfronts with tourist resorts and coconut trees. (*Parts a, e, f, g from Joe Garrison, Garrison Photographic, Cambodia. Part b from Norman C. Duke, University of Queensland, Australia. Part c from Farid Dahdouh-Guebas. Parts d and h from Nico Koedam, Vrije Universiteit Brussel.*)

a climate and pollution record, and they provide educational and scientific information from which we can learn. They also serve as a location for habitation and recreation for indigenous people, sustaining their livelihood as well as their cultural, spiritual, and artistic values.

Past human-mangrove interactions. The history of many mangrove sites in Latin America, Africa, Asia, and Oceania is characterized by detrimental human impacts such as deforestation, conversion to shrimp farms, mangrove land reclamation for tourist resorts, and fragmentation of mangrove populations by

Mangrove loss rates or estimates and the mangrove forest surface area for some select countries

Country or region	Time period	Mangrove loss, %	Mangrove forest surface area, km^2 (year)
Latin America			
Jamaica	xxxx*–1997	30	106 (1993)
Puerto Rico	xxxx–1979	75	92 (1997)
Mexico	1970–1992	65	5315 (1992)
Guatemala	1965–1997	31	161 (1997)
Ecuador	1969–1991	21	2469 (1997)
Peru	1943–1992	68	51 (1997)
Africa			
Gambia	1982–1995	17	497 (1995)
Guinea-Bissau	1973–1995	20	2484 (1995)
Asia			
India	1963–1977	50	6700 (1997)
Malaysia	1980–1990	12	6424 (1997)
Singapore	1822–1997	92	6 (1997)
Thailand	1961–1996	48	2641 (1997)
Vietnam	1943–1999	61	1560 (1999)
Philippines	1920–1990	97	1607 (1997)
Oceania			
Fiji	1869–1986	>9	385 (1993)

*Indicates that the time period during which the mangrove loss occurred is uncertain.
SOURCE: Based on data compiled from M. Jaffar, 1993; M. Spalding et al., 1997; C. D. Field, 2000; D. M. Alongi, 2002; E. Barbier and M. Cox, 2003.

urbanization. This has resulted in tremendous loss of mangrove forests in some countries (see **table**). It has been estimated that the total mangrove surface area has decreased from 198,090 km^2 in 1980 to 148,530 km^2 in 2000, a 25% loss. This impact has been documented through retrospective methods—such as sequential remote sensing, interviews with local people, and archive research—which all indicate that major mangrove functions have been lost (**Fig. 2**). In turn, this loss of natural functions considerably increases the sensitivity of the natural, as well as the human, environment to stochastic oceanic or climatic events such as cyclones and tsunamis.

South-East Asia tsunami. Until December 25, 2004, the most destructive tsunami that ever occurred in the Indian Ocean resulted from the eruption of the Krakatau (Krakatoa) volcano in 1883. The Krakatau explosion and resulting tsunami claimed around 36,000 human lives on Java, Sumatra, and smaller islands scattered over the Sunda Strait (Indonesia). But on December 26, the South-East Asia tsunami struck, resulting in tenfold more deaths, than the Krakatau volcanic eruption and leaving millions homeless. Near the Sumatran island of Nias, a seabed earthquake measuring 9 on the Richter scale generated a huge tsunami wave that spread in all directions, discharging its energy on thousands of kilometers of coast around the Indian Ocean. Waves up to 30 m (100 ft) high were reported to have stripped beaches with tourist resorts, local houses, roads, railways, and other human infrastructures and settlements up to several hundred meters or even several kilometers inland (**Fig. 3**). Recurrent high waves and the receding waves caused a massive flow of water and debris that ravaged the coastal zone. There were victims in Indonesia, Thailand, Burma, Malaysia, Bangladesh, India, Sri Lanka, Maldives, Seychelles, Somalia, Kenya, and Tanzania. The death toll for Indonesia and the Indian subcontinent alone amounted to about 170,000, and this figure does not account for deaths among tourists and other visitors from abroad. In total about 295,000 deaths and 130,000 injuries were recorded. The victims originated from 53 countries.

Fig. 2. Destroyed mangrove stand in Godavari delta, Andhra Pradesh, India. Fishing boats are shown searching for marine species that probably depend on the mangroves as nursery grounds. Apart from loss of function as nursery grounds, these mangroves have lost their ability to buffer oceanic influence on the land such as tidal activity, sea-level rise, and tsunamis. (*Courtesy of Nico Koedam, Vrije Universiteit Brussel, Belgium*)

Barrier function of mangroves. Unfortunately the role of mangroves as living barriers was underappreciated prior to the tsunami event of December 2004, and many mangrove forests had already been destroyed or damaged. This was the case for many mangrove sites in East Africa, Thailand, Indonesia (for example, Banda Aceh), India, and Sri Lanka (for example, the southwest coast)—areas badly affected by the tsunami tragedy. Scientists have repeatedly highlighted that mangrove forests provide goods and services to local communities; however, short-term economic gains have often been perceived to be more important. In Sri Lanka, for instance,

Fig. 3. Pre- and posttsunami coastline in Khao Lak, Thailand. (*a*)These reduced-resolution images were taken by Space Imaging's *IKONOS* satellite on January 13, 2003 (pretsunami), and (*b*) on December 29, 2004, just 3 days after the devastating tsunami hit the area. The images show that most of the lush vegetation, beaches, and resorts on the coast were destroyed by the tsunami. Breaches to the coastline are apparent, and new inlets have been carved into the shoreline. One resort near the newly carved inlet has virtually disappeared. Vegetation has been washed away, and there is standing water in low-lying areas. (*Courtesy of Space Imaging/CRISP-Singapore*)

despite existing regulations from the Forest Department and national and international recommendations for protection of selected mangrove areas, aquaculture ponds were created at the expense of mangrove forests. In addition, less drastic degradation has led to an increase in mangrove sensitivity to anthropogenic and natural hazards. For example, it has recently been reported that mangroves suffering from cryptic ecological degradation proved less resistant than unaltered mangroves during the recent tsunami. Cryptic ecological degradation is the introgression of nonmangrove vegetation into a true mangrove forest, which gives the false impression that the mangrove formation is rejuvenating in a healthy way. The fact that even such subtle changes in species composition (which do not necessarily result in a reduction in mangrove area) have had a profound impact on the damage the tsunami was able to inflict on the coastal zone, makes clear that the clearing of mangroves (or mangrove-shrimp farm conversions in other areas) will dramatically increase the vulnerability of shoreline areas.

Most evidence about the impact of the recent tsunami has come from media-interviewed witnesses who survived the natural catastrophe. Testimonies on the "power of mangroves" were reported from Indonesia, Thailand, Malaysia, Sri Lanka, and India, including the Andaman Islands, a low-lying area with extensive virtually pristine mangrove forests. Of the 418 villages hit by the tsunami along the Andaman coast, only 30, or 7%, were severely devastated. In areas where mangroves have been degraded by the aquaculture or the tourist industries, this percentage reaches an estimated 80 to 100%.

Future research. In addition to tsunamis, a healthy mangrove forest can offer protection against tidal erosion, sea-level rise, the El Niño Southern Oscillation, and associated heavy rains and tropical cyclones. However, the functions of mangrove forests during these extreme meteorologic events have not been investigated in detail. Determining whether mangrove forests play a protective role against these severe weather events requires the collection of data. The typology of the vegetation and of the geomorphologic settings in which mangroves thrive (for example, zoned forests fringing rivers or lagoons versus patchy basin forests), the species composition (that is, major species versus minor species versus associated mangrove species), and the spatial changes over time in the vegetation assemblages (for example, little change in zonation versus strong shifts in mosaic patches) can greatly differ between mangrove forests, as can the degree to which they are able to protect the coast. The assumed buffer function of mangroves has never been studied and compared across these many contexts. More research is necessary to understand the specific role of mangrove ecosystems (or similar vegetation types) in different environmental settings and under different impact types and intensities, information which would be of great value in planning human settlement and land management policy.

Acknowledgement. The author is a postdoctoral researcher of the Fund for Scientific Research (FWO-Vlaanderen) working under the objectives of the International Geosphere-Biosphere Programme (IGBP), Past Global Changes (PAGES) Focus 5: Past Ecosystem Processes and Human-Environment Interactions.

For background information *see* ECOLOGICAL COMMUNITIES; FOREST MANAGEMENT; MANGROVE; TSUNAMI in the McGraw-Hill Encyclopedia of Science & Technology. F. Dahdouh-Guebas

Bibliography. E. Barbier and M. Cox, Does economic development lead to mangrove loss? A cross-country analysis, *Contemp. Econ. Policy*, 21(4):418–432, 2003; E. Barbier and M. Cox, Economic and demographic factors affecting mangrove loss in the coastal provinces of Thailand, 1979–1996, *Ambio*, 31(4):351–357, 2002; R. Costanza et al., The value of the world's ecosystem services and natural capital, *Nature*, 387:253–260, 1997; F. Dahdouh-Guebas et al., How effective were mangroves as a defence against the recent tsunami?, *Curr. Biol.*, 15(12), R443–447, 2005; F. Dahdouh-Guebas et al., Transitions in ancient inland freshwater resource management in Sri Lanka affect biota and human populations in and around coastal lagoons, *Curr. Biol.*, 15(6):579–586, 2005; K. H. G. M. De Silva and S. Balasubramaniam, Some ecological aspects of the mangroves on the west coast of Sri Lanka, *Ceylon*

J. Sci. (Bio. Sci.), 17-18:22-40, 1984-1985; E. J. Farnsworth and A. M. Ellison, The global conservation status of mangroves, *Ambio*, 26(6):328-334, 1997; C. D. Field, Mangroves, in C. Sheppard (ed.), *Seas at the Millennium: An Environmental Evaluation*, vol. 3: *Global Issues and Processes*, Pergamon, Amsterdam, 2000; M. Jaffar, Country report of Fiji on the economic and environmental value of mangrove forests and present state of conservation, in *The Economic and Environment Value of Mangrove Forests and Their Present State of Conservation*, International Tropical Timber Organisation/Japan International Association for Mangroves/International Society for Mangrove Ecosystems, Japan, pp. 161-197, 1993; F. Moberg and P. Rönnbäck, Ecosystem services of the tropical seascape: Interactions, substitutions and restoration, *Ocean Coastal Manag.*, 46:27-46, 2003; P. J. Mumby et al., Mangroves enhance the biomass of coral reef fish communities in the Caribbean, *Nature*, 427:533-536, 2004; P. Rönnbäck, The ecological basis for economic value of seafood production supported by mangrove ecosystems, *Ecol. Econ.*, 29:235-252, 1999; M. Spalding, F. Blasco, and C. Field, *World Mangrove Atlas*, International Society for Mangrove Ecosystems, Okinawa, 1997; M. L. Wilkie and S. Fortuna, *Status and Trends in Mangrove Area Extent Worldwide*, Forest Resources Assessment Working Paper 63, Forest Resources Division, Food and Agricultural Organisation, Rome, Italy, 2003.

Mars Rovers

In June and July 2003, the National Aeronautics and Space Administration's Jet Propulsion Laboratory launched twin *Mars Exploration Rovers*, *Spirit* and *Opportunity*, on a 6-month journey to the Martian surface. Launched 3 weeks apart, both spacecraft successfully landed on Mars, *Spirit* on January 4, 2004, and *Opportunity* on January 25. *Spirit* has roamed a geological depression known as Gusev Crater, while *Opportunity* has explored an area known as Meridiani Planum, one of the flattest locations on Mars. At both landing sites, the twin rovers have made exciting scientific discoveries regarding a past and significant water history on the Martian surface. The *Mars Exploration Rovers* continue the NASA exploration strategy associated with understanding the Martian climate history and the possibility that Mars was once a warm and wet planet that could have been a habitat for life.

Entry, descent, and landing. During the cruise portion of the mission, each rover was contained within a spacecraft configuration that was first developed for the *Mars Pathfinder* mission that landed on July 4, 1997, and deployed the *Sojourner Rover*. For the *Mars Exploration Rovers*, the cruise configuration consisted of a cruise stage, a backshell, a tetrahedral lander structure which contained the rover, and an aeroshell that protected the lander and rover as the spacecraft descended through the Martian atmosphere. Both spacecraft traveled safely on their interplanetary journey from the Earth to Mars (over 450 million kilometers or 280 million miles) and arrived within an atmospheric entry corridor that was only 10 km (6 mi) wide.

Approximately 15 min prior to atmospheric entry, the cruise stage was jettisoned, leaving the lander and rover cocooned within the backshell and aeroshell (**Fig. 1**). Upon arrival at the top of the Martian atmosphere, the lander and rover navigated through what has been called the "6 minutes of terror" as the spacecraft slowed from a maximum speed of 5.4 km/s (12,000 mi/h) to a dead stop on the surface of Mars. During this descent, the thermal protection system on the aeroshell burned away and dissipated the majority of the spacecraft's kinetic energy. A subsonic parachute then opened to further slow the lander. The aeroshell was jettisoned, and the lander was lowered on a tether that separated it from the backshell. A radar sensor was used to determine the altitude of the lander as it continued the decent to the surface. The altitude solution then dictated the timing of the remaining landing events, including the opening of the lander's airbags, the firing of retro rockets to slow the lander to nearly zero velocity, and the cut of the tether that then released the lander to freely bounce on the surface until it came to a complete rest.

Once safely on the surface of Mars, the lander petals opened up revealing the rover stowed inside on what is known as the base petal of the lander. In the event that the lander came to rest on one of the side petals, the petals opened up in a specific order that allowed the lander to end up in a base-petal-down configuration. The *Spirit* lander came to rest in the base-petal-down position, while the *Opportunity* lander came to rest on one of its side panels prior to lander petal opening. A number of single-use mechanisms were then used to deploy the rover's solar panels, deploy the remote-sensing mast, lift the rover off of the lander base petal, and deploy the rocker-bogie mobility suspension hardware. Cable cutters were also used to sever a number of cable harnesses that connected the rover to the lander. These deployments took place over a number of sols (or solar days, whose duration is about 24 h 40 min) on the Martian surface. Finally, 11 sols after arrival, the *Spirit* rover drove off of its lander and onto the surface of Gusev Crater. For *Opportunity*, the egress of the rover occurred 8 sols after the initial landing event.

Instrumentation. The *Spirit* and *Opportunity* rovers (**Fig. 2**) carry identical scientific instrument suites. The remote-sensing instrument suite consists of the multispectral panoramic imaging system known as the Pancam and a miniature thermal emission spectrometer known as the mini-TES. These two instruments are located on a mast that stands approximately 1.3 m (4.3 ft) above the surface. The Pancam is configured as a stereo camera pair and achieves its multispectral capabilities through the use of a filter wheel which places one of eight narrow-band interference filters in front of a black-and-white charge-coupled-device (CCD) imager. The Pancam

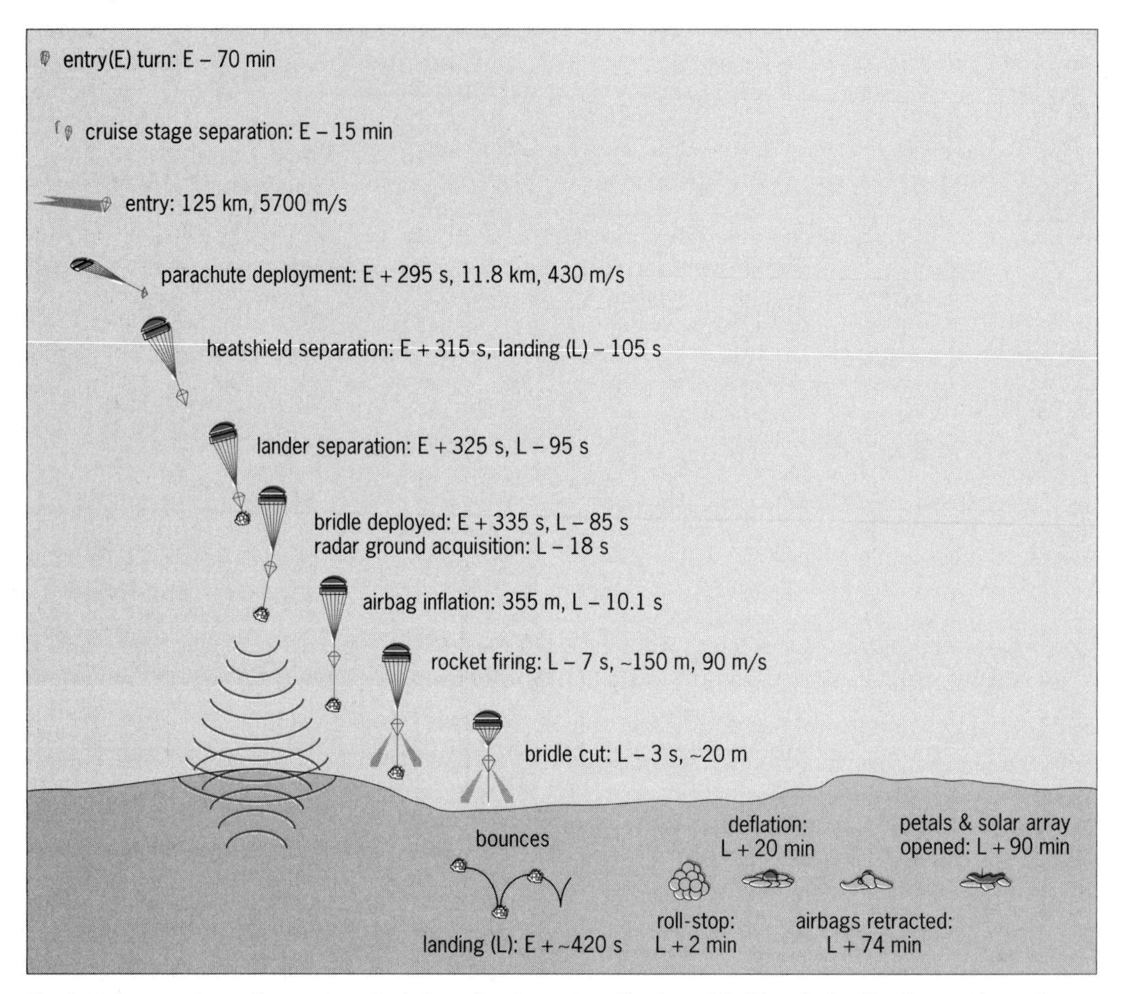

Fig. 1. Sequence of operations and events during entry, descent, and landing of the *Mars Exploration Rovers.* Times of events before or after entry (E) or landing (L) are given, and for some events the altitude above the Martian surface in kilometers and the vehicle speed in meters per second are also given. 1 km = 0.62 mi; 1 m/s = 2.24 mi/h. (*NASA/JPL*)

is a high-resolution imager that provides full-color images of the surrounding terrain and is also used to study the spectral response of rocks and soil. The mini-TES is an interferometer that operates at infrared wavelengths and provides compositional information about the rocks and soil that surround the rover. The mini-TES can also be used to measure the temperature of the atmosphere, and its data can be correlated to the TES instrument that orbits above Mars on the *Mars Global Surveyor* spacecraft. These remote-sensing instruments are used by the rover's science team on Earth to survey the surrounding terrain and determine which science targets should be approached and investigated in more detail.

To perform the detailed investigation of the rocks and soil, the rover carries a set of in-situ instruments that are deployed onto the Martian surface using a robotic arm. Two of the arm-mounted instruments are used to measure the chemical composition of the target: the Alpha-Particle X-ray Spectrometer (APXS) and the Mössbauer spectrometer. The APXS measures the chemical elements of the sample, while the Mössbauer spectrometer determines the target's iron chemistry. The fine-scale details of the rock and soil are studied using the Microscopic Imager, which consists of a close-up imager that covers a 3×3 cm^2 (1.2×1.2 in.^{2}) patch and has a depth of field of approximately 3 mm (0.12 in.). Since the Microscopic Imager is a fixed-focus imaging system, the degrees of freedom of the robotic arm are used to place it within the Imager's best focus region and capture images at and around this region. Finally, the robotic arm carries a Rock Abrasion Tool (RAT) to remove approximately 5–10 mm (0.2–0.4 in.) of the surface of a rock in order to expose the rock's interior. The RAT mechanism includes a grinding wheel that rotates rapidly while mounted to a slowly revolving wheel. It is thought that the Martian rocks have been weathered or oxidized through interaction with the atmosphere that results in a chemically altered rock surface. Therefore, the ability of the RAT to wear away the top layers of a rock surface are very important to the APXS since the depth of penetration of this instrument is very small.

In addition to the science instruments, the rovers also carry three additional stereo camera pairs that are used for navigation purposes and targeting of the robotic arm. Each engineering camera uses the same CCD imager as the Pancam and Microscopic Imager

Fig. 2. Artist's simulation of the *Mars Exploration Rover* on the Martian surface. The remote-sensing instruments on the mast are the Pancam, miniature thermal emission spectrometer (mini-TES), and Navcams. The in-situ instruments are the Mössbauer spectrometer, Alpha-Particle X-ray Spectrometer (APXS), Microscopic Imager, and Rock Abrasion Tool (RAT). (*NASA/JPL/Cornell*)

cameras. The Navcam stereo pair is located on the mast and has a 45° horizontal and vertical field-of-view. It is used primarily for targeting the remote-sensing instruments at rocks and soil in the rover's near-field (less than 30 m or 100 ft from the rover) as well as providing rover drivers (mission operations personnel who are responsible for developing detailed driving and robotic arm sequences) with local terrain maps that are used to navigate the rover through the near-field terrain. The final two stereo camera pairs are known as the Hazcams and are mounted to the front of the rover and the rear of the rover. The front and rear Hazcams are used for autonomous driving on the surface of Mars by providing the flight software (the software code that is executed on the rover's single computer processor and handles all aspects of its activities) with local terrain and navigation maps up to 2 m (6.6 ft) surrounding the rover. Finally, the front Hazcam is used to provide range and surface normal data for the placement of the in-situ instruments onto rock and soil targets using the rover's robotic arm.

Six wheels allow the rover to traverse the Martian surface, with each wheel driven independently. Mounted to the outside four wheels are steering actuators that are used to provide the rover with the ability to turn in place or turn and drive along arc trajectories. The drive and steering wheels are attached to what is known as the rocker-bogie suspension system, which is a passive suspension system that allows the rover to traverse over natural terrain and obstacles that are up to 30 cm (11.8 in.) in height. Communication between the rover and mission controllers at the Jet Propulsion Laboratory occurs using

Fig. 3. View from the *Opportunity* rover, looking back at the landing site known as Eagle Crater. (*NASA/JPL/Cornell*)

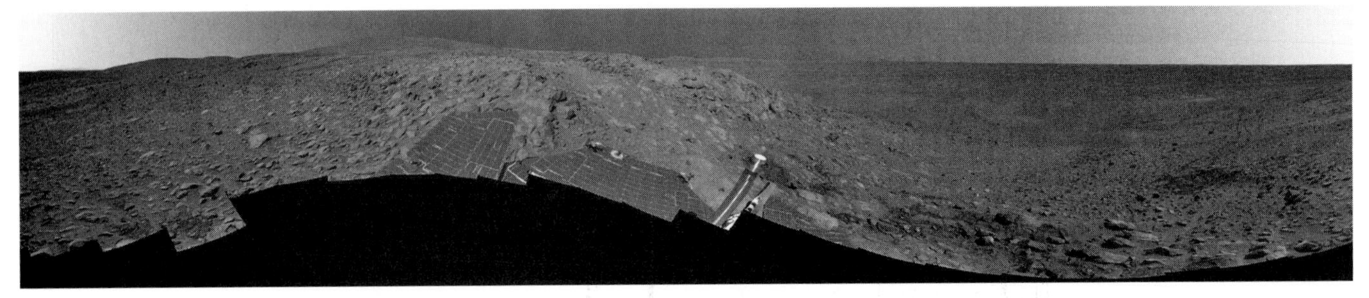

Fig. 4. *Spirit*'s panoramic view of the Columbia Hills and the plains of Gusev Crater. (*NASA/JPL/Cornell*)

an X-band radio that allows the rover to communicate directly to and from Earth on either the low-gain (omni) or high-gain (directional) antennas. A redundant communication path allows the rover to send data over an ultrahigh-frequency (UHF) link to spacecraft that orbit Mars, including the *Mars Global Surveyor*, *Mars Odyssey*, and *Mars Express*. Throughout the *Spirit* and *Opportunity* missions, the vast majority of the science and engineering data returned by the rovers has been sent up to these orbiting spacecraft and then relayed to Earth. *See* MARS EXPRESS.

Surface missions. As of June 2005, both the *Spirit* and *Opportunity* rovers had operated for over 500 sols on Mars, which is well over the 90 sols for which the rovers were designed to last. *Opportunity* had traversed over 5 km (3 mi), and *Spirit* was fast approaching the 5-km traverse milestone. Both rovers have uncovered scientific evidence that points to a significant water history on Mars. The *Opportunity* rover initially landed in a small crater known as Eagle Crater that was 22 m (72 ft) in diameter and 3 m (10 ft) deep (**Fig. 3**). Eagle Crater included exposed Martian bedrock, and the *Opportunity* rover spent the first 60 sols on Mars investigating the rock outcrop. The science instruments on *Opportunity* discovered small gray spheres that were composed of the mineral hematite and were most likely formed as concretions within sedimentary deposits. In addition, the rocks were found to be rich in sulfate salts. Finally, the Microscopic Imager revealed small ripple patterns in the rocks that are consistent with sedimentary layers that are formed by flowing water. After leaving Eagle Crater, *Opportunity* visited a larger crater known as Endurance Crater, which continued to point to an extensive water history at Meridiani Planum. Next, *Opportunity* headed south to investigate an even larger crater known as Victoria Crater. *See* MINERALOGY OF MARS.

The *Spirit* rover landed on the plains of Gusev Crater and, during its first 90 sols on Mars, found little evidence for a water history at this site. However, in the distance, scientists spied a hill complex that was elevated above the Gusev Crater plain and, most likely, represented an older geological unit that was not covered by a volcanic basalt flow that had subsequently filled Gusev Crater (**Fig. 4**). *Spirit* embarked on a 3-km (1.8-mi) journey from its landing site to the base of this formation, named Columbia Hills. Once the rover climbed a short distance up the hills, the geology and rock chemistry changed radically from the basalt rocks investigated on the plains of Gusev. Instead, scientists found evidence for a heavy alteration of the rocks by water processes, including the discovery of rocks with a higher salt content than even those rocks investigated by *Opportunity* on the other side of the planet. *Spirit* continued to survey the Columbia Hills, including a difficult climb to the crest of the hill complex.

In summary, the *Spirit* and *Opportunity* rovers have continued to work well beyond their design life and to provide their ground controllers and scientists on Earth with notable panoramic images and in-situ science data of the Martian surface. These science data have proven the importance of studying the water history of Mars and its transition from a warm and wet environment to the dry, dusty desert that it is today, and have led to a greater understanding of the water processes that were at work on Mars some 4 billion years ago.

For background information *see* CHARGE-COUPLED DEVICES; HEMATITE; MARS; MÖSSBAUER EFFECT; SPACE PROBE; X-RAY SPECTROMETRY in the McGraw-Hill Encyclopedia of Science & Technology.

Eric T. Baumgartner

Bibliography. S. Squyres, *Roving Mars: Spirit, Opportunity, and the Exploration of the Red Planet*, Hyperion, August 2005; *Spirit* at Gusev Crater, *Science* (special issue), 305:793–845, 2004; *Opportunity* at Meridiani Planum, *Science* (special issue), 306:1697–1756, 2004; S. W. Aquyres et al., *Athena* Mars rover science investigation, *J. Geophys. Res.*, 108(E12), 8062, doi:10.1029/2003JE002121, 2003.

Metagenomics

Genomics is the study of all of the genetic material in an organism. Metagenomics is the analysis of many genomes simultaneously, usually from a group of microorganisms (bacteria and archaea as well as eukaryotic microorganisms: fungi, algae, and protozoa). It involves extracting deoxyribonucleic acid (DNA) from a collection of microorganisms, cloning the DNA into a suitable vector, and analyzing the resulting clones. Metagenomic analysis has led to the discovery of new genes, enzymes, and antibiotics. The collective microbial metagenome of the world represents one of the greatest untapped resources for biotechnology as well as for understanding the workings of the natural world.

Origin of metagenomics. The microbial world represents a great frontier of biology. Microorganisms are the oldest forms of life on Earth, having originated over 3 billion years ago. During this time, they have shaped the biogeochemistry of the Earth, creating an atmosphere with sufficient oxygen to support aerobic forms of life and generating the ozone layer, which provided sufficient protection from ultraviolet radiation to make life on land possible. They manage the carbon and nitrogen cycles, transform the chemical state of metals, and convert toxic chemicals to nutrients to support their own growth. Microorganisms have also provided more industrial enzymes and pharmaceuticals than any other group of organisms. The soil bacteria, for example, are the source of most antibiotics currently in clinical use.

Microbiologists have been observing the microbial world for more than three centuries, but they increased the intensity of their efforts during the late nineteenth century with the development of cultivation techniques. These methods were predicated on the assumption that microorganisms would grow on a solid surface or in liquid in the absence of other organisms. The field of microbiology has generated a veritable explosion of knowledge in the last century based on experiments on microorganisms in pure culture. This knowledge is, however, focused entirely on microorganisms grown in biological isolation on the media provided by microbiologists.

The past few decades have illuminated the enormity of the microbial world that is recalcitrant to cultivation by standard methods. Most environments on Earth are dominated by microorganisms that cannot be grown in pure culture using current techniques. For example, fewer than 1% of the bacteria in soil and less than 50% of the bacteria in the human intestinal tract can be cultured. The uncultured microorganisms are highly diverse, representing large divisions of life—phyla or kingdoms—that are entirely unknown to modern microbiology. Metagenomics is distinguished from most other microbiological methods because it is not predicated on cultivation, and so can be used to study the uncultured majority of microorganisms on Earth.

Phylogeny determination via 16S rRNA genes. During the 1980s, the pioneering work of Carl Woese demonstrated that the 16S ribosomal ribonucleic acid (rRNA) gene provides a tool to determine the phylogeny, or evolutionary relationships, among prokaryotes. All organisms contain a related form of this gene (a homolog), and its variable regions provide a molecular signature for the organism. With this tool, Woese defined the three domains of life, Bacteria, Archaea, and Eukarya, which form the main branches of the Tree of Life (**Fig. 1**). Bacteria and Archaea are prokaryotes (single-celled microorganisms that lack nuclei), and eukaryotes are single-celled or multicellular organisms whose cells contain nuclei separated from the rest of the cell by a membrane.

16S rRNA gene analysis of microbial communities. Norman Pace built on Woese's discovery, demonstrating that 16S rRNA genes could be amplified directly from the environment using the polymerase chain reaction (PCR), which synthesizes many copies of each gene, making it possible to clone and sequence them. In order to amplify the 16S rRNA gene from environmental samples, he designed primers directed toward the regions of the gene that are identical in all organisms. Once the genes were amplified and cloned, he determined the complete nucleotide sequence of each, including the regions that vary among organisms and thus provide their molecular signatures. Pace and others built collections of 16S rRNA genes to create portraits of microbial communities in various environments, including hot springs, sediments, soil, seawater, and surfaces of plants and animals. These community portraits include both the readily cultured and as-yet-uncultured microorganisms.

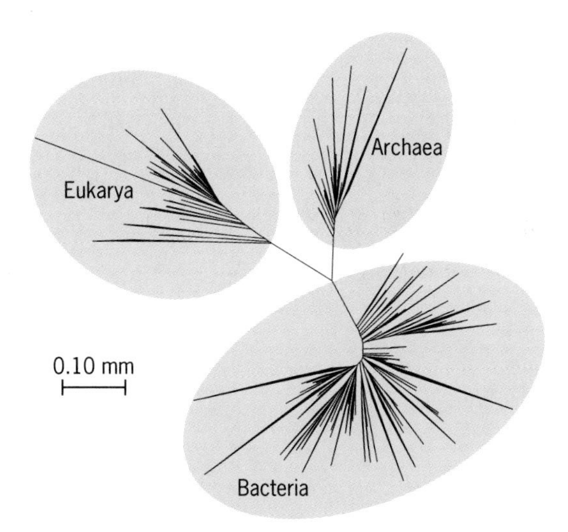

Fig. 1. Tree of Life, giving a modern view of the relationships among the major groups of organisms. These relationships are discerned by sequencing the 16S rRNA genes from organisms and aligning and comparing the sequences to infer evolutionary distances. Note that the two major domains (largest divisions of life), Bacteria and Archaea, are prokaryotic. Eukarya, which appeared more recently on Earth, contain less diversity. Branch length implies the degree of divergence from a common ancestor.

Analyses of 16S rRNA genes in many environments revealed new lineages of the domains Bacteria and Archaea and demonstrated that most species discovered by culture-independent analysis were new. Fifty-six phyla of Bacteria have been defined thus far, and of these, only half contain cultured members; the others are known only by their 16S rRNA gene signature. These results indicate the enormity of what we have yet to learn about the organisms around us, revealing the signatures of thousands of species that have not been cultured and are only distantly related to those that have been.

The limitation of this type of analysis is that PCR amplifies only the targeted gene from an organism, leaving the rest of its genome inaccessible. Therefore, microbiologists sought approaches to study the genetics and physiology of the organisms that were revealed by their 16S rRNA genes. It was from this quest that metagenomics arose. The strategy is to construct a metagenomic library by extracting DNA

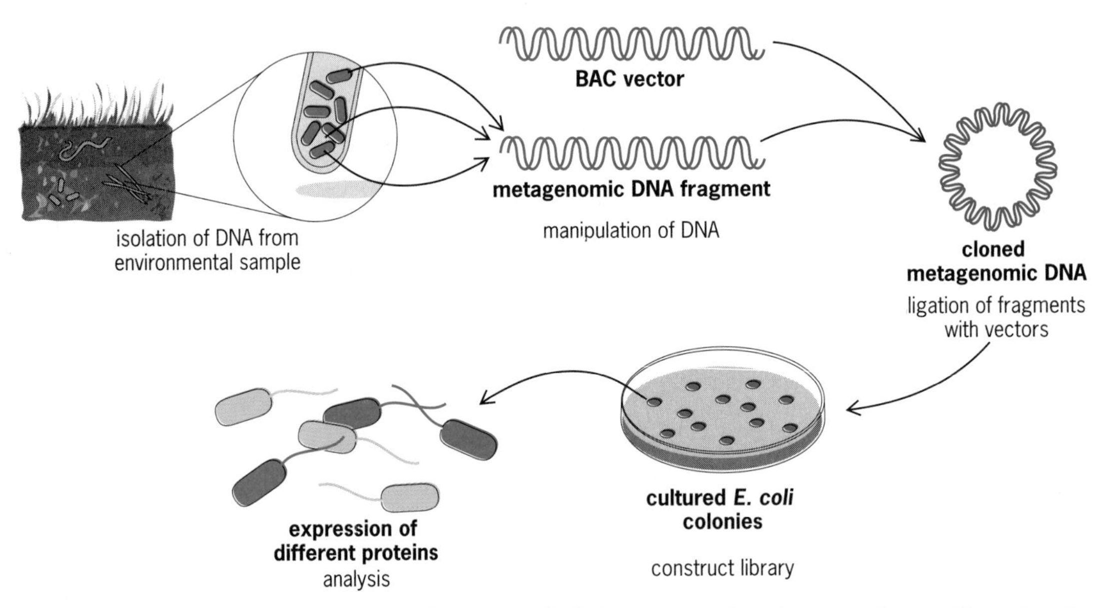

Fig. 2. Construction of metagenomic libraries. The process of isolating microorganisms from an environmental sample and extracting their DNA and cloning it into a culturable bacterium so that it can be studied by sequencing or functional analysis. (*Reprinted with permission from BioTeach, www.bioteach.ubc.ca*)

directly from an environmental sample containing a mixture of organisms, cloning the DNA in fragments into a culturable bacterium, such as *Escherichia coli*, and then studying the cloned DNA or metagenomic library (**Fig. 2**).

Approaches to studying metagenomic libraries. Once a metagenomic library has been constructed, the challenge is to analyze the contents of the library and discover new genes or gene patterns. Two approaches have emerged: sequence-based analysis and functional analysis.

Sequence-based analysis. This approach involves the analysis of the nucleotide sequence of the clones. When sequence-based analysis was used to study the community in seawater, many interesting patterns emerged. For example, a new light-absorbing molecule, a bacteriorhodopsin, was found in clones that came from Bacteria. This was exciting because similar molecules had been found previously only in Archaea.

One of the challenges of metagenomics is to identify the microorganism from which a fragment of DNA originated. One sequence-based analysis technique to deal with this is to find "phylogenetic anchors," such as 16S rRNA genes, which reveal the identity of the microorganism that contributed the DNA. Other genes found on the same fragment, and the functions associated with these genes, provide insight into the species identified by the 16S rRNA gene. Hypotheses about the physiology and ecology of the organism can be developed, and these may lead to culturing strategies. This approach was used to analyze a very simple community found in the drainage from an old copper mine, which had become highly acidic due to microbial activity. Some of the tailings from the mine had a pH of zero or even below. Bacteria and Archaea not only create this extreme environment, they also survive in it. Acid mine drainage contains only a few microorganisms, but most of them cannot be cultured. A metagenomic analysis using phylogenetic anchors suggested which community member was likely to fix nitrogen from the atmosphere. This led directly to the design of a successful strategy to culture this member, which had not been cultured previously, based on its ability to grow using atmospheric nitrogen as its sole source of nitrogen.

Functional analysis. The second approach to metagenomic analysis is functional analysis. This involves screening each clone for an activity or characteristic conferred on the host bacterium by the cloned DNA. Functional analysis has led to discovery of new antibiotics, antibiotic resistance genes, metabolic enzymes, and signal molecules. The power of the approach is that it can detect genes that could not be detected based on the DNA sequence; thus it has tremendous potential to enable the discovery of entirely new classes of genes and proteins. The limitations are that the genes of interest must be expressed in the bacterial host cell used to maintain the metagenomic library, and the assay used to find active clones must be sufficiently sensitive to detect the clone. Since many of the microorganisms whose DNA is represented in metagenomic libraries are very distantly related to the commonly used host *E. coli*, their genes are not likely to be expressed in the cells of this species. Therefore, an essential aspect of functional metagenomic analysis is to test clones for activity in various host bacterial species.

Future directions. Metagenomics has the potential to revolutionize our knowledge of the natural world. It will provide insight into the vast diversity of uncharacterized species of Bacteria and Archaea that can be accessed only through their genomes because they cannot be cultured. New methods for isolating and cloning DNA from environmental sources; rapid sequencing; improvements in computational tools to handle the large datasets generated by metagenomic analysis; and rapid, sensitive, functional screens will accelerate the rate of discovery.

There are many environments that are ripe for metagenomic study. For example, little is known about the community of microorganisms in the air, and these may be ideal subjects for metagenomic analysis, as culturing them may be difficult. Paleometagenomics, or the metagenomic analysis of ancient samples, is particularly appealing because the communities preserved in these samples are unlikely to be alive, but their DNA may be sufficiently intact to be cloned and studied. However, the small amounts of DNA recovered from samples of air and paleocommunities will present significant technical challenges. Nevertheless, the potential of metagenomics to uncover fundamental knowledge about extant and extinct communities and to lead to the discovery of useful drugs and industrial enzymes offers complementary reasons to pursue metagenomic analysis of natural communities.

For background information *see* ARCHAEBACTERIA; BACTERIA; DEOXYRIBONUCLEIC ACID (DNA); GENE AMPLIFICATION; GENETIC MAPPING; MICROBIOLOGY in the McGraw-Hill Encyclopedia of Science & Technology.

J. Handelsman

Bibliography. A. T. Bull (ed.), *Microbial Diversity Bioprospecting*, American Society for Microbiology Press, Washington, DC, 2003; E. F. DeLong, Microbial population genomics and ecology, *Curr. Opin. Microbiol.*, 5:520–524, 2002; M. T. Madigan, J. M. Martinko, and J. Parker, *Brock Biology of Microorganisms*, 10th ed., Prentice Hall/Pearson Education, Upper Saddle River, NJ, 2002; N. R. Pace, A molecular view of microbial diversity and the biosphere, *Science*, 276:734–740, 1997; J. Handelsman, Metagenomics: Application of genomics to uncultured microorganisms, *Microbiol. Mol. Biol. Rev.*, 68(4): 669–685, 2004.

Microbial dechlorination

Halogenated hydrocarbons such as chlorinated ethylenes are commonly found in the environment. For many years it was believed that most of these compounds originated from anthropogenic (human-made) sources. However, it was recently discovered that over 3500 organohalides are naturally occurring. For instance, some researchers have found that organohalides are sometimes used as a part of microbial defense mechanisms against predators. It is now an accepted fact that these compounds are generated from both anthropogenic and biogenic sources.

Trichloroethylene (TCE) and perchloroethylene (PCE) are two common contaminants found in ground water that are usually linked to anthropogenic sources. Most contamination has been linked to their industrial use in dry-cleaning and semiconductor manufacture. Their popularity stemmed from their excellent solvent and low-flammability properties. Over the last several decades, chlorinated derivatives of industrial chemicals have leached into ground water as a result of improper disposal. Because of their relatively high solubility in water and their persistence, these compounds tend to accumulate in soil and ground water.

The presence in the natural environment of halogenated hydrocarbons derived from biogenic sources suggests that there should be many microbial species that could survive in their presence and even metabolize them (that is, use organohalide compounds as food or carbon source). It is now known that these compounds can be broken down by microorganisms in the presence or absence of oxygen using a variety of mechanisms. Many environmental cleanup technologies rely on these organisms for biodegradation.

Degradation mechanisms. Microbial transformation of chlorinated ethylenes has been observed under two sets of conditions. Under anaerobic conditions (in the absence of oxygen), all the molecules in the degradation pathway can be transformed to benign products, although the process happens slowly. In contrast, under aerobic conditions (in the presence of oxygen), only the smaller molecules can be transformed efficiently. Larger molecules can be transformed but not fully metabolized under these conditions.

Anaerobic conditions. Under anaerobic conditions, chlorinated ethylenes can be transformed biologically to ethylene and ethane, which are benign compounds. This chemical transformation process is called reductive dechlorination. Anaerobic degradation occurs through the sequential removal of single chlorine atoms from the molecule and their replacement with hydrogen. This transformation occurs from PCE, to TCE, to *trans*-1,2-dichloroethylene, *cis*-1,2-dichloroethylene (*cis*-1,2-DCE), or 1,1-dichloroethylene, to vinyl chloride (VC), to ethylene, and finally to ethane (**Fig. 1**). The most prevalent transformation product from TCE is *cis*-1,2-DCE. Under anaerobic conditions, chlorinated hydrocarbons can only be used as electron acceptors. For cells to metabolize chemicals (that is,

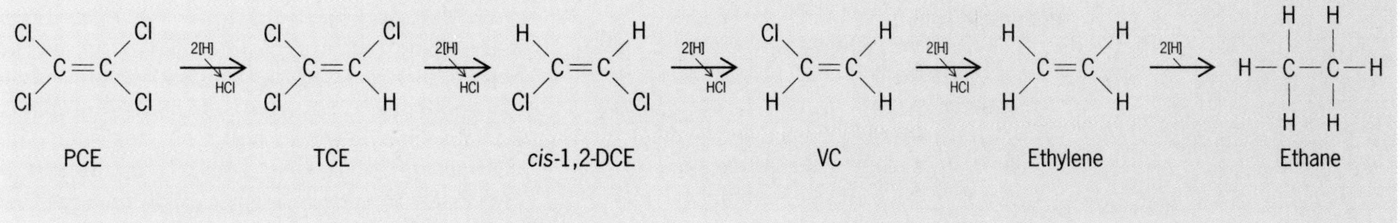

Fig. 1. Anaerobic microbial dechlorination of perchloroethylene (PCE).

$$2\ \mathrm{ClHC{=}CH_2} + 4O_2 \longrightarrow 4CO_2 + Cl_2 + 3H_2$$

Fig. 2. Transformation of vinyl chloride (VC) under aerobic conditions.

degrade a compound and use it to produce energy), they require both electron acceptors and donors. Because the chlorinated hydrocarbon is not used as a source of carbon (it is not an electron donor), reductive dechlorination can only occur in the presence of an additional carbon source. Generally, the rate of dechlorination decreases as the number of chlorine atoms on the molecule decreases. For this reason, since VC is a major intermediate product in TCE and PCE biotransformation, VC tends to accumulate in the environment. Because VC is a known carcinogen, it is important to pay close attention to VC concentrations in anaerobic treatment schemes.

Aerobic conditions. Under aerobic conditions, TCE and PCE cannot be directly metabolized. Their breakdown compounds (*cis*-1,2-DCE and VC), however, can be further broken down. Under these conditions, oxygen is used as the electron acceptor and the chlorinated compound is oxidized (**Fig. 2**) instead of reduced. The final products of aerobic metabolism are carbon dioxide, chloride, and hydrogen. VC transformation occurs much more rapidly under aerobic as compared to anaerobic conditions.

Microorganisms capable of dechlorination. Several different microbes have been identified as capable of anaerobically degrading PCE and TCE. Species such as *Dehalobacter* and *Desulfitobacterium* are able to use PCE or TCE as electron acceptors; however, they are unable to reduce these compounds completely to the benign compound ethane. These microbes are usually capable of transformation only up to the *cis*-1,2-DCE product. Some of the specific strains capable of incomplete dechlorination of PCE and/or TCE include *Dehalobacter restrictus, Desulfitobacterium dehalogens*, and *Desulfitobacterium hafniense*. Microorganisms within the *Dehalococcoides* genus are the only microbial species that have been associated with complete dechlorination of halogenated hydrocarbons. In particular, *Dehalococcoides ethenogenes* strain 195 is, to date, the only microbial species that has been characterized as capable of complete dechlorination of PCE. In this process, only the first three steps are metabolic, and the further dechlorination of VC proceeds at extremely slow rates causing this intermediary to accumulate. Other microbial species closely related to *D. ethenogenes* strain 195 have been identified that can carry out only one or two steps in the overall dechlorination. Under aerobic conditions, several mixed and pure microbial cultures including *Pseudomonas* and *Mycobacterium* species have been identified that are capable of degrading VC. These organisms are capable of transforming VC at much higher rates than their anaerobic counterparts.

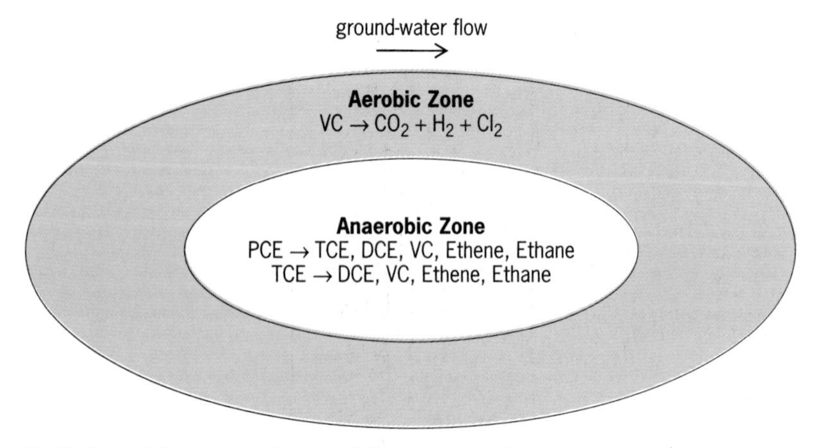

Fig. 3. Degradation scheme for natural attenuation in soil.

Remediation. Because chlorinated hydrocarbons are hazardous to human health and the environment, it is important that they be efficiently removed. The need for rapid treatment is based on the likelihood these compounds will come into human contact, such as through the use of ground-water wells for drinking water supply. Typically, once a chlorinated hydrocarbon plume (a collection of air, soil, and ground water contaminated by a single source) is detected monitoring wells are placed in the immediate area surrounding the known contamination and the progress of contaminant spread is monitored over time. Frequently, natural attenuation of the contamination is sufficient. The natural attenuation process relies on native microbial species in the soil to degrade the chlorinated compounds. When conditions are either not optimal or no natural degradation is observed, additional treatment may be necessary. Several treatment technologies, both in situ and ex situ, exist which can be applied in those situations.

In situ treatment. When chlorinated ethylene plumes are located far from any ground-water sources that are likely to come into human contact, chlorinated hydrocarbons can typically be removed by natural attenuation. In the case of chlorinated ethylenes, the breakdown products are ethylene, ethane, hydrogen, and chloride under anaerobic conditions and carbon dioxide, chloride, and water under aerobic conditions. The presence of naturally occurring anaerobic and aerobic pockets in the soil will dictate the transformation patterns (**Fig. 3**). When native species or local conditions are not favorable for efficient degradation, techniques called bioventing and biosparging may be used to increase the rate of biodegradation. For instance, when insufficient *in situ* oxygen is present, air can be introduced into the unsaturated zone above the water table level (bioventing) or into the saturated zone below the water table level (biosparging). It may also be necessary to introduce other nutrients and/or carbon sources (such as sugarcane molasses or other electron donor) to stimulate *in situ* biological growth and enhance biodegradation.

Ex situ treatment. In some cases, *ex situ* treatment is necessary to entirely degrade environmental contaminants. Some of the most common *ex situ* treatment technologies include pump-and-treat and soil vapor extraction (SVE). In pump-and-treat technology, ground water is removed from the subsurface by pumping, and then treated at the surface prior to reinjection. The treatment could be biological, by mixing the contaminated ground water in a bioreactor with microbes, or it could be physical/chemical, such as by adsorption of contaminants to activated carbon. Typically treatment times are long, thus this alternative is losing popularity. SVE is an alternative that can be used to remove chlorinated hydrocarbons from the subsurface. In SVE, a vacuum is applied to the soil matrix and the contaminants are extracted in their gaseous form. SVE is a powerful tool for removing volatile compounds such as TCE and PCE. Once the volatile compounds are extracted, they are collected at the surface where the vapors are then treated either biologically, such as by using a biofilter, or physically/chemically using activated carbon. Frequently, air injection wells are installed to increase airflow and removal rates. The presence of additional air also serves to further stimulate microbial growth.

For background information *see* BACTERIAL PHYSIOLOGY AND METABOLISM; BIODEGRADATION; CHEMICAL ECOLOGY; GROUND-WATER HYDROLOGY; HALOGENATED HYDROCARBON; HALOGENATION; MICROBIAL ECOLOGY and; WATER POLLUTION in the McGraw-Hill Encyclopedia of Science and Technology. Claudia K. Gunsch

Bibliography. S. Fetzner, Bacterial dehalogenation, *Appl. Microbiol. Biotechnol.*, 50:633–657, 1998; M. D. Lee, J. M. Odom, and R. J. Buchanan, Jr., New perspectives on microbial dehalogenation of chlorinated solvents: Insights from the field, *Annu. Rev. Microbiol.*, 52:423–452; X. Maymo-Gatell et al., Isolation of a bacterium that reductively dechlorinates tetrachloroethene to ethene, *Science*, 276:1568–1571, 1997; H. Smidt and W. M. de Vos, Anaerobic microbial dehalogenation, *Annu. Rev. Microbiol.*, 58:43–73, 2004; L. P. Wackett and C. D. Hershberger (eds.), *Biocatalysis and Biodegradation: Microbial Transformation of Organic Compounds*, ASM Press, Washington, DC, 2001.

Microbial forensics

Forensic science is the identification, collection, preservation, analysis, and interpretation of physical evidence that is relevant to a criminal act or process of interest. Forensic science and its component disciplines often provide crucial information on aspects of a crime beyond what can be derived by traditional means of investigation and intelligence gathering. For example, the advent of DNA technology enabled the forensic community to analyze biological material found at crime scenes to precisely determine the source. The threat of attack by a biological weapon will likely make microbial forensics the next major focus in forensic biology.

Microbial forensics integrates scientific disciplines, such as microbiology, molecular biology, and chemistry, with those traditionally associated with forensic science to assist in the investigation and attribution of threats and hoaxes, as well as acts of bioterrorism, biocrime, and biological warfare.

Microbial forensics is closely related to the field of epidemiology, which is used in public health and agriculture to determine the cause and source of a disease outbreak. Investigators in these fields seek accurate and timely answers to questions, such as whether an outbreak was natural or intentional and what the source and origin of the outbreak was.

Based on how the event presents itself, certain decisions can be made as to how to treat the event and how the investigation should proceed, whether public health (or agricultural) agencies or law enforcement takes the lead. In some instances, it is obvious that an outbreak is natural in origin and not a bioterrorism event. However, if these questions cannot be answered early on, both law enforcement and public health personnel will need to investigate in parallel using their respective approaches. If the outbreak is determined to be natural, public health officials will focus on treatment and prevention. However, if the outbreak is determined to be intentional, public health's role will continue, but law enforcement will manage the investigation and will focus its efforts to develop answers to key questions for a successful prosecution.

The threat. Bioterrorism is the use of disease microorganisms or toxins of biological origin as weapons to threaten or cause harm or death. In recent years, concern has increased that certain countries and terrorist groups have or are seeking to acquire, develop, and use biological weapons against humans, animals, crops, and food. An example of bioterrorism is the 2001 anthrax attacks against people and various facilities in the United States through the mail system.

Because of the nature of bioterrorism and related events and the relative ease for acquiring, developing, or disseminating bioweapons, traditional investigative or intelligence methods alone are often insufficient to develop the breadth and depth of information required. Microbial forensics is intended to fill those gaps, focus an investigation, and support a successful prosecution in a court of law or through national policy decisions for political, economic, diplomatic, or military action.

The response. In 1996, the Federal Bureau of Investigation's (FBI) forensic laboratory created the Hazardous Materials Response Unit (HMRU) as the focal point in the United States for forensic investigations involving biological, chemical, radiological, and nuclear weapons. Based in Quantico, Virginia, HMRU has primarily focused on bioterrorism. HMRU has established working partnerships with federal, state, and local agencies to better address bioterrorism preparedness and response. Very recently, the Department of Homeland Security has established its National Biodefense Analysis and Countermeasures Center (NBACC) at Fort Detrick, MD, which has a major emerging program in the forensic analysis of threat microorganisms and toxins. This program is closely linked to HMRU and its partner, the Chem-Bio Sciences Unit. NBACC supports the FBI and other agencies and is, in turn, supported

Current analytic approaches used in microbial forensics for microbial identification and typing

Approaches	Specific technology
Clinical methods	Culture, standard media, and protocols Microscopy, morphology
Ultrastructure*	Atomic force microscopy
Metabolic profiling	Selective biochemistry
Protein-based methods	Serotyping Antibody capture Antigen capture Enzyme-linked immunosorbent assay Multilocus enzyme electrophoresis
DNA typing	Restriction enzyme methodologies Restriction fragment length polymorphisms Pulse-field gel electrophoresis Amplified fragment length polymorphisms Polymerase chain reaction Intergenic spacer regions Amplified ribosomal DNA restriction analysis Randomly amplified polymorphic DNA Variable number of tandem repeats Insertion sequence elements TIGER™ (mass spectroscopy) Hybridization Subtractive hybridization Microarray Single nucleotide polymorphisms Resequencing Gene expression* DNA sequencing Full genome Multilocus sequence typing
Agent, culture processes, and residues	Mass spectroscopy Accelerator mass spectroscopy Stable isotope analysis* Single-particle stable isotope analysis* Nano secondary ion mass spectroscopy

*Currently under investigation.

scientifically largely by the Department of Energy National Laboratories. In addition, the FBI Laboratory, a number of scientists from federal agencies, national laboratories, and academic institutions meet regularly to identify gaps, develop research agendas, discuss and validate applications, and develop policies for microbial forensics.

Methods. To solve crimes involving bioweapons, attention has turned to applying both traditional and new forensic methods to the analysis and interpretation of physical evidence, with special attention on the unique "signatures" of the microorganisms or toxins used as weapons. Microbial forensics draws on fields such as microbiology, genomics, analytical chemistry, analytical biochemistry, and informatics to characterize the biological threat agent with as much specificity as possible. In each discipline and the associated method applied, information is sought so that the evidence can be described as precisely as possible to compare reference (known, or K) microbe samples with evidence (questioned, or Q) microbe samples, just as in traditional forensic science applications. Based on the power of the available methods, a forensic scientist could state that Q could not have originated from K (exclusion), Q could have originated from K (association, from weak to strong based on the power of the analyses), Q and K could have a common history, or Q absolutely did come from K (attribution). Attribution is the ultimate goal of microbial forensics, though much scientific inquiry and validation must take place before this can be achieved. Available technology and the quality and quantity of the evidentiary material may limit what can be stated.

Genetic-based methods. Much effort in microbial forensics has focused on genetic-based analytical methods (see **table**). The intent is to identify specific genetic features that occur in a particular genus, species, strain, or isolate of a disease organism to develop characteristic genetic "profiles" or "fingerprints" by which identifications and comparisons can be made. Various approaches have been developed, including restriction fragment length polymorphisms, variable-number tandem repeats, insertion sequence typing, multilocus sequence typing, single-nucleotide polymorphisms, and whole-genome-sequencing. Microorganisms such as *Bacillus anthracis* (anthrax), *Yersinia pestis* (plague), and *Francisella tularensis* (tularemia) were among the first to be studied for forensic purposes because of their importance as biological weapons.

New methods. Microbial forensics currently faces considerable challenges to its development and maturation as a discipline and the contributions it can make to investigations and prosecutions. With traditional forensics, DNA and other forensic methods have been characterized, tested, and are well understood. The methods currently applied to microbial forensics have not yet been rigorously and independently accepted. Microorganisms are astoundingly diverse, tend to evolve (mutate) rapidly over time, and interact with their environments in ways that are not well understood. Adding to this challenge, microorganisms are thought to freely exchange genetic material between genera, species, and strains, making the use of specific genetic markers as species identifiers perhaps somewhat limited. How this complexity affects the interpretation of analyses is only beginning to be realized. Because of this, approaches that rely on chemical, elemental, and structural analysis of microbial particles and residues of the weaponization process are also being investigated to further scientists' capabilities to glean information and make comparisons with forensic samples (see table). At present, the use of mass spectroscopy and atomic force microscopy is being investigated in this regard. The capabilities and limits of all microbial forensic methods need to be defined so that they can be applied effectively and accepted into courts of law and by policy makers.

For background information *see* ANALYTICAL CHEMISTRY; ANTHRAX; BACTERIAL GENETICS; CRIMINALISTICS; DEOXYRIBONUCLEIC ACID (DNA); EPIDEMIOLOGY; FORENSIC BIOLOGY; FORENSIC CHEMISTRY; FORENSIC MEDICINE; GENETIC MAPPING; INFECTIOUS DISEASE; MEDICAL BACTERIOLOGY; PLAGUE; TOXIN; TULAREMIA in the McGraw-Hill Encyclopedia of Science & Technology. Randall S. Murch

Bibliography. R. G. Breeze, B. Budowle, and S. E. Schutzer (eds.), *Microbial Forensics*, Academic Press, 2005; W. S. Carus, *Bioterrorism and Biocrimes: The Illicit Use of Biological Agents Since 1900*, Fredonia Books, Netherlands, 2002; National Research Council, *Making the Nation Safer: The Role of Science and Technology in Countering Terrorism*, National Academy Press, Washington, DC, 2002; National Research Council, *Countering Agricultural Bioterrorism*, National Academy Press, Washington, DC, 2002; P. Rogers, S. Whitby, and M. Dando, Biological warfare against crops, *Sci. Amer.*, 280:70–75, 1999; White House Press Secretary, *Presidential Directive/HSPD-10: Biodefense for the 21st Century*, White House, Washington, DC, 2004.

Mineralogy of Mars

Planetary geologists and geochemists are extremely interested in finding out what specific kinds of elements and minerals occur on the surface of Mars, because, just like on Earth, rocks and minerals provide clues about the past climate and history of a planetary surface. Mars today is a very cold, dry, and (as far as we know) lifeless world where wind is the most active process modifying the surface. However, there is evidence from robotic spacecraft images and other measurements that liquid water once flowed on Mars, and that the climate may have been much more Earth-like in the distant past. If Mars really was warmer and wetter and more like the Earth, could life have formed or evolved there as it has on Earth? And how could the climate have changed so dramatically to become as cold and dry as it is today? These are the fundamental questions that planetary scientists are trying to answer. Many believe that the key to solving this mystery is recorded in the composition and mineralogy of the planet's rocks and soils.

Fig. 1. View of bright and dark markings on Mars, acquired by the *Hubble Space Telescope* in August 2003. This is the best-resolution view of Mars possible from the Earth, obtained from a telescope high above the Earth's atmosphere and at a time when Mars was closer to the Earth than any other time in recorded history. The bright regions are parts of the surface covered by tiny particles of oxidized, iron-rich dust, and the dark regions are less dusty parts of the surface covered by unoxidized volcanic minerals. The whitish south polar cap is composed of both carbon dioxide ice (dry ice) and water ice. Thin water ice cloud hazes can be seen along the edges of the planet.

Telescopic measurements. Observations of the surface of Mars reveal bright and dark markings, some of which have changed dramatically over the decades and some of which have remained remarkably constant (**Fig. 1**). Telescopic measurements have revealed that the bright, reddish areas on the planet contain very fine grained, oxidized iron minerals. These areas appear to be covered by the same micrometer-sized dust particles that are lifted up and carried around the planet during occasional dust storms. The specific kinds of oxidized iron minerals covering these dusty regions have been hard to determine precisely from telescopic measurements, but astronomers speculated that microscopic (also called "nanophase") to macroscopic grains of the mineral hematite (α-Fe_2O_3) could explain many of the observations in the bright regions. The darker, less red areas seen in telescopic images were interpreted to contain less oxidized iron-bearing minerals, such as either pyroxene [typically $(Fe,Mg,Ca)Si_2O_6$] or olivine [typically $(Mg,Fe)_2SiO_4$], both of which are common minerals found in unweathered volcanic rocks called basalts on the Earth. Some details about the specific calcium or magnesium or iron contents of these volcanic minerals were inferred from telescopic observations in order to try to constrain the detailed history of Mars volcanism. Telescopic measurements also revealed that most of the surface contains a few percent by weight of a "hydrated" mineral, or a mineral containing OH and/or H_2O in its mineral structure (on Earth, clays are examples of hydrated minerals). The exact mineral or minerals responsible for the observed hydration signature could not be determined. In general, determining the exact mineralogy from telescope measurements is difficult or impossible because the observations cover many hundreds of kilometers at a time and are sometimes blurred by the effects of the terrestrial or Martian atmosphere.

First measurements from the surface. A much more accurate way to determine which minerals occur on the surface of another planet is to send a lander or rover to make direct measurements. The first successful landers sent to Mars were the *Vikings*, which operated from 1976 to 1980. The *Vikings* provided the first detailed color images of the surface (**Fig. 2**) as well as a suite of chemical measurements of the dusty soil at landing sites within two of the brighter regions seen in telescope pictures. The *Vikings* did not carry instruments that could directly determine the mineralogy of the Martian soil, but data they obtained on the elemental composition (amount of iron, silicon, calcium, magnesium, sulfur, and so on) and chemical and magnetic properties of the soils were used to infer the presence of certain classes of minerals. For example, the way in which the soils stuck to various magnets on the lander was used to infer the presence of the iron-bearing minerals magnetite (Fe_3O_4) or maghemite (γ-Fe_2O_3) in the soils. And models of the way the different elements measured by the landers could be combined into different minerals were used to infer the presence of secondary weathering product minerals, such as hematite or maghemite, hydrated clay silicate minerals like nontronite [$Na_{0.3}Fe_2(Si,Al)_4O_{10}(OH)_2 \cdot n(H_2O)$] or montmorillonite [$(Na,Ca)_{0.3}(Al,Mg)_2Si_4O_{10}(OH)_2 \cdot n(H_2O)$], and other possible phases like magnetite or perhaps even

Fig. 2. *Viking 1* image of rocks and dusty drifts at the landing site in Chryse Planitia. For scale, the large split rock on the left is approximately 1.5 m (5 ft) wide.

Fig. 3. Mars *Pathfinder* image of the *Sojourner* rover deploying its elemental chemistry analysis instrument on the rock called Moe on Sol 65 of the mission. For scale, the rover is about 30 cm (12 in.) tall.

carbonates. However, the identification of these specific minerals was highly dependent upon models of the elemental composition, weathering pathways, and magnetic properties, so there was considerable uncertainty about the accuracy of these mineral assignments.

Similar magnetic and elemental chemistry measurements were made by the next mission sent to the surface of the planet, the Mars *Pathfinder* lander and *Sojourner* microrover. *Pathfinder* could also measure rocks, which could not be measured by the *Vikings* (**Fig. 3**). Camera images in many different wavelengths confirmed the presence of oxidized iron minerals in the bright dusty soils and more pristine volcanic minerals in the darker rocks, but specific minerals could not be uniquely identified. High sulfur and chlorine were measured by the elemental chemistry instrument in the dusty soils, just like at the *Viking* sites, implying that sulfates and chloride salts are probably a common surface mineral component distributed by the wind almost everywhere on the planet. The composition of the dark rocks proved to be a bit surprising, as they were found to have higher silicon content than what was expected based on telescopic and other measurements that indicated the presence of basaltic minerals. One possibility is that the rocks at the *Pathfinder* site are a different kind of volcanic rock called andesite, which consists of higher-silicon-content minerals that sometimes form from more complex volcanic processes than simple basalts. As exciting as these implications were for the history of Mars, they were also controversial because, as for *Viking*, the mineralogy could only be inferred or modeled from *Pathfinder* compositional data but not directly measured.

Mineralogy from orbit. Many of the spacecraft that have flown past or orbited Mars have acquired remote sensing measurements that have identified specific minerals on the surface or that can provide new constraints on the kinds of minerals that may be present. The *Mariner 6* and 7 spacecraft carried spectrometers that measured the infrared energy emitted by the planet as the spacecraft flew by Mars in 1969. These data confirmed the presence of hydrated minerals on the surface at the few percent level, but could not identify the specific minerals present.

The *Phobos 2* orbiter mapped part of the planet using its infrared spectrometer in 1988, and confirmed the telescopic identification of oxidized iron-bearing minerals in the bright regions, unoxidized volcanic pyroxene in the dark regions, and minor amounts of hydrated minerals almost everywhere. The high resolution of the observations from orbit meant that some mineral deposits could be associated with specific geologic terrains. For example, the dark volcanic region Syrtis Major (the prominent dark region in Fig. 1) was found to be rich in pyroxene, and the chemistry of the pyroxenes was found to vary from region to region across the volcano. Evidence for nanophase and fine-grained crystalline hematite was seen in some of the bright regions; and possible, though inconclusive, evidence for the oxidized, hydrated iron-bearing mineral goethite (FeOOH) was found. It was not clear if goethite was the mystery hydrated mineral seen in previous data, but *Phobos 2* found that the abundance of hydrated minerals varied slightly from region to region and was higher in the brighter regions. Unfortunately, the spacecraft failed after only a small part of the planet had been mapped, so more extensive follow-on measurements could not be obtained.

The Mars *Global Surveyor* spacecraft began systematically mapping the planet in 1997 with its thermal emission spectrometer (TES), with the goal of detecting the presence of specific minerals and determining their distribution and abundance globally. The mission has been a great success, and a number of important mineral discoveries have been made. For example, TES was able to confirm the presence of pyroxene-bearing basaltic rocks in the dark regions and to reveal that there appear to be two different classes of these basalts. Some researchers believe that TES may be seeing "typical" basalts as one type and *Pathfinder*-like andesite volcanic rocks as a second type, while others believe that the second type of volcanic rock is an oxidized version of the first. There is still significant debate on this issue. TES also discovered small deposits of olivine in some volcanic regions, and small deposits of very coarse grained hematite in several other locations (Fig. 4). The discovery of coarse-grained hematite was an exciting find, because on the Earth this kind of hematite is often formed by the action of liquid water. Partly to test whether this was also true on Mars, the largest hematite-rich region, near the equator at 0° longitude, was ultimately chosen as the landing site for the Mars exploration rover *Opportunity*. Without the high resolution made possible by operating these kinds of instruments so close to Mars, many of these small local deposits of interesting minerals would never have been detected.

Two other orbiters are making continuing contributions to the Mars mineralogy studies. The Mars *Odyssey* orbiter began globally mapping the elemental composition of Mars in 2002. The elemental data are still being acquired and analyzed, but one early and exciting result was the discovery of large abundances of hydrogen in the near surface. At high latitudes, the hydrogen signature is thought to be from water ice in the subsurface, an important finding when considering that water is needed to create many oxidized and weathered minerals. Near the equator, water ice is not stable in the near surface (it would evaporate away), so the hydrogen signature there may be caused by still-unidentified hydrated minerals detected from previous data. In 2004, the Mars *Express* orbiter began systematic mapping of the planet using its infrared spectrometer. Initial results confirm many of the previous findings about the presence of olivine, pyroxene, and nanophase and fine-grained hematite. However, exciting new discoveries have also been made for the presence of specific sulfur-bearing minerals such as gypsum [$CaSO_4 \cdot 2(H_2O)$] and kieserite [$MgSO_4 \cdot H_2O$] and specific clay silicate minerals like nontronite, all associated with specific geologic terrains. Currently, only a small fraction of the surface has been measured at high resolution, so continuing observations are likely to reveal new mineralogic surprises.

Roving mineralogist robots. Perhaps the most exciting and specific information yet about the mineralogy of Mars has come from instrumental measurements by the Mars exploration rovers *Spirit* and *Opportunity* during 2004 and 2005. These identical rovers carry a package of instruments that can either infer or directly identify specific minerals, including a color camera, a microscope, magnets, a rock-grinding tool to remove dust coverings and coatings from rocks, an infrared spectrometer, an elemental composition spectrometer, and an iron-bearing mineral detection instrument called a Mössbauer spectrometer. The rovers' color cameras confirmed the presence of oxidized iron-bearing minerals in the bright soils and basaltic volcanic minerals in the dark soils and rocks. Perhaps more importantly, they provided geologic and color information critical to deciding where to place the other instruments for detailed mineralogic measurements.

At the *Spirit* landing site inside a large impact crater called Gusev, the rover has been searching for evidence that the crater may preserve lakebed or other sedimentary deposits, a hypothesis based on images from the *Viking* and *Mars Global Surveyor* orbiters that show a large water-carved channel flowing into Gusev. No unique evidence for such deposits has been found. Instead, the rover discovered extensive deposits of dust-covered but unoxidized olivine- and pyroxene-bearing basaltic volcanic rocks at the landing site. The dust itself was found to consist of essentially the same relative abundance of elements as dust measured by similar instruments at the *Viking* and *Pathfinder* sites, further supporting the concept of a "global dust" component occurring almost everywhere on Mars. The nanophase

Fig. 4. *Spirit* rover panoramic camera image of sulfur-rich deposits just below the surface in the Columbia Hills region of Gusev Crater. The deposits contain Fe-, Mg-, and Ca-rich sulfate minerals that were exposed by the action of the rover's wheels as it scrambled up the slopes of the hills.

hematite inferred from previous measurements of dust was identified by the Mössbauer spectrometer as a mineral component of this dust, but surprisingly the bulk mineralogy of the dust was discovered to consist primarily of unoxidized minerals like olivine, pyroxene, and magnetite. Apparently, the "oxidized" nature of Mars dust is simply an outer veneer or coating of oxidized particles that hide a relatively pristine unaltered composition from the view of remote sensing instruments. After traversing more than 3 km (2 mi) across the dusty volcanic plains, *Spirit* began climbing into a set of low hills that showed evidence for a much different mineral inventory. Specifically, *Spirit* discovered ilmenite ($FeTiO_3$) in some of the volcanic rocks, implying a significantly different magma evolution history than the volcanic rocks seen on the plains, as well as finely layered deposits of heavily oxidized rocks containing hematite and goethite. Most recently, sulfur-rich deposits were discovered just below the surface in one region of the hills (**Fig. 4**), and infrared and Mössbauer spectra and elemental composition measurements of this material imply the presence of significant deposits of hydrated Fe-, Mg-, and Ca-bearing sulfate minerals in this region, though the specific minerals present have not yet been identified.

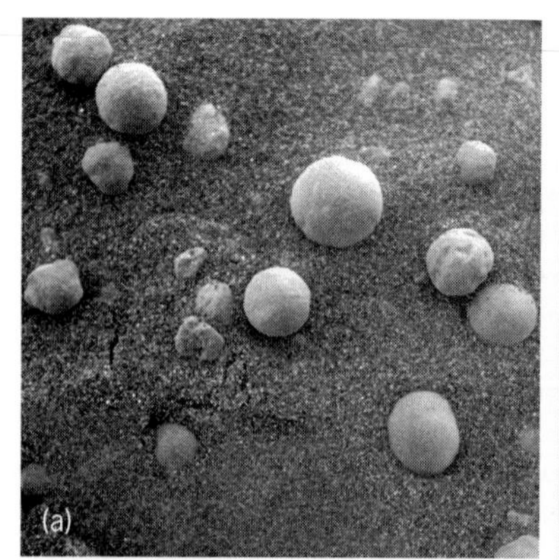

Fig. 5. *Opportunity* rover microscopic imager mosaic of 2- to 5-mm spherical granules discovered (*a*) on the plains and (*b*) embedded within the reddish sulfur-rich outcrop rocks of Meridiani Planum. One of the spherules in *a* has been partially buried by the contact plate of the rover's Mössbauer spectrometer, and in *b* one of the spherules still barely embedded in the outcrop rock on the right has split in half.

Meanwhile, at the second landing site on the other side of the planet in the flat plains of the dark region called Sinus Meridiani, the *Opportunity* rover discovered stunning and unambiguous mineral evidence for the past presence of liquid water on Mars. The dark Meridiani plains are covered by basaltic, olivine-rich dust, sand, and cobbles. But mixed in with these minerals, the rover discovered countless millions of spherical, 2- to 5-mm-sized, hematite-rich granules littering the plains (**Fig. 5**). These spherules appear to be the source of the coarse-grained hematite signature detected in Meridiani by the thermal emission spectrometer on the *Mars Global Surveyor* orbiter. And just below the surface of the plains, exposed in small impact craters or rifts, the rover discovered bright, reddish, finely layered outcrop rocks containing high concentrations of sulfur, chlorine, bromine, and hematite. Images from the rover's microscope showed the spherules to be eroding from the outcrop rocks (Fig. 5), which appear to be soft, layered sedimentary "evaporite" rocks containing sulfates, nanophase iron oxides, and surprisingly the hydrated ferric sulfate jarosite $[(K,Na,H_3O)Fe_{3-x}(SO_4)_2(OH)_6]$. Hematite occurs in the outcrop as finely dispersed nanophase hematite as well as coarse-grained hematite in the spherules. Additional study of rocks and soils by the rover as it traversed the plains revealed evidence for pyroxene-rich impact ejecta rock fragments, and a kamacite-rich (α-Fe,Ni) iron meteorite.

Meteorites from Mars. Additional information on the mineralogy of Mars comes from a small number of meteorites that have fallen to Earth and that are thought to have come from Mars based on their chemical and petrologic signatures. These rocks were blasted off Mars by asteroid or comet impacts and are likely to be samples from deeper in the crust and not the top surface that is directly measured by telescopes and spacecraft. Nonetheless, many of these rocks contain the same kinds of basaltic volcanic minerals, such as olivine and pyroxene, detected on Mars from other measurements. Some contain oxidized and hydrated minerals or mineral assemblages like iddingsite [typically $(MgO,Fe_2O_3,3SiO_2) \cdot 4(H_2O)$], a common alteration product of olivine, that supports the idea that liquid water has interacted extensively with some of the rocks on Mars, forming diagnostic minerals that indicate different past climate conditions.

Implications for the history of Mars. The inventory of minerals and mineral assemblages detected so far on Mars points to an interesting and still not completely understood environmental history for the planet. The presence of jarosite, goethite, hydrated sulfates, evaporitic sedimentary rocks, and coarse-grained hematite supports the argument for the presence of stable liquid water on the surface of Mars, at least in some places and for significant periods of geologic time. How much water and for how long remains the subject of intense debate. At the same time, the presence of putatively younger and relatively unaltered volcanic rocks and soils containing olivine, pyroxene, magnetite, and ilmenite indicate that conditions on Mars have been extremely cold and dry for most of the planet's history. If Mars was Earth-like, it was early on, 3 to 4 billion years ago, and probably only for a relatively short fraction of the planet's history. Regardless of the duration that liquid water was present, the high sulfur content of the altered rocks and soils indicates that the water was probably acidic. That fact may help explain the glaring absence of abundant carbonate and clay silicate minerals on Mars, and presents a challenge to planetary biologists interested in the study of the possible origin and evolution of life on Mars. Could life have developed or thrived in (possible episodic) short-lived environments with acidic ground water? This is a topic of intense current debate, and many biologists are looking to the geologic record of exotic life forms on Earth for clues to this puzzle.

Future measurements. While our knowledge and understanding of the mineralogy of Mars has increased dramatically over the past decade, there is still a critical need for additional measurements to address many outstanding questions. Some new information will come from global multispectral imaging, near-infrared spectrometer, and subsurface radar measurements to be obtained by instruments on the *Mars Reconnaissance Orbiter* spacecraft, launched in 2005 and landed successfully on August 12. Other local-scale compositional and mineralogic information will hopefully be provided by instruments on the planned 2007 *Phoenix* polar lander mission and the 2009 *Mars Science Laboratory* rover mission. Beyond these planned missions, our next quantum leaps in knowledge of the surface mineralogy and past climate/environment history of Mars are likely to come from robotic sample-return missions currently planned for sometime between 2015 and 2020, and then, perhaps a decade or so after that, from the first human exploration missions to Mars. Ultimately, it may take trained astronaut-geologists to unravel the complex and potentially Earth-like history locked within the rocks and minerals of the red planet.

For background information *see* ANDESITE; BASALT; CARBONATE MINERALS; CLAY MINERALS GEOCHEMISTRY; GOETHITE; GYPSUM; HEMATITE; ILMENITE INFRARED SPECTROSCOPY; MAGNETITE; MAGNETOSPHERE; MARS; METEORITE; MINERALOGY; MONTMORILLONITE; MÖSSBAUER EFFECT; OLIVINE; PYROXENE; REMOTE SENSING; ROCK MAGNETISM; SEDIMENTARY ROCKS; SEDIMENTOLOGY; SILICATE MINERALS; SPACE PROBE; SPACE TELESCOPE, HUBBLE; TELESCOPE; WEATHERING PROCESSES in the McGraw-Hill Encyclopedia of Science & Technology. Jim Bell

Bibliography. J. Bell, In search of Martian seas, *Sky Telesc.*, pp. 40–47, March 2005; J. Bell, Mineral mysteries and planetary paradoxes, *Sky Telesc.*, pp. 34–40, December 2003; M. Carr, *The Surface of Mars*, Yale University Press, 1981; B. Jakosky, *Search for Life on Other Planets*, Cambridge University Press, 1998; H. McSween and S. Murchie, Rocks at the Mars Pathfinder landing site, *Amer. Scientist*, pp. 36–45, January/February 1999.

Nanometer magnets

It is interesting to imagine what would happen to a bar magnet if it were broken up into smaller and smaller pieces. Eventually one would end up with very small bar magnets, yet their properties could be vastly different from those of large-scale permanent magnets. Even if a macroscopic object made of a ferromagnetic material such as iron or cobalt may have no magnetic polarization, its nanometer-scale counterparts can be made into perfect bar magnets. The size and shape of these nanometer-scale magnets (nanomagnets) determine their strength, polarization direction, and the ways in which this polarization can be changed by external magnetic fields. The applications of such nanoscale magnets range from mass data storage to cancer treatment and medical imaging.

Basic physics. Ferromagnets can be regarded as a collection of magnetic dipoles that interact like tiny bar magnets whose positions are fixed in space, yet are free to rotate as they please. As a result, these tiny bar magnets, in the absence of other forces, would align themselves in such a way that the overall magnetization of an object would be zero, because the north pole of each tiny bar magnet would be attracted to the south pole of its neighbor. Such attraction is known as magnetostatic interaction. (**Fig. 1***a*)

If this were the only interactive force inside ferromagnets, there would be no permanent magnets, but there is an additional force that acts to align the neighboring dipoles with each other (Fig. 1*b*). This force, known as the exchange interaction, has no analog in everyday life since it arises from a quantum-mechanical interaction between spins located at neighboring atomic sites. In materials such as iron, nickel, and cobalt, the overlap of electron wave functions yields a very strong but very short ranged aligning force, leading to the formation of quasiuniformly magnetized regions within any ferromagnet. These regions are known as magnetic domains, and the lines separating various domains are domain walls. The magnetization orientation can vary from domain to domain, leading to a very complex magnetization distribution in large samples.

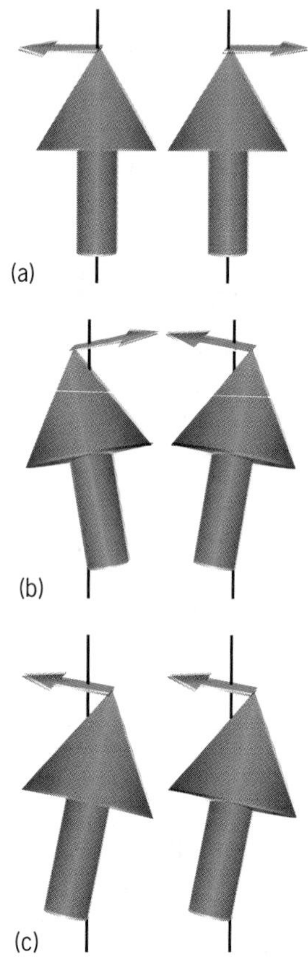

Fig. 1. Diagrams of two neighboring magnetic dipoles, illustrating fundamental forces inside a ferromagnet. The large arrows show the orientations of the dipoles, and the smaller arrows show the forces (more precisely, the torques) acting on them. (*a*) Magnetostatic interaction. The neighboring dipoles have their north poles close together, which produces a strong torque on them and tends to rotate them so the north pole of one dipole is next to the south pole of the other. In this case, the dipoles would end up being antiparallel (one pointing up and the other down). (*b*) Exchange interaction. (*c*) Anisotropy.

Figure 2 shows what happens when a large sample of ferromagnetic material (in this case, a thin ferromagnetic film whose magnetization is confined to lie in plane of the film) is divided into smaller and smaller pieces. First, a complex magnetic domain structure is observed. Machining a little piece out of that pattern gives a flux-closing configuration. The net magnetization of the sample is still zero. If the size of the sample is reduced further, however, the magnetostatic contribution is overcome by the aligning effects of the exchange interaction. Thus the magnetization of the sample is nearly uniform and it has a very strong magnetic moment. This type of nanomagnet is known to be in a single-domain state, and has a vast range of industrial applications. Such nanomagnets act as giant spins.

Magnetization orientation can also be affected by the crystallographic symmetry of the material, making some directions preferable to others. The inequality in orientation of magnetization is known as magnetic anisotropy and is a property of the ferromagnetic material (Fig. 1*c*). It is also possible to induce magnetic anisotropy by varying the shape of a magnetic object. Elongating a magnetic object in one direction induces shape anisotropy, which tends to align the magnetic moments along the longest axis. In the late 1990s it was experimentally shown that the shape and size of nanomagnets strongly affect their magnetic properties such as their magnetization distribution and the resulting anisotropy. The effect was observed in square Permalloy nanomagnets smaller than 500 nm with no elongation. Competition between the exchange and magnetostatic interactions rendered these nanomagnets highly anisotropic, which was in complete contradiction to a theory based on a uniform magnetization approximation. This type of anisotropy is called configurational anisotropy, and stems from the nonuniformity of the magnetization in nanomagnets.

Data-storage applications. The way that digital information is stored on a magnetic hard disk has barely changed since 1955, when the first hard disk drive was manufactured. Regions of ferromagnetic thin film are magnetized in one direction or the other and represent ones and zeros. The magnetization orientation is changed during the write process by a write head and is read back by a stray field sensor, also known as the read head (**Fig. 3**). However, this is a simplistic picture. In reality, all magnetic films are made of small grains which act as small magnets (nanomagnets). The size of these grains is 5–8 nm, and from the size of the data bit one can find that each bit consists of 100–1000 such single-domain grains. The current miniaturization trend is to reduce the size of each data bit. This results in reduction in the grain size, since the number of grains per data bit does not change much. Eventually the magnetic grains would become so small that their polarization could be changed by thermal fluctuations at room temperature, which would lead to data corruption. This is known as the superparamagnetic limit of data storage. Ever since the late 1990s, researchers have been predicting doomsday scenarios, in which the superparamagnetic limit would terminate the increase in data storage density and send the magnetic storage industry into decline. Most of these predictions have already been beaten by a factor of 20 or more.

Ever more innovative magnetic data storage solutions have been proposed, such as patterned magnetic media or self-assembled media, consisting of nanometer-sized magnetic particles. Large companies are also trying to develop magnetic random access memory (MRAM) devices. Such solid-state devices would use nanomagnets to store data. A common function of such devices is reading and writing magnetic data. In order to read data, the MRAM cell is designed as a layered sandwich with two magnetic layers separated by a layer of insulating material such as aluminum oxide (Al_2O_3). The current passing though one metallic magnetic layer acquires some degree of spin orientation. The insulator layer is tailored so that the current can tunnel through into the next layer. It turns out that the scattering probability of tunneling electrons strongly depends on the difference in the polarization direction of the electrons and that of the magnetic material through which they are passing. Since electrons are polarized in the same direction as the first layer, the combined resistance of the sandwich will be small if the magnetization in both layers is aligned and large if it is misaligned. This effect is known as tunneling magnetoresistance, and the device is known as a spin valve and is extensively used in field sensing and read heads as well as in MRAM design.

A curious thing happens when the spin-polarized current is increased. It has been shown that one can switch the bottom layer of the sandwich structure without applying any external magnetic field. The spin-polarized current exerts enough torque on the second layer to change its magnetic polarization. In other words, the magnetization orientation is completely controlled and sensed by electric currents passing through nanomagnet structures. Such use of electron spins for reading, writing, and storing digital information is at the heart of a new area of electronics known as spintronics.

Biomedical applications. Nanomagnets are increasingly used in biomedicine, in applications such as enhancing the signal from magnetic resonance imaging (MRI), targeted drug delivery, tissue manipulation, and hyperthermia treatment of cancerous cells.

One way of boosting the MRI signal is to use contrast agents made of magnetic nanoparticles that vary in size from 10 to 500 nm. These agents are based on superparamagnetic iron-oxide particles which have been coated with a suitable chemically neutral material to prevent them from reacting with body fluids. The size of these particles determines where they accumulate in the body after injection into the bloodstream. This allows a selective study of specific parts of the body.

Nanomagnets can also be used for drug delivery. They are first laced with drug molecules and then steered by external magnetic field gradients until they reach the desired parts of the human body. This targeted drug-delivery technique limits the amount of healthy tissue that is exposed to the drug and is one of the most active areas in cancer research, where it is currently the subject of clinical trials. Magnetic particles can also be attached to biological tissue. An application of an external magnetic field would exert a force on the tissue, allowing some degree of manipulation.

It has also been reported that cancerous cells could be treated thermally, based on the fact that some cancer cells are more susceptible to high temperatures than healthy ones. Therefore, by increasing the body temperature above 42°C (107.6°F), these cells could be selectively destroyed. To achieve this, it is proposed to inject a dose of magnetic nanoparticles into the malignant region and to apply an alternating magnetic field to the particles. If the field is sufficiently strong and of optimum frequency, the particles would absorb energy and heat up the surrounding tissue, thereby affecting only the infected cells. An overheating of tissue is prevented by tailoring the Curie temperature of nanomagnets to be 42°C, above which power absorption stops. Hyperthermia treatment has been demonstrated to work,

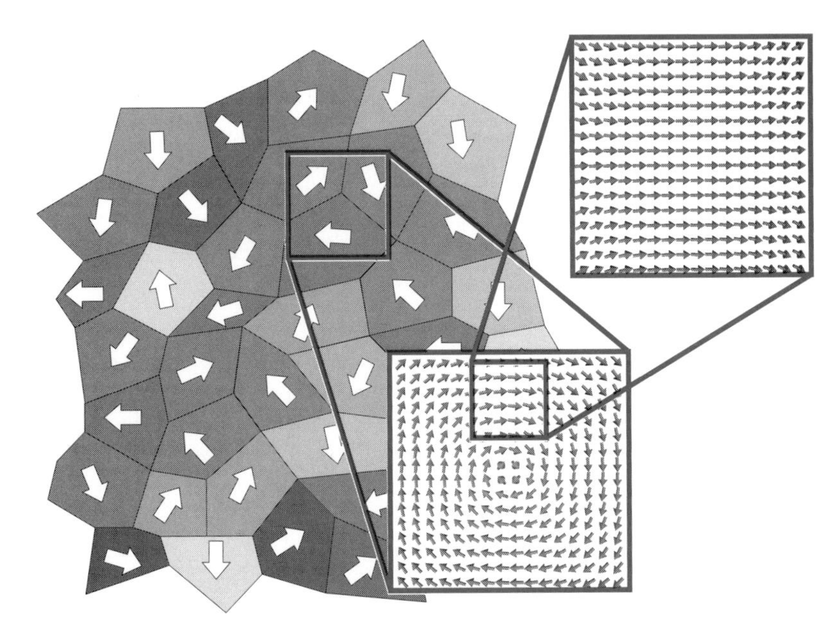

Fig. 2. Multidomain thin ferromagnetic film. The insets demonstrate the effect of cutting smaller and smaller pieces of this film. Arrow plots are results of micromagnetic simulations on a 10-nm-thick Permalloy structures.

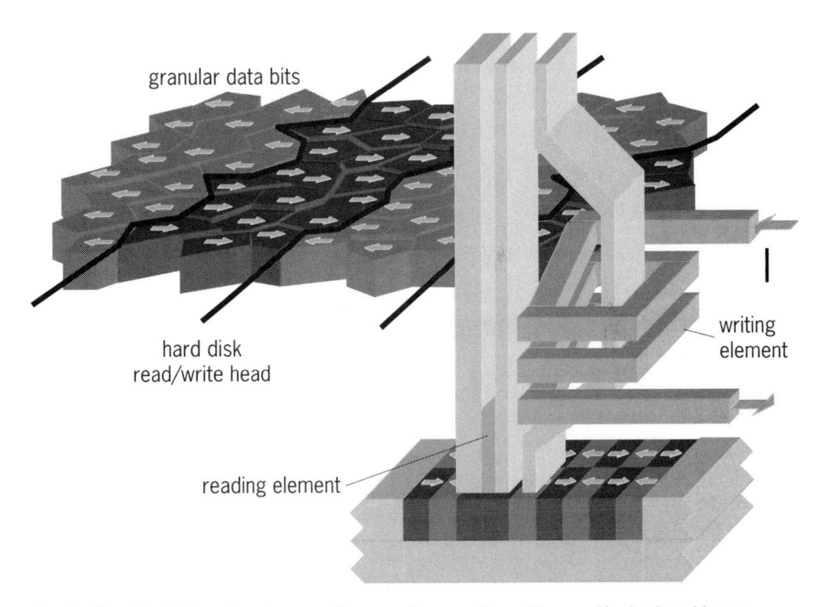

Fig. 3. Read/write head and magnetic recording medium. The read/write head has a writing element with a magnetic coil in it and a reading element. A representation of a granular magnetic medium depicts the effect of grains on the degree of definition of the written data bits.

but still remains problematic for use on humans because of the high magnetic fields required for this technique to be effective.

Despite the many applications of nanomagnets in biomedicine, research in nanomagnet technology is still driven by the search for faster, cheaper, and higher-density magnetic storage devices and sensors. However, applications of nanomagnets in health care are likely to have a more important impact on people's lives in the near future.

For background informations *see* DIPOLE-DIPOLE INTERACTION; DOMAIN (ELECTRICITY AND MAGNETISM); EXCHANGE INTERACTION; FERROMAGNETISM; MAGNET; MAGNETIC RECORDING; MAGNETISM; MAGNETIZATION; MAGNETORESISTANCE; MEDICAL

IMAGING in the McGraw-Hill Encyclopedia of Science & Technology. Denis Koltsov

Bibliography. C. C. Berry and A. S. G. Curtis, Functionalisation of magnetic nanoparticles for applications in biomedicine. *J. Phys. D—Appl. Phys.*, 36(13):R198-R206, 2003; R. P. Cowburn, Property variation with shape in magnetic nanoelements, *J. Phys. D—Appl. Phys.*, 33:R1-R6, 2000; R. P. Cowburn et al., Designing nanostructured magnetic materials by symmetry, *Europhysics Lett.*, 48(2):221-227, 1999; R. P. Cowburn, A. O. Adeyeye, and M. E. Welland, Configurational anisotropy in nanomagnets. *Phys. Rev. Lett.*, 81(24):5414-5417, 1998; S. Parkin et al., Magnetically engineered spintronic sensors and memory, *Proc. IEEE*, 91(5):661-680, 2003.

Nitrogen deposition and forest degradation

Nitrogen is an element of contrasts. In ecological systems, nitrogen is both abundant and limiting, is an essential nutrient and a pollutant, and is driven by natural and human-induced ecological processes. This makes nitrogen an important yet complex element at the center of a number of environmental issues, and one that is of increasing concern in forest conservation.

Nitrogen is the most abundant element in the Earth's atmosphere (78%). It is essential for living organisms, because it is a chemical component of all proteins and nucleic acids [ribonucleic acid (RNA) and deoxyribonucleic acid (DNA)], as well as of chlorophyll, which is essential for photosynthesis. However, due to its strong chemical bonds, most nitrogen is unavailable to plants and animals. Through its limited availability yet essential role, nitrogen exerts a controlling effect on many biological systems. Human actions, particularly over the past 50 years, are dramatically increasing the amount of nitrogen available in the environment with serious consequences. Since 1960, the flow of available nitrogen has doubled, and the rate at which available nitrogen is added to the environment continues to increase. This accumulation of available nitrogen is altering the nitrogen cycle at local and global scales. Nitrogen deposition and accumulation have been identified as critical factors affecting ecosystem composition and services worldwide. These effects are particularly evident in the forests of the northeastern United States.

Nitrogen cycle. Nitrogen is needed by all organisms; it is also a major factor limiting terrestrial plant growth. To be available to living organisms, atmospheric nitrogen must be converted from its molecular form (N_2) to a reactive form (Nr) such as ammonium (NH_4^+) or nitrate (NO_3^-). In this conversion process, called nitrogen fixation, the strong triple bond between two nitrogen atoms is broken, and the atoms are bonded or "fixed" to hydrogen or oxygen. Nitrogen fixation occurs naturally via specialized organisms and bacteria, lightning, and decomposition. However, human-driven processes, such as nitrogen fertilizer production and fossil fuel combustion, now play a dominant role in the nitrogen cycle.

Once converted to Nr, nitrogen can move through the environment having sequential negative impacts; this is termed the nitrogen cascade. For example, imagine an atom of Nr in the atmosphere that contributes to smog and particulate matter. Through precipitation, that atom may fall on a forest or grassland, where, if nitrogen exists in a high enough concentration, it may play a role in decreased ecosystem health. The same atom of Nr can next cycle into nearby lakes and streams, contributing to acidification. Alternatively, Nr may be stored in forests or grasslands for long periods, creating nitrogen sinks, the effects of which can persist for centuries. For example, 50-100% of Nr may be retained in the forests and their soils, as Nr is cycled among a number of various forms (such as ammonium, nitrate, and organic N).

The nitrogen cycle is completed when Nr is converted back to N_2 through denitrification. The denitrification process requires the presence of nitrate, organic matter, and low- or no-oxygen (anoxic) conditions. The most active denitrification occurs in aquatic systems where these conditions are found in bottom sediments all along the wetland-stream-river-estuary continuum. Lakes, rivers, and estuaries all have a similarly wide range of effectiveness in removing Nr through denitrification—from less than 10% to greater than 80%. As much as 30-70% of the total Nr inputs to a river system can be removed along this continuum, although any particular reach or portion (for example, wetlands or small streams) are likely to play a relatively small role (1-20%). The greater the residence time of water in any of these areas, the greater the amount or efficiency of denitrification. The Nr that is not removed ultimately ends up in the ocean on the continental shelf, where it is believed that most of the residual Nr in surface waters is denitrified as N_2.

Humans and nitrogen. People have been responsible for doubling the rate of Nr entering the nitrogen cycle over the last half century. Human activity now produces as much Nr as natural processes. The primary sources of anthropogenic (human-made) Nr are the production of synthetic nitrogen fertilizer, expansion of land under nitrogen-fixing crops, and fossil fuel combustion.

Nitrogen is an essential element in most synthetic fertilizers, and fertilizer production and other industrial uses account for more Nr production than all other sources combined. Fertilizers are used on agricultural crops as well as on lawns, golf courses, parks, and gardens. Up to 50% of nitrogen fertilizer may be lost to the environment largely by being washed into streams and lakes.

The expansion of land dedicated to the cultivation of particular crops, such as legumes, has also contributed directly to increased available Nr. Legumes and certain other plants have symbiotic relationships with specialized bacteria that fix nitrogen from the atmosphere. As human populations grow, the need for increased nutrition is met, in part, by increased production of these plants. In addition, the consumption of nitrogen-containing foods by humans and livestock leads to production of nitrogen-rich waste that typically finds its way into rivers, lakes, and estuaries.

Previously fixed nitrogen stored in fossil fuels such as coal and oil is released into the atmosphere as Nr as these fuels are burned. This Nr is returned to the land—often hundreds of miles from its source—through both wet and dry atmospheric deposition often referred to as acid rain. In many forested areas, atmospheric deposition is the primary or sole source of the Nr in the system.

Nitrogen in northeastern forests. The input of nitrogen to temperate forests is accelerating globally, but the impacts of nitrogen pollution are especially concentrated in certain regions. In particular, weather patterns combined with regional variations in emissions have led to severely elevated deposition of Nr in the northeastern United States, where Nr has increased 5–10 times over preindustrial levels.

Nr contributes to a variety of environmental problems, including altered productivity of forests and grasslands, acidification of lakes and streams, loss of biodiversity, habitat degradation, increased ground-level ozone, global climate change, and decreased human health. Several specific impacts of elevated Nr in northeastern forests have been identified.

Excess Nr. Excess Nr leads to changes in cycling of other key nutrients. The nature and extent of these changes is complex and related to a number of interacting factors, including tree species composition, soil depth, surficial and bedrock geology, elevation, and land-use history. Acidification of soil is the most readily identified problem, as it negatively affects the growth and health of terrestrial forest species. Acidification results from the input of acids—nitric acid, sulfuric acid—and reactions with ammonium from the atmospheric deposition of nitrogen oxides (NO_x), sulfur dioxide (SO_2), and ammonia (NH_3), respectively. Input of these compounds comes in many forms, including not only rain and snow but also clouds and fog, gases, and as dry particles—all of which are considered acid deposition.

Acid deposition can cause some of the essential base nutrients, such as calcium and magnesium, to be depleted or leached from the soil. This has resulted in aluminum—a metal which can be toxic to plants and animals—becoming more available or mobilized in the soil for absorption by living organisms. In some tree species, aluminum absorption can result in decreased root function, inhibiting the uptake of important nutrients. For example, increased mortality and decreased tree vigor in some sugar maples—an economically important species—have been linked to the loss of calcium and magnesium from the soils due to acidification. At low levels of nutrient depletion, aluminum absorption by tree roots exacerbates nutrient imbalances, making the sugar maples less likely to respond to other stresses, such as insect damage or drought. At higher levels of nutrient depletion, the combination of nutrient imbalances coupled with insect defoliation and drought appears to explain the substantial sugar maple mortality in areas of Pennsylvania. Similarly, acid deposition and soil acidification appear to have played a major role in the extensive mortality of red spruce across the Northeast. As with sugar maple, elevated aluminum concentrations in the soil potentially limit the red spruce's ability to absorb water and essential nutrients through its roots. However, acid deposition onto the spruce needles also results in the leaching of calcium directly from the needle membranes. During harsh winters at higher elevations, these needles are less capable of averting damage from freezing—a problem associated with the death of approximately half of the large canopy of red spruce in the Adirondack Mountains of New York since the 1960s, and a problem also observed in the Catskill Mountains in New York and the White Mountains in New Hampshire.

Excess nitrogen. Excess nitrogen can stress plants. Initially, Nr in forest ecosystems will enhance plant growth. However, at some point plants can no longer utilize all of the available Nr, and ecosystems reach nitrogen saturation. Plants then may become stressed from additional Nr. There is a great deal of uncertainty and variability associated with estimates of specific forests' abilities to retain nitrogen, including how long it takes to reach a saturation point.

Stress. Stress from increasing Nr flows increases the opportunity for damage from native and exotic pests and pathogens. Healthy plants may be able to defend themselves against herbivores and disease, but plants that are physiologically stressed by excess Nr are much more vulnerable to attack. Therefore, nutrient depletion and susceptibility to pest and pathogen outbreaks are problems potentially exacerbated by increased Nr inputs. The increased stress from naturally occurring pests and pathogens is further exacerbated by the fact that forests must contend with more new, exotic pests and pathogens than in the past. One consequence of improved human transportation systems has been an increase in the number of potentially damaging forest pests and pathogens now threatening northeastern forests.

Invasive species. Increased Nr appears to favor many exotic plants. Increased Nr levels make it more difficult for forests to retain their natural resilience to invasion, particularly following natural disturbance processes such as wind throws, ice storms, or fire.

Ozone. Nitrous oxide (NO_3) emissions can lead to increased ground-level ozone (O_3) that further degrades forest health. Nitrous oxide can break down and release ground-level ozone (O_3) harmful to humans and the environment. Ground-level ozone in the Northeast is formed primarily from nitrogen oxide emissions when they combine with volatile organic compounds (VOCs) at high temperatures with sunlight. Increasing ozone levels may decrease forest growth by decreasing photosynthetic capacity, as ozone enters the plant through the leaf or needle stomates. Visible signs of damage include the death or discoloration of portions of the leaf or needle called necrotic spotting or foliar stippling. Ozone damage was estimated to reduce forest growth 4–12% across a number of sites in the Northeast. Ozone impacts are not uniform, and more work is needed to expand our understanding of its role in forest health. To date, several tree species have been identified as bioindicators, meaning that they are widespread, easy to identify, and exhibit sensitivity to ozone damage in the field, and damage from ozone has been confirmed in laboratory settings. These bioindicators include red alder, speckled alder, yellow-poplar, white ash, American sycamore, quaking aspen, and black cherry. It is expected that damage is more widespread than just these bioindicator species, as several other tree and forest understory species have been identified as, or suspected of being, ozone-sensitive, including a number of maple, birch, and willow species. The net impact of this secondary effect of nitrogen deposition has yet to be fully understood.

Acidification. Nitrogen saturation may lead to acidification of lakes and streams, degrading aquatic biota. Much recent research has described the flow of Nr from forests and their headwater streams to estuaries and into the oceans. The deleterious impacts are felt throughout these hydrologic systems, significantly impairing lakes and streams across the Northeast. The diversity and abundance of aquatic biota have declined, largely due to increased acidity (pH decreases) promoting the increased mobilization of aluminum that becomes toxic to many aquatic organisms, especially fish. Almost one-quarter (346) of the 1469 Adirondack lakes surveyed no longer support fish, which has been associated with lower pH, higher aluminum, and compromised acid-neutralizing capacity. Although fish have received much of the focus, entire aquatic food webs have been disrupted by changes in species composition and relative abundance. There has been little research on the potential negative-feedback loop that altering the composition and abundance of aquatic biota may have on the forests themselves. For example, nutrient pulses from migrating salmon in the Pacific Northwest are now considered to have played an important role in the nutrient-poor forested systems that contain the headwaters of these salmon streams. Many forests in the Northeast occur on nutrient-poor soils with wetlands, streams, and lakes embedded within them. Dramatic decreases in aquatic biomass could reduce the buildup of detritus in aquatic benthic sediments, altering their ability to denitrify Nr along the wetland-stream-river aquatic continuum.

Overall, increasing rates of Nr in forest systems, combined with increasing pressure from exotic plants, pests, and pathogens, are escalating stress levels in northeastern forests. Nr plays a central role in these complex interactions. Determining what can be done to mitigate the altered nitrogen status of northeast forests in crucial. *See* INVASIVE FOREST SPECIES.

Management options. Concentrations of atmospheric ammonium and nitrates deposited in the northeastern United States have not changed appreciably since the 1960s, despite the Clean Air Act Amendments. The primary need is to reduce the levels of nitrogen deposited from the air, which will require dramatic reductions in nitrogen emissions from utilities and transportation systems. In addition, efforts are needed to limit point and non-point sources of Nr, particularly in the distribution of agricultural inputs and products. National and regional initiatives to reduce the production and control the distribution of Nr are essential for the protection of forests throughout the northeastern United States. Excess nitrogen is having multiple negative impacts on forests, and methods for reducing the flow of reactive nitrogen must be found to ensure adequate forest conservation.

For background information *see* ACID RAIN; AIR POLLUTION; ATMOSPHERIC CHEMISTRY; FOREST ECOSYSTEM; NITROGEN; NITROGEN CYCLE; NITROGEN FIXATION; NITROGEN OXIDE in the Mcgraw-Hill Encyclopedia of Science & Technology. Timothy H. Tear

Bibliography. J. D. Aber et al., Is nitrogen deposition altering the nitrogen status of northeastern forests?, *BioScience*, 53(4):375–389, 2003; C. T. Driscoll et al., *Acid Rain Revisited: Advances in Scientific Understanding since the Passage of the 1970 and 1990 Clean Air Act Amendments*, Hubbard Brook Research Foundation, Science Links™ Publication, vol. 1, no. 1, 2001; C. T. Driscoll et al., *Nitrogen Pollution: From the Sources to the Sea*, Hubbard Brook Research Foundation, Science Links™ Publication, vol. 1, no. 2, 2003; C. T. Driscoll et al., Nitrogen pollution in the northeastern United States: Sources, effects, and management options, *BioScience*, 53:357–374, 2003; J. N. Galloway et al., The nitrogen cascade, *BioScience*, 53:341–356, 2003; Millennium Ecosystem Assessment, *Ecosystems and Human Well-Being: Synthesis*, Island Press, Washington, DC, 2005; P. Vitousek et al., Human alteration of the global nitrogen cycle: Causes and consequences, *Issues Ecol.*, no. 1, 1997.

Nobel prizes

The Nobel prizes for 2004 included the following awards for scientific disciplines.

Chemistry. The chemistry prize was awarded to Aaron Ciechanover and Avram Hershko, both of Technion (Israel Institute of Technology), Haifa, and Irwin Rose of the University of California, Irvine, for the discovery of ubiquitin-mediated protein degradation in cells.

Ubiquitin is a small, 76-amino-acid protein found in all eukaryotic cells. In the cell, damaged, misfolded, or mutated proteins are tagged with multiple ubiquitin molecules and then recognized and processed at the proteasome, a complex tubular structure where proteins are broken down into smaller peptides. The ubiquitin–proteasome system is important in breaking down the proteins that regulate cell division, repair DNA, and control immune responses, to name some. If the system fails due to either under- or overactivity, disease, such as cancer, results.

In the early 1980s, the three researchers made a number of important discoveries that led to the understanding of how ubiquitin selects and identifies proteins for degradation, beginning with the identification of the ubiquitin molecule. This was followed by finding that multiple ubiquitin molecules were covalently bonded to the target protein by an enzyme system, known as the E1–E3 enzymes, and that the process required energy in the form of adenosine triphosphate (ATP).

In ubiquitin-mediated protein degradation, the E1 enzyme activates ubiquitin by a reaction that requires energy (ATP). Next, ubiquitin is transferred to an E2 enzyme. An E3 enzyme then determines which protein is to be marked and catalyzes the transfer of the ubiquitin molecules from the E2 enzyme to the target protein and repeats this step until all the necessary ubiquitin molecules are bonded to the protein. The ubiquitinated protein is recognized at the proteasome, where ubiquitin is removed and recycled, and the protein is unfolded and chopped up.

The discovery of the ubiquitin-mediated protein degradation mechanism is important in medicine for developing proteasome inhibitors to treat human diseases caused by unwanted proteins or to prevent the destruction of specific proteins. For example, if proteasome inhibitors can be synthesized which, in cancer cells, disrupt protein regulation and cause programmed cell death (apoptosis), then cancer cells can be destroyed.

For background information *see* AMINO ACIDS; BIOCHEMISTRY; CELL (BIOLOGY); CELL CYCLE; ENZYME; MOLECULAR BIOLOGY; PEPTIDE; PROTEIN in the McGraw-Hill Encyclopedia of Science & Technology.

Physics. David J. Gross (Kavli Institute for Theoretical Physics at the University of California, Santa Barbara), H. David Politzer (California Institute of Technology), and Frank Wilczek (Massachusetts Institute of Technology) were awarded the prize for the discovery of asymptotic freedom in the theory of the strong interaction. This work was published in 1973 in two papers, one by Gross, who was then at Princeton University, and Wilczek, his graduate student, and the other by Politzer, then a graduate student at Harvard University. Both teams made the same discoveries independently. The strong interaction acts between the quarks inside the protons and neutrons (nucleons) in the atomic nucleus, and the property of asymptotic freedom explains why quarks are never seen in isolation but appear to behave almost as free particles inside the nucleons in experiments in which nucleons are bombarded with high-energy electrons.

The strong interaction is one of the four fundamental forces of nature, the others being gravitation (which is extremely weak in microscopic interactions among particles), the electromagnetic interaction, and the weak interaction. Quantum electrodynamics (QED), an extremely successful theory of the electromagnetic interaction, was developed in the late 1940s. In the early 1970s, a quantum theory was developed that united the electromagnetic and weak interactions into what was called the electroweak interaction. Common to all quantum theories of fundamental interactions is the existence of a force carrier, that is, a particle which carries the force between other particles which have a particular type of charge. For example, in quantum electrodynamics, the photon carries the electromagnetic force between electrically charged particles. In the 1960s, it was suggested that, in order to preserve the Pauli exclusion principle, quarks might have a property called "color," and it was realized that color might function as a charge in a theory of the strong interaction.

However, for some time physicists doubted that it would be possible to develop such a theory or to calculate its effects in the same way as could be done for the electromagnetic and weak interactions. The coupling constant characterizing the electromagnetic interaction, known as the fine structure constant, is about 1/137 at low energies, and this small value makes it possible to calculate the effects of the electromagnetic interaction as a series expansion in this parameter, a technique known as perturbation theory. However, the coupling constant characterizing the strong interaction between two protons in a nucleus is larger than 1, making this method impossible in that context.

The situation seemed to be even worse at high energies. The fine structure constant varies with energy, increasing from about 1/137 at low energies to about 1/128 at energies of 100 GeV. In mathematical terms, its logarithmic derivative with respect to energy, a quantity known as beta, is positive. If the strong interaction also had this property, its coupling constant would be even larger at high energies. The physicist Kurt Symanzik argued that any reasonable theory of the strong interactions must have a negative beta function, which would imply asymptotic freedom of the quarks, since the strong interaction would be weaker when the quarks were close to each other, corresponding to high energies. However, many physicists began to despair of finding such a theory.

However, a way of constructing a theory with a negative beta function was discovered by Gross and Wilczek and by Politzer. A unique characteristic of their theories is that not only the quarks but also the force carriers, called gluons, carry the color charge, and therefore the gluons interact not only with quarks but also with each other. Gross, Wilczek, and others followed with a proposal for a quantum theory of the strong interactions, which was named quantum chromodynamics (QCD) after the color charge, in analogy to QED. Although it is difficult to test, QCD agrees with experiment within 1%, often much better, wherever it can be checked. QCD complemented the electroweak theory, and they were brought together in a model of the electromagnetic, weak, and strong interactions that is known as the standard model of particle physics. The standard model remains the basis for understanding a very broad range of physical phenomena, and is also a starting point for more general theories that unify these interactions and raise the hope that they can be unified with the gravitational interaction as well.

For background information *see* GLUONS; QUANTUM CHROMODYNAMICS; STANDARD MODEL in the McGraw-Hill Encyclopedia of Science & Technology.

Physiology or medicine. The Nobel Prize for Physiology or Medicine was awarded jointly to Richard Axel (Columbia University, New York, and Howard Hughes Medical Institute) and Linda B. Buck (Howard Hughes Medical Institute and University of Washington; Fred Hutchinson Cancer Research Center Seattle, Washington) for their discoveries of odorant receptors and the organization of the olfactory system.

In 1991, Axel and Buck jointly published a fundamental research paper in which they described the large family of 1000 genes that encode odorant receptors in mice. Working independently, they went on to reveal the molecular mechanisms and the cellular organization of the mammalian olfactory system, which is able to distinguish among thousands of odors.

The olfactory system is responsible for detecting and discriminating various odors in the environment, a function critical for the survival of most animals. In the nose, odor molecules in the air bind to a specific odorant receptor on an olfactory receptor cell located in the main olfactory epithelium. Odorant binding triggers G proteins (guanine nucleotide-binding proteins) that are coupled to the odorant receptors to release the chemical messenger 3′,5′-cyclic adenosine monophosphate (cAMP). The release of cAMP opens ion channels, depolarizing (and thus activating) the olfactory receptor cell, causing it to fire action potentials and send nerve impulses along its axons to glomerular cells in the olfactory bulb of the brain. The glomeruli relay these electric signals to projection neurons, known as mitral cells, whose long axons send the olfactory information to various cortical regions of the brain that interpret the information, leading to the conscious recognition of odors.

Working independently, Axel and Buck demonstrated that each olfactory receptor cell expresses only one type of odorant receptor. However, a single olfactory receptor cell can respond to multiple related odorants (having a common motif) with varying levels of intensity. Buck showed that each odorant molecule can be recognized by multiple receptors and that each odorant activates a unique combination of olfactory receptor cells, forming an odorant pattern. This combinatorial coding underlies our ability to discriminate tens of thousands of odors.

Further independent research by Axel and Buck revealed that this combinatorial coding strategy used by olfactory receptors in the nose is carried over into the olfactory bulb. They discovered that a single glomerulus receives convergent input from olfactory receptor cells expressing the same odorant receptor. Thus, a single odorant activates multiple glomeruli, and a single glomerulus can be activated by multiple odorants. This specificity in the flow of information is maintained by mitral cells, each of which is activated by only one glomerulus. Buck went on to show that the mitral cells transmit nerve signals from the glomeruli to defined microregions in the cortex of the brain, where the information from several types of odorant receptors is combined, producing a characteristic pattern that is interpreted as a recognizable odor.

Axel and Buck have made significant contributions to our understanding of the organization and function of the olfactory system, and the general principles revealed by their pioneering work appear to apply to other sensory systems as well.

For background information *see* CHEMORECEPTION; NEUROBIOLOGY; OLFACTION; SENSATION; SIGNAL TRANSDUCTION; SYNAPTIC TRANSMISSION in the McGraw-Hill Encyclopedia of Science & Technology.

Noncoalescence of droplets

Noncoalescence of liquids here indicates a state of hydrodynamic equilibrium in which two liquid bodies (even of the same liquid) apparently refuse to merge when pushed against one another. This state should not be confused with immiscibility. Unlike immiscible liquids, two noncoalescing bodies of liquid never come into molecular contact. Even when they undergo visible deformation while pressed against one another, a thin layer of a third fluid keeps the two liquid bodies apart. In most commonly observed situations, such a state is transient and lasts for just the time needed for the interstitial fluid to be squeezed out of the space between the liquid bodies.

Many examples of temporary noncoalescence are found in everyday life, such as the collision and bouncing of droplets in jets (which is encountered during the watering of a lawn, for example), the retarded coalescence of hot oil drops poured on a cooler oil surface in a pan, and the ephemeral drops of coffee floating in the morning cup. On the other hand, stable noncoalescence is rarely encountered in nature. Indeed, it was recognized as a particular case of lubrication less than a decade ago. Stable inhibition of wetting of a solid surface by a liquid is also possible by means of a similar lubrication mechanism.

While the topic of coalescence retardation is crucial to understanding a number of common phenomena ranging from the stability of emulsions to raindrop formation, the recently learned techniques to achieve stable noncoalescence and nonwetting, as well as those to provoke the rupture of the interstitial film at a well-defined moment, hint at new technologies and applications. Recently, the use of nonwetting drops has been proposed in low-load applications, such as bearings in microgravity environments or micro-electro-mechanical systems (MEMS). The study of noncoalescing and nonwetting liquids is also leading to new insights in lubrication dynamics, in electrohydrodynamics, and in the behavior of gases at the limits of their behavior as continuous media.

Noncoalescence and nonwetting mechanisms. Owing to viscous coupling, a surface (solid or liquid) moving with a nonzero tangential speed is able to drag a layer of a surrounding fluid (possibly air) along with it. Provided that the surface velocity is sufficiently elevated, the layer of dragged fluid may act as a lubricant, preventing the contact of the running surface with another, opposing surface.

One way to set a droplet surface in motion is by means of thermocapillary convection. This kind of convection, also referred to as thermal Marangoni convection, is generated by surface tension gradients due to temperature gradients along the surface. With low-viscosity silicone oils, a temperature difference between the droplets as low as 3°C (5.4°F) is normally sufficient to produce the effect (**Fig. 1**). The higher the oil viscosity, the higher the minimum temperature difference that is necessary to maintain a noncoalescent state.

An alternative way to prevent coalescence is to move the whole body of liquid. For example, a given amount of liquid could be placed in a shallow, cylindrical container, and the container could be set in rotation about its axis. The closer that portions of the surface of the rotating bath are to the container periphery, the higher is their linear velocity. Now, if a droplet, hanging from a support above the bath, is vertically moved toward the bath surface, it may not coalesce with the rotating liquid beneath, provided that the linear bath velocity, directly under the drop, is sufficiently high. If this is the case, the layer of dragged air works as a lubricant, keeping the droplet and the rotating bath surface apart.

Thermocapillary convection can also be exploited to prevent a drop from wetting a clean and smooth, solid surface. In such conditions, a small yet observable amount of liquid may evaporate from the drop and condense upon the colder solid surface (**Fig. 2**). Similar to noncoalescence, nonwetting can also be induced in the absence of temperature gradients: A drop hanging from a fixed support can be pressed against the surface of a rotating disk without wetting, just as a drop slipping upon a rotating bath surface may not coalesce with it. In this case the minimum linear velocity of the solid surface necessary to prevent wetting can be of the order of 10 cm/s (4 in./s).

For nonwetting droplets, the shape and thickness of the lubricating films have been precisely measured by means of interferometry in many different experimental situations (**Fig. 3**). Typically, when the surrounding fluid is air, the minimum film thickness ranges between 1 and a few micrometers.

It has been demonstrated that vapor coming from the involved liquids themselves may contribute to the total lubrication action. However, the relative importance of external gases and vapor contributions can be established only case by case.

In principle, negative van der Waals forces (which are possible in the presence of an interstitial, immiscible liquid) and DLVO interactions (resulting from the combination of van der Waals attraction and double-layer repulsion, and named after B. V. Derjaguin, L. Landau, E. J. Verwey, and J. T. G. Overbeek) can also give rise to permanent noncoalescence of droplets. However, lubrication is the sole

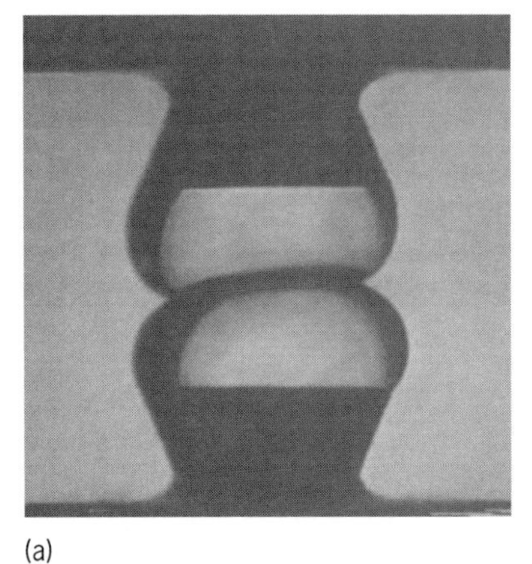
(a)

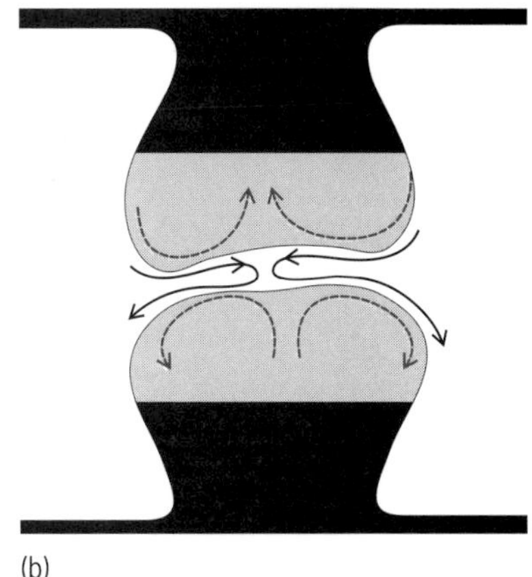
(b)

Fig. 1. Noncoalescing droplets of silicone oils. (*a*) Photograph of the drops attached to copper disks 2.0 mm (0.08 in.) in diameter. (*b*) Thermocapillary convection in the drops and lubricating air film that results from heating the upper disk and cooling the lower disk. (The temperature difference is about 20°C or 36°F, centered at the ambient temperature. The lubricating film is not to scale.)

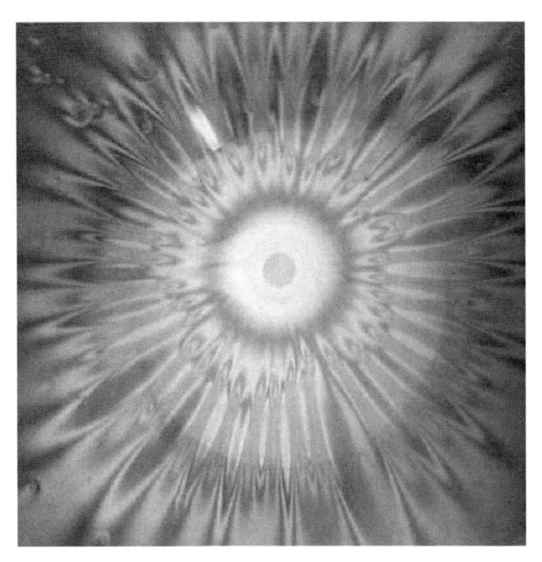

Fig. 2. Interference fringes produced by a white light shed on the oil evaporated from a nonwetting drop in thermocapillary convection and condensed upon a glass. The flowerlike pattern is due to oil migrating from the hotter region beneath the drop toward the colder periphery.

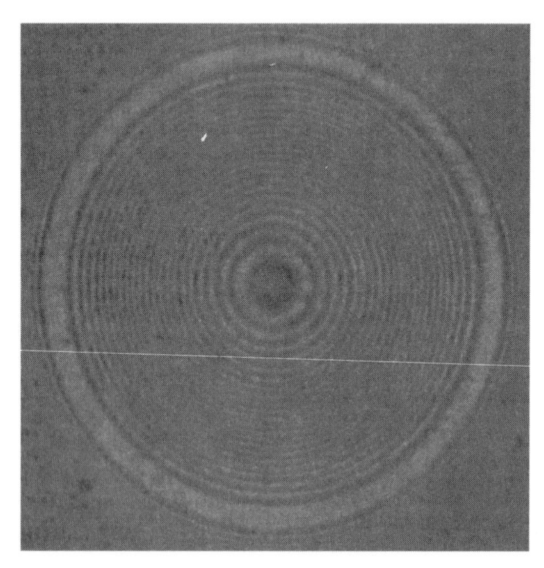

Fig. 3. Interference fringes in monochromatic light, revealing a dimpled shape for a lubricating air film between a nonwetting drop and a glass. Such a shape reflects the symmetry of the thermocapillary flows in the drop.

mechanism that leads to noncoalescence of liquids at a macroscopic scale even when the surrounding medium is a gas.

No matter what particular mechanism is involved, the expressions "noncoalescence" and "nonwetting" are often used to merely represent a macroscopically observable situation where two bodies, at least one of which is a liquid, come into apparent contact without undergoing merging or evident adhesion. Sometimes expressions like noncoalescence and nonwetting are used to the extent that liquids like mercury that undergo very little adhesion or liquids whose surface is coated with some powder may also be referred to as "nonwetting."

Continuum approximation. The parameter that must be considered in order to decide the correct equations to describe a noncoalescing system is the Knudsen number, Kn, that characterizes the problem. For noncoalescing droplets, it is given by the ratio λ/d between the mean free molecular path of the lubrication fluid and the lubrication channel thickness. It is also related to the slip length l by J. C. Maxwell's theory, which suggests the relation $l/d = a\text{Kn}$, where $a = 1.15$, is a slip coefficient. For most practical situations, when Kn is less than about 10^{-2}. Noncoalescing and nonwetting systems can be dealt with rather simply by means of simplified lubrication models derived from continuum equations (the Navier-Stokes equations), applying the no-slip condition at the interface between the lubricant and the lubricated bodies. This can be the case for noncoalescing droplets in normal atmospheric conditions, where the lubrication channel can be several micrometers thick.

However, when dealing with noncoalescing (or nonwetting) droplets in air under reduced pressure conditions, for example, or under relatively large loads, it is necessary to at least account for slip at the interfaces (slip flow). In even more critical conditions, where the film becomes very thin, close to its rupture, the lubrication model must be based on equations, such as the Burnett or Woods equations, which are more suitable to describe a medium (the lubricant) that begins to reveal its molecular nature (transitional flow).

Elastic behavior. Laplace's law, which relates the surface tension σ and the curvature of a liquid surface to the pressure jump across the surface, makes it possible to calculate the reaction F_L a nonwetting drop develops when pressed against a solid, flat surface, $F_L = A\sigma a$. Here, A is a numeric factor accounting for the shape of the drop and a is the radius of the flattened portion of the drop surface.

Apparently the nonwetting drop behaves as an ideal elastic body, since it resumes its original shape, as the load is removed. But, according to Heinrich Hertz's theory for an elastic, solid sphere undergoing a small deformation, the reaction to a load is proportional to a^2 rather than to a. This indicates that a nonwetting drop behaves differently from a common elastic solid: it possesses its own, peculiar characteristics.

It is crucial to bear in mind the differences between noncoalescing or nonwetting liquids and the conventional elastic systems with which most engineers are familiar. This distinction is particularly important when noncoalescing liquids are to be used as mechanical parts, as has been proposed, for instance, for some space applications, described below.

Applications. Applications envisioned for noncoalescing droplets include gas-lubricated liquid bearings and lab-on-a-chip applications.

Gas-lubricated liquid bearings. Noncoalescing drops show that liquids, like solids, can be lubricated and carry a load. Therefore, it has been proposed to use them as bearings. Unlike solid bearings, however, noncoalescing drops can work only in a hydrodynamic lubrication regime, which is a condition where the lubricated surfaces are constantly separated by a continuous gap, without contact. Indeed, at the smallest instance of contact, a gas-lubricated liquid immediately coalesces (or wets) the opposing surface.

On the one hand, noncoalescing droplets have many advantages with respect to solid bearings (high surface compliance, very low friction, no wear, no noise, adjustable stiffness, and the capability to absorb shocks and work as isolators from mechanical disturbances). On the other hand, they are rather fragile and can carry very low loads. Typically, a silicone-oil droplet with a diameter of 3 mm would be limited to carrying a load of about 20 dynes (2×10^{-4} newton or 5×10^{-5} pound-force) or less in order not to risk rupturing the lubricating film. For this reason, microgravity seems a suitable environment to conveniently exploit noncoalescing droplets as bearings.

In conditions of normal gravity, nonwetting liquids provide an opportunity for making easy, direct observations of elasto-hydrodynamic (EHD) deformations of lubricated surfaces, as well as clues for the

Fig. 4. Noncoalescing droplet of silicone oil upon a pool of the same liquid. The drop is heated from a distance by the disk above it. If the disk is displaced laterally, the drop moves accordingly to settle again beneath the disk.

behavior of new compliant-surface bearings for industrial applications.

Lab-on-a-chip applications. Lab-on-a-chip applications are concerned with the transport of small quantities of aqueous solutions (for example, DNA samples) from point to point for testing purposes. A free, noncoalescing droplet can easily be rolled on the surface of a bath of the same (or different) liquid by heating the drop from a distance (**Fig. 4**). The drop can be displaced from one location to another by heating its surface from a side instead of from the top. The jetting flow from the cold side of the droplet provides a momentum flux that results in the droplet being moved in the direction of the heat source. Similar behavior can be expected from nonwetting drops.

It has been proposed that this be done with an aqueous droplet (the sample of interest) encapsulated within an oil droplet. Encapsulation serves three purposes. First, as already noted, it is difficult to drive thermocapillary motion in water. Second, the heating of the aqueous sample would result in some of it evaporating, which would be undesirable. Finally, encapsulation would avoid droplet-to-droplet contamination. In the proposed design, an encapsulated droplet is sandwiched between two optically transparent planes. Infrared radiation is employed to provide localized heating at a point on the front of the droplet, causing thermocapillary flow and driving lubricating gas flows that permit the droplet to be levitated and displaced between the cold planes.

Lubricating film instability. The shapes of the lubrication channels of noncoalescing and nonwetting droplets have been studied, with the aid of interferometry and high-speed photography, both to investigate the many similarities of their dynamics with the EHD deformations of solid bearings and to look into the dynamics of lubricating film rupture. In particular, nonwetting droplets undergoing thermocapillary motion have revealed how complex the rupture mechanism can be, in spite of its being governed by precise rules.

Coalescence or wetting can be induced by decreasing the velocity of the involved surfaces, reducing the pressure of the surrounding gas, increasing the load on the system (that is, increasing drop squeezing), or applying an electric potential. The dynamic behavior of the lubricating film, in the phases that precede its rupture, depends on the specific stress applied to the system (**Fig. 5**). If the drop temperature is decreased, approaching that of the colder solid surface, the drop shrinks (Fig. 5*a*), the thermocapillary motion slows down, and lubricating pressure in the channel is reduced. A decrease in the ambient pressure leads to increased slip at the interfaces and to almost parallel film shrinking (Fig. 5*b*); it may be necessary to lower ambient gas pressures below a few tens of milliibars (a few kilopascals) in order to provoke wetting or coalescence of droplets. Increasing the load on the drop results in more dimpled and more extended films (Fig. 5*c*).

Electric potentials of approximately 10 volts are sufficient to make a droplet wet an electrically

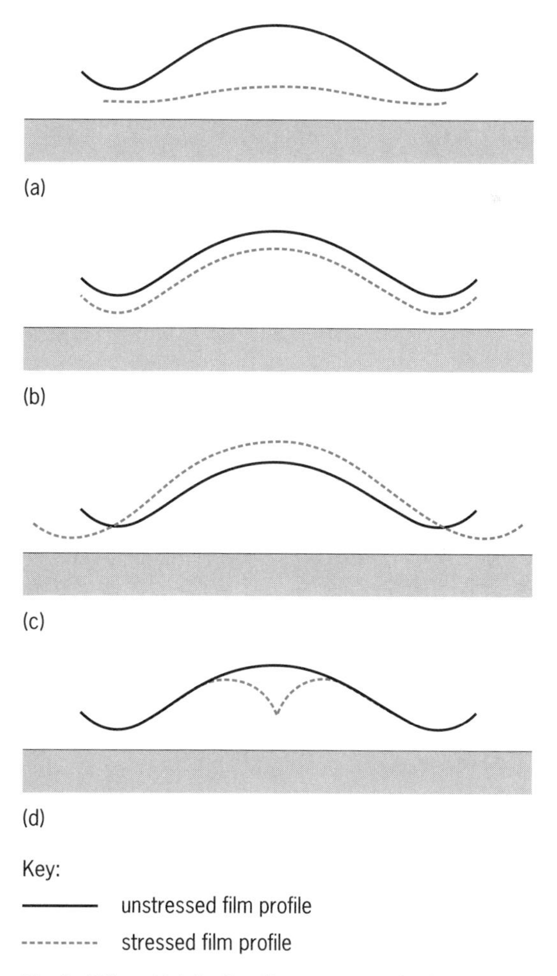

Fig. 5. Different lubricating film rupture modes for a nonwetting drop in response to (*a*) a decrease in drop temperature, toward that of the colder solid surface, (*b*) a decrease in the ambient pressure, (*c*) an increase in the load on the drop, and (*d*) an increase in the electric potential of the solid surface. The vertical scale is greatly expanded to outline the film features.

conductive surface when thermocapillary convection is used, while electrostatic potentials one order of magnitude larger can be necessary to produce wetting in isothermal conditions, with the solid surface running below the droplet. Increasing the electrostatic potential of the solid surface is the sole mechanism that invariably leads to the rupture of the lubricating film at its center, where the film is thicker and the local pressure is maximum. It is thought that the coupling between the axis-symmetric flow field within the drop and the external electrostatic field makes the ions in the liquid accumulate and creates a charge density maximum at the drop vertex. Consequently, it would be there that the electrostatic force is stronger. Eventually, the charge segregation at the drop vertex leads to the formation of a so-called Taylor's cone (Fig. 5*d*).

Prospects. Noncoalescence offers an original perspective to look at liquids, reinforcing the idea that they can be managed as independent bodies, which may not necessarily need a container to stay in or a pipe to flow through. Such a perspective is so new that any envisioned application is still at the design level, including the lab-on-a-chip and liquid-bearing applications.

Rather than being merely considered as a drawback, the intrinsic fragility of noncoalescing and nonwetting systems hints at a new "soft mechanics," where light, compliant components replace heavy, rigid parts, and precision is achieved through self-alignment rather than through low-tolerance machining. "Soft" mechanical parts, inspired by the elastohydrodynamic properties of noncoalescing drops, have already been invented, and their application will not be limited to the microgravity environment.

The previous considerations about nonwetting instability related to electrostatic forces have led to the idea that the coupling between an electric field, perpendicular to the liquid surface, and a centripetal, radial flow along the surface can be exploited in purposely developed systems (different from noncoalescing and nonwetting systems) to yield segregation of ions. This mechanism could be utilized in biomedical applications as an alternate technique to electrophoresis.

For background information *see* BOUNDARY LAYER FLOW; INTERFEROMETRY; INTERMOLECULAR FORCES; KNUDSEN NUMBER; MICRO-ELECTRO-MECHANICAL SYSTEMS (MEMS); NAVIER-STOKES EQUATIONS; SURFACE TENSION; WEIGHTLESSNESS in the McGraw-Hill Encyclopedia of Science & Technology.

Pasquale Dell'Aversana

Bibliography. P. Dell'Aversana and G. P. Neitzel, Behavior of noncoalescing and nonwetting drops in stable and marginally stable states, *Exp. Fluids*, 36:299–308, 2004; M. Gad-el-Hak, *Flow Control: Passive, Active, and Reactive Flow Management*, Cambridge University Press, New York, 2000; G. E. Karniadakis, A. Beskok, and N. Aluru, *Microflows and Nanoflows: Fundamentals and Simulation*, Springer, Berlin, 2005; D. L. Morris, L. Hannon, and A. L. Garcia, Slip length in a dilute gas, *Phys. Rev. A*, 46:5279–5281, 1992; G. P. Neitzel and P. Dell'Aversana, Non-coalescence and non-wetting behavior of liquids, *Annu. Rev. Fluid Mech.*, 34: 267–289, 2002.

Nonconventional aircraft designs

Of the many aviation achievements since the design of the Wright brothers' *Flyer* more than 100 years ago, the development and commissioning of the B-47 bomber in the United States and the Comet airline in Britain in the late 1940s and early 1950s rank among the most important, for they defined the configuration of modern civil aircraft: a cylindrical fuselage with high-aspect-ratio swept wings and a tail. Today's aircraft—for example, the Airbus 380, first flown in 2005—look rather similar to those of over 50 years ago, yet they reflect substantial technological advances in aerodynamic performance, structural integrity, engine fuel efficiency, and flight control systems. An example of such advances is the supercritical wing design for transonic cruise. However, further significant improvement becomes much more difficult under the constraint of this conventional configuration. Alternative, novel designs are now being investigated seriously with the objective of significant improvements in operating costs and environmental issues such as emission of pollutants and noise.

Flying wing. A conventional aircraft design features three primary airframe components: the main wing, the fuselage (central cylindrical body), and the tail, as well as the integration of the engines. These three components play distinct roles. While the fuselage provides the volume for the payload (passengers or cargo), the wing's primary role is to lift the aircraft into the air. The tail is designed to balance the aircraft, and provide the directional control and stability.

A flying wing aircraft design includes neither a fuselage nor a tail. From an aerodynamic point of view, the fuselage provides little lift and large drag. If the entire aircraft is designed to be a single large wing, its aerodynamic efficiency should be significantly higher than that of conventional design. The concept for such flying wings is not new. The earliest example may be traced to the tailless biplane flying wing design by the British aeronautical pioneer John Dunne in 1912. The development of the Northrop flying wing bomber series in the United States, such as the YB49 in 1949, the British flying wing *Avro Vulcan* in the 1950s, and most recently the B-2 bomber, stimulated substantial progress in the flying wing design.

Blended wing body aircraft. Nonconventional aircraft designs, such as the blended wing body (BWB) aircraft, have been proposed, based on earlier flying wing designs, for the revolutionary improvement of future large passenger aircraft. One design is illustrated in **Fig. 1**.

The blended wing body aircraft is a tailless design that integrates the wing and the fuselage. Although there are a number of similarities between a flying

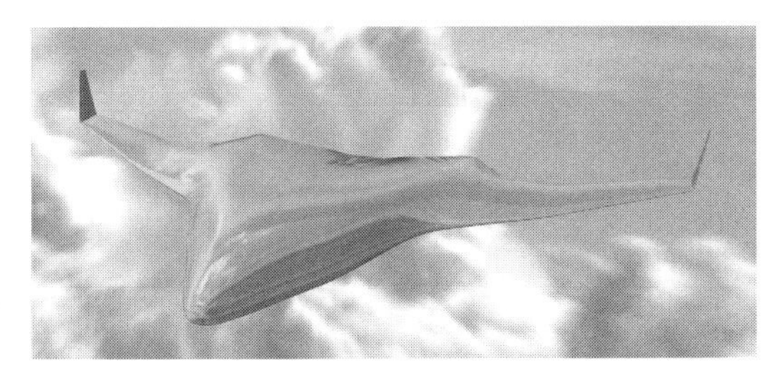

Fig. 1. Blended wing body concept.

wing and a blended wing body with regard to design advantages and challenges, a major design consideration for a blended wing body aircraft is the passenger accommodations. Therefore the central part has a dual role of both a wing and a body, providing both lift and volume for the aircraft.

Design studies were performed in the late 1990s on the blended wing body aircraft as a potential candidate for future large subsonic transport. For the 800-passenger Mach 0.85 design (85% of the speed of sound), the configuration evolves from a 106-m (350-ft) span and a trapezoidal aspect ratio of a 12, to an 85-m (280-ft) span and a trapezoidal aspect ratio of 10, indicating a significant difference in potential aerodynamic performance. It was concluded that a reduction of about 30% fuel burn per seat could be achieved for the 800-passenger blended wing body configurations in comparison with the conventional designs (requiring three instead of four engines).

A group from Cranfield College of Aeronautics in the United Kingdom designed their version of a blended wing body aircraft in 1998. This design is based on a similar payload and performance as the Airbus A380-200, with over 650 passengers accommodated in three classes. It is intended to be compatible with existing airports and facilities, thus limiting the aircraft span to 80 m (260 ft).

A number of blended wing body geometries with a view toward future large transport aircraft configuration designs were reported in 2001 at the Central Aerohydrodynamic Institute (TsAGI) in Russia. In particular, a flying wing, a lifting body, and an integrated wing body were studied in comparison with the conventional design. From the aerodynamic performance aspect, it is noticeable that all the proposed designs have a significantly increased span (100 m or 330 ft) as compared to other designs of about 80 m (260 ft), which also represent the runway capacity of most existing airports. A significant improvement in the aerodynamic performance was promised for the new configurations, with the integrated wing body design performing best at Mach 0.85 cruising speed. Most blended wing body designs have used Mach 0.85 as a cruise design point, as this is consistent with current large transport aircraft operation. A recent study of the aerodynamic behavior of a blended wing body at higher transonic speed concluded that a Mach number of 0.93 is feasible, but with a performance penalty relative to 0.85 Mach designs. In particular, a 10% reduction in ML/D was observed for the design, where M is the cruise Mach number, L the lift, and D the drag of the aircraft. Further increase in the cruise Mach number results in a substantial rise in drag and makes the design unfeasible.

An intensive multidisciplinary design study was carried out for the design of the blended wing body aircraft in Europe in the early 2000s, including multidisciplinary design development, aerodynamic analyses and design, and studies on aeroelasticity and flight mechanics.

Aerodynamic and structural advantages. Conceptually, the main aerodynamic advantage of the novel blended wing body design is its lower wetted area-to-volume ratio. The wetted area of an aircraft is defined as the external area of the aircraft. In other words, for a given volume, blended wing body aircraft should have substantially less surface area as compared with a conventional design. Less surface area implies less friction drag, a substantial part of the aircraft total drag. This aerodynamic benefit can be viewed as a result of the lift-generating central body in the blended wing body design. In addition, without the wing body junctions and the tail, this design should also have much lower interference drag in comparison. The interference drag is the result of the interaction of the wing and the fuselage, which is not accounted for in the wing or body drag as separate components.

Indeed, an increase in the lift-to-drag ratio of about 20% over the conventional design has been estimated for the blended wing body aircraft. However, these benefits can be realized only as an improved aerodynamic performance through careful and detailed aerodynamic shape design. Unlike transonic wing design for conventional aircraft, little is known regarding the best aerodynamic shape for blended wing body aircraft due to a large number of extra design variables and stronger coupling with other disciplines, such as structures and flight dynamics.

On the aerodynamic performance side, the maximum lift-to-drag ratio depends on the ratio of the aircraft span to the square root of the product of the induced drag factor and the zero-lift drag area, which is proportional to the wetted area of the aircraft. From this relation, one can see that larger span, smaller wetted area, lower skin friction (provided, for example, by laminar flow technology), and less induced drag can all potentially provide substantial improvement in aerodynamic performance. As mentioned earlier, a larger span with an increase from the conventional 70–80 m (230–260 ft) to 100 m (330 ft) has been incorporated in some of the blended wing body designs, with obvious aerodynamic advantages, as aerodynamic performance improves with the increase of the aspect ratio. On the other hand, consideration of the current airport capability has motivated other designers to limit the span within an 80-m (260-ft) box.

The blended wing body design provides potential for laminar flow control to reduce skin friction drag. In this respect, it is interesting to note earlier work that advocated the potential benefits of a semi-integrated delta planform with laminar flow control using distributed suction for profile drag reduction for large aircraft design.

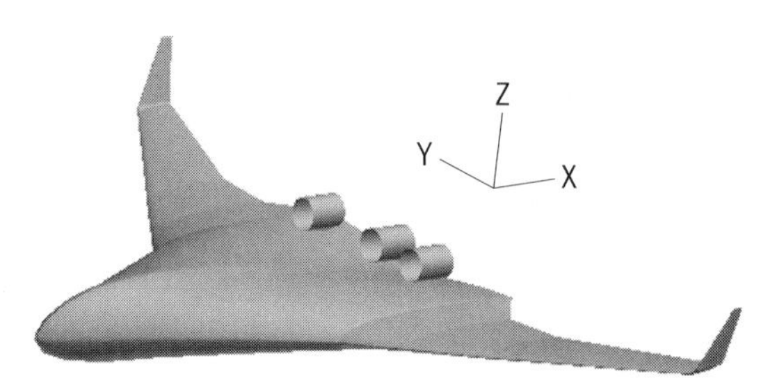

Fig. 2. Blended wing body engine installation concept.

While the primary benefit of the blended wing body design comes from the large aerodynamic gain, there are also obvious structural advantages due to the integration of the wing structure with the thick central body. Peak wing bending moment and shear are estimated to be about half of that for the conventional configuration, implying potential structural weight saving.

Environmental advantages. The aerodynamic and structural advantages provide the benefit of a more environmentally friendly design. Substantial drag reduction implies less thrust from the engines and, in turn, less emission. For example, as mentioned earlier, it is estimated that an 800-passenger blended wing body aircraft requires only three 60,000-lb-class (27,000-kg) engines instead of four for the conventional design.

Most blended wing body designs feature an aft engine location on top of the aircraft as **(Fig. 2)** so that engine noise is naturally shielded by the aircraft. This can substantially reduce the noise level and footprint on the ground at airports during take-off and landing.

Potential problems. Conventional designs use the tail planes to balance the aircraft, which cancels out the moment generated by the main wing. Similar to the earlier flying wing designs, blended wing body aircraft have no tails to balance the aircraft for stability. Therefore, this function has to be fulfilled by the flying wing or the blended wing body itself. One way to achieve stability is through the use of the so-called reflexed camber in the airfoil design, which curves back upward near the trailing edge. Balancing the flying wing or the blended wing body is trickier, and the stability requirement can put a severe constraint on the aerodynamic design. Modern fly-by-wire control systems can contribute to the resolution of the stability problems.

Structurally, the central body has the dual role of the body and the wing as mentioned earlier. This implies the structural need to carry the cabin pressure loads in addition to the bending loads of the wing, which poses a unique design challenge.

Using blended wing bodies for very large transport aircraft may present problems for passenger evacuation in an emergency, and these problems need to be considered at the design stage. Although wing span in some designs has been limited by current airport runways to 80 m (260 ft), airport capacity in accommodating and transporting passengers (potential congestion at check-in, passport control, and arrival of a much larger number of people) in hubs needs to be included in these considerations.

More in-flight entertainment with larger available internal space for passengers should offset the potential disadvantages of the lack of window seats in blended wing body aircraft. However, public acceptance of this revolutionary type of aircraft is a serious issue for commercial success, crucial to both the airlines and the manufacturers, despite its numerous potential technological advantages.

The conventional civil aircraft configuration has matured and nearly reached its performance limit through an evolutionary process over half a century. A revolutionary configuration change is essential to meet increasing demands for direct operational cost reduction and environmental challenges. The blended wing body aircraft offers a significant step toward a more efficient and more environmentally friendly aircraft for future air transportation.

For background information *see* AERODYNAMICS; AIRCRAFT DESIGN; AIRFOIL; AIRFRAME; FLIGHT CHARACTERISTICS; TRANSONIC FLIGHT; WING. N. Qin

Bibliography. A. L. Bolsunovsky et al., Flying wing—problems and decisions, *Aircraft Design*, 4:193–219, 2001; D. Roman, R. Gilmore, and S. Wakayama, *Aerodynamics of High Subsonic Blended-Wing-Body Configurations*, AIAA Pap. 2003–0554, 2003; R. M. Denning, J. E. Allen, and F. W. Armstrong, Future large aircraft design—the delta wing with suction, *Aeronaut. J.*, 101(1005):187–198, 1997; J. E. Green, Greener by design—the technology challenge, *Aeronaut. J.*, 106(1056):57–113, 2002; G. H. Lee, The possibilities of cost reduction with all-wing aircraft, *J. Roy. Aeronaut. Soc.*, 69:744–749, 1965; R. H. Liebeck, Design of the blended wing body subsonic transport, *J. Aircraft*, January–February 2004; A. J. Morris, *MOB: A European Distributed Multidisciplinary Design and Optimisation Project*, AIAA Pap. 2002–5444, 2002; J. K. Northrop, The development of all-wing aircraft, *J. Roy. Aeronaut. Soc.*, 51:481–510, 1947; N. Qin, Aerodynamic considerations for blended wing body aircraft, *Prog. Aerosp. Sci.*, 40(6):321–343, 2004; H. Smith, College of Aeronautics Blended Wing Body Development Programme, *Proceedings of the 22d International Congress of Aeronautical Sciences*, Pap. 1.1.4, 2000.

Ontology (information technology)

From an information technology perspective, ontologies are hierarchical structuring of knowledge about things by subcategorizing them according to their essential or relevant qualities. An ontology can be considered a conceptualization of a domain or subject area typically captured in an abstract model of how people think about things in the domain. Humans have produced ontologies for millennia, from Plato's philosophical framework to modern-day classification systems, for understanding and explaining their rationale and environment.

Ontologies are built very much ad hoc. A terminology is first developed providing a controlled vocabulary for the subject area or domain of interest; then it is organized into a taxonomy where key concepts are identified; and finally these concepts are defined and related to create an ontology.

Ontologies can be built from the ground up by a systematic method based on how we build specialized, faceted classification schemes. The objectives for building ontologies include creating a logical framework, creating a philosophy, creating a classification system, or developing a common understanding in a discipline, to name some. The goal determines the extent and complexity of the process. For example, creating an ontology to provide a basic understanding of a domain might require less effort than creating an ontology to support formal logic arguments and proofs in a domain.

An approach for formalizing the ad-hoc process for building ontologies consists of three steps: ontology capture, which identifies and defines the key concepts and relationships in the domain of interest and the terms that refer to such concepts; ontology coding, which formalizes such definitions and relationships in some formal language; and ontology integration, which associates key concepts and terms in the ontology with concepts and terms of other ontologies—that is, it incorporates concepts and terms from other domains. Although this approach has been used successfully in developing some experimental ontologies, it is strictly top-down requiring extensive domain knowledge and experience in addition to a great effort in consensus building among experts and participants.

Taxonomies and classification. A taxonomy is a structure of categories, and classification is the act of assigning entities to categories within a taxonomy.

A classification scheme in library science is a tool for producing systematic order based on a controlled and structured index vocabulary. This index vocabulary is called the classification schedule; and it consists of a set of names or terms representing concepts, or classes, listed in systematic order to display relationships among them.

A classification scheme and its respective schedule can be considered an extended taxonomy or a reduced ontology. As an extended taxonomy, it goes beyond a mere arrangement of categories since it includes relationships among categories and some brief definitions. As a reduced ontology, it lacks formal definitions of concepts and axioms. Axioms are considered essential components of any formal ontology.

Classification in library science. A classification scheme must be able to express hierarchical relationships as well as relationships that connect two or more concepts belonging to different hierarchies. Hierarchical relationships are based on the principle of subordination or inclusion and are typical in a taxonomy. Relationships among concepts are presented as compounded classes. For example, the compounded class "reproduction of reptiles" relates the term "reproduction" from the class "processes" with the term "reptiles" from the class "taxonomy."

Two types of classification schemes are used in library science: enumerative and faceted. The enumerative (or traditional) method assumes a universe of knowledge divided into successively narrower classes that include all possible subclasses and compounded classes. The Dewey decimal system is a typical example of an enumerative hierarchy, where all possible classes are predefined. These schemes are called enumerative because the predefined classes are listed ready-made in a classification schedule.

The faceted approach was proposed by S. R. Ranganathan in 1939. It works by synthesizing from the subject statements of particular documents. Subject statements are matched against concepts selected from a schedule of elemental classes. The arranged groups of elemental classes making up the scheme are called facets. Facets can be construed as perspectives, viewpoints, or dimensions of a particular domain. Using such a scheme, the classifier expresses a compound class by assembling its elemental classes.

A faceted scheme provides a controlled vocabulary in the form of concepts arranged systematically by facets and a set of rules on how to combine such concepts or terms to define conceptual descriptors, also known as categories.

Deriving faceted classification schemes. Special library collections are typically classified using custom-made classification schemes by a process called literary warrant. The process consists of the following steps: (1) select a random sample of titles from the collection to be classified, (2) list individual terms from the titles, (3) group related terms into common classes, and (4) organize the common classes into a classification scheme. This process requires knowledge of the domain and of the intended use of the collection. The selected concepts and their relationships can be considered as a domain-specific language used to express activities in the domain of the specialized library. Literary warrant assumes that titles capture key concepts in a document, making them representative subject statements.

The following example illustrates this process. Assume we are asked to build a classification for a collection of zoology-related titles (books). The first step is to select a representative sample, such as from the following collection of titles.

"Essays of the physiology of marine fauna"
"Animals of the mountains"
"Amphibious animals"
"Desert reptiles"
"Migratory birds"
"Salt water fish"
"Mammalian reproduction"
"Snakes of the Amazon River"
"Experimental reports on the respiration of vertebrates"
"Tropical leaf moths"

The next step is to group common concepts together.

Faceted classification scheme

By habitat	By element	By taxonomy	By process	By literary form
land	marine	animals/fauna	physiology	essays
tropic	:	invertebrates	respiration	:
desert	:	insects	reproduction	:
mountain	amphibious	moths	:	reports,
:	:	:	:	experimental
:	:	vertebrates	:	:
water		mammals		
sea		:		
river		birds		
Amazon		:		
:		reptiles		
:		snakes		
		:		
		fish		

physiology, reproduction, respiration
tropical, desert, mountains, salt water
marine, amphibious
fauna, animals, vertebrates, reptiles, snakes, birds, fish, moths, mammals
essays, experimental reports

These five groups are the initial facets in our special collection of zoology books. Each group is then named by the general concept it represents.

by process
by habitat
by element
by taxonomy
by literary form

These five facets are ordered by their relevance to the users of the collection, and terms within each group are listed in a logical order. In this example, it is assumed that the users are mainly ecologists, making habitat the most relevant facet. The domain or subject area is animals/fauna. The resulting faceted classification scheme is shown in the **table**.

As new titles enter the collection, new facets may be defined and new concepts added to the scheme, extending and enriching the faceted scheme. This is the core of literary warrant.

This scheme is a structured, controlled vocabulary that can be used systematically to define each title of the collection. Each title can now be reduced to a normal form of terms or concepts from each facet. To describe a title using this scheme, each term in the title is matched to the concept in the scheme by order of relevance.

For example, the title "Essays on the physiology of marine fauna" can be represented (classified) by the following set of concepts selected from the faceted scheme.

/null/marine/animals/physiology/essays/

The first entry is null because there is no concept in the title that corresponds to any concept in the habitat facet. The remaining terms are selected by conceptually matching keywords in the title to facet concepts in the scheme.

Similarly, the titles below are normalized (classified) following the same process.

"Animals of the mountains" →
/mountain/null/animals/null/null/

"Desert reptiles" →
/desert/null/reptiles/null/null/

This is the process of synthetic classification using a faceted classification scheme. The scheme provides a vocabulary and some basic rules for converting titles to a normalized set of concepts. From the bottom-up, we have produced a knowledge structure that can be used to generate normalized descriptors of statements that facilitate their categorization; that is, titles sharing the same facet concept belong to same category. This can also be seen as a primitive domain language.

As another example, function descriptors (instead of book titles) can be used to generate a faceted classification scheme for software components as part of a reuse library system. Function descriptors are, in most cases, "one-liners" defining a software function. In some cases, a function descriptor is the title of the function. Function descriptors are representative subject statements.

Building ontologies. We will describe a method for building the basis for ontologies. The concepts are defined by clusters of phrases and statements extracted from a body of textual experience. The resulting ontologies do not include formal definitions of concepts and axioms.

One advantage of this approach is that it is practical and useful. The ontologies built by this method may not yet be comprehensive or formal enough for some purposes, but provide sufficient information and concepts to facilitate the task of ontology coding and formal documentation.

The proposed approach is based on a combined top-down and bottom-up method borrowed from a domain analysis method (**Fig. 1**). Domain analysis is a process by which information used in developing software systems is identified, captured, and organized for making it reusable. First, a high-level ontology is proposed, and then it is revised and validated based on a bottom-up analysis of existing

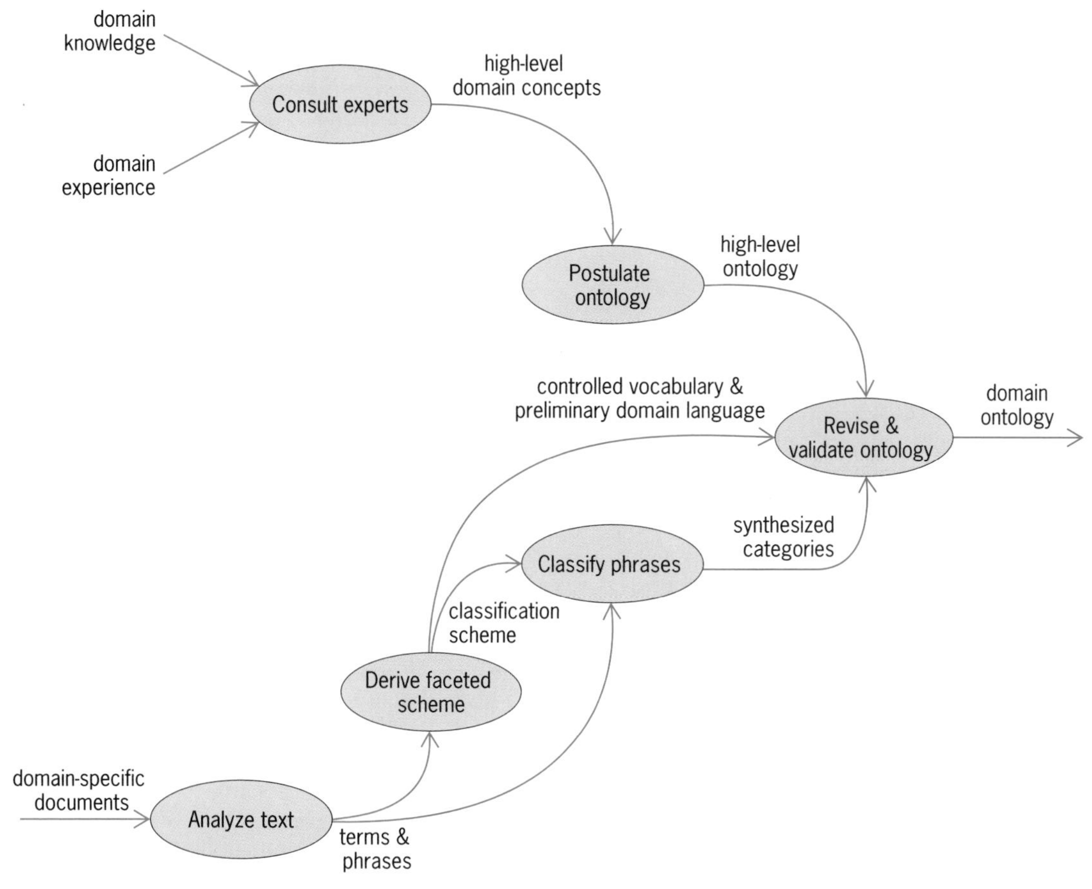

Fig. 1. Process for building ontologies. (*From R. Prieto-Díaz, 2003*)

domain-specific documents. While the top-down method is highly creative, somewhat informal, and manual, the bottom-up part can be made systematic, repeatable, and partially automatic. In the top-down process, experts identify key concepts so that they can propose an initial high-level taxonomy.

In the bottom-up process, key words and phrases are extracted from domain documents using standard text analysis tools. The literary warrant technique is then used to build a domain-specific faceted classification scheme. The resulting scheme is used to group phrases into categories, creating clusters that represent concepts in the domain.

The synthesized clusters are then compared to the proposed concepts, and in iterative process the ontology is modified and adjusted to match the bottom-up clusters. **Figures 2** and **3** illustrate how the method works.

Mapping clusters to proposed concepts, as shown in Fig. 3, may require modifications to the ontology and the concept clusters. Some of the following examples may arise, showing how a proposed ontology is modified and validated by concepts extracted from domain documents.

1. Cluster A maps to concepts s and t. In this case, cluster A can be broken into two separate concepts, one matching s and another matching t, or either s or t can be deleted from the ontology and keep only one link to A from either the parent of t or the parent of s.

2. Clusters B and C map to single-concept u. In this case, clusters B and C can be merged to represent concept u, or concept u can be partitioned into two different concepts.

3. Elements from clusters D and E and cluster F map to concept x. In this case, a new cluster can be

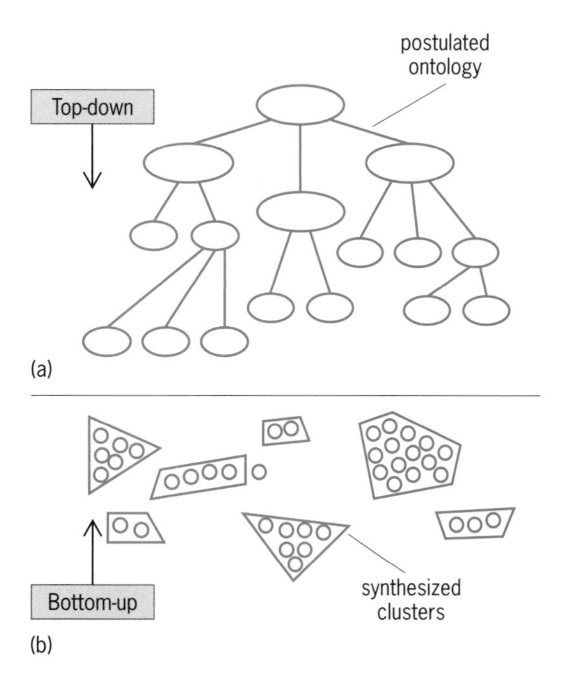

Fig. 2. Elements of the process. (*a*) Top-down. (*b*) Bottom-up. (*From R. Prieto-Díaz, 2003*)

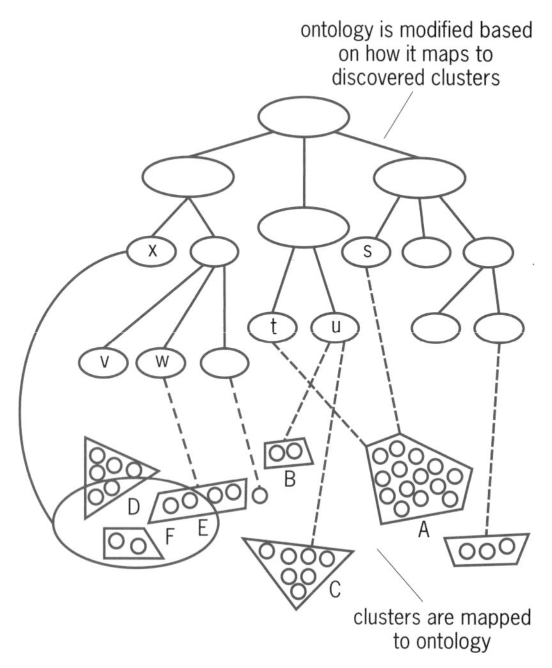

Fig. 3. Revising and validating the initial ontology. (*From R. Prieto-Díaz, 2003*)

created to map to concept x, or concept x can be deleted and cluster F merged into D and E.

For background information *see* INFORMATION MANAGEMENT; INFORMATION TECHNOLOGY; SOFTWARE; SOFTWARE ENGINEERING; TAXONOMY in the McGraw-Hill Encyclopedia of Science & Technology.

Rubén Prieto-Díaz

Bibliography. B. Buchanan, *Theory of Library Classification*, Clive Bingley, London, 1979; W. Frakes, R. Prieto-Díaz, and C. Fox, DARE: Domain analysis and reuse environment, in W. Frakes (ed.), *Ann. Software Eng.*, 5:125–141, Baltzer Science, 1998; R. Prieto-Díaz, Domain analysis: An introduction, *ACM SIGSOFT Software Eng. Notes*, 15(2):47–54, April 1990; R. Prieto-Díaz, A faceted approach to building ontologies, *Proceedings of the 2003 IEEE International Conference on Information Reuse and Integration, Las Vegas*, pp. 458–465, October 27–29, 2003; R. Prieto-Díaz and P. Freeman, Classifying software for reusability, *IEEE Software*, 4(1):6–16, 1987; S. R. Ranganathan, *Prolegomena to Library Classification*, Asian Publishing House, Bombay, 1967; M. Uschold and M. King, *Towards a Methodology for Building Ontologies*, Workshop on Basic Ontological Issues in Knowledge Sharing, IJCAI-95, 1995; V. C. Vickery, *Faceted Classification: A Guide to Construction and Use of Special Schemes*, Aslib, 3 Belgrave Square, London, 1960.

Paleozoic bivalve borings

In the ocean ecosystem, competition for food and living space is often very intense, and some of the adaptations of marine organisms for carving out a spot to live on, or in, the sea floor are remarkable. One very special adaptation is bioerosion, or the biological process of excavating holes in rocks or shells. A wide spectrum of organisms, ranging from bacteria to sponges to barnacles, are able to bioerode quite effectively. Some animals, such as parrot fish, sea urchins, and chitons (polyplacophoran mollusks), bioerode by scraping the surface of rocks or shells in order to eat encrusting algae. Others, such as certain groups of sponges, worms, and bivalves (pelecypods), bioerode by creating holes (borings) in hard substrates, which they occupy permanently.

Bivalves are among the most prominent and widespread bioeroders in modern marine environments. Bioeroding bivalves often produce borings in which they reside permanently. Since the borings continually enlarge as the animals grow, the holes partly mirror the external morphology of the bivalve shells. Typical bivalve borings are flask- or teardrop-shaped holes with a broadly rounded base and a narrow neck leading to a single opening. Such borings are generally known by the genus *Gastrochaenolites*, which is the scientific name of the trace fossil (fossilizable biogenic structure) made by the bivalve rather than the taxonomic name of the bivalve that created the structure.

Bioerosion methods. The ability of an invertebrate animal to penetrate hard substrates is a specialized adaptation, because it involves a significant expenditure of energy. There are only two ways an animal can make a deep hole in a rock or shell: by carving it away via mechanical abrasion or by etching it away via chemical dissolution. Both methods are employed by different groups of bivalves in the oceans today, and the trace fossil record indicates that both mechanisms have been employed by marine bivalves in the distant past.

Mechanical bioerosion. Some bivalve taxa, including certain pholadids (for example, modern *Zirfea* and *Penitella*), excavate their borings mechanically by jostling back and forth, using the coarse ribbing on their shell as a drill bit to bore into a hard substrate. The interior walls of their borings often contain characteristic scratches that indicate mechanical abrasion of the substrate, and in some cases the geometry of these scratches is preserved in such intricate detail as to allow for a precise determination of the genus or species of the bivalve that created the boring.

Chemical bioerosion. Other bivalve taxa, including certain mytilids (for example, modern *Lithophaga* and *Botula*), use corrosive chemical secretions to dissolve rock or shell material. The interior walls of their borings usually have a smooth lining, so identification of the particular type of borer is not so straightforward. The exact chemical nature and composition of the secretions produced by chemically boring bivalves are not well understood for modern borers, let alone for those in the fossil record. In modern borers, it seems clear that the solvents are not simple inorganic acids, but organic secretions of the mantle and/or siphonal tissues that dissolve the calcareous substrate. Dissolution rates are slow—on the order of only a few millimeters per year. The exterior of the bivalve shell typically is covered with a proteinaceous periostracum, which protects the calcareous shell from being dissolved. However, there is no clearcut evidence to indicate the chemical composition of the bioeroding secretions produced by fossil borers, nor is there any direct evidence that a periostracum may have been present on their shells.

Origin of bivalve borings. Because of the special anatomical and behavioral attributes required for an animal to bore holes into rocks and shells, the origin of the lifestyle of borers, especially that of bioeroding bivalves, is an intriguing paleontological mystery. In fact, the geologic record contains evidence that animals have been creating such bioerosion trace fossils since almost the beginning of the Paleozoic Era, 544 million years ago (mya). However, the nature of these ancient borings (mechanical versus chemical) and the identity of their creators are not easy to discern.

Lower Cambrian borings. The oldest reported borings are finger-shaped holes, called *Trypanites*, that were excavated in archeocyathid skeletons (an extinct group of calcareous spongelike reef formers) of Lower (Early) Cambrian age (543 mya) in Labrador, Canada. These ancient trace fossils do not resemble the kinds of borings that are normally attributed to bivalves; furthermore, there are no fossils of bivalve shells preserved in Lower Cambrian rocks to indicate that bivalves may have been the borers. Thus, it may be concluded with confidence that bivalve borers were not around in the Early Cambrian seas.

Upper Ordovician. The oldest borings that can be attributed directly to a particular bivalve producer are the trace fossils, *Petroxestes pera*, of upper (Late) Ordovician age (460 million years old) in Ohio (**Fig. 1**). The creators of these borings apparently belong to the bivalve species *Corralliodomus scobina*, because fossilized shells of these animals actually were found inside the borings, which had been excavated in the skeletons of stromatoporoids (an extinct group of densely laminated calcareous sponges). *Petroxestes* borings look quite different from most bivalve borings today, as they are shallow, rounded, elongate slots rather than deep, flask-shaped holes.

Lower Ordovician. Recent discoveries in marine limestone beds of equivalent age at widely separated sites in western North America (Utah) and northern Europe (Sweden and Norway) have revealed that the oldest borings with a flask- or teardrop-shaped morphology, closely resembling those made by boring bivalves today (such as *Gastrochaenolites*), are in fact of Lower (Early) Ordovician age (505 million years old) [**Fig. 2**]. Unlike the younger Ordovician occurrences in Ohio of *Petroxestes* borings, the older Ordovician *Gastrochaenolites* borings in Utah and Scandinavia do not contain fossilized shells of bivalves (or anything else) that can be interpreted as the producers of these trace fossils. Nevertheless, these Lower Ordovician trace fossils are intriguing, because they obviously represent a major innovation in the evolution of organisms that were capable of creating borings as permanent dwelling structures in shallow marine habitats.

Until these recent reports of Lower Ordovician *Gastrochaenolites* borings came to light, the oldest *Gastrochaenolites* that have been attributed to bioeroding bivalves were described from Lower Pennsylvanian (325-million-year-old) strata in Arkansas. In fact, until now, the Arkansas specimens were the only *Gastrochaenolites* trace fossils known from the Paleozoic Era. The next oldest *Gastrochaenolites* come from rocks of Jurassic age (213–145 mya), when bivalve borings suddenly became abundant worldwide, occurring especially inside the skeletons of colonial corals in large-scale reef systems.

Although the Lower Ordovician *Gastrochaenolites* borings from Utah and Scandinavia look very

Fig. 1. Plain view of top of a bed of Paleozoic borings (*Petroxestes pera*) made by a bioeroding bivalve (*Corallidomus scobina*) in the Kope Formation (Upper Ordovician), southern Ohio. (*Photo courtesy of Leif Tapanila*)

Fig. 2. Vertical cross-sectional view of a Paleozoic boring (*Gastrochaenolites oelandicus*) made by an unknown bioeroding invertebrate in the Bruddesta Formation (Lower Ordovician), Öland, Sweden. Arrows point to margins of the boring. (*Photo courtesy of A. A. Ekdale*)

similar to modern *Gastrochaenolites* that are obviously created by various species of bivalves, there are several lines of evidence to suggest that the Ordovician specimens were actually produced by something else, probably a soft-bodied chemical borer that is not known from the body fossil record, because its lack of anatomical hard parts rendered it essentially unpreservable. Most importantly, there are no bivalve shells directly associated with the Ordovician *Gastrochaenolites*; and the bivalve taxa that are known with certainty to be the producers of *Gastrochaenolites* of Mesozoic (248–65 mya) and Cenozoic (65 mya—present) age, such as *Lithophaga* and *Gastrochaena*, are not known in the body fossil record until the Late Paleozoic (260 mya) at the earliest. And even these taxa are known only from poorly preserved specimens that are subject to question.

Evolution of macroborers. One main reason that the recognition and ecologic interpretation of borings in the Ordovician Period is of great interest to paleontologists is that the development of a lifestyle requiring an animal to expend energy to dig a deep hole into a hard substrate for the purpose of dwelling represents a major evolutionary innovation for making use of a potential habitat that previously had been unexploited by marine animals. That this development occurred during the Ordovician is particularly noteworthy, because this period of geologic time was marked by one of the most significant intervals of biological diversification on Earth in the past half billion years. This is when the bivalves and many other marine taxa began to flourish and diversify.

The boring niche apparently was well established by Middle and Upper Ordovician time, when diverse borings began to be widespread in both rocky sea bottoms and skeletal substrates, such as in the calcareous skeletons of stromatoporoids and colonial corals. The new discoveries of Lower Ordovician *Gastrochaenolites* in Utah and Scandinavia suggest that the evolution of macroborers actually preceded the beginnings of the major phase of global biological diversification, which began to increase in the Middle Ordovician.

It even has been suggested that the advent of macroborers as prominent members of marine communities may have spurred coastal erosion, because the excavating activities of the boring animals would have increased the input of gravel-size debris (bioeroded cobbles and pebbles) from submarine outcrops into the sediment on the sea floor. The potential sedimentologic implications of intensified bioerosion throughout the Ordovician Period are not yet well understood, but they offer some very interesting possibilities for further research to link the living communities of animals on and in the sea floor with their physical environment.

Conclusion. The appearance of borings in Early Paleozoic marine ecosystems was a major evolutionary and ecological development. The fact that bivalves, which are among the most prominent macrobrers in the sea today, began to diversify taxonomically at approximately the same time as borings became established causes one to wonder if bivalves may have included some species that were adapted for a bioeroding lifestyle from the very beginning. The oldest trace fossils that can be attributed to boring bivalves with a reasonable degree of confidence are Upper Ordovician in age. The oldest trace fossils that bear a striking resemblance to modern bivalve borings are Lower Ordovician in age, but there is no tangible evidence to prove a bivalve origin of those borings. Thus, the search continues for direct body fossil evidence of the identity of the producers of the remarkable flask-shaped borings in Lower Ordovician (and older) hard substrates.

For background information *see* BIVALVIA; BORING BIVALVES; FOSSIL; ORDIVICIAN; PALEOECOLOGY; PALEOZOIC; TRACE FOSSILS in the McGraw-Hill Encyclopedia of Science & Technology. A. A. Ekdale

Bibliography. J. S. Benner, A. A. Ekdale, and J. M. de Gibert, Macroborings (*Gastrochaenolites*) in Lower Ordovician hardgrounds of Utah: Sedimentologic, paleoecologic, and evolutionary implications, *Palaios*, 19:543–550, 2004; A. A. Ekdale et al., Bioerosion of Lower Ordovician hardgrounds in southern Scandinavia and western North America, *Acta Geologica Hispanica*, 37:9–13, 2002; A. A. Ekdale and R. G. Bromley, Bioerosional innovation for living in carbonate hardgrounds in the Early Ordovician of Sweden, *Lethaia*, 34:1–12, 2001; L. Tapanila, P. Copper, and E. Edinger, Environmental and substrate control on Paleozoic bioerosion in corals and stromatoporoids, Anticosti Island, eastern Canada, *Palaios*, 19:292–306, 2004.

Permian-Triassic mass extinction

Five mass extinctions, known as the "Big 5," have caused abrupt, drastic decreases in animal biodiversity over the last 540 million years. Each resulted from global-scale environmental perturbations and resulted in the extinction of a significant proportion of the world's biota in a geologically negligible period of time. Numerous studies have focused on the taxonomic changes and patterns that are the hallmark of these mass extinctions. However, mounting research indicates that the most significant results of mass extinctions actually may have been new ecological patterns that arose during their aftermath, rather than the extinctions themselves.

Prolonged biotic crisis. At the end of the Permian Period, approximately 250 million years ago, the Earth experienced the most devastating biotic crisis in the Phanerozoic history of life: over 50% of marine vertebrate and invertebrate families went extinct. The cause(s) of this mass extinction is still hotly debated, but evidence increasingly indicates that it is ultimately rooted in the supercontinent configuration of Pangea and its disassembly by extensive continental flood basalt volcanism. Numerous lines of evidence indicate that these factors led to a cascade of global detrimental environmental consequences, including increased atmospheric CO_2 levels, global warming, a reduced pole-to-equator temperature gradient, and sluggish ocean circulation that facilitated massive release of CH_4 (methane), marine anoxic (no oxygen) and dysoxic (low oxygen) conditions, hypercapnia (CO_2 poisoning), and H_2S (hydrogen sulfide) poisoning. These chemical and physiological stresses pulsed throughout the Early Triassic for 5–6 million years after the end-Permian mass extinction.

Unusual Early Triassic gastropod fauna. Gastropods are a group of marine organisms that survived and thrived during the aftermath of the end-Permian mass extinction despite prolonged environmentally harsh conditions around the world. Their abundance during this time is reflected by the prevalence of Early Triassic rocks comprised solely their shells and gastropod grainstones and packstones (**Fig. 1**). Gastropod-dominated grainstones and packstones are common in strata deposited during the Early Triassic in the present-day western United States, northern Italy, and Japan, indicating that gastropods were common in sea-floor communities around the world during this time (**Fig. 2**).

Gastropod-dominated faunas are common throughout the geologic record and in modern settings, but several characteristics of the Early Triassic gastropod fauna indicate that it is actually unique in the history of life.

Small size. Larger gastropods were very rare in the Early Triassic; the majority were actually "microgastropods," 1 cm or less in height. Only one Lower Triassic section contains larger gastropods or even fragments of larger gastropods (those greater than 1 cm) before the late Early Triassic; even adult gastropods are smaller than 1 cm in height (Fig. 2).

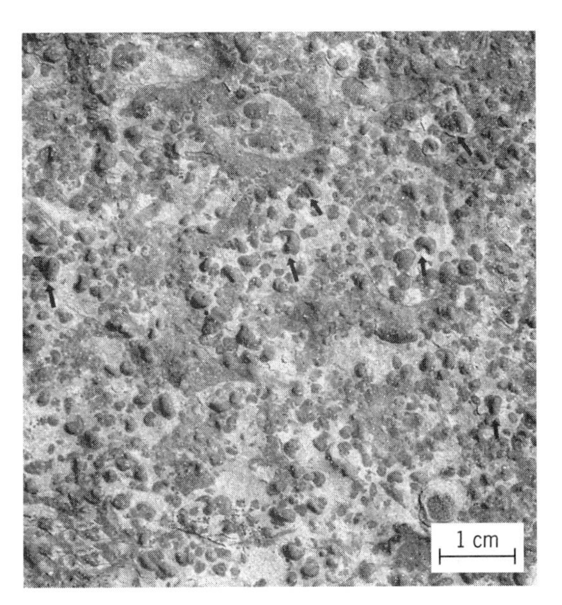

Fig. 1. Sample of Lower Triassic microgastropod grainstone from the Gastropod Oolite Member, Werfen Formation, Punta Rolle, Italy. Arrows point to individual microgastropods.

The Early Triassic microgastropod phenomenon contrasts starkly with preextinction (Late Permian) and post-aftermath (Middle Triassic) gastropod faunas. For example, in Upper Permian and Middle Triassic strata of the western United States, only 28% and 27% of gastropod species, respectively, are microgastropod species, whereas 90% of Early Triassic gastropod species are microgastropod species. In the modern eastern Pacific Ocean off North America, 46% of gastropods are microgastropod species and 54% are larger gastropod species.

Environment. The depositional environments inhabited by Early Triassic microgastropod-dominated faunas were also unusual. Early Triassic microgastropods flourished in open-marine, subtidal environments around the world (Fig. 2). However, today and generally throughout geologic history, microgastropod-dominated marine faunas have been

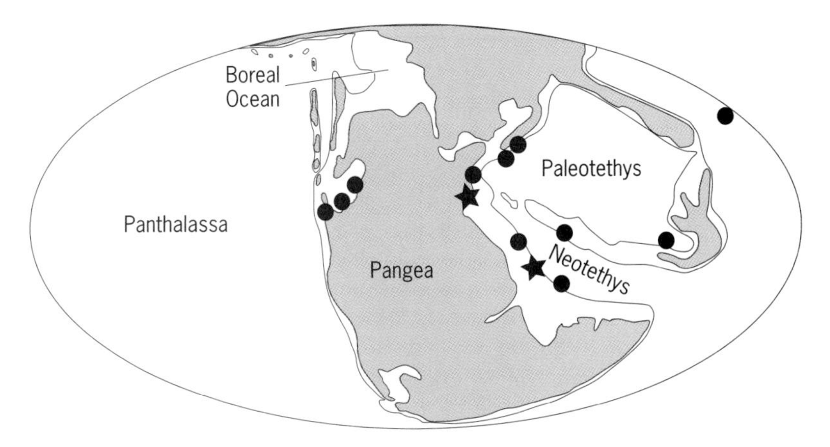

Fig. 2. Paleogeography during the Late Permian-Early Triassic. Circles indicate the approximate locations of microgastropod-dominated faunas during the Early Triassic; stars indicate the approximate locations where larger gastropods (greater than 1 cm) were common.

restricted to locally stressful environments, such as hypersaline (for example, salt marshes, mud flats) and low-oxygen (for example, deep ocean basins) environments. Except for the Early Triassic, global, long-ranging microgastropod-dominated fauna are absent in the geologic record.

Opportunistic Early Triassic microgastropods. Early Triassic microgastropods are considered to be ecological opportunists because they are extremely abundant in Lower Triassic strata around the world in environments in which they do not typically occur under normal conditions. During and after chemically and/or physiologically harsh environmental conditions, opportunistic taxa are the pioneering organisms that rapidly colonize the affected areas and reproduce to prolific numbers in the vacated ecospace. (For example, opportunistic weeds and mice proliferate after localized events such as forest fires.) Opportunistic behavior is due to certain biological attributes of the organisms, including broad environmental tolerance, generalized feeding habits, rapid development, small body size, and high reproduction rates. Opportunistic populations are characterized by low diversity and a high number of individuals.

The prolonged global physiochemically harsh environmental conditions during the Late Permian–Early Triassic facilitated ecological opportunism. Conditions were ideal for surviving generalist gastropods to proliferate and inhabit space that was vacated when other organisms became extinct, including habitats in which they were uncommon or absent throughout the rest of the Phanerozoic. Throughout geologic history, microgastropod-dominated faunas typically characterize harsh environments from which most groups were excluded. The dominance of microgastropods in subtidal marine environments during the aftermath of the end-Permian mass extinction indicates that typical open-shelf faunas could not tolerate the conditions of the world's oceans.

Early Triassic microgastropods behaved opportunistically because they were able to cope with the physiochemically harsh perturbations to the global environment, such as marine anoxia (low O_2) and increased levels of atmospheric CO_2, during the aftermath of the end-Permian mass extinction. Atmospheric CO_2 passively diffuses into the ocean and joins with water to make carbonic acid (H_2CO_3), which breaks down to make bicarbonate (HCO_3^-) and hydrogen ions (H^+). Bicarbonate breaks down to make carbonate (CO_3^{2-}) and more hydrogen ions (H^+). As levels of atmospheric CO_2 increase, more carbonate ions join back to the hydrogen ions to make more bicarbonate, leading to a decrease in the carbonate (CO_3^{2-}) concentration in the surface ocean over time. If there are low levels of carbonate in the ocean, the $CaCO_3$ saturation state will decrease (since the $CaCo_3$ saturation state is dependent on the amount of carbonate in the ocean), and it will become difficult for organisms to secrete a shell. However, early Triassic microgastropods possessed a unique ability to secrete a shell in conditions of low $CaCO_3$ saturation; their small size was likely an adaptation to the increased energetic costs for carbonate secretion.

Other ecological opportunists. Conditions during the Early Triassic facilitated opportunistic behavior by several other types of organisms. For example, the small inarticulated brachiopod *Lingula* is a type of opportunist known as a "disaster taxon," an organism that is geologically long-ranging but becomes unusually abundant during the aftermath of mass extinctions. Other groups of small marine organisms (such as the inarticulate brachiopod *Lingula*, the bivalve *Claraia*, and annelid worms) are found in Lower Triassic strata, too.

Studying the aftermath of mass extinctions. Study of the aftermath of mass extinctions has revealed that global environmental perturbations can affect the Earth's biota for millions of years. Mass extinctions represent just one effect of broad-scale environmentally harsh conditions; their aftermath can be prolonged and characterized by nonactualistic (not like today) ecology, such as global opportunism. An understanding of how the Earth's biota reacts to environmental perturbations will enable the present biodiversity crisis to be better managed. The geologic record is a natural laboratory with clues for the future of life on Earth that should not be ignored.

For background information *see* EXTINCTION (BIOLOGY); GASTROPODA; PALEOECOLOGY; PERMIAN; TRIASSIC in the McGraw-Hill Encyclopedia of Science & Technology. Margaret L. Fraiser

Bibliography. M. J. Benton, *When Life Nearly Died: The Greatest Mass Extinction of All Time*, Thames & Hudson, London, 2003; A. Hallam and P. B. Wignall, *Mass Extinctions and Their Aftermath*, Oxford University Press, New York, 1997; P. D. Taylor (ed.), *Extinctions in the History of Life*, Cambridge University Press, New York, 2004; P. B. Wignall, Large igneous provinces and mass extinctions, *Earth-Sci. Rev.*, 53:1–33, 2001; E. O. Wilson, *The Future of Life*, Vintage Books, New York, 2002.

Pharmaceutical residues in the environment

Because of advancements in the analytical methods used for detecting chemicals in the environment, trace levels of prescription and over-the-counter drugs have been found in waterways. The frequent detection of many pharmaceutical compounds in the environment has been an increasing concern because of their potential to cause undesirable ecological effects, which may range from endocrine disruption in fish and wildlife to antibiotic resistance in pathogenic bacteria. Residues of human and veterinary pharmaceuticals (such as antibiotics, estrogens, and active ingredients of drugs) are introduced into the environment via a number of pathways, but primarily from discharges of wastewater treatment plants or land application of sewage sludge and animal manure. Most active ingredients of

Pharmaceuticals detected in 10 wastewater treatment plants in the United States

Pharmaceutical compound	Maximum concentration, μg/L	Frequency of detection, %	Primary use
Acetaminophen	1.780	50.0	Antipyretic
Albuterol	0.034	32.5	Antiasthmatic
Codeine	0.730	72.5	Analgesic
Diltiazem	0.146	67.5	Antihypertensive
Diphenhydramine	0.387	55.0	Antihistamine
Erythromycin–H_2O	0.610	52.5	Antibiotic
Fluoxetine	0.021	2.5	Antidepressant
Metformin	0.698	17.5	Antidiabetic
Ranitidine	0.295	27.5	Antacid
Sulfamethoxazole	0.763	72.5	Antibiotic
Thiabendazole	0.515	7.5	Antifungal agent
Trimethoprim	0.414	60.0	Antibiotic

pharmaceuticals are transformed only partially in the body and thus are excreted as a mixture of metabolites and bioactive forms into sewage systems.

Treatment plant effluents. Several monitoring studies have shown the presence of a wide range of pharmaceuticals, metabolites, and their conjugates in environmental samples, including surface water, ground water, and drinking water. These findings indicate that while the existing treatment technologies producing water satisfy current regulatory standards, most wastewater treatment plants are not designed to completely remove micropollutants such as pharmaceuticals. A list of the most frequently detected pharmaceuticals in surface waters receiving effluents from wastewater treatment plants in the United States is presented in the **table**. Many other pharmaceuticals not included in this list have also been detected in surface and ground waters from Asia, Europe, and other parts of North America at various concentrations and frequency due to differences in use patterns, treatment regulations, and analytical methodologies.

There has been an increased effort in improving the pharmaceutical removal efficiency of wastewater treatment plants because of the need to provide sustainable water supplies and meet the escalating water consumption associated with population growth, industrialization, and increased agriculture. Pharmaceutical residues have been detected in the effluents of wastewater treatment plants even in places using advanced treatment technologies such as those found in the United States, Canada, and Europe. Conventionally, most wastewater plants use primary and secondary treatment process. In some wastewater facilities, the effluent is subjected to an advanced treatment (such as precipitation/coagulation) and disinfection process (such as chlorination or ultraviolet radiation) to reduce bacteria and odors. While the design of the primary treatment (for example, clarification) stage is common to many wastewater treatment plants, the secondary treatment can differ significantly between plants. The most common secondary treatment used in the United States is the activated-sludge process, in which the operating conditions, such as solids retention time and hydraulic retention time, can have a significant effect on the biodegradation and adsorption of contaminants.

To illustrate the effect of plant design and operation on the removal of pharmaceuticals, consider the concentrations of four commonly detected antibiotics in the environment [ciprofloxacin (CIP), sulfamethoxazole (SMX), tetracycline (TC), and trimethoprim (TRI)] monitored at various stages of three wastewater treatment plants. In Plant 1, primary treatment is followed by a two-stage secondary biological degradation process (**Fig. 1**). The first stage is a conventional activated-sludge process with a hydraulic retention time of 1.05 h for reduction of biological oxygen demand, and the second stage is a nitrification process with a hydraulic retention time of 2.1 h. Figure 1 shows that the second stage is more effective in removing antibiotics than the first stage. In Plant 2, primary treatment is not included, but the plant uses extended aeration during the biodegradation process, which holds the sludge in an aeration tank for 28–31 h, in contrast to the typical aeration time of 1–4 h (**Fig. 2**). In addition, ferrous chloride is added directly to the aeration tank, rather than as a separate metal-salt precipitation/coagulation. The Plant 2 design appears to be more effective in the overall removal of the four antibiotics relative to Plant 1. In Plant 3, a rotating biological contactor is used, containing a set of disks with a film of microorganisms affixed to them (**Fig. 3**). The contactor is submerged halfway into the effluent and is rotated slowly, allowing the microorganisms on the disks to alternate between the effluent (where the microorganisms adsorb or degrade organic matter) and the air (where microorganisms can consume oxygen). This design provides antibiotic removal efficiencies similar to that of the second-stage nitrification tank used in Plant 1. In all three plants, antibiotics were removed only partially, resulting in concentrations generally below 1 μg/L in the final effluents, except for the higher trimethoprim level in Plant 1.

Agricultural waste. It is estimated that more than 50 million pounds (22.5 million kilograms) of antibiotics are produced each year in the United States, with approximately 40% used in animal production to treat diseases (therapeutic use) and promote

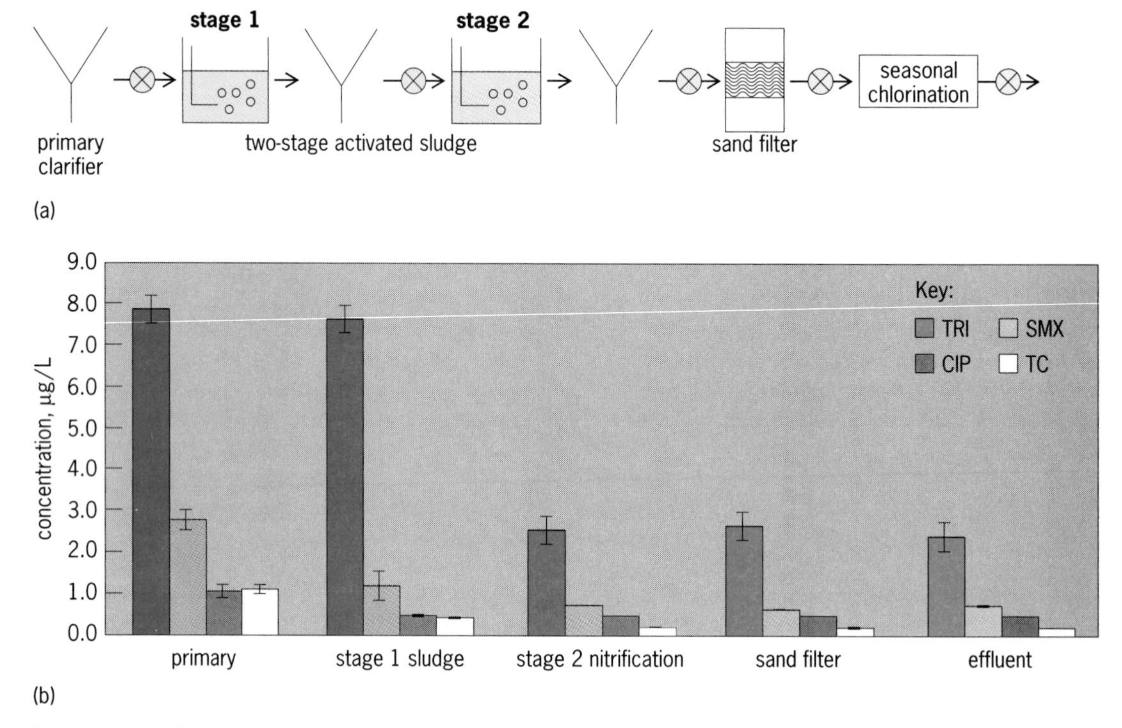

Fig. 1. Plant 1. (*a*) Two-stage activated-sludge process (sampling points indicated by circled X). (*b*) Average concentration of each antibiotic detected throughout the treatment process. (Chlorination is only during May through October.)

growth and weight gain. A significant portion of the antibiotics administered to animals is excreted, which could contaminate soil and water through cropland application of animal wastes. As with human pharmaceuticals, available data on the environmental fate, behavior, and ecological effects of veterinary antibiotics are lacking such that complete risk assessment is difficult.

Although some antibiotics used in animal production are decomposed quickly by soil microbes, others may persist in the environment. For instance, manure-borne oxytetracycline, used as

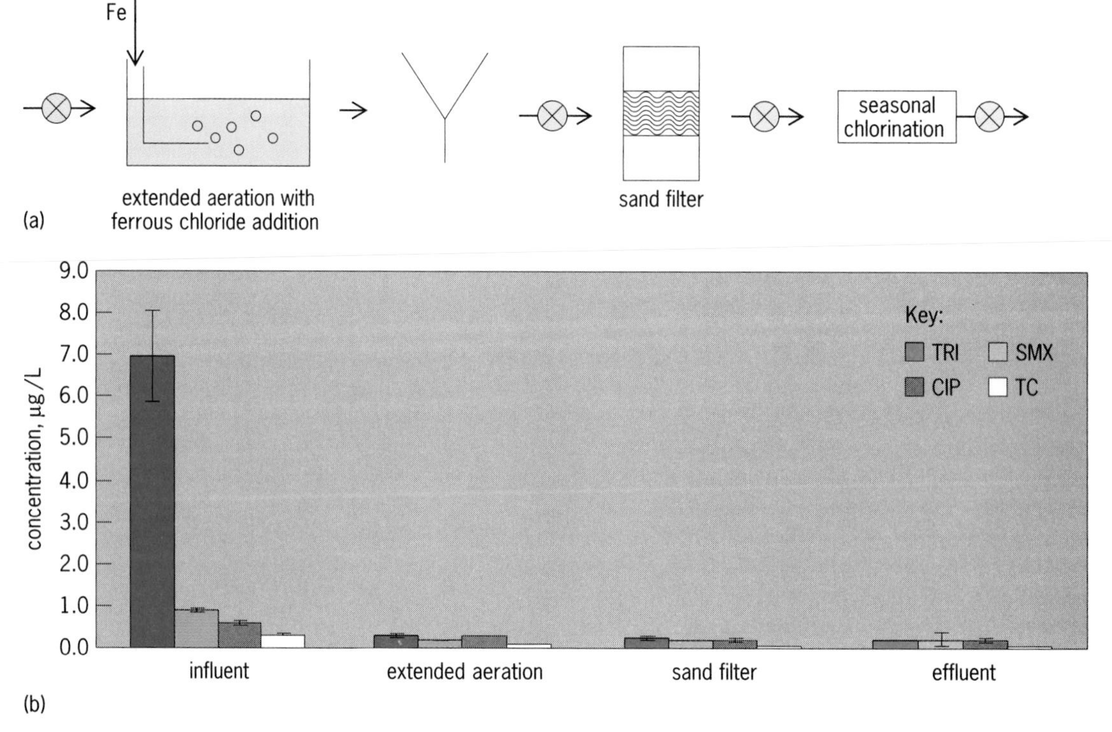

Fig. 2. Plant 2. (*a*) Extended aeration (sampling points indicated by circled X). (*b*) Average concentrations of each antibiotic detected throughout the treatment process. (Chlorination is only during May through October.)

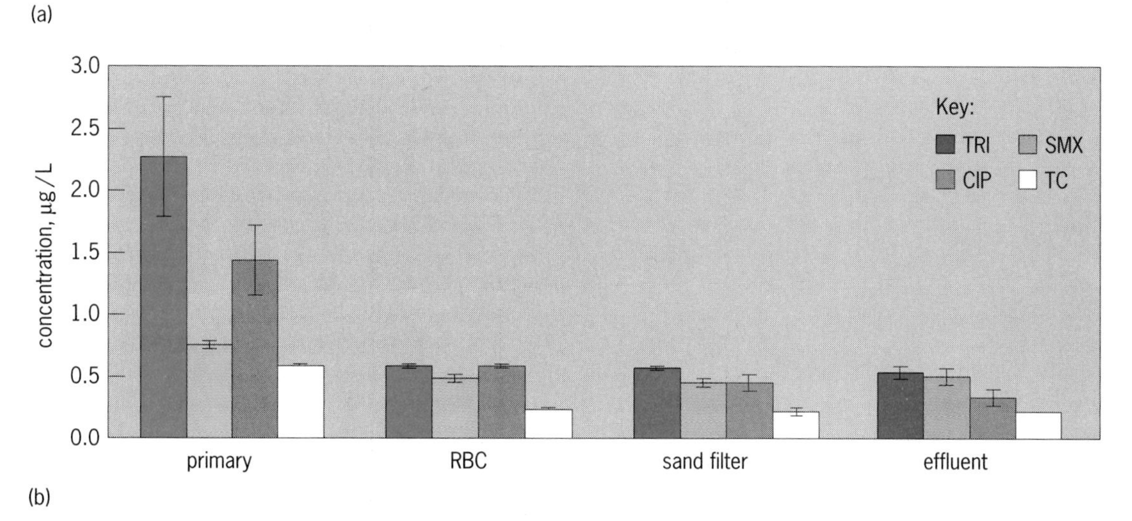

Fig. 3. Plant 3. (*a*) Rotating biological contractors (sampling points indicated by circled X). (*b*) Average concentrations of each antibiotic detected throughout the treatment process. (UV treatment is only during May through October.)

growth-promoting agent in cattle and swine production, can be transformed into derivatives that tend to persist in soil. It is possible that there are many unidentified transformation products of animal antibiotics in soil that are modified only slightly and may still have antibacterial activity that could change the structure of natural soil microbial flora. Thus, it is important to account for the presence of antibiotic analogues and transformation products in soil during risk assessment. It is difficult to completely assess the effect of veterinary pharmaceuticals on the environment because of the lack of available data on their fate, behavior, and ecological effects at environmentally relevant concentrations. This lack of information is largely due to analytical difficulties encountered in detecting trace levels of pharmaceuticals and in identifying unknown degradates in complex matrices such as soil and animal manure.

Analytical and environmental challenges. The recent availability of liquid chromatography coupled to either an ion-trap mass spectrometer (IT-MS) or

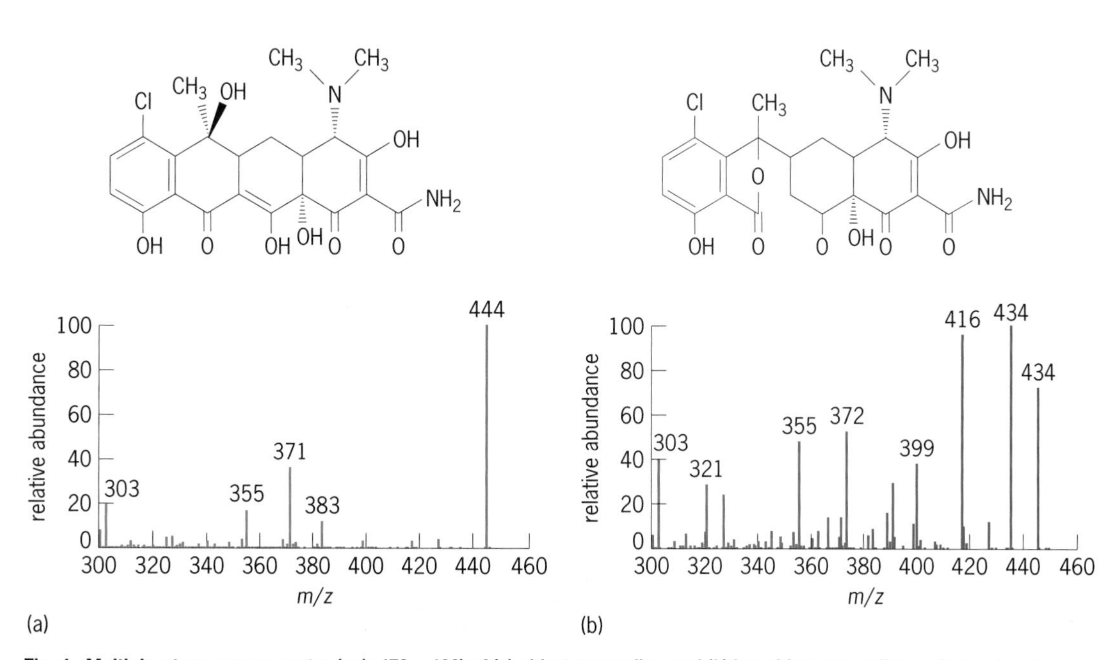

Fig. 4. Multiple-stage mass spectra (*m/z* 479→462) of (*a*) chlortetracycline and (*b*) iso-chlortetracycline, using an ion-trap mass spectrometer.

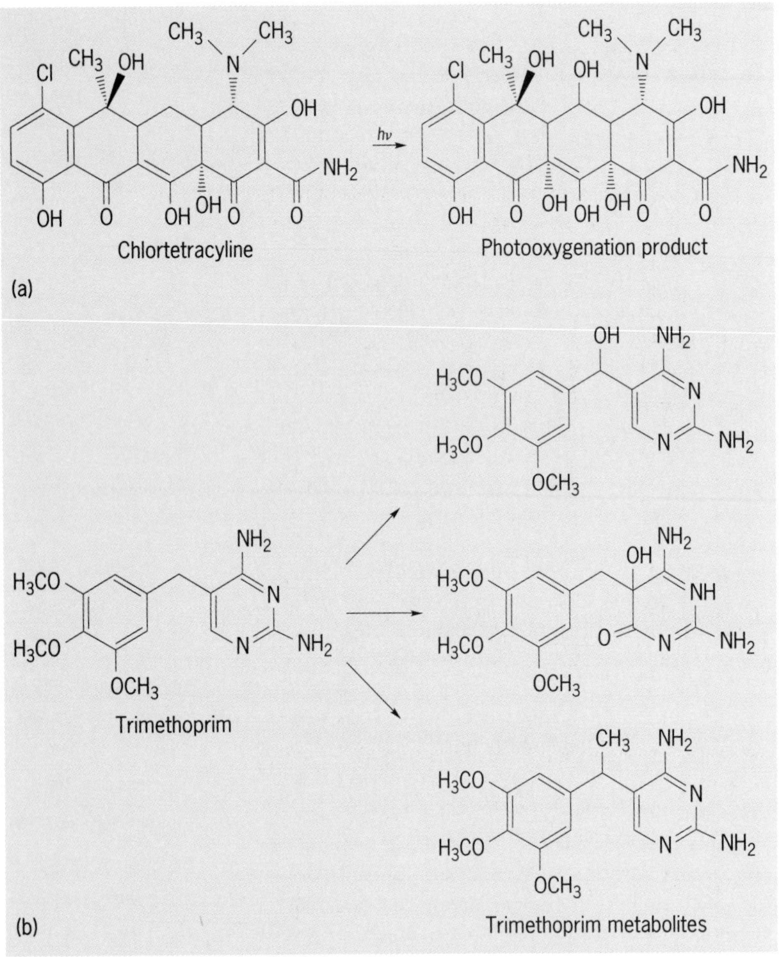

Fig. 5. Reactions showing (*a*) chlortetracycline and its photodegradation product in hog manure, and (*b*) chemical structures of trimethoprim and its metabolites in activated sludge.

time-of-flight mass spectrometer (ToF-MS) has resulted in more degradates and metabolites being identified. These two mass-spectrometric techniques provide complementary data that facilitates structural identification of compounds. For instance, multiple stages of fragmentation in an IT-MS can generate spectra with large amounts of structural information that allow the identification of an unknown degradate. Without multiple stages of mass spectrometry, isomers may not be distinguishable from each other. For example, the protonated molecular ions [mass/charge ratio (m/z) 479] of chlortetracycline and its isomeric conversion product, iso-chlortetracycline, both produce only one fragment ion with m/z 462 corresponding to the neutral loss of NH_3. Isolation and further fragmentation of this ion results in distinct spectra providing valuable structure-specific information (**Fig. 4**).

Further confirmation of the assigned identity or chemical formula of new degradation products can be achieved by accurate mass measurements using ToF-MS. Current benchtop ToF-MS instruments can now achieve low-femtomole-level sensitivity, with high resolving power and mass accuracy. A combination of these mass-spectrometric approaches has been successfully used in the identification of a photooxygenation product of tetracycline in manure and trimethoprim metabolites in activated sludge, with structures shown in **Fig. 5**. Other researchers have also used IT-MS and ToF-MS for effective screening and identification of unknown pharmaceutical contaminants in water.

Recent studies have indicated that some degradation products and metabolites of recalcitrant pharmaceutical compounds are not very different from their parent compounds. For example, the photooxygenation product of chlortetracycline formed at environmentally relevant conditions has a structure that is only slightly modified compared to its parent compound (Fig. 5*a*). Likewise, the metabolites of the antibiotic trimethoprim in activated sludge are also very similar to the parent trimethoprim (Fig. 5*b*), indicating that the absence of the active ingredient of the pharmaceutical from the waste-treatment effluent does not necessarily mean that the compound has been completely eliminated. Knowledge of the nature and quantities of the by-products of degradation or treatment is important because they may also have long-term ecotoxicological effects.

For background information *see* ANTIBIOTIC; BIODEGRADATION; ENVIRONMENTAL TOXICOLOGY; HAZARDOUS WASTE; INDUSTRIAL WASTEWATER TREATMENT; LIQUID CHROMATOGRAPHY; MASS SPECTROMETRY; SEWAGE; SEWAGE SOLIDS; SEWAGE TREATMENT; WATER POLLUTION; WATER TREATMENT in the McGraw-Hill Encyclopedia of Science & Technology.

Diana S. Aga

Bibliography. A. B. A. Boxall et al., Are veterinary medicines causing environmental risks?, *Environ. Sci. Technol.*, 37(15):286A–294A, 2003; P. Eichhorn et al., Application of ion trap-MS with H/D exchange and QqTOF-MS in the identification of microbial degradates of trimethoprim in nitrifying activated sludge, *Anal. Chem.*, 77(13):4176–4184, 2005; P. Eichhorn and D. S. Aga, Identification of a photooxygenation product of chlortetracycline in hog lagoons using LC/ESI-ion trap-MS and LC/ESI-time-of-flight-MS, *Anal. Chem.*, 76(20):6002–6011, 2004; I. Ferrer and E. M. Thurman, Liquid chromatography/time-of-flight/mass spectrometry (LC/TOF/MS) for the analysis of emerging contaminants, *TrAC: Trends in Analytical Chemistry*, 22(10):750–756, 2003; S. T. Glassmeyer et al., Transport of chemical and microbial compounds from known wastewater discharges: Potential for use as indicators of human fecal contamination, *Environ. Sci. Technol.*, 39(14):5157–5169, 2005.

Phononic crystals

Phononic crystals are periodic composite materials made from constituents with different densities and acoustic wave velocities. These synthetic materials, which are analogous to photonic crystals for electromagnetic waves, are of growing interest because they can change the way in which sound or ultrasound travels through matter, leading to a number of novel applications. Phononic crystals may be constructed by arranging identical objects [for example, rods in two dimensions (2D) or spheres in three dimensions (3D)] in a regular periodic array or crystal lattice, and embedding these elementary units in a host material or matrix. The key feature of phononic crystals is their periodicity, which causes the propagation of acoustic or elastic waves to be dramatically modified when length scale of the periodicity is comparable with the wavelength of sound (or ultrasound). The origin of these effects is the interference of waves scattered from the periodically arranged constituents. The overall size of phononic crystals may vary from several meters down to several micrometers, depending on the size of the elementary units (such as the diameter of a sphere or a rod), the lattice constants (the shortest distances over which the structure repeats), and the number of layers. The condition that these characteristic lengths be comparable to the wavelength dictates the operational frequency range of the phononic crystal, which can thus be tuned from hundreds of hertz to gigahertz.

Since about 1990, there has been a growing number of theoretical and experimental studies of phononic crystals, illustrating the wide range of different phononic materials that can be constructed and allowing their remarkable properties to be investigated. Phononic crystals can be used to block sound propagation over certain frequency ranges due to the formation of band gaps, and can cause sound to bend in unusual ways (negative refraction) at other frequencies, leading to a new way of focusing sound using a material with flat surfaces.

Band gaps. One of the main features of phononic crystals that distinguishes them from uniform materials is the existence of the frequency ranges, known as band gaps, in which acoustic waves cannot propagate. The basic physics explaining the existence of these acoustic band gaps is similar to the band theory of solids, which explains the formation of energy bands and band gaps for electrons in metals, semiconductors, and insulators. In atomic crystals, the Bragg scattering of the electron wavefunctions (brought about by the periodic arrangement of the atoms) makes them interfere destructively at certain energies, effectively creating ranges of energy (band gaps) for which no electron states exist. The same Bragg scattering of acoustic waves inside phononic crystals is responsible for the creation of frequency band gaps, with the result that no propagating modes exist at such frequencies. When the gap exists in all directions, it is called a complete band gap, while if there are no modes only in certain directions, the term stop band is used.

The effects of periodic structure on wave propagation are most conveniently represented by a band structure plot, where the frequency is plotted as a function of wave vector along different crystal directions (**Fig. 1*a***). The example shown depicts the

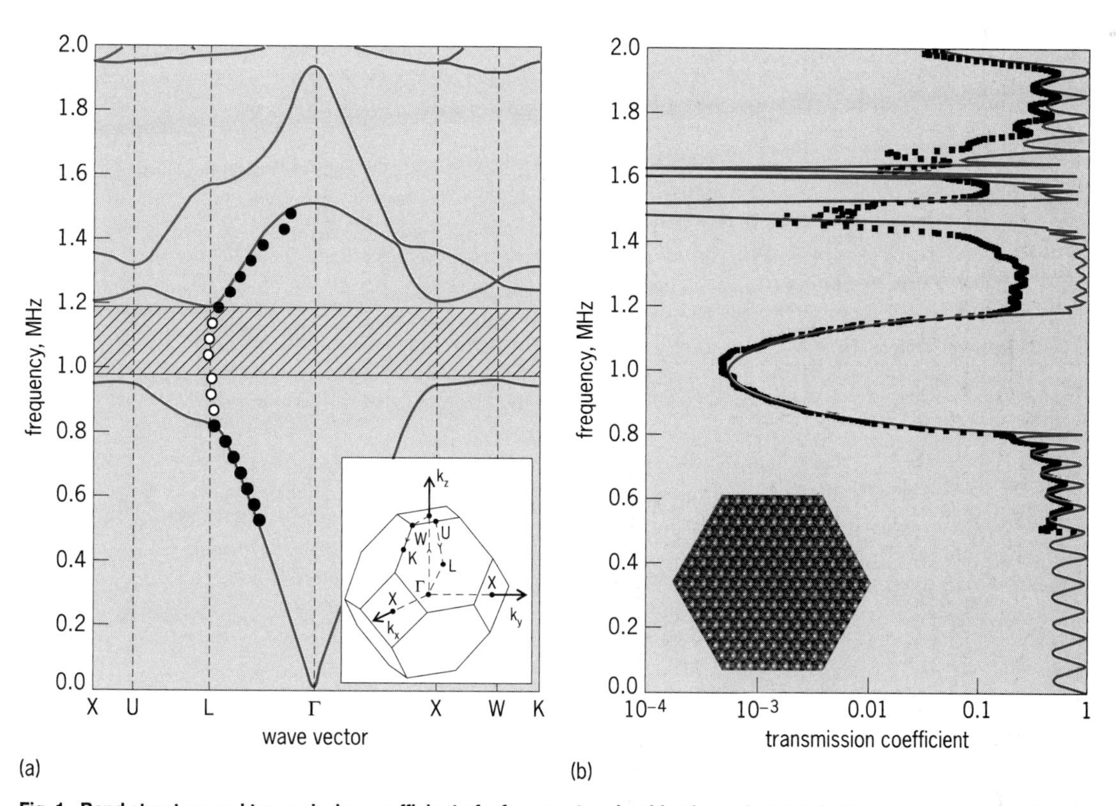

Fig. 1. Band structure and transmission coefficient of a face-centered-cubic phononic crystal of tungsten carbide beads in water. (*a*) Band structure plot. Solid curves give the variation of the frequency with the wave vector along high-symmetry directions of the crystal, as calculated using multiple scattering theory. The different crystallographic directions of the wave vector are represented by the capital letters explained in the inset. Data points give representative experimental results obtained from ultrasonic phase velocity experiments on a 12-layer crystal; solid and open symbols correspond to measurements in the pass and stop bands, respectively. (*b*) Frequency dependence of the ultrasonic transmission coefficient through this 12-layer phononic crystal for waves traveling parallel to a body diagonal of the cubic unit cell. Solid curves give results of multiple scattering theory; data points give experimental results. The inset shows a top view of the crystal.

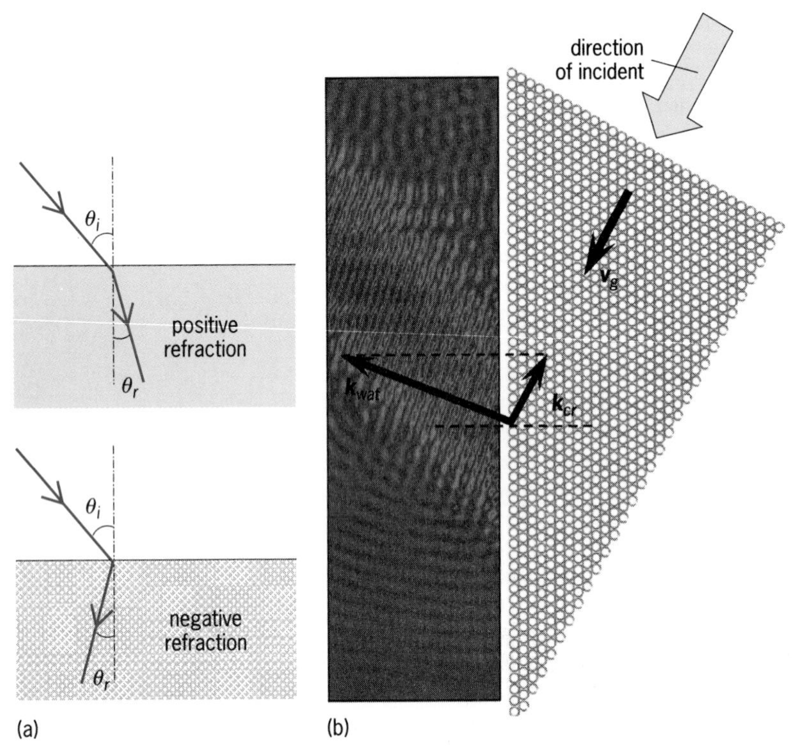

Fig. 2. Phenomenon of negative refraction. (*a*) Comparison of positive and negative refraction at an interface. θ_i = angle of incidence; θ_r = angle of refraction. (*b*) Negative refraction of ultrasonic waves emerging from a 2D phononic crystal prism. The directions of the group velocity, v_g, and of the wave vector, k_{cr}, inside the crystal, and of the wave vector in the water, k_{wat}, are shown by arrows.

band structure, calculated using multiple scattering theory, for a three-dimensional phononic crystal of 0.8-mm-diameter tungsten carbide beads immersed in water. The band structure can be investigated experimentally by measuring the ultrasonic phase velocity as a function of frequency; representative data are shown in Fig. 1*a*. This phononic crystal has a complete band gap near 1 MHz, at which frequency the spacing between adjacent planes of beads is approximately half the ultrasonic wavelength in water. For the frequencies in the band gap, the transmission coefficient or fraction of the incident wave amplitude that is transmitted through the crystal exhibits a deep minimum (Fig. 1*b*).

However, in a band gap the transmitted signal is not zero, although there are no propagating modes. The origin of this signal is explained by the tunneling of ultrasound, an effect that is completely analogous to the tunneling of a quantum-mechanical particle through a potential barrier. As a result, the transmission coefficient decreases exponentially with crystal thickness in a band gap, and is thus very small even for thin crystals. A remarkable signature of the tunneling mechanism is that the velocity of a pulse transmitted through the crystal (the group velocity) can be larger that in any of the constituent materials, and increases with the crystal thickness. This unusual effect has been observed experimentally using ultrasonic waves, in good agreement with theoretical predictions.

Application to noise suppression. Because of the ability of phononic crystals to effectively block sound waves over a range of frequencies, they can be used as acoustic filters, suppressing noise in places where a vibrationless or noise-free environment is desired. Consequently, it is of the great importance to determine and optimize the factors that influence the width of the band gap. In general, for two phononic crystals having the same crystal structure, the one with the greater density and sound velocity mismatch between constituent materials will have the wider band gap. Other factors that control the width of the band gap are the shapes of the scattering elements, the symmetry of the crystal lattice, and the filling fraction, which is the ratio of the volume occupied by the scatterers to the total volume. One practical example that has been proposed for blocking traffic noise along the edges of a highway is a 2D phononic crystal fence made from an array of solid cylinders in air. Another is the use of locally resonant microstructures to construct compact crystals with band gaps in the audible range. By exploiting the low-frequency resonance of heavy spherical scatterers coated with a weak elastic material and embedded in a stiffer matrix material, band gaps can be obtained in phononic crystals with interscatterer separations that are 1/100th the wavelength in air. Such compact structures have considerable potential for noise-proof devices.

Applications of defects. Another promising field with numerous potential applications takes advantage of defects to modify wave transport in phononic crystals. Defects may be any objects breaking the symmetry in the otherwise perfect crystal. For example, in case of a two-dimensional phononic crystal, the defect can be a single rod that differs in its acoustic properties from the rest of the rods (for example, the defect rod may be hollow, of different shape, or made of different material) or simply a missing rod. It has been shown both theoretically and experimentally that, as a result of such modifications,

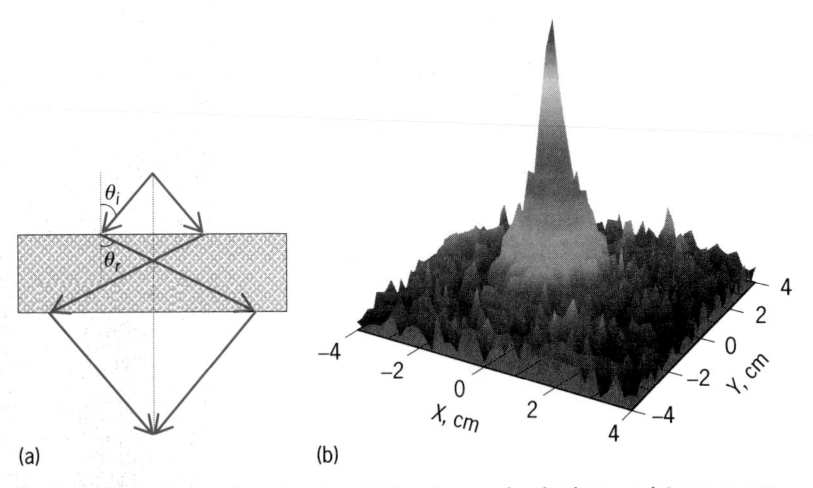

Fig. 3. Focusing by negative refraction. (*a*) Ray diagram showing how a point source may be focused by negative refraction in a phononic crystal. (*b*) Demonstration of focusing by negative refraction in a 3D phononic crystal; this picture shows the focal spot obtained when a diverging beam from a small-diameter transducer is imaged by the crystal. The height of the surface maps the sound intensity across the neighborhood of the focal point. (*After S. Yang et al., Focusing of sound in a 3D phononic crystal, Phys. Rev. Lett., 93:024301:1–4, 2004*)

the transmission spectrum of the phononic crystal with the defect exhibits a narrow peak at a particular frequency inside the band gap. By changing the different characteristics of the defect (for example, changing the inner or outer diameters of a hollow rod or filling the rod with various fluids), the frequency of the transmitted peak can be controlled. Thus, phononic crystals with defects may be used as tunable acoustic filters to select waves of a certain frequency.

By removing an entire row of rods, a line defect inside a 2D phononic crystal can be created. If the crystal has a complete band gap, waves entering the line defect will be confined inside it, creating a waveguide for acoustic waves. The waveguide also can be made tunable to transmit waves of a selected frequency or frequencies. To this end, the line defect can be created by using rods different from the rest of the rods in the phononic crystal rather than removing the rods completely. These applications of defects in 2D crystals can be readily extended to 3D phononic crystals, even though their implementation may be more difficult to engineer.

Negative refraction and sound focusing. Another unusual phenomenon exhibited by phononic crystals is negative refraction. At some frequencies, ultrasonic waves entering or leaving a phononic crystal may bend in a direction that is opposite to the norm for uniform materials, so that the effective angle of refraction at the interface is negative (**Fig. 2***a*). This effect can occur when the group velocity, $\boldsymbol{v}_g$, which determines the direction in which energy is transported through the crystal (and hence the ray direction), points in a different direction than the wave vector, $\boldsymbol{k}$, which describes the direction of travel of the phase oscillations in the wave. In the simplest case of negative refraction, $\boldsymbol{v}_g$ and $\boldsymbol{k}$ are antiparallel.

Figure 2*b* shows an example of negative refraction for ultrasonic waves emerging from a prism-shaped phononic crystal of steel rods in water. The waves are incident from the upper right side of the figure. The directions of the group velocity, $\boldsymbol{v}_g$, and of the wave vector inside the crystal, $\boldsymbol{k}_{cr}$, are shown by arrows. In this frequency range, which corresponds to the second pass band of the crystal, the frequencies of the modes decrease as the wave vector increases (similar to the branches of the 3D band structure shown in Fig. 1*a* for frequencies above 1.2 MHz); hence the group velocity $\boldsymbol{v}_g$, which is given by the derivative of frequency ω with respect to wave vector, has the opposite sign to the wave vector. Since the wave can be transported across the crystal only when the group velocity is positive, the wave vector $\boldsymbol{k}_{cr}$ points backward, as indicated. (In general, the relationship between the direction of the group velocity and the wave vector can be found by determining the equifrequency or slowness surface, which displays the magnitude of the wave vector as a function of its direction for waves of a single frequency. Since $\boldsymbol{v}_g = \nabla_k \omega(\boldsymbol{k})$, $\boldsymbol{v}_g$ is perpendicular to this surface; it points in the opposite direction to $\boldsymbol{k}$ if the equifrequency surface is circular and if the surface shrinks in size as frequency increases, as is the case here.) Consequently, Snell's law, which states that the component of the wave vector parallel to the interface should be the same on both sides of the boundary, predicts that a negatively refracted wave will emerge from the prism crystal as shown. These experimental results are in excellent agreement with predictions based on multiple scattering theory for the magnitudes and directions of the wave vector and the group velocity for this 2D crystal.

The ability of phononic crystals to refract sound waves negatively allows them to be used as lenses for focusing sound. When an initially diverging beam from a source point is incident on the crystal, the rays are bent back by negative refraction toward the axis of the lens, so that when they emerge of the far side of the crystal, where they travel in their original directions, they are brought to a focus (**Fig. 3***a*). The focusing of sound by a flat 3D phononic crystal has been demonstrated experimentally (Fig. 3*b*). A potential advantage of this novel type of focusing is improved resolution compared with traditional lenses, and theoretical predictions of image of a point source as small as one-seventh the wavelength have been made.

For background information *see* ACOUSTIC NOISE; BAND THEORY OF SOLIDS; COMPOSITE MATERIAL; CRYSTAL DEFECTS; GROUP VELOCITY; REFRACTION OF WAVES; SOLID-STATE PHYSICS; TUNNELING IN SOLIDS in the McGraw-Hill Encyclopedia of Science & Technology.

Alexey Sukhovich; John H. Page

Bibliography. Z. Liu et al., Locally resonant sonic materials, *Science*, 289:1734–1736, 2000; J. H. Page et al., Phononic crystals, *Phys. Stat. Sol. (b)*, 241:3454–3462, 2004; Y. Pennec et al., Tunable filtering and demultiplexing in phononic crystals with hollow cylinders. *Phys. Rev. E*, 69:046608:1–6, 2004; I. E. Psarobas (ed.), Phononic crystals: Sonic band-gap materials, a special issue of *Zeitschrift für Kristallographie*, vol. 220, iss. 9–10, 2005; S. Yang et al., Focusing of sound in a 3D phononic crystal, *Phys. Rev. Lett.*, 93:024301:1–4, 2004; S. Yang et al., Ultrasound tunneling through 3D phononic crystals, *Phys. Rev. Lett.*, 88:104301:1–4, 2002.

Phylogenetic diversity and conservation

"Biodiversity" refers to the variety of life on the planet—extending in scale from genes to species to ecosystems. Our understanding of biodiversity conservation, however, faces a twofold knowledge gap. First, we have no complete list of the components of biodiversity at the different levels, and second, it is difficult to judge what their values might be in the future. Therefore a core strategy for biodiversity conservation is to estimate patterns of variation, and then try to conserve as much of that variation as possible, so as to retain the full range of possible future values.

Phylogeny, the evolutionary "tree of life," can play an important role in estimating biodiversity patterns. A phylogenetic pattern for a taxonomic group displays evolutionary relationships among species, or other taxa, in two ways. It reflects the process of cladogenesis, where one lineage splits into two, and it also reflects the process of anagenesis, where evolutionary changes in genetic, morphological, or other features occur along a given lineage. Such a summary of the complex evolutionary history of a group is never revealed to us as the definitive "true" phylogeny. Instead, the discipline of phylogenetic inference uses genetic or other features of organisms to evaluate competing hypotheses about the true phylogenetic pattern.

Phylogeny-based strategies for biodiversity estimation reverse this discovery process. The corresponding biodiversity inference process uses inferred phylogenetic patterns to make a predictive link from the taxon level to general patterns of variation at the lower level of genes or other features. Conservation priorities may then be based on assessment of the potential losses in feature diversity from different scenarios of taxon extinction. Theoretical arguments, based on the simple idea that shared ancestry should account for shared features, suggest that any subset of taxa that has greater phylogenetic diversity will represent greater genetic or feature diversity. A measure of phylogenetic diversity (PD) is defined as the minimum total length of all the phylogenetic branches required to span a given set of taxa on the tree (see **illustration**). Larger PD values imply greater expected feature diversity. The total PD represented by different-sized sets of taxa defines a "features/taxa" curve. It is analogous to the well-known species/area curve, which illustrates the linear relationship between log transformations of number of species and total area. Using PD, the increase in represented feature diversity is initially rapid as we add taxa to a set. As the size of the set grows larger, the rate of gain in features becomes progressively lower.

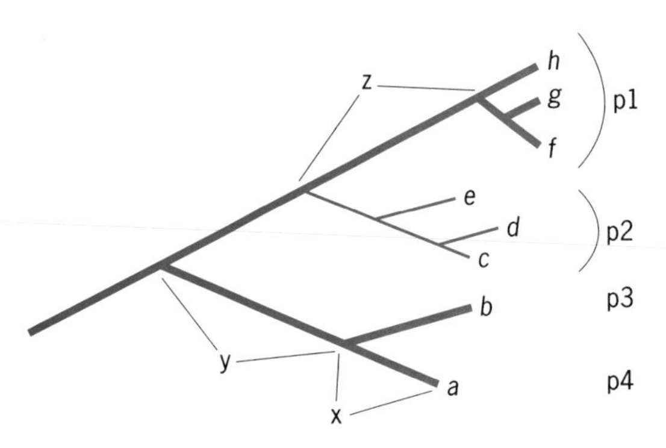

Phylogenetic tree for taxa *a–h*. The branching pattern reflects splitting of lineages and the length of the branches reflects character change along lineages. The total PD of the set of taxa {*a, b, f, g, h*} is represented by the sum of the lengths of the thick branches. Extinction of taxon *a* implies a PD loss of x units. However, if taxon *b* is extinct, then loss of *a* implies a PD loss of an additional y unit represented by the deeper branch. Taxa are restricted in distribution (endemic) to individual places, p1–p4. Phylogenetic clumping of taxa in p1 implies that loss of that place corresponds to loss also of the deeper branch, z. Historical relationships among places imply that phylogenetic patterns for other groups may be concordant, supporting general PD-endemism, for example, of p1.

It has been argued that conservation actions that maximize species/taxon diversity automatically maximize conservation of PD. However, because species are not equal in factors such as extinction vulnerability and branch length contributions, actual gains and losses in PD often are not predicted well by species counts.

PD gains and losses. In practice, PD is typically used to assess marginal gains and losses, such as those implied by the loss of a single species, or other taxon, that is "pruned" by extinction from the phylogenetic tree (see illustration). Similarly, gains may be recorded as additional taxa that are represented in new conservation areas. Unlike traditional metrics using species diversity, PD focuses on the overall number of features that might be lost from a given community with the loss of a particular species. This shift in focus also presents a potential advantage of PD-based assessments, namely that PD sidesteps conventional difficulties in deciding exactly what is, or is not, a "species." Conservation scenarios therefore can be evaluated in terms of gains in feature diversity, ignoring species counts.

The PD loss corresponding to loss of a given taxon is not a fixed value. It depends on which other taxa are extant (not extinct). For example, in the illustration, the PD loss implied by the loss of taxon *a* will be greater if taxon *b* already has been lost. The loss of both taxa does not imply a PD loss simply equal to the sum of the individual losses. Instead, losing both taxa means we also lose the PD contribution from the branch or lineage linking those two taxa. Such phylogenetic "clumping" of losses, and consequent inflation of PD loss, has been documented in some studies. A recent assessment of Indonesian fauna showed that the vulnerability of these species to extinction is phylogenetically clumped. Therefore the amount of PD at risk in Indonesia is significantly greater than it would be for the loss of the same number of species selected at random. Global-scale studies of both birds and carnivores suggest that many extinct and currently threatened species are phylogenetically clumped. Other global-scale studies, such as one for Felids (the "cat" family), suggest that extinction vulnerability of species is distributed more or less randomly across the phylogeny of the group.

Other forms of extinction bias also may inflate the loss of PD. For example, greater extinction vulnerability for taxa at the ends of long branches will increase potential PD loss. Also, greater vulnerability of species with few close relatives will place more PD at risk, because there is a greater chance that deeper branches will not retain any extant representation. Both of these factors have been implicated in extinction patterns for Australian marsupials.

The nonrandom geographic distribution of taxa is a major factor in determining potential PD losses. Often the units for conservation planning are geographic places, and PD assessment reflects possible gains and losses for whole collections of taxa, potentially from many taxonomic groups. The loss of a place may imply taxon loss that is phylogenetically clumped or dispersed depending on the environment (see illustration). Both scenarios may flag conservation opportunities. Numerous places that individually capture high phylogenetic diversity provide flexibility in conservation planning for representative protected areas, and individual places recognized as having a phylogenetic concentration of taxa attract higher conversation priority based on the potential high loss of PD.

PD endemism. PD contributions may be restricted (endemic) to a particular place. While endemism conventionally refers to counts of the species restricted to a place, PD endemism reflects the presence of unique amounts of phylogenetic diversity, and indicates the feature diversity inevitably lost if that place is lost (see illustration). For example, PD endemism based on an estimated phylogeny of amphipods has highlighted the unique biodiversity value of forests in northwestern Tasmania that were not apparent from corresponding species counts.

Global hotspots are regions where high species endemism currently overlaps with a high overall threat to biodiversity. A recent study used estimated phylogenetic trees for carnivores and for primates to argue that the 25 named global hotspots are critical for conservation, as they contain large amounts of PD found nowhere else. The high PD endemism is reflected in the fact that a remarkable one-third of the total PD of primates and carnivores is contained only in these hotspots. Collectively, they correspond to the scenario in the illustration, where the endemic taxa are phylogenetically clumped. If we lost the hotspots, the amount of PD lost would be significantly greater than for a loss of a random selection of the same number of species in other areas.

At the continental scale, an Australian hotspots study revealed that the places having high PD endemism often differ from places of high species endemism. While that comparison was based on limited phylogenetic information, PD endemism patterns for one phylogenetic group may predict more general PD endemism. One factor promoting general predictions is co-evolution, for example, of hosts and parasites. Similarly, phylogenetic patterns for different taxonomic groups may reflect shared historical relationships among areas (see illustration), so that a single area may have PD endemism for many different groups.

Phylogenetic clumping and PD endemism are important at other scales of variation as well. For example, a recent study showed that the phylogenetic diversity of a little known group of bacteria, occurring widely in anaerobic environments, displayed strong phylogenetic clumping linked to different animal hosts (for example, termites and birds). The gut of each animal species corresponds to a phylogenetic radiation, resulting in PD endemism at that scale. The phylogenetic clumping seen there, analogous to that for global hotspots, means that loss of a gut fauna would imply a disproportionate loss in bacterial PD.

Consideration of fine-scale variation extends to applications of PD within species. Here, phylogenetic patterns describe relationships among populations and genetic variants. For example, mitochondrial DNA phylogenies for different vertebrate species from Queensland rainforests show congruent patterns for the PD contributions of different areas. Another recent study has argued that within-species variation generally can be expected to be distributed unevenly among populations. Global samples of *Pseudomonas* bacteria revealed that a disproportionate fraction of the phylogenetically structured diversity was clumped in small subpopulations. Conservation priority for these places could avoid inflated PD loss associated with deep branches (see illustration).

Prospects. The utility of PD for conservation will depend on phylogenetic information being readily available in contexts where biodiversity is under threat. Prospects for this were highlighted in a recent study of the freshwater invertebrate taxa found in headwaters of streams in N.S.W. Australia, which are under threat from mining and new dams. The phylogenetic and distribution patterns for these groups suggested repeated cases of PD endemism and location-based phylogenetic clumping. For example, the "spiny crayfish" had one lineage restricted to a location that could be lost to mining impacts. This suggests a higher conservation priority for another threatened location, uniquely holding a sister lineage, because the combination of losses would imply a large PD loss (see illustration).

Phylogenetic inference for that study was based on a mitochondrial gene called cytochrome-c-oxidase subunit I (*COI*). The sequence of this gene can potentially distinguish between closely related species across a broad range of biodiversity, and rapid *COI* assessment is seen as providing a general "barcoding" approach for assignment of unidentified individuals to particular species. As suggested by the *COI* phylogenetic/spatial patterns for freshwater organisms, future rapid conservation assessments could side-step the associated difficulties in distinguishing species and apply PD to phylogenetic patterns based in part on *COI*. A global bar-coding program integrating PD assessment could provide more rapid identification of places of high PD endemism. *See* DNA BARCODING.

For background information *see* BIODIVERSITY; CONSERVATION OF RESOURCES; EXTINCTION (BIOLOGY); ORGANIC EVOLUTION; PHYLOGENY; ZOOGEOGRAPHY in the McGraw-Hill Encyclopedia of Science and Technology. Daniel P. Faith; Kristen J. Williams

Bibliography. D. P. Faith, Conservation evaluation and phylogenetic diversity, *Biol. Conserv.*, 61:1-10, 1992; G. M. Mace et al., Preserving the tree of life, *Science*, 300:1707-1709, 2003; D. P. Faith, Biodiversity, in E. N. Zalta (ed.), *The Stanford Encyclopedia of Philosophy*, 2003; W. Sechrest et al., Hotspots and the conservation of evolutionary history, *PNAS*, 99(4):2067-2071, 2002.

Phytoremediation

Phytoremediation is the use of plants to clean up environmental contamination of surface soils. It is more cost-effective and environmentally appealing than other currently available methods for soil detoxification. The most common approach for soil cleanup involves the excavation and removal of polluted soil to a chemical treatment facility or a long-term storage landfill facility. This method is very costly for large-scale decontamination and can be destructive to the environment. Phytoremediation, on the other hand, costs significantly less and does not require the same degree of environmental perturbation.

Phytoremediation methods. A variety of phytoremediation methods can be applied depending on the type and concentration of pollution as well as the desired end result (see **illustration**). Phytoremediation can be used to clean up soils with organic contaminants (via phytodegradation) as well as inorganic pollutants (via phytoextraction). Phytodegradation uses plants and the soil microorganisms associated with roots to degrade organic soil pollutants, such as trinitrotoluene (TNT) or trichloroethylene (TCE). Phytoextraction uses plants to accumulate high concentrations of inorganic soil contaminants such as heavy metals [for example, cadmium (Cd), lead (Pb), and zinc (Zn)] into aboveground biomass where it can then be harvested. Phytovolatilization disperses contaminants into the atmosphere by way of plant transpirational processes. Phytostabilization uses plants to reduce the bioavailability or mobility of pollutants in the soil.

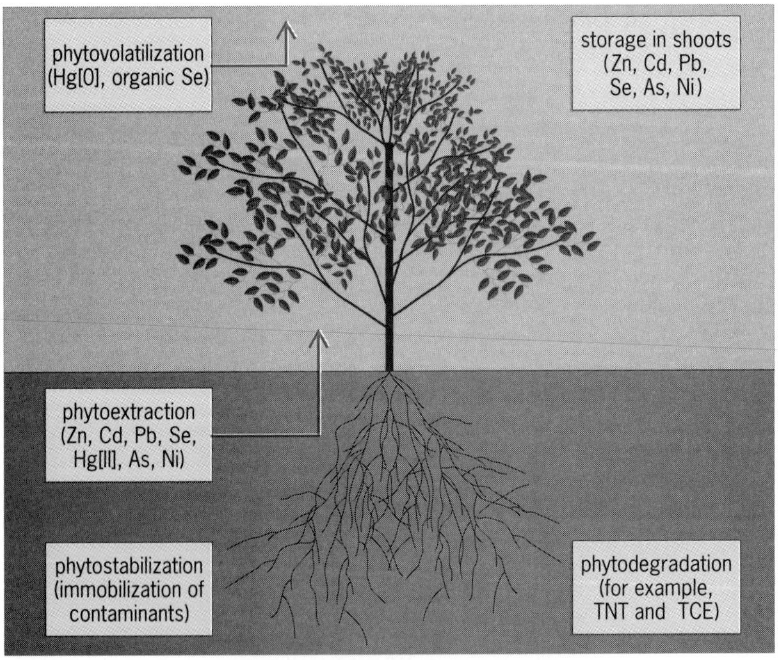

This article will focus on phytoremediation via phytoextraction of heavy metals in the soil and their storage in the easily harvestable aboveground biomass.

Heavy metal toxicity. Soils with high heavy-metal concentrations can occur naturally or due to anthropogenic factors. Human activities such as mining, smelting, and application of pesticides and fertilizers can significantly increase the concentrations of heavy metals in the soils. Some heavy metals taken up by plants, such as Zn and copper (Cu), are essential for plant growth, while others, such as Cd and Pb, have no known biological function. All these elements are extremely toxic for plants and all other organisms at high concentrations. Upon exposure to high concentrations of heavy metals, plants exhibit various toxicity symptoms such as reduced biomass, chlorosis (yellowing of green parts), and necrosis (tissue death). A common environmental contaminant, Cd, has similar physical characteristics to Zn ions and can replace Zn in enzyme-protein complexes requiring Zn, resulting in enzyme inactivation. Further damage caused by heavy metals involves the formation of free radicals and reactive oxygen species, which can cause disruption of biological membranes and damage to the photosynthetic apparatus. However, there are certain plant species, known as hyperaccumulators, that have a unique ability to tolerate and accumulate high concentrations of heavy metals. *See* REACTIVE OXYGEN SPECIES.

Hyperaccumulating plant species. Hyperaccumulators are generally defined as plants that are able to accumulate 100 times higher levels of a specific heavy metal in shoot tissues compared with regular, nonaccumulator plants. To date, more than 400 metal-hyperaccumulating plant species have been identified. Hyperaccumulator plant species for specific toxic metals have been identified, including those that hyperaccumulate nickel (Ni), Zn, Cd, arsenic (As), and selenium (Se) [see **table**]. Unfortunately, most hyperaccumulators are relatively slow-growing and produce small amounts of shoot biomass; thus they are usually not good candidates for commercial-scale phytoremediation projects.

Probably the best-known example of a metal hyperaccumulator is *Thlaspi caerulescens*, which is a small, slow-growing member of the Brassicaceae (cabbage) family; it is able to accumulate up to

Depiction of the various methods of phytoremediation. Heavy-metal phytoremediation involves phytoextraction of the toxic metal from the soil and subsequent transport and storage of the metal in the shoots (for removal). For some toxic metals that form volatile organic forms in the shoot, plants can release these volatile forms (for example, Hg[0] and organic Se) into the air (phytovolatilization). Other uses of plants for remediation involve the degradation of organic contaminants in the soil by microbes associated with plant roots (phytodegradation) and the use of plants to stabilize and reduce the bioavailability of certain contaminants (phytostabilization). TCE, trichloroethylene; TNT, trinitrotoluene.

Selected naturally occurring metal hyperaccumulating plant species

Element	Plant species	Metal accumulation, ppm
Selenium (Se)	*Astragalus bisulcatus*	6,500
Arsenic (As)	*Pteris vittata*	7,500
Zinc (Zn)	*Thlaspi caerulescens*	30,000
	Arabidopsis halleri	30,000
Cadmium (Cd)	*Thlaspi caerulescenes*	14,000
	Arabidopsis halleri	2,700
Nickel (Ni)	*Alyssum lesbiacum*	23,000

30,000 ppm Zn and over 10,000 ppm Cd in its shoots without exhibiting toxicity symptoms. Because of these unique abilities, scientists are studying hyperaccumulators such as *T. caerulescens* in order to identify metal hyperaccumulation genes and the associated mechanisms. The goal is to build a molecular toolbox so that plants can be genetically engineered to have large biomasses and the ability to survive and accumulate high concentrations of heavy metals in their shoots.

Metal uptake and transport. In order for a plant to be a successful remediator, it needs to have two distinct features: (1) effective transport of heavy metals from the roots to the shoots and (2) the ability to convert or store toxic metals in the aboveground biomass.

Nonhyperaccumulators. The uptake of metal ions by plants uses a well-honed regulatory system to accumulate adequate levels of these scarce metals for growth, yet avoid accumulation of toxic concentrations of these metals. This regulation is further complicated in multicellular plants because metal ion requirements and tolerance levels vary in different cells and tissues. The general path of ion uptake by plants starts in the soil solution, where ions are mobilized and taken up by the roots. In the roots, movement of metals starts with transport across the plasma membrane of epidermal or cortex cells and is followed by the radial movement to the center of the root to the endodermis via symplastic (through plasmodesmata) and apoplastic (via cell walls) pathways. Once the metal ion reaches the endodermis, apoplastic movement is stopped by the Casparian strip of endodermal cell walls, and all ions must be taken up symplastically. This allows the plant to regulate the movement of ions into the xylem and limit transport of ions to the rest of the plant if necessary.

Once within the endodermis, metals are transported to xylem parenchyma cells and then actively loaded into xylem vessels for movement to the shoot via the transpirational pull of xylem sap to the top of the plant. Once in the shoot, ions are unloaded into the leaf apoplast. The movement of metal ions through leaves occurs both apoplastically and symplastically, with plasma membrane transporter proteins taking ions into cells to maintain necessary concentrations for enzymatic processes and structural requirements. Unnecessary metal ions and excess metals taken up by cells are usually stored in the vacuole or pumped back out of the cell into the cell wall.

In hyperaccumulating plant species, this system of metal homeostasis is altered at a number of regulatory levels, resulting in plants that can accumulate metals far in excess of normally required concentrations.

Hyperaccumulators. Physiological research comparing *T. caerulescens* with related nonhyperaccumulators (such as *T. arvense*) has shown a number of modifications in metal uptake and transport processes. For example, *T. caerulescens* maintains a much higher metal influx into root epidermal cells and keeps this absorbed metal in a more mobile phase in the root for radial transport to the xylem compared with nonaccumulators. Furthermore, hyperaccumulator plants have a more active loading step into the xylem and thus transport most of the absorbed metal to the shoot; whereas in nonaccumulator plants the majority of the metal is stored in the root. Once the metal is in the shoot, it has been shown that leaf cells of *T. caerulescens* absorb metal more actively. (Interestingly, the heavy metals stored in the *T. caerulescens* shoots are localized to the leaf epidermal cells; thus, the metal concentrations in these cells must be enormous.)

These increases in metal transport at a number of sites within hyperaccumulating plants, along with other findings, suggest that there are alterations in the way *T. caerulescens* recognizes and responds to internal metal concentrations. Further evidence comes from the identification and characterization of genes for a number of micronutrient/heavy-metal transporters, such as ZNT1 and ZTP1, which are Zn/Cd transporter proteins cloned from *T. caerulescens*, whose levels of gene expression remain high even when grown at high Zn concentrations (unlike in nonhyperaccumulators in which expression of these transporters decreases as the plant accumulates high levels of Zn). This suggests that in hyperaccumulators there are alterations in global regulation of micronutrient/heavy-metal transport and accumulation according to the concentration of metal in the plant; however, no transcriptional regulators have yet been identified to explain these differences in hyperaccumulator versus nonhyperaccumulator species.

Using biotechnology to improve mercury tolerance and accumulation. Ultimately, in order to improve the phytoremediation potential of plants it will be necessary to use biotechnology to incorporate novel phytoremediation traits from other species into a target plant species. One well-characterized example is the incorporation of a mercury (Hg) resistance operon (the *merA* and *merB* genes) from gram-negative bacteria into a range of plant species to remove methyl mercury from contaminated soil. The *merA* gene encodes an enzyme (a mercuric ion reductase) that reduces Hg(II) to the volatile Hg(0) form. The *merB* gene encodes an enzyme (an organomercurial lyase) that catalyzes the removal of Hg(II) from organic mercury groups (such as methyl mercury, a highly toxic compound that is efficiently transferred through the food chain). By modifying a number of features of the bacterial genes and the proteins they encode, researchers working with transgenic plants (that is, plants into which the foreign bacterial genes had been transferred) have been able to make the plants significantly more Hg-tolerant and able to accumulate significantly higher levels of Hg into the shoot biomass than wild-type plants.

The first generation of these transgenic plants exhibited high levels of Hg volatilization into the air from the shoots when both the *MerA* and *MerB* genes were highly expressed in plants. In subsequent work, transgenic plants were designed to express just the *merB* gene and produce the *merB* enzyme in only the cell wall or endoplasmic reticulum vesicles. These plants were able to accumulate high concentrations of Hg(II) in the shoot without volatilization. The researchers produced plants that could not only effectively grow at concentrations of organomercurial or elemental Hg(II) that killed wild-type plants, but also remove a significant amount of the Hg from the soil to the shoot for harvest of the toxic heavy metal. This is an excellent example of using biotechnology to produce plants useful for phytoremediation.

Future research. Considerable further work is necessary to understand and ultimately modify the mechanisms and molecular basis of hyperaccumulation before this phenotype (trait) can be transferred via biotechnology to other, larger and faster-growing plant species for commercial phytoremediation work.

For background information *see* ENVIRONMENTAL TOXICOLOGY; GENE ACTION; GENETIC ENGINEERING; PLANT MINERAL NUTRITION; PLANT TISSUE SYSTEMS; SOIL CHEMISTRY in the McGraw-Hill Encyclopedia of Science & Technology.

Melinda A. Klein; Ashot Papoyan; Leon V. Kochian

Bibliography. A. J. M. Baker et al., Metal hyperaccumulator plants: A review of the ecology and physiology of a biological resource for phytoremediation of metal-polluted soils, in N. Terry and G. Banuelos (eds.), *Phytoremediation of Contaminated Soils and Water*, pp. 171–188, CRC Press, Boca Raton, FL, 2000; R. R. Brooks, *Plants That Hyperaccumulate Heavy Metals*, CAB International, United Kingdom, 1998; Environmental Protection Agency, *Introduction to Phytoremediation*, EPA Pub. EPA/600/R-99/107, National Risk Management Research Laboratory, Cincinnati, 1999; S. C. McCutcheon and J. L. Schnoor (eds.), *Phytoremediation: Transformation and Control of Contaminants*, Wiley, Hoboken, NJ, 2003; M. N. V. Prasad and J. Hagemeyer, *Heavy Metal Stress in Plants*, Springer, Berlin, 1999; R. D. Reeves and A. J. M. Baker, Metal accumulating plants, in B. D. Ensley and I. Raskin (eds.), *Phytoremediation of Toxic Metals: Using Plants to Clean Up the Environment*, pp. 193–229, Wiley, New York, 1999.

Plant cell walls

Research over the past two decades has resulted in a major change in how plant cell walls are perceived. Long gone is the early notion of the wall as a rigid static box that acts mostly as a structural support. A newly evolved perspective of the wall sees it as both a highly complex and dynamic structural matrix and as a compartment with many roles. These roles include defining cell size and shape and mediating interactions between the plant cell and its environment.

This article briefly summarizes some recent developments in the areas of cell wall biosynthesis, wall modification and restructuring, secondary cell walls, the role of the wall in defense, and new technologies that are being developed to study wall biology.

Cell wall synthesis. The structural and chemical makeup of plant cell walls results from the coordinated expression of numerous genes involved in the biosynthesis, assembly, and deposition of polysaccharides, structural proteins, waxes, and other hydrophobic compounds. In the case of secondary cell walls, lignin and its precursors are also involved. The synthesis of these macromolecules and their assembly into a mature wall is still not well understood, although much progress has been made recently in identifying the genes and enzymes that contribute to the biosynthesis of specific polysaccharides.

Cellulose is synthesized from uridine diphosphate (UDP)-glucose, an intermediate product in protein glycosylation, by cellulose synthase (CES), an enzyme complex composed of six catalytic subunits that is embedded in the plasma membrane. The *CESA* genes that encode subunits of the CES complex were identified by randomly sequencing a library of cotton deoxyribonucleic acid (DNA) sequences and looking for genes with homology to known bacterial cellulose synthase genes. It has been shown that two sets each of three different subunit types are required to form a functional "rosette" structure, and that the subunits of the CES complex for primary wall synthesis are different from those used in secondary wall synthesis. Noncellulosic wall polysaccharides are synthesized in the Golgi apparatus. The enzymes for synthesis of these polysaccharides are encoded by families of cellulose synthase-like (*CSL*) genes, which are members of the cellulose synthase superfamily and are thus related to the *CESA* genes. Another gene family, referred to as glucan synthase-like (*GSL*) or callose synthase-like (*CalS*), appears to be responsible for callose (β-1,3-glucan) synthesis, although this has not yet been definitively proven.

Cell expansion and wall reorganization. The plant cell wall must be a highly adaptable structure to accommodate the impressive variety of plant cell shapes and sizes. A major structural challenge is how to loosen and disassemble the wall matrix during cell division and directional cell elongation, while still maintaining turgor and tensile strength and simultaneously allowing the incorporation of new wall material. Thus, a delicate balance of wall loosening, disassembly, and biosynthesis must be achieved.

Many models have been proposed to explain the molecular basis of wall loosening. One of the most long-standing, the acid growth hypothesis, purports that the plant hormone auxin induces wall acidification through activation of a plasma membrane–localized proton pump, which in turn promotes the activities of wall loosening enzymes. While polysaccharide hydrolases were originally thought to be major participants in this process, a family of proteins termed expansins has been shown to act as potent inducers of cell expansion. Expansins have no defined catalytic activity, but appear to disrupt the hydrogen bonds between cellulose microfibrils and cross-linking glycans in the cell wall, and current evidence suggests that they mediate acid-induced growth.

Many other classes of wall-modifying proteins that may contribute to cell expansion have also been characterized in the last decade, including endo- and exo-acting glycanohydrolases, pectate lyases, and xyloglucan endotransglucosylase/hydrolases (XTHs). In addition to these enzyme-mediated processes, it has been suggested that free radicals may play a nonenzymatic role in breaking polysaccharide chains during cell expansion. It seems that the various wall-loosening mechanisms occur in parallel, and there is growing support for the idea that the battery of wall-modifying enzymes and agents work in a synergistic manner. For example, the activity of one may facilitate or inhibit the activity of another, thereby keeping the wall modification process in check and allowing wall plasticity without compromising strength.

Wall disassembly. Wall assembly and disassembly should be thought of as simultaneous processes. In some cases the wall loosens transiently to allow rapid expansion, while maintaining tensile strength, and in other cases wall disassembly can result in irreversible disintegration. This is particularly apparent in processes such as the development of phloem cells (the sugar transport vasculature of plants), tracheary elements (part of the xylem which provides physical support and water conduction), organ abscission, and fruit ripening.

The softening of fleshy fruits provides a particularly dramatic example of wall degradation. Typical events in the fruit wall and apoplast (the extracellular space in plants surrounded by cell walls, where water and solutes can flow freely) include changes in the concentration of apoplastic ions, wall acidification, reduction in cell-cell adhesion, depolymerization and solubilization of pectins and hemicelluloses, and elevated expression of genes and activities associated with polysaccharide breakdown.

The tomato has long been used as an experimental model to explore the molecular basis of wall disassembly in ripening fruit. Early studies proposed that tomato fruit softening was due to pectin depolymerization by the enzyme polygalacturonase (PG); however, this model was reevaluated following experiments with transgenic tomatoes. While PG is no longer considered to be the primary enzymatic cause of softening, it does appear to influence fruit texture and fruit processing characteristics.

Searches for other wall proteins that might act as major softening agents revealed that isoforms of many of the enzymes associated with cell expansion are also ripening-related. Over the past few years the functions of several wall proteins, such as expansin, pectin methylesterase, XTH, endo-1,4-β-glucanase, and β-galactosidase, have been tested by reducing their expression in transgenic fruits. While minor effects were seen in some cases, substantial softening still occurred, and it therefore appears that wall disassembly results from the collective actions of a cocktail of proteins, probably acting synergistically. Since many of the same families of wall-modifying proteins that catalyze transient wall loosening are also apparently involved in wall degradation, an important question is what determines the balance between reversible and irreversible wall disassembly.

Secondary walls and their uses. Secondary cell walls are those that are formed once the cell has reached its mature size and function to provide additional structural support and resistance to attack by pathogens and herbivores. They are typically composed of cellulose (40–60%), hemicellulose, and lignin (up to 30%). Many products of enormous commercial importance are derived from secondary wall material—for example, in the form of wood and processed-wood products—such as paper. Moreover, there is an increasing interest in using lignocellulosic biomass for the production of ecologically sustainable fuels and industrial raw materials. There is therefore a major incentive to understand secondary wall synthesis and, ultimately, to manipulate its structure and composition.

While much has been learned recently about polysaccharide synthesis, such as the identification of specific cellulose synthases that contribute to secondary cell wall formation, many aspects of lignin biosynthesis are still not completely understood. Lignin is a polymer of phenolic compounds, especially phenylpropanoids, which can form chemical bonds with structural proteins and other cell wall polysaccharides. Lignin synthesis begins with the formation of monolignols from the aromatic amino acid phenylalanine, then goes through a series of hydroxylation reactions, phenolic *O*-methylation, and finally conversion of the side-chain carboxyl to an alcohol. These lignin precursors are then transported to the wall, where they are oxidized and polymerized to generate the lignin macromolecular complex. Several extracellular proteins have been proposed as catalysts for the oxidation steps, including peroxidases, laccases, polyphenol oxidases, and coniferyl alcohol oxidase; however, it is still not clear which of these enzymes, or combinations thereof, play a major role. It appears that the actual polymerization/lignification process is regulated by the physiochemical, not biochemical, environment of the cell wall, and a relatively new theory suggests that specific proteins, termed dirigent proteins, direct the structure of the polymer. Other structural wall proteins may also act as a molecular scaffold around which the lignin is polymerized.

Plant defense. The plant wall provides a formidable defensive barrier against the armies of microbial pathogens, including bacteria, fungi, and oomycetes, which constantly seek to colonize a new host. The wall and apoplastic environment can be viewed as a molecular battleground in which numerous complex systems for assault, defense, counterdefense, and surveillance are deployed by both sides.

Plant defense systems are typically divided into two categories: preformed and inducible. The first class includes the physical obstruction of the wall itself, the overlying waxy cuticle, and a broad spectrum of wall-localized secondary metabolites and proteins with antimicrobial properties. Upon attack, a second battery of wall-related defenses is mobilized, resulting in, among other things, altered wall synthesis and restructuring and an apoplastic oxidative burst. Additionally, numerous functionally divergent defense-related proteins are secreted into the wall, such as those involved in degradation of pathogen cell wall polysaccharides and inhibitors of various pathogen-derived enzymes.

Plant pathogens attack by secreting numerous peptides and proteins into the host apoplast during infection, including enzymes involved in degradation of the plant wall, nutrient acquisition, and counterdefense, such as the detoxification of host antimicrobial compounds and the inhibition of host defense-related proteins. Many of these secreted pathogen proteins can be detected by the plant, as can certain by-products of their action such as oligosaccharide fragments that are released during plant and pathogen cell wall degradation. Such molecules which activate a spectrum of plant defenses are termed elicitors. A number of protein classes have been identified that are involved in the detection of elicitors, some of which are localized at the plant cell surface and have distinct extracellular domains. In addition to the perception of elicitors, it appears that plants have also evolved mechanisms to sense cell wall integrity or abnormal wall status, although their molecular basis is not currently known.

Wall analysis and emerging technologies. The structure and function of the plant wall typically varies between tissues, cells, and even within microdomains of a single cell face. For this reason, researchers must continuously develop new techniques to fine-tune the study of the unique biochemistry of this cellular compartment. Plant genetics research is now sufficiently advanced that a plant can be created in which virtually any specific gene has been deleted. Techniques are available, such as DNA chips and microarrays, that can associate specific genes with cellular and developmental processes, even on the scale of a single cell type. Such techniques facilitate the selection of potential cell wall-related candidate genes for further study. This can be complemented by efforts to characterize the protein complement, or proteome, of the cell wall, in order to identify new wall proteins and discover novel wall localized processes. Other analytical methods and tools now in development include polysaccharide fingerprinting by Fourier transform infrared (FTIR) microscopy and the production of libraries of polysaccharide-specific antibodies and enzymes that cleave specific sugar linkages. These tools will help to resolve wall structure in much greater detail and will be used to study a growing number of cell wall-related mutants. As novel technologies are brought to bear in studying the wall and apoplast, the complexity of our understanding continues to grow, and future reviews will doubtless describe many new, exciting, and currently unsuspected aspects of wall biology.

For background information *see* ABSCISSION; CELL PERMEABILITY; CELL WALLS (PLANT); CELLULOSE; GENE ACTION; GENETICS; HEMICELLULOSE; LIGNIN; PLANT CELL; PLANT GROWTH; PLANT TRANSPORT OF SOLUTES; REPRODUCTION (PLANT); TOMATO; URIDINE DIPHOSPHOGLUCOSE (UDPG); WOOD ANATOMY; XYLEM in the McGraw-Hill Encyclopedia of Science and Technology.

Breeanna Rae Urbanowicz; Sarah Nell Davidson; Jocelyn Kenneth Campbell Rose

Bibliography. W. Boerjan, J. Ralph, and M. Baucher, Lignin biosynthesis, *Annu. Rev. Plant Biol.*, 54:519-546, 2003; D. A. Brummell and M. Harpster, Cell wall metabolism in fruit softening and quality and its manipulation in transgenic plants, *Plant Mol. Biol.*, 47:311-340, 2001; J. K. C. Rose, M. Saladié, and C. Catalá, The plot thickens: New perspectives of primary cell wall modification, *Curr. Opin. Plant Biol.*, 7:296-301, 2004; T. A. Richmond and C. R. Somerville, The cellulose synthase superfamily, *Plant Physiol.*, 124:495-498, 2000; W. Scheible, S. Bashir, and J. K. C. Rose, Plant cell walls in the post-genomic era, and cell wall disassembly, in *The Plant Cell Wall*, Annual Plant Reviews Series, edited by J. K. C. Rose, pp. 325-375, Blackwell, 2003.

Plume imagery

Convection plumes, anchored in the lower mantle of the Earth, were proposed by W. J. Morgan in 1970 to explain the existence of oceanic islands such as Hawaii, their relative fixity with respect to one another over geological time, and the "tracks" of volcanoes left by the lithospheric plate sliding over such a "hotspot." The basalts found on oceanic islands differ in their isotopic composition from basalts found at midocean ridges, which lent geochemical support to the notion that these islands tap into a different reservoir for basaltic material, presumably the lower mantle, the silicate shell of the Earth below about 660 km (410 mi) depth where the upper-mantle minerals olivine, garnet, and pyroxene break down to very high pressure phases of perovskite and magnesiowuestite. But while the geochemical observations give some strong arguments in favor of the plume hypothesis, only seismology can pinpoint the source location of such plumes. Until recently, direct seismic evidence was missing. This allowed a small but vocal minority to oppose the plume hypothesis, advocating instead shallow lithospheric processes to be responsible for the observed phenomena.

Some plume characteristics. The classical mushroom-shaped plume image in which a large head is fed by a thin tail derives from laboratory studies, but the situation in the Earth is vastly more complicated. There is a large contrast between the viscosity in the upper mantle (10^{20}–10^{21} pascal seconds) and lower mantle (10^{22}–10^{23} Pa · s), with the deeper mantle being more viscous by two orders of magnitude or more. Chemical heterogeneities cause density variations in addition to the thermal perturbations, and the rock undergoes phase transitions as it moves up into regions with lower hydrostatic pressure. The temperature dependence of the phase transition at 660 km depth is such that a rising mantle plume will remain in the perovskite state when crossing this discontinuity, thus delaying

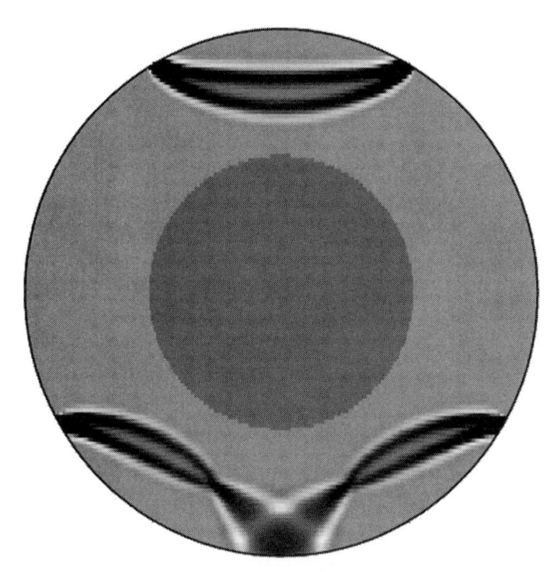

Fig. 1. Example of integral kernels for the travel time delay in the Earth's mantle of a P wave and a surface-reflected PP wave (upper and lower hemispheres, respectively). Dark areas denote negative values, lighter areas are positive. The Earth's core is indicated by the darker disk at the center. (*Adapted from G. Nolet, F. A. Dahlen, and R. Montelli, Travel Times and Amplitudes of Seismic Waves: A Re-assessment, AGU Monog., in press, 2005*)

its transition to a lower density and viscosity, and material may pile up for a while for lack of buoyancy. The reverse is true near 410 km (255 mi) depth, when the olivine to β-spinel transition comes earlier because of the higher-temperature plume. Depending on the geological setting, the plume may encounter a low-viscosity asthenosphere, which may reach as deep as 300 km (190 mi) near slowly spreading oceanic ridges, where the viscosity of the surrounding mantle is less than 10^{19} Pa · s. When the plume reaches the lithosphere and melting sets in, the ascent is very rapid. The presence of an oceanic ridge may strongly influence the behavior of the plume. Spreading at the ridge enhances flow perpendicular to the ridge and explains how plumes may influence spreading even when located far away from the oceanic ridge itself.

Near the Earth's surface, the excess temperature of the mantle plume is thought to be in the range of 160–280 K. This temperature contrast with the surrounding mantle is large enough to cause seismic waves to slow down appreciably. Since the Earth's mantle is solid, it supports two different seismic-wave types: longitudinal or compressional waves (P waves) and transverse or shear waves (S waves). The propagation velocities of these waves as a function of depth are well known, but deviations from the average velocity cause small variations (of the order of a few seconds) in travel times of the order of 1000 seconds. Their arrival time T in seismic stations will then be earlier or later than the known time T_0 predicted by a standard Earth model, and by combining the delays δT from many different ray paths, a mantle temperature anomaly can be localized and imaged. This method is known as seismic tomography. Seismic tomography has been very successful in imaging large features, such as subducting slabs and two major low-velocity regions at the bottom of the mantle, known as the African and Pacific superplumes. Narrow plumes, however, suffer from the fact that the seismic-wave energy easily diffracts

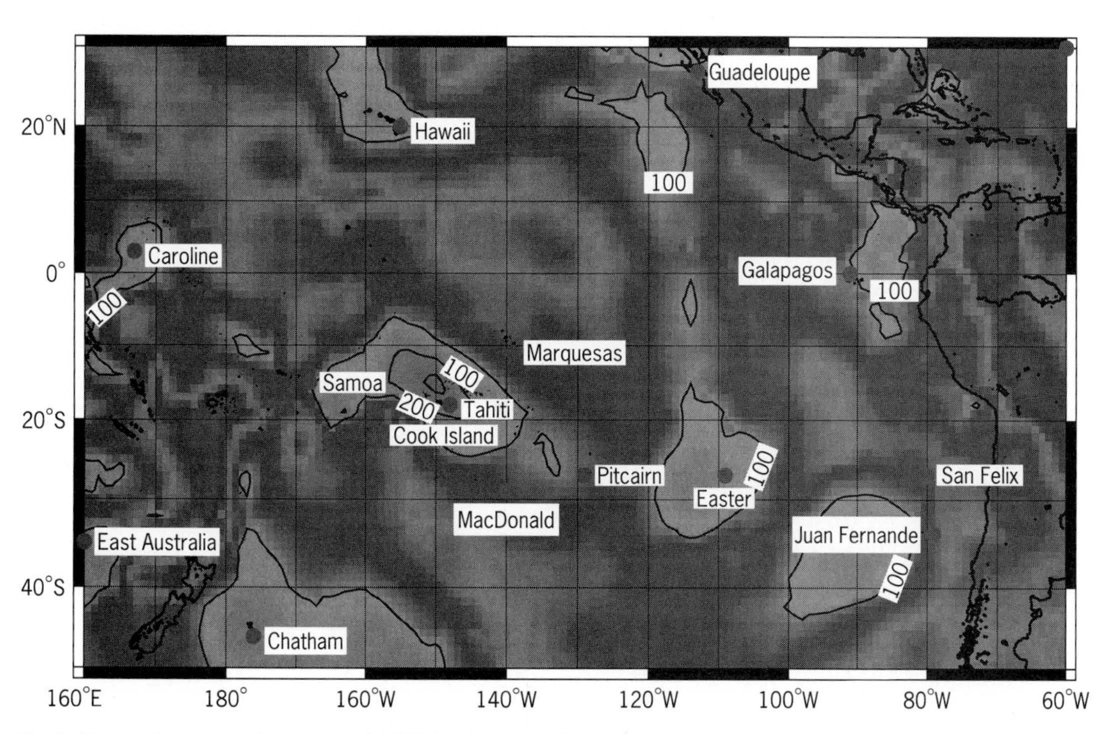

Fig. 2. Temperature anomalies from model PRI-P04 (*data from R. Montelli et al., Science, 303:338–343, 2004*) in the Pacific Ocean at a depth of 800 km. Contour lines for positive temperature anomalies at 100, 200, and 300 K (see Fig. 4) define the gray scale.

around an anomaly, causing diffracted energy to arrive earlier than the delayed, direct wave. Classical seismic tomography uses an infinite-frequency approximation, ray theory, akin to that in geometrical optics, which does not acknowledge the existence of diffraction phenomena.

Ray theory and finite-frequency theory. The travel time $T = T_0 + \delta T$ of a seismic wave in ray theory is given by the integral of the inverse velocity v along a raypath of infinitesimal width for a velocity anomaly δv with respect to a standard (usually radially symmetric) Earth model given by a velocity v_0 according to Eq. (1).

$$T_0 + \delta T = \int_{\text{ray}} \frac{ds}{v_0 + \delta v} \qquad (1)$$

This can be replaced by a better approximation that takes the finite-frequency content of the wave into account, leading to a volume integral, where $K(\mathbf{r})$ is an integral kernel which is defined in the full Earth volume, according to Eq. (2).

$$T_0 + \delta T = T_0 + \iiint_{\text{Earth}} K(\mathbf{r})\delta v(\mathbf{r}) d^3\mathbf{r} \qquad (2)$$

Figure 1 shows two examples for such kernels, for a direct P wave and for the surface-reflected PP wave with a dominant period of 20 s, typical for delay times measured by cross-correlation of seismograms from modern, digital, broadband instruments. The width of these kernels is an indication of the resolving power, and is equal to that of the Fresnel zone known from geometrical optics: $\sqrt{\lambda L}$ for a wave of wavelength λ traveling a raypath of length L. The zero sensitivity, visible at the center of the P kernel, reflects the fact that a very small anomaly may become completely invisible through the effects of diffraction around it.

Finite-frequency tomography. Given a large collection of seismic delay times, each with a linear constraint on the velocity perturbation δv given by Eq. (1) or Eq. (2), we may invert these linear constraints to obtain a three-dimensional tomographic model of the Earth. The first tomographic models using finite-frequency theory [Eq. (2)] have recently been published. A remarkable improvement in the images is immediately visible: the finite-frequency tomographic models indisputably show images of lower-mantle plumes. **Figures 2** and **3** show cross sections through the tomographic model PRI-P04 at a depth of 800 km (500 mi), that is, at the top of the lower mantle. In the figures, the perturbations in compressional-wave velocity v_P and shear-wave

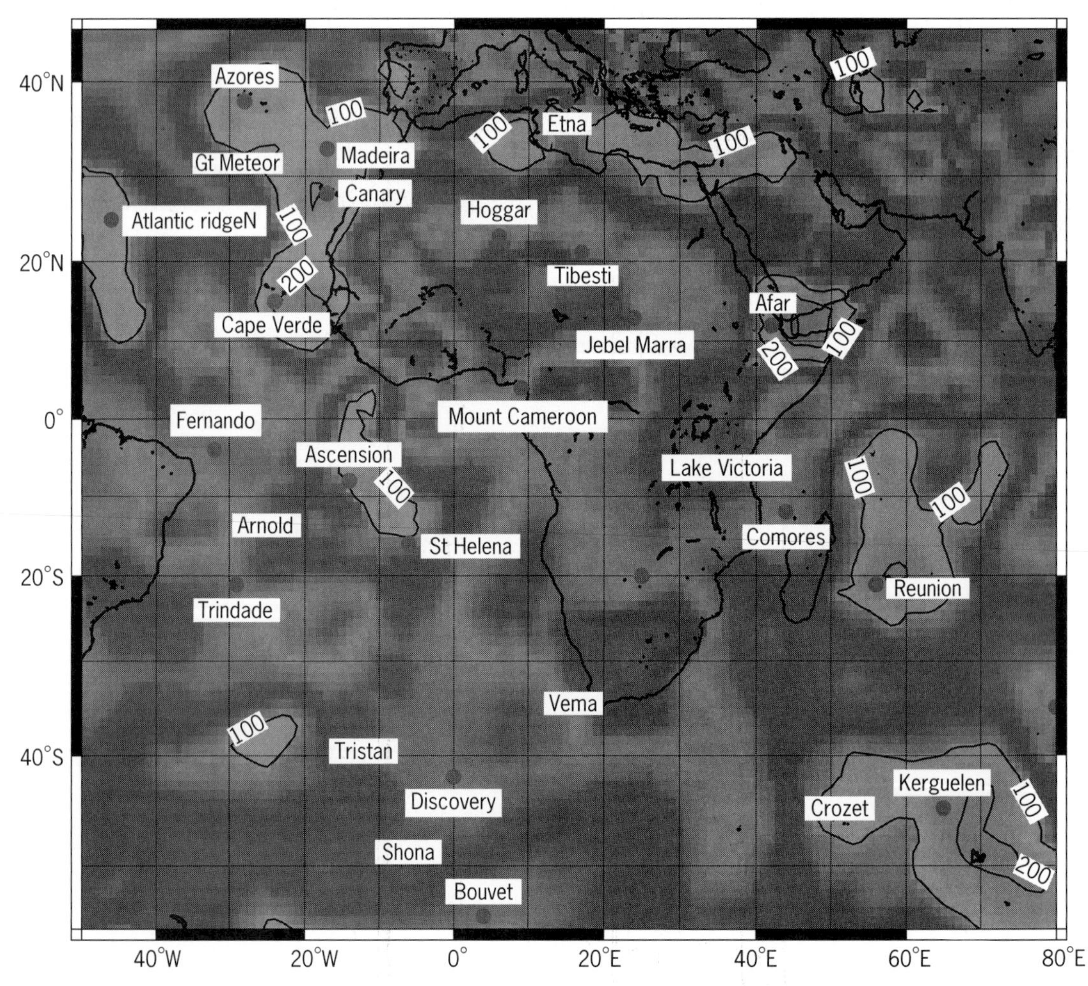

Fig. 3. Temperature anomalies from model PRI-P04 in the Indo-Atlantic region at a depth of 800 km. The gray scale in this and subsequent figures is as in Fig. 2.

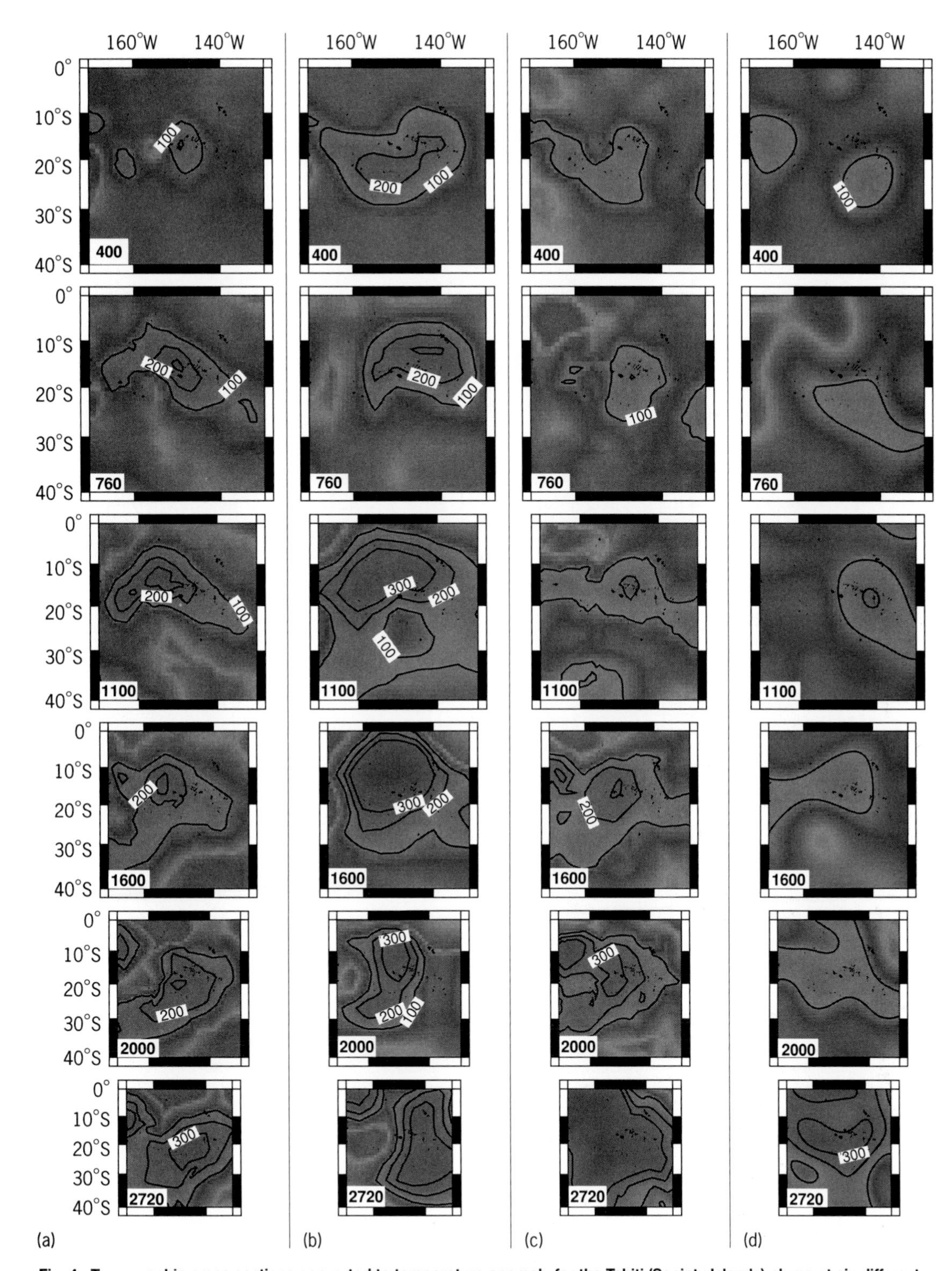

Fig. 4. Tomographic cross sections converted to temperature anomaly for the Tahiti (Society Islands) plume at six different depth levels for models (*a*) PRI-P04, (*b*) HPZ (*data from D. Zhao, Phys. Earth Planet. Int.*, 146:3–34, 2004), (*c*) PRI-S05 (*data from R. Montelli et al., submitted to Geochem. Geophys. Geosys., 2005*), and (*d*) S20RTS (*data from J. Ritsema et al., Science, 286:1925–1928, 1999*). Depth in kilometers is indicated in boldface. The deepest cross section, at 2720 km, is located 170 km above the core–mantle boundary.

velocity v_S have been converted into a temperature anomaly δT using estimates for the derivatives $\partial \ln v_P/\partial T$ and $\partial \ln v_S/\partial T$ listed in the **table**.

The strong correlation of several low-velocity anomalies (the lightly shaded areas) with known hotspots long suspected to cap a mantle plume—such as Tahiti, Hawaii, and Easter Island in the Pacific, as well as Azores, Cape Verde, and Ascension in the Atlantic, and Reunion, Afar, and Kerguelen in the Indian Ocean—are striking and provide strong support for the hypothesis that mantle plumes originate in the lower mantle.

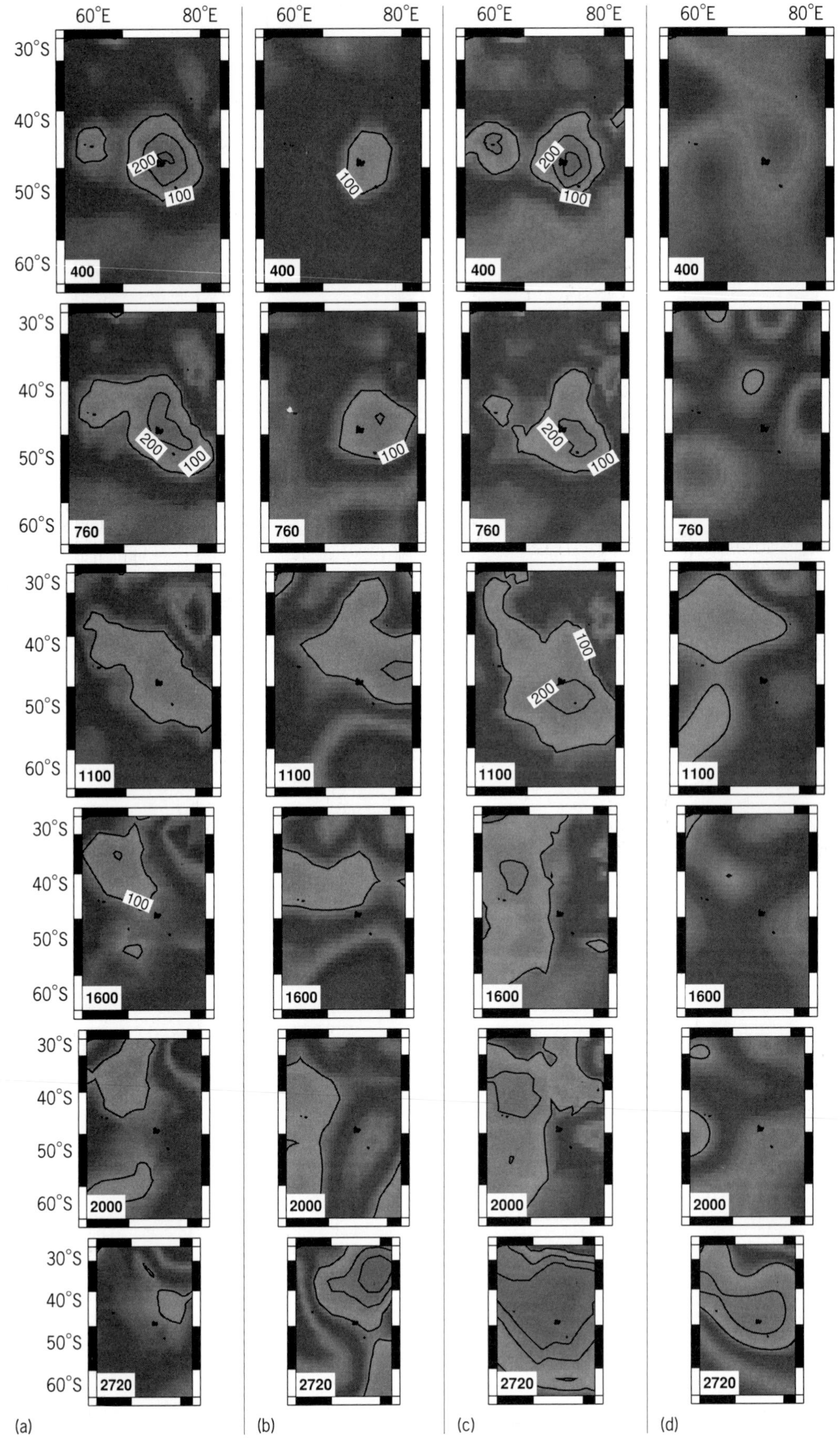

Fig. 5. Tomographic cross sections converted to temperature anomaly for the Kerguelen plume at six different depth levels for models (*a*) PRI-P04, (*b*) HPZ, (*c*) PRI-S05, and (*d*) S20RTS.

Temperature derivatives

Depth, km	$\partial \ln v_P/\partial T$, 10^{-4} m/s/K	$\partial \ln v_S/\partial T$, 10^{-4} m/s/K
400	−0.62	−0.93
760/800	−0.42	−0.77
1100	−0.36	−0.69
1600	−0.29	−0.56
2000	−0.24	−0.48
2720	−0.19	−0.37

Figures 4 and 5 show a direct comparison between four recent models for two selected lower-mantle plumes. Models PRI-P04 and PRI-S05 used finite-frequency theory. P04 used compressional (P) waves in both a high- and a low-frequency band, but only low-frequency shear (S) waves were used for S05. For comparison, we show two other models obtained using ray theory: a high-frequency P-wave model (here named HPZ) and the S-wave model S20RTS constructed with very low frequency shear waves. The latter model has been included to show the effects of wave diffraction around plumes for waves with very long wavelengths.

Tahiti. PRI-P04, PRI-S05, and HPZ agree that Tahiti is a very strong plume in the lower mantle that extends all the way to the core-mantle boundary with a temperature anomaly δT in excess of 200K (Fig. 4). The finite-frequency models provide a sharper image, as expected, and PRI-P04 clearly is the most coherent image with depth. This is a direct consequence of the use of different frequency bands. A high-frequency (thin) ray will acquire a delay proportional to the plume radius R_{plume}, whereas a low-frequency wave, with a Fresnel zone comparable to the width of the plume, will be delayed proportionally to the plume cross section, that is, to R^2_{plume}. This difference in sensitivity greatly contributes to constrain the width of the plume and increases the stability of the image. PRI-S05 does not have the added contribution of high-frequency travel times, and the image seems more distorted. Because of the large dimension of the plume, the diffraction effects suffered by the short-wavelength P waves are minor, and HPZ agrees largely with the PRI models. S20RTS gives a much smoother image, in which the contours of the plume that is visible in Fig. 4*a–c* can still be recognized. Note that the temperature conversion in the lowermost mantle is problematic, since it does not allow for chemical contributions to the seismic velocity perturbation, or for a possible phase transition in perovskite that is suspected to be present, so that we do not expect P and S models to agree near the core-mantle boundary in this simplified mapping.

Kerguelen. Down to 2000 km (1240 mi) depth there is obvious agreement between PRI-P04, PRI-S05, and HPZ (Fig. 5). The plume remains centered below the island of Kerguelen until 1100 km (680 mi) depth, then bends in the northwest direction, toward the African superplume. The plume is weaker and narrower than the Tahiti plume. The very long wavelength S waves used to construct S20RTS easily diffract around it, and in a ray-theoretical interpretation the plume remains almost invisible, with a temperature anomaly mostly below 100 K.

Outlook. Though there are still important limitations to the imaging of plumes—in particular a lack of data caused by the scarcity of seismic stations in the Southern Hemisphere where most important plumes are located—the introduction of finite-frequency theory in the interpretation has already given undisputable evidence for the existence of plumes in the lower mantle. Important improvements in the images are to be expected when seismic delays will systematically be measured in a range of frequency bands.

For background information *see* ASTHENOSPHERE; BASALT; COMPUTERIZED TOMOGRAPHY; EARTH INTERIOR; GEOMETRICAL OPTICS; HOTSPOTS (GEOLOGY); LITHOSPHERE; MID-OCEANIC RIDGE; OLIVINE; PEROVSKITE; PLATE TECTONICS; SEISMOLOGY; SPINEL; SUBDUCTION ZONES; VOLCANO in the McGraw-Hill Encyclopedia of Science & Technology. Guust Nolet

Bibliography. F. A. Dahlen, S.-H. Hung, and G. Nolet, Fréchet kernels for finite-frequency traveltimes, I. Theory, *Geophys. J. Int.*, 141:157–174, 2000; G. F. Davies, *Dynamic Earth*, Cambridge University Press, 1999; R. Montelli et al., Finite frequency tomography reveals a variety of plumes in the mantle, *Science*, 303:338–343, 2004; W. J. Morgan, Convection plumes in the lower mantle, *Nature*, 230:42–43, 1971; H.-C. Nataf, Seismic imaging of mantle plumes, *Annu. Rev. Earth Planet. Sci.*, 28:391–417, 2000; G. Nolet (ed.), *Seismic Tomography*, Reidel, 1987; J. Ritsema, H. van Heijst, and J. Woodhouse, Complex shear wave velocity structure imaged beneath Africa and Iceland, *Science*, 286:1925–1928, 1999.

Polymer stereochemistry and properties

The properties of polymers are significantly influenced by their stereochemistry. Cellulose (structure **1**) and amylose (**2**), the main components of

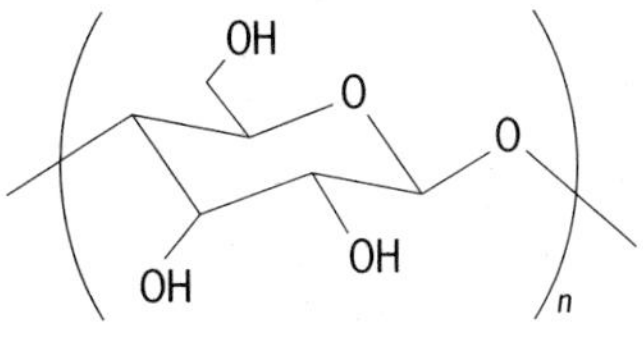

(1)

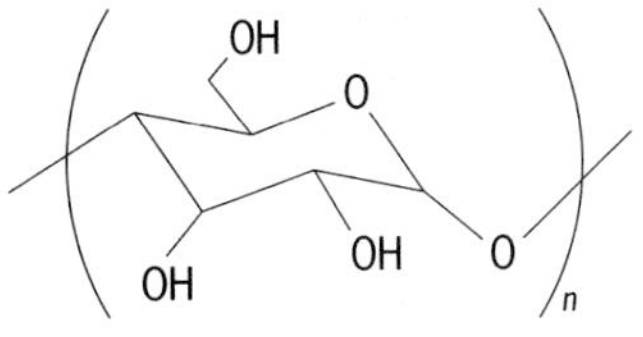

(2)

wood and starch, respectively, are among the most abundant polymers on Earth. Both are produced from D-glucose and are stereoisomers of each other.

These two have different linkages between 1- and 4-carbons; the former has a 1,4-β-structure and the latter a 1,4-α-structure. This difference in the linkage of these polysaccharides leads to different higher-order structures of the polymer chains; cellulose has a stretchable polymer chain, which facilitates the regular arrangement of the polymer chains through the interchain hydrogen bond, whereas amylose has a helical chain, in which about seven glucose residues construct one turn of a helix. Because of their structural differences, their properties are also very different; for instance, cellulose is not soluble in water because of the strong interchain hydrogen bonds, but amylose is. Human beings cannot digest cellulose to D-glucose to use as an energy source, but can digest amylose or starch to D-glucose.

Vinyl polymers. When vinyl monomers (CH_2= CH—X) like propylene (X = CH_3), styrene (X = C_6H_5), vinyl acetate (X = $OCOCH_3$), and *N*-isopropylacrylamide (X = $CONHCH(CH_3)_2$) are polymerized, each step of the monomer addition to a propagating chain end yields an asymmetric *R* or *S* center. The continuing formation of the same center produces an isotactic polymer (**3**), and the alternating formation of the *R* and *S* centers produces a syndiotactic polymer (**4**). This kind of arrangement of side chains

S S S S S
X H X H X H X H X H
Isotactic
(**3**)

S R S R S
X H H X X H H X X H
Syndiotactic
(**4**)

has been called stereoregularity or tacticity. The control of tacticity was first realized by G. Natta in 1956 for the stereospecific coordination polymerization of propylene using the Ziegler-Natta catalyst consisting of $TiCl_3$—$Al(C_2H_5)_3$. The isotactic polymer had properties superior to those of the atactic polypropylene, which has randomly arranged side-chain methyl groups. Isotactic polypropylene is a hard, durable material with a high melting point (~165°C or 330°F) and high mechanical strength, whereas the atactic polypropylene is a greasy material. Isotactic polypropylene is one of the most widely used polymers.

Nearly 100 million tons per year of the polyolefins polypropylene and polyethylene are produced worldwide. The production of these polyolefins is steadily increasing due to their low cost and high performance. They have been used for a number of products, such as grocery bags, containers, gasoline tanks, and automobile bumpers. The Ziegler-Natta catalyst is an insoluble and heterogeneous catalyst, and therefore control of the polymerization reaction with the catalyst is difficult. This defect has been improved by the discovery of new homogeneous catalyst systems composed of metallocenes (**5**, **6**).

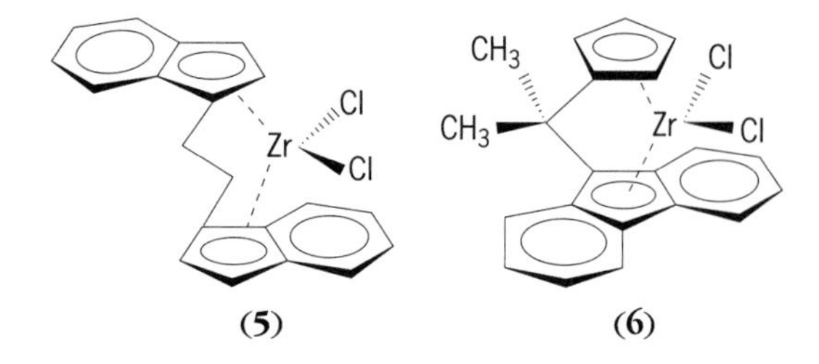

(**5**) (**6**)

By changing the ligand structures on zirconium, the tacticity of polypropylene can be varied from highly isotactic to syndiotactic.

Polystyrene is also a very popular polymer and has been industrially prepared by radical polymerization, which usually produces an atactic polymer. The polymer is used as kitchen utensils, video cassettes, and machine enclosures. Styrene is also polymerized by coordination polymerization using a catalyst containing a half metallocene (**7**), which gives a highly

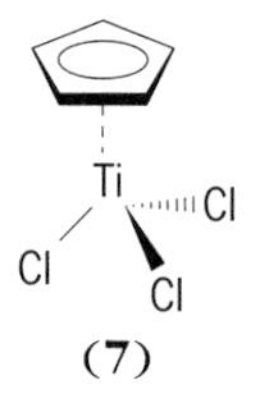

(**7**)

syndiotactic polymer. The properties of this stereoregular polymer significantly differ from those of the atactic polymer. Polystyrene obtained by radical polymerization becomes soft at around 100°C (212°F), while the syndiotactic polymer has high crystallinity and a melting point as high as 270°C (520°F). Syndiotactic polystyrene has both high thermal and chemical stability and therefore can be used as an engineering plastic.

The properties of poly(vinyl alcohol) (**8**) are also

$$-\!\left(CH_2\overset{H}{\underset{OH}{C}}\right)_{n}\!-$$
(**8**)

significantly influenced by its stereoregularity, and both the isotactic and syndiotactic polymers are known to have higher melting points than the atactic polymer. Poly(vinyl alcohol)s with different tacticities have been prepared by cutting the ether linkage of poly(vinyl ether)s obtained by cationic polymerization under various conditions. However, this method is not a practical process. The polymer is manufactured by the saponification (hydrolysis) of poly(vinyl acetate), which is obtainable only by the radical polymerization of vinyl acetate. For this polymerization, there exist almost no temperature and solvent effects on tacticity, making it is difficult to change the properties of poly(vinyl alcohol) by these processes.

N-isopropylacrylamide is also polymerized only by a radical process. The stereochemistry of this polymerization is difficult to change by the reaction conditions, as is the polymerization of vinyl acetate. Because the polymer exhibits a unique phase transition in an aqueous solution and becomes a crosslinked gel at around 32°C (90°F), it has been extensively studied without paying attention to the polymer structure, particularly the tacticity. However, recent progress in the stereocontrol during the radical polymerization of *N*-isopropylacrylamide using a catalytic amount of Lewis acids, such as the trifluoromethanesulfonates of yttrium (Y) and ytterbium (Yb), enables the production of isotactic-rich polymers. The phase-transition temperature of the isotactic-rich polymer decreases with an increase in the isotacticity.

Polyester and polyamide. A polyester, poly(L-lactic acid) **(9)**, is prepared from L-lactic acid, which is manufactured by the fermentation of glucose from starch.

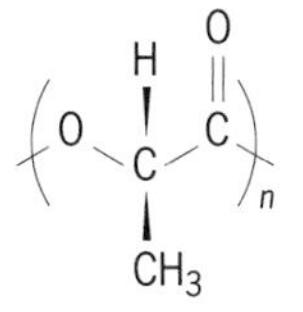

(9)

The polymer has the continuous L-configuration, and is a totally stereoregular and optically active polymer. This polymer has drawn much attention as being an environmentally friendly, biodegradable polymer. Its stereoisomer, poly(D-lactic acid), is prepared from D-lactic acid. Most properties of these two polymers are identical, but they differ in biodegradability. The degradation is catalyzed by enzymes, which often discriminate enantiomers like poly(L-lactic acid) and poly(D-lactic acid). To change the polymer properties, the preparation of the atactic polymer containing both L- and D-isomers is effective.

Protein analogue homopolymers, that is, poly(L-amino acid)s **(10)**, have been synthesized from

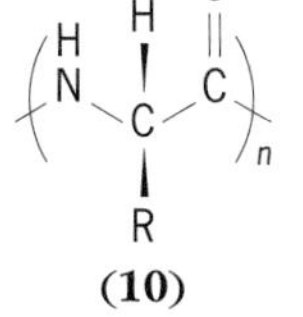

(10)

L-amino acids and are totally stereoregular polyamides. The incorporation of D-amino acids into the L-amino acid sequences yields an atactic polymer, the properties of which significantly differ from those of the corresponding homopolymers.

Higher-order structure: helicity. Most stereoregular polymers have a helical conformation at least in the solid state, and some of them can maintain the helicity in solution. It is also well known that most proteins have a partial helical conformation to exert their sophisticated functions in the living systems, and therefore some poly(L-amino acid)s also have a helical conformation in solution. Polymethacrylate **(11)**, poly(*N*,*N*-disubstituted acrylamide) **(12)**, and polychloral **(13)** are examples that can be synthe-

(11) **(12)**

(13)

sized as one-handed helical polymers using optically active initiators. Due to the steric hindrance of the bulky side chains, these polymers can maintain the helical structure that is constructed via the polymerization processes. Some of the helical polymers exhibit high chiral recognition and have been used as the chiral stationary phases in high-performance liquid chromatography (HPLC) to separate enantiomers.

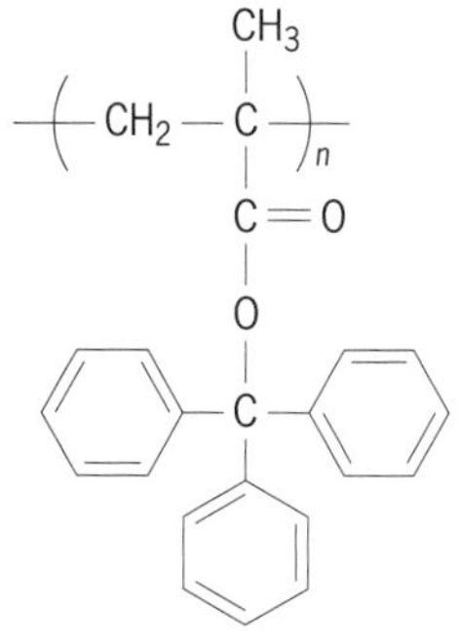

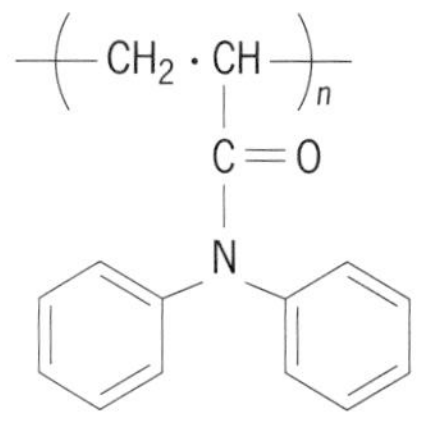

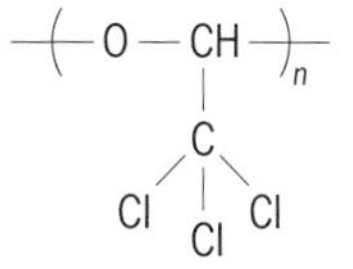

Polyisocyanates **(14)**, stereoregular polyacetylenes **(15)** with the cis-transoid structure, and polysilanes **(16)** have a dynamic conformation,

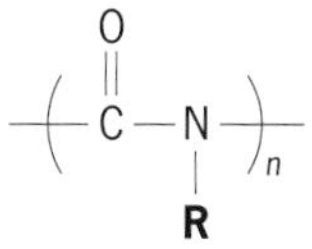

(14) 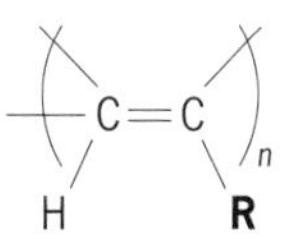 **(15)** 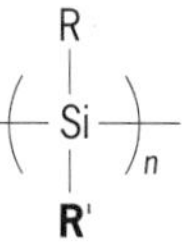 **(16)**

which can be changed through the incorporation of a tiny chiral component or through the interaction with chiral stimuli. For instance, poly(*N*-hexyl isocyanate) (**12**, **R** = n—C_6H_{13}—) takes a prevailing one-handed helical structure by introducing a 1% optically active R group. Poly(4-carboxyphenylacetylene) (**13**, **R** = —C_6H_4—COOH) can change its helicity through the interaction with chiral amines. This change is readily detected by circular dichroism (CD) spectroscopy, which shows intense CD peaks in the ultraviolet (UV) and visible regions above 300 nanometers depending on the chirality of the amines. Because the amines with analogous structures induce the same CD pattern, the CD measurement can identify the absolute configuration of the chiral amines. The behavior of polysilanes **(16)** is similar to that of polyisocyanates **(14)**, and the introduction of a very small amount of a chiral side chain forces the entire polymer chain to have the same helicity.

For background information *see* METALLOCENE CATALYST; POLYACRYLATE RESIN; POLYESTER RESINS; POLYMER; POLYMERIZATION; POLYOLEFIN RESINS;

POLYSTYRENE RESIN; POLYURETHANE RESINS; POLYVINYL RESINS; STEREOCHEMISTRY in the McGraw-Hill Encyclopedia of Science & Technology.

Yoshio Okamoto

Bibliography. M. Farina, Stereochemistry of linear macromolecules, *Top. Stereochem.*, vol. 17, no. 1, 1987; T. Nakano and Y. Okamoto, Synthetic helical polymers: Conformation and function, *Chem. Rev.*, 101:4013, 2001; Y. Okamoto et al., *Lewis Acid-catalyzed Tacticity Control during Radical Polymerization of (Meth)acrylamides*, ACS Symp. Ser. 854, ed. by K. Matyjaszewski, vol. 59, 2003.

Polymers for microelectronics

The fabrication of the first integrated circuits (ICs) in the 1960s sparked a revolution that has profoundly influenced modern society. Subsequently, the microelectronics industry in general has grown at an explosive rate due to constant technological innovations that have decreased the cost and size of each chip, while increasing performance. A large number of these improvements are a direct result of technical advances in the polymeric materials used to manufacture the devices. Currently, polymers are used as photoresists in lithographic applications for fabricating microprocessors and memory chips, in photovoltaic and other applications that require conjugated polymers, and in polymeric dielectrics used for charge storage. Because of the high demands placed on the microelectronics industry, new polymeric materials will be required with improved physical properties such as ultralow dielectric constants, low dissipation factors, increased moisture and oxygen compatibilities, high thermal stabilities, and enhanced conductivities. This article examines the use and importance of polymers in leading areas of microelectronics as well as some opportunities and trends in the field.

Photolithography. Photolithography, the process of transferring or writing an image with light, is used to create the fine features and patterns on microprocessors and other types of circuit boards, liquid-crystal displays, magnetic recording heads, and micro-electro-mechanical systems (MEMS). In photolithography, silicon wafers are first coated with a conducting, insulating, or semiconducting material, and then a thin film containing a light-sensitive polymer, called a photoresist, is applied **(Fig. 1)**. When photoresists are irradiated through a quartz mask containing the desired pattern, chemical reactions occur that alter their solubility characteristics. These reactions include molecular-weight reduction through chain scission, mass loss through degradation, and macromolecular rearrangement. Since non-irradiated areas of the photoresist exhibit different solubilities when compared to those areas that have been exposed, selective removal (development) of either is possible. After development, any area not covered by the photoresist is removed through etching, the process by which the pattern is permanently transferred. After etching, the remaining photoresist material is stripped from the wafer. This process is repeated multiple times to fabricate the three-dimensional architecture of an integrated circuit.

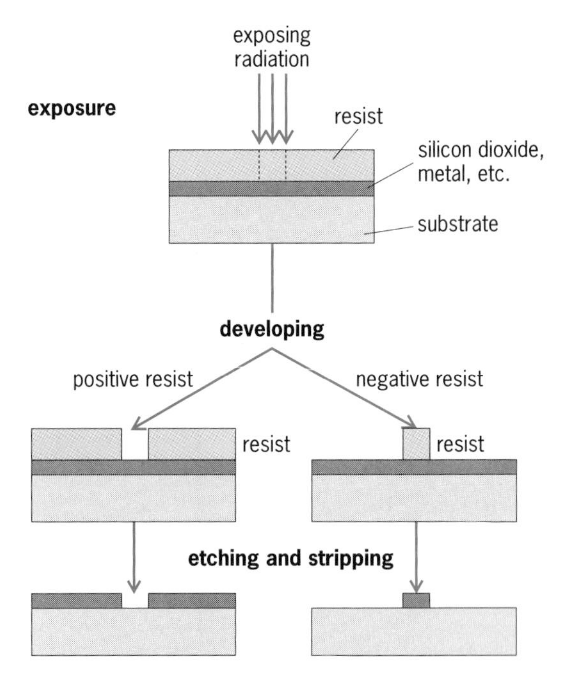

Fig. 1. Photolithographic process.

In 1965, Gordon Moore (cofounder of the Intel Corporation) predicted that the average transistor density of semiconductor devices would double every 18 months. This became known as Moore's law; and to meet it, a continual reduction in the minimum feature sizes on a microprocessor or a memory chip was needed. Since resolution is inversely related to the exposure wavelength, using shorter wavelengths of light was an obvious solution. Over the years, the exposure sources for photoresist patterning have shifted from 436- and 365-nanometer emitting high-pressure mercury lamps to 248- and 193-nm excimer lasers. In the late 1990s, the semiconducting industry was preparing to use 157-nm radiation. Initially this caused a major setback as nearly all chemical functionalities (including the C-C bonds that make up the photoresist backbones) strongly absorb at this wavelength. This is highly undesirable as it leads to signal attenuation, diffusion issues, and ultimately poor image quality. In the pursuit of polymers suitable for 157-nm-based photolithography, an empirical observation was made that integrating fluorine atoms and fluorine-containing groups into aliphatic polymeric materials with high degrees of hydrogen deficiency results in greatly reduced absorbance at 157 nm.

Although a solution was eminent, the semiconductor industry abruptly skipped 157-nm photolithography and decided to exploit the immersion lithography technique at 193 nm. In this way, the existing mask, resist, and material infrastructures could be used. Immersion as practiced in microlithography involves adding water into the gap between the final lens element and the wafer. This improves the depth of focus but does not improve resolution directly. However, it does enable an increase in the lenses' numerical aperture, which can be increased to the point where the lenses would not function if air were in the gap. Increasing the index of refraction of the fluid (over that of water) and increasing the index of refraction of the resist would enable fabrication of lenses with even higher numerical apertures, which ultimately would result in higher resolution. It is anticipated that new, high-refractive-index polymeric imaging materials will be needed to support the growth of this technology.

The emerging worldwide focus on nanotechnology will undoubtedly influence the future of photolithography. For example, the smallest physical dimensions of microprocessors currently in production are rapidly approaching the size of the individual polymer chains of the photoresist materials. Therefore, it will become important to accurately understand how photoresist structure affects performance. Ultimately, this will require a new level of control in the synthesis and characterization of polymeric materials. Until recently, the principal polymers used in commercial photoresists were based on novolac resins or poly(4-hydroxystyrene) copolymers with varying chain lengths. To help understand the effect of varying polymers' chain lengths, molecular weights, and architectures on their dissolution rates, a range of well-defined photoresists were prepared by synthetic techniques developed in the last 10 years (**Fig. 2**). This allowed experimental confirmation that dissolution rates of photoresist materials vary exponentially with molecular weight. In addition, these investigations showed that random copolymers exhibited faster dissolution rates than analogous block or graft copolymers. It is expected that the dependence on advances in polymer synthesis will be increasingly important as feature sizes are pushed to 100 nm and below.

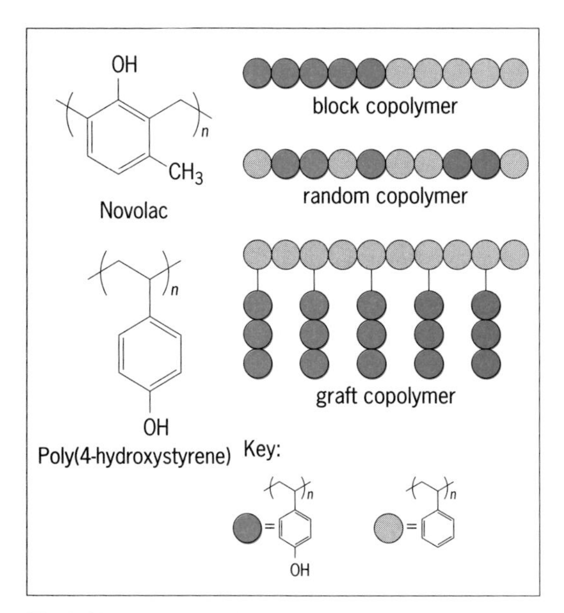

Fig. 2. Various photoresist materials with advanced architectures.

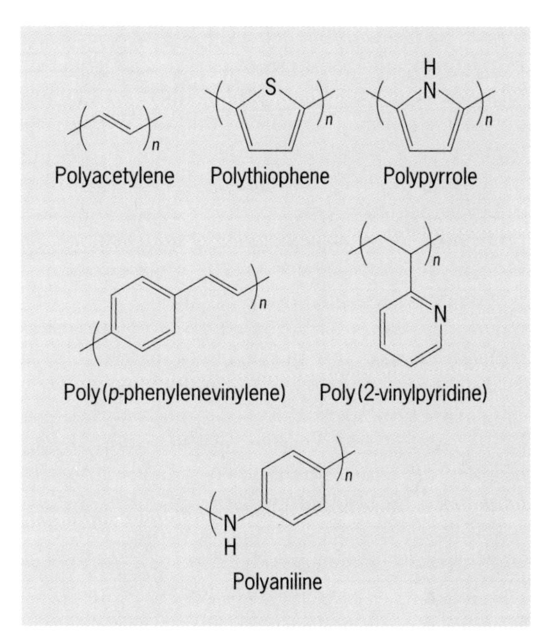

Fig. 3. Various conjugated polymers used in photovoltaics.

Applications of conjugated polymers. A. G. MacDiarmid, H. Shirakawa, and A. J. Heeger discovered that polymers whose backbones contain a series of alternating single (σ) and double (π) bonds (conjugated polymers) exhibit dramatically increased electronic conductivities when chemically doped with oxidants, reductants, or acids. Since their discovery (the 2001 Nobel prize was awarded to the discoverers), these conjugated polymers have been used in a variety of devices, including sensors, light-emitting diodes, polymer actuators, corrosion protection coatings, rechargeable batteries, and electromagnetic shields. Conducting polymers also offer a unique combination of properties that makes them attractive alternatives to the metallic alloys that currently are used in microelectronics. Most polymeric materials are lightweight, easily processed, flexible, and generally inexpensive. Their physical properties, including their inherent conductivity, can be tuned through structural manipulation of the backbone, by the nature or amount of the dopant, or through blending or copolymerization.

Certain conjugated polymers are ideally suited for use as optoelectronic devices. Effort has been devoted to using conjugated polymers, such as polyacetylene, polythiophene, poly(2-vinylpyridine), poly(*p*-phenylenevinylene), polypyrrole, and polyaniline, in photovoltaics (for example, solar cells) [**Fig. 3**]. The overall efficiency of photovoltaic devices is determined by their ability to generate highly energetic charge-separated intermediates (excitons) from incoming radiation. Since excitons typically recombine when they are at distances of 10 nm or less, new polymeric materials are needed for increased efficiencies. Leading efforts involve the development of interpenetrating networks containing donor (electron-donating/hole-accepting) and acceptor (electron-accepting/hole-donating) polymers. Currently, the best polymers for photovoltaics produced are less efficient at harvesting light and transferring charge than their silicon counterparts. To overcome these limitations, polymers are being developed that use dyes to enhance their harvesting efficiencies, and techniques are being devised that increase their structural order, which may enhance their charge-transport properties. Despite these drawbacks, the major advantage of photovoltaics based upon conjugated polymers is their high open-circuit voltages. Combined with the other desirable physical properties typically associated with polymeric materials, such as their processibility and use on flexible devices, the range of photovoltaic-based applications should greatly expand. For example, imagine painting your house or car with a material that not only enhances its esthetics but also produces energy during the day.

Polymeric dielectrics. The push to produce logic and memory circuits with faster clock speeds has been so successful that signal propagation through the interconnecting wiring is becoming the speed-limiting factor. The speed of signal propagation through this wiring depends on the dielectric constant of the insulating medium. Historically, thermally stable polymeric materials, such as polyimides, have been widely used in the semiconductor industry as interlayer dielectrics. However, because of their aromatic hydrocarbon backbones and carbonyl moieties, these polymers possess relatively little potential for further lowering of their inherent dielectric constants. During the 1980s and 1990s, Dow Chemical Co. developed and commercialized a new class of polymers based on biscyclo[4.2.0]octa-1,3,5-triene (benzocyclobutene or BCB). The advantages of BCB-based polymers are that, during the polymerization process, they do not emit any volatiles (detrimental in ultraclean circuit manufacturing environments) and show minimal shrinkage (**Fig. 4**). More recently, fluorinated polymers, such as polytetrafluoroethylene and fluorinated ethylene-propylene copolymers, polysiloxanes, and polysilsesquioxanes, have been explored as dielectrics in integrated circuits. In addition to their low dielectric constants, these polymers are very resistant to water uptake and show high chemical durability. Since most fluorinated polymers are crystalline and therefore exhibit poor solubility, recent efforts have been focused on the developing fluorinated derivatives with enhanced processing and solubility characteristics.

Outlook. Over the past 40 years, the role of polymers in the expansion of the microelectronics industry has steadily increased. Ultimately, this has been a direct result of overcoming limitations or obstacles through advancements and innovations in polymer synthesis. Regardless of the field, it is envisioned that the development of new polymeric materials with advanced architectures and unprecedented properties will be needed to sustain such growth.

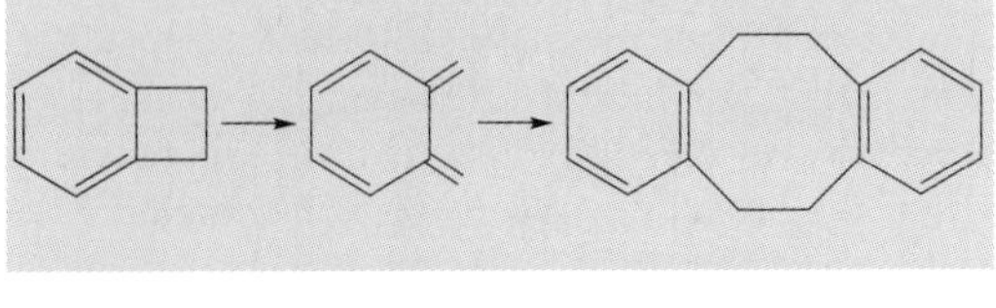

Fig. 4. Dimerization of benzocyclobutene (BCB).

For background information *see* CONJUGATION AND HYPERCONJUGATION; COPOLYMER; DIELECTRIC MATERIALS; ELECTRON-HOLE RECOMBINATION; EXCITON; INTEGRATED CIRCUITS; LASER; MERCURY-VAPOR LAMP; MICRO-ELECTRO-MECHANICAL SYSTEMS (MEMS); MICROPROCESSOR; ORGANIC CONDUCTOR; PHENOLIC RESIN; PHOTOVOLTAIC CELL; POLYMER; PRINTED CIRCUIT; SEMICONDUCTOR MEMORIES in the McGraw-Hill Encyclopedia of Science & Technology.

C. W. Bielawski; C. G. Willson

Bibliography. T. Blyte and D. Bloor, *Electrical Properties of Polymers*, Cambridge University Press, 2005; B. J. Lin, Immersion lithography and its impact on semiconductor manufacturing, *J. Microlith. Microfab. Microsys.*, 3:377–395, 2004; S. A. MacDonald, C. G. Willson, and J. M. J. Fréchet, Chemical amplification in high-resolution imaging systems, *Acc. Chem. Res.*, 27:150–158, 1994; C. A. Mack, The lithography expert: Immersion lithography, *Microlithog. World*, May 2004; K. Matyjaszewski, *Advances in Controlled/Living Radical Polymerization*, Oxford University Press, 2003; A. Reiser, *Photoreactive Polymers: The Science and Technology of Resists*, Wiley, New York, 1989; L. F. Thompson, C. G. Willson, and M. J. Bowden, *Introduction to Microlithography*, American Chemical Society, Washington, DC, 1994.

Post-perovskite

Rocks in the Earth's crust and mantle are composed of minerals: crystalline structures having elements in ordered lattices. For a common mineral, such as olivine [$(Mg,Fe)_2SiO_4$], there is a stable crystal form that exists for the pressure and temperature conditions present near the Earth's surface. The physical properties of a mineral like olivine are determined by its composition and crystal structure. At higher pressures and temperatures, below 410 km (255 mi) deep in the mantle, olivine is not stable in its near-surface form, and $(Mg,Fe)_2SiO_4$ occurs in different mineral polymorphs, with denser packing of the elements. An olivine-composition mineral form, called wadsleyite, exists below 410 km deep in the mantle. Below 520 km (320 mi) depth, there exists an even more densely packed mineral form of olivine called ringwoodite. If an olivine-bearing rock sinks in the mantle in a subducting lithospheric plate, the olivine minerals undergo phase transitions over narrow pressure (depth) ranges, transforming from one mineral form to the next with increasing pressure and temperature conditions. At 650 km (400 mi) depth, olivine composition in the ringwoodite structure undergoes a different type of phase transition, called a disassociative transition, which forms two distinct minerals—magnesium-silicate perovskite [$(Mg,Fe)SiO_3$] and ferropericlase [$(Mg,Fe)O$]. Rising mantle material undergoes the reverse sequence of phase transformations. At each phase transition, physical properties of the olivine-bearing rock, such as the density, bulk modulus (incompressibility), and shear modulus (rigidity), abruptly change. This causes corresponding rapid increases in elastic (seismic) wave velocities at depths of 410, 520, and 650 km.

The predominant upper-mantle minerals all transform to the remarkably stable magnesium-silicate perovskite form at lower-mantle conditions, with no further phase changes occurring over a several-thousand-kilometer depth range. Thus, $(Mg,Fe)SiO_3$ perovskite is believed to be the most abundant mineral in the Earth, making up 70–80% of the vast lower mantle. The physical properties of magnesium-silicate perovskite have been extensively studied by experiments and theory. In 2004, high-pressure experiments in Japan first demonstrated that for pressures greater than about 120 gigapascals, corresponding to depths in the lowermost mantle within a few hundred kilometers of the core–mantle boundary, magnesium-silicate perovskite undergoes a transition to a new mineral structure, assigned the inelegant name post-perovskite.

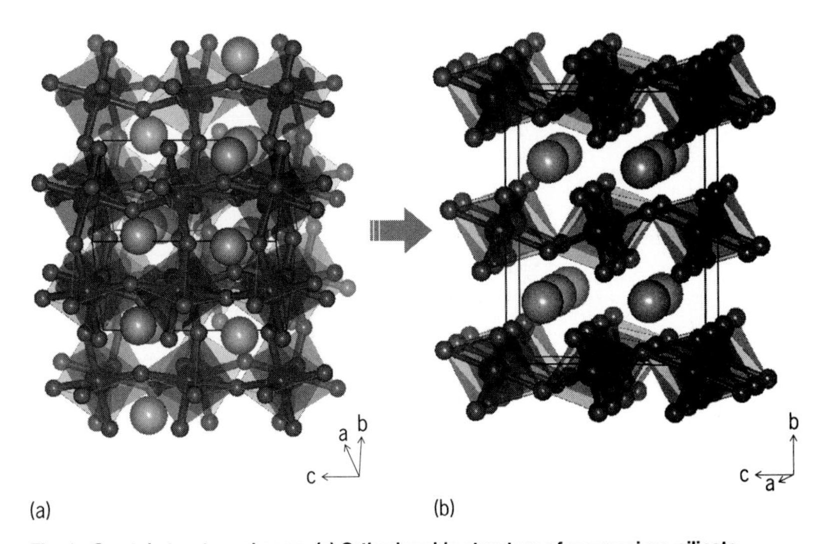

Fig. 1. Crystal structure change. (*a*) Orthorhombic structure of magnesium–silicate perovskite and (*b*) base-centered orthorhombic polymorph post-perovskite structure. Large spheres represent Mg ions, and octahedrals represent SiO_6 units. There is about a 1.0–1.2% density increase from perovskite to post-perovskite. (*From T. Lay et al., 2005; reproduced by permission of American Geophysical Union*)

Physical properties. The phase transition from perovskite to post-perovskite does not affect mineral composition, but the post-perovskite phase is 1–1.2% denser and has higher shear modulus than perovskite. X-ray diffraction studies of experimental samples at high pressure established the existence and the change in volume of the post-perovskite phase and provided constraints on the atomic lattice for theoretical modeling of the precise crystal structure of the mineral. **Figure 1** shows the change in crystal structure from perovskite to post-perovskite, based on molecular dynamic modeling. The theoretical calculations provide many important physical characteristics of post-perovskite that have not yet been directly confirmed by experiments. This includes prediction of the slope of the phase boundary in pressure–temperature (P–T) space: the Clapeyron slope. The computed Clapeyron slope for the pure magnesium (Mg) end-member composition, $MgSiO_3$, is about 7.5 MPa/K, a rather large positive value. This indicates that the phase boundary should occur at lower pressure (shallower in the mantle) in regions that are relatively lower in temperature. Large variations in the depth of the phase transition could thus result from the strong thermal heterogeneity expected to exist in the vicinity of the thermal boundary layer in the lowermost mantle.

Numerical calculations also predict the crystal elasticity of post-perovskite at P–T conditions likely to exist in the deep mantle, providing estimates of the seismic pressure-wave (P-wave) and shear-wave (S-wave) velocities. The P-wave velocity changes little relative to that for perovskite as a result of competing effects of increasing shear modulus, increasing density, and decreasing bulk modulus; however, the S-wave velocity is about 2% faster than for perovskite. If the transition from perovskite to post-perovskite is confirmed to occur over a small pressure (depth) range, the resulting rapid increase in S-wave velocity is expected to produce a velocity discontinuity that can reflect shear-wave energy. The theoretical models of elasticity also predict anisotropic properties (such as the directional dependence of the wave velocity) of the post-perovskite crystals. These differ significantly from those for perovskite in low-temperature calculations, with increasing temperature reducing the differences but still allowing a contrast in anisotropic properties under deep-mantle conditions.

The pioneering experimental and theoretical work on post-perovskite was performed for the pure Mg end-member, but effects of the presence of iron (Fe) and aluminum (Al) have recently been explored experimentally and theoretically. It is believed that lower-mantle silicates probably contain 10–15% iron substitution for magnesium. Initial work suggested that having iron in the post-perovskite mineral should reduce the pressure of the phase transition, such that it may occur hundreds of kilometers shallower in the mantle than for an iron-free mineral. These results have been contested in very recent experiments that find less pressure effect due to inclusion of Fe. Theoretical predictions of the effects of Al substitution for both Mg and silicon (Si) in the crystal lattice suggest that there may a significant depth range (a few hundred kilometers) over which perovskite and post-perovskite can coexist, which would reduce any velocity discontinuity, weakening seismic-wave reflections from the transition.

Lower-mantle rocks, like all rocks in the Earth, will involve an assemblage of mineral phases with variations in crystal size and rock fabric because of solid-state convection. While there has been some experimental work done on real rock samples at lower-mantle pressures and temperatures (with the

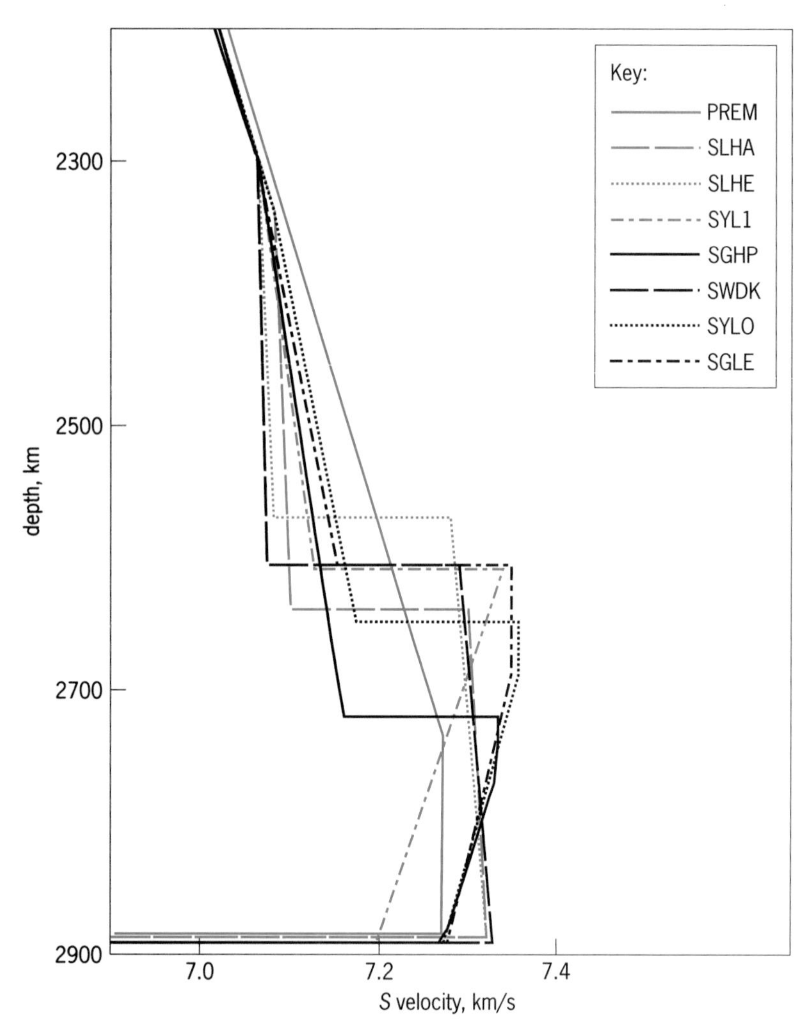

Fig. 2. Models of seismic S-wave velocity in the deep mantle. PREM is an average Earth model. The other models, determined for the localized regions shown on the map in Fig. 3 by analysis of seismic waves, all indicate the presence of a 2–3% shear velocity discontinuity 200–300 km (120–190 mi) above the core–mantle boundary (2891 km or 1796 mi deep). This is well explained by the presence of post-perovskite in the lowermost mantle in these regions.

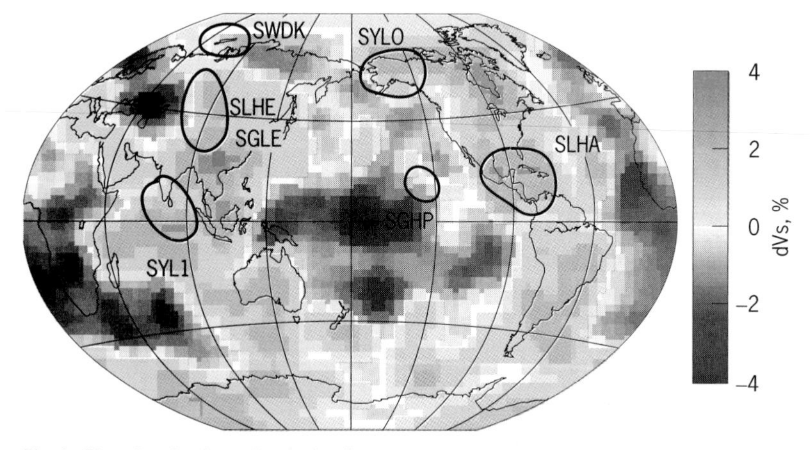

Fig. 3. Map showing laterally varying S-wave velocities at a depth of about 2800 km (1740 mi) deep in the mantle. Color regions indicate faster-than-average velocity, while gray regions indicate slower-than-average velocity, where average is defined by the PREM model in Fig. 2. The labeled subregions correspond to areas where an S-wave-velocity discontinuity has been observed a few hundred kilometers above the core–mantle boundary. This usually corresponds to regions of faster-than-average velocity, which are likely to be regions of lower-than-average temperature, with a post-perovskite transition at shallower depths than in low-velocity (hotter) regions.

post-perovskite phase being observed), full assessment of coexisting multiple phases is just beginning. For example, the properties of ferropericlase [(Mg,Fe)O] are important, especially the partitioning coefficient of iron between perovskite and ferropericlase. Recent experimental work indicates that at high pressure, Fe, normally in its high-spin state (Fe^{3+}) in the lower mantle, will prefer to be in a low-spin state (Fe^{2+}) in the lowermost mantle, which will favor iron partitioning into ferropericlase rather than perovskite. This Fe spin transition may occur at depths similar to the post-perovksite phase boundary, so iron partitioning may affect the post-perovskite composition. Thermal, electrical, and mechanical transport properties of lower-mantle rocks will be influenced by iron distribution. Thus, future work on realistic assemblages under high pressure-temperature conditions is very important for assessing the effects of the precise composition of post-perovskite in the Earth.

Seismological evidence for post-perovskite in deep mantle. The experimental discovery of the perovskite-to-post-perovskite phase transition may offer an explanation for a long-standing seismological observation. In 1983, S-wave reflections from a velocity increase in the deep mantle were first observed in several regions, and there has been extensive mapping of the reflecting interface in subsequent studies. **Figure 2** shows examples of seismological S-wave-velocity models for the lowermost mantle, indicating the presence of an abrupt 2–3% increase in shear velocity about 200–300 km (120–190 mi) above the core-mantle boundary. Global mapping of three-dimensional S-wave velocity variations in the deep mantle indicates that the lowermost 300 km of the mantle has strong, large-scale patterns of heterogeneity, with relatively high S-wave velocities underlying the margins of the Pacific Ocean (**Fig. 3**). This coincides with the regions where large volumes of oceanic lithosphere have subducted during the past 200 million years, and there are indications of slab material extending downward throughout the mantle, possibly connecting to large provinces of relatively high-S-wave-velocity material at the base of the mantle.

Subducted oceanic slab material should be relatively low in temperature, compared to surrounding ambient mantle, so that at lowermost mantle pressures this material may preferentially undergo transition to post-perovskite, resulting in a reflecting surface at the phase change. Figure 3 indicates that the models in Fig. 2 are found for regions with higher-than-average S-wave velocity in the lowermost mantle, where it is plausible that slab material has descended. If the patterns of heterogeneity in Fig. 3 truly indicate relative temperatures (rather than a compositional change), low-velocity regions should be hotter, and therefore any post-perovskite phase transition may occur at greater depth or not at all in the hotter regions. Seismologists are seeking to establish whether the lowest-velocity regions have any S-wave-velocity discontinuity, but so far there is little evidence for this. Thus, post-perovskite may exist in large patches of lower-mantle material that have been cooled by recently subducted slab material, and as it heats up over time the material may change back to perovskite. The core–mantle boundary is likely to be at a temperature too high for post-perovskite to be stable, so there may be a thin basal layer, with rapidly increasing temperature below regions cooled by slab material, in which the minerals transform back to perovskite.

The anisotropic properties of post-perovskite may further explain why seismologists find a strong association between regions with a lower-mantle S-wave-velocity discontinuity and regions with strong S-wave splitting, which is not observed in lower-velocity regions. S-wave splitting into two waves (a fast and a slow) is caused by wave propagation through a region with anisotropic rock properties, which can result from alignment of anisotropic post-perovskite crystals by shearing flow of the rock in the lower-mantle boundary layer.

Dynamical consequences. The large positive Clapeyron slope of the post-perovskite phase boundary in the presence of lateral temperature differences at the base of the convecting mantle may influence the generation of boundary layer instabilities. Warmer regions of the boundary layer will have a thinner layer of dense post-perovskite mineralogy, while colder regions of the boundary layer will have a thicker layer of the denser material. This thermally induced topography on the phase boundary is like that near the 410-km olivine–wadsleyite phase transition. And in both cases, the pattern promotes flow of material across the boundary layer (as the elevated dense material sinks and pulls down overlying material that transforms to the denser phase). Because the Clapeyron slope for the deep-mantle transition is about twice that of the upper-mantle transition, the effect is enhanced. Convection models that include the post-perovskite transition have quite unstable lower thermal boundary layers that tend to generate vigorous deep-mantle flow. Seismological mapping of the phase boundary can thus provide a probe of the thermal and dynamical processes in the deep mantle.

For background information *see* EARTH INTERIOR; ELASTICITY; GEOPHYSICS; HIGH-PRESSURE MINERAL SYNTHESIS; MINERAL; OLIVINE; PEROVSKITE; SEISMOLOGY; SUBDUCTION ZONES in the McGraw-Hill Encyclopedia of Science & Technology. Thorne Lay

Bibliography. T. Iitaka et al., The elasticity of the $MgSiO_3$ post-perovskite phase in the Earth's lowermost mantle, *Nature*, 430:442–444, 2004; T. Lay et al., Multidisciplinary impact of the deep mantle phase transition in perovskite structure, *EOS, Trans. AGU*, 86:1,5, 2005; M. Murakami et al., Post-perovskite phase transition in $MgSiO_3$, *Science*, 304:855–858, 2004; A. R. Oganov and S. Ono, Theoretical and experimental evidence for a post-perovskite phase of $MgSiO_3$ in Earth's D″ layer, *Nature*, 430:445–448, 2004; T. Tsuchiya et al., Elasticity of post-perovskite $MgSiO_3$, *Geophys. Res. Lett.*, 31:L14603, doi:10.1029/2004GL020278, 2004.

Primate communication

Primates are our closest biological relatives. Therefore, we expect to find many similarities between primate and human communication. Primates primarily use vocal and visual modalities, as do humans, and some primate communication demonstrates impressive complexity. Nonetheless, primate communication differs from human communication, especially language, in important ways. This article focuses on natural communication processes in primates rather than artificial training to use sign language or computer symbols to mimic human language.

Modalities. Primates use many sensory modalities for communication.

Vocal communication. Vocal communication is important for species living in trees or dense vegetation and for communication over long distances (**Fig. 1**). Some primates have large vocal repertoires with up to 40 different types of calls. Slight variations of call structure that may not be detectable to human ears indicate subtle differences in context and motivation. Some pair-bonded species engage in complex duet songs to which each mate contributes its own calls. Sequences of calls in some species follow simple grammatical rules.

Visual communication. Visual communication is important to species living in open areas and to close interactions in all species. There are many types of visual signals. The origins of human smiles, laughter, and aggressive and sad facial expressions can be seen in monkeys and apes. Hand and arm gestures are used for greeting, recruiting others, and pointing in ways that appear similar to human gestures. Other types of visual cues include the ability to erect hair on the head or entire body to make an animal look larger; changes in the position of the tail; changes in the coloration of the face or genital areas; periodic swellings of the genital area, which correspond to sexual receptivity in females; and rapid tongue flicking in sexual and aggressive contexts.

Fig. 1. Cotton-top tamarin vocalizing. (*Photo by Carla Y. Boe*)

Facial expressions often correlate with specific vocalizations, an example of cross-modal matching, which is thought to be an important prerequisite for language. Monkeys and apes can match a vocalization with the appropriate facial expression and will look longer at a picture of a face that matches a vocalization.

Olfactory communication. Olfactory communication is challenging to study because odors are difficult for humans to detect. Careful observation can suggest whether olfactory signals are being used, and behavioral experiments can determine if primates can discriminate between odors. Ring-tailed lemurs mark their tails with secretions from glands in their forearms and then engage in tail-waving "stink-fights" at territorial boundaries. Many species of marmosets and tamarins (small monkeys from the Neotropics that tend to live in extended family groups) have scent glands in the anogenital region, on the lower abdomen, and sometimes on the chest. Female odors change in quality during their ovarian cycles. Males respond to these odors with increased sexual arousal and display a rapid increase in testosterone after brief exposure to odors of an ovulating female. Species, individual identity, sex, and social status are communicated by differences in odors.

Touch. Touch is an important mode of communication. Primates groom each other extensively. Although grooming can help keep fur clean and free of parasites, it also serves an important social function by helping maintain relationships (**Fig. 2**). Grooming releases endogenous opioids in the brain of the receiver, providing an important way to reward a social partner. Primates spend much time in physical contact with close social companions; pair-bonded titi monkeys that sit together with their tails entwined are the most striking example.

Vocal development. One significant difference between primates and humans is that primates do not appear able to learn new vocalizations. Virtually every individual within a primate species has a similar vocal repertoire. Development of communication has three distinct components: production of signals, using signals in appropriate contexts, and understanding signals.

Fig. 2. Grooming between tamarins. (*Photo by Carla Y. Boe*)

Signal production. In general, primates show little flexibility in vocal production, although development of adult vocal structure may occur over several months or even years. Adult call forms often do not appear until after individuals become reproductively active. For example, in cooperatively breeding tamarins in which adult offspring are reproductively suppressed, the young animals give imperfect versions of adult calls. In pair-bonded species, both males and females can adjust their call structure to match those of their mates or can coordinate singing behavior with that of their mate.

Appropriate use of signals. Primates appear to need some experience to learn how to use signals appropriately. For example, infant vervet monkeys give aerial alarm calls to falling leaves and small, nonthreatening birds, only gradually learning to use calls to bird species that are actual threats. Young tamarins give food calls to small manipulatable objects as well as to food.

Understanding signals. Learning appears to be most important in understanding the calls of others. Macaque mothers that cross-foster infants of a different species appear to understand the calls of their foster infants even though the infants cannot learn to produce the calls of their foster mothers. Vervet monkeys, lemurs, and Diana's monkeys can learn the meaning of alarm calls of other species (both primates and birds), and use the calls of other species to avoid predation.

Vocalization variations. Dialects are uncommon in primate vocalizations, but some evidence for dialects has been found in marmosets, macaques, and chimpanzees. Whether these differences occur as local adaptations to habitat acoustics, as a result of genetic drift, or from social learning is not known with certainty, but there is some support for each mechanism.

Babbling is a common feature of human vocal development, but vocal play is rare in primates. However, pygmy marmosets and related species demonstrate babbling. Infant pygmy marmosets produce long sequences of calls similar in structure to adult calls, but unlike adult calls, infant babbling contains calls of different functional types juxtaposed. Thus, an affiliative call may be followed by a threat, an alarm call, and a food call. Infants can call for minutes at a time, making them conspicuous to parents and predators. Babbling infants are more likely to receive social interaction from other group members than nonbabbling infants. Infants that babble more appear to develop adult call structure sooner, suggesting that babbling may provide vocal practice. It is not known why marmosets show this "babbling" and other primate species do not.

Signal specificity and flexibility. Some primates have predator-specific alarm calls. Lemurs and capuchin monkeys have distinct calls for aerial versus terrestrial predators. Vervet monkeys have distinct calls for eagles, snakes, and leopards, and Diana's monkeys also have distinct calls for eagles and leopards. When these calls are played back to an undisturbed group, animals behave as though the call represents the actual predator. Vervet monkeys, on hearing an aerial alarm call, run to the ground and take cover. If they hear a snake alarm, they climb a tree and look down. Diana's monkeys respond equally to a natural call from a predator and to a specific alarm call of either their own species or another monkey species in the same habitat. Interestingly, when Diana's monkeys give a leopard alarm call, the leopard leaves the area. However, Diana's monkeys do not call when they hear chimpanzees, as chimpanzees will pursue a calling monkey.

Some primates have calls that appear specific to discovering food, and infants in some species use different forms of screams in agonistic situations that relate to the status of the threatening animal and the degree of aggression. These studies suggest that primates use "referential" calls that appear to refer explicitly to objects or events in the environment, as words do.

Primates are also very flexible in using signals. They can use their signals selectively in a socially strategic fashion and appear able to apply calls to novel contexts. Wild pygmy marmosets have three types of calls used to contact other group members, and these calls differ in how well they transmit through the environment. Marmosets use the most cryptic calls when they are close to each other, and more readily localized calls when farther apart. Many primates use food calls selectively, giving more calls to food that can be shared with others than to food that can be monopolized. Captive tamarins give alarm calls toward familiar food made noxious and give mobbing calls toward veterinarians, but not toward natural predators, such as snakes.

Relationship to human communication and language. Although primates appear to have complex communication skills, primate signals are more similar to human nonverbal signals than to language. Primates have small vocabularies and cannot recombine signals to create new ones. Some primates exhibit simple rules governing sequences of calls, but these are extremely limited compared with human grammar. Semantic signaling is limited to a few species that communicate about predators or food. There is little evidence that primates can take the perspective of others, an important factor in successful human conversation.

Conservation. Humans are the greatest threat to primates. In some environments, monkeys give terrestrial predator alarms to humans as well as to lions and leopards. In other environments, primates vocalize less often and with less complexity close to human settlements or after disturbance by humans compared with primates in less disturbed environments. Most primate signals are adapted to physical features of natural habitats. If those habitats are disrupted, communication efficiency is affected. In order to continue to study and understand primate communication, their habitats must be protected.

For background information *see* ANIMAL COMMUNICATION; LINGUISTICS; PRIMATES; SOCIOBIOLOGY in the McGraw-Hill Encyclopedia of Science & Technology.

Charles T. Snowdon

Bibliography. M. D. Hauser and M. Konishi (eds.), *The Design of Animal Communication*, MIT Press, 1999; D. K. Oller and U. Griebel, *Evolution of Communication Systems: A Comparative Approach*, MIT Press, 2004; D. H. Owings and E. S. Morton, *Animal Vocal Communication: A New Approach*, Cambridge University Press, 1998; C. T. Snowdon, Expression of emotion in nonhuman animals, pp. 457–480 in *Handbook of Affective Science*, ed. by R. J. Davidson, K. Scherer, and H. H. Goldsmith, Oxford University Press, 2002; C. T. Snowdon and M. Hausberger (eds.), *Social Influences on Vocal Development*, Cambridge University Press, 1997.

Primate origins

By the midtwentieth century most scientists shared the view that the primate order consisted of the living and fossil tree shrews, lemurs, galagos, lorises, tarsiers, monkeys, and apes, including humans. The fossils of three extinct radiations known primarily from the Paleocene and Eocene of western Europe and North America were also included: the Plesiadapiformes (sometimes referred to as Archaic Primates), the Adapiformes, and the Omomyiformes. Primates were seen as the archetypal arboreal mammals and were consequently thought to have evolved most of their characteristics as a direct result of the demands of arboreal life.

During the second half of the last century, new fossils and the growing influence of cladistic systematics (the classification of organisms based on common ancestry) combined to gradually erode this view. Plesiadapiforms and tree shrews were each relegated to a separate order (Plesiadapiformes and Scandentia, respectively), and new explanations emerged for the evolution of a newly defined characteristic set of primate traits: forward-facing eyes (resulting in a wide field of stereoscopic vision), grasping hands and feet with digits topped by nails rather than claws, and increased brain size.

In a new consensus, the pruned primate order was thought to have originated at some point during the Paleocene (65 to 55 million years ago) as small, nocturnal animals adapted for foraging in the small terminal branches of trees and bushes. In the two main conflicting hypotheses, the characteristic traits were interpreted as evidence that either the ancestral primate was a visual predator, adapted for stalking and grasping insects and other prey (Matt Cartmill's visual predation hypothesis), or the ancestral primate evolved in parallel with Paleocene/Eocene angiosperms, exploiting their fruits, flowers, and nectar (R. W. Sussman's angiosperm co-evolution hypothesis). While these views still dominate discussions of primate origins today, new evidence from the fields of molecular biology, paleontology, and statistics, as well as novel insights from comparative studies of modern species, have recently combined to reopen a debate that had largely stalled.

Ancestral diet. Overall, comparative data remain ambiguous regarding ancestral primate dietary adaptations. However, some recent research aimed at reconstructing ancestral primate body mass suggests that ancestral primates may have weighed around 1 kg (2.2 lb), twice as much as the highest previous estimates. If correct, this would significantly impact traditional interpretations of ancestral primate adaptations because of the pervasive influence that body mass has on the biology of a species. Most notably, comparative data from modern primates show that it would be highly unlikely for an animal of that size to be able to rely on a primarily insectivorous diet.

Fossil evidence. One of the attractions of reconstructing ancestral character states from comparative analyses of modern primates comes from the fact that, to this day, the fossil record has doggedly refused to release any direct evidence of primate evolution prior to the Paleocene-Eocene boundary, some 55 million years ago (see **illustration**). A recent find from the Late Paleocene of Morocco (*Altiatlasius koulchii*) was initially described as an omomyiform primate but later reclassified as a plesiadapiform. The oldest undoubted fossil primates come from basal Eocene rocks in Europe, North America and Asia. Their sudden, virtually simultaneous appearance across the Northern Hemisphere (a major unit of the Earth's surface extending from 30–45°N latitude and characterized by faunal homogeneity) clearly suggests that they did not evolve in situ, but immigrated from an as yet unspecified part of the globe. Until recently, research on early primates was strongly focused on western Europe and North America and the two groups of extinct primates known from these regions, the Adapiformes and Omomyiformes. Even today, despite a surge of activity in other parts of the world, nearly 80% of known Eocene species come from restricted areas of North America and western Europe. The most intriguing recent finds, however, have come from Asia.

Dispersal from Asia to Europe. In 2004, the description of a well-preserved skull of an omomyiform primate (*Teilhardina asiatica*) from the earliest Eocene of China markedly affected our knowledge of early primate evolution. It was long assumed that direct dispersal of mammals into Europe from Asia at the beginning of the Eocene was prevented by the presence of a marine barrier formed by the epicontinental West Siberian Sea and the Turgai Straits. As Europe was also cut off from Africa by the presence of the Tethys Sea to the south, the only remaining route for mammalian, and hence primate, dispersal into Europe appeared to be via North America and Greenland. Significantly, the close affinity between Asian *T. asiatica* and European *T. belgica* to the exclusion of North American *T. americana* now adds important support to recent evidence that direct dispersal from Asia to Europe must have been possible. The antiquity of the *T. asiatica* deposits and the biogeographic implications of their phylogenetic affinities

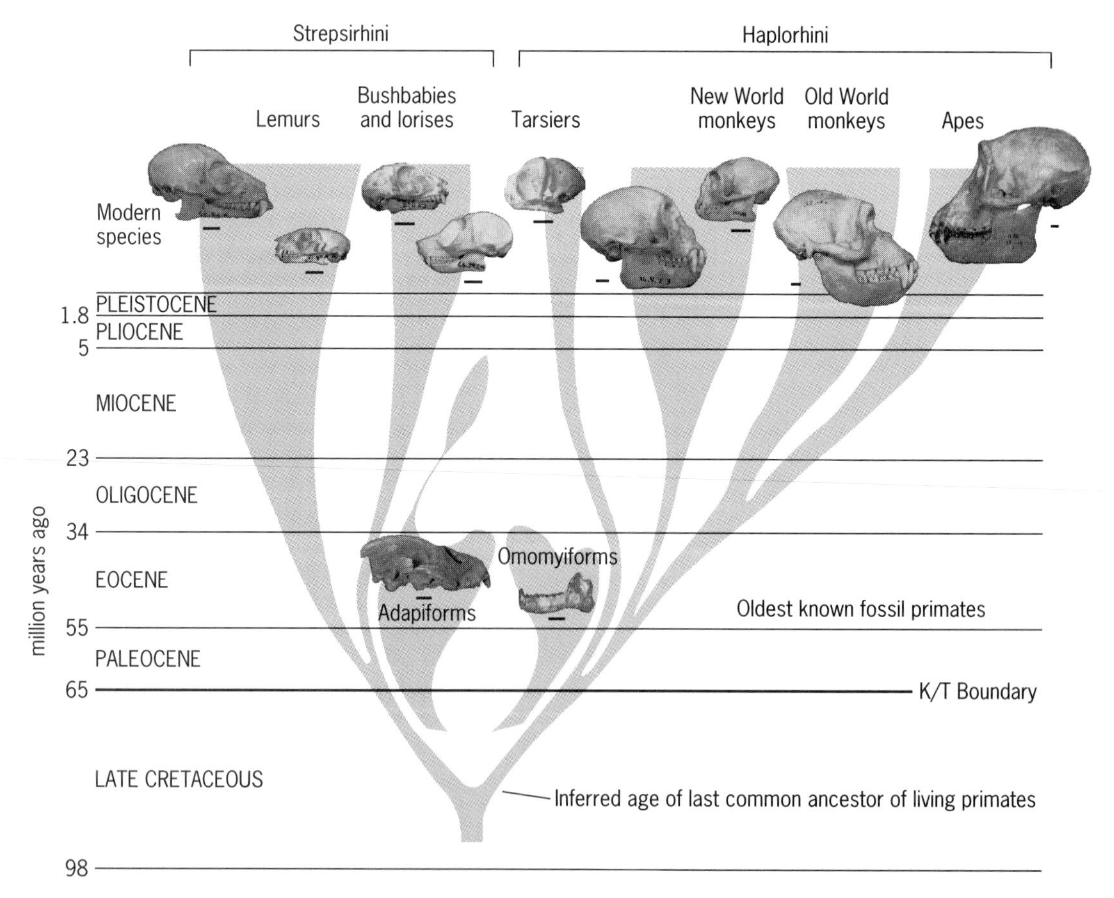

Outline evolutionary history of the Primates. Skulls of modern species (top): *Lemur catta, Cheirogaleus medius, Galago senegalensis, Loris tardigradus, Tarsius bancanus, Cebus apella, Callithrix humeralifer, Maccaca sylvanus, Pan troglodytes.* Fossil species (bottom): skull of *Adapis parisiensis*, lower jaw of *Microchoerus erinaceus*. Scale bars: 1 cm.

have confirmed Asia as a key player in early primate evolution.

Lemur origins. Possibly the most intriguing fossil of recent years has come from Pakistan. Today, lemurs form a diverse taxonomic unit that is restricted to the island of Madagascar. Because lemurs are the closest relatives (sister group) of the bushbabies and lorises (the Lorisiformes; see illustration), their lineage must be as old as that of the lorisiforms. In 2003, the documented antiquity of lorisiforms was dramatically extended from about 20 to about 40 million years with the description of two new genera, *Karanisia* and *Saharagalago*, from late Middle Eocene deposits in Egypt. In contrast, until very recently not a single fossil lemur—with the exception of some recently extinct members of the modern Malagasy fauna—had been found. In the face of an evident gap in the fossil record of more than 40 million years, the origin and evolution of lemurs remains shrouded in mystery, although most scientists currently think that lemurs must at some point have invaded Madagascar from mainland Africa. It was therefore remarkable when, in 2001, the first fossil primate with direct affinities to lemurs was not described from mainland Africa or Madagascar, but from Early Oligocene deposits in Pakistan. The new species, *Bugtilemur mathesoni*, is said to show close dental similarities to modern dwarf lemurs, members of the lemur family Cheirogaleidae. *Bugtilemur* has raised a number of questions, notably in relation to the biogeographic origin of lemurs, but also of primates in general.

Molecular and statistical evidence. A fundamental issue in contextualizing the evolution of the first primates is to determine their time and place of origin. A traditional approach for deriving the time of origin of an evolutionary group of organisms has been to assume that the group originated not long before the appearance of its oldest known fossils. For primates, this means an origin some time during the Paleocene epoch, in line with the general view that most modern mammalian orders evolved as a consequence of the fundamental changes brought upon terrestrial biota by mass extinction at the K/T boundary, which marked the end of the Cretaceous Period (65 million years ago).

However, as the fossil record provides no direct evidence of where primates originated it also provides no direct evidence of how long primates may have existed before they appear in the fossil record. A direct reading of the fossil record may accurately reflect the evolution of groups whose fossil record is very dense, but where substantial gaps exist it may result in a serious underestimate of the actual time of origin of a group. A point that has long been argued by R. D. Martin is that the overall quality of the primate fossil record is likely to be so poor as to render any direct inferences of the time when various groups of primates originated highly unreliable. In line with this argument, most molecular data have estimated the divergence of primates from other mammals, as well as the initial divergence within primates (the divergence between Strepsirhini and Haplorhini), to have occurred during the Cretaceous (see illustration). A recently developed statistical approach that estimates chronological range extensions of evolutionary groups based on their fossil record and models of their patterns of diversification has added to the evidence by estimating that the last common ancestor of living primates most likely lived over 80 million years ago. The most significant consequence of an ancient origin of primates is that it cannot be considered a fall-out of K/T-boundary events. Instead, both the emergence of the primate lineage and the initial diversification of primates would have coincided with a period during which ancient landmasses were broken up by continental drift and the epicontinental seas formed, suggesting vicariance (geographical isolation of populations) as a possible mechanism driving early primate evolution.

Although less conclusive, some evidence regarding the place of origin of primates has also come from recent molecular analyses. The important question of which modern mammalian order is most closely related to primates has yet to be settled, but the availability of substantial molecular datasets is beginning to result in a clearer overall picture of mammalian interrelationships. A significant development has been the identification of four supergroups of Eutherian orders: Afrotheria (a group of primarily African mammals that includes elephants, sea cows, hyraxes, tenrecs, golden moles, aardvarks, and elephant shrews), Xenarthra (armadillos, sloths, and the true anteaters), Euarchontoglires (rodents, rabbits, hares, pikas, tree shrews, flying lemurs, and primates), and Laurasiatheria (hoofed mammals, whales and dolphins, carnivores, pangolins, bats, hedgehogs, moles, and shrews). Continental drift is thought to have played a major role in the divergence of these groups, with Afrotheria becoming isolated on Africa at an early stage. The fact that primates do not belong to Afrotheria now seems to provide good evidence against the notion that ancestral primates occurred on Africa. An alternative hypothesis that has been advanced is that primates, and perhaps ancestors to all Euarchontoglires, became isolated on the combined landmasses of India and Madagascar as these separated from Africa. The subsequent separation of India from Madagascar and India's onward journey toward Asia may help explain both the existence of lemurs on Madagascar and the presence of *Bugtilemur* in the Oligocene of Pakistan. It is clear, however, that there are no easy scenarios either way.

For background information *see* ANIMAL EVOLUTION; BIOGEOGRAPHY; EUTHERIA; FOSSIL PRIMATES; MAMMALIA; PRIMATES in the McGraw-Hill Encyclopedia of Science & Technology.

Christophe Soligo

Bibliography. M. Cartmil, New views on primate origins, *Evol. Anthropol.*, 1:105–111, 1992; W. C. Hartwig, *The Primate Fossil Record*, Cambridge University Press, 2003; R. D. Martin, New light on primate evolution, *Ber. Abhandl. Berlin-Brandenburg Akad. Wiss.* 12, in press; R. D. Martin, *Primate Origins and Evolution: A Phylogenetic Reconstruction*, Princeton University Press, 1990; M. J. Ravosa and M. Dagosto (eds.), *Primate Origins and Adaptations*, Kluwer, Academic, 2005.

Printing on demand

Printing on demand uses digital content and reproduction technology to print a digital page on paper.

The importance of printing on demand, also known as digital printing, on-demand printing, print on demand, and, more narrowly, books on demand, is not the technology itself, as xerographic and inkjet printers are not new. The importance is how this technology has been combined with computing technology to change not only the printing industry but also industries that use the technology, such as the publishing industry.

The availability of on-demand printing technology has fueled growth in the United States publishing industry. According to R. R. Bowker, in 1993 there were 104,124 new books published, and in 2003 there were 164,609 new books published. This growth coincides with the launch of the first commercially available digital printers in the mid-1990s.

Digital printers. The printing industry's increasing use of digital printers, which input digital data streams and output pages printed on paper, is fueling on-demand printing. Digital printers include electrophotographic and inkjet. Electrophotographic printers are laser printers that fuse toner on paper. They are the most common type of digital printer used for on-demand printing. Inkjet printers are used for wide-format paper, such as posters and other content.

One reason for the growing popularity of digital printers is that the average number of copies printed per job dropped by 25 percent during the 1990s. Shorter press runs mean less efficiency for traditional offset lithography because the setup or make-ready time and paper waste are essentially the same for all traditional offset print jobs. Shorter press runs also mean less profit per job. Shorter press runs require that a greater percentage of time is spent scheduling jobs.

Digital printing systems do not require make-ready time and significantly reduce paper waste. From start to finish, production time is quicker on digital printers because time does not have to be allocated for the ink to dry before postpress processing, such as binding, can begin. Companies that publish books using offset lithography are now adding digital printers to their facilities.

Digital workflow. On-demand printing is popular with publishers because there are clear cost savings. The workflow for producing a book printed on an offset press is the same as that for producing a book printed on a digital printer. As a result, a publisher could have the first edition of a book printed on an offset press and then have subsequent, smaller runs printed on a digital printer, without having to recreate the source files. Workflow standards have been created by a consortium of corporations and trade groups, including publishing software developers, printers, and publishers, to create Extensible Markup Language (XML)-based standards such as Job Definition Format (JDF) to manage print jobs and Personalized Print Markup Language (PPML) to define and describe variable (for example, personalized) content within print jobs.

Book publishing. A key use of on-demand printing technology is to create books on demand, with the use of a digital printer to print quantities ranging from one to hundreds of copies.

Typically, the book is stored in digital format, usually as a collection of PDF (portable data format), PostScript®, or TIFF (tag image file format) files that can be printed with a digital printer, packaged (that is, bound or shrink wrapped), and shipped to the customer directly from the printer. Not only can a book be printed one at a time, it can be printed in black and white or color. A single copy of a 300-page book can be printed in less than 1 minute and bound in less than 5 minutes.

Minimizing risk. An important element of on-demand printing is understanding the typical life-cycle of a book. R. Voytko, an expert in printing-on-demand, described it as a bell curve. At the beginning, sales are typically slow as marketing efforts and reviews build demand. Once demand has increased, sales increase and eventually peak. From the peak, sales decrease until the book goes out of print.

First-edition books are usually produced in quantities large enough to justify traditional offset printing. After printing and binding, the books are stored in warehouses or shipped to distributors. Publishers accept a great deal of risk when they commit to the length of press runs for first editions, based on their estimates of potential sales. If book sales fail to sell as projected, the publisher faces returns, which must be warehoused, sold to discounters, or destroyed. If sales increase quicker than projected, the publisher may decide to do a second printing, which entails even more risk.

When demand slackens and a book goes out of print, it may not be because there is a lack of customers or back orders for the book, but because it is too expensive to print a small quantity of books using an offset press. Even if there are customers and back orders for a book, the number must be large enough to justify offset printing. This is where digital printing can extend the lifecycle of a book indefinitely.

According to Voytko, books-on-demand benefits publishers by letting them better manage a book's life-cycle, while reducing their risk and enabling them to gain incremental sales that would have ended if book went out of print.

Technological benefits. Printers, publishers, and consumers benefit from on-demand printing because the technology provides a cost-effective method for printing small quantities of books, newsletters, or any other content.

Build to order. Publishers do not have to estimate how many books to print, since they can print the books as orders come in. In essence, publishers now have just-in-time manufacturing, which was not possible with offset printing.

Reduces warehouse costs. Most business models for books on demand are based on centralized print facilities, where orders come in for books, and books are printed and shipped immediately. This reduces or eliminates the costs of carrying inventories.

Enables niche publishing. This is because the fixed costs of printing on-demand books are less than the fixed costs of offset printing, resulting in far less costs for printing small quantities of books.

Eliminates out-of-print books. Printing books when they are needed makes it possible for established authors to offer their books again to their audience and allows publication of books that are in the public domain.

Enables variable printing. This allows the content to be customized. For example, a publisher might want to personalize a newsletter for its customers by including their names or, based on their preferences, a reference to an article a customer might find interesting.

Pick and choose. Content can be selected for a book or package of information. For example, a textbook publisher might choose to publish a complete course package that includes the textbook, student notes, and quizzes, or it might choose to publish chapters from several textbooks as a collection printed specifically for a single course.

For background information *see* BOOK MANUFACTURE; INK; INKJET PRINTING; INVENTORY CONTROL; PHOTOCOPYING PROCESSES; PRINTING; PRODUCTION PLANNING; TYPE (PRINTING) in the McGraw-Hill Encyclopedia of Science & Technology. Harold Henke

Bibliography. H. Henke, *Electronic Books and ePublishing: A Practical Guide for Authors*, Springer, 2001; H. Henke, The global impact of eBooks on ePublishing, *Proceedings of the 19th Annual International Conference on Computer Documentation*, SIGDOC, ACM, 172–179, 2001; *One for the Books: Barnes and Noble Launches Print-On-Demand Operations*, IBM, G563-0139-00, 2000; F. J. Romano, *An Investigation into Printing Industry Trends*, pp. 5–7, Rochester Institute of Technology, 2004.

Process improvement methodologies

Modern product development faces many challenges, including competitive forces from the global marketplace, highly constrained resources, and the need to keep pace with rapid technological change. A common response to the challenges is a formal process improvement program to ensure that product development processes are continuously improved in order to maximize both efficiency and effectiveness. At least six categories of such process improvement methodologies are in wide use today. Each category takes a unique perspective and includes specific methodologies that are applied to improve the product development processes. Contemporary improvement programs most often invoke two or more methodologies and are customized to the individual business culture and needs.

Strategy-based methodologies. Strategy-based methods take a holistic view of the business enterprise, as exemplified by the Balanced Scorecard methodology. The Balanced Scorecard is a management system that enables organizations to clarify their vision and strategy and translate them into action. It provides feedback related to both internal business processes and external outcomes to continuously improve strategic performance and results. The Balanced Scorecard method includes the development of strategy maps, which show how value is created at every level of the enterprise. This method can help increase understanding of how product development processes contribute to the overall strategic intent of the enterprise. The results demonstrated by strategy-based methods may not relate directly to cost and schedule reduction, but are manifest in customer satisfaction with products, better return on investment, and greater revenues.

Customer-based methodologies. Customer-based improvement methodologies focus on identifying the intersection of critical customer needs with existing or planned attributes of products. In general, customer-based methods focus on capturing and understanding the Voice of the Customer (VOC), that is, the identification of the "unstated" needs and desires of customers. An example of a customer-based method is Quality Function Deployment (QFD), a structured approach to defining customer requirements and translating them into specific plans to make products to meet those requirements. While QFD is typically applied to products themselves, some organizations are applying it to identifying customer needs and desires that can be addressed by process improvement. A QFD-like approach provides the ability to do a self-assessment and identify the development practices that provide the greatest improvement within their organization, and thus form the basis of an improvement action plan.

Quality-based methodologies. There are various quality-based methods, Total Quality Management (TQM) being the most recognized. TQM is an overall business (quality) improvement system, encompassing leadership, strategic planning, human resources, and process improvement. Six Sigma is another quality-based method that is in widespread use and is driving improvement in product development processes. Invented by Motorola, Inc., in the mid-1980s as a metric for measuring defects and improving quality, Six Sigma has evolved into a robust business improvement methodology that focuses an organization on customer requirements, process alignment, analytical rigor, and timely execution. Six Sigma is a structured statistical analysis method to identify and reduce production defects to the target of 3.4 defects per million within a specific process. Whereas traditional quality programs have focused on detecting and correcting defects, Six Sigma provides methods to redesign the process with the goal of seeing that defects are never produced in the first place. Although Six Sigma brings a somewhat new direction to quality and productivity improvement, its underlying tools and philosophy are based on fundamental principles of total quality and continuous improvement that have long been in use. Design for Six Sigma (DFSS) is a comprehensive approach to product development that links business and customer needs to critical product attributes, product functions, detailed designs, and verification. Like Six Sigma, DFSS uses a structured set of steps to ensure repeatability and continuous improvement, but focuses specifically on the product design activities.

Process-enhancement-based methodologies. The process-based methodologies generally fall into the category of business process reengineering and have been in widespread use for more than a decade. Business process reengineering involves the redesign of business processes and the associated systems and organizational structures to achieve a dramatic improvement in business performance. A specific technique that has been resulting in improvements to product development processes (for example, cycle-time reduction) uses Dependency Structure Matrices (DSMs). The DSM visually represents process interactions in a matrix format and involves specific techniques for understanding how to improve these interactions to achieve optimal efficiencies.

Capability-maturity-based methodologies. Capability-based process improvement methods use Capability Maturity Models that describe general and specific best practices for development organizations. The Capability Maturity Model is a method for evaluating the maturity of development processes of organizations on a scale of 1 to 5. The latest model, CMM®I, stands for Capability Maturity Model Integration and includes a comprehensive set of practices for evolving process capability, with five defined levels of capability maturity. Many commercial, defense, and government organizations are using the CMM®I to drive improvements in product development, particularly in software products and software-intensive systems. The developer of CMM®I, the Software Engineering Institute (SEI), maintains a website with a growing database of documented improvement outcomes. One of the benefits of capability models is that they help an organization to prioritize and focus on where to apply improvement resources given the current state of the organization's processes.

Lean-based methodologies. Lean principles focus on the creation of value through the elimination of waste and the insertion of practices that contribute to overall value. The desired outcome is reducing cost and schedule while improving performance of products. The contribution of lean practices was first recognized on the manufacturing production floor as used by Toyota Motor Company. The basic lean principles and practices have since been extended to focus at the enterprise level and have been applied to a full set of processes, including product development. An essential element in the lean approach is identification of the value stream, the set of all actions required to bring a specific product through its crucial tasks to the desired outcome. In analyzing the value stream, non-value-added activities and waste are identified, and thus can be eliminated through improvement of the relevant processes and activities. In product development, the waste is often related to ineffective information handling and to waiting time between tasks. Recent studies point to as much as 40% of time charged to product development as waste, with up to 60% of tasks being in an idle state at any given time. In the lean-based approach, it is recognized that each product may be unique, but product development processes are typically repetitive and can be continuously improved to yield higher value while seeking to eliminate waste.

Methodology improvement criteria. Organizations seeking to improve their product development processes are faced with the difficult choice of selecting an appropriate methodology. This selection requires careful consideration of business culture, desired outcomes, and specific improvement goals. Each of the categories of methods has demonstrated successful outcomes in regard to improvement of product development processes. In many cases today, the effective strategy involves the synergistic use of multiple methodologies, for example, a Lean Six Sigma program along with a Capability Maturity Model.

Improvement of product development processes has traditionally involved streamlining of work processes and the creation of efficiencies in process interactions to drive quality up and costs and cycle time down. This will continue to be a valuable endeavor for product development organizations in satisfying customers, meeting marketplace demands, and navigating the competitive environment. However, the next generation of process improvement methodologies will focus not only on traditional quality and efficiency but also on evolving specific organizational process capabilities to realize products with advanced properties such as flexibility, adaptability, and robustness. Desired product design attributes are a reflection of the environment in which these are created; thus products with these advanced properties will be obtained only through an enabling process environment.

Outlook. The challenges related to product development in the contemporary environment drive the need for highly effective process improvement programs. There are many improvement methodologies that are highly suited for this task. There is increasing evidence to demonstrate the positive results of applying each of the various types of improvement methods. Given the complexity of modern enterprises, particularly those that have a history of mergers and acquisitions, the most effective strategy may be to develop a customized improvement approach through a synergistic use of multiple improvement methods. The critical need that transcends the selection of the improvement method is that every endeavor must begin with understanding the desired outcomes and specific goals for the product development improvement in the context of the particular organization.

For background information *see* MANUFACTURING ENGINEERING; PROCESS ENGINEERING; PRODUCT ENGINEERING; PRODUCT QUALITY; QUALITY CONTROL in the McGraw-Hill Encyclopedia of Science & Technology. Donna H. Rhodes

Bibliography. M. B. Crissis et al., *CMMI: Guidelines for Process Integration and Product Improvement*, Addison-Wesley Professional, 2003; M. George, *Lean Six Sigma: Combining Six Sigma Quality with Lean Production Speed*, McGraw-Hill, 2002; V. D. Hunt, *Reengineering: Leveraging the Power of Integrated Product Development*, Wiley, 1995; R. Kaplan and D. Norton, *The Strategy-Focused Organization: How Balanced Scorecard Companies Thrive in the New Business Environment*, HBS Publishing, 2001; E. Murman et al., *Lean Enterprise Value: Insights from MIT's Lean Aerospace Initiative*, Palgrave, 2002.

Proteomics

Proteomics is the study of the proteome, the total protein complement of a cell, a tissue, or an entire organism. Proteomics involves comprehensive approaches to study protein expression, posttranslational modification, interactions, organization, and function. Analysis of protein derangements on a proteome-wide scale is expected to reveal insights into deregulated pathways and networks involved in the pathogenesis of disease. Additionally, it is anticipated that proteomic studies will lead to identification of biomarkers that are important for the early detection, diagnosis, prognosis, and treatment of human diseases.

Principles and tools. There are three main components to the study of the proteome: (1) the biologic sample, (2) analytical approaches for the extraction, separation, and identification of proteins, and (3) bioinformatics analysis. Most studies require proteins extracted from fresh or frozen archived cellular material or biological fluids, such as serum, urine, or tissue samples. Cellular proteins have to be isolated from samples containing other biological molecules, including carbohydrates, lipids, and nucleic acids. Highly enriched tissues can be obtained using laser capture microdissection of specific cell types, and can be successfully used for proteomic analysis.

Several different modalities are commonly used for separation of proteins from complex mixtures. These include one-dimensional gel electrophoresis, which achieves resolution of proteins based on molecular weight; two-dimensional gel electrophoresis, which involves separation of proteins based on isoelectric point followed by separation based on size; high-performance liquid chromatography; ion exchange; and different types of affinity chromatography.

While recent developments in protein/peptide array technology hold promise for widespread future applications in proteomics, mass spectrometry is currently the principal technology for the high-throughput analysis of peptides and proteins. Mass spectrometers measure the mass-to-charge ratio of the smallest molecules with high accuracy, and have the ability to detect low-abundance proteins at sub-picomolar concentrations. In essence, peptide sequence-based protein identification by tandem mass spectrometry (MS/MS) centers on the fact that peptide sequences of 6–30 amino acid residues or greater may be sufficiently unique to provide unequivocal identification of the parent proteins by matching them with those in databases that contain genomic sequences translated to their protein counterparts.

Mass spectrometers. Mass spectrometers typically consist of three components: an ionization source, a mass analyzer, and a detector. The ionization source creates ions from the sample to be analyzed. The mass analyzer resolves the ions by their mass-to-charge ratio (m/z). The detector records the m/z and abundance of the ions. Matrix-assisted laser desorption/ionization (MALDI) and electrospray (ESI) techniques are the common ionization techniques utilized in biological mass spectrometry. In MALDI, the sample is incorporated into a chemical matrix. Laser activation of the target leads to the release of peptide/protein ions into gas phase. Surface-enhanced laser desorption/ionization (SELDI) is a variation on the MALDI concept and is embodied in the Ciphergen Chip™ system. It contains several chip matrices that separate proteins based on their preferential binding to different chromatographic surfaces. ESI involves the generation of peptide ions from solutions with liberation of ionized peptides as a spray of droplets before sampling by the mass analyzer.

Mass analyzers measure the mass-to-charge ratio of gas-phase ions generated from the ionization source. Examples include quadrupole, ion traps, time-of-flight (TOF), and Fourier-transform ion cyclotron resonance-type analyzers. In addition, hybrid instruments incorporating one type of analyzer with another (for example, quadrupole-TOF spectrometers) also have been utilized with great success.

Protein identification by mass spectrometry. Peptide mass fingerprinting (PMF) is the simplest method for protein identification. Sequence-specific enzymes (such as trypsin) are used to generate a series of peptides from a protein of interest. The peptides generated are analyzed by mass spectrometry, and the masses obtained are compared with theoretical mass spectra of proteins in a sequence database. By comparison, peptide sequencing by tandem mass spectrometry is a more definitive method for protein identification based on the random cleavage of the peptide bonds between adjacent amino acid residues in a peptide sequence achieved by collisional-induced dissociation (CID) with an inert gas. CID of the peptides yields ion series that are important for the identification of the amino acid residues in a peptide. Numerous software algorithms, such as SEQUEST™ and MASCOT™, utilize statistical modeling algorithms that permit assignment of confidence values for each identification.

Protein microarrays. Protein and antibody microarrays are synthesized by immobilization of hundreds to thousands of peptides, proteins, or antibodies on a solid matrix. Incubation with biologic fluids, such as serum, followed by a visualization step leads to the identification of antibodies that are reactive with spotted protein. Similarly, microarrays composed of thousands of antibodies can be used to screen for disease-related antigenic proteins in biologic fluids or tissue extracts. Protein microarrays are ideal for studying receptor–ligand interactions, enzyme activity, and antibody–antigen interactions in a high-throughput manner. They have the potential to detect a protein with sensitivity a thousandfold greater than an enzyme-linked immunosorbent assay, and do not require a mass spectrometer.

Medical applications of proteomics. The application of proteomics to the identification of disease markers from body fluids and tissue has received considerable attention. The most investigated areas are cancer, infectious pathogens, and inflammatory conditions such as cardiovascular and autoimmune disorders. The potential of obtaining mass-spectral profiles of peptides and proteins without the need to carry out laborious and time-consuming protein separation could be advantageous for biomarker discovery. SELDI-TOF is a popular method that requires only small quantities of sample. The mass-spectral patterns reflect the protein and peptide content of the samples, and subsequent bioinformatics analyses of mass spectral patterns have been utilized to distinguish normal patients from those with cancer, cardiovascular diseases, and autoimmune diseases. Proteomic patterns of nipple aspirate fluids, cytologic specimens, and tissue biopsies have also revealed a potential for discovery of novel biomarkers that aid in diagnosis. Additionally, the proteomes of several human infectious pathogens, such as *Plasmodium, Bacillus subtilis, Helicobacter pylori*, and *Candida albicans*, have been elucidated by mass spectrometry.

Imaging mass spectrometry. Mass spectrometry has been used for the in-situ analyses of proteins in tissue sections, thereby allowing imaging and comparison of protein expression between normal and disease tissues. In this strategy, frozen tissue sections are applied to a MALDI plate and analyzed at regular spatial intervals. The mass-spectral data obtained at different intervals are compared to yield a spatial distribution of masses (proteins) across the tissue section. Using this approach, investigators have been able to distinguish brain tumors from benign brain tissues.

Quantitative proteomics. Quantitative proteomic studies are designed to determine the "relative" proteomic differences between one cellular state and another. Two-dimensional gel electrophoresis has been utilized extensively with great success, although it is a relatively low-throughput approach which requires a large amount of starting material.

Stable isotope labeling methods also have been used with great success for quantitative proteomics. Two major approaches include stable isotope labeling with amino acids in culture (SILAC) and isotope-coded affinity tags (ICAT™). Both involve sample labeling with an isotope/tag that is either metabolically (SILAC) or chemically (ICAT™) incorporated. More recently, isobaric tags for relative assessment of quantitation (iTRAQ™) permitting multiplex comparison of up to four samples have been developed. The proteins from each labeled sample are isolated, mixed at a 1:1 ratio, and subjected to mass-spectrometric analysis. Relative quantification is achieved by comparison of the peak height or areas of the isotope/tag pairs for each peptide distinguished by the mass difference of the isotope or tag.

Subcellular subproteomics. One of the major initiatives of the Human Proteome Organization (HUPO) is the comprehensive characterization of the complete subproteome of each cell type. Defining the global fingerprint of proteins expressed in a given cell type will aid in the identification of deregulated proteins that are characteristic of disease states and aid in the diagnosis and prognosis. Protein secretion by diseased cell types may provide a means for earlier detection of disease states, including cancer. Systematic approaches to purify and identify secreted proteins from a variety of cell types have been reported and thus far include the identification of secreted proteins during differentiation of adipocytes, osteoclasts, and astrocytes. It is anticipated that these proteins represent candidate biomarkers of human disease.

Drug development. Proteomics has been applied to the rational design of drugs and vaccines, and investigations into their mode of action and markers of toxicity. The interaction of potential drugs with particular proteins can be evaluated with the use of activity-based probes placed on microarrays. Furthermore, preclinical evaluation of potential drug toxicities can be carried out by proteomic approaches.

Thus, recent advances in protein separation techniques, mass spectrometry, and completion of the genome sequences of several organisms are critical developments that facilitate proteomics. Substantial progress has also been made in the development of protein and antibody arrays. The challenges for the future lie in the archiving, integration, and, analysis of the vast amounts of data derived from mass-spectrometry experiments into cohesive information relevant to physiologic and disease processes. These studies will have to occur in concert with large-scale validation studies on clinical samples obtained from well-controlled patient populations before they can be considered for potential diagnostic, prognostic, or therapeutic applications. In the near future, proteomics will impact the discovery of novel biomarkers and targets for the practical diagnosis and treatment of various diseases.

For background information *see* ANALYTICAL CHEMISTRY; CHROMATOGRAPHY; DEOXYRIBONU-

CLEIC ACID (DNA); ELECTROPHORESIS; MASS SPECTROMETRY; PROTEIN; RIBONUCLEIC ACID (RNA) in the McGraw-Hill Encyclopedia of Science & Technology.

Megan S. Lim; Kojo S. J. Elenitoba-Johnson

Bibliography. J. B. Fenn et al., Electrospray ionization for mass spectrometry of large biomolecules, *Science*, 246:64–71, 1989; S. P. Gygi et al., Evaluation of two-dimensional gel electrophoresis-based proteome analysis technology, *Proc. Nat. Acad. Sci. USA*, 97:9390–9395, 2000; M. S. Lim and K. S. Elenitoba-Johnson, Proteomics in pathology research, *Lab. Invest.*, 84:1227–1244, 2004; M. Mann, R. C. Hendrickson, and A. Pandey, Analysis of proteins and proteomes by mass spectrometry, *Annu. Rev. Biochem.*, 70:437–473, 2001; E. F. Petricoin et al., Clinical proteomics: Translating benchside promise into bedside reality, *Nat. Rev. Drug Discovery*, 1:683–695, 2002; M. Stoeckli et al., Imaging mass spectrometry: A new technology for the analysis of protein expression in mammalian tissues, *Nat. Med.*, 7:493–496, 2001; J. R. Yates III, Mass spectrometry and the age of the proteome, *J. Mass Spectrom.*, 33:1–19, 1998.

Proximal probe molecular analysis

Optical microscopy remains one of the most powerful research tools in the biological and material sciences. The main advantage of optical techniques is that they enable noninvasive studies on micrometer length scales under native environmental conditions. Their major drawback is that the diffraction of light limits the spatial resolution to about 100 nanometers, preventing visualization of smaller species such as single molecules and molecular assemblies. For many years, scientists searched for a complementary technique with the high resolution of electron microscopy that would operate in a physiologically relevant environment. With the development of scanning probe or proximal probe techniques, and in particular atomic force microscopy (AFM) in 1986, a new era of microscopy began.

Proximal probe techniques are based on a local probe scanning near the substrate. By moving the probe across the surface, one is able to map out surface features. In the case of AFM, an extremely sharp tip is used to probe the interactions between the tip and the surface. As the tip approaches a surface feature with a different height, the force of the interaction changes. By mapping out the changes in the interaction, a topographic image of the surface is obtained, with a lateral resolution on the order of nanometers.

AFM enabled three-dimensional (3D) imaging of untreated surface structures and probing their physical properties on the subnanometer scale. In addition, AFM allowed tracking of individual molecules in real time and real space. The role of molecular visualization has become especially important in recent years with the synthesis of complex molecules whose structures are difficult to confirm using conventional light-scattering techniques. Currently, molecular visualization is making the first steps toward "molecular metrology" by providing unambiguous proof of the molecular architecture, along with accurate analysis of size, conformation, and ordering of molecules on surfaces.

Verifying molecular architecture. The first molecular images appeared shortly after the development of AFM. In the early 1990s, many scientific journals covered the visualization of various molecular species, most of which were biological objects such as chromosomes, DNA, viruses, and proteins due to their well-defined and easily recognizable tertiary structures. In recent years, AFM has been used for visualizing synthetic molecules. The clearest images were obtained for hyperbranched macromolecules possessing a well-defined 3D shape,

Fig. 1. High-resolution AFM height images of different molecules with individual architectures. (*a*) Carbosilane dendrimers of ninth generation (*courtesy S. S. Sheiko and A. M. Muzafarov*), (*b*) Four-arm starlike molecular brushes (*reprinted with permission from K. Matyjaszewski et al., Effect of initiation conditions on the uniformity of three-arm star molecular brushes, Macromolecules, 36(6):1843–1849, 2003; © 2003 American Chemical Society*). (*c*) λ-DNA molecules endcapped with streptavidin molecules (*reprinted from Rivetti et al., Scanning force microscopy of DNA deposited onto mica: Equilibration versus kinetic trapping studied by statistical polymer chain analysis, J. Mol. Biol., 264(9):919–932, 1996, with permission from Elsevier*). (*d*) Individual fetal bovine aggrecan monomer adsorbed on mica, showing ∼100 chondroitin sulfate glycosaminoglycan (CS-GAG) chains that are covalently bound at extremely high densities (2–4 nm molecular separation distance) to a 250-kDa protein core (*reprinted from Ng et al., Individual cartilage aggrecan macromolecules and their constituent glycosaminoglycans visualized via atomic force microscopy, J. Struc. Biol., 13(3):242–257, 2003, with permission from Elsevier*).

such as spherical dendrimers, cylindrical "bottle-brushes," and monodendron-jacketed chains.

Figure 1 shows four prominent examples of designer and biological molecules visualized by AFM. Carbosilane dendrimers (Fig. 1*a*) represent one of the most challenging materials because the molecules are small (3–9 nm in diameter) and form a disordered fluid at room temperature. Only recently has it been possible to attain clear resolution of individual dendrimers using ultrasharp AFM probes. Figure 1*b* demonstrates a peculiar example of bottlebrush molecules—starlike brushes synthesized by K. Matyjaszewski and coworkers. As designed, most of the molecules possess four arms; however, molecules with two and three arms are also observed, indicating possible termination reactions. Figure 1*c* shows DNA molecules wherein both biotinylated ends anchor streptavidin-horseradish peroxidase fusion protein. The image obtained by C. Bustamante and coworkers resolves the DNA chain and the molecular recognition groups at the chain ends. Figure 1*d* shows the biological macromonomer, aggrecan, consisting of a protein core backbone with densely spaced polysaccharide side chains, which further self-assembles with link protein and hyaluronan to form larger proteoglycan aggregates in cartilage tissue. The image obtained by C. Ortiz and coworkers shows a "bottle-brush" type morphology which is thought to be largely responsible for the viscolelastic properties of mucus layers in lung airways and the exceptional stiffness, toughness, and resilience of articular cartilage.

Molecular size and shape. In addition to the verification of synthetic strategies, molecular visualization enables accurate measurements of molecular weight, size, and conformation. The unique advantage of AFM is that one obtains molecular dimensions in direct space, affording more opportunities for statistical analysis. The pictorial resolution allows fractionation of the visualized molecules by size, branching topology, and chemical composition, as well as sorting out irrelevant species. For example, molecular images similar to those in Fig. 1*b* enable characterization of the branching topology of polymer chains. Branching, which either can be introduced purposely or can occur spontaneously during polymer synthesis, is known to strongly affect the flow properties of polymer melts. Characterization of individual branches of the entire molecule can be done only by molecular visualization.

Molecular weight distribution was determined by S. S. Sheiko and coworkers using a combination of

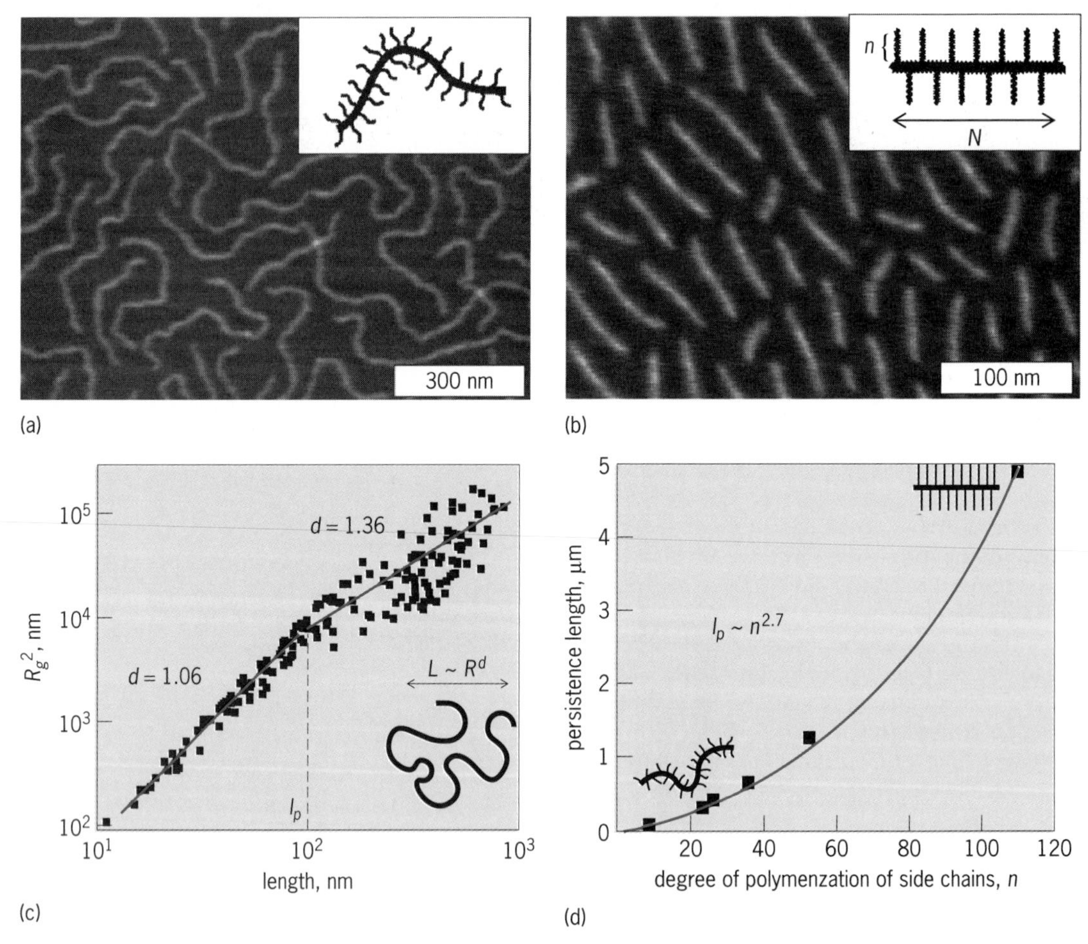

Fig. 2. AFM enables measurements of the fractal dimension and the persistence length of molecular brushes as a function of the side-chain length. Molecules change from (*a*) flexible to (*b*) rodlike with increasing side-chain length. (*c*) Persistence length l_p is determined as a distance along the chain at which the fractal dimension changes from $d \cong 1$ to $d \cong 1.33$. (*d*) Persistence length was shown to depend on the side-chain length as $l_p \sim n^{2.7}$. (*Courtesy of S. S. Sheiko*)

the Langmuir-Blodget (LB) monolayer film-forming technique and AFM. The LB technique provided mass density information, while visualization of monolayer by AFM enabled accurate measurements of the number of molecules per unit area. From the ratio of the mass density to the molecular density, the number average molecular weight was determined. This approach can be applied to a large variety of molecular and colloidal species. In particular, the visualization-based approach is useful for relatively large ($>10^7$ daltons) species, which are difficult to study by light scattering and chromatography. The AFM–LB method also uses smaller sample amounts compared to conventional techniques.

Polymer chains are fractal objects whose dimensionality is determined by interactions between the monomeric units and solvent molecules. **Figure 2*a*** and *b* shows typical AFM images of brush molecules, making direct analysis of their conformation in two dimensions possible. Through molecular visualization, one can measure the persistence length, as well as the fractal dimensionality (Fig. 2*c* and *d*). In addition, the micrographs provide a unique opportunity to measure the local curvature of the molecular contour and verify theoretical models of chain conformation in dilute and semidilute solutions.

Molecules in motion. Currently, there are very limited data available on the dynamics of polymer chains on surfaces. Most significant experiments were performed on DNA molecules using fluorescence microscopy, whose applicability is restricted by the optical resolution limit. Even less data are available on the motion of molecules within dense monolayers. In this respect, one can regard AFM as a powerful and appropriate experimental technique for molecular studies because it provides real space information about the translational, rotational, and bending motions of molecules at nanometer length scales. These applications of AFM techniques are in their infancy and constitute the priority targets of current research.

Recently, H. Xu and coworkers have succeeded in monitoring polymer flows with molecular resolution by AFM. For the first time, it is possible to measure simultaneously the displacement of the contact line and the movement of individual molecules within the film (**Fig. 3*a***) and thus shed light on the mechanism of the mass transport. The film length

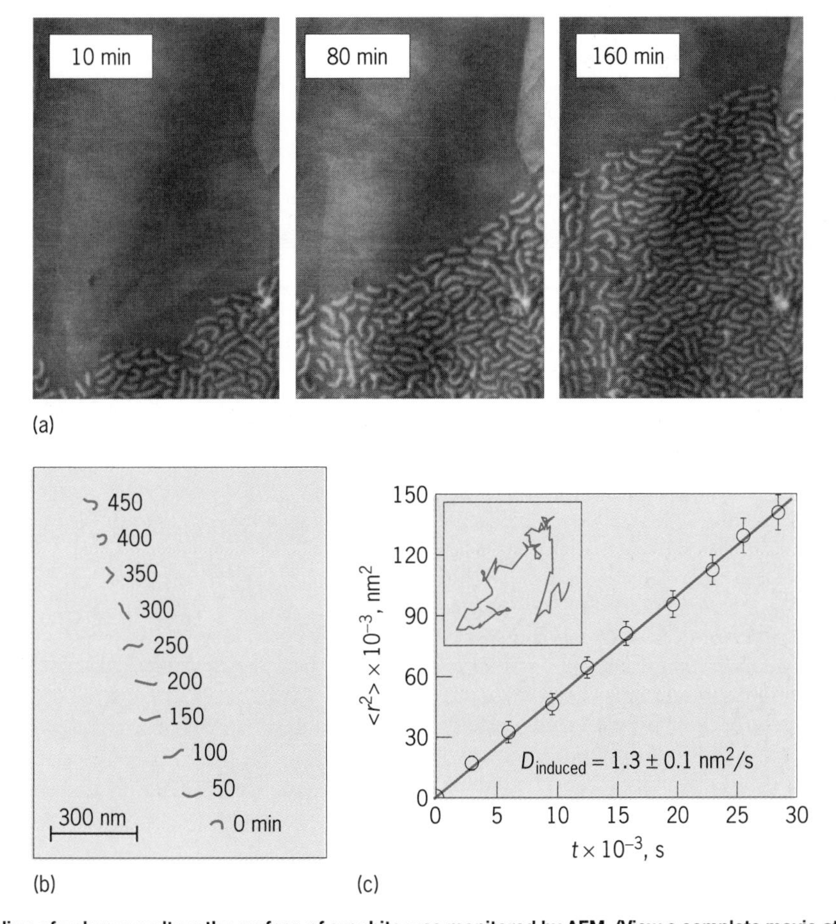

Fig. 3. Spreading of polymer melt on the surface of graphite was monitored by AFM. (View a complete movie at http://pubs.acs.org/cen/news/8246/8246molecules.html.) The melt is composed of brush molecules with poly(*n*-butyl acrylate) side chains. (*a*) Images captured at different times of the spreading process show the displacement of the film edge. (*b*) Animation of one of the flowing molecules demonstrates different modes of the molecular motion, including translation of the center of mass, chain rotation, and fluctuations in the backbone curvature. (*c*) Molecules within the film exhibit diffusive motion. Inset shows molecular path in the frame of the flowing monolayer. (*Reprinted with permission from H. Xu et al., Molecular motion in a spreading precursor film, Phys. Rev. Lett., 93:206103, 2004, copyright 2004 by the American Physical Society*)

follows the classical dependence $L = \sqrt{D_{spread}t}$ with a spreading rate of $D_{spread} = (3.9 \pm 0.2) \times 10^3$ nm^2/s. Along with the position of the contact line, the position of the center of mass, orientation, and the conformation for every molecule were measured, each in its unique environment (Fig. 3*b*). Within the frame of the moving film, one obtained the molecular diffusion coefficient (Fig. 3*c*). Since $D_{spread} \gg D_{induced}$, plug flow of the polymer chains, with insignificant contribution from the molecular diffusion, was identified as the main mass-transport mechanism. The origin of the diffusion within the flowing monolayer is still a subject for debate. Probably, the diffusion is induced by the flow and originates from the heterogeneous structure of the substrate, resembling the flow behavior of granular fluids. The understanding of what controls molecular diffusion in monolayers will provide opportunities for the enhanced mixing of molecules, which is imperative for chemical reactions and the ordering of molecules on surfaces.

Monitoring chemical reactions and self-assembly processes. Even more challenging is observation of molecular assembly processes, such as crystallization, micelle formation, and chemical reactions. It is predicted that traditional top-down lithographic methods for manufacturing microelectronics will soon reach their resolution limits. Anticipating this problem, researchers have begun developing bottom-up techniques based on self-assembly of molecules into a desired pattern. One potential approach, developed by N. C. Seeman, E. Winfree, J. H. Reif, and others, is to use DNA molecules as an engineering material for the programmable construction of molecular devices and nanometer scaffolds. Recently, multifunctional DNA tiles have been readily assembled into nanotubes, ribbons, and grids. As shown in **Fig. 4**, T. H. LaBean and coworkers used AFM to confirm the programmable self-assembly of 4 × 4 tiles into ordered lattices with periodic 20-nm^2 cavities. One can further use these nanogrids as templates for targeted adsorption of proteins and metal nanoclusters, leading to uniform-width metallic nanowires.

Complementary to biomolecules, Matyaszewski and Sheiko view molecular brushes as a powerful platform for designing elementary building blocks of controllable length, stiffness, and branching functionality. The molecular structure also allows "sticky" groups at the chain ends to support association of the blocks into larger complexes. **Figure 5** shows one of the first examples of cylindrical brushes with "sticky" ends due to crystallization of the octadecyl moieties in ABA block brushes, wherein A-blocks are short poly(octadecyl methacrylate) chains. The molecules spontaneously associate to form multimers resulting in chains and branches. As such, the images highlight the existence of competitive reactions and multiple paths of the molecular assembly which, in the present case, are ascribed to the lack of selectivity of the crystallizing chain ends. The advantage of the crystalline stickers is that they dissociate upon melting, leading to reversible association.

Currently, scientists are working on taking molecular visualization even further in order to observe and control chemical reactions and assembly processes on the molecular level directly and in situ. In addition to temperature and pressure, which are the usual variables chemists use to control a chemical reaction, proximal probe techniques offer an additional variable—force. As pointed by E. Evans and Bustamante, the force applied to a single molecule directly affects reaction pathways, and might require reformulation of the expressions from classical thermodynamics and kinetics of chemical reactions to describe their dependence on the external force. Although very young, the field of mechanochemistry

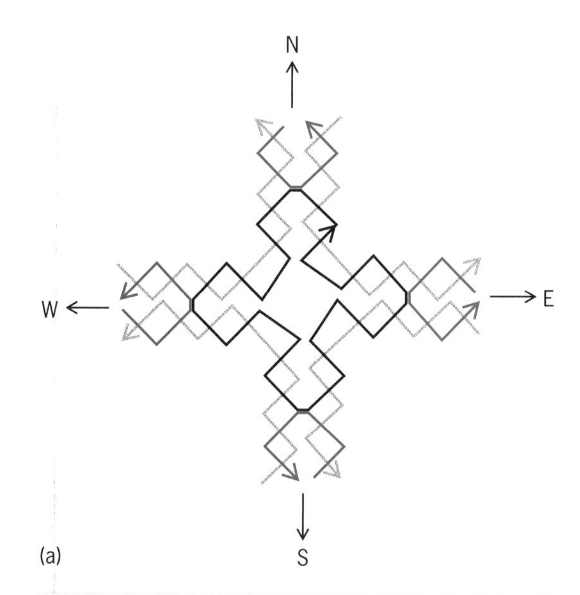

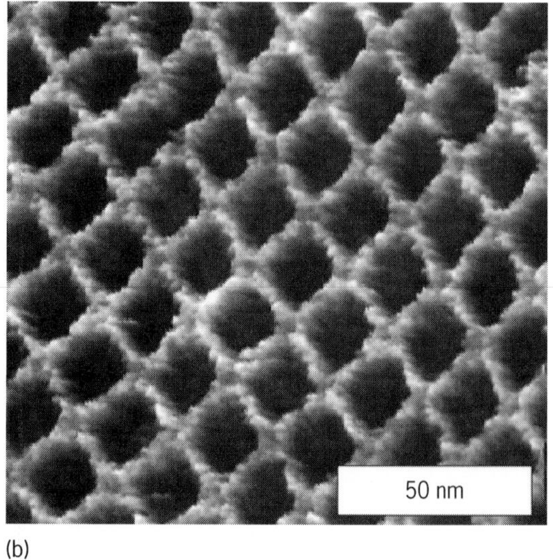

Fig. 4. Self-assembled DNA nanogrids using 4 × 4 DNA tiles. (*a*) The 4 × 4 DNA tile strand structure points in four directions. It is composed of nine strands, with one of the strands participating in all four junctions. The structure is sufficiently rigid to act as a building block in larger superstructures. (*b*) An example is grids. The periodic cavity size is about 15 nm. The smaller cavity (6 × 6 nm) inside the tiles can be also distinguished. (*Reprinted with permission from H. Yan et al., DNA-templated self-assembly of protein arrays and highly conductive nanowires, Science, 301:1882–1884, 2003, copyright 2003, AAAS*)

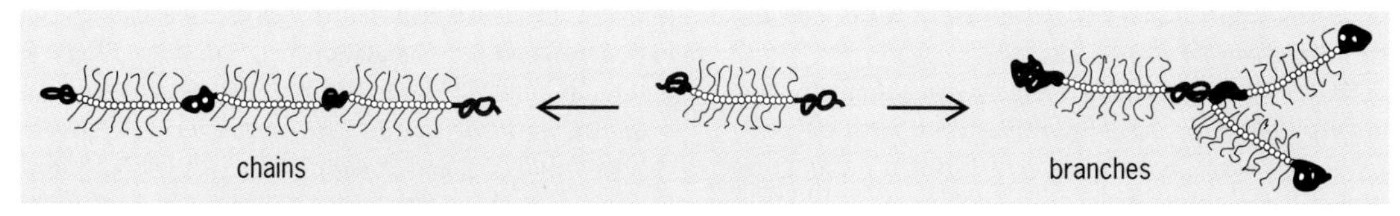

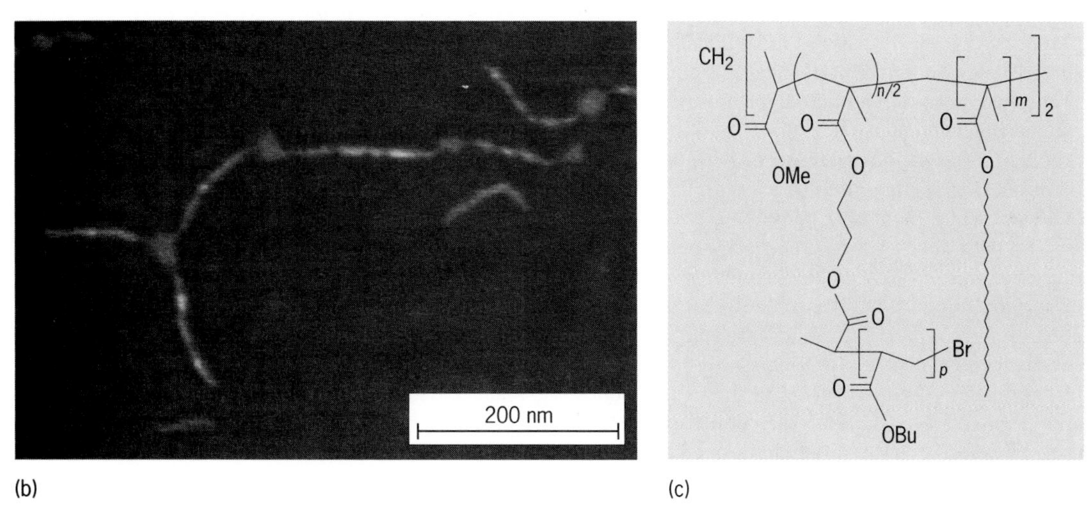

Fig. 5. End-chain association of poly(*n*-butyl acrylate) brushes with crystallizing poly(octadecyl methacrylate) chains on both ends. Leads to spontaneous formation of linear and branched structures observed by AFM. (*a*) Diagram. (*b*) Image. (*c*) Chemical structure. (*Courtesy of S. S. Sheiko*)

is already showing new ways and unlimited possibilities of controlling reactions. The gained understanding of the kinetics involved in elementary reactions will aid in the design of functional nanostructures and molecular devices.

Through molecular visualization, one can directly confirm synthetic strategies, characterize molecular conformation, and monitor chemical reactions, as well as physical association processes. For many systems, such as branched macromolecules, the contribution of molecular visualization is unique and cannot be reproduced by other techniques.

For background information *see* BRANCHED POLYMER; COPOLYMER; DIFFUSION; MOLECULAR MACHINE; MOLECULAR WEIGHT; MONOMOLECULAR FILM; NANOSTRUCTURE; NANOTECHNOLOGY; POLYMER; SUPRAMOLECULAR CHEMISTRY; SURFACE PHYSICS in the McGraw-Hill Encyclopedia of Science & Technology. S. S. Sheiko

Bibliography. C. Bustamante et al., Mechanical processes in biochemistry, *Annu. Rev. Biochem.*, 73:705, 2004; E. Evans, Probing the relation between force-lifetime-and chemistry in single molecular bonds, *Annu. Rev. Biophys. Biomol. Struc.*, 30:105–128, 2001; S. Qin et al., Synthesis and visualization of densely grafted molecular brushes with crystallizable poly(octadecyl methacrylate) block segments, *Macromolecules*, 36(3):605–612, 2003; S. S. Sheiko et al., Visualization of molecules: A first step to manipulation and controlled response, *Chem. Rev.*, 101:4099, 2001; E. Winfree et al., Design and self-assembly of two-dimensional DNA crystals, *Nature*, 394:539, 1998; H. Xu et al., Molecular motion in a spreading precursor film, *Phys. Rev. Lett.*, 93:206103, 2004; H. Yan et al., DNA-templated self-assembly of protein arrays and highly conductive nanowires, *Science*, 301:1882, 2003.

Psychopathology and the Human Genome Project

The prevalence of psychopathology in the general population is surprisingly high and appears to be rising. A Harvard University study of the global burden of disease, sponsored by the World Health Organization (WHO) and the World Bank, estimated that by 2020 depression will be exceeded only by heart disease in worldwide economic importance. Although it is now widely accepted that our genes contribute to the risk of becoming psychiatrically unwell, it is commonly misunderstood how genes have such an effect. Frequently the media report that the "gene for" a specific disorder or behavior has been discovered, leading to the public perception that psychopathology follows a straightforward single-gene–single-disorder model. In fact, few psychiatric disorders follow this simple model, and those that do tend to be rare, such as Huntington's disease or some early-onset forms of Alzheimer's disease. The majority of psychiatric disorders follow a more complex pattern, involving multiple genes that may interact with one another as well as showing interplay with environmental factors. The genes contributing to such complex genetic systems are often referred to as quantitative trait loci (QTLs), the idea being that what is inherited is not the certainty of illness or health but a liability to disorder, where only those individuals who at some point in their lives exceed a certain threshold can be classified as ill.

Nature, nurture, or both? The starting point for believing that genes have anything to do with psychopathology is the age-old observation that mental disorders show a tendency to run in families. However, we share environments as well as genes with members of our families, and a key question is whether psychopathology is clustered in families because of shared genes, shared experiences, or a combination of the two. Adoption studies are one form of "natural experiment" that can help provide an answer. The rationale is that adoptive parents provide an environment but do not share their genes with the adopted children, while biological parents of the adoptees pass on their genes but not their upbringing. Adoption study results have been particularly influential in schizophrenia, where as recently as the 1970s debates were still raging as to whether this disorder resulted from faulty upbringing. Adoption studies, particularly by researchers in the United States and Denmark, showed conclusively that the risk of schizophrenia is elevated only in biological and not in adoptive relatives of schizophrenics.

Studies of twins provide a second experiment of nature. Monozygotic twins (MZ; identical) have all their genes in common, while dizygotic (DZ; fraternal) twins share an average of 50% of their genes. If it is assumed that both MZ and DZ twins share the environment to roughly the same extent, any greater resemblance in MZ compared with DZ pairs for a particular disorder or trait is likely to indicate that genes are having an effect. Modern twin studies suggest that the effects of genes on behavior are pervasive, influencing normal traits, such as personality and intelligence, as well as a whole range of psychiatric disorders, including schizophrenia, depression, bipolar disorder (manic depression), and childhood autism. However, in none of these is there 100% resemblance between MZ twin pairs, even though they are 100% genetic clones of each other, providing strong evidence that such behaviors result from environmental as well as genetic influences.

Such quantitative genetic studies therefore have been critical in establishing that genes play an important role in the development of psychopathology, but they cannot tell us which genes. The ability to identify genes depends on various types of information derived from the Human Genome Project, the decoding of the entire sequence of deoxyribonucleic acid (DNA) carried on our 23 pairs of chromosomes.

Impact of Human Genome Project. Genetics is a comparatively young science. Its statistical foundations were laid by the English polymath Francis Galton in the second half of the nineteenth century, but he and the rest of the scientific community were ignorant of the basic rules of inheritance discovered by his near-contemporary Gregor Mendel, who experimented on pea plants in virtual obscurity in a Moravian monastery garden. Consequently, genetics did not become fully established as a branch of biology until the rediscovery of Mendel's laws of heredity in the early twentieth century. Molecular genetics began in earnest a half-century later with the discovery of the double-helix structure of DNA by J. Watson and F. Crick in 1953, and the genetic code, written in sequences of just four bases—adenine (A), cytosine (C), guanine (G), and thymine (T)—was deciphered in 1966. By 2001, we had a working draft of most of the human genome, consisting of the sequence of approximately 3 billion DNA bases, and by the end of 2004 had filled in nearly all of the gaps. One surprising finding was that there are far fewer genes than had been predicted—only about 25,000—which to many people seemed a small number to provide the basic code for something as complex as a human being. Nevertheless, tracking down precisely those genes that influence common diseases and traits remains a difficult task. Major tools that the human genome project has provided to complete the task are more and better DNA markers, effectively a system of signposts that help us to navigate our way around the genome and tag the locations of QTLs that contribute to the liability to disease.

Finding "psychopathology genes." Broadly speaking, there are two approaches to finding the genes that contribute to psychopathology (or any other common form of disease): the candidate gene approach and positional cloning.

Candidate genes. Potential candidate genes are those that code for proteins that might plausibly be involved in the cause of the disorder. Since about half of all the genes are expressed in the brain, the number of potential candidates for "psychopathology genes" is enormous. One way the list can be narrowed down is by using knowledge of the action of therapeutic drugs. An interesting clue is provided by antidepressants, many of which act by augmenting the effect of the chemical messenger serotonin, also known as 5-HT (5-hydroxytryptamine), in the brain. A common class of antidepressants, the selective serotonin reuptake inhibitors (SSRIs), such as fluoxetine (Prozac®), do this by slowing down the rate at which 5-HT is reabsorbed by the brain cells that secrete it. Normally 5-HT reuptake is effected by a protein on the cell surface called the serotonin transporter (5-HTT). 5-HTT is coded for by a gene that has a common variation in its promoter region, effectively the gene's "dimmer switch," that causes the 5-HTT to have high or low activity. The high-activity version of the promoter variant is known as the long form (l) and the low activity version as the short form (s). Since everyone inherits genes in pairs, one from each parent, there are three possible types (genotypes) of individuals with respect to the 5-HTT promoter, ll, ls, and ss, with ll showing the highest transporter activity, ss the lowest, and ls being in between.

There is some evidence that 5-HTT genotype is associated with personality traits affecting anxiety and depression. Moreover, there are converging lines of evidence pointing to the 5-HTT promoter variant affecting susceptibility to environmental stress. For example, a study of approximately 1000 people from a birth cohort in Dunedin, New Zealand, found an interaction between 5-HTT promoter genotype and the development of depressive symptoms in response to unpleasant life events. Other studies in the United Kingdom and the United States were able to replicate the effect. The 5-HTT promoter genotype also influences the extent to which depressed mood can be induced by depleting research subjects of the amino acid tryptophan, as well as the extent to which a "fear center" in the brain, the amygdala, "lights up" in response to fearful stimuli in functional magnetic resonance imaging (fMRI) studies. It should be emphasized that the 5HTT promoter variant is not the "gene for" depression but the first gene to be confidently identified of what will probably be the fairly large number that can contribute to the liability of getting depressed. In discovering these other genes, positional cloning approaches will be important.

Positional cloning. The starting point in positional cloning is a search of the genome, using hundreds of evenly spaced DNA markers, within families in which two or more individuals are affected by the disorder. The aim is to detect genetic linkage, a phenomenon that occurs when two genes are close together on the same chromosome. Usually pairs of genes do not show any tendency to be co-inherited when passed from parents to offspring. This is because they either are on different chromosomes or are sufficiently far apart on the same chromosome to be separated by the process called crossing over, effectively a shuffling of the genetic deck of cards, which occurs during meiosis (the type of cell division involved in the production of sperm and eggs). Mendel, although he had no idea about the cellular mechanisms involved, observed this in his pea plants and called it the law of independent assortment. Finding departure from independent assortment—for example, the tendency for pairs of siblings affected by the disorder to inherit one or more marker genotypes at a level greater than chance—suggests that the marker and a disease gene are close together. Once linkage between a disorder and marker, or a set of markers, is detected, the region of the genome "signposted" by the markers can be explored in greater detail using an allied approach called allelic association or linkage disequilibrium mapping. This effectively treats an entire population as if it were a family, searching for association between a particular marker genotype and the disease by comparing frequencies in cases with healthy controls. Association will be found only if the marker and the disease QTL are extremely close together, because for populations there will have been very many generations of genetic shuffling, or crossing over, since the population was founded. Hence association studies have much higher resolution than linkage studies and can be used to narrow down linkage regions.

This approach, linkage followed by association mapping, has begun to bear fruit in discovering genes involved in susceptibility to schizophrenia. Two such genes, dysbindin on chromosome 6 and neuregulin on chromosome 8, have now been independently found to be schizophrenia susceptibility genes, and several other genes are beginning to look promising. A novel gene, G72 on chromosome 13, now looks to be involved in susceptibility to both schizophrenia and bipolar disorder, contradicting the traditional view that the two conditions are quite separate. It must be emphasized, however, that as with depression and 5HTT each of these genes on its own has quite small effects, and the disorders result from an interplay with environmental factors and multiple other genes as yet undiscovered.

Postgenomic psychiatry. Purely in terms of scientific discovery, the future for the study of psychopathology in the postgenomic era is looking bright, but what does all this mean for patients and for clinical practice? First, identification of relevant genes will improve understanding of the molecular neurobiology and basic causation of psychiatric disorders. This should lead to the development of more efficacious and more specific medications. It also seems likely that DNA testing will be useful in predicting response to treatment and susceptibility to unwanted effects of therapeutic drugs. Although DNA testing is unlikely to be useful for population screening, it probably will inform the counseling of the relatives of patients at high risk of heritable disease. Finally, although it has sometimes been suggested, that there is a danger of genetic connotations increasing stigma, it is in the authors' view more likely that improved understanding of the causes and mechanisms of disease will help demystify psychiatric disorders in the public perception and actually reduce stigma.

For background information *see* AFFECTIVE DISORDERS; ALZHEIMER'S DISEASE; GENETIC CODE; HUMAN GENETICS; HUMAN GENOME PROJECT; HUNTINGTON'S DISEASE; SCHIZOPHRENIA in the McGraw-Hill Encyclopedia of Science & Technology.

Sarah Cohen; Peter McGuffin

Bibliography. A. Caspi et al., Influence of life stress on depression: Moderation by a polymorphism in the 5-HTT gene, *Science*, 301:386–389, 2003; T. Eley et al., Gene-environment interaction analysis of serotonin system markers with adolescent depression, *Mol. Psychiat.*, 9:908–915, 2004; A. Elkin, S. Kalidindi, and P. McGuffin, Have schizophrenia genes been found?, *Curr. Opin. Psychiat.*, 17(2):107–113, 2004; P. McGuffin, B. Riley, and R. Plomin, Genomics and behavior: Toward behavioral genomics, *Science*, 291(5507):1232–1233, 2001; R. Plomin, M. J. Owen, and P. McGuffin, The genetic basis of complex human behaviours, *Science*, 264:1733–1739, 1994.

Pterosaur

Pterosaurs were the first vertebrates capable of flapping flight (that is, able to use their wings to produce thrust as well as lift) arising some time before the end of the Triassic Period (248 to 220 million years ago), when they first appear in the fossil record. They died out at the end of the Cretaceous Period (65 million years ago), leaving no descendants. They included the largest flying animals that have ever lived, *Quetzalcoatlus northropi*, the largest species yet known, with a wingspan of about 12 m (40 ft). The pterosaur wing skeleton was dominated by the enormously

Fig. 1. Computerized reconstruction of the Early Cretaceous pterodactyloid pterosaur *Anhanguera santanae*, with an estimated wingspan of about 4 m (13 ft). The reconstruction is based on three-dimensional preserved fossils from the Santana Formation (Lower Cretaceous) in Brazil, one of the most important pterosaur localities.

elongated fourth finger, which supported a superficially saillike wing membrane reinforced by a radiating array of structural fibers (actinofibrils). The other three digits (the fifth is absent) were free of the membrane and remained small.

Pterosaurs can be broadly classified into two groups: the earlier rhamphorhynchoids, which went extinct at the end of the Jurassic and bore long, stiff tails; and the later pterodactyloids, which first appeared in the Late Jurassic and whose tails were reduced (**Fig. 1**). The rhamphorhynchoids do not constitute a true group in the strict taxonomic sense, because the derived forms are more closely related to the pterodactyloids than to the most basal forms. Many aspects of pterosaur biology—from their anatomy to their ecology—have been controversial subjects for decades. In the last few years, however, the discovery of a number of spectacular fossils, and the application of novel techniques to their study, have begun to resolve some of these old debates, and have uncovered some startling new information about this enigmatic group.

Wing membrane anatomy. The suggestion that the pterosaurs' elongate fourth fingers supported wings of skin was first made by Georges Cuvier in the early nineteenth century. He was proved correct by the discovery of many beautifully preserved fossils, mostly from the Upper Jurassic Solnhofen Limestone of Germany, with which impressions of the flight membranes were associated. Unfortunately, in the vast majority of these specimens the membrane nearest to the body could not be discerned. The equivocal nature of the evidence led to a polarization of opinion about pterosaur wing shape, with some workers arguing that the trailing edge of the wing ran to the ankle (the so-called batlike reconstruction), and others that it ran to the hip, leaving the legs free (the birdlike reconstruction). A reappraisal of the Late Jurassic rhamphorhynchoid *Sordes pilosus* from Kazakhstan indicated that, in this species at least, the membrane did run all the way to the ankle. Nevertheless, many workers remained unconvinced that the construction of the wing in this isolated example was shared by all pterosaurs. In recent years, however, a number of additional specimens have been found in the Solnhofen Limestone, and also in the Lower Cretaceous Yixian Formation of China and Crato Formation of Brazil, in which the wing membrane clearly attaches to the lower leg. Moreover, these fossils represent four distinct pterosaur groups: two rhamphorhynchoids and two pterodactyloids. Given the absence of firm evidence to the contrary, it therefore seems increasingly likely that the condition reported in *Sordes* is universal for pterosaurs.

These soft tissue specimens have also provided a wealth of new information about wing membrane microanatomy. Particularly significant is the so-called dark-winged *Rhamphorhynchus* from Solnhofen. This specimen is one of the most superbly preserved pterosaurs ever discovered, consisting of a nearly complete, articulated, three-dimensional skeleton with associated wing membranes. The system of actinofibrils is clearly visible, but it was only when scientists illuminated the fossil with ultraviolet light (which makes phosphatized soft tissues fluoresce) that the truly remarkable extent of preservation was revealed. Under the ultraviolet lamp the delicate tracery of two previously unseen tissues appeared. The first was a network of tubular structures that are probably the remains of blood vessels. The second was a system of fibers that, unlike the densely packed, regularly arranged actinofibrils, were wavy in appearance and formed an irregular latticework. These may be the remains of intrinsic membrane muscles, a suggestion consistent with the earlier discovery of muscle fibers in a three-dimensionally preserved fragment of wing membrane from Brazil.

Head crests. It has long been known that several pterodactyloids possessed head crests, the most famous being the backwardly directed bladelike crest of *Pteranodon*. Over the years, more examples have been found, including the bizarre saillike soft tissue crests of the tapejarids, the more modest bony ridges adorning the jaws of the ornithocheirids, and the huge sparlike extension of the skull of *Nyctosaurus*. However, recent finds have shown that they were more widespread than has generally been appreciated. Once again, the use of ultraviolet light has proved critical in this regard. Subjecting a Solnhofen specimen of *Pterodactylus* to this treatment revealed a hitherto unknown soft tissue crest running along the midline of the skull, with a rearward extension supported by a curious cone-shaped structure. It now seems probable that most, if not all, pterodactyloids had some kind of head crest. It is doubtful, given the great variability in crest form, that these structures had a specific aerodynamic function, although they would certainly have had an aerodynamic effect. A more likely explanation is that they played a role in species recognition or sexual display.

Neural anatomy. Our knowledge of pterosaur brains was for a long time dependent either on the fortuitous discovery of natural endocasts, which form when sediment fills the hollow brain cavity, or on the use of latex to make artificial endocasts—a potentially destructive technique. Recently, however, x-ray computed tomography (CT) has been used to scan the three-dimensionally preserved skulls of a representative rhamphorhynchoid (*Rhamphorhynchus* from the Solnhofen Limestone) and pterodactyloid (*Anhanguera* from the Santana Formation). The scans were used to reconstruct high-resolution virtual endocasts of the brain cavities (**Fig. 2**).

The virtual reconstructions confirmed that the pterosaur brain was somewhat similar to that of birds, with reduced olfactory lobes and an expanded cerebrum and cerebellum. There were, however, some surprising differences. Most significantly, the floccular lobes, on either side of the cerebellum, were comparatively enormous, accounting for 7.5% of total brain mass—about five times the relative size of this part of the brain in birds. In birds and mammals, the floccular lobes process sensory inputs from the skin and muscles; therefore, it seems likely that in pterosaurs the primary function of these enlarged regions was to integrate the great quantity of sensory information relayed from the flight membranes. Thus, pterosaurs could have obtained a detailed, constantly updated picture of the local forces acting on the wings, and the putative network of muscle fibers within the membrane would have enabled them to make rapid fine adjustments to wing shape in response to any changes in local aerodynamic loading. This highly responsive system would have given pterosaurs superb flight control capabilities, perhaps even surpassing those of birds.

Reproduction. The manner of pterosaur reproduction was, until very recently, a mystery. Many presumed that pterosaurs laid eggs: after all, their closest living relatives—birds and crocodiles—are both obligate egg-layers. However, the continued failure to find any fossil evidence of eggs or nests caused some workers to question this belief; they suggested that pterosaurs gave birth to live young. This opinion was swept away in 2004 by the first discovery of a pterosaur egg in the Yixian Formation of China. The egg can be instantly identified as pterosaurian because, within the outline of the shell, the embryo is preserved in exquisite detail, displaying the hallmark feature of the group—the elongate wing finger. Remarkably, soon after this fossil was unearthed, two other pterosaur eggs were discovered: another in China and one in the Lower Cretaceous Lagarcito Formation of Argentina. Microscopic examination revealed that the shell of each egg was extremely thin—between 0.25 and 0.3 mm thick. In addition, while a single layer of calcite crystals was identified in the Argentinian specimen using a scanning electron microscope, no calcareous remains were found in the Chinese eggshell. This cannot simply be the result of dissolution prior to burial, for several calcareous shells of mollusks were found in the immediate vicinity. It can be inferred, therefore, that pterosaur eggs were rather soft and leathery like those of turtles—and completely unlike bird eggs. Being soft, pterosaur eggs could not have been directly incubated by the parents. Instead, like turtles

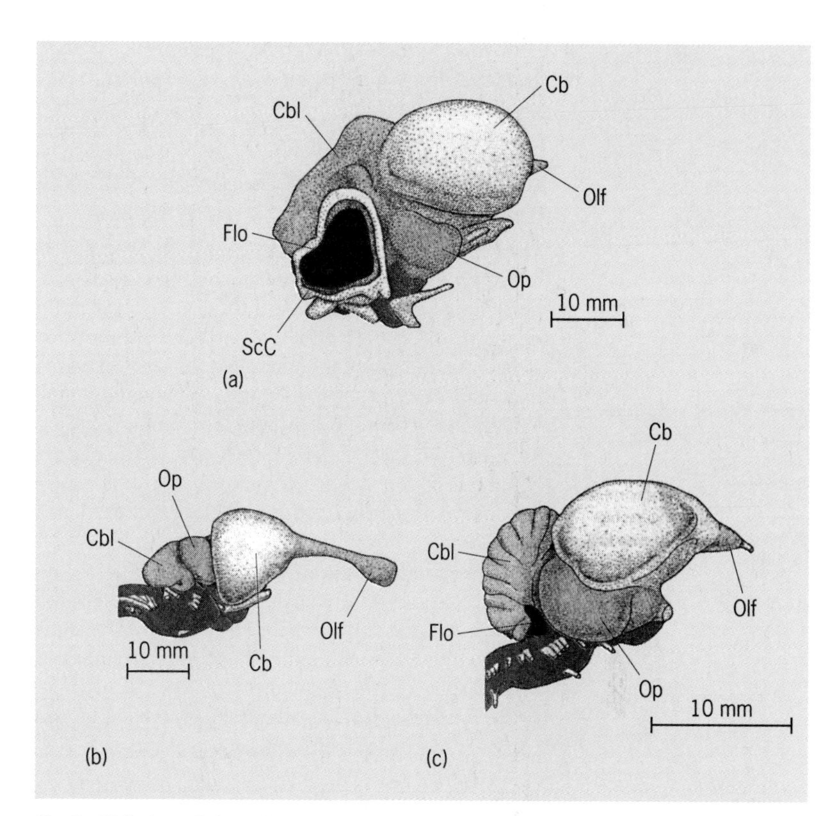

Fig. 2. Right lateral view of the brains of pterosaurs, which were both crocodilians and birds. (*a*) Reconstructed brain of *Anhanguera*, based on a virtual endocast generated by CT. (*b*) Brain of an alligator (*Alligator mississippiensis*). (*c*) Brain of a pigeon (*Columba livia*). Relative to a typical reptile brain, the cerebrum, cerebellum and optic lobes were enlarged in pterosaurs, as they are in birds, whereas the olfactory lobes were reduced. The floccular lobes, however, were much larger in pterosaurs than in birds. Cb, cerebrum; Cbl, cerebellum; Flo, floccular lobes; Olf, olfactory lobes; Op, optic lobes; ScC, semicircular canals. (*Reprinted with permission from L. M. Witmer et al., Neuroanatomy of flying reptiles and implications for flight, posture, and behaviour, Nature, 425:950–953, 2003, Nature Publishing Group*)

and crocodiles, pterosaurs may have buried their eggs in the ground, in which case parental care is likely to have been minimal or nonexistent. This scenario is in marked contrast to the traditional view of doting pterosaur parents diligently feeding their offspring.

Conclusion. The discoveries described in this article have greatly increased the understanding of pterosaur biology. It is now clear that these animals were not ineffectual gliders, but neither were they the birdlike creatures envisioned in the 1980s. Instead, it is increasingly obvious that the pterosaur flight apparatus and the neural hardware that controlled it were unique and highly sophisticated. The challenge now is to use the newly acquired anatomical information to address the aerodynamics of pterosaur flight—this being the most conspicuous, but still the most poorly understood aspect of the way of life of these animals.

For additional information *see* AVES; COMPUTERIZED TOMOGRAPHY; DINOSAUR; FLIGHT; REPTILIA in the McGraw-Hill Encyclopedia of Science & Technology.

Matthew Wilkinson

Bibliography. E. Buffetaut and J.-M. Mazin (eds.), *Evolution and Palaeobiology of Pterosaurs*, Geol. Soc. Lond. Spec. Publ. 217, 2003; D. M. Unwin, Pterosaurs: Back to the traditional model?, *Trends*

Ecol. Evol., 14:263–268, 1999; P. Wellnhofer, *The Illustrated Encyclopedia of Pterosaurs*, Salamander Books, London, 1991.

Quantum chaos in a three-body system

The behavior of a quantum system whose classical motion could involve chaotic behavior is called quantum chaos. Chaos can arise in a classical system—a system of objects which are large compared to atomic scale—when the equations governing its behavior are nonlinear. Chaos in this classical sense generally cannot be found in a system fully described by quantum mechanics, since the Schrödinger equation, which describes quantum systems, is linear. The correspondence principle requires that a result obtained from quantum mechanics approach the classical result in the classical limit, that is, the limit of large size. The primary issues in quantum chaos in a three-body system are what are the signs of chaos, and how large the quantum system has to be to display chaotic behavior.

Chaotic dynamics essentially means that the outcomes of a repeated experiment are exquisitely sensitive to the initial conditions of the experiment, and that they can diverge with time. Chaotic motion can be observed for very simple systems, an example of which is a double pendulum.

Three-body systems. Humans have observed the motion of the Moon for thousands of years. Knowledge of the lunar calendar and the seasons has played an important role in the life of agrarian people. The notion that the Sun is the center of the solar system and that the Moon revolves around the Earth, along with the development of classical mechanics, allows detailed understanding of the orbits and behavior of this three-body gravitational system.

The classical problem of three bodies interacting under their mutual gravitational force has been known for more than a century—since the work of the mathematician Henri Poincaré in the late nineteenth century—to exhibit a mixture of regular and chaotic dynamics. The possibility of chaotic motion arises because the differential equations governing the dynamics of a three-body gravitational system are nonlinear. We are accustomed to celestial dynamics manifesting regular (as opposed to chaotic) dynamics: The behavior of the Earth, Sun, and Moon are regular and predictable. However, since the equations of motion of the Earth-Sun-Moon system allow chaotic dynamics, it would be possible, for example, for the Moon to be ejected from Earth's orbit.

Three bodies interacting under the influence of their mutual electric force should exhibit the same dynamical behavior as planets, because the gravitational force and the electric force both obey the same inverse-square power law with distance between the objects. However, an atomic-scale three-body electrical system—the helium atom, which consists of a nucleus and two electrons—is also governed by quantum mechanics. The underlying chaotic classical behavior of the three-body problem should manifest itself in a similar quantum system. An atom in the limit of large size (electrons far from the nucleus) should show the underlying classical chaotic behavior. Quantum mechanics is governed by the Schrödinger equation, which is linear, and thus does not admit chaos in the classical sense. A recent experiment performed using an ultrabright beam of photons from a synchrotron light source has allowed study of the doubly excited autoionizing states of the helium atom, in which the electrons are both far from the nucleus and the atom approaches the size for which it should be described by classical dynamics.

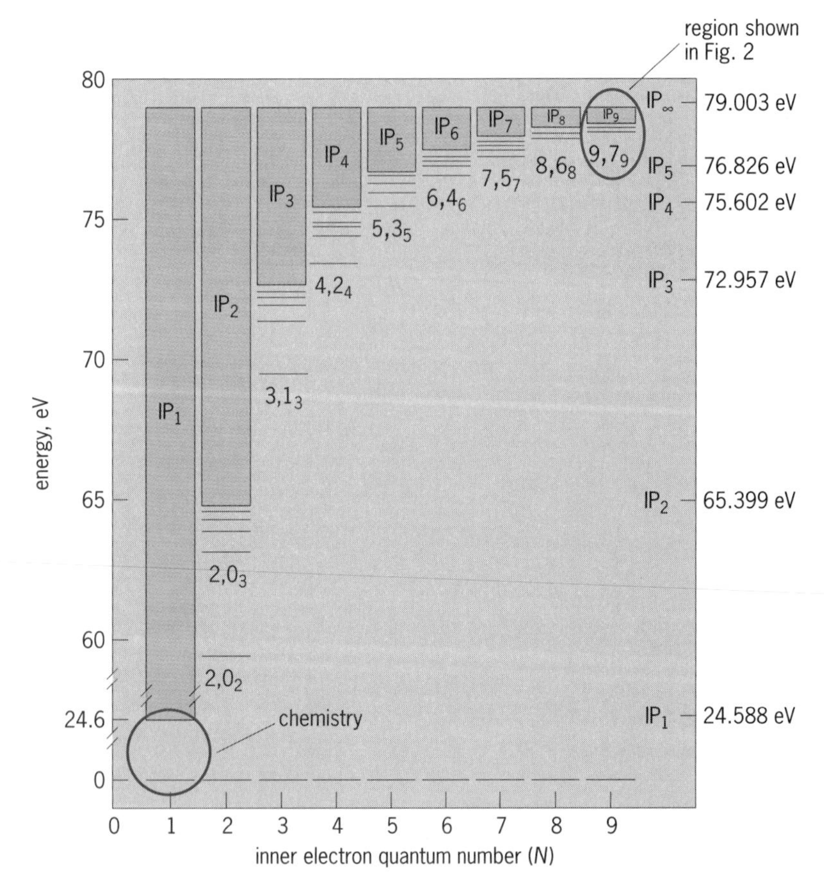

Fig. 1. Energy-level diagram for the helium atom. The horizontal axis is the "inner-electron" quantum number. N,K_n means the state of a helium atom with the inner electron having quantum number N and the outer electron quantum number n. (K refers to an angular-correlation quantum number.) Chemistry generally takes place with only one electron excited ($N = 1$), which is in the lower left corner of the diagram. The notation IP_N means the ionization potential (energy) to remove the outer electron and excite the inner electron to state N. A shaded area indicates an ionization continuum (outer electron removed). (*After M. Domke et al., High-resolution study of* $^1P^o$*double-excitation states in helium, Phys. Rev. A, 53:1424–1438, 1996*)

Helium energy levels. The helium atom has been studied for many years, and its energy-level diagram or spectrum of excited states is known to beginning chemistry students. States in which only one electron is excited from the ground state give the familiar singlet and triplet series of the helium atom. The states are completely specified by a complete set of quantum numbers. The complete set of quantum numbers provides a complete description of the helium atom. As long as a complete set of quantum numbers can be identified for a state of an atom, chaos is not present.

The description of an atomic state by a complete set of quantum numbers, and the absence of chaos, is due to an averaging influence of the Heisenberg uncertainty principle on the classical chaotic motion. This applies to the atom in low states of excitation (at least one electron near the nucleus). However, at higher excitation energies—for both electrons far from the nucleus—the averaging influence of the uncertainty principle decreases and leads to a loss of the complete set of quantum numbers. As a consequence, the quantum-mechanical description is increasingly influenced by the underlying classical chaos.

An energy-level diagram for the helium atom is shown in **Fig. 1**. The vertical scale shows energy: 0 eV is the ground state of the helium atom; the atom will be doubly ionized (lose both electrons) for an excitation greater than 79.003 eV. The columns are labeled by the quantum number N of the "inner" electron, meaning the electron which is the less excited of the two. A few states are shown, labeled by the quantum numbers of both electrons, N and n, along with an additional quantum number, K, describing angular correlation of the electrons. Ionization potential is shown for each series of excitations. The shaded region above each series is an ionization continuum. All atoms with an excitation energy above 24.6 eV, thus with both electrons excited, will autoionize, resulting in a helium ion and an electron.

Excitation of doubly excited states. A single x-ray photon can excite both electrons in a helium atom. The doubly excited states so produced decay by autoionization to an ion and an electron, and can be detected in a suitable photoabsorption experiment. Doubly excited states of helium have been studied since R. P. Madden and K. Codling performed one of the first experiments with synchrotron radiation. Their 1963 publication demonstrated the complexities of low-lying doubly excited states of helium, where the two excited electrons couple to each other and to the nucleus to produce resonance series which have subsequently been described by complete sets of quantum numbers.

More recent experiments by Günter Kaindl and coworkers in Berlin and Berkeley have utilized the greatly increased brightness of synchrotron-radiation sources to measure the energy levels of doubly excited states of the helium atom closer to the limit of double ionization. (Both electrons would be removed for photon energies above the double ionization limit.) A beam of low-energy x-rays (extreme ultraviolet light) with very high spectral (energy) resolution was directed onto a gas of helium atoms. Ions produced by electron ejection from the doubly excited helium atoms (autoionization) were measured as the photon energy was varied.

Figure 2 shows the photoabsorption spectrum of doubly excited autoionizing resonances in helium for very highly excited states lying just below the double ionization threshold. The notation is the same as in Fig. 1. The upper curve shows a fit to the experimental data; the lower curve is a theoretical model calculation. The experiment and theory, plus a theoretical model extrapolated to electrons farther from the nucleus (larger values of the principal quantum numbers of the two electrons), provide the basis for determing whether there is an indication of chaos in the results.

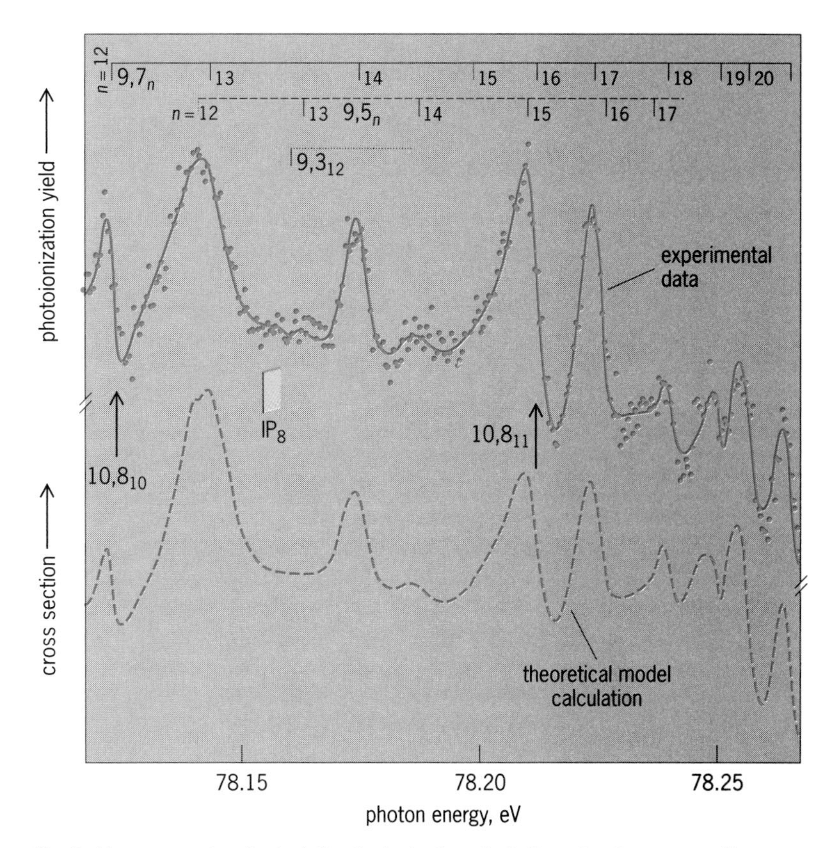

Fig. 2. Measurement and calculation for ionization of a helium atom by an x-ray. Upper curve and data points represent the photoabsorption spectrum of helium gas for exciting two electrons. The solid line through the data points indicates the best fit. Lower curve shows the spectrum calculated from theoretical model. (*Advanced Light Source, Lawrence Berkeley National Laboratory*)

Evidence for chaotic behavior. The measurements and model calculations of resonance energies were tested for their chaotic nature by use of a statistical measure in research by R. Püttner and colleagues combining theoretical and experimental results. It is known from random matrix theory that the energy-level spacing distribution is different for systems with regular and chaotic behavior. Thus a statistical test can be applied to data to determine the degree to which a spectrum is characteristic of chaotic behavior. Analysis shows that the results of Püttner and colleagues would characterize chaos in the helium atom for atoms with both electrons excited to approximately N and n having values of 17 or greater. This result can be understood by a loss of the quantum number N as a good quantum number; that is, the identification of the precise quantum number is lost. A helium atom with both electrons having quantum numbers approximately 17 is a very large atom, as both electrons are very far from the nucleus.

The classical helium atom exhibits chaotic behavior if the distances of both electrons from the nucleus is approximately the same, that is, $n \cong N$. In this case the electrons can come close to each other so that the repulsion is strong, resulting in significant chaotic behavior. In the case where one electron is much farther away from the nucleus than the other ($n \gg N$), the interaction between the electrons is weak and the mutual disturbance of the electronic motion becomes small. Therefore, the most interesting states in the context of quantum chaos are those with $n \cong N$; it will be assumed that $n = N$ for this discussion. The size of such an atom scales as the square of the principal quantum number, $N = n$. The use of a photon to simultaneously excite both electrons in the helium atoms places the atom temporarily in a state with both electrons far from the nucleus. The result of Püttner and colleagues indicates that a helium atom approaches a sufficient size to have classical properties when the principal quantum number of both electrons is greater than $N = 17$. An atom with both electrons having $N = 20$ would be 400 times its usual size. A helium atom this large shows classical behavior, and the classical underlying chaotic nature of the dynamics is manifested by the nearest-neighbor spacing distribution. An atom made even larger (one with electrons with even higher values of the principal quantum number) should behave fully classically; that is, the quantum-mechanical and the classical solutions of the problem should agree with each other according to the correspondence principle. At present, it is not experimentally feasible to observe this behavior.

For background information *see* ATOMIC STRUCTURE AND SPECTRA; CHAOS; CORRESPONDENCE PRINCIPLE; IONIZATION POTENTIAL; NONLINEAR PHYSICS; QUANTUM MECHANICS; RANDOM MATRICES; RESONANCE (QUANTUM MECHANICS); RYDBERG ATOM; SCHRÖDINGER'S WAVE EQUATION; SYNCHROTRON RADIATION; UNCERTAINTY PRINCIPLE in the McGraw-Hill Encyclopedia of Science & Technology.

Alfred S. Schlachter

Bibliography. M. Altarelli, F. Schlachter, and J. Cross, Making ultrabright x-rays, *Sci. Amer.*, 279(6):2–9, December 1998; G. L. Baker and J. P. Gollub, *Chaotic Dynamics: An Introduction*, Cambridge University Press, 1996; R. Blümel and W. P. Reinhardt, *Chaos in Atomic Physics*, Cambridge University Press, 1997; M. C. Gutzwiller, Moon-Earth-Sun: The oldest three-body problem, *Rev. Mod. Phys.*, 70:589–639, 1998; R. Püttner et al., Statistical properties of inter-series mixing in helium: From integrability to chaos, *Phys. Rev. Lett.*, 86:3747–3750, 2001.

Queuing theory (health care)

A queue (or a waiting line) is a familiar concept. We experience queues at the checkout counter of the grocery store, at the movie theater, in traffic, or in other daily activities. Queues also exist in the provision of health care. Large health-care concerns, such as clinics, hospitals, or entire health-care systems, are most appropriately viewed as a network of queues.

Health care is time-sensitive. When we are ill or in pain, our physical, emotional, psychological, and cognitive functions may be handicapped. We may not be able to perform daily activities of living, or our ability to enjoy life may be compromised. Indeed, life itself may be in jeopardy. Timeliness in health-care delivery is crucial.

Health care is also extremely costly. Health-care spending in the United States in 2004 was estimated by the OECD (Organisation for Economic Co-operation and Development) to be about $1.7 trillion, or roughly 15% of the gross domestic product (GDP). This amounted to an estimated $5635 for every man, woman, and child in the United States. Health care expenditures for other industrialized nations indicate that health care is costly, though generally less so than in the United States. Switzerland and Germany spend about 11% of their GDP on health, while Canada, France, the Netherlands, and Belgium spend about 10%, and the United Kingdom about 7.7%.

Definition of a queue. A queue is an ordered list of objects (people, cars, or information packets) waiting to be processed by a resource or set of resources (doctor, traffic lane, or switch) having a finite capacity. Queues arise when the rate at which objects arrive for service exceeds the rate at which they can be processed. The study of queues belongs to a branch of operational research called queuing theory. Queuing theory has its origins in the 1920s and 1930s at Bell Laboratories, where it was used to study telephone networks. Queuing theory has been used to study health-care and hospital operations since the 1950s.

Queuing theory provides a compact framework for describing a queuing system. Six elements are sufficient to describe most queuing systems: arrival process, service mechanism, number of servers, queuing discipline, system capacity, and the calling population.

The mechanism by which the customers arrive at the system is known as the arrival process and is defined by an arrival distribution that provides information about both the arrival rate λ and the time between arrivals. The service process is the mechanism by which customers are served and then depart the system. Like the arrival process, the service process is defined by a service-time distribution that provides information about the service rate μ and process variability. The number of servers s describes the number of resources, each with unit capacity, which are available to serve customers. Queue discipline describes the mechanism by which customers queue for service. While it is common to assume that customers are serviced on a first-in-first-out (FIFO) basis, other disciplines, such as last-in-first-out (LIFO) or service-in-random order (SIRO), are possible. System capacity is the upper limit on the number of customers that may be resident in the system at any time, and this limit can be physical or logical. The calling population represents the pool of available customers who might use the system. In general, it is assumed that the pool of customers and the system capacity is infinitely large.

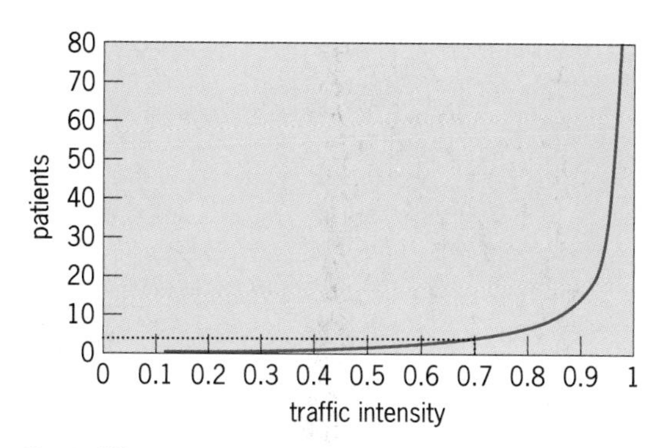

Fig. 1. Effect of traffic intensity ρ on queue size.

Important results. Queuing theory is particularly useful for health care because it provides a compact framework for understanding and addressing process issues. One of the most important results coming from queuing theory is the idea of traffic intensity or service use. Traffic intensity ρ is defined as the arrival rate/total service rate, $\rho = \lambda/s\mu$. As ρ approaches 1, the arrival rate begins to equal the system service rate and patients experience increasingly long waits for service. If the traffic intensity of a queuing system meets or exceeds 1, the system will usually become unstable and the queue will grow without bound. In such a system, the number of patients waiting for service increases without limit, as does the time a patient spends in the system. In all queuing systems, it is necessary to ensure traffic intensity < 1. Because health care is a time-sensitive service, issues tend to arise long before ρ approaches 1 (**Fig. 1**).

Using queuing theory as a framework, there are four broad strategies that can be used to change the performance of a waiting-line system. (1) Change the arrival process by either speeding up or slowing down patient arrivals (that is, by changing the mean of the arrival distribution) or changing the shape of the interarrival distribution (that is, by changing the variance of the distribution). (2) Change the number of servers. (3) Change the service process by altering either the service rate or the variance of the process. (4) Change the manner in which patients line up for service.

Optimizing health-care operations. In health care, queuing issues are usually about lines that are too long rather than too short. Thus, most work in health-care operations is focused on methods that minimize wait times, subject to constraints on cost, resource availability, clinical requirements, patient desires, and regulatory or legal obligations.

If it were possible to ignore cost constraints, optimizing health-care operations would be a relatively simple matter. A decision maker would simply increase the number of servers s such that ρ, which equals $\lambda/s\mu$, is lowered sufficiently to ensure an acceptable wait-time level. This is illustrated as the dotted line in Fig. 1, which has been set to correspond to $\rho = 0.7$, the target threshold used in many non-health-care service operations. However, increasing the number of servers also increases the overall cost of providing health care. Given the high cost of health care in most countries, this alternative is not attractive.

Accordingly, a great deal of operational research has focused on making health-care delivery more efficient. Efficiency efforts often concentrate on the elimination of non-value-added work, such as delay, rework, or rescheduling, in the service process. Eliminating non-value-added work changes the service process by either increasing the service rate μ which makes the service faster, or reducing the variation of service delivery which makes the service delivery more predictable. As process variability decreases, queue length and patient waiting time decrease, even if the overall number of resources available in the system remains constant (**Fig. 2**).

A substantial body of research has also been devoted to modifying aspects of the patient arrival rate. Typically, efforts focus not on reducing the number of arrivals but on modifying the arrival process to make the overall process variability decrease. Appointment scheduling is the most common form of arrival process control. Process designers may also use incentives to encourage patients to use services at periods of low use or avoid peak periods. For example, some institutions mount a yearly publicity campaign to discourage the unnecessary use of emergency services during flu season.

Networks of queues. Although queuing theory provides a convenient framework for analyzing wait-time issues in health-care delivery systems, it does have limitations. Queuing models are valid only for systems involving a single queue. Queuing models also make a number of restrictive assumptions about the operating characteristics of the systems they represent to preserve tractability. In some cases, it is possible to represent a health-care system (such as an outpatient clinic or a small emergency room) as a network of single-queue systems that can be solved analytically. However, as the system represented grows in size and complexity, the tractability and validity of queuing models decrease. Accordingly, most researchers rely on simulation methods to represent, analyze, and improve health-care operations. Simulation is a numerical approximation of a real-world system over time. In a simulation, a model is used to generate an artificial history of the system, which is analyzed to draw conclusions about how the real

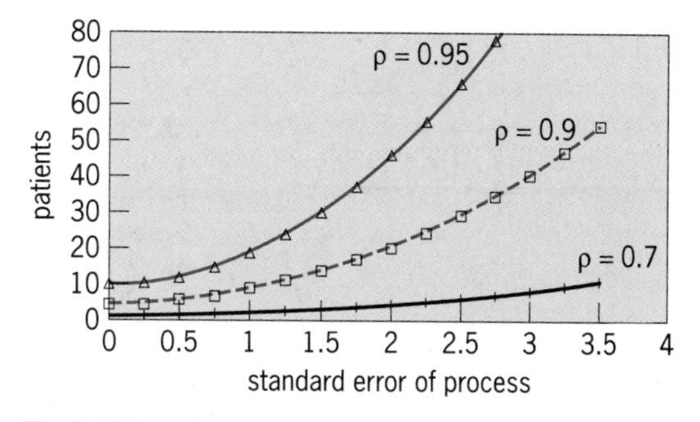

Fig. 2. Effect of process variability on queue size for varying traffic intensities.

system might function under similar circumstances. Simulation provides a method for analyzing networks of queues that is flexible, robust, and accurate. Simulation models are particularly appropriate for health care, since they allow decisions to be made about the structure of operations without endangering the safety of patients.

Simulation models have been applied in a broad range of studies in hospital and health systems operations. Recent applications include surgical wait-list management, cardiac treatment, allocation of organs to donors, emergency-room operations, bed planning and use, operating-room function, and intensive-care-unit flow.

For background information *see* INDUSTRIAL ENGINEERING; METHODS ENGINEERING; OPERATIONS RESEARCH; OPTIMIZATION; PUBLIC HEALTH; QUEUEING THEORY; SIMULATION; SYSTEMS ENGINEERING in the McGraw-Hill Encyclopedia of Science & Technology.

John Blake

Bibliography. M. Carter, Diagnosis: Mismanagement of resources, *OR/MS Today*, 29(2):26–32, 2002; D. Larson, Perspectives on queues: Social justice and the psychology of queuing, *Operat. Res.*, 35(6):895–905, 1987; J. Preater, Queues in health, *Health Care Manag. Sci.*, 5(4):283, 2002; W. Winston, *Operations Research: Applications and Algorithms*, 3d ed., Duxbury Press, 1994; D. Worthington, Hospital waiting list management models, *J. Operat. Res. Soc.*, 42(10):833–843, 1991.

Fig. 1. Chilbolton Advanced Meteorological Radar (CAMRa), near Andover in the United Kingdom. With its 25-m-diameter (82-ft) dish antenna, it is the world's largest fully steerable meteorological research radar. In 1978, it became the first dual-polarization weather radar to implement the Z_{DR} measurement parameter. Since 1994, this S-band radar has been capable of measuring the full set of polarimetric-Doppler parameters.

Radar meteorology

The science of radar meteorology has its origins in the early experiments using military radar equipment during the period immediately following World War II. This work demonstrated the potential of radar techniques for use in the detection and mapping of clouds and precipitation. Over the years, the sensitivity of the radar systems evolved to the point whereby all significant precipitation could be detected at ranges sufficiently remote (about 100 km or 60 mi) so as to permit short-term forecasting of rainfall conditions at ground level several hours in advance. National weather services realized the potential of ground-based radar remote sensing to provide wide-area, real-time precipitation data to forecasters. In many countries, this led to initiatives to develop and deploy coordinated networks of identical weather radars, so as to achieve national coverage. These radar networks proved to be very capable in determining the location of areas of precipitation, but were found to be much less effective in assessing the amount, type, and rate of precipitation.

Until recently, the radars' ability to perform quantitative precipitation estimation, an important capability for such practical applications as flood forecasting, river-level management, and hazardous weather prediction, was limited, for fundamental scientific reasons, by their use of single-polarization transmitting and receiving systems. Modern, dual-polarization meteorological radar techniques have solved several operationally important problems.

Need for dual polarization. Ground-based weather radars, such as the Chilbolton research radar shown in **Fig. 1**, generally operate at frequencies in the S-band (around 3 GHz), or sometimes in the C-band (around 5.5 GHz). At these frequencies, the microwave power backscattered from a volume of atmosphere illuminated by the radar beam, filled with raindrops having an exponential distribution of drop sizes, is dependent on both the number of drops per unit volume, N_0, and the median drop diameter, D_0. With appropriate normalization, this quantity is known as the radar reflectivity, denoted Z. It is measured on a logarithmic scale, in decibels (dBZ), and ranges in value from 10 dBZ in drizzle to 55 dBZ in torrential rain and hail. Since the power received at the radar due to scattering from a single drop is proportional to the sixth power of that drop's diameter, it follows that the radar cannot distinguish between 10^6 drops of 1-mm (0.04-in.) diameter and a single drop of 10-mm (0.4-in.) diameter. However, the volume of liquid water contained in each of these cases differs by a factor of 10^3.

Improving precipitation estimation. The resolution of this dilemma makes use of the fact that raindrops are generally not spherical; rather, they are oblate to an extent which depends on their size. Oblate drops have a larger radar cross section for horizontally polarized microwave radiation than for the vertically polarized case. Consequently, a radar which uses dual-linear polarization can measure D_0 by alternately transmitting and receiving with horizontal, then vertical, polarization in a repeating sequence. Taking the ratio of the received co-polar powers yields the differential reflectivity, Z_{DR}, from which D_0 may be inferred. At frequencies in the S-band, Z_{DR} typically ranges from 0 to 3 dB for $D_0 \leq 0.5$ mm to $D_0 = 2.5$ mm. Measurements of Z and Z_{DR} enable N_0 to be determined. In turn, knowledge of the parameters N_0 and D_0 leads to an unambiguous estimate of the volume of water in the radar's sample volume.

Distinguishing the melting layer. Further benefits of the dual-polarization technique are its ability to detect the melting layer, to discriminate between different phases of precipitation (water, ice, melting particles), and to estimate the variety of precipitation particle shapes and sizes. The melting layer is commonly observed as a narrow horizontal band, around the 0°C (32°F) isotherm, where falling snowflakes and ice crystals melt to form liquid raindrops. Due to the sharp increase in water's dielectric constant on melting and the formation of a thin liquid water coating on melting snowflakes (which then appear as large-diameter water particles at microwave wavelengths), the melting layer has a very high radar reflectivity. To discriminate between the melting layer and regions of genuinely very high rainfall, dual polarization techniques have proved invaluable. The asymmetric melting process strongly depolarizes the radar echo. This depolarization signature may be detected by transmitting horizontal polarization while receiving vertical (cross) polarization, and comparing this measurement with co-polar reception. The ratio between the co-polar and cross-polar received powers is known as the linear depolarization ratio (LDR). Its magnitude can be used to readily distinguish regions of snow and ice, melting particles, and rain (with typical LDRs at frequencies in the S-band of ≤ -25, $\cong 20$, and ≤ -30 dB, respectively).

Identifying data contamination. Under enhanced but frequently occurring propagation conditions in the troposphere, the radar beam may be refracted to approximately follow the Earth's curvature. This can result in returns from distant ground targets, well beyond the radar's normal operational range, which are apt to be misinterpreted as echoes from areas of intense precipitation. The co-polar cross-correlation coefficient, $\rho_{HV}(0)$, often abbreviated to "co-polar correlation," is a useful discriminatory parameter in these circumstances. It is the complex correlation between the time series of horizontally and vertically polarized co-polar returns. This correlation is defined by the equation

$$\rho_{HV}(0) = \langle HV^* \rangle / (\langle |H|^2 |V|^2 \rangle)^{1/2}$$

where H and V are signal voltages representing the horizontally and vertically polarized returns, (0) denotes simultaneous sampling of H and V, and the angle brackets indicate an average over a set of samples. The co-polar cross-correlation coefficient is a measure of the variety of particle shapes and orientations present in the radar's sample volume. The quantity $|\rho|^2$ exhibits values around 0.99 in rain and 0.9 in the melting layer, but near 0 in both local ground clutter and in distant clutter echoes arriving via anomalous propagation.

Improved rainfall-rate measurements. Many modern weather radars are Doppler systems, with the ability to measure both the radial velocity of precipitation (from the frequency shift that arises from the Doppler effect) and the width of the frequency spectrum of radar echoes (an indicator of turbulence). When both Doppler and polarimetric capabilities are combined in a radar, an important new parameter may be measured. This is the differential phase shift, denoted ϕ_{DP}, which is a measure of the extra phase delay suffered by horizontally polarized waves, relative to that for vertically polarized waves, as the radar beam passes through regions containing oblate raindrops. Oblate raindrops have a larger cross section in the horizontal plane than in the vertical. Consequently, the large refractive index of water at microwave wavelengths causes a wave propagating through rain to exhibit a gradually increasing phase lag of the horizontally polarized component with respect to the phase of the vertically polarized component. In order to measure the relative phase of the horizontal and vertical waves, the radar must employ both polarimetric and Doppler (that is, phase-measuring) techniques. It has been shown that the specific differential phase shift, K_{DP} (the derivative of ϕ_{DP} with respect to range), is an excellent quantitative estimator of rainfall rate.

Figure 2 shows an example of multiparameter Chilbolton radar data from October 24, 1995, for a squall-line event. Exploitation of the ϕ_{DP} and Z_{DR} data reduce the uncertainty in rainfall-rate estimation from around a factor of 2 to within 25%. The LDR data confirm that the high values of Z measured are due to the presence of heavy rain, rather than the melting layer. The Z_{DR} data indicate median raindrop diameters of around 2 mm (0.08 in.). There is a strong gradient in ϕ_{DP} with range (that is, a large value of K_{DP}) as the radar beam passes through the squall line. Substitution of the measured Z, Z_{DR}, and K_{DP} data into quantitative precipitation prediction algorithms yield rainfall rates of about 80 mm/h (3 in./h).

A further benefit of ϕ_{DP} measurements is that the radar can determine its own Z calibration by making multiparameter measurements in heavy rain.

Status of operational radar meteorology. Currently, extensive operational weather radar networks exist in the United States, Europe, and parts of Asia. Many of these radars already employ Doppler techniques. However, so far, relatively few incorporate dual polarization. This situation will change in the near future, since it is intended that the NEXRAD radar network in the United States be upgraded to dual-polarization capability. In Europe, several national weather services are experimenting with dual-polarization radars as adjuncts to their existing single-polarization networks. As recommended by the conclusions of the European Union's COST-75 project, it is likely that these radars will gradually be replaced or

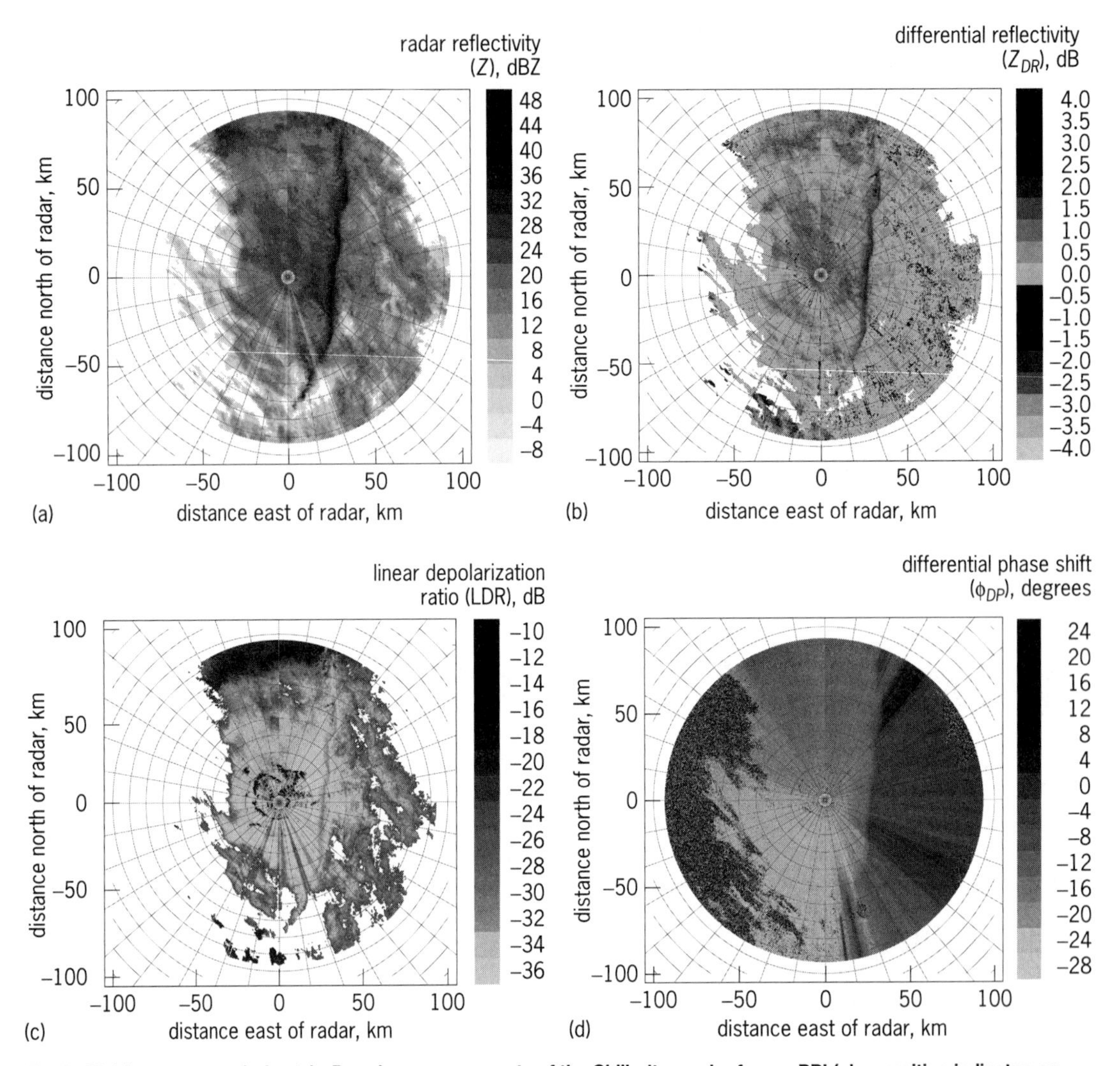

Fig. 2. Multiparameter polarimetric-Doppler measurements of the Chilbolton radar from a PPI (plan position indicator, or horizontal-plane) scan through an active squall line about 30 km (19 mi) east of Chilbolton, bearing heavy rainfall. 1 km = 0.62 mi. (*a*) Radar reflectivity (*Z*). (*b*) Differential reflectivity (Z_{DR}). (*c*) Linear depolarization ratio (LDR). (*d*) Differential phase shift (ϕ_{DP}).

upgraded to dual-polarization operation over the course of the next decade. The impetus to obtain improved meteorological and hydrological data, which dual-polarization radars can provide, is driven by concerns regarding the increased frequency of severe weather events, and the consequent economic and human costs.

Areas of research. In the research domain, several countries currently operate dual-polarization weather radar facilities. Much effort continues to be devoted to assessing the performance of the various polarimetric and polarimetric-Doppler parameters described above in operational scenarios. Whereas large, multiparameter, research radars operating at frequencies in the S-band have achieved promising results in improved quantitative precipitation measurement, it is not clear whether this performance can be fully realized in operational radars at frequencies in the C-band. Issues such as the effects of wider antenna beamwidth, higher scanning rates, and the more complex interpretation of dual-polarization observations at the C-band as compared to the S-band, are the subject of current studies. The incorporation of data from bistatic radar receivers into radar network observations is also a promising area of research. Besides precipitation radars, very active research programs exist worldwide in the use of millimeter-wave radar for cloud characterization, and in exploiting ultrahigh-frequency (UHF, 0.3–3 GHz) radar for clear-air weather observation.

For background information *see* DOPPLER EFFECT; DOPPLER RADAR; METEOROLOGICAL RADAR; POLARIZATION OF WAVES; PRECIPITATION (METEROLOGY); RADAR METEOROLOGY in the McGraw-Hill Encyclopedia of Science & Technology. Jon Eastment

Bibliography. D. Atlas (ed.), *Radar in Meteorology*, American Meteorological Society, Boston, 1990; L. J. Battan, *Radar Observation of the Atmosphere*, University of Chicago Press, London, 1973; V. N. Bringi and V. Chandrasekar, *Polarimetric Doppler Weather Radar: Principles and Applications*, Cambridge University Press, 2001; R. J. Doviak and D. S. Zrnic, *Doppler Radar and Weather Observations*, 2d ed., Academic Press, 1993; P. Meischner (ed.), *Weather*

Radar: Principles and Advanced Applications (Physics of Earth and Space Environments), Springer-Verlag, Berlin, 2004.

Reactive oxygen species

Free radicals are chemicals with an unpaired electron (denoted as •). Depending on the chemical structure, they can be highly reactive due to the tendency of electrons to pair. Reactive oxygen species (ROS) are oxygen-containing molecules which may be free radicals, for example, superoxide ($O_2^{\bullet -}$) and hydroxyl radical (•OH), or nonradical, for example hydrogen peroxide (H_2O_2) and singlet oxygen ($^1O_2$); (see **table**). Normal cellular metabolism can generate ROS when electrons "leak" from reactions in the cell and combine with oxygen (O_2). The deliberate generation of ROS is useful for the cell, as ROS can have an antimicrobial function, as well as act as signals within and between cells. Due to their reactive nature, ROS have the potential to damage important cellular molecules. As a consequence, their production is tightly controlled, for example by various antioxidant defenses such as vitamins C and E, and antioxidant enzymes such as catalase and glutathione peroxidase. However, a proportion of the ROS evade these defenses and can result in the oxidative modification of cellular molecules [lipids, deoxyribonucleic acid (DNA) and proteins], which can be detected even in normal tissue (see **illustration**).

Oxidative stress and oxidative damage. Various external events, such as exposure to ionizing or ultraviolet radiation, or the metabolism of certain chemicals (for example; paracetamol overdose) can also lead to an increase in the generation of ROS. When the generation of oxidants is excessive and antioxidant defenses are overwhelmed, or antioxidant defenses are inherently poor, the resulting disturbance of the pro-oxidant/antioxidant balance in favor of the former leads to a condition called oxidative stress. This can result in increased oxidation of cellular components, perturbation of cell signaling pathways, alterations in gene expression, and subsequent impacts on cell metabolism and function, perhaps leading to cell death or disease. An important target for oxidation is DNA as damage may lead to mutation. Consequently, DNA damage has been well studied, and several classes of product identified. By far the most studied product of oxidative modification of DNA is 8-hydroxyguanine (8-OHGua) and its deoxynucleoside equivalent, 8-hydroxy-2′-deoxyguanosine (8-OH-dG). In addition to mutation, failure to protect or repair the genome may lead to other forms of genetic instability, such as loss of heterozygosity and alterations in methylation. Consequently, oxidative damage to DNA has been associated with a number of pathologies.

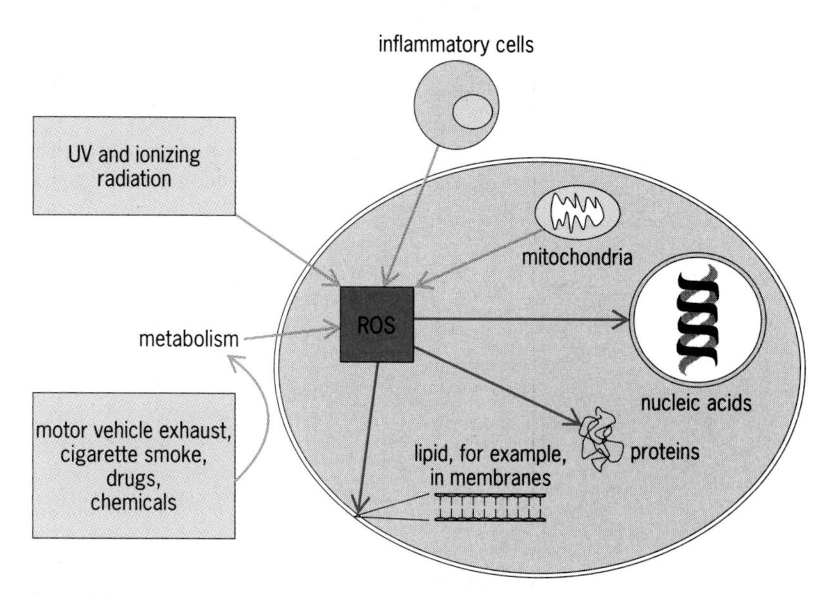

Potential sources and targets for ROS.

To combat the deleterious biological effects of ROS, cells have developed specific protective mechanisms. The most highly developed and intensively studied defense mechanism is DNA repair. Defective repair may lead to increased levels of damage, as

Precursors and nature of reactive oxygen species

Precursors	Process	ROS product
O_2	Visible light plus a chemical photosensitizer to transfer energy to oxygen	Singlet oxygen ($^1O_2$)
O_2	Electron from enzymes, electron transport chain (for example, mitochondria, inflammatory cells)	Superoxide ($O_2^{\bullet -}$)
$O_2^{\bullet -}$, H^+	Spontaneous production or catalyzed by the enzyme superoxide dismutase	Hydrogen peroxide (H_2O_2)
H_2O_2, reduced metal ions (such as Fe^{2+})	One electron reduction	Hydroxyl radical (HO•)
H_2O_2, Cl^-, Br^-	Enzyme catalysis (myeloperoxidase; eosinophil peroxidase)	Hypohalous acids (HOCl; HOBr)
Lipid, oxygen	Free-radical initiated lipid degradation (lipid peroxidation)	Lipid oxy radicals (LO•, $LO_2^{\bullet}$) Lipid hydroperoxide (LOOH)
Arginine, oxygen	Nitric oxide synthase in neurons, macrophages, endothelial cells	Nitric oxide (NO•)
Superoxide, nitric oxide	Direct chemical reaction	Peroxynitrite (OONO)

might lowered antioxidant capacity, each with detrimental consequences for the cell.

Oxidative damage and disease. Elevated levels of oxidative DNA damage have been reported in the tissues of many diseases, with the accompanying proposal that this damage may be important in the etiology of the disease.

Cancer. One consequence of oxidized base lesions persisting in DNA is mutation, which is a crucial step in carcinogenesis. Elevated levels of oxidative DNA lesions have been noted in many tumors, strongly implicating such damage in the etiology of cancer. Such increases in damage are understood to arise as a consequence of (1) an environment in the tumor low in antioxidant enzymes and high in ROS generation, or (2) reduced DNA repair. Furthermore, mutations characteristic of oxidative DNA damage have been observed in key genes in lung and liver cancer. Increased ROS production may lead to an increase in oxidative damage to DNA, but also may decrease the expression or activity of the enzymes which prevent the persistence of such damage. There are numerous experimental data which suggest that decreased repair activity may be linked with cancer development, and that many types of DNA repair pathways are reduced in cancer patients. Subtle inherited alterations in the repair of oxidative DNA damage known as polymorphisms have the potential to affect lesion levels, although whether this is of significance in terms of predisposition to disease remains to be established. Taken together, the data suggest that, while the role of oxidative stress in carcinogenesis is well established, the extent to which oxidative DNA damage contributes has not been well defined, but at the very least it is a pertinent marker of oxidative stress.

Brain diseases. The neurodegenerative conditions, Alzheimer's disease, Huntington's disease, and Parkinson's disease, have oxidative stress implicated in their pathogenesis. This is supported by in-vitro studies demonstrating that neurotransmitters such as dopamine and serotonin can generate free radicals. The role of oxidative stress and oxidative damage to biomolecules other than DNA in the pathogenesis of neurodegenerative disease—and Alzheimer's disease specifically—has been reiterated in several recent reviews of the subject, although the greatest significance for the pathogenesis of the disease has been placed upon lipid and protein oxidation.

Inflammation and infection. Oxidative stress is an important consequence of inflammation, with a number of studies reporting elevated levels of 8-OH-dG with hepatitis, HIV infection, *Helicobacter pylori* infection (for example, peptic ulcer), and atopic dermatitis. The inflammatory response can lead to the recruitment of activated white blood cells, which may give rise to a "respiratory burst"—an increased uptake of oxygen that causes the release of high quantities of ROS, with subsequent DNA damage. Chronic inflammation and hence oxidative stress have been closely linked to the pathogenesis of certain autoimmune diseases, such as rheumatoid arthritis and systemic lupus erythematosus, with radical production resulting not only in connective tissue damage but also in oxidative DNA damage, which, upon exposure to the systemic circulation, can help produce the autoantibodies frequently seen in these diseases.

Mechanistically, chronic inflammation can be closely linked to carcinogenesis according to the hypothesis by which ROS generated during chronic inflammation initiates and promotes events important in carcinogenesis. However, there is little evidence to suggest that patients with chronic inflammatory diseases, such as systemic lupus erythematosus, have an increased incidence of cancer development. Nevertheless, it has been estimated that chronic inflammation may be involved in the development of about one-third of all cancer cases worldwide.

Cardiovascular disease. Atherosclerosis may be viewed as a chronic inflammatory disease, with growing evidence for the involvement of ROS in atherosclerotic plaque development, although the actual role of DNA damage in this lesion is less clear. While there are relatively few reports examining levels of oxidative DNA damage in cardiovascular disease, one striking result was that of a strong association between premature coronary heart disease in men and elevated 8-OH-dG levels in white blood cells. Interestingly, one component of dietary fat (oxidized low-density lipoprotein), an established risk factor for heart disease, has been linked to reduced DNA repair, which could have important consequences for oxidative DNA damage.

Aging. The "free radical theory of aging" suggests that aging occurs through the gradual accumulation of free radical damage to cellular components. Failure of antioxidant defenses to scavenge all radical species, and decreasing DNA repair efficiency—both evident from the increasing "background" levels of damage with age—will result in the insidious accumulation of damage and gradual loss of function. Numerous studies have reported the accumulation of 8-OH-dG, and by implication other lesions, with age, in the DNA of both the nucleus and mitochondria. Overall, it can be concluded that high metabolic rate equates to shorter maximum life-span potential and faster aging. While the experimental evidence is not conclusive, the hypothesis for the involvement of free radicals in aging remains compelling.

Conclusions. Despite the many diseases in which levels of oxidative DNA damage are elevated, demonstration of a causal link between elevations in damage and disease development remains elusive. Nevertheless, the proposal that oxidative DNA damage has an important role in disease remains credible, and is supported by the presence of multiple, highly redundant pathways for its repair. Clearly it is vital for the cell not to allow this damage to persist. However, it is important to appreciate that the elevated DNA damage levels present under oxidative stress may have more effects upon the cell than simply mutation, and that simply increased levels of ROS can affect cellular function. The wide-ranging consequences of oxidative stress may account for the breadth of diseases which appear to have oxidative stress in their etiology. Equally, it must not be forgotten that biomolecules other than DNA may be oxidatively modified and this may also have a significant effect. Both possibilities, taken together, suggest that a better understanding of ROS and oxidative DNA damage may lead to intervention strategies which may prevent, or at least ameliorate, many diseases.

For background information *see* AGING; ALZHEIMER'S DISEASE; ANTIOXIDANT; FREE RADICAL; INFLAMMATION; ONCOLOGY in the McGraw-Hill Encyclopedia of Science & Technology.

Marcus S. Cooke; Mark D. Evans

Bibliography. M. F. Beal, A. Lang, and A. Ludolph (eds.), *Textbook of Neurodegenerative Diseases*, Cambridge University Press, in press, 2005; R. G. Cutler and H. Rodriguez (eds.), *Oxidative-Stress and Aging: Advances in Basic Science, Diagnostics and Intervention*, World Scientific Publishing, 2003; B. Halliwell and O. I. Aruoma (eds.), *DNA and Free Radicals*, Ellis Horwood, Chichester, 1993; K. K. Singh (ed.), *Oxidative Stress, Disease and Cancer*, Imperial College Press, London, in press, 2005.

Regional climate modeling

Both regional climate models and general circulation models include time-dependent partial differential equations and algebraic equations to represent the motion and physics of the atmosphere and surface properties in the climate system (such as wind, pressure, temperature, clouds, radiation, rainfall, and soil hydrology). Most regional climate models cover an area of 10^5–10^7 km^2 (10^4–10^6 mi^2) with a typical resolution of 10–100 km (6–60 mi), compared to a few hundred kilometers resolution for most general circulation models, which cover the global atmosphere. Integration times for regional climate models range from a month to more than 10 years. Most regional climate models use the observed sea surface temperature or the mixed-layer ocean model for the ocean. The coupled climate model system, including the regional atmospheric model, ocean–sea ice model, and hydrology model, has also been successfully applied for studying some extreme weather events in Germany. At the lateral boundaries, a regional climate model is driven by atmospheric wind, temperature, and humidity generated from a general circulation model output or global meteorological reanalysis, which is derived from the combination of the general circulation model output and observations. A regional climate model is, in principle, similar to the operational limited-area numerical weather forecasting models. But a regional climate model is integrated continuously over time intervals much longer than those used in numerical weather forecasting are. Consequently, lateral and surface boundary conditions become dominant in a regional climate model, while limited-area numerical weather forecasting is more dependent on the initial data.

Capability. When the global meteorological reanalysis (a set of observational data compiled and edited by the National Centers for Environmental Prediction) is applied at boundary and initial conditions, a regional climate model can reproduce the synoptic patterns in both spatial and temporal evolutions (**Figs. 1** and **2**). In addition, a good regional climate model has greater capability of resolving details and providing more information, including condensational heating of clouds or sub-grid-scale fluxes, than observations do. (Subgrid scale indicates that the length scale involved is too small to be resolved explicitly.) If a regional climate model can reasonably reproduce the reanalyses in both spatial and temporal resolutions, it may be used to simulate extreme events and study the effects of variations in physics, aerosol concentration, and surface conditions on weather/climate, as long as those changes do not

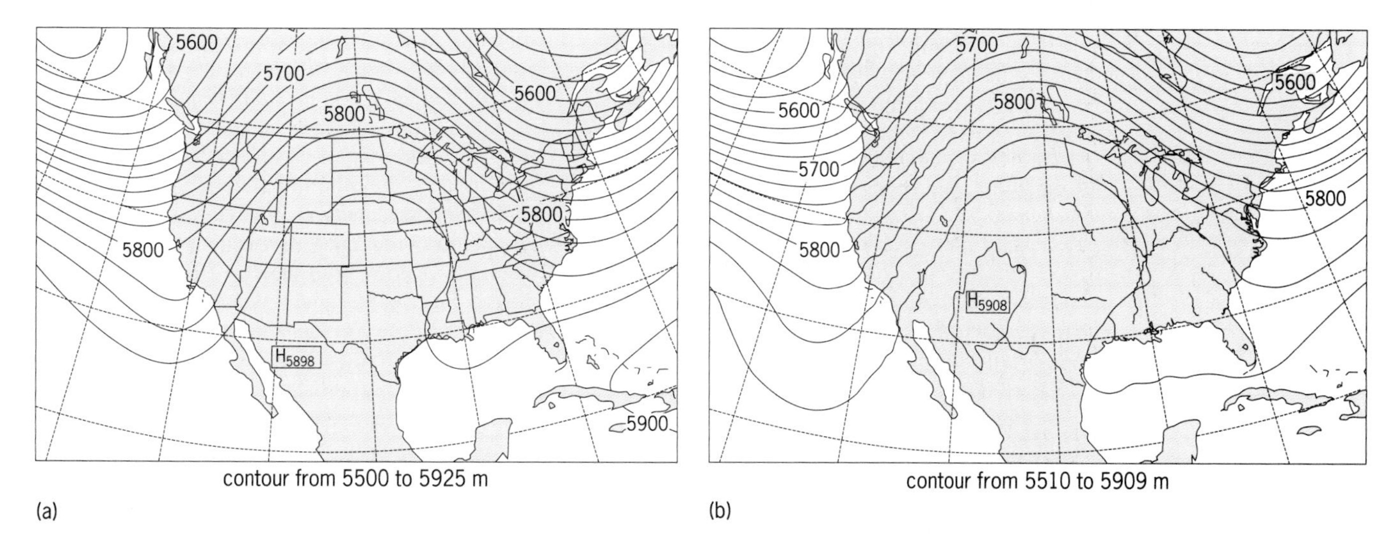

Fig. 1. Regional climate models. (*a*) European Center for Medium-Range Weather Forecasts (ECMWF) reanalysis and (*b*) Purdue Regional Model (PRM) simulation of monthly mean of geopotential at 500 millibars in June 1988 when a warm ridge dominated the midwestern United States during severe drought. (*After W.-Y. Sun et al., 2004*)

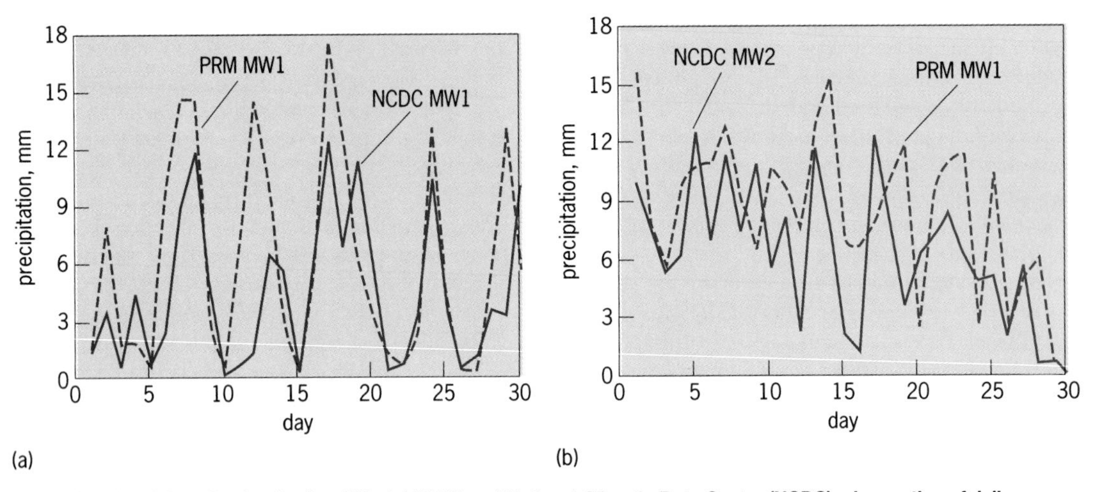

Fig. 2. Simulated from Purdue Regional Model (PRM) and National Climatic Data Center (NCDC), observation of daily integrated precipitation over flooding area in the midwestern United States for (*a*) June and (*b*) July 1993. (*After M. G. Bosilovich and W.-Y. Sun, 1999*)

significantly alter the imposed lateral boundary. The domain of a regional climate model should be large enough to allow the synoptic and mesoscale disturbances to develop. But if the domain is too large, the regional climate model results can be drifted away from the large-scale pattern imposed by reanalysis or a general circulation model through the lateral boundary. Occasionally, the regional climate model has been applied beyond its limitation. For example, S. I. Seneviratne and coworkers applied the National Center for Atmospheric Research's Regional Climate Model (RegCM) to study the climate of 1986, 1988, 1990, and 1993 (that is, control cases) over the United States. They also simulated the made-up, warm cases by increasing the atmospheric and sea surface temperatures uniformly by 3 K, while the relative humidity remained unchanged and all other initial and boundary fields were identical for both sets of simulations. However, it is unlikely that the flow pattern and lateral boundary conditions would remain the same in a warm episode as in the control cases.

Nested models. Instead of using analysis, a regional climate model can be nested in a general circulation model, which provides both initial and boundary conditions for the regional climate model. When the chemistry and physics of trace gases/aerosols are included, the models can be used to predict the transport and deposition of aerosols and greenhouse gases, as well as their effects and other forcings on weather, climate, and hydrology. The nested regional climate model can also be applied to simulate the past or future climates when the composition of the atmosphere or land-ocean distribution is different from the current situation. The nested regional climate model is a better tool than the statistical downscaling method for simulating the local climate/weather. The statistical downscaling is based on the assumption that the relationship between the large-scale climate variables (such as surface pressure and atmospheric temperature) and the surface climate at point locations (for example, precipitation measured by a rain gauge) remains the same under different climate systems. *See* DOWNSCALING CLIMATE MODELS.

Climate prediction. A fine-resolution regional climate model can simulate the structure of the frontal system, mesoscale convective systems, hurricanes, typhoons, land sea breezes, downslope winds, lee-vortices, and orographic precipitation, which are usually not well resolved by a general circulation model or conventional observational systems. A regional climate model can provide climate information with useful local details, including realistic models of extreme events. Predictions using regional climate models will lead to substantially improved assessments, for example, of the vulnerability of a region to climate change and possible adaptation strategies. Regional modeling is a relatively affordable means of bringing climate change closer to impact research, the public, and decision makers. However, the bias of a general circulation model or reanalysis can be passed to a regional climate model. The quality of a regional climate model depends upon the quality of the reanalysis or general circulation model. Compared to the driving general circulation model, a regional climate model generally produces more realistic regional detailed climate. Validation experiments also show that a regional climate model can both improve and degrade aspects of the regional climate, compared to driving general circulation model runs, especially when regionally averaged. Overall, both general circulation models and regional climate models performed better at midlatitudes than they did in tropical regions.

Limitations. In addition to the errors of regional climate models from initial and boundary conditions, errors came from numerical schemes, physics, and parameterizations of the models. It is very difficult to formulate realistic sub-grid-scale parameterizations for convective clouds and turbulence in a regional climate model, because the scale separation (in both time and space) in a regional climate model is more difficult than in a general circulation model. In the popular one-way nesting technique, a regional climate model is driven by a general circulation model output through the lateral boundary. However, the result of the regional climate model cannot be fed back to the general circulation model. In this way, the horizontal fluxes of heat, moisture, and momentum over the entire domain of the regional climate model are completely controlled by the general circulation model at any time. In a real atmosphere, a large-scale system can affect a small-scale system, just as a severe storm or a small-scale disturbance can influence the large-scale circulation. Therefore, a two-way nest should be applied to couple the general circulation model and the regional climate model. But more computing resources and further studies are required in this area.

Outlook. It has been proven that a regional climate model can simulate the real climate using reanalysis as initial and boundary conditions. In climate research, some scientists are still concerned about the validity of nesting a regional climate model in a general circulation model. Specifically, they question whether a regional climate model can really provide realistic regional climate information for global-change research and can improve seasonal climate prediction, or if a regional climate model just adds spatial details that are intricately tied to the general circulation model (Wang, 2004). A Japanese general circulation model with resolution of about 20 km (12 mi) and 60 levels vertical will become an operational global model soon. Japanese researchers have also developed a regional cloud-resolving model on the Earth Simulator. It is expected that very high resolution general circulation models and regional climate model will become popular within 10 years. In addition to extra computing resources, new physics, parameterizations, and numerical schemes are required for those models. Three-dimensional observational data, including soil properties and sea surface temperatures, are also needed as the initial and boundary conditions, especially for the high-resolution regional climate models.

For background information *see* ATMOSPHERE; ATMOSPHERIC GENERAL CIRCULATION; CLIMATE MODELING; CLIMATIC PREDICTION; CLIMATOLOGY; METEOROLOGY; WEATHER FORECASTING AND PREDICTION in the McGraw-Hill Encyclopedia of Science & Technology.

Wen-Yih Sun

Bibliography. M. G. Bosilovich and W.-Y. Sun, Numerical simulation of the 1993 Midwestern flood: Land-atmosphere interactions, *J. Climate*, 12:1490–1505, 1999; D. Jacob and P. Lorenz, The Baltic Sea Coupled Regional Modeling System: BALTIMOS, *Geophys. Res. Abstr.*, vol. 7, 08342, 2005; S. I. Seneviratne et al., Summer dryness in a warmer climate: A process study with a regional climate model (RegCM), *Climate Dynam.*, 20:1:69–85 2002; W.-Y. Sun, J. D. Chern, and M. Bosilovich, Numerical study of the 1988 drought in the United States, *J. Meteorol. Soc. Jap.*, pp. 1667–1678, 2004; Y. Wang et al., Regional climate modeling: Progress, challenges, and prospects, *J. Meteorol. Soc. Jap.*, 1599–1627, 2004.

Risk analysis (homeland security)

Applying risk analysis to strategies against terrorism requires an innovative combination of approaches. Assessing the risk of terrorism (which is dynamic and adaptable) is fundamentally different from assessing the risk of accidents or acts of nature. In particular, terrorists can adopt different strategies in response to our defenses. Game theory, an economic tool for mathematically modeling optimal strategies, provides a way of accounting for this. Counterterrorism can benefit from the combination of risk analysis and game theory. While game theory quickly becomes mathematically complex and may not be applicable to problems of realistic scale and complexity, the results of such analyses can yield useful qualitative insights into the nature of optimal defensive investments.

Introduction. Since the September 11, 2001, terrorist attacks, there has been increased interest in methods for analyzing the risks of terrorism and identifying cost-effective defenses. However, the development of such methods poses two challenges: the relative scarcity of empirical data on severe terrorist attacks and the intentional nature of such attacks.

Risk analysis and applications to terrorism. In dealing with events for which empirical data are sparse, risk analysis is used to break down complex systems into individual components for which larger amounts of empirical data may be available. Quantification of risk analyses generally relies on some combination of expert judgment and Bayesian statistics to estimate the parameters of interest.

In Bayesian statistics, a prior probability distribution for an unknown quantity of interest (such as the failure rate of a key component) is updated using empirical data to derive a posterior probability distribution. Unlike in classical statistics, the prior probability distribution explicitly allows for the use of subjective judgment, which can be obtained from expert opinion. In assessing the risks of a new technology, the initial estimate could be based on expert opinion, with empirical data being used to update the assessment as it becomes available.

Following September 11, numerous researchers have proposed the use of risk analysis for homeland security. Because terrorist threats can span such a wide range, the emphasis has been on using risk analysis to target security investments at the most important threats (such as threats against critical infrastructure).

In most applications of risk analysis, the decision maker can review a list of possible actions, rank them by the magnitude of risk reduction per unit cost, and choose the most cost-effective ones. This does not work so well in the context of terrorism (especially if the potential attacker can readily observe system defenses), since terrorist strategies can change in response to our defenses. Rerunning an analysis with some assumed security improvements but the same threat can overestimate the effectiveness of those improvements. Game theory provides a way of accounting for this.

Applications of game theory to terrorism. Due to its value in understanding conflict, game theory has been applied to many security problems, such as military strategy, arms control, and computer security. With respect to terrorism, most applications have been by economists. Much of this work has been designed to provide policy insights (such as the relative merits of deterrence versus other protective measures). There has also been interest in using game theory to help determine which assets to protect.

For example, the Brookings Institution has recommended a "weakest-link" model, where defensive investment is allocated only to the targets that would cause the most damage if attacked, such as sites of high national significance and targets involving large numbers of casualties or significant economic costs. However, weakest-link models tend to be unrealistic. Instead, real-world decision makers will generally "hedge" by investing in the defense of additional targets to cover contingencies such as whether they have correctly estimated which targets will be most attractive to attackers.

It is also important to go beyond the general guideline of protecting only the most valuable assets and consider the success probabilities of attacks against various possible targets. Terrorists appear to take the probability of success into account in their choice of targets. Moreover, defending one set of targets could deflect attacks to alternative targets, which are less attractive to the attackers (for example, more costly or difficult to attack) but could be equally or more damaging to the defenders. Past economic analysis documents the existence of substitution effects in terrorism.

Security as a game between defenders. In addition to viewing security as a game between an attacker and a defender, it is instructive to consider the effects of defensive strategies on the incentives faced by other defenders. Some defensive actions (such as visible hardening of facilities) may increase the risk of terrorism against other targets that have not been similarly hardened. This can lead to overinvestment in security, because the payoff to any one organization from investing in security is greater than the net payoff to society as a whole (since the risk has not been eliminated but merely deflected). Conversely, other types of defensive actions, such as vaccination against smallpox, decrease the risk even to individuals that have not taken such actions. This can result in underinvestment in security, if some individuals attempt to "free ride" on the investments of others. In such cases, where investments in security have either positive or negative benefits, there may be a role for the federal government or other "coordinating mechanisms" to help ensure socially desirable levels of investment in security.

Combining risk analysis and game theory. Many applications of risk analysis to security do not explicitly model the response of potential attackers to defensive investments and may overstate the effectiveness and cost-effectiveness of those investments. Similarly, much of the existing work in game theory for security focuses on nonprobabilistic games and considers individual assets in isolation, failing to account for their role in ensuring the functionality of a larger system, such as the transportation system. Combining risk analysis with game theory could be a fruitful research direction for identifying cost-effective ways of protecting against terrorist threats to complex systems. In this type of approach, possible protective actions would be evaluated in light of terrorists' ability to adapt and alter their tactics, making it possible to capture aspects of the problem that might not be well modeled using either game theory or risk analysis alone.

Preliminary results emphasize the value of redundancy as a defensive strategy, since redundancy reduces the flexibility available to a potential attacker and increases the flexibility available to defenders. Other principles in the design for reliability (such as spatial separation and functional diversity of key components) can be important in defensive strategies to ensure that the same type of attack cannot disable all redundant components. Finally, secrecy and even deception can be strategies for improving security from terrorism, although identifying the situations in which secrecy and deception are likely to be helpful requires further research.

For background information *see* BAYESIAN STATISTICS; DECISION ANALYSIS; DECISION THEORY; GAME THEORY; OPERATIONS RESEARCH; PROBABILITY; RISK ASSESSMENT AND MANAGEMENT in the McGraw-Hill Encyclopedia of Science & Technology. Vicki M. Bier

Bibliography. T. Bedford and R. Cooke, *Probabilistic Risk Analysis: Foundations and Methods*, Cambridge University Press, 2001; G. Hart, W. B. Rudman, and S. E. Flynn, *America: Still Unprepared, Still in Danger*, Council on Foreign Relations, Washington, DC, 2002; B. H. Lindqvist and K. A. Doksum (eds.), *Mathematical and Statistical Methods in Reliability*, World Scientific Publishing, 2005; M. O'Hanlon et al., *Protecting the American Homeland*, Brookings Institution, Washington, DC, 2002; J. Von Neumann, *Theory of Games and Economic Behavior*, Princeton University Press, 1953.

Salmon farming

The natural range of the Atlantic salmon (*Salmo salar*) stretches from Portugal and the northern coast of New England in the United States northward to subarctic Norway, Russia, and Canada. Salmon are anadromous, whereby they breed in freshwater but migrate to sea to feed. Within their range, adult salmon spawn in cool and well-oxygenated streams. Juveniles spend 1–5 years in freshwater before transforming into silvered smolts. At this stage they migrate to sea, following ocean currents to feeding grounds in the Atlantic. Once in the sea, salmon grow rapidly, feeding on small fish and crustaceans. After 1–3 more years, the fish reach sexual maturity and migrate with great accuracy to their natal rivers, often spawning in the same tributary where they were born. As a result, salmon stocks have evolved into genetically distinct local populations.

Throughout its range, adult Atlantic salmon are highly valued for food. Prior to the eighteenth century, subsistence fisheries existed wherever salmon were found. After about 1750, commercial net and trap fisheries were established to meet the demands of growing urban populations. However, during the late twentieth century the cumulative effects of declining marine productivity and human impacts such as dams, pollution, and overfishing began to limit the abundance of stocks, and by the 1980s commercial fisheries were becoming unviable; wild salmon had become a luxury food item.

During the 1960s and 1970s, trials were completed in Norway and Scotland to artificially propagate wild salmon in freshwater hatcheries and transfer smolts to saltwater net cages for on-growing. Salmon were successfully fed pellets manufactured from industrial fish, and selective breeding for late- maturation maximized growth rates. Since the 1980s, the salmon aquaculture industry has expanded rapidly within the species' range, with production dominated by Norway, Scotland, Ireland, and eastern Canada. In Scotland, annual production increased from negligible amounts in 1979 to 150,000 metric tons in 2002 (see **illustration**). Salmon farming has also been successfully established outside the species' range in suitable regions of Tasmania, Chile, and western Canada.

By exploiting the "fish gap" between the limited supply and increasing demand for wild salmon, salmon farming has become an economic success story. Because suitable areas are often within remote, underdeveloped coastal regions, the industry has encouraged socioeconomic development, encouraging investment and stemming population drift to urban areas. For example, in western Scotland salmon farming is estimated to provide 30% of rural employment, and the industry exceeds the export value of traditional beef and sheep agriculture combined.

Despite this, there are concerns about the industry's long-term economic and environmental sustainability. The global growth in salmon farming has flooded most markets, forcing the prices of salmon down and pressuring multinational companies to expand further to exploit economies of scale and mechanize production. In many areas the industry's growth has outstripped planning frameworks and understanding of its ecological "footprint." The levels of protection of wild salmon populations have increased due to their depleted status, for example under the Endangered Species Act in the United States, and the European Union Habitats Directive. The impacts of salmon aquaculture on wild stocks also are of increasing concern. Consequently, the North Atlantic Salmon Conservation Organisation passed resolutions in 1994 and 2003 to minimize the impact of aquaculture on wild salmon.

Impacts on wild salmonids. When the industry began, few predicted the potential impacts of farmed fish on neighboring populations of wild salmonids. Marine cages were often sited in sheltered fiords and bays with easy shore access, usually close to river mouths, and freshwater hatcheries and cages were placed in areas of catchments accessible to wild fish. Consequently sites were often adjacent to migration and rearing areas for wild salmonids, and since the late 1980s the following impacts on wild fish have become evident.

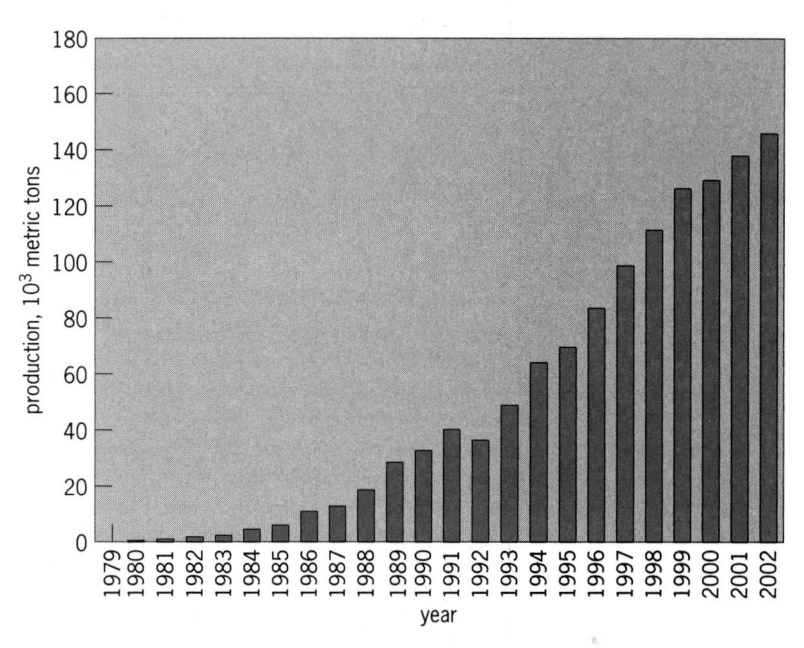

Annual production of farmed Atlantic salmon in Scotland, 1979–2001. (*Scottish Executive data*)

Escapes. The escape of farmed salmon from marine and freshwater sites occurs regularly due to storm and predator damage to net cages and operational accidents. Studies have proven that farmed salmon successfully breed with wild salmon, diluting unique local genetic characteristics and reducing population fitness. Due to the relatively small size of wild stocks and the large size of farms, even a small percentage of farmed fish can vastly outnumber wild fish. As a result, in some Norwegian rivers over 80% of spawning adults are escapees. Escapes are also an issue for biodiversity conservation in countries where Atlantic salmon are nonnative, such as the west coast of Canada and Chile.

Parasites. Parasites proliferate in situations where hosts are enclosed in farms; they can then be transmitted back to wild stocks at elevated levels, causing abnormal mortality. The sea louse *Lepeoptheirus salmonis*, a marine copepod specific to salmonids, is the most serious parasite, and epidemics have caused significant losses of farmed and wild fish. The freshwater fluke *Gyrodactylus salaris* is native to salmon populations in Baltic rivers, and the fish have evolved resistance to it. However, through the movement of farmed fish it has spread to countries where salmon populations have no resistance, and mass mortalities have resulted, notably in Norway.

Disease. As with parasites, bacterial and viral pathogens also proliferate in farmed salmon and can be transmitted at elevated levels to local wild salmonids. Diseases such as furunculosis, infectious pancreatic necrosis, and infectious salmon anemia are examples.

Impacts on aquatic environment. Salmon aquaculture also has a wider footprint on the aquatic environment, which has not been fully quantified.

Local benthic impact. The growing of fish in cages results in the accumulation of waste under the site, with related changes in the local habitat. Although some of this pollution is dispersed by currents, the deposition of matter with high levels of nutrients can lead to oxygen-poor conditions, depleting biodiversity of local flora and fauna.

Regional ecosystem impacts. The dispersal of nutrients, wastes, and chemical compounds from sites is cumulative over time. Consequently there are concerns that salmon farming is altering ecosystems through nutrient enrichment and bioaccumulation of waste. For example, toxic algal blooms may have increased in frequency in salmon farming areas, with impacts on water supplies and also shellfish aquaculture. The structure of ecological communities, including fish, may also be altered by increased aquatic productivity. Establishing ecological carrying capacities for marine and fresh waters is a major research priority for government regulators.

Marine mammals. Fish-eating predators are attracted to fish farms. Although birds and small mammals such as otters can be a problem, seals are the major source of economic loss to fish farms. Problem seals are shot, with potential impacts on local populations. Acoustic deterrents are also deployed to scare seals, which can affect porpoises, dolphins, and whales.

Feed supplies. The feed for farmed salmon is obtained from industrial fisheries for species such as capelin. However, the food conversion ratio of feed to salmon is approximately 3:1, and this inefficient use of marine productivity is controversial, particularly when the sustainability of industrial fisheries is under question.

Addressing impacts on wild salmonids. In collaboration with government and wild salmonid stakeholders, the salmon farming industry is developing measures to minimize impacts on wild salmonids. The rationale is often that solutions can be of mutual benefit to farmed and wild fish.

Siting. New fish farm sites can be located away from catchments and river mouths where wild salmonid populations of high conservation value occur. In some cases, existing sites are being relocated to less sensitive areas. Minimum separation distances from river mouths are also being introduced in some areas.

Escapes. Escapes are being reduced through codes of practice recommending minimum standards for cage design. In some countries, freshwater farms are permitted only in areas of catchments inaccessible to wild salmon. Tagging of all farmed fish and reporting of escapes are also being explored as methods for tracking the problem. The breeding of sterile salmon which cannot breed in the wild is also under consideration, but as for tagging, the resulting increased costs of production may be prohibitive.

Parasite and disease management. Advances in sea lice control are being made, including more efficacious chemical treatments. Integrated with management techniques such as synchronized treatments between farms, single-year class production, and coordinated fallow periods, lice infestations are being brought under control. Controls on fish movement also reduce the risks of parasite and disease transfer.

Addressing impacts on aquatic environment. The wider impacts of salmon farming are increasingly being addressed through more effective planning based on integrated coastal zone management. For example, in Europe the Water Framework Directive is developing river basin planning, where aquaculture pollution is considered in the context of other anthropogenic sources. Further mitigations for the wider impacts of salmon farming include the following.

Diversification. As with terrestrial agriculture, diversification is being developed as a method of increasing productivity while reducing environmental impact. Cod and halibut are two species which offer realistic opportunities.

Land-based systems. The rearing of farmed salmon in land-based tanks would solve many of the environmental problems associated with the industry. Contained units would nullify the possibility of escapes and allow the treatment of water for parasites and pollution. However, the costs of pumping seawater currently make this option economically unviable. Instead, the use of closed containment sites at sea, both in coastal and offshore locations, is more feasible.

Alternative feed sources. The issue of industrial fisheries as a source of salmon feed is being addressed by the development of vegetable oils as a substitute.

For background information *see* AQUACULTURE; FISHERIES ECOLOGY; MARINE CONSERVATION; MARINE FISHERIES; SALMONIFORMES in the McGraw-Hill Encyclopedia of Science & Technology.

James R. A. Butler

Bibliography. M. Esmark, S. Stensland, and M. S. Lilleeng, *On the Run: Escaped Farmed Fish in Norwegian Waters*, WWF-Norway Rep. 2/2005, Oslo, 2005; S. J. Northcott and A. F. Walker, Farming salmon, saving sea trout: A cool look at a hot issue, pp. 72–81 in *Aquaculture and Sea Lochs: Proceedings of a Joint Meeting of the Scottish Association for Marine Science and the Challenger Society at Dunstaffnage Marine Laboratory, U.K.*, June 1996; G. H. Porter, *Protecting Wild Atlantic Salmon from Impacts of Salmon Aquaculture: A Country-by-Country Progress Report*, 2d ed., World Wildlife Fund (Washington, DC) and Atlantic Salmon Federation (St. Andrews, Canada), 2005; A. F. Youngson, L. P. Hansen, and M. L. Windsor, *Interactions between Salmon Culture and Wild Stocks of Atlantic Salmon: The Scientific and Management Issues*, Report by a symposium organized by the International Council for the Exploration of the Sea (ICES) and the North Atlantic Salmon Conservation Organization (NASCO), held at Bath, England, April 18–22, 1997, NINA, Trondheim, Norway, 1998.

Sedna

In November 2003 the known solar system grew significantly larger with the discovery of a new object, Sedna, 90 astronomical units from the Sun. (1 AU = 150×10^6 km = 93×10^6 mi, the average distance from the Earth to the Sun.) For comparison, Jupiter orbits 5 AU and Neptune 30 AU from the Sun. Objects in the Kuiper Belt, discovered beginning in 1992, orbit beyond Neptune at typical distances of around 40 AU, and are likely to be the closest relatives to Sedna.

Discovery. Michael Brown, Chadwick Trujillo, and David Rabinowitz reported the discovery based on images taken at the Palomar Observatory 1.2-m (48-in.) Oschin Schmidt telescope. They discovered Sedna using the same method Clyde Tombaugh used to discover Pluto in 1930, and indeed the same method used to discover nearly all of the over 100,000 solar system objects known today. All of these objects move against the background field of stars. Most of the motion relative to the stars is apparent, being caused by the orbital motion of the Earth and an effect called parallax (**Fig. 1**). If two pictures of the same patch of sky are taken at different times and then compared, solar system objects will appear to have moved relative to the background stars. The exact position of any moving targets seen in the images is then compared to the predicted position of all of the known solar system objects. If the positions do not match, the observer has discovered a new member of the solar system.

Sedna's apparent motion in the discovery images was very small: it moved a total of 4.6″ (or 0.0013°) in 3 hours. The amount of motion is proportional to the distance to the target. In this case the distance to the target was estimated to be 100 AU; the discoverers knew immediately that they had discovered something much farther from the Sun than anyone previously had. The immediate problem was that their single night of data gave only a rough idea of where Sedna would be in the coming weeks: without more data their newly discovered object could be lost. Beginning one week later and continuing through December, they and collaborators took more images at other observatories (including the giant Keck telescope in Hawaii, and at Cerro Tololo in Chile). By the end of December they had measured Sedna's motion well enough to predict where Sedna would be months in the future and, more importantly, where it had been months previously. Extrapolating Sedna's position into the past, they were able to find archival images from August and September 2003 that showed Sedna. Using those positions, they were then able to extrapolate farther into the past, and found archival data from late 2002 and then late 2001 that also showed Sedna. By March 2004 they had measured Sedna's position over a time baseline of 2.5 years, and were able to confidently predict its position in the sky through June 2005.

In March 2004, Sedna was given the provisional name 2003 VB_{12} by the Minor Planet Center of the International Astronomical Union (IAU). By

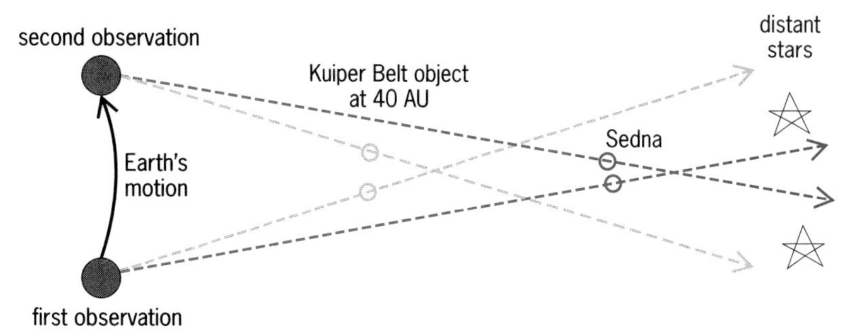

Fig. 1. Effect of parallax. The apparent position of a foreground object changes relative to the stars as the Earth moves in its orbit. More distant objects such as Sedna appear to move less, or more slowly, than closer objects, such as a typical Kuiper Belt object.

September 2004, 2003 VB_{12} had been promoted to a numbered minor planet (90377) and given the name Sedna. The name, picked by the discoverers, is the Inuit name for the goddess of the ocean, who lives at the bottom of the Arctic Ocean and is the creator of all sea creatures.

Current orbit. Besides being the most distant observable solar system object at the time of its discovery, Sedna has an unusual orbit: the semimajor axis of its orbital ellipse is 480 AU, the eccentricity is 0.84 (the eccentricity is 0 for a circle, and equal to or greater than 1 for an object that is unbound from the Sun), and the plane of the orbit is inclined 11° to the plane of the orbits of most of the major and minor planets (the ecliptic) [**Fig. 2**].

Relation to Kuiper Belt objects. Kuiper Belt objects (KBOs) are the closest orbital relatives to Sedna. These objects currently fall into three broad orbital classifications: classical, resonant, and scattered.

Classical Kuiper Belt objects have nearly circular, low-inclination orbits. It is thought that the classical Kuiper Belt objects are remnant's from the time that planetesimals were forming in the outer solar system, about 4.6×10^9 years ago.

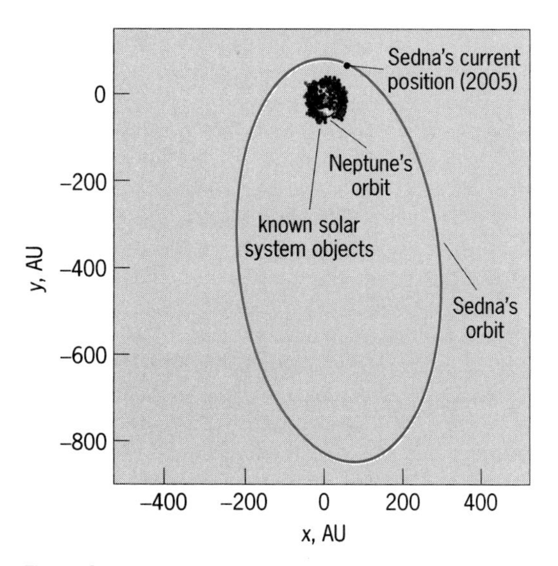

Fig. 2. Comparison of Sedna's orbit and current position with the positions of all other known objects in the solar system (indicated by dots, many of which overlap). The horizontal and vertical scales are *x* and *y* coordinates of a Cartesian system whose origin lies at the Sun.

Resonant objects have moderately eccentric (elliptical) orbits that are inclined by 10–20° to the ecliptic, and have orbital periods that are "resonant" with the orbital period of Neptune. These periods are related to Neptune's period by a ratio of small integers (3:2, 2:1, 4:3, 5:2, 5:3, and so forth). Pluto falls into this orbital class and has an orbital period that is 3:2 that of Neptune's (as do many of the resonant Kuiper Belt objects). The presence of many Kuiper Belt objects in orbital resonances with Neptune is thought to be evidence that Neptune formed not more than 25 AU from the Sun, and subsequently migrated outward to its current 30-AU distance. As it did so, its orbital resonances migrated outward with it, sweeping through the original Kuiper Belt. Resonant Kuiper Belt objects became trapped in the resonances and had their orbital eccentricity and inclination pumped up as they were swept outward along with Neptune.

Scattered objects have the most eccentric, inclined orbits of the three classes; these orbits most resemble Sedna's. These objects also probably started out on "classical" orbits, but rather than getting caught into orbits resonant with Neptune's, they suffered a close encounter with Neptune itself. The resulting scattered object orbits have large eccentricities and inclinations.

Uniqueness. Before the discovery of Sedna, all known Kuiper Belt objects spent at least a part of their orbit in the 30–50-AU region, regardless of their orbital classification. In particular, the perihelia (the closest points of their orbits to the Sun) are all less than 47 AU. This spacing was consistent with a primordial disk of classical Kuiper Belt objects from 25 to 50 AU, which was then sculpted into its current form by Neptune's migration. Sedna never comes closer to the Sun than 76 AU; its orbit cannot be explained by any of the mechanisms just described.

Origin of the orbit. Because gravitational interactions with the giant planets, which are interior to the orbits of Kuiper Belt objects, cannot permanently raise the perihelia of Kuiper Belt objects beyond 50 AU, Sedna's orbit requires a strong perturbing force with a source farther away than Neptune. Dynamicists have explored several hypothetical candidates for perturbing Sedna. Each of these hypotheses has strengths and weaknesses, but an overarching uncertainty is due to the fact that there is only one Sedna. The discovery of more such objects would go far toward testing the hypotheses and revealing details of the dynamics of the Kuiper Belt and beyond.

Perturbation by a large, distant, unknown planet. A massive planet, Earth-mass or more massive, at 70 AU could explain Sedna's orbit, but would almost certainly have been detected by now. Just the survey being conducted by Brown and his colleagues, which is only one of many, has covered 80% of the area of sky where such a planet would be seen. Such a massive and presumably large planet would also be considerably brighter than the Kuiper Belt objects that are regularly being detected, and so would be even easier to discover. Also, if the planet were much more massive than Earth, it would noticeably perturb the orbits of the giant planets, and no such perturbations have been seen. Thus, this hypothesis, while not impossible, is extremely unlikely.

Perihelion fluctuations due to perturbations by known planets. Another possibility is that Sedna's orbit (like those of some of the scattered Kuiper Belt objects) currently has a large perihelion distance, but that the current state is temporary and the orbit will evolve back to having a smaller perihelion, less than 50 AU. Dynamical models predict such fluctuations of perihelion distance due to the gravitational influence of the known giant planets, so that no distant undiscovered planet is required to explain large perihelia. A weakness of this hypothesis is that when objects are in their large-perihelion state they also have very high inclinations, probably inconsistent with the moderate inclination of Sedna's orbit. Another consideration is that if the large-perihelion state is temporary, the likelihood of discovering an object with a large perihelion is proportional to the fraction of time it remains in that state. While this hypothesis is promising, more work is needed to see if it can account for Sedna and for any new Sedna-like objects that may be discovered.

Formation at around 70 AU. Computer models of the formation of bodies in the 50–100-AU region indicate that large objects like Sedna could have formed early in solar-system history. (Another mechanism would still be required to explain the high-eccentricity, high-inclination orbit.) However, if Sedna formed there, other objects should have too, and we should be finding them. Instead, there are no known classical Kuiper Belt objects beyond 50 AU. This observation suggests either that the disk of material from which the planets formed ended at about 50 AU, or that objects that formed beyond that distance have since disappeared. In the first case, Sedna would not have formed because there was too little material. In the second, Sedna would be one of a small population of outer Kuiper Belt objects that survived the event that removed all the others.

Perturbation by passage of a massive extrasolar object. The apparent edge to the classical Kuiper Belt at around 50 AU has fueled speculation that perhaps Kuiper Belt objects did originally form in that region but were stripped away from the solar system by a close encounter with another star. Such close encounters may seem unlikely, given the great distance to even the nearest stars. But the solar system was probably formed contemporaneously with many other star systems, all coalescing from a giant cloud of gas and dust. In such star-forming cluster environments, systems are much closer together and are relatively likely to encounter one another. Computer simulations of such encounters between the early solar system and another star system show that if the two stars passed within about 200 AU, the classical Kuiper Belt beyond 50 AU would have largely been disrupted, and that some objects inside that distance would have ended up on Sedna-like orbits. The outcomes of these simulations depend on the detailed initial assumptions about the two systems and the geometry of the encounter, and so do not prove that Sedna's orbit resulted from such an encounter.

Fig. 3. Artist's rendering of the solar system as seen from the surface of Sedna. The Sun is in the center in the middle of the cloud of light scattered from dust in the Kuiper Belt. (*NASA, ESA, A. Schaller*)

However, the scenario is very plausible, Sedna's orbit is easily reproduced, and the observed edge of the Kuiper Belt is an additional natural outcome. Also, if more objects are found with large perihelia, inclinations, and eccentricities, their orbits will provide a direct test of whether such an encounter occurred.

Physical nature. In brief, very little is known about Sedna except where it is in the sky, how bright it appears, its color, and its orbit. For now, it is necessary to rely in part on the imagination (**Fig. 3**). While Sedna, with an apparent visual magnitude of about 21, is not particularly faint by the standards of Kuiper Belt objects, it is still so faint that detailed study is extremely difficult. Its color is very red, like some, but by no means all, Kuiper Belt objects. The surface composition could be determined in more detail using near-infrared spectra at wavelengths of 1–2.5 μm, but so far even the largest telescopes, such as Keck, Gemini, and Very Large Telescope, have yielded spectra that only hint at the composition. (There may be methane ice, which would make Sedna the first Kuiper Belt object known to have methane, although two others have subsequently been shown to have ice on their surfaces.) Similarly, attempts to measure Sedna's diameter, using the both the *Hubble Space Telescope* and the *Spitzer Space Telescope*, have failed, although those data indicate that it is no larger than 1800 km (1120 mi) in diameter, compared to Pluto's diameter of 2400 km (1490 mi), and it could be as small as 900 km (560 mi) in diameter, which would still be as large as the largest asteroid, Ceres, and comparable to the largest Kuiper Belt objects.

Many Kuiper Belt objects have moons, and there was an initial suggestion that Sedna might have a large moon as well, based on an erroneous observation that Sedna's rotation period was very long. Moons interact with their primary via tides, slowing the rotation rate of the primary. For example, Pluto's rotation period is 6 days because its large moon, Charon, has slowed its rotation. It now seems clear that Sedna rotates in about 10 hours, which is fairly typical for Kuiper Belt objects and asteroids, and *Hubble Space Telescope* observations failed to show a moon.

Sedna and the Kuiper Belt objects are likely to remain enigmatic for years, since only the most advanced telescopes are capable of revealing information about their size, reflectivity, and composition. Meanwhile, Sedna's orbit will be refined through continuing observations of its position, and perhaps new objects on orbits like Sedna's will be discovered. Such discoveries may help to probe the earliest history of the solar system. New insights could include whether close encounters with another star system may have truncated our planetary system, and whether Sedna (and its brethren, if they are discovered) are really members of the Oort Cloud of comets, most of which are at 50,000 AU from the Sun, or are a new class of Kuiper Belt object.

For background information *see* ASTRONOMICAL SPECTROSCOPY; COMET; KUIPER BELT; NEPTUNE; PARALLAX (ASTRONOMY); PLUTO; SOLAR SYSTEM in the McGraw-Hill Encyclopedia of Science & Technology.

John Stansberry

Bibliography. M. E. Brown, C. Trujillo, and D. Rabinowitz, Discovery of a candidate inner Oort cloud planetoid, *Astrophys. J.*, 617:645–649, 2004; S. Ida, J. Larwood, and A. Berkert, Evidence for early stellar encounters in the orbital distribution of Edgeworth-Kuiper belt objects, *Astrophys. J.*, 528: 351–356, 2000; D. C. Jewitt and J. X. Luu, Discovery of the candidate Kuiper belt object 1992 QB_1, *Nature*, 362:730–732, 1992; S. J. Kenyon and B. C. Bromley, Stellar encounters as the origin of distant solar system objects in highly eccentric orbits, *Nature*, 432:598–602, 2004; R. Malhotra, The origin of Pluto's peculiar orbit, *Nature*, 365:819–821, 1993; A. Morbidelli and H. F. Levison, Scenarios for the origin of the orbits of the trans-Neptunian Objects 2000 CR_{105} and 2003 VB_{12} (Sedna), *Astron. J.*, 128:2564–2576, 2004; S. A. Stern, Regarding the accretion of 2003 VB_{12} (Sedna) and like bodies in distant heliocentric orbits, *Astron. J.*, 129:526–529, 2005.

Self-recognition

One of the first things most people do upon awakening in the morning is to look in the mirror. We instantly equate the identity of the reflection with ourselves. This, in essence, is self-recognition. Self-face recognition has been extensively studied by Gordon Gallup for more than 30 years by placing nonhuman primates in front of a mirror. Gallup observed that while most animals emit "other-directed" social responses in front of a mirror (for example, aggression), some primates engage in "self-directed" nonsocial behavior (such as grooming) that suggests an understanding that what they perceive is themselves and not another organism.

Mark test. Gallup also devised the mark test, a more rigorous procedure demonstrating self-oriented behavior (and self-recognition) in front of a mirror. Using this technique, Gallup and others have been able to show that only humans, chimpanzees, bonobos, and orangutans are capable of self-recognition. In nonhuman primates, a red dye is applied on the right eyebrow ridge and upper left ear of anesthetized subjects; once awake they are placed in front of a mirror, and their behavior is monitored. Subjects who touch the marks on their face ("self"), as opposed to the mirror ("other"), pass the self-recognition test. Human infants (on whom a mark is placed on the nose when asleep) pass the mark test between ages 18 and 24 months. However, chimpanzees develop self-recognition much later in life (between 4.5 and 8 years of age), with a greater variation in its onset, as well as individual differences (that is, not all chimpanzees come to self-recognize).

All other groups of primates (for example, baboons, gibbons, rhesus macaques) and animals (for example, elephants) persist in emitting other-directed responses in front of a mirror even after extensive exposure and/or failure of the mark test. Fish and birds tend to engage in aggressive display in front of a mirror; cats seem uninterested by their reflection, while dogs typically run behind the mirror to meet the "other" dog; sea lions emit underwater clicking-type vocalizations when seeing their reflection, and attempt to bite the mirror image and slap it with their flippers.

The evidence is more controversial in gorillas and dolphins. Mirror self-recognition (using the mark test) has been demonstrated in only two home-reared gorillas. For obvious anatomical reasons, self-directed behavior in dolphins is more difficult to establish, but experiments have shown that they tend to spend more time in front of a mirror following the application of a dot on their body (as opposed to a dot on their face), seemingly examining a previously nonexistent mark on the self.

Patterns of self-recognition in humans. Recent work on the onset of self-recognition in humans has studied precursory behavior of infants confronted with various specular images. For instance, infants from 4 months of age can discriminate between an image of themselves and that of a mimicking other. That is, they tend to smile and look more at the mimicking other compared with the self. This discrimination gets even more obvious by age 9 months, when infants produce additional social behaviors directed toward the image of a mimicking adult. This seems to indicate that at this age infants perceive the image of the imitating other—and not the self—as a social partner. Antecedents of self-recognition in toddlers have been investigated as well. Current studies suggest that a high-quality relationship between the mother and the child (for example, a relationship in which the mother displays a readiness to accept and reinforce emotional expressions) leads to earlier mirror self-recognition. Second-born children, as opposed to first- and third-born children, show slight delays in self-recognition. Gender does not seem to be significantly correlated with self-recognition.

Clinical populations have also been tested for self-recognition. Profoundly developmentally disabled adults and most schizophrenic patients fail the mark test. The latter actually tend to spend extended periods of time gazing at their image—a behavior believed to indicate that these patients are "strangers to themselves." Most autistic children pass the test; some individuals with Alzheimer's disease lose the ability to recognize themselves in mirrors.

Other forms of self-recognition. While most research in this area has focused on self-face recognition, other forms of self-recognition have also been investigated in humans. People not only can recognize their own face—they are also efficient at correctly identifying their body, including limbs (such as hands) and shadows produced by the body. People can also readily recognize self-odors, such as perspiration (olfactory self-recognition), their spoken name (auditory name-recognition), and seen name (visual name-recognition). These different sensory modalities actually interact to generate "cross-modal self-recognition": a more rapid self-face recognition following exposure to visual, auditory, or olfactory self-primes.

Neuroanatomy of self-recognition. Recent advances in brain-imaging techniques have led to a major shift in self-recognition research: the quest for the identification of brain areas mediating this ability. A typical experiment involves having healthy participants process self- and other-faces while scanning their brains with functional magnetic resonance imaging (fMRI) or positron emission tomography (PET) equipment. One variation of this approach, known as the intracarotid amobarbital (WADA) test, involves presenting self- and other-face stimuli during selective anesthesia of the right and left hemispheres to observe interferences with recall of the "self" face. Another technique is to measure right-hand (which is controlled by the left hemisphere) and left-hand (controlled by the right hemisphere) advantages in self-face recognition tasks.

Early reports suggested that the right prefrontal lobe represents the anatomical seat of self-recognition. More recent studies suggest a much more distributed network, implicating areas such as the left prefrontal cortex, left inferior parietal lobe, left fusiform gyrus, hippocampus, and putamen. Two studies examining epileptic patients who had the commissures connecting their cerebral hemispheres surgically removed (to reduce the severity of their convulsions) showed that both the right and left disconnected hemispheres were capable of self-face recognition.

Hypotheses on self-recognition and self-awareness. Although the mark test has been criticized on technical grounds, overall, testing for self-recognition in visually competent organisms represents a fairly straightforward venture. What has been much more challenging is the interpretation of self-recognition. Two core views have been put forward: the self-awareness explanation by Gallup and the kinesthetic-visual matching hypothesis by Robert Mitchell.

Self-awareness hypothesis. Gallup's proposal is that self-recognition indicates the presence of complete self-awareness that includes introspective access to one's own mental states. Recognizing oneself in a mirror means that one can become the object of one's attention; it also presupposes a self-concept because one first has to know who one is in order to self-recognize. Gallup and others rely on converging evidence from developmental psychology, comparative psychology, and cognitive neuroscience to substantiate their claim. For instance, self-recognition in human infants correlates with the emergence of theory-of-mind ability (that is, making inferences about others' mental states), prosocial behavior, self-conscious emotions, and the use of personal pronouns. In addition, Gallup asserts that primates who pass the mark test also seem capable of engaging in theory of mind.

Another argument is that self-awareness, self-recognition, and theory of mind all depend on the same neurological structure—the right prefrontal lobe. Critics argue that other right- and left-hemispheric areas are involved in not only self-recognition but also self-awareness and theory of mind. For example, various self-awareness tasks (such as self-description, self-evaluation, and autobiographical memory) recruit the medial prefrontal cortex, posterior cingulate, left inferior temporal lobe, and/or right inferior parietal lobe; theory-of-mind tasks reliably activate the anterior paracingulate cortex bilaterally.

Kinesthetic-visual matching hypothesis. Mitchell proposes that self-recognition does not require an awareness of one's mental states. One problem with Gallup's view is that there is no obvious connection between being aware of one's mental states and recognizing oneself in a mirror. A more plausible link exists between possessing a mental representation of one's body (that could be contrasted with what one sees in the mirror) and self-recognition. Mitchell suggests all that is needed for an organism to recognize itself in a mirror is a kinesthetic representation of its own body. The organism "matches" what it sees in the mirror with an internal image of its own body and concludes that the specular image is the self—hence the term "kinesthetic-visual matching hypothesis." Experimental evidence for this view has been provided by researchers who have been studying the underlying mechanisms of self-recognition. One of these mechanisms involves the matching of visual (that is, what is seen in the mirror), tactile, and proprioceptive signals coming from the same body, creating an intermodal sensory body image.

Conclusion. Initial reports of animals' reactions to mirrors date back more than 100 years. In addition to identifying the neuroanatomical basis of self-recognition, current research is focused on the use of increasingly sophisticated designs and measures to carefully reexamine the claim that self-recognizing primates can engage in theory of mind. It is hoped that these studies will eventually provide psychologists with the information needed to decide which interpretation of mirror self-recognition—Gallup's self-awareness hypothesis or Mitchell's kinesthetic-visual matching hypothesis—is the more plausible.

For background information *see* BRAIN; COGNITION; PERCEPTION; PRIMATES; SOCIAL MAMMALS in the McGraw-Hill Encyclopedia of Science & Technology.

Alain Morin

Bibliography. G. G. Gallup, Jr., Chimpanzees: Self-recognition, *Science*, 167:86–87, 1970; G. G. Gallup, Jr., J. L. Anderson, and D. P. Shillito, The mirror test, in M. Bekoff, C. Allen, and G. M. Burghardt (eds.), *The Cognitive Animal: Empirical and Theoretical Perspectives on Animal Cognition*, pp. 325–333, University of Chicago Press, 2002; J. P. Keenan, *The Face in the Mirror: The Search for the Origins of Consciousness*, Harper Collins, New York, 2003; R. W. Mitchell, Kinesthetic-visual matching and the self-concept as explanations of mirror-self-recognition, *J. Theory Soc. Behav.*, 27(1):18–39, 1997; R. W. Mitchell, Subjectivity and self-recognition in animals, in M. R. Leary and J. P. Tangney (eds.), *Handbook of Self and Identity*, pp. 3–15, Guilford Press, New York, 2002; D. J. Turk et al., Mike or me? Self-recognition in a split-brain patient, *Nat. Neurosci.*, 5:841–842, 2002.

Size effect on strength (civil engineering)

Various types of structures used in civil, mechanical, naval, aerospace, and biomedical engineering must be designed such that they can safely resist loads expected to act upon them during their lifetime. Assessment of the load-carrying capacity of a given structure is the fundamental task of structural analysis. The structure is usually considered as a solid body occupying a given spatial domain and formed by material with given properties. In the traditional approach, it is assumed that the material is a continuum infinitely divisible into arbitrarily small volume elements that have the essential properties of the whole. This fundamental assumption allows decoupling material behavior from the effects of structural geometry.

On the global level of the whole structure, changes in size and shape are characterized by displacements, and loads acting on the structure are described as body forces distributed throughout the volume (for example, gravity forces) and surface forces distributed along a part of the boundary (for example, wind pressure on a building). On the local level of an infinitesimal material volume, changes in size and shape are characterized by the strain components, and the forces between a particular volume element and its immediate neighbors are characterized by the stress components.

Size effect. In the simple case of a prismatic bar of length L and cross-sectional area A clamped at one end and stretched in the longitudinal direction by a force F applied at the other end (**Fig. 1**), the global characteristics are the elongation (change of length) ΔL and force F, while the local ones are strain $\varepsilon = \Delta L/L$ and stress $\sigma = F/A$. The relationship between stress and strain is described by the constitutive equations, which depend on the specific material but should be independent of the structural size and geometry. If the material fails at a certain stress level σ_{max}, the structure fails when the

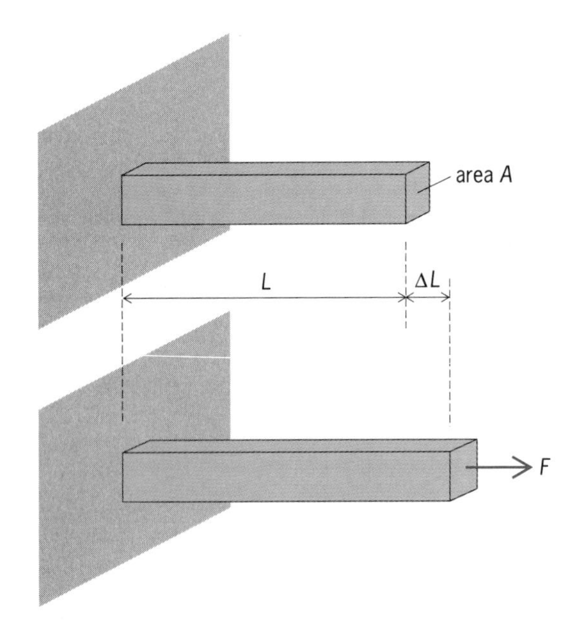

Fig. 1. Bar under uniaxial tension.

applied force attains its ultimate value $F_{max} = A\sigma_{max}$. So, the maximum force resisted by a wire or rope should be proportional to the area of the cross section and independent of the length. However, experimental evidence shows that this is not always the case. Leonardo da Vinci noted that long ropes are usually weaker than short ones. Later it was found that the maximum stress that can be sustained by very thin glass fibers dramatically increases with decreasing fiber diameter. In such cases, the nominal strength $\sigma_N = F_{max}/A$, defined as the ratio between the measured failure load and the cross-sectional area, is no longer an intrinsic material property but depends on the size of the structure. This phenomenon is referred to as the size effect on nominal strength.

The foregoing considerations can be extended to more complicated structures for which the stress state is multiaxial and nonuniform. Stress and strain are in general described by symmetric, second-order tensors that vary in space. Depending on the type of material (for example, brittle or ductile), structural failure can occur right away when the stress at one point attains the critical level, or later after some stress redistribution. Dimensional analysis indicates that, according to all models dealing with local constitutive equations and stress-based failure criteria, the failure load F_{max} on geometrically similar structures should be proportional to the square of the characteristic structural dimension D. In other words, the nominal strength $\sigma_N = F_{max}/D^2$ should depend only on the material and on the geometry (shape of the structure and boundary conditions) but not on D. In many real tests, size dependence of the nominal strength is observed, and typically σ_N decreases with increasing D. The fact that larger specimens appear to have lower strength complicates the development of design rules and formulas. Ideally, one should be able to extract the material properties from laboratory tests and then use those properties in safety assessment of large structures. If the particular failure mode exhibits size effect, the actual failure load would be smaller than the predicted one, and the design may not be safe. Galileo pointed out that a size effect is manifested by the fact that large animals have bulkier bones than small ones, which he called the "weakness of giants."

Weakest-link (Weibull) theory. Different explanations for the size dependence of strength have been advanced, and nowadays it is clear that size effects have multiple sources. Depending on the specific conditions, one of these sources becomes dominant or several are combined. The fact that longer ropes fail at lower load levels than shorter ones with the same diameter can be explained by the existence of random flaws that locally weaken the material. The rope fails when the strength of the weakest section is exhausted. This type of behavior is characteristic of structures in which failure of one element leads to failure of the whole structure, and mathematically it is described by the weakest-link theory developed by W. Weibull. He showed that the tail of the statistical distribution in the region of extremely small failure probabilities can be described only by a power law with a threshold. Once the statistical distribution of local strength is established, the probability that the whole structure fails under a given load can be evaluated. For chainlike structures under uniform stress, such as ropes or wires, the probability of failure is simply the probability that there exists at least one link in the chain whose strength is lower than the applied stress. Extension of this concept to structures under nonuniform stress reveals that, for geometrically similar structures of different sizes D, the nominal strength is a power function of D with a negative exponent: $\sigma_N = \sigma_{N0}(D/D_0)^{-n/m}$, where D_0 is a fixed reference size, σ_{N0} is the nominal strength for structure of size D_0, $n = 3$ for three-dimensional similarity, and m is the Weibull modulus characterizing the strength distribution for the particular type of material, with typical values between 5 and 50.

Weibull theory explains the size effect on strength by the random distribution of local material properties. In a larger structure, the probability of finding a weak spot is higher, and therefore the structure fails at a lower overall stress level than a smaller one. This statistical size effect certainly exists, but it is not the only reason for size dependence of nominal strength.

Energetic size effects. Another, very strong size effect arises if the failure is related to crack propagation. The energy needed to propagate a crack is released from the energy stored in elastic deformation of the cracking body. For geometrically similar bodies of different sizes D, the energy spent by formation of new crack surfaces grows with D^2 while the energy released from the volume surrounding the crack grows with D^3. Therefore, the crack can propagate in larger bodies at lower stress levels than in smaller ones. Detailed analysis shows that the nominal strength of an elastic body with an initial crack is inversely proportional to $\sqrt{D}$. This very strong size effect is applicable to bodies with a notch or initial crack proportional to D (**Fig. 2**). The derivation is also based on the assumption that the material around the crack remains linear elastic; that is, stress is proportional to strain. It can be shown that the linear elastic solution leads to a stress singularity at the crack tip, where stress tends to infinity. Since no material can resist arbitrarily large stresses, there is always a certain zone around the crack tip in which the material response is inelastic. If the size of this process zone is negligible compared to the size of the crack and of the ligament (unbroken part of the cross section), the solution based on linear elastic fracture mechanics is acceptable and a strong energetic size effect can be expected. This is the case for sufficiently large structures. If the structure is small, the inelastic process zone around the crack tip cannot be neglected and the formula describing the size effect must be modified. For quasibrittle materials, such as concrete, fiber-polymer composites, stiff foams, toughened ceramics, many types of rock, snow, bones, filled elastomers, and various other materials, Z. P. Bažant proposed a two-parameter size effect formula that describes the transition from ductile behavior with no size effect for small sizes to brittle behavior with strong size effect for large sizes (**Fig. 3**). According to Bažant's formula, the nominal strength is given by $\sigma_N = \sigma_0/\sqrt{1+D/D_0}$, where D is the size of the structure and σ_0 and D_0 are two parameters. For small sizes D, σ_N is almost constant and approaches σ_0 as D tends to zero. For large sizes D, the foregoing expression for σ_N asymptotically approaches a power law with exponent $-1/2$, valid for linear elastic fracture mechanics.

Size-effect sources. This article has discussed two principal sources of size effect on nominal strength, namely the stochastic size effect originating from random distribution of local material properties, and the energetic (deterministic) size effect originating from the balance between elastic energy released from the volume and fracture energy dissipated on the crack surface. The former type of size effect is described by the Weibull theory while the latter can be described by the scaling law of linear elastic fracture mechanics if the material is brittle, or by Bažant's extension of that law if the material is quasibrittle. However, each of the simple scaling laws mentioned here applies only to a specific class of problems. More general formulas that combine both sources of size effect and cover different types of failure have been derived by Bažant and coworkers, based on asymptotic matching.

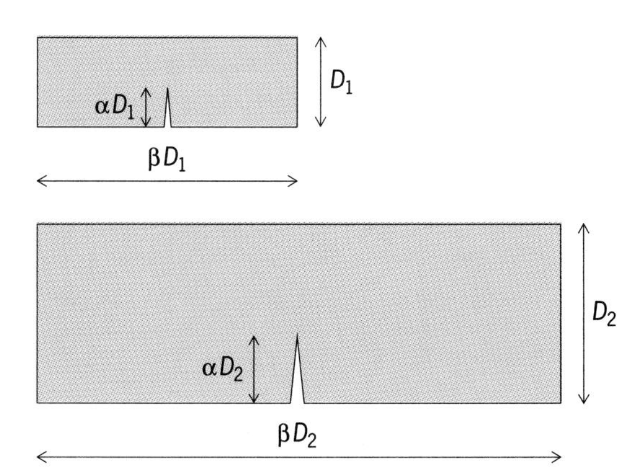

Fig. 2. Geometrically similar specimens with an initial notch.

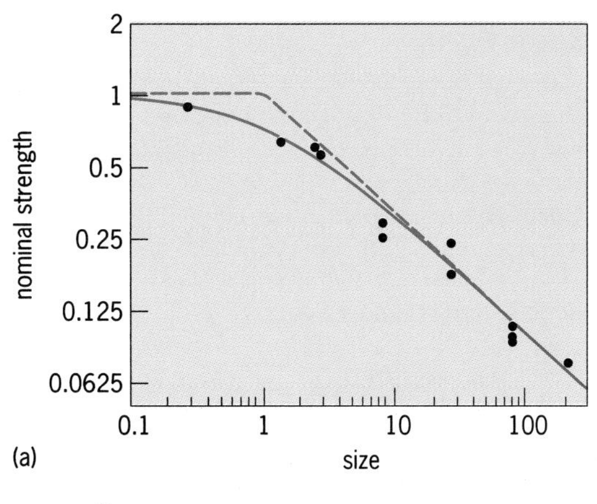

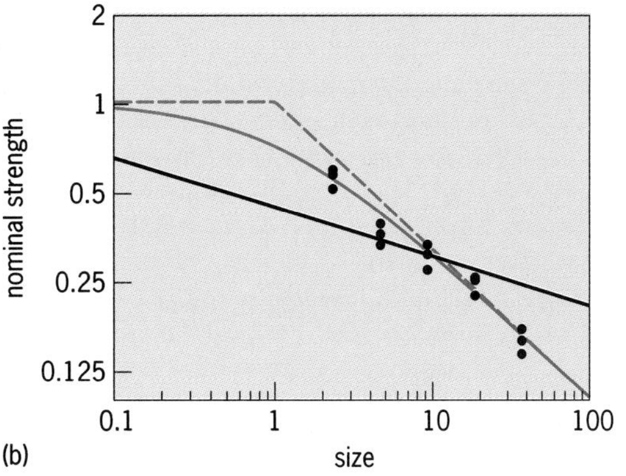

Fig. 3. Plots of dimensionless nominal strength σ_N/σ_0 versus dimensionless size D/D_0 in logarithmic scale. (*a*) Failure of notched sea-ice blocks in the size range 1:800. (*b*) Diagonal shear failure of reinforced concrete beams in the size range 1:16. Points correspond to experimental results, color curve is the best fit by Bažant size effect formula, dashed lines are the asymptotes for small and large sizes, and black line in *b* corresponds to Weibull size effect law with $n = 2$ and $m = 12$. (*Experimental data taken from J. P. Dempsey et al., 1999, and Z. P. Bažant and M. T. Kazemi, 1991*)

Another potential source of size effect resides in the fractal nature of the fracture surface and perhaps of the entire fracture process zone, as recently suggested by A. Carpinteri.

For background information *see* BEAM; BRITTLENESS; COMPOSITE MATERIAL; CONCRETE BEAM; LOADS, DYNAMIC; LOADS, TRANSVERSE; SHEAR; STRESS AND STRAIN; STRUCTURAL ANALYSIS; STRUCTURAL DESIGN; STRUCTURAL MATERIALS; STRUCTURAL MECHANICS in the McGraw-Hill Encyclopedia of Science & Technology.

Milan Jirasek

Bibliography. Z. P. Bažant, *Scaling of Structural Strength*, Hermes Penton Science, 2002; Z. P. Bažant, Size effect, *Int. J. Solids Struc.*, 37:69–80, 1999; Z. P. Bažant and J. Planas, *Fracture and Size Effect in Concrete and Other Quasibrittle Materials*, CRC Press, 1998; A. Carpinteri (ed.), *Size-Scale Effects in the Failure Mechanisms of Materials and Structures*, E&FN Spon, 1996; B. B. Mandelbrot, *The Fractal Geometry of Nature*, Freeman, 1983.

Space flight

For space flight, 2004 was a year of strong contrasts. Human and robotic space activities set new marks with a lineup of unique accomplishments, preparing the stage for new developments that will contribute greatly to human exploration beyond the Earth. At the same time, 2004 saw a decline in the utilization of space, as the number of launches to orbit plus the number of satellites reached the lowest levels since 1961.

While the United States space budget managed to stay at a relatively stable level, the continuing standdown of the space shuttle kept human flights limited to Russian launch services. Commercial flights dropped below the 2003 level, and international space activities continued their trends of reduced public spending and modest launch services. A total of 53 successful launches worldwide carried 73 payloads (2003: 86), compared to 60 launches in 2003 (2002: 61; 2001: 57). There also were two rocket failures (down from three in 2003), a Tsyklon 3 (Russia), and a Shaviyt (Israel).

In its traditional leadership role among world space organizations, the National Aeronautics and Space Administration (NASA) started the year on a positive note, when President George W. Bush announced the Vision for Space Exploration on January 14, 2004, charging the space agency with a robust space exploration program to advance national scientific, security, and economic interests. Also in January, NASA successfully landed the mobile geology labs *Spirit* and *Opportunity* on the planet Mars. Other milestones of 2004 were the arrival of the *Cassini-Huygens* spacecraft at Saturn and its moon Titan; the *Stardust* deep-space probe's flyby of Comet Wild 2 within 235 km (146 mi); the launch of the *Swift* satellite to look for the birth of black holes; strange cosmic sights unveiled by the *Spitzer Space Telescope*; the deepest portrait of the visible universe taken by the *Hubble Space Telescope*; and the crash landing of the solar-sample return mission of *Genesis* in the Utah desert, still containing preserved and usable samples of the Sun.

In 2004, the commercial space market continued its decline, begun in 2003 after a surprising recovery in 2002 from the dramatic slump of previous years. Of the 53 successful launches worldwide, about 19 (36%) were commercial launches (carrying 27 commercial payloads), compared to 20 (33%) in 2003 and 28 (46%) in 2002. In the civil science satellite area, worldwide launches totaled nine, down four from the preceding year. But in 2004, the dream of opening space to the general public was greatly advanced by *SpaceShipOne's* prize-winning suborbital flight and congressional legislation to help establish a space travel industry in the United States. *See* SCIENTIFIC SATELLITES.

Russia's space program, despite chronic shortage of state funding, showed continued dependable participation in the buildup of the *International Space Station*. This partnership had become particularly important after the shuttle standdown caused by the loss of *Columbia* on February 1, 2003. Europe's space activities in 2004 dropped below the previous years (of four missions), with only three missions of the Ariane 5 heavy-lift launch vehicle, which brought the number of successes of this vehicle to 16.

During 2004, two crewed flights from the two major space-faring nations (down from three in 2003) carried six humans into space. This brought the total number of people launched into space since 1958 (counting repeaters) to 977, including 100 women, or 438 individuals (38 women). Some significant space events in 2004 are listed in **Table 1**,

TABLE 1. Significant space events in 2004

Designation	Date	Country	Event
Progress M1-11/13P	January 29	Russia	Crewless logistics cargo/resupply mission to the *International Space Station (ISS)* on a Soyuz-U rocket
Rosetta	March 2	Europe	Successful launch of ESA's intercept mission on an Ariane-5G to Comet 67 P/Churyumov-Gerasimenko, to arrive in 2014
Soyuz TMA-4/ISS-8S	April 19	Russia	Launch of the third *ISS* crew rotation flight on a Soyuz-FG, bringing the Expedition 9 crew of Gennady Padalka and Michael Fincke, plus 10-day visitor André Kuipers from ESA
Gravity Probe-B (GP-B)	April 20	United States	Successful launch of NASA probe on a Delta 2 to test two predictions of Einstein's theory of general relativity
Progress M-49/14P	May 25	Russia	Crewless logistics cargo/resupply mission to the *ISS* on a Soyuz-U rocket
Aura	July 15	United States	Launch of the third major Earth Observing System (EOS) platform on a Delta 2, joining Terra and Aqua
Messenger	August 3	United States	Successful Delta 2 launch of NASA's Mercury orbiter probe, to arrive at the Sun's innermost planet in March 2011
Progress M-50/15P	August 11	Russia	Crewless logistics cargo/resupply mission to the *ISS* on a Soyuz-U rocket
EDUSAT	September 20	India	Successful launch of India's first operational GSLV rocket carrier, with the country's first educational services satellite
Soyuz TMA-5/ISS-9S	October 13	Russia	Launch of the fourth *ISS* crew rotation flight on a Soyuz-FG, bringing the Expedition 10 crew of Leroy Chiao and Salizhan Sharipov, plus 10-day visitor Yuri Shargin from Russia.
Swift	November 20	United States	Successful Delta 2 launch of NASA satellite to research interstellar gamma-ray bursts (GRBs)
Progress M-51/16P	December 24	Russia	Crewless logistics cargo/resupply mission to the *ISS* on a Soyuz-U rocket

TABLE 2. Successful launches in 2004 (Earth-orbit and beyond)

Country	Number of launches (and attempts)
United States (NASA/DOD/ commercial)	19 (19)
Russia	22 (23)
People's Republic of China	8 (8)
Europe (ESA/Arianespace)	3 (3)
India	1 (1)
Israel	0 (1)
TOTAL	53 (55)

and the launches and attempts are enumerated by country in **Table 2**.

International Space Station

One continuing partnership issue regarding the *International Space Station (ISS)* in 2004 was the debate about the provision of assured crew return capability after the Russian obligation to supply Soyuz lifeboats to the *ISS* expires in April 2006. In NASA's space transportation planning, under revision (August 2005) after the unveiling of a long-range space exploration strategy by President Bush on January 14, 2004, which includes retirement of the space shuttle by 2010, a U.S. crew rescue capability would be available only by 2014. Partnership efforts continue to work out a solution for dealing with the gap. Of much greater significance to the continuation of *ISS* assembly and operation was Russia's shouldering the burden of providing crew rotation and consumables resupply flights to the station after the loss of space shuttle *Columbia* in 2003. This accident brought shuttle operations to a standstill that lasted until July 2005.

After the *Columbia* loss, there was an immediate consequence of the unavoidable reduction in resupply missions to the station which now could be supported only by Russian crewless automated Progress cargo ships. Station crew size was reduced from a three- to a two-person "caretaker" crew per expedition (also known as increment), except for brief 10-day stays by visiting cosmonaut/researchers arriving and departing on the third seat of Soyuz spacecraft.

Three crews lived on the station during 2004 as *ISS* entered its fifth year of operations as a staffed facility. Each two-person crew, working with ground teams, did its part to keep the station safely operating while accumulating knowledge ("lessons learned") for future deep-space missions of the new Vision for Space Exploration. Crews made unprecedented repairs to an oxygen generator, a crucial piece of exercise equipment, and a U.S. spacesuit. They also performed an extravehicular activity (EVA), or spacewalk, to restore power to a gyroscope.

After appropriate training for the temporary two-person situation in the United States and Russia, Expedition 9 was launched to the *ISS* in April 2004 with Russian Commander (CDR) Gennady Padalka and U.S. Flight Engineer (FE) Michael Fincke on a Soyuz TMA spacecraft. Two members of Expedition 8, CDR Michael Foale and FE Alexander Kaleri, plus visiting cosmonaut/researcher André Kuipers from the Netherlands, who flew for the European Space Agency (ESA), returned on the previous Soyuz that had served as a contingency crew return vehicle (CRV) for almost the duration of its certified lifetime of 200 days. The replacement crew, Expedition 10, came 6 months later in a fresh Soyuz TMA, consisting of U.S. CDR Leroy Chiao and Russian FE Salizhan Sharipov, to continue station operations into 2005. All three U.S. crew members had personal milestones in 2004: With an endurance time of 374 days 11 h 19 min accumulated in his six space flights, CDR Foale has become the first U.S. astronaut to exceed 1 year of "space time," breaking Carl Walz's record of 231 days (four flights) and moving to position number 16 of the overall record list of international space fliers.

Following the recommendations of an independent advisory panel of biological and physical research scientists called Remap (Research Maximization and Prioritization), NASA in 2002 had established the formal position of a Science Officer (SO) for one crew member aboard the *ISS*, responsible for expanding scientific endeavors on the station. After FE Dr. Peggy Whitson of Expedition 5 became NASA's first Science Officer, Expedition 8 CDR Foale, Expedition 9 FE Fincke, and Expedition 10 CDR Chiao were the fourth, fifth, and sixth Science Officers in 2004.

By the end of 2004, 44 carriers had been launched to the *ISS*: 16 shuttles, 2 heavy Protons (FGB/*Zarya*, SM/*Zvezda*), and 26 Soyuz rockets (16 crewless Progress cargo ships, the DC-1 docking module, and 9 crewed Soyuz spaceships).

Progress M1-11 (no. 260). Designated ISS-13P, the first of four crewless cargo ships to the *ISS* in 2004 lifted off on a Soyuz-U rocket at the Baikonur Cosmodrome in Kazakhstan on January 29. As all Progress transports do, it carried about 1.8 metric tons (2 tons) of resupply for the station, including maneuver propellants, water, food, science payloads, equipment, and spares.

Soyuz TMA-4 (no. 214). The *Soyuz TMA-4*, ISS Mission 8S (April 19–October 24), was once again a flawless success of the Soyuz launcher. The third crew rotation flight by a Soyuz because of the shuttle standdown, it carried Expedition 9, the two-man station crew of Padalka and Fincke, plus visiting cosmonaut/researcher Kuipers. *TMA-4* docked to the *ISS* on April 21, replacing the previous CRV, *Soyuz TMA-3*/7S, which, when it undocked on April 29, had reached an in-space time of 194 days. In *TMA-3*, Expedition 8 crew members Foale and Kaleri plus Kuipers landed in Kazakhstan on the morning of April 30. Expedition 8 had spent 194 days 18 h 35 min in space (192 days onboard the *ISS*).

Progress M-49 (no. 249). ISS-14P was the next crewless cargo ship, launched in Baikonur on a Soyuz-U on May 25 and arriving at the station with fresh supplies on May 27.

Progress M-50 (no. 250). ISS-15P, the third automated logistics transport in 2004, lifted off on its Soyuz-U on August 11, docking at the *ISS* on August 14.

Soyuz TMA-5 (no. 215). *Soyuz TMA-5*, ISS mission 9S (October 13, 2004–April 24, 2005) was the fourth *ISS* crew rotation flight by a Soyuz. Its crew constituted Expedition 10, the fourth caretaker crew of Chiao and Sharipov, plus visiting Russian cosmonaut/researcher Yuri Shargin. With its perfect liftoff, the Soyuz-U launch vehicle racked up another success in its history of—at that point—436 flights (426 successes; there was one more in 2004 for a total of seven launches). On October 16, *TMA-3* docked smoothly to the *ISS* at the DC-1 docking compartment's nadir (downward)-pointing port, achieving successful contact and capture. On October 23 the previous CRV, *Soyuz TMA-4*/8S, undocked from the FGB nadir port, where it had stayed for 186 days (187 days in space), and landed safely in Kazakhstan with Padalka, Fincke, and Shargin. Expedition 9's total mission elapsed time (MET), from launch to landing, was 187 days 21 h 17 min in space (186 days onboard the station).

United States Space Activities

Launch activities in the United States in 2004 showed a sizable decrease from the relatively elevated level of the previous year. There were 19 NASA, Department of Defense (DOD), and commercial launches out of 19 attempts (in 2003 there were 26 of 27 attempts [loss of *Columbia*]; in 2002 there were 18 of 18).

Space shuttle. After the loss of orbiter *Columbia* on the first shuttle mission in 2003, operations with the reusable shuttle vehicles of the U.S. Space Transportation System (STS) were halted until July 2005, as NASA and its contractors labored on intensive return-to-flight (RTF) efforts. Resupply and crew rotation flights to the *ISS* were taken over solely by Russian Soyuz and Progress vehicles.

Advanced transportation systems activities. NASA's study activities of the original 5-year Space Launch Initiative (SLI) project, announced in 2001, for developing the technologies to be used to build an operational reusable launch vehicle (RLV) before 2015 were terminated when in 2004 President Bush announced NASA's long-range Vision for Space Exploration. New study activities would focus on concepts of the Vision's Crew Exploration Vehicle (CEV) and heavy-cargo lifters, with the goal to retire the space shuttle by 2010.

Space sciences and astronomy. In 2004, the United States launched four civil science spacecraft, one less than in the previous year: *Gravity Probe-B*, *Aura*, *Messenger*, and *Swift*.

Gravity Probe-B. *Gravity Probe-B* (*GP-B*) is a NASA mission to test two predictions of Albert Einstein's theory of general relativity. Launched on April 20, 2004, on a Delta 2 rocket, the 3100-kg (6834-lb) spacecraft, orbiting 644 km (400 mi) above Earth, uses four ultraprecise gyroscopes to test Einstein's theory that space and time are distorted by the presence of massive objects. To accomplish this, the mission measures two factors: how space and time are warped by the presence of the Earth, and how the Earth's rotation drags space-time around with it. In early September, the probe achieved a milestone with the completion of the Initialization and Orbit Calibration (IOC) phase of its mission and the transition into the science phase, bringing the GP-B mission one step closer to shedding new light on the fundamental properties of the universe. NASA's Marshall Space Flight Center manages the GP-B program, with Stanford University, the prime mission contractor which conceived the experiment, responsible for the design and integration of the science instrument as well as for mission operations and data analysis. Lockheed Martin, a major subcontractor, designed, integrated, and tested the space vehicle and built some of its major payload components.

MESSENGER. NASA's *MESSENGER* (Mercury Surface, Space Environment, Geochemistry, and Ranging), set to become the first spacecraft to orbit the planet Mercury, was launched on August 3, 2004, aboard a Delta 2 rocket from Cape Canaveral Air Force Station, Florida. The approximately 1100-kg (1.2-ton) spacecraft, designed and built by the Johns Hopkins University Applied Physics Laboratory (APL) in Laurel, Maryland, was placed into a solar orbit 57 min after launch, targeted to begin orbiting Mercury in 2011. During a 7.9-billion-kilometer (4.9-billion-mile) journey that includes 15 trips around the Sun, *MESSENGER* will fly past Earth once, Venus twice, and Mercury three times before easing into orbit around its target planet. The Earth flyby, on August 2, 2005, and the Venus flybys, in October 2006 and June 2007, are using the pull of the planets' gravity to guide *MESSENGER* toward Mercury's orbit. The Mercury flybys in January 2008, October 2008, and September 2009 will help the probe match the planet's speed and location for an orbit insertion maneuver in March 2011. The flybys will also allow the spacecraft to gather data critical to planning a year-long orbit phase. Since *MESSENGER* is only the second spacecraft sent to Mercury—*Mariner 10* flew past it three times in 1974–1975 and gathered detailed data on less than half the surface—the mission has an ambitious science plan. With a package of seven science instruments, *MESSENGER* will determine Mercury's composition; image its surface globally and in color; map its magnetic field and measure the properties of its core; explore the mysterious polar deposits to learn whether ice lurks in permanently shadowed regions; and characterize Mercury's tenuous atmosphere and Earth-like magnetosphere.

Swift. NASA's *Swift* satellite was successfully launched on November 20, 2004, aboard a Delta 2 rocket from Launch Complex 17A at the Cape Canaveral Air Force Station. The satellite was designed and built with international participation (England, Italy) to solve the 35-year-old mystery of the origin of gamma-ray bursts. Scientists believe the bursts, distant yet fleeting explosions, are related to the formation of black holes throughout the universe—the birth cries of black holes. Each gamma-ray burst is a short-lived event, lasting only a few milliseconds to a few minutes, never to appear again. The bursts occur several times daily somewhere in the universe, and *Swift* should detect several weekly. To track these mysterious bursts, *Swift* carries a suite of three main instruments. The Burst Alert Telescope (BAT) instrument, built by Goddard Space Flight Center, detects and locates about two gamma-ray bursts weekly, relaying a rough position to the ground within 20 s. The satellite swiftly repoints itself to bring the burst area into the narrower fields of view of the onboard X-ray Telescope (XRT) and the UltraViolet/Optical Telescope (UVOT). These telescopes study the afterglow of the burst produced by the cooling ashes that remain from the original explosion. During its 2-year mission, *Swift* is expected to observe more than 200 gamma-ray bursts—the most comprehensive study of gamma-ray burst afterglows to date.

GALEX. *GALEX* (Galaxy Evolution Explorer), launched by NASA on April 28, 2003, on a Pegasus XL rocket from an L-1011 aircraft into a nearly circular Earth orbit, is an orbiting space telescope for observing tens of millions of star-forming galaxies in ultraviolet light across 10 billion years of cosmic history. Its telescope has a basic design similar to the *Hubble Space Telescope* (*HST*), but while *HST* views the sky in exquisite detail in a narrow field—like a grain of sand held at arm's length—the *GALEX* telescope is tailored to view hundreds of galaxies in each observation. Thus, it requires a large field of view, rather than high resolution, in order to efficiently perform the mission's surveys. During 2004, astronomers using *GALEX*'s sensitive ultraviolet detectors detected three dozen bright, compact galaxies that greatly resemble the youthful galaxies of more than 10 billion years ago. These new galaxies are relatively close to Earth, ranging from 2 to 4 billion light-years away. They may be as young as 100 million to 1 billion years old (the Milky Way is approximately 10 billion years old). This discovery suggests the aging universe is still alive with youth. It also offers astronomers their first close-up glimpse at what our galaxy probably looked like in its infancy.

Spitzer Space Telescope (SST). This instrument, formerly known as *SIRTF* (*Space Infrared Telescope Facility*), was launched on August 24, 2003, as the fourth and final element in NASA's family of Great Observatories, and represents an important scientific and technical bridge to NASA's Astronomical Search for Origins program. The observatory carries an 85-cm (33-in.) cryogenic telescope and three cryogenically cooled science instruments capable of performing imaging and spectroscopy in the 3.6–160-micrometer range. Its supply of liquid helium for radiative-cryogenic cooling was estimated post-launch to last for about 5.8 years, assuming optimized operation.

In 2004, *Spitzer* discovered for the first time dusty discs around mature, Sun-like stars known to have planets, and the *Hubble Space Telescope* captured the most detailed image ever of a brighter disc circling a much younger Sun-like star. The findings offer "snapshots" of the process by which our own solar system evolved, from its dusty and chaotic beginnings to its more settled present state. The young star observed by *Hubble* is 50–250 million years old. This is old enough to theoretically have gas planets, but young enough that rocky planets like Earth may still be forming. The six older stars studied by *Spitzer* average 4 billion years old, nearly the same age as the Sun. They are known to have gas planets, and rocky planets may also be present. Prior to these findings, rings of planetary debris, or "debris discs," around stars the size of the Sun had rarely been observed, because they are fainter and more difficult to see than those around more massive stars. Debris discs around older stars the same size and age as our Sun, including those hosting known planets, are even harder to detect. These discs are 10–100 times thinner than the ones around young stars. *Spitzer*'s highly sensitive infrared detectors were able to sense their warm glow for the first time.

RHESSI. *RHESSI* (Reuven Ramaty High Energy Solar Spectroscopic Imager) was launched in 2002 and named in honor of the late NASA scientist who pioneered the fields of solar-flare physics, gamma-ray astronomy, and cosmic-ray research. In 2004, *RHESSI* continued its operation in Earth orbit, providing advanced images and spectra to explore the basic physics of particle acceleration and explosive energy release in solar flares. Since its launch the spacecraft has been very successful in observing solar flares, which are capable of releasing as much energy as a billion 1-megaton nuclear bombs. In 2004, *RHESSI* was taken off the Sun to point at the Crab Nebula, to obtain the finest imaging ever done of a cosmic source in the hard x-ray/soft gamma-ray range.

Hubble Space Telescope. Fourteen years after it was placed in orbit, the *Hubble Space Telescope* continued to probe far beyond the solar system, producing imagery and data useful across a range of astronomical disciplines to expand knowledge of the universe. *Hubble* is making discoveries at a rate that is unprecedented for a single observatory, and its contributions to astronomy and cosmology are wide-ranging. In 2004, scientists using the *HST* measured the age of what may be the youngest galaxy ever seen in the universe. Called I Zwicky 18 (**Fig. 1**) at a distance of 45 million light-years, it may be only 500 million years old (so recent an epoch that complex life had already begun to appear on Earth). The Milky Way Galaxy by contrast is over 20 times older or about 12 billion years old, the typical age of galaxies across the universe. This "late-life" galaxy offers a rare glimpse into what the first diminutive galaxies in the early universe looked like. On Jupiter, *HST* spotted a rare triple eclipse: an alignment of three of Jupiter's largest moons—Io, Ganymede, and Callisto—across the planet's face. Another accomplishment of the *HST* in 2004 was observing the explosion of a massive star blazing with the light of 200 million Suns, called a supernova. The

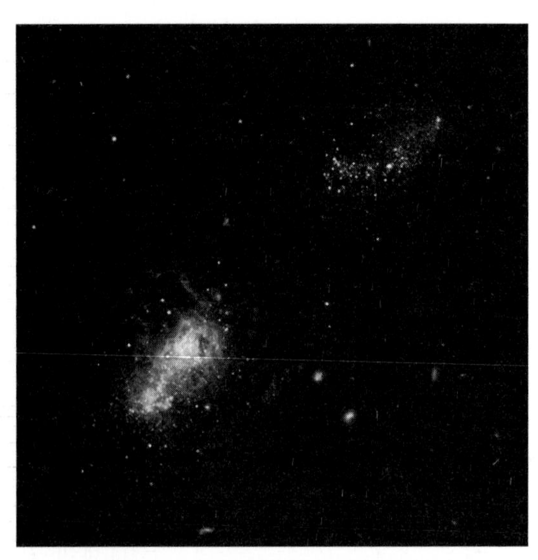

Fig. 1. ***Hubble Space Telescope*** **image of the galaxy I Zwicky 18, which may be only 500 million years old and the youngest galaxy ever seen in the universe. The galaxy may have gone through several sudden bursts of star formation, resulting in the concentrated knots embedded in the heart of the galaxy. The recent star formation may have been triggered by interaction with a companion galaxy just above and to the right of the dwarf galaxy. (*NASA; ESA; Y. Izotov, Main Astronomical Observatory, Kyiv, UA; T. Thuan, University of Virginia*)**

supernova is so bright that it easily could be mistaken for a foreground star in the Milky Way, but in reality this supernova, called SN 2004dj, resides far beyond our galaxy—in the outskirts of NGC 2403, a galaxy located 11 million light-years from Earth (the closest stellar explosion to Earth discovered in more than a decade).

In 2004, design activities continued on the *HST*'s successor, the *James Webb Space Telescope* (*JWST*), by a contracting team headed by Northrop Grumman Space Technology. Plans are to launch the giant cosmic telescope (5400 kg/11,880 lb) in 2011 on a European Ariane 5 toward the second Lagrangian point (L2), 1.5 million kilometers (930,000 mi) beyond Earth's orbit on the Sun–Earth line, where effects of their light on its optics are minimized and gravitational pull is relatively well balanced.

Chandra Observatory. Launched on shuttle mission STS-93 in 1999, the massive (5870-kg/12,930-lb) *Chandra* X-ray Observatory uses a high-resolution camera, high-resolution mirrors, and a charge-coupled-detector (CCD) imaging spectrometer to observe x-rays of some of the most violent phenomena in the universe which cannot be seen by the *Hubble*'s visual-range telescope. Throughout its fifth year of operation, *Chandra* continued to provide scientists with views of the high-energy universe never seen before which potentially revolutionize astronomical and cosmological concepts. In 2004, *Chandra*'s most popular image was a spectacular view of Cassiopeia A (**Fig. 2**) that had nearly 200 times more data than the "First Light" *Chandra* image of this object made in 1999. The new image reveals clues that the initial explosion caused by the collapse of a massive star was far more complicated than suspected. Also in 2004, an international team of scientists used *Chandra* data to measure the temperature of the pulsar at the center of 3C58, the remains of a star observed to explode in the year 1181. *Chandra*'s image of 3C58 (**Fig. 3**) also showed spectacular jets, rings, and magnetized loops of ionized particles. Data indicated that the surface of the 3C58 pulsar has cooled to a temperature of slightly less than a million degrees Celsius, which is extremely cool for a young neutron star. Pulsars are formed when the central core of a massive star collapses to create a dense object about 24 km (15 mi) across that is composed almost entirely of neutrons. Collisions between neutrons and other subatomic particles in the interior of the star produce neutrinos that carry away energy as they escape from the star. This cooling process depends critically on the density and type of particles in the interior, so measurements of the surface temperature of pulsars provide a way to probe extreme conditions where densities are so high that current understanding of how particles interact with one another is limited. They represent the maximum densities that can be attained before the star collapses to form a black hole.

Cassini/Huygens. NASA's 5.4-metric-ton (6-ton) spacecraft *Cassini* continued its epic 6.7-year, 3.2-billion-kilometer (2-billion-mile) journey to the planet Saturn. During 2004, the spacecraft remained in excellent health and successfully entered orbit around Saturn on June 30, when flight controllers received confirmation that *Cassini* had completed the engine burn needed to place the spacecraft into the correct orbit. This began a 4-year study of the giant planet, its majestic rings, and 31 known moons. Already in August, the probe discovered two new moons, approximately 3 km (2 mi) and 4 km (2.5 mi) across. The moons, located 194,000 km (120,000 mi) and 211,000 km (131,000 mi) from

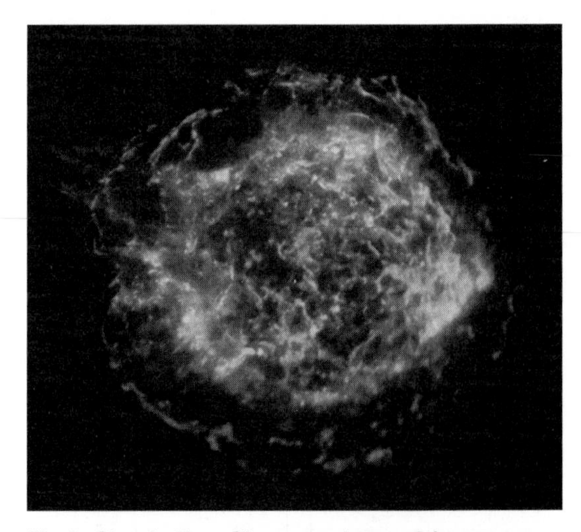

Fig. 2. ***Chandra X-ray Observatory*** **image of the supernova remnant Cassiopeia A, taken in nine observations totaling 10^6 s (more than 11 days) in 2004. The most detailed image ever made of the remains of an exploded star shows a bright outer ring 10 light-years in diameter that marks the location of a shock wave generated by the supernova explosion. A large jetlike structure protrudes beyond the shock wave in the upper left. (*NASA, Chandra X-ray Center, Goddard Space Flight Center, U. Hwang et al.*)**

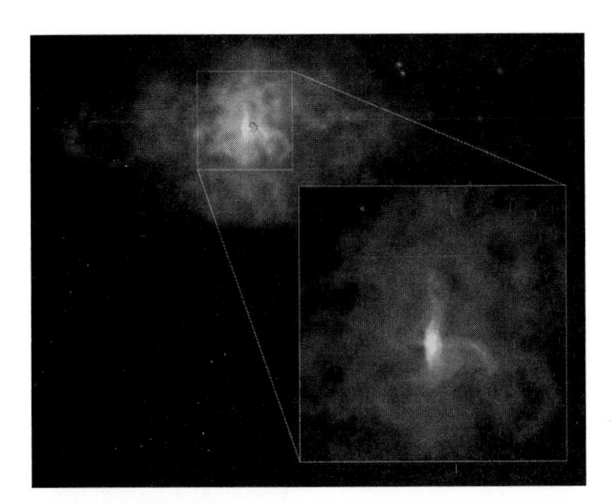

Fig. 3. *Chandra X-ray Observatory* **image of the supernova remnant 3C58. A bright torus of x-ray emission surrounds the central pulsar (not visible). An x-ray jet erupts in both directions from the center of the torus and extends over a distance of a few light-years. Farther out, an intricate web of x-ray loops can be seen. These features are due to radiation from extremely high energy particles moving in a magnetic field, and show a strong resemblance to the rings, jets, and loops around the Crab pulsar. (*NASA, Chandra X-ray Center, Smithsonian Astrophysical Observatory, P. Slane et al.*)**

the planet's center, are between the orbits of two other Saturnian moons, Mimas and Enceladus. They were provisionally named S/2004 S1 and S/2004 S2. In January 2005, they were given the permanent names XXII Methone and XXIII Pallene. XXII Methone may be an object spotted in a single image taken by NASA's *Voyager* spacecraft in 1981, called at that time S/1981 S14. Pictures and data taken during the first close flyby of Saturn's moon Titan by *Cassini* revealed greater surface detail than ever before and showed that Titan has lost much of its original atmosphere over time. The surface appears to have been shaped by multiple geologic processes. Although a few circular features can be seen, none can be definitively identified as impact craters. On December 24, 2004, the *Huygens* probe successfully detached from NASA's *Cassini* orbiter to begin a 3-week journey to Titan. The *Huygens* probe, built and managed by ESA, was bolted to *Cassini* and had been riding along during the nearly 7-year journey to Saturn largely in a "sleep" mode. Set to touchdown on Titan on January 14, 2005, *Huygens* was the first human-made object to explore on-site the unique environment of this moon, whose chemistry is assumed to be very similar to that of early Earth before life formed. By end-2004, its mother ship had found Saturn roiled with storms, detected lighting, discovered a new radiation belt, found four new moons and a new ring around the planet, and mapped the composition of the planet's rings. On December 31, *Cassini* encountered Saturn's "yin-yang" moon Iapetus, on the probe's closest pass yet by one of Saturn's smaller icy satellites since its arrival around the ringed giant. The next close flyby of Iapetus, a world of sharp contrasts, is not until 2007. Its leading hemisphere is as dark as a freshly tarred street, and the white, trailing hemisphere resembles freshly fallen snow, like the yin-yang symbol. *See* CASSINI-HUYGENS MISSION.

WMAP. NASA's *Wilkinson Microwave Anisotropy Probe* (formerly called the *Microwave Anisotropy Mission, MAP*) was launched in 2001 on a Delta 2. Now with the probe located in an orbit around the second Lagrangian libration point L2, its differential radiometers measure the temperature fluctuations of the cosmic microwave background radiation (CMBR) with unprecedented accuracy. The CMBR is the light left over from the big bang, bathing the whole universe in this afterglow light. It is the oldest light in the universe, having traveled across the cosmos for 14 billion years, and the patterns in this light across the sky encode a wealth of details about the history, shape, content, and ultimate fate of the universe. Since the start of *WMAP* operations, scientists have produced the first version of a full sky map of the faint anisotropy or variations in the CMBR's temperature (now averaging a frigid 2.73 K). Results to date indicate that the universe is 13.7 billion years old, with a margin of error of close to 1%; the first stars ignited 200 million years after the big bang; light gathered in revealing *WMAP* pictures is from 379,000 years after the big bang; and the universe consists of 4% atoms, 23% cold dark matter, and 73% dark energy. The data place new constraints on the dark energy, which now seems more like a "cosmological constant" than a negative-pressure energy field called quintessence (the latter, however, is not ruled out). Fast-moving neutrinos do not play any major role in the evolution of structure in the universe (they would have prevented the early clumping of gas in the universe, delaying the emergence of the first stars, in conflict with the new *WMAP* data). The expansion rate of the universe, called the Hubble constant, is $H_0 = 71$ km/s/Mpc (megaparsecs) with a margin of error of about 5%. There is new evidence for inflation (in the polarized signal), and for the theory that fits all the data, the universe will expand forever. (But the nature of the dark energy remains a mystery. If it changes with time, or if other unknown and unexpected things happen in the universe, this conclusion could change.)

Genesis. The solar probe *Genesis* was launched on August 8, 2001, on a Delta 2 rocket, and went into a perfect orbit about the first Earth–Sun Lagrangian libration point L1, about 1.5 million kilometers (932,000 mi) from Earth and 148.5 million kilometers (92.3 million miles) from the Sun, on November 16, 2001. After an unconventional Lissajous orbit insertion (LOI), *Genesis* began the first of five "halo" loops around L1, lasting about 30 months. Collection of samples of solar wind material expelled from the Sun started on October 21, 2001. One year later, on December 10, 2002, with the spacecraft in overall good health and spinning at 1.6 rotations per minute, its orbit around L1 was fine-tuned with the seventh of 15 planned station-keeping maneuvers during the lifetime of the mission. Throughout 2003, *Genesis* continued its mission of collecting solar wind material, with all spacecraft subsystems still reported in excellent health. In April 2004, the sample collectors were deactivated and stowed, and the spacecraft returned to Earth, where the sample return capsule was to be recovered in midair by helicopter over the Utah Test and Training Range on September 8, 2004. However, *Genesis*' return did not go according to plan. The vessel, which had spent 27 months collecting data and samples of the solar wind, entered Earth's atmosphere as scheduled on September 8, but its parachutes failed to deploy and the capsule crashed into the Utah desert at nearly 322 km/h (200 mi/h). After the crash, the 181-kg (400-lb) capsule was recovered and transported by helicopter to a nearby Army base equipped with a clean room for analysis. In October, scientists reported that a large amount of material within the *Genesis* scientific collectors had remained intact and will provide useful information about the beginning and development of the solar system.

ACE. In 2004, the *Advanced Composition Explorer* (*ACE*) continued to observe, determine, and compare the isotopic and elemental composition of several distinct samples of matter, including the solar corona, the interplanetary medium, the local interstellar medium, and galactic matter. By end-2004, *ACE* has been positioned in a halo orbit around L1 for more than 7 years, and things were still working very well, with the exception of the SEPICA (Solar Energetic Particle Ionic Charge Analyzer) instrument. SEPICA had trouble with the gas regulation of its proportional counters and with a high-voltage power supply. Two-thirds of the instrument was nonfunctional, but the third counter was returning good science data. The problems are still under investigation. As of spring 2004, over 350 peer-reviewed papers had been published by *ACE* science team members.

Stardust. In January 2004, having weathered a strong "sandblasting" by cometary particles hurtling toward it at about six times the speed of a rifle bullet, NASA's comet probe *Stardust*, launched on February 3, 1999, on a Delta 2, passed by Comet P/Wild 2, collected particles, and began its 2-year, 1.14-billion-kilometer (708-million-mile) trek back to Earth. The probe had entered the comet's coma—the vast cloud of dust and gas that surrounds a comet's nucleus—on December 31, 2003. From that point on, it kept its defensive shielding between it and what scientists hoped would be the caustic stream of particles it would fly through. The encounter with Comet P/Wild 2 occurred on January 2, 2004, with a closest approach of about 150 km (93 mi) at a relative velocity of about 6.1 km/s (3.8 mi/s), at 277 million kilometers (172 million miles) from the Sun and 389 million kilometers (242 million miles) from Earth. The sample collector was deployed in late December 2003, and was retracted, stowed, and sealed in the vault of the sample reentry capsule after the Wild flyby. Images of the comet nucleus were also obtained (**Fig. 4**), with coverage of the entire sunlit side at a resolution of 30 m (98 ft) or better. On January 15, 2006, the capsule is scheduled to separate from the main craft (with a stabilizing spin of 1.5 rpm) and reenter Earth's atmosphere. A parachute will be deployed, and a chase aircraft will "snatch" and recover the descending capsule over the U.S. Air Force Test and Training Range in the Utah desert. Special engineering analyses were performed to ensure that *Stardust* will not suffer the same fate as *Genesis*.

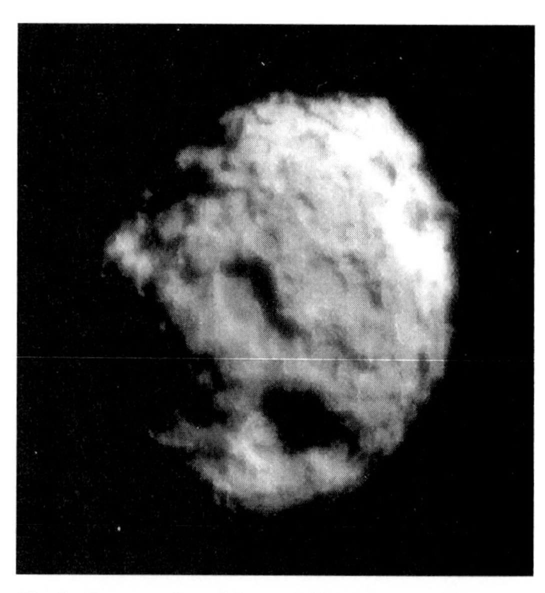

Fig. 4. Close-up view of Comet P/Wild 2 from the *Stardust* spacecraft. This is only the third close look at a comet, after Comet Halley and Comet Borelly. Unlike the smoother features of the other two, this comet core consisted of unique rocky, jagged terrain, craters, and ice, a possible result of its greater amount of time spent in deep space and remaining unaffected by solar heating. (*Courtesy of NASA/ JPL-Caltech*)

Ulysses. In February 2004, ESA's Science Program Committee unanimously approved a proposal to continue operating the highly successful *Ulysses* spacecraft until March 2008. This latest extension, the third in the history of the joint ESA–NASA mission, will enable *Ulysses* to add an important chapter to its survey of the high-latitude heliosphere. In 2007 and 2008, the space probe will fly over the poles of the Sun for a third time. Unlike the high-latitude passes in 2000 and 2001 that brought *Ulysses* over the solar poles near the maximum of the Sun's activity cycle, conditions for the third set of polar passes are expected to be much quieter. In fact, they are likely to be similar to those in 1994/1995 when *Ulysses* first visited the Sun's poles.

Voyager. The Voyager mission, begun in 1977, continues its quest to push the bounds of space exploration. The twin *Voyager 1* and *2* spacecraft opened new vistas in space by greatly expanding our knowledge of Jupiter and Saturn. *Voyager 2* then extended the planetary adventure when it flew by Uranus and Neptune, becoming the only spacecraft ever to visit these worlds. *Voyager 1* is the most distant human-made object in the universe. At end-2004, the spacecraft was about 94 times as far from the Sun as is Earth (14 billion kilometers or 8.7 billion miles). It was deflected northward above the plane of the planets' orbits when it swung by Saturn in 1980 and is now speeding outward from the Sun at nearly 1.6 million kilometers/day (1 million miles/day). The only spacecraft to have made measurements in the solar wind from such a great distance from the source of the dynamic solar environment, it has entered the solar system's final frontier, the so-called heliosheath beyond the termination shock of the solar wind, a vast turbulent expanse where the Sun's influence ends and the solar wind crashes into the thin gas between stars. *Voyager 2* also continues this ground-breaking journey. In July 2003, *Voyager 2* had reached a distance from the Sun of 10.6 billion kilometers (6.6 million miles). It too continues to return valuable science data. Each *Voyager*'s cosmic-ray detector, magnetometer, plasma-wave detector, and low-energy charged particle detector are still operational. In addition, the ultraviolet spectrometer on *Voyager 1* and the plasma science instrument on *Voyager 2* are producing and transmitting data. Both spacecraft are expected to continue to operate and send back valuable data until at least the year 2020, when *Voyager 1* will be more than 21 billion kilometers (13 billion miles) from Earth and may have reached interstellar space.

Mars exploration. The year 2004 began with the successful landing and deployment of the twin Mars Exploration Rovers *Spirit* and *Opportunity*, after their launches in 2003, at near-equatorial locations on opposite sides of Mars. The interplanetary navigation systems enabled exceptionally accurate achievement of the desired atmospheric entry conditions for both mobile geology laboratories, and the actual surface landing points differed from the targets by only 10 km (6.2 mi) for *Spirit* and 25 km (15.5 mi) for *Opportunity*. In April 2004, both rovers successfully completed their primary 3-month missions.

The six-wheeled rover vehicle *Spirit*, launched on June 10, 2003, on a Delta 2 Heavy rocket, landed on January 3, 2004, in Gusev Crater, in excellent condition. During 2004, *Spirit* completed a 3.2-km (2-mi) trek to a formation called Columbia Hills (after the lost shuttle), where it found a water-signature mineral called goethite in bedrock, one of the mission's surest indicators yet for a wet history on *Spirit*'s side of Mars. *Spirit*, during its primary mission, explored a plain strewn with volcanic rocks and pocked with impact craters. It found indications that small amounts of water may have gotten into cracks in the rocks and may also have affected some of the rocks' surfaces. This did not indicate a particularly favorable past environment for life. *Spirit*'s Extended Mission began with the rover starting a long trek toward the Columbia Hills on the horizon whose rocks might have come from an earlier and wetter era of the region's past. In late September 2004, NASA approved a second extension of the rovers' missions. The solar-powered machines were still in good health, though beginning to show signs of aging. They had come through the worst days of the Martian year from a solar-energy standpoint. Also, they had resumed full operations after about 2 weeks of not driving in mid-September while communications were unreliable because Mars was passing nearly behind the Sun. *Spirit* had driven 3.6 km (2.3 mi), six times the goal set in advance as a criterion for a successful mission. It was climbing hills where its examinations of exposed bedrock found more extensive alteration by water than what the rover had seen in rocks on the younger plain. During the long trek, *Spirit*'s right front wheel developed excessive friction. Controllers found a way to press on with the exploration by sometimes driving the rover in reverse with the balky wheel dragging.

Opportunity, launched on July 7, also on a Delta 2 Heavy, touched down on January 25, 2004, on Meridiani Planum, halfway around the planet from the Gusev Crater site of its twin, also in excellent condition. By the end of 2004, *Opportunity* had driven about 1.6 km (1 mi). It was studying rocks and soils inside a crater named Endurance, about 130 m (430 ft) wide and 22 m (72 ft) deep. The rover entered this crater in June 2004 after careful analysis of its ability to climb back out. Inside, *Opportunity* examined layers of bedrock with characteristics similar to those of the outcrop inside the smaller crater where it landed. This indicated a much longer duration for the watery portion of the region's ancient past. The rover also found some features unlike any it had seen before, evidence of changes in the environment over time. The rovers' life spans were unpredictable, but by October 2005 they had already racked up successes beyond the high expectations set for them when the Mars Exploration Rover project began. *See* MARS ROVERS.

The *Mars Odyssey* probe, launched April 7, 2001, successfully reached Mars on October 24, after a 6-month and 460-million-kilometer (286-million-mile) journey. Entering a highly elliptical polar orbit around the planet, it began to change orbit parameters by aerobraking, reducing its ellipticity to a circular orbit at 400 km (249 mi) by the end of January 2002. The orbiter is circling Mars with the objective of conducting a detailed mineralogical analysis of the planet's surface from space and measuring the radiation environment. On August 25, 2004, it began working overtime after completing a prime mission that discovered vast supplies of frozen water, ran a safety check for future astronauts, and mapped surface textures and minerals all over Mars. *Odyssey*'s camera system obtained the most detailed complete global maps of Mars ever, with daytime and nighttime infrared images at a resolution of 100 m (328 ft). The spacecraft has been examining Mars in detail since February 2002 (more than a full Mars year of about 23 Earth months) and was approved for an extended mission through September 2006. About 85% of the images and other data from NASA's twin Mars rovers, *Spirit* and *Opportunity*, have reached Earth via communications relay by *Odyssey*, which receives transmissions from both rovers every day. The orbiter helped analyze potential landing sites for the rovers and is to do the same for NASA's *Phoenix* mission, scheduled to land on Mars in 2008. Plans also call for *Odyssey* to aid NASA's *Mars Reconnaissance Orbiter*, due to reach Mars in March 2006, by monitoring atmospheric conditions during months when the newly arrived orbiter uses calculated dips into the atmosphere to alter its orbit into the desired shape. *See* SPACE PROBE COMMUNICATIONS.

Mars Global Surveyor (*MGS*) had completed its primary mission at the end of January 2001 and entered an extended mission. *MGS* has returned more data about Mars than all other missions combined. In September 2004, *MGS* started its third mission extension after 7 years of orbiting Mars, using an innovative technique to capture pictures even sharper than most of the more than 170,000 it has already produced. One dramatic example from the spacecraft's Mars Orbiter Camera showed actual wheel tracks of the Mars Exploration Rover *Spirit* and the rover itself. Another told scientists that no boulders bigger than about 1–2 m (3–7 ft) are exposed in giant ripples created by a catastrophic flood. The new technique involves rolling the entire spacecraft so that the camera compensates image motion while scanning; in this way it is able to show details at three times the resolution normally obtained.

Earth science. In 2004, NASA launched one Earth science satellite, the *Aura*. The *ICESat* observation system and *GRACE* satellites continued gathering information.

Aura. *Aura* (Latin for "breeze"), launched from Vandenberg Air Force Base on July 15 on a Delta 2 rocket, is NASA's third major Earth Observing System (EOS) platform, joining its sister satellites *Terra* and *Aqua*, to provide global data on the state of the atmosphere, land, and oceans, as well as their interactions with solar radiation and each other. *Aura*'s design life is 5 years with an operational goal of 6 years. The satellite flies in formation about 15 min behind *Aqua*. During 2004, observations from *Aura* showed that ozone destroyed chemically in the Arctic during the winter in near-record levels was restored by other atmospheric processes to near-average amounts, stopping high levels of harmful ultraviolet radiation from reaching Earth's surface. The reason for this appeared to lie in the unusual Arctic atmospheric conditions, which caused polar ozone to be replenished by shifted stratospheric winds which transported ozone-rich air from Earth's middle latitudes into the Arctic polar region.

ICESat. *ICESat* (Ice, Cloud, and land Elevation Satellite) is also an EOS spacecraft and the benchmark mission for measuring ice sheet mass balance, and cloud and aerosol heights, as well as land topography and vegetation characteristics. Launched on January 12, 2003, on a Delta 2 Expendable Launch Vehicle (ELV) into a near-polar orbit at an altitude of 600 km (373 mi) with an inclination of 94°, the spacecraft carries only one instrument, the Geoscience Laser Altimeter System (GLAS). Scientists trying to understand the dynamics of the Earth are using the lasers of *ICESat* to measure the height of ice sheets, glaciers, forests, rivers, clouds, and atmospheric pollutants from space with unprecedented accuracy, providing a new way of understanding the changing planet. For example, in winter 2004 *ICESat* showed thicker sea ice grouped together in its usual place near the Canadian Arctic than it was in 2003. It also showed a larger area of thinner ice in the Beaufort and Chukchi seas, where the summer ice cover has been rapidly decreasing. The location and amount of ice are important to climatologists and also ships that travel those seas.

GRACE. Launched on March 17, 2002, on a Russian Rockot carrier, the twin satellites *GRACE* (Gravity Recovery and Climate Experiment), named "Tom" and "Jerry," are mapping the Earth's gravity fields by taking accurate measurements of the distance between the two satellites, using the Global Positioning System (GPS) and a microwave ranging system. Its science data are being used to estimate global models for the mean and time-variable Earth gravity field approximately every 30 days for the 5-year lifetime of the mission. The science data from *GRACE* consist of the intersatellite range change measurements, and the accelerometer, GPS, and attitude measurements from each satellite. The project is a joint partnership between NASA and the German DLR (Deutsches Zentrum für Luft- und Raumfahrt).

Department of Defense (DOD) space activities. In 2004, there were five military space launches (2003: 11), carrying five payloads: one Titan 4B/IUS vehicle from Cape Canaveral, with the DSP-022 early warning satellite, three Delta 2's with GPS navigation satellites, and one Atlas 2AS with a National Reconnaissance Office (NRO) communications satellite.

Commercial space activities. In 2004, commercial space activities in the United States exhibited a sluggish increase over prior years, after the 2001/2002 downturn in the communications space market caused by failures of satellite constellations for mobile telephony and a slight recovery in 2003.

In addition to the financial problems, some difficulties remained due to the export restrictions on sensitive technologies imposed on U.S. industry. In general, commercial ventures continue to play a relatively minor role in U.S. space activities, about as in 2001 (50%), but more than the 26% in 2002 and 31% in 2003, of commercial satellites and associated launch services worldwide.

Of the 19 total launch attempts by the United States in 2004 (26 in 2003), 10 were commercial missions (NASA: 4; military: 5). In the launch services area, Boeing sold seven Delta 2 vehicles, while competitor ILS/Lockheed Martin flew 4 Atlas 2AS and one Atlas 3B (with Russian engines). Both companies also had successful launches of their next-generation EELV (evolved expendable launch vehicle) rockets, Lockheed Martin with the fourth Atlas 5/Centaur (comsat *AMC-16*), and Boeing with the first Delta 4H (heavy) launcher (Demosat, plus imaging test satellites *3CS-1* and *3CS-2*), while the partnership of Boeing, RSC-Energia (Russia, 25% share), NPO Yushnoye (Ukraine), and Kvaerner Group (Norway) successfully launched three Russian Zenit 3SL rockets carrying Brazil's *Estrela do Sul* (Telstar 14), the United States US DirecTV-7S, and China's Telstar 18 comsats from the *Odyssey* sea launch platform floating at the Equator (first launch 1999).

The year 2004 was a historic one for privately funded personal space travel. On June 21, *SpaceShipOne*, a joint venture between Vulcan and the Scaled Composites Company of Burt Rutan, became the first commercial spacecraft when it rocketed beyond the 100-km (62-mi) threshold of space, launched from its piloted mother ship/airplane *White Knight*, reaching 100,124 m (328,491 ft) with pilot Mike Melvill. After this test flight, on September 29, *SpaceShipOne* made the first of the two flights for the Ansari X-Prize of $10 million, again piloted by Melvill (who had to control over 25 unscheduled rolls) and carrying ballast representing a second passenger, to 102.9 km (64 mi). This flight was followed on October 4 by the second flight, with Brian Binnie at the helm, to 112 km (69.6 mi).

Russian Space Activities

With financial support by a slowly improving national economy only slightly increasing over previous years, Russia in 2004 showed relatively unchanged activity in space operations from 2003. Its total of 22 successful launches (of 23 attempts) was one more than the previous year's 21 (of 21 attempts): five Soyuz-U, two Soyuz-FG (both crewed), one Soyuz-2-1A (test), eight Protons, one Zenit-2, three Zenit-3SL (sea launch, counted above under United States Activities), one Molniya-M, two Kosmos-3M, one Tsiklon 2, one Tsiklon 3 (failed), and one Dnepr. The upgraded Soyuz-FG rocket's new fuel injection system provides a 5% increase in thrust over the Soyuz-U, enhancing its lift capability by 200 kg (441 lb) and enabling it to carry the new *Soyuz-TMA* spacecraft, which is heavier than the Soyuz-TM ship used previously to ferry crews to the *ISS*. *Soyuz-TMA* was flown for the first time on October 30, 2002, as *ISS* mission 5S. It was followed in 2003 by *Soyuz TMA-2* (6S) and *TMA-3* (7S), and in 2004 by *TMA-4* (8S) and *TMA-5* (9S).

The Russian space program's major push to enter the world's commercial arena by promoting its space products on the external market, driven by the need to survive in an era of severe reductions of public financing, increased in 2004. First launched in July 1965, the Proton heavy lifter, originally intended as a ballistic missile (UR500), by end-2004 had flown 231 times since 1980, with 14 failures (reliability: 0.94). Its launch rate in recent years has been as high as 13 per year. Of the eight Protons launched in 2004 (2003: five), six were for commercial customers (*Eutelsat W3A*, *Ekspress-AM11* and *-AM-1*, *Intelsat 10-02*, *Amazonas* [Spain], *AMC-15*), the other two for the state/military (*Raduga 1* comsat, three *GLONASS* navsats). From 1985 to 2004, 179 Proton and 403 Soyuz rockets were launched, with 10 failures of the Proton and 10 of the Soyuz, giving a combined reliability index of 0.966. Until a launch failure on October 15, 2002, the Soyuz rocket had flown 74 consecutive successful missions, including 12 with crews; subsequently, another 15 successful flights were added, including four carrying a total of 11 humans.

European Space Activities

Europe's efforts to reinvigorate its faltering space activities after the long decline since the mid-1990s took a considerable step back in 2004, compared to astronautics activities of NASA, DOD, Russia, and China. Ongoing efforts by the European Union (EU) on an emerging new European space strategy for ESA to achieve an autonomous Europe in space under Europe's new constitution that makes space and defense an EU responsibility remained unresolved.

Commercial activities. After 2003 did not bring the much-needed breakthrough of Europe's commercial space industry in its flagging attempts at recovery, given particular emphasis by the last flight of the Ariane 4 workhorse on February 15, 2003, the year 2004 continued the decline of Europe's commercial activities in space. As in 2003, the heavy-lift Ariane 5G (generic) was launched only three times (compared to four times in 2002), bringing its total to 20. After the December 2002 failure of the new EC version of the Ariane 5, designed to lift 9 metric tons (10 tons) to geostationary transfer orbit, enough for two big communications satellites at once, European industry quickly developed an Ariane 5 recovery plan; and on June 20, 2003, Arianespace concluded a preliminary contract with European aerospace conglomerate EADS Space Transportation for a batch of 30 Ariane 5 launchers, which was signed in May 2004.

Space science. In the space science area, there was only one new European launch in 2004, the international comet mission *Rosetta*.

Rosetta. The mission was approved in November 1993 by ESA's Science Program Committee as the Planetary Cornerstone Mission in ESA's long-term space science program. The mission goal was initially set for a rendezvous with Comet 46 P/Wirtanen. After postponement of the initial launch, a new target was set: Comet 67 P/Churyumov-Gerasimenko. The launch of *Rosetta* on an Ariane 5 occurred on March 2, 2004. After its rendezvous with the comet in 2014, *Rosetta* is to release a landing craft named *Philae*. It is hoped that on its 10-year journey to the comet, the spacecraft will pass by at least one asteroid.

Envisat. In 2004, ESA's operational environmental satellite *Envisat*, the largest Earth observation spacecraft ever built, continued its observations after its launch on March 1, 2002, on the eleventh Ariane 5. During September 6–10, ESA held the 2004 ENVISAT & ERS Symposium in Salzburg, Austria, open to all interested parties from scientists to operational users and covering both ENVISAT and Earth Resource Satellite (ERS) missions.

SPOT 5. Launched on May 4, 2002, by an Ariane 4 from Kourou (French Guiana), the fifth imaging satellite of the commercial Spot Image Company in 2004 continued operations in its polar Sun-synchronous orbit of 813 km (505 mi) altitude. Unique features of the *SPOT* system are high resolution, stereo imaging, and revisit capability. The *SPOT* satellite Earth observation system was designed by CNES, the French Space Agency, and developed with the participation of Sweden and Belgium.

INTEGRAL. ESA's *INTEGRAL* (International Gamma-Ray Astrophysics Laboratory), a cooperative project with Russia and the United States, continued successful operations in 2004. In 2004, a gamma-ray burst detected by *INTEGRAL* on December 3, 2003, was thoroughly studied for months by an armada of space and ground-based observatories. Astronomers then concluded that this event, called GRB 031203, was the closest cosmic gamma-ray burst on record, but also the faintest, which suggests that an entire population of subenergetic gamma-ray bursts has so far gone unnoticed.

XMM-Newton. Europe's *XMM* (X-ray Multi Mirror)–*Newton* observatory, launched on December 10, 1999, on an Ariane 5, is the largest European science research satellite ever built. By December 10, 2004, the observatory had achieved oversubscription factors in observing time of seven or more, and had performed about 3800 scientific observations, and spawned about 700 publications in the refereed literature, with 290 observations published.

Smart 1. *Smart 1* (Small Missions for Advanced Research in Technology) is Europe's first lunar spacecraft. The 370-kg (816-lb) spacecraft was launched on September 27, 2003, with two commercial communications satellites (*Insat 3E*, *e-Bird*) on an Ariane 5G. Built by Swedish Space Corp., it is intended to demonstrate new technologies for future missions, in this case the use of solar-electric propulsion as the primary power source for its ion engine, fueled by xenon gas. The single engine was fired for the first time on September 30, 2003. On November 15, 2004, the spacecraft encountered its first perilune after 332 orbits around the Earth. The ion drive was fired on that day to brake the spacecraft into lunar orbit, after which, over several months, its engine was to be fired repeatedly to lower the spacecraft into an operational orbit whose height ranged from approximately 300 to 3000 km (200 to 2000 mi). This was achieved in January 2005, heralding the beginning of its science program, executed with spectrometers for x-rays and near-infrared radiaton as well as a camera for color imaging.

Mars Express. *Mars Express* is Europe's entry into the slowly expanding robotic exploration of Mars as a precursor to later missions by humans. The probe was launched on June 2, 2003, from the Baikonur launch site by a Russian Soyuz/Fregat rocket. After a 6-month journey, it arrived at Mars in December. Six days before arrival, *Mars Express* ejected the *Beagle 2* lander, which was to have made its own way to the correct landing site on the surface, but was lost, failing to make contact with orbiting spacecraft and Earth-based radio telescopes. The orbiter entered Martian orbit on December 25, first maneuvering into a highly elliptical capture orbit, from which it moved into its operational near-polar orbit later in January 2004. Highly successful operations and stunning close-up imagery of the Mars surface continued during 2004. On February 6, while *Mars Express* was flying over the area that NASA's Mars Exploration Rover *Spirit* was examining, the orbiter transferred commands from Earth to the rover and relayed data from the rover back to Earth, a successful pioneering demonstration of communications between ESA's orbiter and NASA's rover and the first working international communications network around another planet. *Mars Express* continues to remotely explore the planet with a sophisticated instrument package comprising the High Resolution Stereo Camera (HRSC), Energetic Neutral Atoms Analyzer (ASPERA), Planetary Fourier Spectrometer (PFS), Visible and Infrared Mineralogical Mapping Spectrometer (OMEGA), Sub-Surface Sounding Radar Altimeter (MARSIS), Mars Radio Science Experiment (MaRS), and the Ultraviolet and Infrared Atmospheric Spectrometer (SPICAM).

Asian Space Activities

China, India, and Japan have space programs capable of launch and satellite development and operations.

China. With a total of eight launches in 2004, China remained solidly in third place of spacefaring nations after Russia and the United States, following its successful orbital launch in 2003 of the first Chinese "taikonaut," 38-year-old Lt. Col. Liwei of the People's Liberation Army in the spacecraft *Shenzou 5* ("Divine Vessel 5").

The launch vehicle of the *Shenzhou* spaceships is the new human-rated Long March 2F rocket. China's Long March (Chang Zheng, CZ) series of launch vehicles consists of 12 versions, which by the end of 2004 had made 83 flights, sending 95 payloads (satellites and spacecraft) into space, with a 90% success rate. China has three modern (but land-locked, thus azimuth-restricted) launch facilities: at Jiuquan (Base 20, also known as Shuang Cheng-Tzu/East Wind) for low-Earth-orbit (LEO) missions, Taiyuan (Base 25) for Sun-synchronous missions, and Xichang (Base 27) for geostationary missions.

Its eight major launches in 2004 served to demonstrate China's growing space maturity (six successful launches in 2003, four in 2002, and one in 2001). On April 18, a CZ-2C launched the *Shiyan 1* and *Naxing 1* imaging and technology satellites, on July 25 the *Tan Ce-2* science satellite, on August 29 the recoverable science research satellite *FSW 19* which returned to Earth 27 days later, on September 8 the science satellites *SJ-6A* and *-6B*, on September 27 the twentieth *FSW*, on October 19 the weather satellite *Feng Yu-2C*, on November 6 the imaging satellite *ZY-2C*, and on November 18 the *Shiyan 2* remote sensing satellite. This wound up a record of 41 consecutive launch successes for the Long March, which in its two-stage 2C version has a lift-off weight of 222 metric tons (245 tons), a total length of 41.9 m (137.5 ft), a diameter of the rocket and payload fairing of 3.4 m (11 ft), and a low-Earth-orbit launching capacity of 3.5 metric tons (3.9 tons).

In 2004, the China National Space Administration (CNSA), through its Lunar Exploration Program Center (LEPC), initiated the project of a lunar satellite mission called *Chang'e*, regarded as a symbolic project in China like the Human Spaceflight Program, with the objectives to serve as a community-wide exercise to bolster advanced research and development, and to improve project planning capabilities and master engineering system management of a complex program.

Also in 2004, CNSA under its new administrator Sun Laiyan (since April) strongly increased its efforts in engaging in international cooperation, as exemplified by the cooperative project DSP (Double Star Program), a joint ESA/Chinese project to study the effects of the Sun on Earth's environment, and in particular the "magnetotail," where storms of high-energy particles are generated, and the collaboration on ESA's *Galileo* navigation satellite system. CNSA improved relationships with Russia, Ukraine, France, Germany, Italy, United Kingdom, and other European countries and also started promoting cooperation with Asia Pacific countries as well as with "south-south" countries such as Brazil, Argentina, Nigeria, and Venezuela.

India. India's main launchers are the PSLV (Polar Space Launch Vehicle) and the Delta 2 class GSLV (Geosynchronous Satellite Launch Vehicle). In 2004, India conducted only one launch, but it was important: On September 20, a GSLV on its first operational flight demonstrated the reliability of the vehicle with the successful launch of *EDUSAT*, India's first thematic satellite dedicated to educational services. Also in 2004, India saw the inauguration of the first cluster of Village Resource Centers and further expansion of its Telemedicine network, reiterating India's commitment to use space technology for societal applications.

India has announced that it plans to explore the Moon and to send a crewless probe there in 2007. The Indian Space Research Organization (ISRO) calls the Moon flight project *Chandrayan Pratham*, which has been translated as "First Journey to the Moon" or "Moonshot One." The 525-kg (1157-lb) *Chandrayan 1* would be launched on one of India's own PSLV rockets. At first, the spacecraft would circle Earth in a geosynchronous transfer orbit (GTO). From there, it would fly out into a polar orbit of the Moon some 97 km (60 mi) above the surface, carrying x-ray and gamma-ray spectrometers and sending back data to Earth for producing a high-resolution digital map of the lunar surface. The project's main objectives are high-resolution photography of the lunar surface using remote-sensing instruments sensitive to visible light, near-infrared light, and low-energy and high-energy x-rays. Space aboard the satellite also will be available for instruments from scientists in other countries. *Chandrayan 1* is expected to be the forerunner of more ambitious planetary missions in the years to come, including landing robots on the Moon and visits by Indian spacecraft to other planets in the solar system.

Japan. In 2004, efforts were underway to improve the H-2A launch vehicle, following the failure of its sixth launch on November 29, 2003. In its longer-range view, the new space agency JAXA (Japan Aerospace Exploration Agency) is studying versions of a "new generation" launch vehicle, essentially a version of the H-2A with 10–20% greater lift capacity, which would put it into the Delta 4 class.

The Space Engineering Spacecraft *Hayabusa* (*MUSES-C*) was launched at the Kagoshima Space Center in southern Japan on May 9, 2003, on a solid-propellant M-5 rocket. In 2004, it was in a heliocentric orbit using its ion engines. Japan's first asteroid sample return mission, *Hayabusa* ("Falcon") continued on a trajectory toward the asteroid 25143 Itokawa/1998 SF36. On May 19, 2004, *Hayabusa* came close to the Earth at a distance of approximately 3700 km (2300 mi) and successfully carried out an Earth swingby to place it in a new elliptical orbit toward Itokawa. The Earth swingby is a technique to significantly change direction of an orbit and/or speed by using the Earth's gravity without consuming onboard propellant. After its flyby, the probe restarted its ion engines. By the end of 2004, the three microwave discharge engines had accumulated 20,000 h of operational time. The deep-space probe arrived at Itokawa on September 14, 2005. It was to spend 5 months at the asteroid, conducting observations as well as gathering a tiny sample of the asteroid. It was then to depart the asteroid in late 2005 and return to Earth in mid-2007, with the sample to be recovered in a reentry capsule parachuted to the surface in the Australian outback.

Other Countries' Space Activities

Brazil, Chile, and Argentina haltingly continued small efforts to develop their own space launch and operations capability for garnering a share of future satellite markets. In 2004, there were no launches from these countries. Brazil has the most advanced space program in Latin America, with significant capabilities in launch vehicles, launch sites, and satellite manufacturing. In 2004, Brazil and Russia agreed to expand their cooperation in space, including the joint development and production of launch vehicles, the launch of geostationary satellites, and the joint development and utilization of Brazil's Alcântara Launching Center (Centro de Lançamento de Alcântara, CLA) in Maranhão.

The Canadian Space Agency (CSA) continued supporting work on its contribution to the ISS partnership, the Mobile Service System (MSS), consisting of the 1800-kg (3960-lb) Space Station Remote Manipulator System (SSRMS) *Canadarm2*, the Mobile Base System (MBS), and the Special Purpose Dexterous Manipulator (SPDM). Canada also has an active Canadian Astronaut Program.

On July 17, 2004, Telesat Canada's innovative, high-speed Ka-band multimedia telecommunications satellite *Anik-F2* was launched to geostationary orbit in a perfect liftoff by an Ariane 5G+ heavy rocket. At 5900 kg (13,000 lb), the Boeing-built *Anik-F2* is one of the most powerful communications satellites ever built and the fifteenth satellite launched by Telesat.

Also in 2004, Canada's *Radarsat 1* "eye-in-the-sky" completed its ninth year of operation, well beyond its 5-year nominal lifetime. Over the years, it has delivered precision images and garnered 15% of the world's Earth observation market for Canada. Development of *Radarsat 2* is underway for launch in 2005.

For background information *see* COMET; COMMUNICATIONS SATELLITE; COSMIC BACKGROUND

RADIATION; GALAXY, EXTERNAL; INFRARED ASTRONOMY; MARS; MERCURY (PLANET); METEOROLOGICAL SATELLITES; MILITARY SATELLITES; MOON; REMOTE SENSING; SATELLITE ASTRONOMY; SATELLITE NAVIGATION SYSTEMS; SCIENTIFIC SATELLITES; SOLAR WIND; SPACE FLIGHT; SPACE PROBE; SPACE STATION; SPACE TECHNOLOGY; SPACE TELESCOPE, HUBBLE; SUN; X-RAY ASTRONOMY in the McGraw-Hill Encyclopedia of Science & Technology. Jesco von Puttkamer

Bibliography. *AEROSPACE AMERICA*, November 2004; *Aerospace Daily*; *AIAA Aviation Week & Space Technology (AW&ST*, various 2004 issues); ESA *Press Releases*; NASA Public Affairs Office *News Releases*; *SPACE NEWS*.

Space probe communications

One of the biggest challenges of deep space exploration is communicating over links spanning distances of solar system scale. Compared to a typical commercial link between a geostationary satellite and a user on the Earth's surface, which spans a distance of roughly 40,000 km (25,000 mi), a link between Mars and Earth can be up to 4×10^8 km (2.5×10^8 mi), and links to a Jupiter spacecraft can approach 10^9 km (6×10^8 mi). Because the performance of a communications system scales inversely as the square of link distance, these deep-space links pose unique problems and, in many cases, fundamentally limit the quantity of scientific data that can be returned.

These challenges are particularly taxing for small space probes, with highly constrained mass and power. Such probes are limited in the amount of radio-frequency power they can transmit, and in the size of the antenna they can use to focus their radiated signal toward Earth. For such missions, relay communications offers an approach to significantly increase data return.

In a relay scenario, a small space probe relays its telemetry to a nearby relay spacecraft over an energy-efficient, short-range communications link. The relay spacecraft, with much greater mass and power availability, can then take on the task of communicating the collected data back to Earth, utilizing higher transmit power and larger, higher-gain antennas. In addition to the advantages of energy efficiency and increased data rates, communications via a relay spacecraft allows telemetry capture even at times when the Earth is not in view (for example, from the far side of Mars).

Two recent planetary missions have successfully demonstrated the benefits of relay communications for returning data from small space probes. The *Mars Exploration Rovers*, *Spirit* and *Opportunity*, developed by the National Aeronautics and Space Administration (NASA), have returned the vast majority of their science data via relay through three Mars science orbiters that have been equipped with relay capabilities. And the *Huygens* probe, developed by the European Space Agency (ESA), relayed the data it collected during its descent through the atmosphere of Saturn's moon, Titan, via NASA's *Cassini* spacecraft, which carried *Huygens* on the long journey from Earth to the Saturnian system. An examination of each of these scenarios illustrates how large data return can be achieved, even for small probes at other planets. *See* CASSINI-HUYGENS MISSION; MARS ROVERS.

Relay support for Mars Rovers. On January 4, 2004, the first of the *Mars Exploration Rovers*, *Spirit*, began its 6-min descent through the Martian atmosphere toward touchdown in Gusev Crater. During this critical period, it was essential to capture engineering telemetry characterizing the lander's status so that, if the landing failed, NASA could diagnose the failure in time to modify the entry, descent, and landing (EDL) scenario for the second rover, *Opportunity*, scheduled to arrive 3 weeks later. However, the lander's X-band (8.4-GHz) direct-to-Earth link could support an effective data rate of only about 1 bit per second, due to the enormous distance to Earth and the fact that, during EDL, only a low-gain, omnidirectional antenna could be used. While this was sufficient to provide some high-level information on spacecraft state during EDL, more detailed information was strongly desired. Fortunately, *Spirit*'s landing was within view of the *Mars Global Surveyor* spacecraft, a NASA science orbiter launched in 1996 and equipped with an ultrahigh-frequency (UHF, 400-MHz) relay communications system for use by future Mars landers. With an orbit altitude of only 400 km (250 mi), *Mars Global Surveyor* offered a much shorter communications path; over this UHF link, *Spirit* was able to relay data during EDL at a rate of 8000 bits per second, providing a much more comprehensive engineering assessment.

Spirit landed safely and commenced a long and highly successful program of robotic exploration of the geology of the Gusev Crater. The sister spacecraft, *Opportunity*, touched down 3 weeks later, also relaying critical engineering telemetry through *Mars Global Surveyor* during the descent to the Meridiani plains. Once on the surface, each 170-kg (375-lb) rover was able to deploy its steerable X-band high-gain antenna, allowing typical data rates of 1 to 10 kilobits per second on the direct-to-Earth link to the large, 70-m (230-ft) receiving antennas of NASA's Deep Space Network on Earth. But once again, relay communications offered a more efficient data return path. Both *Mars Global Surveyor* and *Odyssey*, another NASA Mars science orbiter, launched in 2001 and also equipped with relay capability, offered high-rate, energy-efficient UHF relay links through a simple, low-gain antenna on the rover. In their low-altitude polar orbits, each orbiter would typically overfly each lander at least twice per sol (a sol is a Martian day, about 24 h 40 min)—once in the Martian afternoon and once in the early Martian morning. Because of the low altitude of the orbiters, these passes were short—typically only 10–15 min—but the short link distance allowed the relay links to operate at a high data rate of 128 kbit/s; the link to *Odyssey* occasionally supported data rates as high as 256 kbit/s. Once onboard the orbiters, the scientific data from the *Mars Exploration Rovers* were relayed back to Earth using the orbiters' much more capable direct-to-Earth links.

Using the UHF relay, each rover is able to return an average of more than 100 megabits of data every sol. As of June 2005, more than 130 gigabits of data had been returned from *Spirit* and *Opportunity*, with over 97% of that data volume having been sent via the UHF relays through *Mars Global Surveyor* and *Odyssey*; only 3% has been sent on the X-band direct-to-Earth links. This telecommunications capability has greatly benefited the science return from the rovers' data-rich instrument suite. The stereo cameras on *Spirit* and *Opportunity* can generate 3D color panoramas of the Martian surface, with a full panorama representing a data volume of several hundred megabits, even after lossy image compression. Other rover instruments, including the microscopic imager and the infrared spectrometer, also generate large volumes of data. The availability of the relay links has enabled *Spirit* and *Opportunity* to vastly increase the quantity and quality of science data, and enhance the fidelity of humans' virtual presence on Mars.

In addition to *Mars Global Surveyor* and *Odyssey*, a third science orbiter—the European Space Agency's *Mars Express Orbiter*—arrived at Mars in December 2003, just prior to the arrival of the *Mars Exploration Rovers*. It, too, is equipped with a UHF relay radio and utilizes the same communications protocol as the NASA spacecraft, allowing interoperability between NASA and ESA spacecraft. While not used in day-to-day operations for *Spirit* and *Opportunity* data return, *Mars Express* has successfully conducted several demonstration relay passes with the rovers. Thus, a truly international telecommunications network is orbiting Mars, prepared to provide efficient, high-bandwidth telecom services to future Mars explorers. *See* MARS EXPRESS.

Return of data from Huygens probe. On January 14, 2005, the European Space Agency's *Huygens* probe descended through the methane-rich atmosphere of Saturn's moon, Titan, providing a first glimpse of the surface of this fascinating world. Twenty-two days earlier, the *Huygens* probe had been released by NASA's *Cassini* spacecraft, which had carried the probe from Earth to the Saturnian system, 12×10^9 km (7.5×10^9 mi) distant. *Cassini* was launched on October 15, 1997, and entered orbit around Saturn on June 30, 2004, where the craft is carrying out extended observations of Saturn and its moons, in addition to fulfilling the role of delivering *Huygens* to Titan.

The *Huygens* probe had no capability to transmit high-rate data back to Earth as it descended to Titan's surface. Instead, the probe relayed its data to the *Cassini* spacecraft as it flew by just 60,000 km (37,000 mi) overhead. The probe transmitted its telemetry at a rate of 8192 bits/s via a 10-watt S-band (2-GHz) signal radiated from the probe's low-gain antenna, and captured by the *Cassini* spacecraft through its large, 4-m (13-ft) high-gain antenna, which was pointed directly toward the probe throughout its descent. Redundant radio systems were utilized on the probe and on *Cassini* to increase the robustness of the overall relay data return and mitigate against the failure of any single radio component.

The *Huygens* probe spent roughly 2 h 15 min descending through Titan's atmosphere, all the while returning scientific data from its suite of instruments and acquiring descent imagery, chemical analysis of the constituents of Titan's atmosphere, temperature and pressure profiles, and information on atmospheric winds. Designed to last only a few minutes on the surface, the probe in fact continued to transmit for another 2 h, at which time the *Cassini* spacecraft reoriented its high-gain antenna toward Earth and relayed the probe data back to one of NASA's Deep Space Network antennas. In all, over 100 megabits of probe telemetry were relayed back to Earth for analysis.

Future interplanetary relay scenarios. Relay communications will continue to be used to enable small deep-space probes to return science data from distant sites throughout the solar system. At Mars, NASA plans to launch the *Mars Reconnaissance Orbiter*, another science orbiter equipped with relay communications capability, in August 2005, further enhancing the orbital telecommunications infrastructure that will support landers slated for launch in 2007 and 2009. Ultimately, dedicated telecommunications orbiters are envisioned, with higher-altitude orbits and enhanced radio systems designed to provide one to two orders of magnitude greater communications capability. These capabilities will greatly increase the science return from next-generation landers and rovers, and when humans ultimately walk on the Red Planet, they will look to a robust telecommunications and navigation infrastructure to be in place to support their exploration activities.

For background information *see* MARS; SATURN; SPACE COMMUNICATIONS; SPACE PROBE; SPACECRAFT GROUND INSTRUMENTATION in the McGraw-Hill Encyclopedia of Science & Technology.

Charles D. Edwards

Bibliography. C. D. Edwards et al., Strategies for telecommunications and navigation in support of Mars exploration, *Acta Aatronautica*, 48(5-12): 661-668, 2001; D. J. Mudgway, *Uplink-Downlink: A History of the Deep Space Network 1957-1997*, NASA SP-2001-4227, NASA History Series, NASA Office of External Relations, Washington, DC, 2001; *Opportunity* at Meridiani Planum, Special Issue, *Science*, 306(5702):1697-1756, Dec. 3, 2004; *Spirit* at Gusev Crater, Special Issue, *Science*, 305(5685): 793-845, Aug. 6, 2004.

Spider silk

Spider silks are fine fibers produced from fibrous-protein solutions for use by spiders in the various functions required for their survival. Spiders are the only animals that use silk for a variety of functions

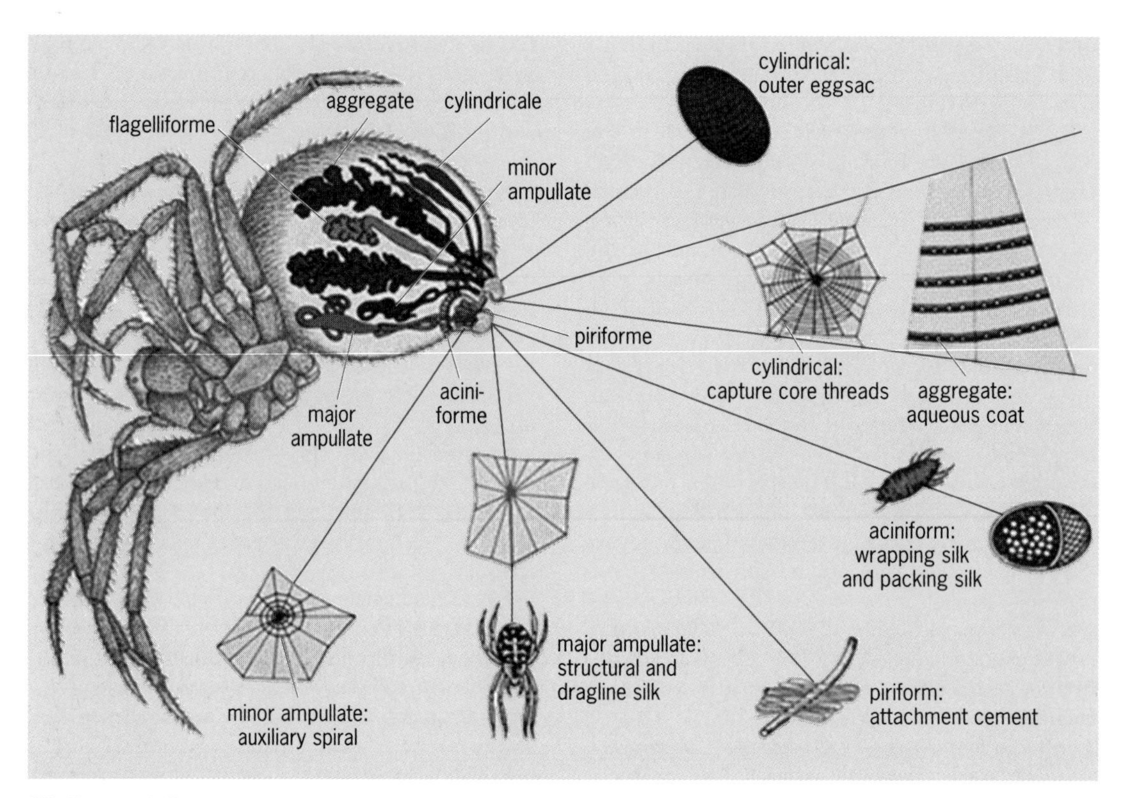

Silk fibers and silk glands in *Araneus diadematus*. (*Reprinted from F. Vollrath, Strength and structure, Rev. Mol. Biotechnol., 74(2):67–83, 2000*)

in their daily lives, including reproduction, food capture, and construction. Silk fibers produced by silkworms to build their cocoons have been used in the textile industry for more than 5000 years. But in recent years, spider silk, particularly silk from the major ampullate glands used to make elaborate "orb" webs, has attracted the attention of scientists and engineers in many fields from evolutionary biology to mechanical engineering and material science. The exceptional toughness, stiffness, and extensibility of spider-silk fibers outperform even the best synthetic materials. With the recent and expected advances in molecular genetics and proteomics, as well as in the understanding of material properties in the nanoscale size range, orb-web-weaving spiders are ideal organisms for studying the relationship among the physical properties of protein-based materials in terms of their function and evolution.

Mechanical properties. Over the past 400 million years, many different kinds of silk have evolved in the nearly 37,000 known spider species. For every kind of spider silk that has been characterized in any detail, there are more than 1000 that have not been characterized. Silk from only a few species, such as *Nephila clavipes* (the golden orb-weaving spider) and *Araneus diadematus* (common garden spider), have been examined in detail. Orb-weaving spiders produce silk for different functions from as many as seven different glands (see **illustration** and **Table 1**). Dragline and viscid silk fibers have been the focus of recent research because they are among the toughest materials known. Toughness is a measure of a material's energy absorption capacity. A single silk fiber is only 1–10 micrometers in diameter, compared to a human hair which typically is close to 50 μm. Dragline silk will support the spider's weight and is used to make the radial threads in webs. Viscid silk is the sticky, spiral threads in webs used for capturing prey. **Table 2** lists the characteristic mechanical properties of dragline and viscid silk, compared to other materials. Most of the other materials are either very strong but not tough and

TABLE 1. Types of common spider silk and their function

Silk	Gland	Function
Dragline	Major ampullate	Radial threads of web frame safety line
Viscid	Flagelliform	Prey capture
Gluelike	Aggregate	Prey capture, attachment
Minor ampullate	Minor ampullate	Orb-web frame construction
Cocoon	Cylindriform	Reproduction
Wrapping	Aciniform	Wrapping captured prey
Attachment	Pyriform	Attachment to environmental substrates

TABLE 2. Tensile mechanical properties of spider silk and other materials

Material*	Stiffness or Young's modulus, GPa (kilo psi)	Strength or breaking stress GPa (kilo psi)	Extensibility or maximum strain, %	Toughness or maximum energy absorption, MJ/m^3 ($Btu/in.^3$)
AD dragline silk	10 (1450)	1.1 (159)	27	160 (2.48)
AD viscid silk	0.003 (0.43)	0.5 (72)	270	150 (2.32)
NC dragline silk	22 (3190)	1.1 (160)	9	150 (2.32)
NE dragline	8 (1160)	1.1 (160)	28	165 (2.56)
Silkworm silk	7 (1015)	0.6 (87)	18	70 (1.08)
Tendon, collagen	1.5 (217)	0.15 (22)	12	7.5 (0.12)
Bone	20 (2900)	0.16 (23)	3	4 (0.06)
Elastin	0.001 (0.145)	0.002 (0.29)	150	2 (0.03)
Synthetic rubber	0.001 (0.145)	0.05 (7.25)	850	100 (1.55)
Nylon fiber	5 (725)	0.95 (137)	18	80 (1.24)
Kevlar 49	130 (18850)	3.6 (522)	2.7	50 (0.77)
Carbon fiber	300 (43500)	4 (580)	1.3	25 (0.39)
High tensile steel	200 (29000)	1.5 (217)	0.8	6 (0.09)

*AD: *Araneus diadematus*; NC: *Nephila clavipes*; NE: *Nephila edulis*.

extensible (such as steel) or are highly extensible but not strong (such as rubber). The extraordinary characteristic of dragline silk is that it has an exceptional combination of high stiffness, strength, extensibility, and toughness. The breaking stress or strength of dragline silk is comparable to steel, Kevlar® (used in bullet-proof clothing), and carbon fibers, but it is more extensible and tougher. Viscid silk is even more extensible, with stiffness comparable to rubber, but it is much stronger than rubber. The high stiffness of dragline and the rubberlike extensibility of viscid silk with the exceptional ability to absorb energy appear to be optimized for their required functions. Scientists seek to understand how spiders are able to spin silk fibers from an aqueous protein solution with the desired properties. Using spider silk as a guide, it should be possible to make the next-generation high-performance (very strong and yet very extensible) materials.

Structure. All types of silk have the same basic building blocks—protein polymers made of amino acids. Although the different silk glands in an evolutionarily advanced orb-weaving spider, such as *N. clavipes*, possibly evolved from a single type of gland, silk glands now differ significantly from one another in morphology and composition, and produce different kinds of silk. The best-characterized silk protein and associated fiber is the dragline silk of *N. clavipes*. Dragline-silk fibers are composed primarily of two protein polymers: spidroin I and spidroin II. Both spidroin I and II consist of repeated alternating protein sequences rich in alanine (Ala) and glycine (Gly) amino acids. These proteins are stored in the glands as concentrated solutions (30–50 wt %). A combination of physicochemical and rheological processes (including ion exchange and elongational flow) cause the protein molecules to self-assemble and transform from an unaggregated and soluble state in the gland, through an intermediate liquid-crystalline state, into an insoluble fiber, while passing through the spinning canal.

After spinning, the dry silk fibers consist of stiff nanometer-scale crystals embedded in a softer protein matrix. The nanocrystals are formed by the aggregation of hydrophobic polyalanine regions into a highly ordered crystalline β-sheet via self-assembly, while the glycine-rich protein regions form the softer rubbery matrix. This combination of hard crystallites embedded in a soft matrix is considered the primary reason for the exceptional mechanical strength and extensibility of dragline fibers. It is still not clear how spiders and silkworms prevent and regulate the formation of β-sheet aggregates during the spinning process, but understanding of the mechanism is of great relevance for protein folding. An understanding of this mechanism could help in developing therapies for protein-folding diseases, such as Alzheimer's, which result from the uncontrolled aggregation of proteins into β-sheet aggregates known as amyloid plaques.

Mimicking spider silk. Although spidroin I and spidroin II have been sequenced, a complete understanding of the various physicochemical processes during the spinning of silk fibers is still lacking. Before we can completely mimic spider silk, we must fully understand the biochemistry, processing strategy, and molecular basis of the mechanical properties of silk. Advances are being made on different fronts to emulate different aspects of spider silk. Three main approaches are being pursued to make materials with mechanical properties comparable to dragline silk. The first approach is to use recombinant DNA methods to produce proteins with structure similar to spidroin I and II in other organisms, including potato, tobacco, *Escherichia coli*, and goat's milk. The second approach is to synthesize high-molecular-weight block copolymers that mimic the molecular structure of silk proteins. The third approach is to develop strong, yet extensible synthetic composites by reinforcing elastomeric polymers with inorganic nanoparticles. To date, these approaches have had limited success. One resulting application is the spinning of fibers from recombinant silk proteins.

In addition to the material aspects, the silk-spinning process is far superior and more sophisticated than industrial spinning processes for making polymeric fibers. Conventional industrial processes, such as wet spinning or dry spinning, operate only under very controlled conditions during which the polymer filaments are rapidly solidified or "quenched" to provide sufficient mechanical strength to be drawn into fibers. Spiders are able to spin silk fibers under widely different ambient conditions, and the silk filaments rarely break during spinning. Even in the wet state during spinning, the silk filaments are strong enough to support the weight of the spider. Microscopic analysis shows that the fibers have radially graded mechanical properties, with aligned, stiffer molecules surrounding a softer elastic core. Spiders also produce different silk fibers using the same apparatus through highly regulated protein self-assembly. A complete understanding of spiders' spinning processes would help in developing new technologies that produce nanometer- and micrometer-scale fibers from self-assembling polymers and nanocomposite materials.

For background information *see* AMINO ACIDS; ARANEAE; ELASTICITY; FIBROUS PROTEIN; MANUFACTURED FIBER; NANOSTRUCTURE; NATURAL FIBER; POLYMER; PROTEIN; SILK; STRENGTH OF MATERIALS; STRESS AND STRAIN; YOUNG'S MODULUS in the McGraw-Hill Encyclopedia of Science & Technology.

Nitin Kumar; Gareth H. McKinley

Bibliography. C. L. Craig, *Spiderwebs and Silk: Tracing Evolution from Molecules to Gene to Phenotypes*, Oxford University Press, New York, 2003; J. Gosline et al., The mechanical design of spider silks: From fibroin sequence to mechanical function, *J. Exp. Biol.*, 202:3295–3303, 1999; D. Kaplan et al., *Silk Polymers: Material Sciences and Biotechnology*, American Chemical Society, Washington DC, 1994; F. Vollrath, Strength and structure of spiders' silks, *Rev. Mol. Biotechnol.*, 74:67–83, 2000; F. Vollrath and D. P. Knight, Liquid crystalline spinning of spider silk, *Nature*, 410:541–548, 2001.

Steel construction

Steel is produced in many forms, and its composition varies greatly from one application to the next. Structural steel is the common name for the material that is used for buildings, bridges, and structures in general. With properties that allow it to withstand the effects of high winds, earthquakes, and other types of loads, it has been the construction material for a very large number of structures since 1874. That was the year of completion of the Eads Bridge in St. Louis, Missouri, recognized worldwide as the first application of what is called structural steel today. Since then, the number and types of projects where steel has been the key material has increased exponentially, including such historical examples as the Eiffel Tower in Paris, the Empire State Building in New York, and the Golden Gate Bridge in San Francisco.

Over the past 10 to 15 years, the material itself and how it is used in major projects have undergone significant changes. Production processes have become very efficient, making the material more versatile and economical. Through the use of scrap as the main ingredient in the production, steel is now highly environmentally friendly ("green"). Steel is therefore a material that satisfies the most stringent "green" construction demands.

Some of the greatest advances have taken place in the areas of analysis and design of structures. Sophisticated software has allowed structural engineers to design structures that could not be analyzed, much less fabricated and built, even as recently as 10 years ago. Building codes and standards have followed suit, recognizing that safety and economy are critical to the success of any project.

Improved structural steel. The steels of the past needed relatively high contents of carbon to reach the necessary strength. They also tended to exhibit significant variations in the amounts of chemical components that make up the matrix of the material. In addition, nonmetallic inclusions, such as manganese sulfides (MnS), sometimes created problems for fabrication operations such as welding. These steels were not uniform as commonly assumed and applied by engineers in design and construction. In most cases, the structures were built without problems; in other cases, small or large-scale failures took place.

Today's structural steel is produced in electric arc furnaces, using mostly scrap. The iron ore, limestone, and coke of the past are now used only to a very limited extent in steel mills in the United States and many other countries. Instead of developing the steel products via casting into very large blocks (ingots), now the production is based on continuous casting, a process that yields products that are much closer in size and shape to the final version. As a result, today the steel has much less variation in the characteristics that are central to the success of any construction project. The microscopic structure of the steel is nearly uniform, impurities occur rarely and are much smaller, and the major strength properties (mechanical properties) that are essential to any design are much less variable. The appropriate steel strength is reached through the use of alloying elements, such as manganese, molybdenum, and columbium, and the amount of carbon is typically less than half of what the steels of 15 years ago used. Steel is therefore easier to fabricate, especially in welding, it offers increased resistance to brittle fracture (toughness), and the designer can count on a material that exhibits much less strength variability. These characteristics are essential for structures that will be subjected to earthquakes, high winds, or the effects of the highly variable (cyclic) loads associated with trucks on bridges.

An important example of an advanced metallic material is high-performance steel, which exhibits the strength and performance requirements that are needed to withstand all types of environmental effects, including corrosion. Originally developed for bridges as an outgrowth of research intended to provide advanced steel for the hulls of submarines and other naval vessels, such steels are now commonly used in buildings. They are available in a range of strengths.

The structural steel of today is a much better defined material than what was used even as recently as 10 to 15 years ago, with near uniformity in all its properties. As a result, structures are being built that reflect great artistry, complexity, and innovative uses of the material, and they are safer and more reliable, environmentally advanced, and more economical than those of the past were.

Analysis techniques. Analyzing a structure involves determining the forces and deformations that are developed in the elements that make up the framework, as a result of weight of the structure, its occupants, and similar effects of gravity (gravity loads), as well as the effects of the wind and earthquakes that typically act transversely to the structure (lateral loads). Disastrous loads in the form of the effects of blast and fire are now also taken into account. *See* BUILDING SAFETY DESIGN.

Because of the limitations of early computational equipment, such as slide rules, elementary calculators, and early forms of computers, structures in the past were modeled as two-dimensional assemblies of elements connected to each other through simple joints. Using fixed (static) loads, this approach worked well for many years, although it could not be used efficiently for some complicated structures, and not at all for others. As a result, the buildings and bridges that were analyzed in this fashion were not as economical as they could have been.

The advances that have been made for computer software over the past 10 years are very impressive. Structures are now commonly analyzed as three-dimensional assemblies being subjected to the cyclic effects of wind, earthquakes, and other forms of loads. The limitations of having to treat a material as elastic (that is, where all deformations revert to zero once the loads are removed) are now gone, since full inelastic response is built into the advanced computer programs. This is particularly important when considering the effects of large-scale hurricanes, earthquakes, and other disasters. It also allows the designers to assess the energy absorption capacity of the structure. The ability of steel to absorb extreme environmental demands is a critical feature of the material, since large deformations will not necessarily lead to failure, given that the material is capable of deforming without fracturing. Referred to as the ductility of the steel, this characteristic translates into structures that are highly efficient in resisting the most demanding load effects. Ductility is a unique feature of steel and one of the key reasons for the excellent performance of steel-framed structures of all kinds.

Figures 1 and **2** show structures that have been designed for the demanding conditions of Seattle, Washington, and Taipei, Taiwan, respectively. The analyses and design details reflect state-of-the-art applications; that is, the structures are very complex and yet highly satisfying examples of innovative architecture and engineering in steel.

Fig. 1. Seattle Public Library, completed in 2005. (*Photo courtesy of Michael Dickter/ Magnusson Klemencic Associates*)

State-of-the-art design codes. The demands for safety, reliability, and economy of the large number of structures that are built for service in locations and environments of great variety necessitate uniformity of the criteria that are used by design engineers. At the same time, it is critical to avoid stifling innovative concepts and structural applications. For example, a simple warehouse does not require advanced analysis techniques; whereas a multistory structure, a building covering a large area (large footprint), or a major bridge need the most advanced design evaluations that address all the details of the performance of the structures and their construction materials.

This is where codes and standards enter the design process. Many types of criteria apply to all structures, and some are relevant only to special cases. But above all, the public users of the structures must be assured that safety is paramount, and this is achieved with codes and standards that apply to all design situations. It is simply impossible to achieve safety and at the same time to allow a designer complete freedom to do whatever is regarded as "necessary." The design process is so complicated that it is not possible for any individual to be aware of all the needs that must be satisfied to meet the necessary safety and reliability requirements.

Seventy-five years ago, design codes for steel structures were published by individual cities, counties, or states, creating a complicated system that was prone to failure. The American Institute of Steel Construction (AISC) was founded in 1921 specifically to develop a set of uniform design criteria for application throughout the United States. The first code was issued in 1923. Since that time, a number of editions of increasingly accurate and practical codes have been issued. In 2005, a milestone was reached for steel design codes: a unified, highly advanced, and practical set of design criteria was published. The 2005 AISC code (specification) is the most advanced

Fig. 2. Taipei 101, the tallest building in the world. (*Courtesy of Taipei Financial Center Corporation and Thornton-Tomasetti Group, Inc.*)

in the world today. It treats loads, strengths, and materials in the most realistic fashion, recognizing that nothing is fixed: the design parameters include random variations that must be addressed in order to ensure satisfactory safety and reliability. The 2005 AISC code ensures that steel structures in the United States will continue to be among the most advanced in the world.

For background information *see* ARCHITECTURAL ENGINEERING; METAL, MECHANICAL PROPERTIES OF; METAL CASTING; METALLURGY; PLASTIC DEFORMATION OF METAL; STEEL; STEEL MANUFACTURE; STRUCTURAL ANALYSIS; STRUCTURAL DESIGN; STRUCTURAL STEEL; WELDING AND CUTTING OF MATERIALS in the McGraw-Hill Encyclopedia of Science & Technology.

Reidar Bjorhovde

Bibliography. R. Englekirk, *Steel Structures: Controlling Behavior through Design*, Wiley, 1994; J. E. Gordon, *Structures, or Why Things Don't Fall Down*, Dacapo Paperbacks, 1978; H. Petroski, *To Engineer Is Human*, Vintage Books, 1992; M. Salvadori, *Why Buildings Stand Up: The Strength of Architecture*, Norton, 1980.

Sum-frequency vibrational spectroscopy

Interfaces play an important role in the physical world. The air-water interface is involved in evaporation from the seas and the condensation of pollutants on rain droplets, causing acid rain. The air-water interface is also crucial to the functioning of our lungs. Liquid-liquid interfaces are used in chemical extractions and are an integral part of oil-water emulsions found in food products. Solid-liquid interfaces are of great technological importance in fields such as lubrication and catalysis. The ability to examine the properties of interfaces in biology is also important. This is particularly true of biological membranes, which represent the interface between the cell interior and the extracellular environment. Analyzing the composition, structure, and dynamics of molecules at an interface is not an easy task due to the extreme difficulty in isolating the interfacial molecules or atoms from those of the bulk material (solid, liquid, or gas).

Spectroscopic techniques can be used to obtain information about the composition and structure of matter, as well as reveal information on the dynamics of molecules. All spectroscopic methods use radiation of different wavelengths to analyze a sample. For example, infrared (IR) spectroscopy uses radiation in the wavelength region of 2500 to 25,000 nanometers (or 4000 to 400 cm^{-1}). In this region of the light spectrum, the energy of the photons corresponds to the vibrational motions of atoms within molecules. The motion of two atoms linked by a covalent bond will have a characteristic vibrational frequency dependent upon the mass of the nuclei. As a result, IR spectroscopy can be used to characterize the composition of organic and inorganic molecules by measuring the absorption of light as a function of wavelength (or frequency), providing a unique "fingerprint" of the substance. IR spectroscopy is considered a linear spectroscopic technique; that is, only one photon is absorbed per vibration in a single molecule. IR absorbance spectroscopy is extremely well suited for the analysis of bulk materials; however, an increasing number of technological and biological problems require molecular-level information on the structure, composition, and dynamics of molecules at a surface or at the interface of two materials.

Interface specific vibrational spectroscopy. Sum-frequency vibrational spectroscopy (SFVS) is a coherent spectroscopic technique which has the chemical sensitivity of IR spectroscopy, yet is also surface-selective. The theoretical foundations of SFVS are well established in the literature and will not be thoroughly discussed here. SFVS is a nonlinear optical spectroscopy which, like IR spectroscopy, can provide information on the structure and organization of molecules. It is the nonlinear aspect of SFVS which gives the method its surface specificity. Experimentally, SFVS is performed by overlapping, both spatially and temporally, a visible (ω_{vis}) and tunable IR (ω_{IR}) laser source on a surface where they combine to produce a third photon at the sum of their respective frequencies ($\omega_{sum} = \omega_{vis} + \omega_{IR}$) [**Fig. 1**]. A sum-frequency spectrum is obtained by tuning the IR frequency through the vibrational resonance of the molecules that make up the interface and measuring the intensity of the resulting light at ω_{sum}. A laser source is used due to the low conversion efficiency of the process; that is, about 2 in 100,000,000 photons will combine to produce a photon at ω_{sum}. In addition, SFVS is a coherent spectroscopy which requires a coherent light source, such as a laser, to be observed. A coherent light source produces light in which the phases of the photons (or light waves) are in unison (**Fig. 2**).

The interface specificity of SFVS arises from the fact that an interface (or surface) represents a break in the otherwise continuous distribution of molecules (or atoms), thus producing an asymmetry in the arrangement of molecules. SFVS is a spontaneous optical process which does not require the adsorption of a photon. Instead, the molecules experience the oscillating electric field and respond to its presence by reemitting a photon which is at the sum of the two incident frequencies. This process can be approximated by an electric dipole responding to an applied oscillating electric field. Consider, for example, bulk water with its associated dipole moment (Fig. 2). When a coherent optical field is applied to the sample, the dipoles will oscillate in response to this stimulus. If two coherent light fields are applied, as with SFVS, some dipoles will oscillate at the sum of the two frequencies, emitting a photon at (ω_{sum}). The key to the surface specificity of SFVS is that in a bulk material the emitted photons at ω_{sum} are not coherent, which is to say they are not in phase with each other. When the various light waves generated from the individual molecules (of water in this example) combine, they destructively interfere, resulting in no detectable SFVS emission from the sample. Another way of looking at this is to say that for every water molecule pointed in a particular direction (*A* in Fig. 2*a*), there will always be another water molecule which has the opposite orientation (B in Fig. 2*a*). The combined effect of these oppositely oriented dipoles is that they will generate SFVS light of opposite phases, which will cancel each other out. At an interface, molecules tend to orient preferentially in a given direction, breaking the random arrangement found in the bulk (Fig. 2*b*). It is this loss of symmetry which restricts SFVS to an interface, where the inversion symmetry of the bulk phases is broken, making the technique surface-specific.

In the past 10 years, SFVS has proven to be a powerful and versatile surface analytical tool. Applications of the technique include the characterization of numerous liquid–air interfaces. Most notable are

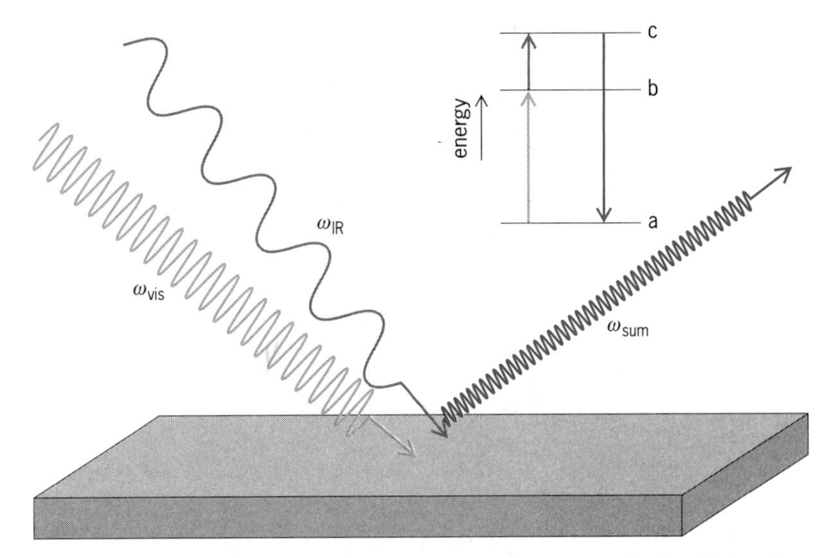

Fig. 1. Representation of an SFVS experiment in which a visible laser (ω_{vis}) and infrared laser (ω_{IR}) are directed at a surface, producing sum-frequency light (ω_{sum}). An energy-level diagram is also depicted (longer arrows represent greater energy) showing the addition of the visible and infrared photons to produce a sum-frequency photon. If the energy gap between b and c corresponds to a vibrational frequency in a molecule, the intensity of the transition from c to a will be greatly enhanced, increasing the intensity of the light emitted at ω_{sum}.

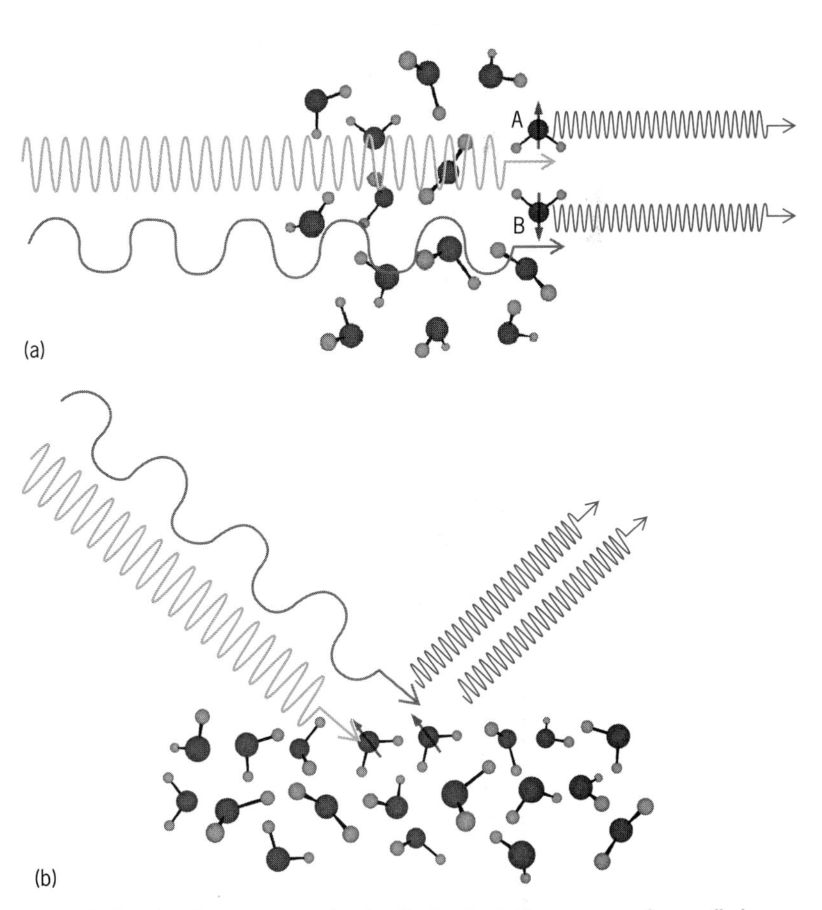

Fig. 2. Bulk and surface response of water dipoles due to the presence of an applied visible (light color) and infrared (gray) coherent light source. (*a*) In the bulk water sample, the antiparallel orientation of the water dipole moments (A and B) causes the generated sum-frequency to be out of phase, and thus cancel. (*b*) At a surface, the break in the random distribution of water molecules induces a preferential alignment of the individual water molecules which can then give rise to a coherent sum-frequency (dark color) response.

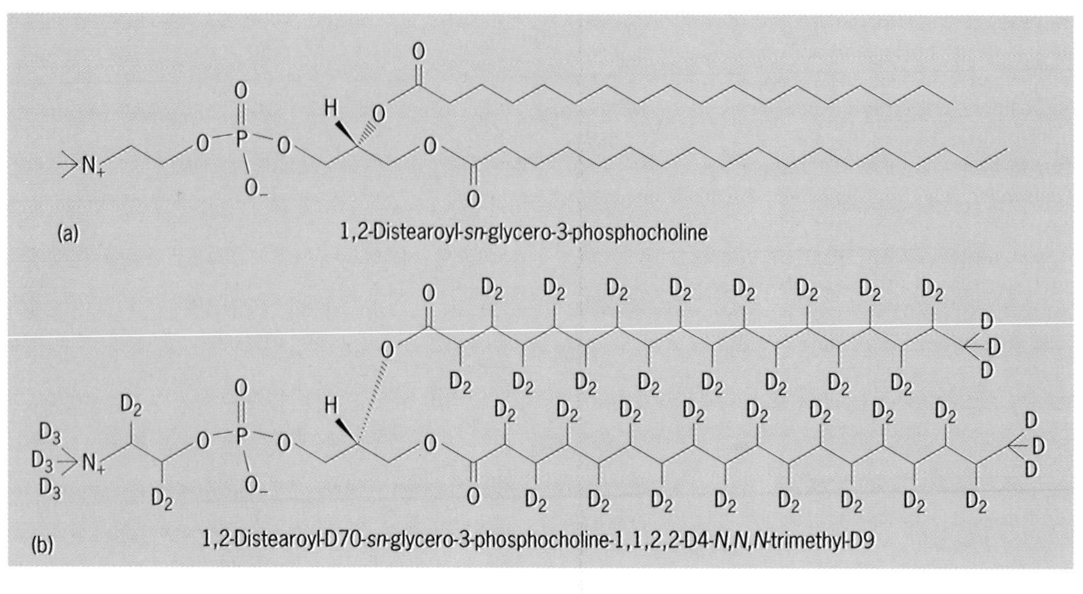

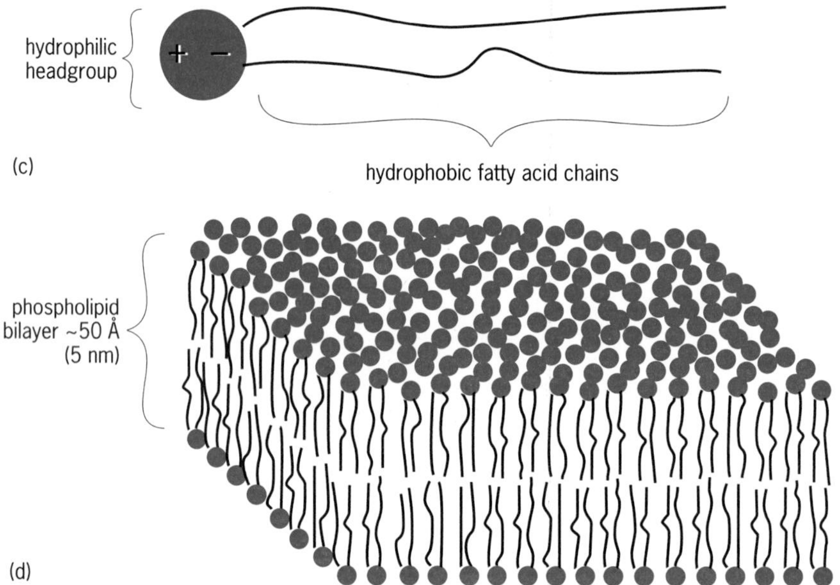

Fig. 3. Structure of (*a*) a typical phosphocholine lipid and (*b*) a perdeuterated analog of the same lipid. (*c*) The hydrophilic headgroup and hydrophobic tail of a typical phospholipid. (*d*) The bilayer structure of a phospholipid membrane in which the hydrophobic tails are in contact with each other and the hydrophilic headgroups are in contact with the aqueous environments of the cell interior and exterior.

detailed studies of the air–water interface, which have provided new insights into the structure and hydrogen-bonding interactions of this ubiquitous interface. In many instances, such as the analysis of the neat liquid–air interface, SFVS has proven to be the only analytical tool capable of providing vibrational spectroscopic information. Detailed studies of molecular adsorption to a variety of surfaces have been done, including surfactant adsorption at the air–water and oil–water interfaces, providing new insight into the action of soaps and detergents. The most recent applications of SFVS have been in the characterization of biological systems, in particular lipid membranes.

Measuring lipid dynamics. In biological systems, the most important interfaces involve the cell membrane. The cell membrane is a fluid barrier which partitions the cell from its environment and subdivides and compartmentalizes functions within a cell. The membrane is primarily composed of proteins, cholesterol, and phospholipids, of which 1,2-diacyl-*sn*-glycero-3-phosphocholine is the most prevalent lipid species found in eukaryotic cells. These lipids are arranged in a bilayer structure composed of an inner (cytosolic) and outer (extracellular) leaflet (**Fig. 3**). Understanding the movement of lipid

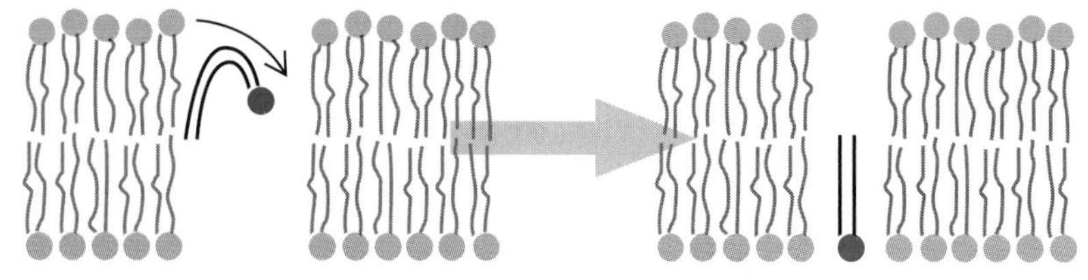

Fig. 4. Lipid translocation (flip-flop) across a phospholipid bilayer.

species across cellular membranes is a fundamental goal in molecular biology. Since the biosynthesis of phospholipids is generally confined to the cytosolic leaflet of a membrane, the continual growth of cells requires the alteration of the outer leaflet by the translocation of phospholipids from the inner leaflet, commonly called flip-flop (**Fig. 4**). Lipid flip-flop has direct relevance to cell apoptosis (programmed cell death), the viral infection of living cells, the functioning of antibiotics and other drugs, and the regulation and growth of cells. As a result, the ability to measure lipid flip-flop is of great importance.

The symmetry constraints imposed on SFVS provide a highly sensitive method for determining the asymmetry in lipid bilayers and means to directly measure lipid flip-flop. Since SFVS is a coherent vibrational spectroscopy, the arrangement of the transition moments of the molecules at the interface has a dramatic effect on the measured response. For example, looking at the symmetric stretch (ν_s) of the terminal CH_3 groups on the lipid fatty acid chains of a typical phospholipid, the transition dipole moment lies along the C-C bond (**Fig. 5***a*). If another lipid chain is placed antiparallel to the first, in an arrangement found in a typical lipid bilayer, the sum of the two transitions will cancel each other due to their opposing phases. The result is that no SFVS response will be observed from such an assembly.

For a lipid bilayer composed of identical lipids in the inner and outer leaflets, nearly complete destructive interference of the terminal fatty acid CH_3 ν_s is observed due to the antiparallel orientation of the transition dipole moments. If an artificially asymmetric structure is constructed in which a lipid monolayer is placed in contact with a lipid monolayer

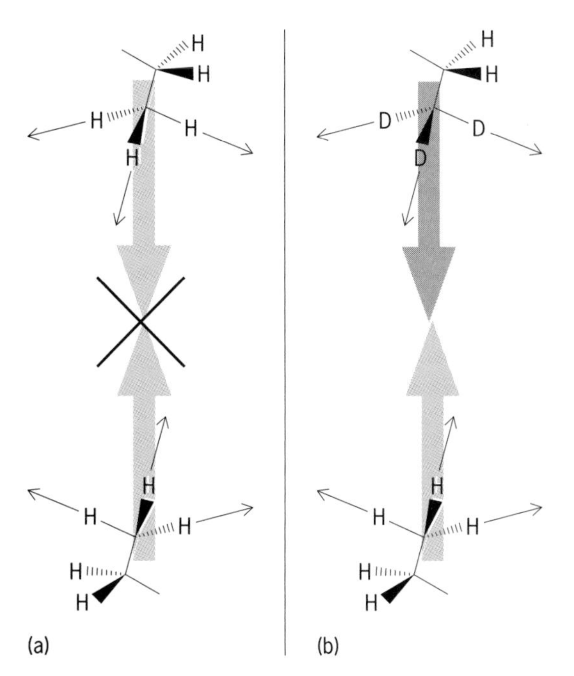

Fig. 5. Representation of (*a*) the dipole cancellation of the CH_3 ν_s on the terminus of the lipid chains of a phospholipid in a bilayer structure and (*b*) the absence of interference when one of the lipids is deuterated (CD_3), altering the vibrational frequency.

of perdeuterated lipids, the CH_3 ν_s resonance will no longer experience destructive interference, due to the $\sim$760 cm^{-1} shift to lower frequency of CD_3 (Fig. 5*b*). By measuring the intensity of the CH_3 ν_s with time, changes in the membrane lipid composition due to direct exchange (flip-flop) between leaflets can be followed. These experiments have provided new information on the dynamics of lipids in membranes and the energetics of lipid flip-flop.

Outlook. SFVS is a powerful and versatile analytical tool for characterizing interfaces. In some cases, SFVS is the only analytical spectroscopic technique available that can provide information on the structure, composition, or dynamics of molecules at an interface. The growing use of the technique will no doubt make an impact in many areas of surface science. SFVS will also continue to play an important role in expanding our understanding of interfacial phenomena in biological systems and in the biosciences in general.

For background information *see* CELL MEMBRANES; INTERFACE OF PHASES; LASER SPECTROSCOPY; LIPID; MOLECULAR BIOLOGY; PHOSPHOLIPID; SPECTROSCOPY; SURFACE AND INTERFACIAL CHEMISTRY in the McGraw-Hill Encyclopedia of Science & Technology.

John C. Conboy

Bibliography. K. B. Eisenthal, Liquid interfaces probed by second-harmonic and sum-frequency spectroscopy, *Chem. Rev.*, 96(4):1343–1360, 1996; J. Liu and J. C. Conboy, Direct measurement of the transbilayer movement of phospholipids by sum-frequency vibrational spectroscopy, *J. Amer. Chem. Soc.*, 126(27):8376–8377, 2004; P. B. Miranda and Y. R. Shen, Liquid interfaces: A study by sum-frequency vibrational spectroscopy, *J. Phys. Chem. B*, 103(17):3292–3307, 1999; R. J. Raggers et al., Lipid traffic: The ABC of transbilayer movement, *Traffic*, 1(3):226–234, 2000; G. L. Richmond, Vibrational spectroscopy of molecules at liquid surfaces and interfaces, *Anal. Chem.*, 69:536A.2P, 1997; Y. R. Shen, *Nonlinear Spectroscopy for Molecular Structure Determination*, chap. 10 in R. Field et al. (eds.), *IUPAC Chemical Data Series*, Blackwell Science, Oxford, UK, 1998.

Sumatra-Andaman earthquake

On December 26, 2004, a 1300-km (800-mi) length of sea-floor boundary between two tectonic plates ruptured in the Sumatra-Andaman earthquake. This earthquake was one of the five largest earthquakes of the past 100 years and the largest since 1964. Seafloor deflection from this massive earthquake created one of the largest and most far-reaching tsunamis observed by humans. Estimated deaths along the coastlines of Indian Ocean nations approached 300,000, marking this as one of the most lethal natural disasters in human history.

Analysis. Sumatra-Andaman was a "megathrust" earthquake, in which a large portion of one tectonic plate (Indian Ocean plate) thrust beneath an adjoining plate (Burma microplate) in meter-scale

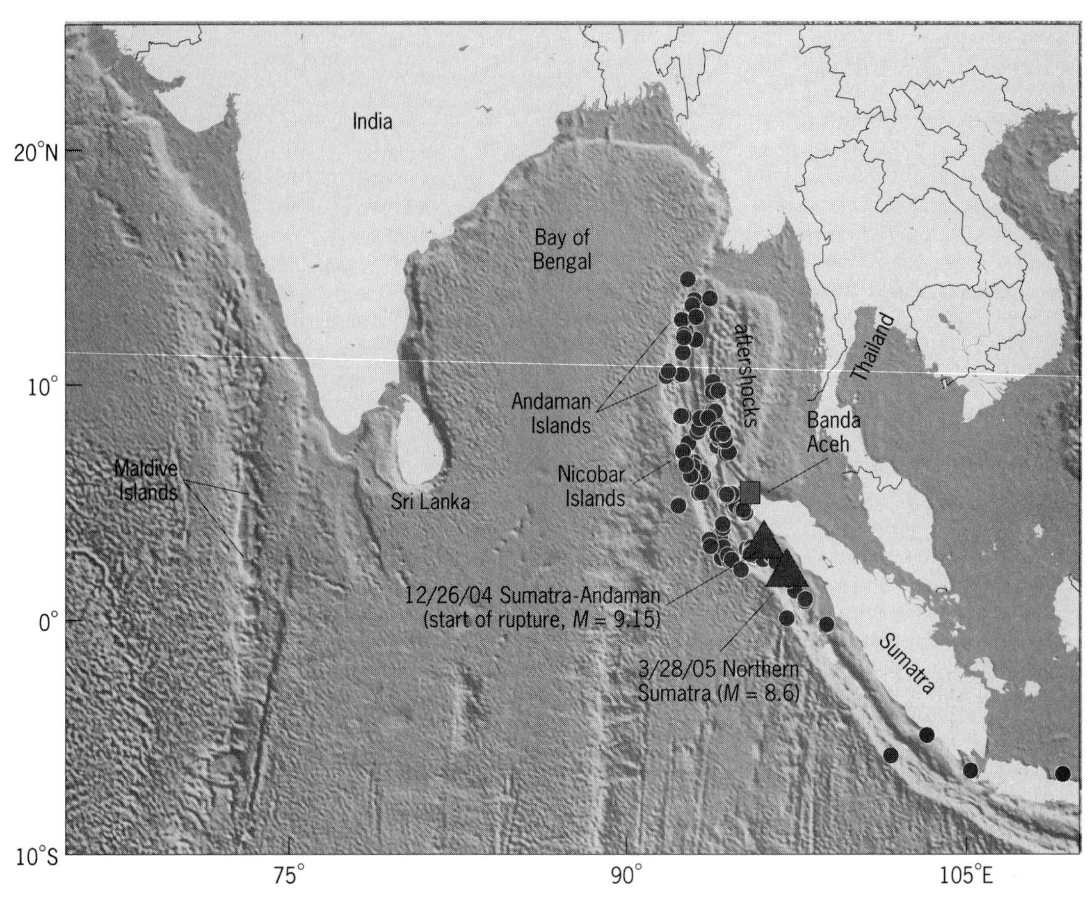

Fig. 1. Map of the region most affected by the Sumatra-Andaman earthquake and tsunami. Triangles indicate initial rupture locations of the great earthquakes on 12/26/04 and 3/28/05. Circles indicate aftershocks of the 12/26/04 earthquake. The aftershocks outline the area of the fault rupture for the main shock, extending over a 1300-km (780-mi) segment of the plate boundary along the Indian Ocean coastline of Sumatra.

jerks, grinding forward on its downward journey into Earth's mantle. Seismic waves from this event were recorded at a worldwide network of seismometers. Seismologists used data from thousands of seismometers in the following weeks and months to reconstruct how the fault rupture progressed. Starting offshore of northern Sumatra, one of the islands that make up Indonesia, the rupture raced northward at rates between 2 and 3 km/s (1.2 and 1.9 mi/s) for more than 500 s, eventually extending beneath the Nicobar and Andaman island chains (**Fig. 1**). The fault dip (angle) was shallow (8-15°), so most fault motion occurred at depths less than 30 km (19 mi). Scientists interpret large pulses within the rupture to suggest patches of the fault zone that moved up to 20 m (66 ft). On average, 7 m (23 ft) of fault motion occurred offshore of Sumatra, decreasing progressively to an average of 2 m (6.6 ft) beneath the Andaman Islands. Additional fault motion likely occurred more slowly in the north, because permanent displacements of the Earth's surface in the surrounding landmasses, as measured with Global Positioning Satellite (GPS) systems, are best explained with an average of 5 m (16 ft) of total fault motion under the Andaman Islands.

Earthquake signals are recorded as a sequence of pulses. P (pressure) waves arrive first, and S (shear) waves arrive next. Typically, the seismic surface waves arrive some minutes later, like large ripples on a pond of water. Every point on Earth's surface moved back and forth at least 1 cm (0.4 in.) as the largest surface waves passed by. Earthquake size scales with seismic moment M_0, which measures, in a rough sense, the force on the fault multiplied by how far the fault slips (moves relative to a point on the opposite side of the fault). Initial estimates for M_0 varied from 4.0×10^{22} to 10.0×10^{22} newton-meters, depending on which seismic waves were measured. Source estimates, which allow for variable slip and variable dip on the fault, suggest that $M_0 = 6.5 \times 10^{22}$ N · m, equivalent to a 1300 × 200 km (800 × 120 mi) fault rectangle slipping more than 8 m (26 ft). This seismic moment corresponds to Richter magnitude 9.15. The largest earthquake previously measured with modern digital seismometers was the 2001 Peru earthquake, with magnitude 8.4. Each unit on the logarithmic Richter scale corresponds roughly to a factor of 30 in seismic moment, so the 2004 Sumatra-Andaman earthquake was more than 10 times larger than the 2001 Peru earthquake. In fact, its moment exceeded the total moment of all earthquakes in the previous 15 years.

The 2004 Sumatra-Andaman earthquake produced the clearest, most detailed seismographic data set in history. To detect the progression of fault rupture from south to north, researchers studied subtle differences in P waves from different recording locations, direction-dependent Doppler shifts in the period of the largest S waves, and the migration of ocean-acoustic sources caused by sea-floor vibrations along the trace of the fault. At long periods, the earthquake caused the Earth to vibrate like a ringing bell, setting off an unprecedented multitude of observable free oscillations. The "breathing mode" oscillation, in which the Earth expands and contracts every 20.5 min, was raised to an initial amplitude of 49×10^{-6} m or 49 micrometers, roughly twice the diameter of a pollen grain, and remained visible in seismic data for months. The breathing-mode oscillation persisted until March 28, 2005, when the Northern Sumatra earthquake ($M = 8.7$), the second largest earthquake since 1964, excited it again. A southward transfer of stress from the 2004 Sumatra-Andaman fault zone may have triggered this later quake.

The last megathrust earthquake, on March 28, 1964, in the Gulf of Alaska, destroyed the nearby city of Anchorage and jammed seismographs worldwide for hours, leaving many sensitive galvanometer springs actually broken. By contrast, on December 26, 2004, most seismometers in global networks recorded on scale. Data telemetered by satellite and the Internet allowed the Pacific Tsunami Warning Center (PTWC) to estimate the Richter magnitude within 11 min of the earthquake onset, when the P waves had barely reached halfway around the Earth. Although the first PTWC magnitude estimate was too low, it triggered tsunami-warning alerts. Tragically, these alerts were directed toward Pacific Ocean nations, not those bordering the Indian Ocean, because international tsunami warning protocols had been established only in the Pacific region. A more accurate magnitude estimate was available within hours from surface waves recorded at dozens of stations.

Tsunami. The Sumatra-Andaman earthquake unleashed a devastating tsunami on the coastlines of Sumatra, Thailand, India, Sri Lanka, and nations as far away as Kenya in eastern Africa. Even though fault motion was largely lateral, it caused a meter or more deflection of the sea floor, which displaced the ocean water and sea surface. Tugged by gravity, the sea surface distortion became a tsunami, propelled laterally at speeds of 650 km/h (400 mi/h) in the open ocean. By chance, a satellite fly-by measured sea-surface displacements of 1 m (3.3 ft) while the tsunami raced across the Bay of Bengal (**Fig. 2**).

Wind-driven waves involve water motion only in the top layer of the ocean, but a tsunami involves motion from the sea surface to the sea floor. As the tsunami approached the coastlines of Sumatra, Thailand, India, and Sri Lanka, the energy distributed throughout the water column became concentrated in shallow coastal waters, amplifying the wave to heights of 3–30 m (10–100 ft), depending on the location. The largest tsunami struck the northwest coast of Sumatra within minutes of the earthquake, washing away numerous coastal villages and much of the city of Banda Aceh. Nearly 2 h after the earthquake, the tsunami struck the western coast of Thailand and the eastern coasts of India and Sri Lanka, inundating thousands of kilometers of coastline. Passing over the low-lying Maldive Islands without great amplification, the tsunami retained sufficient amplitude to cause damage along the coasts of Tanzania and Somalia in Africa.

Close to the earthquake's shaking, most coastal residents of northwest Sumatra had little time for warning and evacuation. Farther from the source, there was more time to react, even in the absence of a formal tsunami warning. Tsunamis typically consist of several peaks and troughs, which alternately inundate and expose the shoreline. The times between peaks and troughs varies between 15 and 30 min, depending on the shape and depth of the ocean basin and the earthquake fault geometry. If the approaching tsunami begins with a wave trough, the first sign of impending danger is, paradoxically, an inexplicable retreat of the sea. On the beaches of Thailand, India, and Sri Lanka, many unsuspecting villagers and tourists followed the retreating water out of curiosity, because it revealed rocks and sea creatures that were otherwise hidden by water. The proper response to the sea's retreat was to run inland to high ground. Only a handful of people on that fateful morning recognized the danger and warned those around them.

Tsunamis in the deep ocean have wavelengths of hundreds of kilometers, too broad to have wave crests noticeable to the human eye. When a tsunami approaches land, it slows drastically in shallow water and onrushing water builds into a wave face. Ordinary wind-driven waves develop crests and breakers, but a tsunami is the leading edge of a thick, wide slab of water, arriving with a force sufficient to propel boats, cars, trees, and small structures far inland. Many tsunami victims drowned, but many were crushed by onrushing debris or hurled into structures by the advancing water. Casualties and damage were greatest in Sumatra (230,000 deaths estimated), numerous in Thailand (8500 deaths), India (16,500 deaths), and Sri Lanka (36,600 deaths), and less numerous in more distant affected areas (Somalia and Tanzania, 150 and 10 deaths, respectively). The steep undersea flanks of some midocean islands inhibit tsunami amplification, so damage and casualties on such islands were less severe; for example, 82 deaths were reported in the Maldives.

The Nicobar and Andaman islands lay directly above the slipping fault zone. The 2004 quake tilted many islands because of undersea plate flexure. Subsidence of roughly 1 m (3.3 ft) on the western coasts was accompanied by uplift of roughly 1 m on eastern coasts. On coastlines that lay above or immediately beside the rupture zone, nearshore regions dropped below sea level or rose above it, with large permanent excursions in some cases. The lighthouse at the southern tip of Great Nicobar Island subsided more than 4 m (13 ft), its base falling into the surf zone.

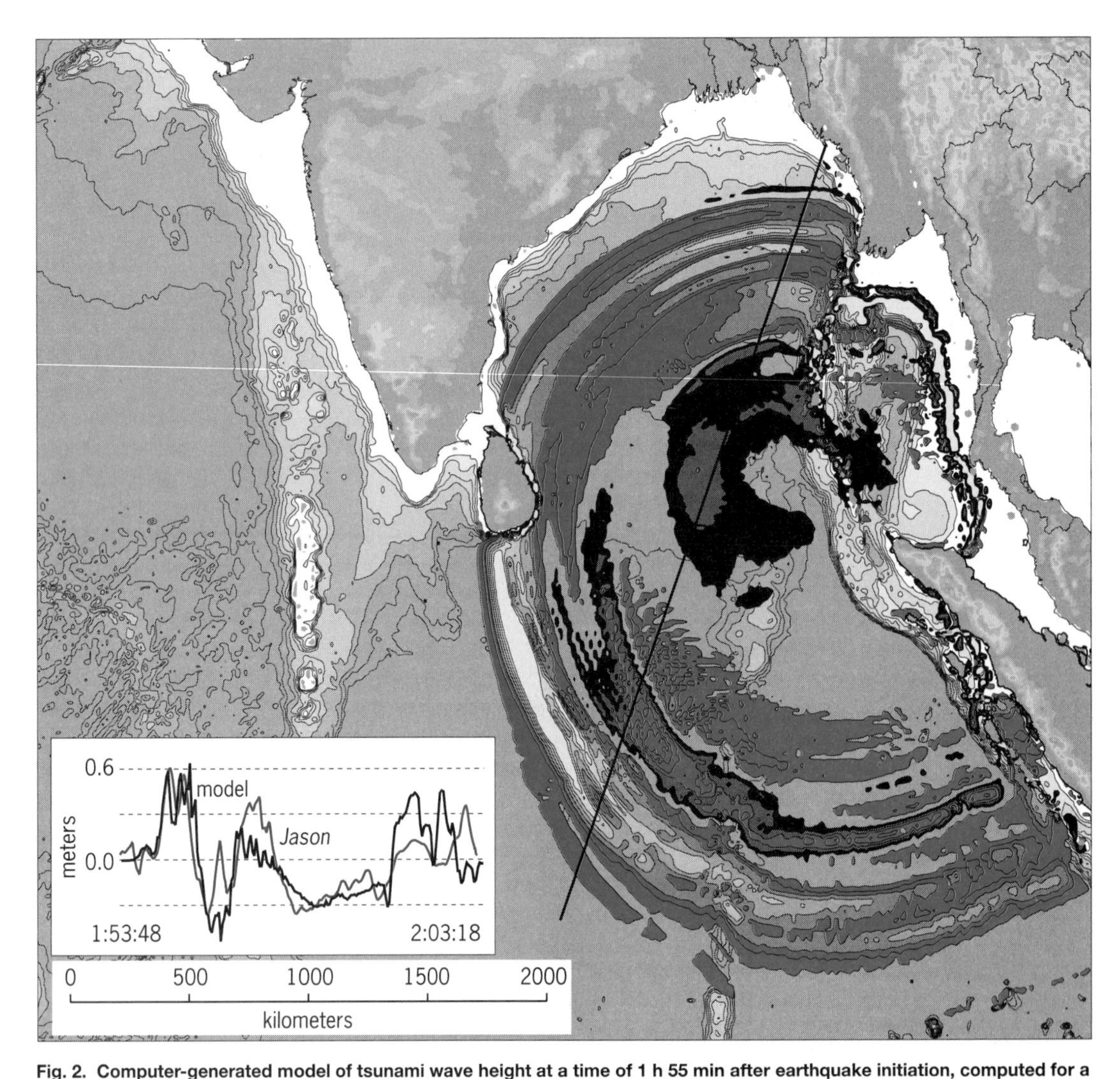

Fig. 2. Computer-generated model of tsunami wave height at a time of 1 h 55 min after earthquake initiation, computed for a fault rupture model derived from seismic data recorded worldwide. Color in the map indicates positive ocean wave height, and black, negative. The diagonal line is the track of the *Jason* satellite that passed over the region at about this time (10 min of actual transit time along the profile). Inset shows the predicted (black) and observed (color) tsunami waves. (*Courtesy of Steven Ward, University of California, Santa Cruz*)

Far from the fault zone, the coastline remained stable after the tsunami's inundation retreated.

Outlook. The Sumatra-Andaman earthquake released hundreds of years of pent-up plate motion, which accumulates more slowly there than in most subduction zones. In the Sunda Trench offshore central and southern Sumatra, plate motion is faster and great earthquakes may be more frequent. In 1833 and 1861, great subduction-zone earthquakes rocked central Sumatra and generated substantial tsunamis. Similar earthquakes in the twenty-first century are a threat, especially to Indonesia and Australia. This hazard motivates the extension of the Pacific Ocean tsunami warning system, now based in Hawaii, to the Indian Ocean. However, all the Earth's oceans present a tsunami hazard at some level. Scientists have proposed a worldwide network of oceanic motion sensors as part of an environmental monitoring infrastructure called GEOSS, the Global Earth Observing System of Systems. The present Pacific system uses seismometers to detect and locate large earthquakes and undersea pressure sensors, cabled to land, to detect the passage of a tsunami near the shore. GEOSS would extend and improve this system by using seismometers on the sea floor to detect earthquake waves closer to their source, and floating buoys to monitor tsunami in the open ocean.

For background information *see* AFRICA; ASIA; EARTHQUAKE; FAULT AND FAULT STRUCTURES; INDIAN OCEAN; NEARSHORE PROCESSES; OCEAN WAVES; PLATE TECTONICS; SEISMIC RISK; SEISMOLOGY; TSUNAMI in the McGraw-Hill Encyclopedia of Science & Technology. Jeffrey J. Park

Bibliography. B. Bolt, *Earthquakes*, 5th ed., W. H. Freeman, New York, 2003; W. Dudley and M. Lee, *Tsunami!*, 2d ed., University of Hawaii Press, Honolulu, 1998; S. E. Hough, *Earthshaking Science: What We Know (and Don't Know) about Earthquakes*, Princeton University Press, 2002; K. Sieh and S. LeVay, *The Earth in Turmoil: Earthquakes, Volcanoes, and Their Impact on Humankind*, W. H. Freeman, New York, 1999.

Superconductor wires

High-temperature superconductor (HTS) wires are so efficient that they can carry up to 140 times the power of conventional copper wires of the same size. The potential uses for HTS wires in electric power applications include underground transmission cables, oil-free transformers, superconducting magnetic-energy storage (SMES) units, fault-current limiters, high-efficiency motors, and compact generators. Electricity grid losses have grown to be more than 10% of all electricity generated due to resistance and other losses in conventional equipment, and blackouts highlighted transmission bottlenecks in the United States in 2003. Efficiency and reliability will be enhanced when new transmission technologies are used that have reduced line losses and the capability to carry more current for a given size of conductor. Superconductors have virtually no electrical resistance; therefore, they can carry current with no electrical energy loss. Superconductors are of two types: low- and high-temperature. Low-temperature superconductors (LTS) work at only very frigid temperatures, near 4 K (−452°F). Equipment made with LTS wires can be expensive to operate because they need to be chilled with liquid helium. Equipment with HTS wires, however, works at relatively warmer temperatures, near 77 K (−320°F), and requires liquid nitrogen to operate.

High-temperature oxide superconductors, discovered in the late 1980s, are moving into the second generation of development. The first generation relied on bismuth strontium calcium copper oxide (BSCCO), and the second generation is based on yttrium barium copper oxide (YBCO), which has the potential to be less expensive and perform better under magnetic fields. One of the main challenges in developing high-performance superconductors has been the brittleness of many of the most promising materials and drawing them into wires that can carry current. Yet, recent developments allow fabrication of superconductor wires and tapes.

Fabrication methods. Following the discovery of high-temperature superconductors, notably $(Bi,Pb)_2Sr_2Ca_2Cu_3O_{10}$ (BSCCO or 2223) and $YBa_2Cu_3O_{7-x}$ (YBCO or 123), researchers worldwide have searched for ways to fabricate affordable, flexible wires that will carry high current density. The U.S. Department of Energy's target cost for the conductor is close to the current copper wire cost of $10/kA-meter (where 1 m carries 1000 amperes; it is defined as kA-meter). The near-term goal is to achieve HTS wire 100 m (330 ft) in length, with current-carrying capacity of 300 amperes. Robust, high-performance HTS wire will revolutionize the electric power grid and various other electric power equipment. The American Superconductor Corporation (AMSC) and Sumitomo (Japan) are the leading manufacturers of first-generation (1G) HTS wire, based on BSCCO materials using the oxide-powder-in-tube (OPIT) process. Typically, first-generation HTS wire carries critical currents I_c of over 125 amperes in piece lengths of several hundreds of meters at the standard 0.41-cm (0.16-in.) width and ~210 micrometer thickness. However, due to the higher cost of first-generation wire and the intrinsic properties of YBCO, researchers shifted their effort toward the development of YBCO (second-generation or 2G) wires and tapes. One main obstacle to the manufacture of commercial lengths of YBCO wire has been the phenomenon of weak links; that is, grain boundaries formed by the misalignment of neighboring YBCO grains are known to form obstacles to current flow. By carefully aligning the grains, low-angle boundaries between superconducting YBCO grains allow more current to flow. In fact, below a critical misalignment angle of 4°, the critical current density approaches that of YBCO films grown on single-crystal substrates. First- and second- generation HTS wire architectures are shown in the **illustration**. Typically, second-generation HTS wires have three components: flexible metal substrate, buffer layers, and YBCO superconductor layers.

Several methods have been developed to obtain biaxially textured metal substrates suitable for fabricating high-performance YBCO-coated conductors. They are ion-beam-assisted deposition (IBAD), rolling-assisted biaxially textured substrates (RABiTS), and inclined-substrate deposition (ISD). The industry standard for characterizing the second-generation wire is to divide the current by the width of the wire. With either a 3-μm-thick YBCO layer carrying a critical current density J_c of 1 MA/cm^2 or 1-μm-thick YBCO layer carrying a J_c of 3 MA/cm^2, the electrical performance would be 300 A/cm-width. Converting these numbers to the industry standard of a 0.4-cm-wide HTS wire would correspond to 120 A, which is comparable to that of the commercial first-generation wire available in the market. Increasing the YBCO film thickness or critical current density, or finding a way to incorporate two layers of YBCO (either double-sided coating or joining two YBCO tapes face-to-face) in a single-wire architecture would result in performance exceeding the first-generation materials, that is, a high overall engineering critical current density J_E at 77 K (−320°F). The other important advantages of second-generation wires are that YBCO has better in-field electrical performance at higher temperatures, potentially lower cost, and low alternating-current (ac) losses.

Ion-beam-assisted deposition. In the ion-beam-assisted deposition (IBAD) process, an ion beam is used to grow textured buffer layers onto a flexible but untextured metal, typically a nickel-based superalloy. After the initial announcement of the ion-beam-assisted deposition process to grow textured yttria-stabilized zirconia (YSZ) layers by Y. Iijima and coworkers, researchers at the Los Alamos National Laboratory in 1995 perfected the process and achieved high-performance YBCO films on IBAD-YSZ templates. To date, three IBAD templates, namely YSZ, gadolinium

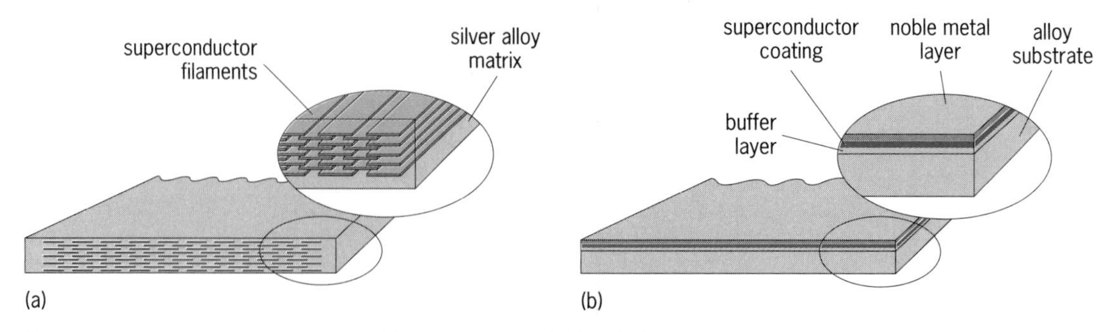

High-temperature superconductor wires. (*a*) First-generation (1G) BSCCO wire multifilamentary composite. (*b*) Second-generation (2G) YBCO-coated conductor tape. (*Courtesy American Superconductor*)

zirconium oxide [$Gd_2Zr_2O_7$ (GZO)], or magnesium oxide (MgO), are being used to make YBCO tapes.

Rolling-assisted biaxially textured substrates. The rolling-assisted biaxially textured substrates (RABiTS) process, developed at Oak Ridge National Laboratory in 1996, uses thermomechanical processing to obtain flexible, cube-textured nickel-alloy substrates. Both buffers and YBCO superconductors are then deposited epitaxially on textured nickel-alloy substrates. The starting substrate serves as a structural template for the YBCO layer, which has substantially fewer weak links. For comparison, wires made by the oxide-powder-in-tube (OPIT) process (1G wire) are limited to a higher cost since the majority component is high-purity silver. However, in the RABiTS or IBAD process, silver is replaced by a low-cost nickel alloy, which allows for fabricating less expensive HTS wires.

Inclined-substrate deposition. In the inclined-substrate deposition (ISD) process, the textured buffer layers are produced by vacuum-deposited material at a particular angle on an untextured nickel alloy substrate. After the discovery of the ISD-YSZ process by K. Hasegawa and coworkers, Sumitomo, Japan, in 1996, THEVA/Technical University of Munich, Germany, in 2003 perfected the reel-to-reel MgO buffer-layer texturing by ISD on Hastelloy® (nickel-based alloy) tape.

Recently, several industries have demonstrated that they can produce second-generation wires 10–100 m (33–330 ft) in length with I_c ranging 50–300 A/cm-width, based on either IBAD or RABiTS technology. In 2005, the THEVA group from Germany produced its first 40-m-class (132-ft) YBCO tape, based on ISD-MgO technology. The 37-m (121-ft), 10-mm-wide wire, based on a nonmagnetic Hastelloy C276 steel tape with an ISD-aligned MgO buffer, exhibited an average critical current of 158 A. This demonstration would correspond to 5846 (158 A × 37 m) A-m.

Present status of wires and tapes. Methods for producing textured templates for growing high-performance YBCO-coated conductor wires include IBAD-YSZ, IBAD-MgO, IBAD-GZO, ISD-MgO, and RABiTS. Using these five templates, high-deposition-rate YBCO processes such as trifluoroacetate-based metal-organic deposition (MOD), metal-organic chemical vapor deposition (MOCVD), and high-rate pulsed laser deposition (PLD) are being used to deposit the superconductor films. The main challenge is to combine the oriented template concept and superconductor deposition process and fabricate high-temperature superconductor tapes in kilometer lengths. Industries from the United States and Japan are leading this area, while industries from Europe, Korea, and China are trying to catch up with them. The present status of the second-generation high-temperature superconductor wires is summarized in the **table**.

Choices that enhance 2G attributes include the substrate and its mechanical properties (especially critical tensile stress); ac loss due to ferromagnetic loss; ferromagnetism; eddy-current loss; resistivity, hysteretic loss, and grain size; HTS deposition process and its throughput, availability of long piece lengths, in-field performance (especially performance in field perpendicular to tape), thick films, small aspect ratio for less hysteretic loss; and stabilizer and its surround stabilizer fully encapsulating wire or stabilizer on one side, rounded edges for direct dielectric integration, or sharp edges.

Per the Coated Conductor Development Roadmapping Workshop II, conducted by U.S. Department of Energy in Washington, DC (July 28- 29, 2003), the following issues related to the development of second-generation HTS wires were identified; and the following geometry of conductor design and engineering issues toward power applications in 2010 were recommended: face-to-face, neutral axis; alternate conductor designs, conducting substrates, two-sided coating, current-carrying capacity of stabilizer, low ac loss; piece length of 100–1000 m (3300–33,000 ft); physical size $\leq$4.1 mm in width; and price (large volume) of $\sim$\$10/kA-m. For 2010, the recommended performance and operating specifications are J_E of 10,000–20,000 A/cm^2 at 30–65 K (-406 to -343°F) and 3 T (or at operating conditions); I_c of 100–200 A at operating conditions; stabilizer design; 200 MPa stress (300 MPa) at 77 K; irreversible strain limit 0.4–0.6% tension; 0.3–1% compression (for magnets); 2–3.5-cm bend diameter; ac loss of 0.25 W/kA-m; n value $\geq$14; and I_c of 1000 A/cm-width at 77 K and self-field.

In summary, five different templates comprising IBAD-YSZ, IBAD-GZO, IBAD-MgO, ISD-MgO, and RABiTS have been developed, and superconductivity

Status of the 2G HTS wire technology (September 2005)

Country/organization	Length, m	Critical current at 77 K, A/cm width	Substrate/HTS deposition process*
U.S./American Superconductor	85	175	RABiTS/MOD
	10	272	RABiTS/MOD
	Short samples	~400	RABiTS/MOD
U.S./SuperPower	100	70	IBAD/PLD
	207	107	IBAD/MOCVD
	Short samples	407	IBAD/MOCVD
Japan/Sumitomo	117	110	RABiTS/PLD
	Short samples	357	RABiTS/PLD
	10	130	RABiTS/MOD
	Short samples	196	RABiTS/MOD
	105	NA	RABiTS
Japan/Showa Electric	6	69	IBAD-GZO/MOD
	230	NA	RABiTS (Ni-W)
Japan/Fujikura	105	126	IBAD-GZO/PLD
	Short samples	~300	IBAD-GZO/PLD
	255	NA	IBAD-GZO
Japan/ISTEC	100	159	IBAD-GZO/PLD
	25	100	IBAD-GZO/MOD
	Short samples	413	IBAD-GZO/MOD
	220	NA	IBAD-GZO
Europe/THEVA	37	158	ISD-MgO/Evap.
	5	237	ISD-MgO/Evap.
	1	422	ISD-MgO/Evap.
Europe/Edison Spa	2	120	RABiTS/Coevap.
	Short samples	220	RABiTS/Coevap.
Korea/KERI	4	97	RABiTS/Coevap.
	1	107	RABiTS/PLD

*RABiTS = rolling-assisted biaxially textured substrates; MOD = metal organic deposition; IBAD = ion-beam assisted deposition; PLD = pulsed laser deposition; MOCVD = metal organic chemical vapor deposition; GZO = gadolinium zirconium oxide, $Gd_2Zr_2O_7$; ISD = inclined-substrate deposition; MgO = magnesium oxide; Evap. = electron beam evaporation; Coevap. = electron beam co-evaporation. Data from Oak Ridge National Laboratory and Los Alamos National Laboratory.

companies around the world are in the process of taking the technology to the pilot scale to produce commercially acceptable 100-m lengths. In addition, three different methods including metal-organic deposition, metal-organic chemical vapor deposition, and high-rate pulsed laser deposition have been used to demonstrate high I_c in 100-m lengths of YBCO-coated conductors. The current research in the area of HTS wire technology is to increase the flux pinning properties of YBCO superconductor wires and to reduce the ac loss in these wires for various military applications.

Applications. The U.S. Department of Energy has funded three superconductivity partnership initiatives projects to demonstrate the use of HTS power cables for electric transmission and distribution. They are the Long Island HTS power cable, Albany HTS power cable, and Columbus HTS power cable. The goal of the Long Island project is to demonstrate a 610-m (2000-ft), 600-megawatt HTS power transmission cable operating at 138 kV in the Long Island Power Grid—the first-ever installation of a superconductor cable in a live grid at transmission voltages. The goal of the Albany project is to demonstrate the technical and commercial viability of HTS cables by operating a 350-m (1150-ft) superconducting cable, including a 30-m (100-ft) section made from second-generation HTS wire, between two Niagara Mohawk substations. The goal of the Columbus project is to complete the development, installation, and testing of a 200-m (660-ft), three-phase HTS power cable at a substation in Columbus, Ohio. This project will demonstrate how a triaxial HTS cable may be used in the future to replace existing oil-filled underground copper cables and greatly increase the capacity of power link. The future of HTS wire technology research depends heavily on the successful demonstration of these power cables.

For background information *see* ALLOY; CURRENT DENSITY; ELECTRIC CURRENT; ELECTRIC POWER SYSTEMS; ELECTRICAL RESISTANCE; EPITAXIAL STRUCTURES; GRAIN BOUNDARIES; LASER-SOLID INTERACTIONS; SUPERCONDUCTING DEVICES; SUPERCONDUCTIVITY; VAPOR DEPOSITION in the McGraw-Hill Encyclopedia of Science & Technology.

M. Parans Paranthaman

Bibliography. A. Goyal et al., High critical current density superconducting tapes by epitaxial deposition of $YBa_2Cu_3O_x$ thick films on biaxially textured metals, *Appl. Phys. Lett.*, 69(12):1795–1797, 1996; K. Hasegawa et al., Biaxially aligned YBCO film tapes fabricated by all pulsed laser deposition, *Appl. Supercond.*, 4(10-11):487–493, 1996; Y. Iijima et al., In-plane aligned $YBa2Cu3O_{7-x}$ thin films deposited on polycrystalline metallic substrates, *Appl. Phys. Lett.*, 60(6):769–771, 1992; J. L. Macmanus-Driscoll et al., Strongly enhanced current densities in superconducting coated conductors of $YBa2Cu3O_{7-x}$ + $BaZrO_3$, *Nat. Mater.*, 3:439–443, July 2004; M. Parans Paranthaman and T. Izumi (eds.), High-performance YBCO-coated superconductor wires, *MRS Bull.*, vol. 29, no. 8, August 2004; W. Prusseit et al., Evaporation—the way to commercial coated conductor fabrication, *Physica C*, 392: 801–805, 2003; X. D. Wu et al., Properties of $YBa_2Cu_3O_{7-x}$ thick films on flexible buffered metallic substrates, *Appl. Phys. Lett.*, 67(16):2397–2399, 1995.

Telerobotics (mining)

The field of robotics explores the potential for machines and control systems to perform the tasks of a human operator. The subfield of telerobotics allows a human operator to remain in control of a semirobotic machine incorporating limited automation. This permits the operator to focus on only the work requiring human skill to increase productivity, significantly increasing the output of a process. Telerobotics is an emerging industry with potential applications in manufacturing, underwater mining, outer-space construction, and the military. In mining, telerobotic technology is used in the discovery, delineation, development, extraction, and closure of ore bodies. It offers the potential for enhancing safety, productivity, cycle time, quality, value, and costs. Teleremote mining is the operation of mining equipment over a network from any location out of the line-of-sight of the machine. Teleremote mining combines remote sensing, remote control, and limited automation of mining equipment and systems with the process of mining mineral ores. It is, essentially, the application of advanced manufacturing systems to the process of mining, the main difference being that the machines used in the mining process must be mobile.

Telerobotic technology. Telerobotics in the mining industry was introduced to production environments in the early 1970s, with the implementation of line-of-sight radio remote controls in underground mines and truck dispatch systems in open-pit mines. As the technology improved, line-of-sight remote control was extended to longer and longer distances underground. Today, many underground mining companies and suppliers are developing teleoperation systems using operators that are hundreds of kilometers from the units they control, and some methods depend on this technology on a day-to-day basis.

Long-distance teleoperation has enabled operators on the surface to run multiple load-haul-dump machines and drills from comfortable control rooms. This style of operation removes operators from the mine environment, offering dramatic benefits in safety. In addition, the support staff enters the mining area only when no mining work is occurring, thereby maximizing safety for the whole operation. Teleoperation offers many additional benefits, including increased labor productivity because of instantaneous connection to many machines, reduced maintenance requirements by ensuring that machines are operated within their capability, and more intimate control of the mining process through direct measurement and communication of real-time information. Profitability of a mining operation increases as the combination of these benefits alters the traditional economics involved in mining a deposit.

The elements needed for teleremote mining include advanced underground mobile computer networks, underground positioning and navigation systems, mining process monitoring and control software systems, mining methods designed specifically for teleremote mining, and advanced mining equipment. An underground mobile computer network connects the underground mine to surface operations centers with virtual control stations. This high-capacity network allows the operation of mobile telephones, handheld computers, mobile computers onboard machines, and multiple video channels needed to run many pieces of mining equipment. Accurate positioning systems are also critical in the application of mobile robotics to mining. Positioning systems consisting of ring laser gyros (RLG) and accelerometers function like Global Positioning Systems. These positioning units can be mounted on all types of drilling machines, allowing operators to position the equipment remotely and without the need for surveying. In addition, mine planning, simulation, and process control systems are critical elements in successfully applying telerobotic technology to mining. Mine planning systems supply data directly to the working machines, which then feed information back to the mine planning systems, which are then coordinated with the overall strategic mine plan.

Ongoing cost increases are rendering current mining methods prohibitively cost-restrictive and limiting the feasibility of using lower-grade ore bodies. Using a simulation model to evaluate the impact of teleremote technology on mine operations, substantial benefits appear possible. While teleremote crew settings are not currently included in the model, the ability to increase operating hours and flexibility in process schedules results in reduced mine life, increased throughput, and maximized resource use. Specifically, teleremote mining simulations show a 38% reduction in mine life and allow optimization to smooth out production peaks.

Sudbury research facility. A research and testing facility in Sudbury, Ontario, Canada, has been established to support the development of telerobotics for mining and other industrial environments, cooperative telerobotics, and multimine teleoperation. The facility consists of two sites 20 km (12.4 mi) apart [**Fig. 1**]. Site 1 consists of a teleoperation chair and a telecommunications control room. A teleoperation chair has a telehaptic feedback system, which simulates the movements of a telerobotic machine and enables the operator to feel the motion and orientation of the machine. The teleoperation chair incorporates

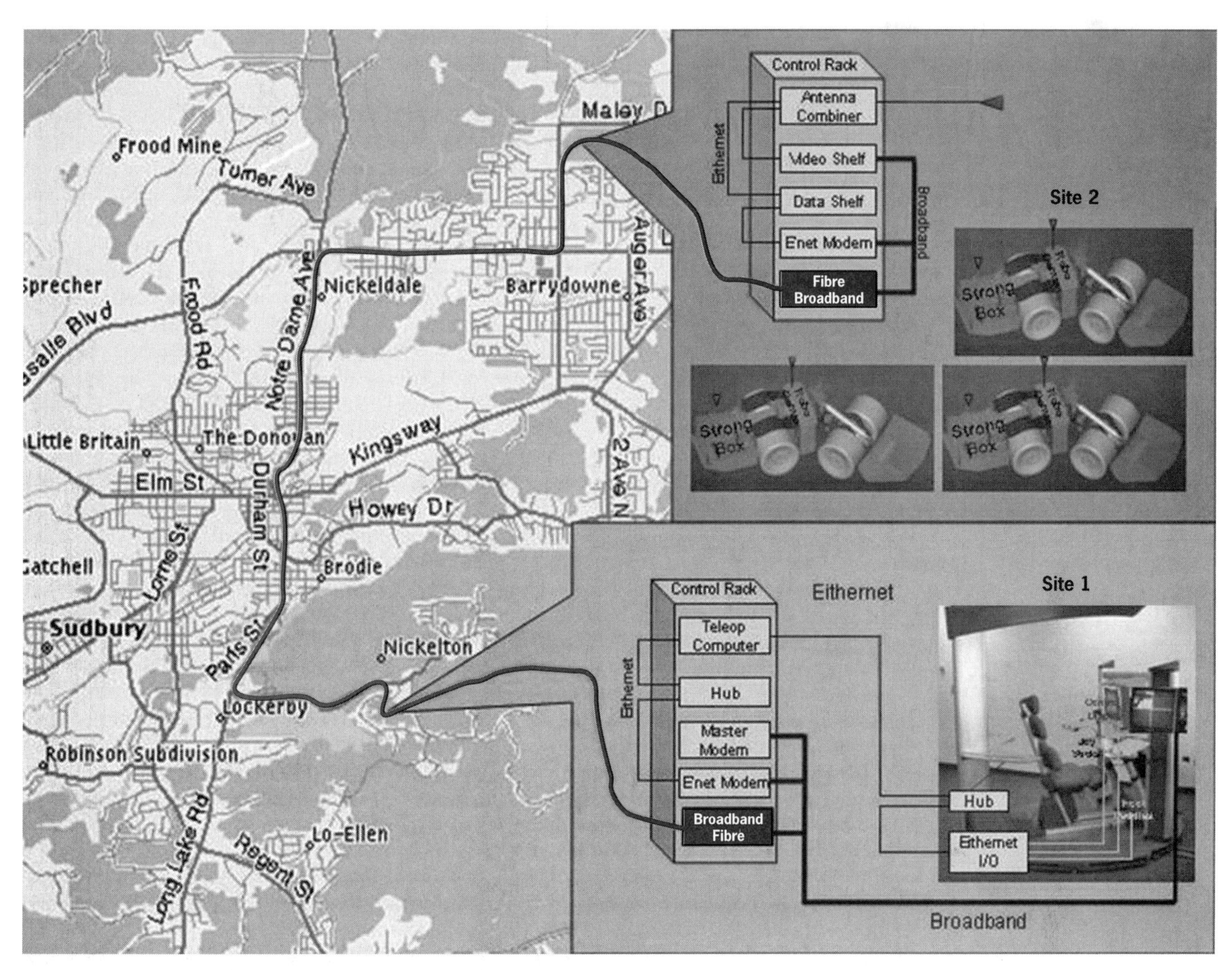

Fig. 1. Telerobotic test facility, showing the two sites and wireless network.

an ergonomically designed seat with joysticks and a display to allow teleoperation of a machine. Site 2 supports mobile data communications and mobile video, as well as laboratories for designing and building quarter-scale models to full-sized mining machines and full-mine simulations. The wireless network between these two sites provides the opportunity to study mobile and multimachine operation and to investigate many other telerobotics-related issues such as the impact of multiple levels of automation and the economic and safety implications of multiple mine operation scenarios using telerobotics systems. Building the Sudbury facility was an important step in the development of telerobotic technology and its applications, and the facility will continue to be a significant resource for research and development. The units developed here were designed with electronic and telecommunication systems identical to those that are used on current robotic machines working in the field, which allows experimentation with new systems on scale models both at the teleoperation labs and in mining facilities.

Creating a virtual environment. Teleoperation in industrial, military, and space applications is extremely challenging. Because of the lack of feedback, operators have great difficulty understanding the environment in which their remote machines are working. Workstations provide very limited video feedback and typically offer only joystick and footpedal control. Underground and open-pit mining companies around the world are beginning to implement operations from surface control rooms, but are limited by the requirement for constant and substantial feedback from the mobile equipment. In one such mining operation's excavation cycle, ore is collected from a rock-loading bin, transported some distance in the vehicle's bucket, and then deposited at a dump for further processing. However, the teleremote operator is unable to monitor either the machine's condition or the mine's infrastructure without some feedback from the machine's environment. Teleoperation in other situations faces the same difficulty, such as industrial and military applications where machine rollovers occur due to steep slopes. In addition, a teleoperator cannot feel or see poor road conditions, and therefore cannot judge how to maneuver the equipment to compensate for these conditions.

Fig. 2. Teleoperator's virtual reality workstation.

A workstation has been proposed that will dramatically enhance the teleoperation of equipment by creating a virtual environment using vestibular (balance and equilibrium) and haptic (touch) feedback, as well as an immersive visual display (**Fig. 2**). The control station will include a suite of sensors to provide critical information for the operator to remotely see, hear, and feel the machine orientation and conditions. To accomplish these goals, the workstation will support three-dimensional cameras; pitch, yaw, and roll sensors and actuators for the entire machine; and force feedback for accelerator and brake pedals, and bucket and arm controls. A control software environment will provide the operator with additional information needed to work at a distance, which is processed and communicated back to the workstation. Actuators connected to a movable base, a joystick, and foot pedals will provide the operator with force-feedback for hands and feet. In order for the operator to feel sideways tilting, inclines, and declines, the chair and video screen will mimic the movements of the operating equipment. The chair's movements will be proportional and scaled to the actions of the equipment and will be sufficient to enable the operator to understand where the equipment is and what position it is in. For full-body feedback, the movable base will allow the operator to sense balance while accelerating, braking, and controlling the bucket or arm of a machine. The proposed teleoperator workstation will also provide the operator with information about floor conditions that might degrade or damage the tires or vehicle, jolts that the machine receives as it is driven through the tunnel, and problems with the vehicle manifested by increased vibration which, without feedback, would not be noticed or repaired.

Aboveground, underwater, and space environments. The bulk of telerobotics research, development, and application to date has been applied to underground mining operations. This is largely because of the difficulty in implementing telecommunications technology within surface applications such as open pits. Limited communication bandwidth within the open-pit environment presents a considerable challenge. The introduction of optical wireless communications systems offers a possible solution for the development of a high-bandwidth communication system for open-pit mining and other aboveground applications. The underwater medium is the most difficult environment for high-bandwidth communication and therefore offers the optimal environment in which to develop this technology. This capability is also relevant to space construction because of the many characteristics that space shares with the underwater environment.

Teleoperation in underwater applications has not been successful to date due to the limitations inherent in the methods being used. A long, trailing antenna can be used to allow a small submarine to transmit and receive information at the surface, but such antennas are many kilometers in length, which presents a significant problem with tangling when trying to run multiple machines. Long-wave communication can be used; however, this method suffers from a lack of bandwidth and allows only paging type of communications when used in an underwater environment. Direct-cabled connection, which is used today, provides the necessary bandwidth, but it presents the same drawback of tangling in multimachine operation.

A high-capacity, underwater telecommunications system for teleoperating underwater machines has been developed and is being enhanced. A positioning system for the unstructured underwater environment is under development. A fiber-optic light buoy system and corresponding onboard telerobotic system to optically transmit and receive the information required to operate a telerobot has also been developed. Methods are being developed for buoys and underwater robots to communicate with sufficient bandwidth to transmit video information that will enable full teleoperation of the units using an inertial navigation system that provides accurate positioning and orientation in three-dimensional space for mobile, untethered equipment (**Fig. 3**).

High-speed, high-capacity, underwater, and nonconductive communications systems will enable wireless underwater operation of teleremote vehicles, creating many new opportunities in both mining and nonmining industries. Some examples are mining underwater while avoiding the expense of building dams to mine a lakebed, exploiting flooded mines without pumping the water out, exploring and mining the ocean depths, robotic sewer repairs, maintenance work in flooded waste fill sites, and communications to submersible vehicles leading to a multitude of military applications.

For background information *see* ACCELEROMETER; BANDWIDTH REQUIREMENTS (COMMUNICATIONS);

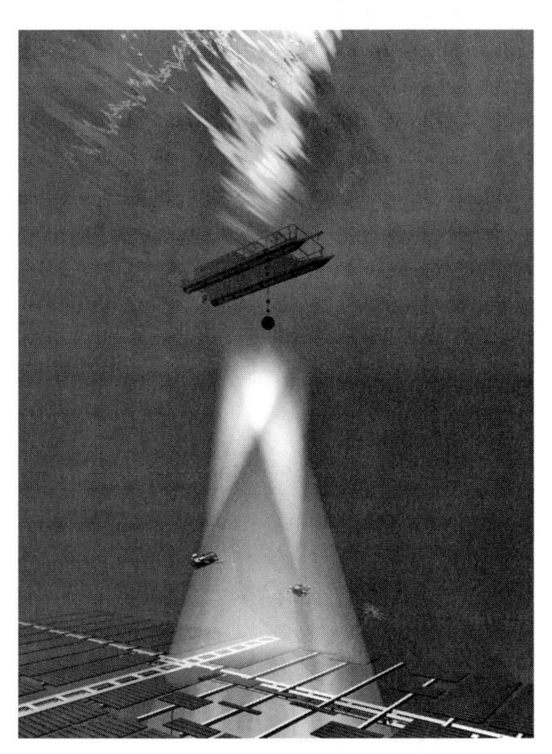

Fig. 3. Underwater telerobotics, showing pontoon and miniature submersible.

CONTROL SYSTEMS; GYROSCOPE; MINING; OPEN-PIT MINING; OPTICAL COMMUNICATIONS; OPTICAL FIBERS; PROPRIOCEPTION; REMOTE-CONTROL SYSTEM; REMOTE MANIPULATORS; ROBOTICS; SIMULATION; UNDERGROUND MINING; VESTIBULAR SYSTEM; VIRTUAL REALITY in the McGraw-Hill Encyclopedia of Science & Technology. Gregory R. Baiden

Bibliography. G. R. Baiden, *MAP: Mining Automation Programme Results: Final Presentation*, Copper Cliff, Ontario, Canada, Oct. 25, 2000.

Tetrapod origins

Tetrapods may be defined as vertebrates with four limbs and digits (that is, arms and legs bearing fingers and toes). The term includes not only the extinct forms described here, but all their relatives and descendants, including dinosaurs, snakes, frogs and salamanders, and even mammals like ourselves. It has, however, become increasingly difficult to draw a line between some of the earliest fossil tetrapods and their closest relatives with fins, in part as a result of the increase in the number of transitional fossils that are being discovered. Thus, the definition of the word "tetrapod" has become almost as murky as the swamps from which the first four-legged animals emerged.

Evidence relating to the origin of tetrapods has expanded remarkably in the recent past. The number of recorded tetrapod taxa from the Late Devonian Period, the time segment during which tetrapods first evolved (between about 382 and 362 million years ago), has increased exponentially over the last few decades, doubling in the last 10 years.

Acanthostega studies. In the early 1990s, discovery and description of the Devonian tetrapod *Acanthostega*, from East Greenland, changed the way people thought about the so-called fish-tetrapod transition. This animal showed us that the earliest limbs equipped with fingers and toes probably evolved before the animals began to use them for walking on land. The limbs of *Acanthostega* were unsuited to weight bearing and would have been incapable of the complex movements required in walking. The first use of such limbs was likely to have been for pushing through swampy weed-choked waters where fins would have been a disadvantage. At the time that *Acanthostega* was being studied, it was one of only three known genera of Devonian tetrapod. The first recognized genus was *Ichthyostega*, also from East Greenland and discovered in the early 1930s, and the other was *Tulerpeton*, from Russia, discovered in 1984 (see **illustration**).

One of the most exciting things to have emerged from the study of these animals was that they all possessed more than five digits on their limbs, unlike the previously assumed primitive pattern of five. These discoveries came at around the same time that developmental studies on living animals renewed interest in the origin of limbs, and all combined to produce new ideas about how digits arose, their initial function or functions, and the genetic control mechanisms that lay behind them.

One of the unexpected outcomes of the study of *Acanthostega* was that in learning more about an apparently obscure part of its anatomy, namely the inner face of its lower jaw, paleontologists were able to identify other jaw fragments as more similar to *Acanthostega* and other tetrapods than they were to any fish relatives. This has led to the remarkable increase in known Devonian tetrapod taxa from all over the world. Previously unidentified fragments in museum collections suddenly made sense, and gave clues to where further Devonian tetrapod finds could be made. Among the first of the newly identified taxa was one from Latvia, named *Ventastega* (see illustration). At a site initially identified from a lower jaw, renewed collecting in the area has produced some wonderful new material. An almost complete skull has recently been found, in addition to limb and girdle elements and many more lower jaws. The preservation of this material is exquisite—it is three-dimensional and uncrushed. It is encased only in very soft sediments, in sharp contrast to the hard sandstones in which *Acanthostega* and *Ichthyostega* specimens are embedded. This material promises to produce some great surprises for studies of the origin of tetrapods.

Red Hill findings. A previously disputed lower jaw from Australia was confirmed as that of a Devonian tetrapod, and, subsequently, newly discovered specimens have come from the United States, Scotland, China, Belgium, and Russian localities (see illustration). Further collecting at most of these sites is actively taking place or is planned for the future. From the point of view of the richness of the fauna and the information it provides about the context in which

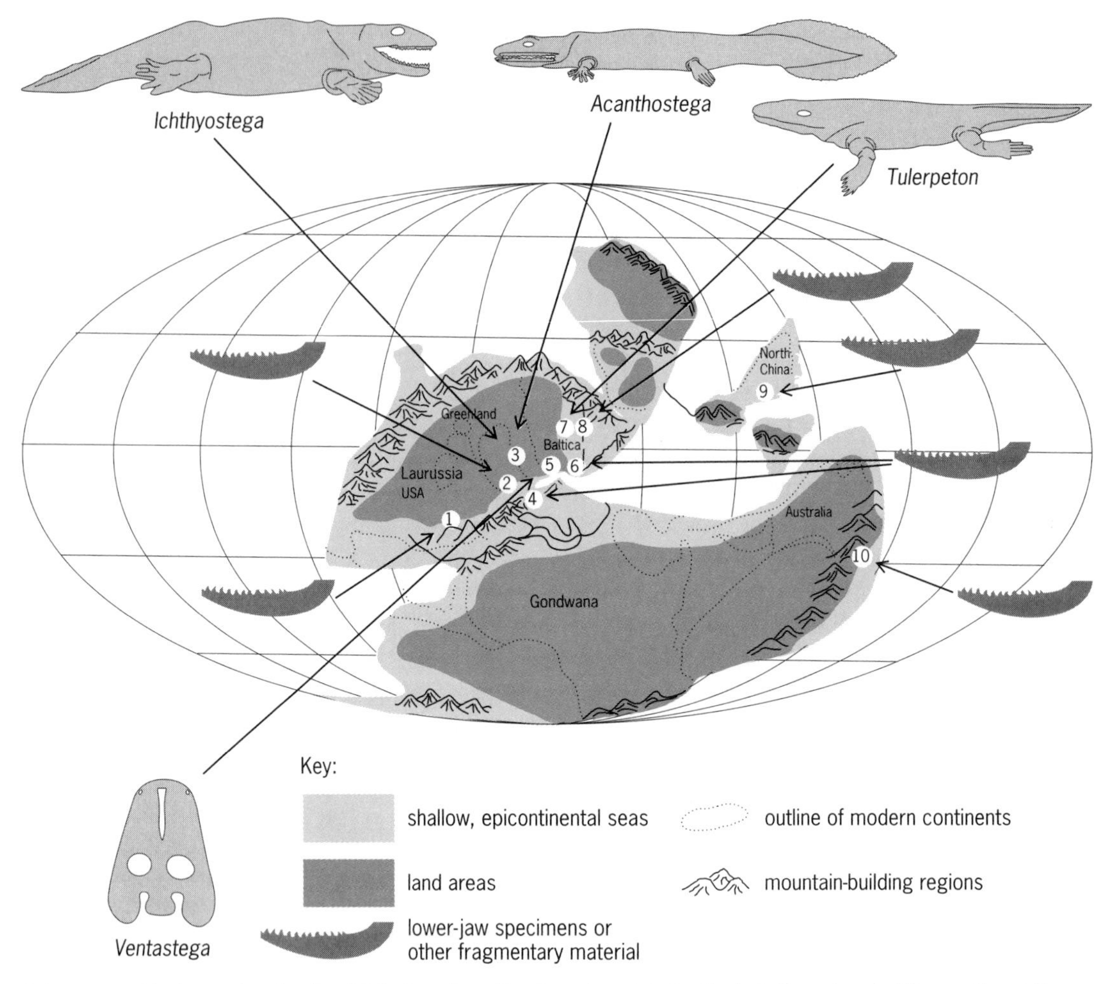

Paleogeographical map showing the distribution of continents and seas during the Late Devonian, and the position of the localities that have produced tetrapod remains. 1, Pennsylvanian Red Hill (U.S.); 2, Scat Craig (Scotland); 3, East Greenland; 4, Strud (Belgium); 5, Pavari/Ketleri (Latvia); 6, Novgorod (Russia); 7, Andreyevka (Russia); 8, Gornostayevka (Russia); 9, Ningxia (China); 10, Jemalong (Australia). The manus (forefoot or hand) of *Ichthyostega* is still unknown; *Tulerpeton* is known from limbs, girdles, and part of the snout, and other parts are reconstructed; the skull of *Ventastega* is still not fully described so is indicated by a generalized outline. Laurussia, Baltica, North China, and Gondwana are ancient continental masses. Greenland, United States, and Australia are modern regions indicated for clarity.

the animals lived, perhaps the most significant locality is the site known as Red Hill in Pennsylvania. Two taxa of tetrapod have been identified from this locality. One, *Hynerpeton*, is known from two partial shoulder girdles and an associated lower jaw, and the other, *Densignathus*, is known from two lower jaws. Other tetrapod material has also been recovered from the site and is awaiting study. One limb bone, a humerus (upper arm bone), was particularly unusual, though the kind of limb that it came from is open to debate. This humerus combined some obviously primitive and fishlike features, some that are more closely associated with known tetrapod humeri, including evidence that the arm it belonged to was operated by strong and complex musculature. It appears to have functioned as some kind of support structure, though not necessarily for walking. There is at present no way of knowing whether the limb bore a fin or tetrapodlike digits, and, if the latter, then how many. It does, however, give us clues to the sequence in which tetrapodlike features arose. There is some evidence to suggest that supportive features of the forelimb arose in concert with an increase in air-breathing capabilities, and may initially have been employed in lifting the head out of water.

Another important feature of the Red Hill locality is that other aspects of the flora and fauna are also well preserved. Current studies have found evidence of a warm subtropical climate, rich vegetation-lined river channels, and that tetrapods shared their habitats with many invertebrates such as myriapods, scorpions, and their relatives, as well as a diversity of fishes. Other early tetrapod localities represent very different environments. For example, *Ventastega* was found to live in marginal marine conditions formed in a low-tidal near-shore environment such as that which might be found in a lagoon or estuary. At the other end of the spectrum, *Tulerpeton* was found in deposits representing shallow marine conditions in which the salinity varied quite widely (see illustration). Organisms that usually occur only in marine conditions were found alongside *Tulerpeton* in a location that seems to have lain at least 200 km (125 mi) from the nearest landmass.

New technology and changing interpretations. In recent years, not only have ideas about the origin of limbs and digits had to be revised, but some new interpretations of the first-known Devonian tetrapod, *Ichthyostega*, are challenging views long taken for granted. The first such challenge came with the discovery that its hind limb was paddlelike and bore seven digits in a unique pattern, but further surprises followed. Its ear and braincase region (the braincase houses the brain and ear capsules) proved to be utterly unlike those of any other known vertebrate, living or extinct. They show a highly modified morphology that has only yielded to interpretation following the application of the new technique of computer-assisted tomography to newly collected specimens. This peculiar morphology has been hypothesized to be a highly specialized ear that was used under water. More recent work has shown that *Ichthyostega's* postcranial skeleton was also highly specialized in several ways, in particular the vertebral column. The picture that is being put together shows an animal that did not walk in the conventional sense. Although it certainly seems to have been more land-capable than *Acanthostega*, its unique ear suggests that it must have spent a large part of its time in the water. Work in progress is expected to fill in the puzzle more fully, to reveal an image quite different from the old view of a clumsy salamanderlike animal.

The whole picture of tetrapod origins is becoming more complex. We now see that Devonian tetrapods were scattered widely throughout the world, ranging across a supercontinent that stretched from what is now the east coast of North America to Australia and northern China (see illustration). We also see that they inhabited a wide variety of ecological habitats, from apparently marine, or at least brackish, to freshwater rivers and lakes. The animals found there show a wide variety of morphologies, seen even among the fragmentary remains that are all that has been found of some of them. The two animals known from the most complete remains, *Ichthyostega* and *Acanthostega*, are very different in skeletal morphology and must have had very different lifestyles, even though they lived in nearby environments.

Just as our knowledge of *Acanthostega* changed ideas about early tetrapods and their evolution, showing us a tetrapod that was more fishlike than any previously known, so some recent discoveries are set to fill in yet more of the puzzle by showing us fish that are more tetrapodlike than any previously known. Finds of fossil fish in the Canadian Arctic made over the past couple of years promise a wealth of new information about the other side of the divide—information likely to blur the distinction between fish and tetrapods even further.

For background information *see* ANIMAL EVOLUTION; ANIMAL SYSTEMATICS; DEPOSITIONAL SYSTEMS AND ENVIRONMENTS; DEVONIAN; ICHTHYOSTEGA; SKELETAL SYSTEM; TETRAPODA in the McGraw-Hill Encyclopedia of Science and Technology.

Jennifer A. Clack

Bibliography. J. A. Clack, Fossil vertebrates—Palaeozoic non-amniote tetrapods, in *Encyclopaedia of Geology*, vol. 2, pp. 468–497, Elsevier, 2005; J. A. Clack, *Gaining Ground: The Origin and Evolution of Tetrapods*, Indiana University Press, Bloomington, 2002; J. A. Clack, The origin of Tetrapods, chap. 2 in H. Heatwole and R. L. Caroll (eds.), *American Biology*, vol. 4: *Paleontology*, pp. 973–1079, Surrey Beatty, 2000; J. A. Clack et al., A uniquely specialised ear in a very early tetrapod, *Nature*, 425:65–69, 2003; N. H. Shubin, E. B. Daeschler, and M. I. Coates, The early evolution of the tetrapod humerus, *Science*, 304:90–93, 2004.

3D inkjet printing

The ability to form a raised image using a desktop printer makes it possible to use this technology for a variety of specialized print products, such as business cards, postcards, and certain children's books. A very recent application includes printing of micro-electro-mechanical machined structures (MEMS). If the raised image attains the required height and firmness, it could be used for Braille characters (see **illustration**). One technology under development is the use of modern inkjet systems to print 3D structures.

Inkjet printing. There are two types of inkjet printing, continuous inkjet (CIJ) and drop-on-demand (DOD). In continuous inkjet, a high voltage is applied to an electrically responsive crystal in contact with ink. A stream of uniformly sized and spaced droplets is generated by applying pressure pulses at a suitable frequency. Droplets are directed to a moving surface, and a line of dots is created. Droplets are charged with varying voltage as they leave the nozzle. They pass by an oppositely charged plate, and are deflected to a degree proportional to the charge carried. Creation of droplets is a continuous process; some droplets will not be printed because no voltage is charged to them, and instead they are collected in a trap, filtered, and returned to the ink reservoir. Drop-on-demand is a simpler system; no electrostatic charge is applied to the droplet, thus deflection and recovery are not part of the process. An ink droplet is generated when it is needed for printing. DOD is capable of producing excellent print quality on suitable, usually absorbent substrates.

DOD ink formulations are simpler than CIJ. The ink must be designed for the particular device used.

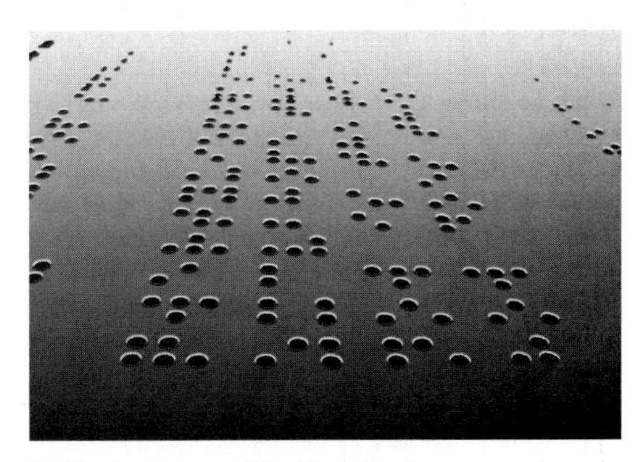

Braille characters as 3D printed structures.

The principal types of DOD printers are valve-jet, piezoelectric, thermal or bubble jet, and hot-melt ink printers. Hot-melt (phase-change) ink formulations are frequently used for inkjet printers. Phase-change inks are solid at room temperature. The ink melts into a reservoir where it is kept fluid by a heating element. This hot, liquid ink is pumped through a nozzle, using thermal DOD technology. The ink is jetted from the printing head as a molten liquid at elevated temperature. Upon hitting a recording surface, the molten ink drop solidifies immediately, thus preventing the ink from spreading or penetrating the printed media. Quick solidification ensures that image quality is good on a wide variety of media.

3D printing techniques. One approach to 3D printing is to use hot-melt inks, combined with various additives, and commercial inkjet printers equipped with a thermal print head. Customized inks have to be created, because none of the commercially available hot-melt inks is suitable for printing raised images. (Existing inkjet printers are designed to produce a thin, flat ink film, employing a fuser to flatten the film.) Another approach is to use a DOD nonimpact printing technique with multiple droplet-deposition nozzles. Ink droplets are ejected and have to be cured immediately to avoid ink penetration into substrate, ink spreading, and ink leveling. After curing, the second, third, fourth, and fifth layers of the ink are deposited on previously printed areas in order to create a 3D image. Adjoining portions are formed from different ink formulations, such as hot-melt and UV-curable. This creates a three-dimensional ink film, layer by layer, at the same position on the substrate.

Ingredients of 3D inks. The ink for forming raised images contains thermoplastic polymers along with particular chemical substances, such as blowing agents (also called foaming agents), which will decompose at an elevated temperature, releasing gas bubbles. The gas bubbles are entrapped in the solidifying ink, creating raised dots. The blowing agent and thermoplastic polymers are selected according to the temperature in the print head. Blowing agents are thermally unstable materials, such as diammonium phosphate, urea, ammonium oxalate, azo compounds, sulfonylhydrazides, sulfonylcarbazides, or azides of acids. The polymer formulation will serve to trap the gas bubbles until drying or curing are completed. Examples of suitable polymers for ink compositions include alkyd resins, amides, acrylic polymers, benzoate esters, citrate plasticizers, coumarone-indene resins, dimer fatty acids, epoxy resins, fatty acids, ketone resins, maleate plasticizers, long-chain alcohols, olefin resins, petroleum resins, phenolic resins, phthalate plasticizers, polyesters, polyvinyl alcohol resins, rosins, styrene resins, sulfones, sulfonamides, terpene resins, urethanes, vinyl resins, their derivatives, and their combinations. No limitation is placed on the type or the amount of the polymer that is present in the ink. Generally, polymers that are used for hot-melt inks have melting points in the range 40–200°C (104–392°F). In a molten state, the polymer should be stable so that there is no formation of gaseous products or deposits on the printer device.

Nanoparticles can be dispersed into the ink formulation, along with suitable blowing agents and binders, to act as a gas barrier. These nanostructured polymer composites can be made from carbon nanotubes, nanocoils, nanosilicates, and other structures. Nanostructured polymer composites provide very effective barriers to gases and liquids. They are characterized by a combination of chemical, physical, and mechanical properties that cannot be obtained with macro- or microscopic dispersions of inorganic fillers. Moreover, the materials exhibit the properties at very low filler level, usually less than 5% by weight.

Binder selection is critical to secure the nanoparticles to the surface and to form an ink film of sufficient strength to withstand, for example, the repeated touching of 3D Braille letters. Selection of thermoplastic resins and waxes is essential. The resins assure adhesion of the ink to the printing substrate. At the same time, resins control the viscosity of the ink at the melting point, inhibiting crystallization of the wax (which would impart transparency to the ink).

Waxes are used alone or in the form of a mixture. If the wax component in the ink formulation is less than 5% by weight, the properties of other additives may have a higher or unsettled melting point, which will negatively influence the ink composition (ink will not melt sharply at the inkjetting temperature). If the wax component in the ink formulation is more than 95% by weight, the ink composition may have an insufficient melt viscosity, and adhesion to the substrate can be accompanied with various problems. There is a wide selection of waxes, such as plant waxes, animal waxes, synthetic hydrocarbon waxes, higher fatty acids, higher alcohols, and their derivatives. Petroleum wax consists of paraffin wax and microcrystalline wax. The most essential synthetic hydrocarbon waxes are a polyethylene wax and a Fisher-Tropsch wax.

An alternative to using hot-melt inks in combination with nanoparticles and blowing agents is to use UV-curable hot-melt ink, which is crosslinked by ultraviolet radiation either in-flight or immediately after ink droplet deposition. UV-curable inks are actually composite materials, containing a UV-curable low-viscosity monomer, higher-viscosity prepolymer or oligomer, and a photoinitiator. Chemically, they can be acrylic resins, epoxy resins, urethane resins, urethane acrylics, unsaturated polyester resins, or epoxy acrylics. The advantage of this composition is that the ink is harder and does not crumble, which can sometimes occur in hot-melt 3D printed ink. The amount of wax is sufficient to permit a phase change of the UV-curable composition after jetting. The UV-curable hot-melt inks have excellent heat and abrasion resistance and good adhesion to a substrate.

Three-dimensional ink may contain various kinds of pigments and dyes to color it, or just fillers such as silicone. The silicone loading decreases the surface energy of the print material and increases nonwetting properties, which forces ink droplets to create rounder, higher droplets.

Outlook. Printing with a digital inkjet printer using thermal wax is much faster and quieter than the current technology for Braille printing, which is done by embossing individual dots using low-speed desktop embossers. In addition to being noisy and slow, these embossers are typically very expensive (up to $10,000, compared to $2000 for the average inkjet printer) and require expensive specialty substrates to print on. Therefore, although the use of inkjet printers and thermal wax to print 3D structures is still in the development stage, prospects are good that this technology will soon be commonly used in Braille printing and similar specialty printing processes.

For background information *see* INK; PRINTING; POLYMERIZATION; ULTRAVIOLET RADIATION in the McGraw-Hill Encyclopedia of Science and Technology.

Alexandra Pekarovicova; Jan Pekarovic

Bibliography. A. Pekarovicova, H. Bhide, and J. Pekarovic, Phase change inks, *J. Coatings Technol.*, 75(936):65–72, 2003; K. A. Schmidt, Selective deposition modeling with curable phase change materials, US Patent 6,841,116, January 11, 2005; S. Speakman, 3D printing and forming of structures, US Patent 6,164,850, December 26, 2000; C. Williams, *Printing Ink Technology*, Pira International, Leatherhead, UK, 2001.

Tissue and organ printing

Tissue and organ printing is an evolving technology that uses a special-delivery device (the bioprinter) to deposit self-assembling biological material (the bio-ink) into a three-dimensional biocompatible scaffold (the biopaper) with high spatial accuracy. Delivery of the bio-ink takes place layer by layer according to the anatomical blueprint (that is, the shape and histological composition) of the desired structure. This "blueprint" is obtained by bioimaging—for example, through magnetic resonance imaging (MRI) or ultrasound—the structure to be reproduced and stored in the computer controlling the operation of the printer (**Fig. 1**). Accordingly, the technology has three main steps: preprocessing, or computer-aided design (CAD) of the organ; processing, the actual printing and fast solidification of tissue and/or organ construct; and postprocessing, accelerated tissue and organ maturation. The overall objective of this technology is to mitigate the chronic shortage of replacement organs by building three-dimensional functional biological structures, or organ modules, of specific shape in the laboratory. Organ printing relies on classical tissue engineering, which is based on the process of seeding cells into biodegradable polymer scaffolds, culturing and expanding the cells in bioreactors for several weeks, and implanting the resulting tissue into the recipient organism. At the same time, the new technology represents a departure from the classical approach in that the seeding of cells is performed by printers and proceeds according to a CAD.

Bioprinter. To date, delivery devices used for bioprinting have been either inkjet printers or sophisticated pressure-operated mechanical extruders. Inkjet printers use either piezoelectric ejectors or large and rapid variation in temperature (thermoprinters) to expel ink particles. Thomas Boland and coworkers at Clemson University have redesigned commercially available desktop inkjet printers (**Fig. 2*a***) to accommodate cartridges filled with live cells or extracellular matrix material and perform three-dimensional printing. According to the specific application, a printer can be outfitted with several nozzles to deliver cells of different types. These particular authors have already printed branching cellular tubes and organ rudiments. Others have employed inkjet technology to fabricate scaffolds possessing predefined internal structure.

Mechanical extruders (Fig. 2*b*, *c*) are more versatile than inkjet printers, since they can be operated with more complex bio-ink particles—in particular, inks made from spherical aggregates composed of several thousands cells. This method has also been used to build tubelike structures, as well as sheets of cells several hundred micrometers in thickness.

Each of the two printing schemes has their advantages and disadvantages. Inkjet printers are inexpensive and easy to operate. On the other hand, during printing, cells are exposed to harsh conditions (such as high temperature and strong shear forces) and, similar to what has been seen in the more traditional tissue engineering approaches, maturation of the printed structure into functional tissue or organ is slow, putting cell survival at risk. Mechanical extruders provide a gentler environment. Furthermore, when used in combination with cell aggregates (prebuilt small tissue blocks), printing time can be drastically reduced, structure formation accelerated, and thus cell survival enhanced. The major disadvantage of extruders is their high cost.

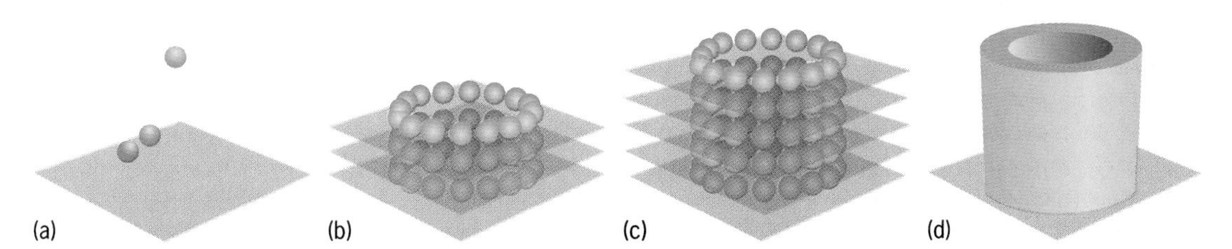

Fig. 1. Schematic representation of organ printing. (*a*) Spherical bio-ink particles are deposited by the printer one by one into the biopaper (itself printed). (*b, c*) Deposition occurs layer by layer. (*d*) During postprocessing, the bio-ink particles fuse into the desired three-dimensional organ—here, a lumen-containing tube.

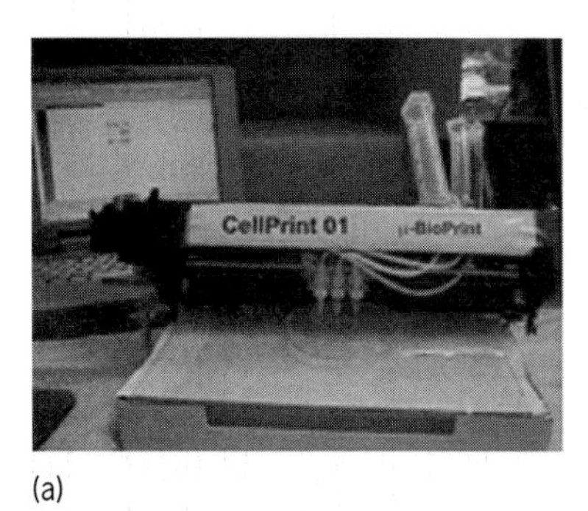

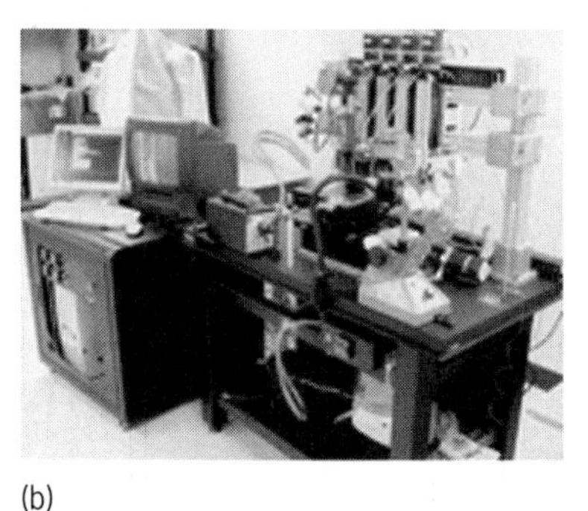

(a) (b) (c)

Fig. 2. Bioprinters. (*a*) Commercially available inkjet printer redesigned for organ printing by Thomas Boland and coworkers at Clemson University. (*b*) Pressure-operated bioprinter with four nozzles, constructed by VaXDesign Sciperio, Inc. (*c*) Pressure-operated bioprinter with two nozzles, constructed by a joint effort of three companies: Roland, Fishman, and Naetco.

Biopaper. As with any tissue engineering approach, bioprinting utilizes biocompatible scaffolds to accommodate the bio-ink particles. These scaffolds, the biopaper, are typically hydrogels (polymer networks that swell in water). A unique feature of bioprinting is that the scaffold-biopaper itself is printed by the printer (Fig. 1). To successfully deliver the bio-ink particles into the biopaper, the latter must be in the "right phase." It can be neither a fully rigidified gel nor a free-flowing solution (sol): it must be near the sol-gel transition.

The ideal hydrogel for tissue printing must both allow cells to survive and provide favorable conditions for postprinting self-assembly. Furthermore, it should be degradable, either by the cells themselves or by external manipulations, so that it can be replaced by the extracellular material secreted by the cells. There already exist several candidates for such gels: thermoreversible gels, photosensitive gels, pH-sensitive gels, and gels sensitive to specific molecular entities. Detailed studies must be performed on these "intelligent materials" to identify the optimal ones for specific cells and their mixtures. In the case of a thermoreversible material, for example, printing takes place at a temperature at which the biopaper is in the gel phase. Once postprinting structure formation is complete, the system is brought to a temperature at which the hydrogel turns back into a solution and can easily be removed. In the case of a light-sensitive material, the role of temperature in the previous example is assumed by the wavelength of an external light source.

Bio-ink. In inkjet printing, the bio-ink is made mostly of individual cells. Cartridges of different "color" contain different cell types. Drops of these bio-inks are delivered on demand, at the right time and the right location.

Bio-ink particles that are made of aggregates of cells, which are used with mechanical extruders, may be composed of single or multiple cell types. They can be prefabricated according to histotypical requirements with specific internal structure, or they may contain specific molecules to enhance postprinting structure formation. These aggregates vary in size, typically in the range 100–500 μm, and are printed one by one. The use of spherical cell aggregates, composed of motile and adhering cells, as bio-ink particles is made possible by the fact that in many respects they mimic liquid drops. They are prepared from multicellular fragments that spontaneously round into spheres to minimize their surface area with the surroundings (interfacial area), just as liquid drops do. When an aggregate contains two distinct cell types with differing adhesive properties, sorting takes place: the more adhesive cells locate themselves in the center of the sphere surrounded by the less adhesive cells. This configuration is analogous to the one assumed by two intermixed immiscible liquids (such as oil and water), with the one of higher surface tension (water) being surrounded by the one of lower surface tension (oil). Finally, when two identical cellular aggregates come into contact they fuse into one spheroid, similar to the coalescence of liquid drops. The time course of aggregate fusion follows quantitatively the outcome predicted by the theory of liquids.

Postprocessing: biological structure formation. Bioprinting accomplishes the deposition of biological materials, but in itself does not assure the development of a functional tissue or organ. This is accomplished during postprocessing in a controlled environment (bioreactor), where the printed structure rapidly self-assembles. Postprinting structure evolution depends on a number of factors. Using inkjet technology, the initial shape of the structure can be controlled with high precision, but the final pattern depends on the ability of individual cells to properly adhere, interact, and differentiate in histotypical and organotypical fashion. In the case of cell aggregate printing, cells are in adhesive contact right away and thus in a more physiological environment. However, the printed pattern initially represents a collection of discrete aggregates. It is the self-organizing and liquidlike properties of these aggregates that assure that, under the right conditions, the aggregates will fuse or coalesce into the correct anatomical structure. Whereas the self-organizing capacity of these aggregates is due to their biological nature, the analogy with liquids is the consequence of their biophysical attributes. In either method, the outcome of postprocessing is to

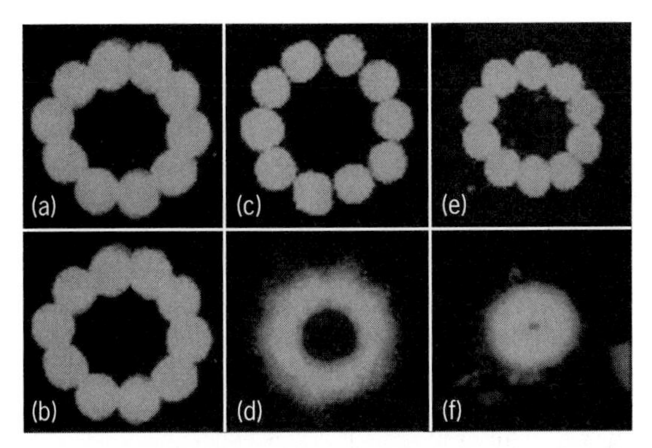

Fig. 3. Postprinting structure formation. Upper row: initial configuration of 10 aggregates (approx. 480 micrometers in diameter) of Chinese Hamster ovary cells, printed into (*a*) agarose gel, (*c*) 1.0-mg/ml collagen gel, and (*e*) 1.7-mg/ml collagen gel. Cells are fluorescently labeled. Lower row: final configuration of cells after 140 h. (*b*) Agarose does not favor aggregate fusion. (*f*) Collagen at 1.7 mg/ml is too permissive for cell movement: the structure eventually collapses into a single sphere, which has the lowest gel-cell interfacial energy. (*d*) Collagen at 1.0 mg/ml favors structure formation: a long-lived cellular toroid forms—in this case the layer-by-layer deposition of rings of aggregates—results in a cellular tube.

a large extent controlled by the properties of the biopaper (**Fig. 3**).

It is unlikely that it will ever be possible to engineer the minute details of such complex naturally occurring biological structures as organs, details of which are the result of millions of years of evolution. Fortunately, self-organization makes such attempts unnecessary. Therefore, it is the most crucial element of tissue engineering in general and organ printing in particular. Once the right initial cues are provided (including appropriate deposition pattern, and bioink and biopaper of the right composition), self-organization assures the development of the correct anatomical structure.

Perspective. Organ transplantation has saved the lives of countless patients and become a routine surgical procedure. However, there is a chronic shortage of available organs. Solutions presently at hand all have serious deficiencies. Implantable artificial organs, mostly mechanical devices, risk malfunction and from time to time have to be replaced. Transplantation of animal organs (xenotransplantation) raises serious ethical questions. Both methods also have immunological implications. It is clear that in the long run the fast and reproducible development of functional human organs under laboratory conditions is the most promising alternative to harvested organs. Organ printing could use the patient's own cells (autologous tissue engineering) or stem cells, providing a technology that could satisfy the increasing demand. Besides applications in regenerative medicine, the ability to construct functional biological structures would have far-reaching consequences in other areas as well. For example, in basic science it would make it possible to study questions such as biological pattern formation or tumor growth. In pharmaceutical applications it would allow the development of methods for patient-tailored drug testing.

For background information *see* BIOMEDICAL CHEMICAL ENGINEERING; CELL ORGANIZATION; COMPUTER-AIDED DESIGN AND MANUFACTURING; MEDICAL IMAGING; STEM CELLS; SURFACE AND INTERFACIAL CHEMISTRY; TISSUE CULTURE; TRANSPLANTATION BIOLOGY in the McGraw-Hill Encyclopedia of Science and Technology.

Gabor Forgacs; Vladimir Mironov

Bibliography. K. Jakab et al., Engineering biological structures of prescribed shape using self-assembling multicellular systems, *Proc. Nat. Acad. Sci. USA*, 101: 2864–2869, 2004; R. Langer and J. Vacanti, Tissue engineering, *Science*, 260:920–926, 1993; R. S. Langer and J. P. Vacanti, Tissue engineering: The challenges ahead, *Sci. Amer.*, 280:86–89, 1999; V. Mironov et al., Organ printing: Cell assembling cell aggregates as "bioink," *Sci., Med.*, 9:69–71, 2003; V. Mironov et al., Organ printing: Computer-aided jet-based three-dimensional tissue engineering, *Trends Biotechnol.*, 21:157–161, 2003; F. Oberpenning et al., De novo reconstitution of a functional mammalian urinary bladder by tissue engineering, *Nat. Biotechnol.*, 17:149–155, 1999; E. A. Roth et al., Inkjet printing for high-throughput cell patterning, *Biomaterials*, 25:3707–3715, 2004; M. S. Steinberg and T. J. Poole, Liquid behavior of embryonic tissues, in R. Bellairs, A. S. G. Curtis, and G. Dunn (eds.), *Cell Behavior*, pp. 583–697, Cambridge University Press, 1982; G. M. Whitesides and B. Grzybowski, Self-assembly at all scales, *Science*, 295:2418–2421, 2002; T. Xu et al., Inkjet printing of viable mammalian cells, *Biomaterials*, 26:93–99, 2005.

Transgressive segregation (plant breeding)

Extraordinary variation can be found within the plant kingdom, but how do extreme new forms arise? Although new mutations are generally thought of as the primary source of variation, they are not always necessary. In many cases, simply crossing two relatively similar varieties of plants will yield at least a few offspring with dramatically different traits. This phenomenon is called transgressive segregation.

Highly uniform inbred lines are generally used in plant breeding to maintain consistency over multiple generations. When two different inbred lines are crossed, the first-generation (F1) hybrids are also highly uniform, but are sometimes superior to both parents because of favorable allele combinations at each locus. However, when these plants produce seed, the resultant second generation (F2) is often extremely variable due to the varying assortment ("segregation") of alleles from both grandparents. Offspring with a majority of alleles from a single grandparent will be quite like that grandparent, but most offspring will have a mixture of alleles from both grandparents. In these offspring, many traits will be different from either grandparent—either developing at an intermediate level between the grandparents' traits or becoming more extreme ("transgressive") than either (see **illustration**). Transgressive segregation describes

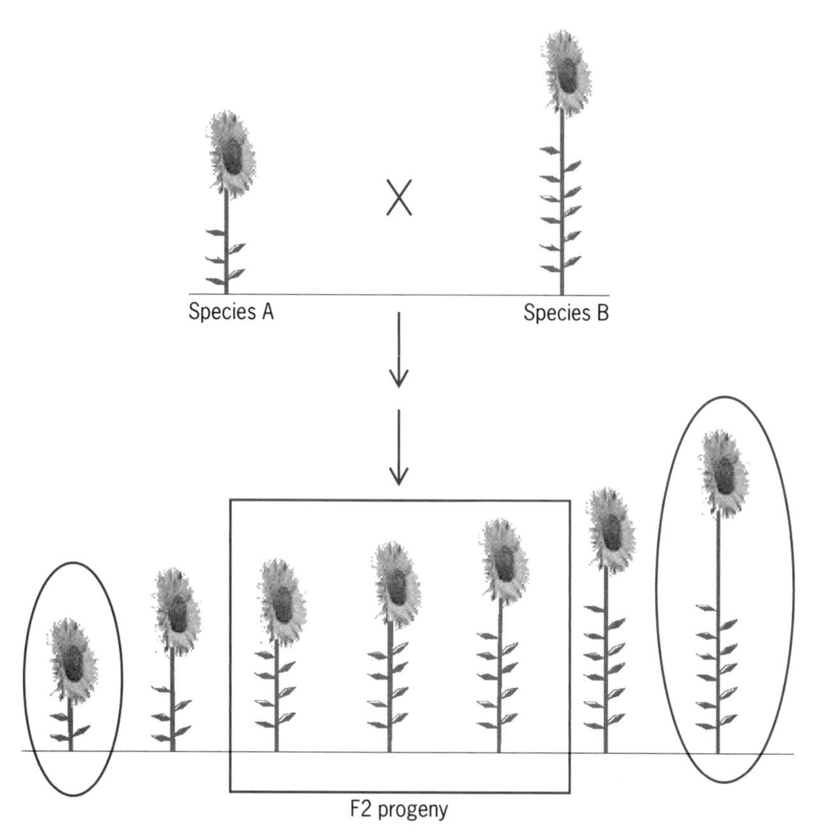

Example of transgressive segregation for height. Second-generation (F2) progeny in ellipses are transgressive for height, while those in a box are intermediate between the two parental lines. The remaining two lines resemble the parental lines.

the presence of phenotypic variation in F2 or later-generation hybrid offspring that exceeds the phenotypes of both original parents. Progeny are considered to be transgressive whether they are extreme in a positive or a negative direction (for example, much taller or much shorter than both parents). These extreme variations are often highly heritable, thus distinguishing transgressive segregation from the "hybrid vigor" of F1 generation hybrids, which is not transmitted uniformly to offspring. Transgressive segregation is commonly seen both in the progeny of crosses made between lines within a species (intraspecific crosses) and in the progeny crosses made between species (interspecific crosses).

Effect of multiple loci. Many explanations have been offered to account for the extreme phenotypes often found in F2 or later-generation hybrids, including elevated mutation rate, developmental instability, epistasis, overdominance, recessive alleles unmasked in the hybrid, and complementary action of alleles from both parents in the hybrid. Evidence based on crop breeding and genetic mapping suggests that complementary effects of alleles from both parents are the most common cause of transgressive segregation in hybrid progeny.

For genes from two individuals to combine to produce offspring that exceed parental means, the traits must be complex and determined by multiple genetic loci. Many traits, such as height, are controlled by a large number of loci. The different loci are frequently fixed for alleles with effects in opposing directions; some result in an increase in trait value and others result in a decrease in trait value. The presence of loci with opposing effects within the same genome may seem counterintuitive, but it is a logical outcome of the selective pressures experienced by many species. A combination of alleles within the genome of a given species affords a population the genetic flexibility to respond rapidly to selective pressures in changing environments. (For example, an environment that favors tall plants may change to favor short plants.) Because a trait such as height is achieved by combining loci fixed for alleles with positive effects (making an organism tall) and loci fixed for alleles with negative effects (making an organism short), two species with the same height might have very different ways of achieving this trait value genetically. Both might have alleles with opposing effects, but it is unlikely that the positive and negative alleles will be distributed identically across the genetic loci. Thus, a cross between individuals of these two species could result in some progeny with positive alleles at all the loci contributing to height determination (making them very tall) and some with negative alleles at all the loci contributing to height determination (making them very short) [see **table**]. In phenotypic terms, this is manifest as transgressive segregation.

Plant breeding. The fact that crosses between two strains can yield hybrids with extreme phenotypes is often used by plant breeders to improve crop species, and has been used to enhance a wide variety of important plants. For example, crosses between two strains of tomatoes, both of which were vulnerable to bacterial canker infection, produced a number of transgressive lines which were resistant. In another study, crosses between two semidwarf early-flowering rice strains yielded some transgressive lines which were both much shorter and earlier flowering than either parent. Crosses between barley strains have shown transgressive segregation for resistance to disease or stalk breakage, changes in height or yield, and many other characteristics.

More recent efforts to cross domesticated plants with wild or weedy relatives promise to enrich diversity even further. Untapped genetic variation may exist in wild populations, even for traits such as fruit size or yield that have already been vastly increased during domestication. The objective of these crosses is to identify genetic regions from wild plants that may enhance domestic traits, and then introduce those wild alleles into domestic genetic backgrounds. One of the best examples involves a cross of the domesticated tomato (*Solanum lycopersicum*) and its wild relative *S. pennellii*, which has much smaller fruit. In most cases, alleles from *S. pennellii* are associated with smaller fruit size compared to alleles from *S. lycopersicum*. However, because fruit size is controlled by many genes, it was not surprising to find some combinations with wild alleles that lead to bigger fruit than the *S. lycopersicum* alleles alone. Once these and other wild alleles were identified, they were introgressed into domesticated

Example of transgressive segregation for height due to the complementary action of alleles at many loci with additive effects

Locus	Effect of alleles at each locus on height			
	Species A	Species B	Transgressive F2*	Transgressive F2*
1	+1	−1	+1 (A)	−1 (B)
2	+1	−1	+1 (A)	−1 (B)
3	+1	−1	+1 (A)	−1 (B)
4	−1	+1	+1 (B)	−1 (A)
5	−1	+1	+1 (B)	−1 (A)
Resulting height	+1	−1	+5	−5

*Letters in parentheses indicate the species of origin for the alleles at each locus in the second generation crosses (F2's).

backgrounds using molecular genetic markers, leading to further improvements in fruit size and quality.

The great promise offered by transgressive segregation in crop breeding is not without challenges. One primary problem is that the transgressive segregants frequently develop in the wrong direction. Extreme phenotypes are often found to be far outside the range of the parental lines, but not in a useful way—fruits or yields are smaller, rather than larger, than either parent. Furthermore, even when alleles are found that improve traits in a hybrid, there is no guarantee that those alleles will improve domestic crops. The new allele may not function properly in popular domestic genotypes, and thus would provide no benefit (epistasis). In other words, an allele that depends on the products of many other genes may function well in one variety, but could be nonfunctional in a different variety unless compatible alleles are present at the other necessary genes. This is commonly true for biosynthetic pathways, which rely on multiple proteins, each of which is involved in one step toward the final product (for example, pigments for fruit or flower coloration). Another problem may occur if the new allele is associated with undesirable effects on other important traits (pleiotropy). This is the case for several perennial varieties, which can be harvested without replanting for several years. While these varieties provide the advantage of reduced labor and soil erosion, they also typically have lower yields than annuals because resources go to root growth rather than fruits or leaves. Nonetheless, transgressive segregation provides an important source of crop improvement and continues to be used to great advantage in plant breeding.

Role in the wild. The inter- and intraspecific hybridization techniques used by humans for crop breeding also occur naturally in the wild, often with transgressive segregation as a result. It has been suggested that the extreme traits found in transgressive hybrids might be advantageous in extreme environments that would otherwise be inaccessible, allowing the colonization of new habitats.

The importance of transgressive segregation in the wild has been well explored in the annual sunflowers (*Helianthus*) native to North America. The genus *Helianthus* contains three hybrid species that occur in widely divergent habitats, despite the fact that all three hybrids share the same two parental species. Both parental species are found in fairly moderate habitats; *H. annuus* is found in moderately wet, clay-based soils and *H. petiolaris* occurs in slightly dryer, sandier soils. The hybrid species, in contrast, are found in extreme habitats in the desert southwest; *H. anomalus* is found on active sand dunes, *H. deserticola* occupies extremely dry habitats in the Great Basin Desert, and *H. paradoxus* inhabits desert salt marshes with salinity often exceeding that of typical seawater. Greenhouse comparisons have identified a suite of transgressive traits that differentiate each taxon from its parental species and that seem likely to confer a fitness advantage to each hybrid species in its native habitat. For example, *H. deserticola* has early flowering and small leaves compared to the parental species; both traits are typical of desert annuals. *Helianthus anomalus* has evolved an extremely long taproot to obtain water deep underneath sand dunes, and can grow out from beneath burying sand. *Helianthus paradoxus* tolerates water saltier than the ocean, thriving in places where few other organisms can grow. Multiple experiments conducted in the greenhouse and in the natural habitats of the three hybrid species revealed that the hybrid species extreme trait values can be fully replicated in at least a few individuals by making hybrids between the parental species. These newly derived hybrids could then grow and thrive in their extreme habitats. Overall, the experiments have shown that extreme or transgressive phenotypes for ecologically relevant traits can be created via hybridization. This work was extended to determine the genetic mechanism underlying the generation of extreme trait values in the synthetic hybrids, and mapping quantitative trait loci revealed that extreme or transgressive trait values were almost certainly generated by complementary gene action rather than new mutation.

Summary. The nature of the genetic mechanisms underlying transgressive segregation has implications for its importance in both applied and natural settings. Because transgressive segregation relies on additive genetic variation, extreme traits resulting from complementary effects of alleles from two parents can be maintained and fixed through natural or artificial selection, leading to stable new strains with traits quite different from either parent. Transgressive segregation is also interesting in that its mechanism yields novel variation rapidly without depending on potentially detrimental mutagenic events. Because of these two features, transgressive segregation can be extremely important in generating heritable novel variation in both domestic and wild settings.

For background information *see* ALLELE; BREEDING (PLANT); GENE ACTION; GENETICS; SPECIATION; SUNFLOWER; TOMATO in the McGraw-Hill Encyclopedia of Science & Technology.

Nolan C. Kane; Briana L. Gross; Loren H. Rieseberg

Bibliography. M. L. Arnold, *Natural Hybridization and Evolution*, Oxford University Press, New York, 1997; J. A. Coyne and H. A. Orr, *Speciation*, Sinauer Associates, Sunderland, 2004; M. Lynch and B. Walsh, *Genetics and Analysis of Quantitative Traits*, Sinauer Associates, Sunderland, 1998; L. H. Rieseberg et al., Major ecological transitions in wild sunflowers facilitated by hybridization, *Science*, 301(5637):1211–1216, 2003; N. W. Simmonds and J. Smartt, *Principles of Crop Improvement*, Blackwell Science, Malden, 1999.

Trapped-ion optical clocks

Increasingly stringent demands on atomic timekeeping, driven by applications such as global navigation satellite systems (GNSS), communications, and very long baseline interferometry (VBLI) radio astronomy, motivate the development of improved time and frequency standards. This article discusses frequency standards and clocks based on optical transitions in trapped, singly ionized atomic systems. The year 2005 marked the fiftieth anniversary of the first working cesium atomic clock. Cesium microwave frequency standards have since improved in accuracy by an order of magnitude per decade. Today the best primary standards can realize the second, as defined in the International System of Units (SI), with an accuracy of better than 1 part in 10^{15}. However, this is near the limit to which a standard based on the cesium hyperfine transition can be developed. Standards based on optical transitions are anticipated to reach two or more orders of magnitude greater accuracy and stability. It is widely expected that the development of such optical clocks will lead to the redefinition of the SI second in terms of an optical transition in place of the cesium hyperfine transition.

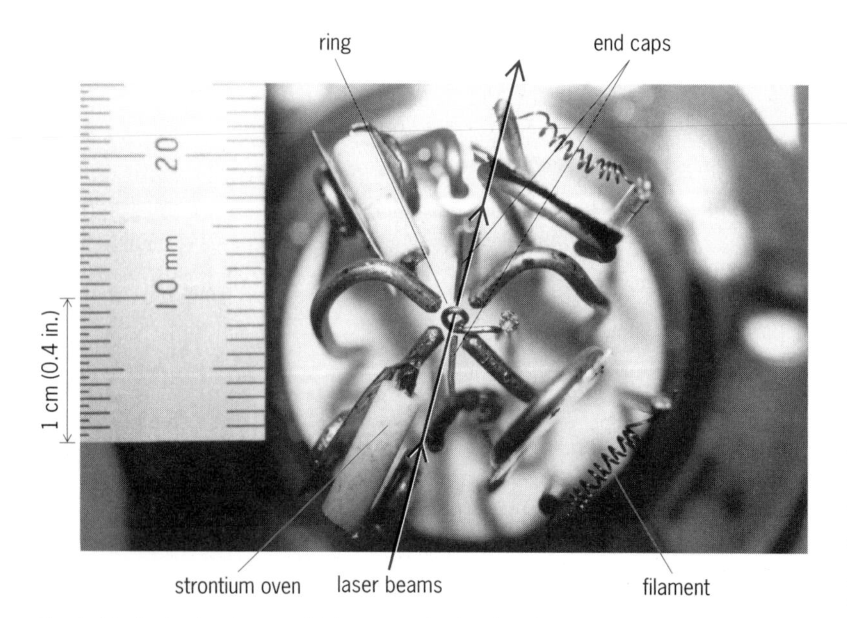

Fig. 1. Electrode structure used to trap a single strontium ion.

The accuracy and stability of any frequency standard is closely related to its quality factor Q, the ratio of its resonant frequency f to the frequency width Δf of the resonant feature. A clock based on a weak optical transition between a long-lived metastable state and a lower ground state can have at least 10^5 times higher Q than microwave standards. The best cesium primary standards have a Q of about 10^{10}; to achieve a precision better than one part in 10^{15}, the center of the resonant feature has to be found to better than one part in 10^5. This is achieved using the signal from many atoms. In contrast, various atoms and ions possess metastable states and hence "clock" transitions with natural linewidths at the subhertz level, corresponding to a Q of around 10^{15}. Thus improved precision can be obtained using only a single ion. Ion-based clocks have the advantage that a single ion can be confined in a nearly perturbation-free environment, whereas atom-based clocks may have the advantage that by using a large ensemble of atoms they may display better short-term stability. The principal features of transitions currently being studied are presented in the **table**.

The use of optical frequency standards as optical clocks has become possible as a result of the development of optical frequency "combs" of accurately and equally spaced frequencies, based on femtosecond lasers. The award of the 2005 Nobel Prize for Physics to T. Hänsch and J. Hall recognizes the importance of these devices. Such femtosecond combs enable the stability of an optical frequency standard to be transferred to the microwave region for comparison with the cesium primary frequency standard; they made possible the first demonstration of a trapped ion optical clock, at the laboratories of the National Institute for Standards and Technology (NIST) in Boulder, Colorado, in 2001. This NIST mercury ion standard and others worldwide have since been measured relative to cesium using femtosecond combs (see table). In 2004, for example, a group at the United Kingdom National Physical Laboratory (NPL) demonstrated a fractional frequency accuracy of 3.4 parts in 10^{15}.

Ion traps. For an atomic transition to be used as a frequency standard, it must be free of any perturbation that could change its frequency. A single ion, confined in the dynamic quadrupole electric field of a radio-frequency trap, is a good approximation to such a perturbation-free environment. Variations on this type of trap, invented by W. Paul, are in use, differing in their electrode geometry. **Figure 1** illustrates a design consisting of two grounded electrodes and a ring electrode, to which an oscillating electric potential with a frequency of several megahertz and an amplitude of a few hundred volts is applied, generating a pseudo-potential with a trap well depth of a

Candidates for trapped ion optical clocks and their published experimental status. The observed transition linewidth is limited by the probe laser linewidth and the interrogation time*

	Transition		Current status		
				Frequency uncertainty	
Species	Wavelength	Natural linewidth	Observed linewidth	Absolute	Fractional
$^{27}Al^+$	266 nm	0.5 mHz	—	—	—
$^{40}Ca^+$	729 nm	0.14 Hz	1 kHz	—	—
$^{88}Sr^+$	674 nm	0.4 Hz	70 Hz	1.5 Hz	3.4×10^{-15}
$^{115}In^+$	236 nm	1 Hz	170 Hz	230 Hz	2×10^{-13}
$^{171}Yb^+$	435 nm	3 Hz	30 Hz	6 Hz	9×10^{-15}
$^{171}Yb^+$	467 nm	0.5 nHz	180 Hz	600 Hz	9×10^{-13}
$^{199}Hg^+$	282 nm	1.7 Hz	7 Hz	11 Hz*	1.0×10^{-14}

* An uncertainty of 1.5 Hz was reported in 2005.

few electronvolts. Ions are loaded into the trap either by electron bombardment or by photoionization of an atomic beam, the latter technique having the advantage of avoiding contamination of the electrodes, which can lead to small changes in the trapping potential and hence a perturbation of the transition frequency. By laser cooling on its fast resonance transition (**Fig. 2***a*), the ions' motion is confined to a region smaller than the clock transition wavelength, further reducing its sensitivity to external perturbations of the transition frequency, as discussed below.

Probe laser. The oscillator used to interrogate the clock transition is a stable laser that has a linewidth comparable with the natural linewidth of the clock transition. This can be achieved by stabilizing the laser frequency to a highly stable, narrow-linewidth optical cavity. High stability is achieved by making the cavity out of a material that has a very low thermal expansivity and by minimizing its sensitivity to vibrations. The cavity is mounted in a vacuum chamber on a vibration-isolated platform. Subhertz linewidth probe lasers have been demonstrated with drift rates of fractions of a hertz per second.

Quantum jumps. The clock transition is interrogated using the "electron shelving" technique first proposed by H. Dehmelt. The laser radiation, tuned to the cooling or resonance transition, causes the ion to oscillate between the ground state and a short-lived excited state, scattering photons at a high rate. The probe laser induces transitions from the ground state to the long-lived metastable state (Fig. 2*a*). An ion in the metastable state no longer scatters cooling transition photons, giving rise to a "quantum jump" in the number of scattered photons (Fig. 2*b*). The rate of quantum jumps provides a signal for stabilizing the probe laser frequency to the optical clock transition.

Femtosecond comb. The optical frequency stability of the probe laser is transferred to the radio-frequency or microwave region using a femtosecond comb (**Fig. 3**). In the frequency domain, the train of femtosecond pulses emitted by a mode-locked laser appears as a comb of optical frequency modes. These "teeth" of the comb are equally spaced by the pulse repetition rate f_R, which is typically in the range from 100 MHz to 1 GHz. A radio-frequency signal is obtained at f_R by detecting the optical pulse train on a fast photodiode. The comb modes are offset from integer multiples of f_R by a frequency f_0, which arises from the difference in phase of the optical wave relative to the pulse envelope for successive pulses.

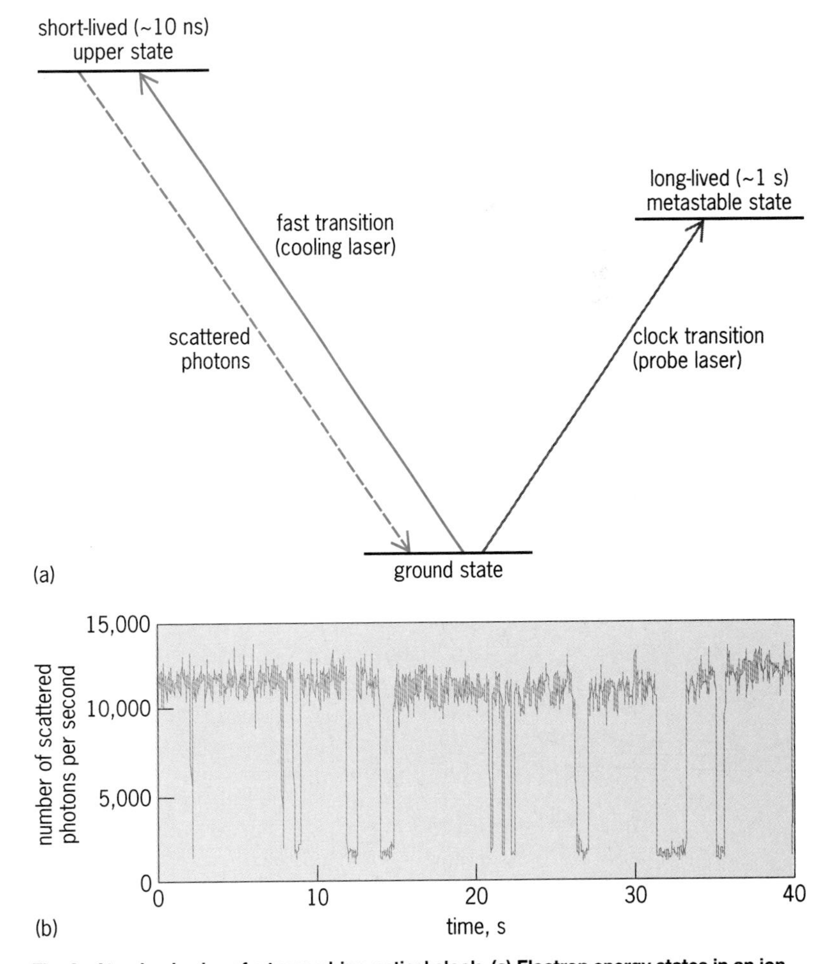

Fig. 2. Atomic physics of a trapped-ion optical clock. (*a*) Electron energy states in an ion having a weak clock transition, which serves as the optical reference, and a strong transition for laser cooling and state detection. (*b*) Quantum jumps in the number of cooling transition photons induced by transfer of the ion from the ground state to the metastable state. The variable widths of the jumps reflect the random times that the ion remains in the metastable state, and can be used to measure the natural linewidth of the clock transition.

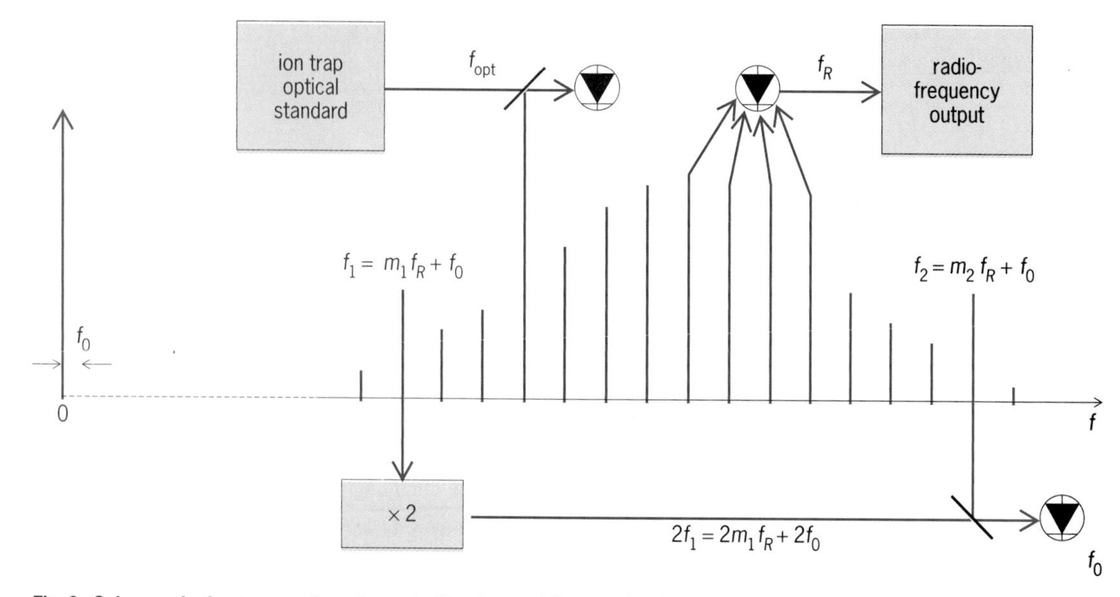

Fig. 3. Scheme of a femtosecond comb used with a trapped-ion standard to make an optical clock, illustrating the self-referencing technique used to measure the carrier-envelope offset frequency f_0.

The comb spectrum of the laser, which typically spans a wavelength range of 30 nm at around 800 nm, can be broadened to cover at least an octave of optical frequency by self-phase modulation of the pulses in a highly nonlinear optical fiber. This greatly increases the utility of the comb, enabling it to address a wide range of optical frequency standards. It also enables f_0 and hence the absolute frequency of the comb modes to be determined by a self-referencing technique. Modes from the left-hand (infrared) end of the comb are frequency-doubled by second harmonic generation in a nonlinear crystal. The beat frequency of these doubled modes with modes from the right-hand end of the comb is the offset frequency f_0 for modes for which $m_2 = 2m_1$.

When the mth comb mode, of frequency $f_{opt} = mf_R + f_0$, is stabilized to the optical standard, the repetition-rate beat frequency f_R is equal to the optical frequency f_{opt} divided by m (apart from a small correction equal to f_0/m); f_R then reproduces the accuracy and frequency stability of the optical standard. Femtosecond combs based on mode-locked Ti:sapphire lasers of differing design have already been demonstrated to agree at the level of 1.4×10^{-19} in fractional frequency uncertainty.

Accuracy and stability. Confinement of the ion in a region smaller than the clock transition wavelength at the center of the trapping potential (called the Lamb-Dicke regime) enables shifts of the transition frequency due to the ion's motion to be suppressed. In particular, the Doppler shift is eliminated to first order. Moreover, by constraining the ion to the center of the trap, the quadrupole electric field and its gradient are minimized. The quadrupole shift, due to the interaction of the ion's quadrupole moment with the residual gradient of the trapping field, can be eliminated by methods that depend on the shift's symmetry properties. Frequency shifts due to static and time-varying magnetic fields can be reduced by the use of an isotope with a magnetic-field-insensitive transition. Currently, these shifts have been measured and controlled at the parts-in-10^{15} level or better; with improved experimental techniques, reduction to the parts-in-10^{18} level is anticipated.

For an ion trap operated at room temperature, a correction has to be applied for the blackbody frequency shift due to the thermal radiation field of the trap environment. The magnitude of this correction has to be calculated from atomic physics theory or determined by experiment; it may be that the uncertainty in this blackbody shift due to nonuniformity in the thermal emissivity of the trap components and the uncertainty in the ambient temperature will limit the accuracy of optical clocks at the parts-in-10^{18} level.

The fractional frequency stability of a trapped-ion optical clock based on a hertz-wide optical clock transition is expected to be at the part-in-10^{-16} level at 100 seconds averaging time. Experimentally, a stability of 10^{-15} has been demonstrated with a single mercury ion (at 100 s) and a ytterbium ion (at 1000 s); this is significantly better than microwave cesium fountain and hydrogen maser capability. Projected stabilities for clocks based on subhertz wide transitions, such as the 467-nm octupole transition in the ytterbium ion, are several orders of magnitude below these values.

Navigation applications. Global navigation satellite systems (GNSS), such as GPS and the future European Galileo system, rely on accurate microwave atomic clocks both on the ground and in space. Optical clocks could improve positional accuracy and reduce reliance on regular ground-to-satellite time transfer; they may eventually be used in the Deep Space Network, which guides spacecraft through the solar system.

Scientific applications. Fundamental science experiments such as tests of Einstein's relativity theories, searches for gravitational waves, and tests of the stability of the fundamental constants rely on ultrastable clocks and frequency standards. Attempts to unify gravitation with the strong and electroweak interactions predict violation of Einstein's equivalence principle. One manifestation of this would be a time variation of fundamental constants such as the fine-structure constant, α. The accuracy now obtained with trapped-ion optical frequency standards enables very precise limits to be placed on the present-day variation of α. By combining repeated measurements, at an interval of a few years, of different optical frequency standards relative to cesium primary standards, a limit of $(-0.3 \pm 2.0) \times 10^{-15}$ per year can currently be placed on the fractional time variation of α.

Optical clocks. Trapped-ion optical clocks are leading the way to a new generation of time standards based on an optical rather than a microwave atomic reference. More accurate clocks will aid improved terrestrial navigation and surveying, as well as space travel to distant planets. It is widely anticipated that, once measurement of the accuracy of optical clocks becomes limited by the present definition of the SI second in terms of the cesium hyperfine transition frequency, the second will be redefined in terms of an optical transition—a new era of time measurement will have been born.

For background information *see* ATOMIC CLOCK; ATOMIC STRUCTURE AND SPECTRA; FREQUENCY MEASUREMENT; FUNDAMENTAL CONSTANTS; LASER; LASER COOLING; LASER SPECTROSCOPY; NONLINEAR OPTICS; PARTICLE TRAP; PHYSICAL MEASUREMENT; Q (ELECTRICITY); SATELLITE NAVIGATION SYSTEMS in the McGraw-Hill Encyclopedia of Science & Technology. Hugh Klein; Stephen Lea

Bibliography. S. A. Diddams et al., An optical clock based on a single trapped $^{199}Hg^{+}$ ion, *Science*, 293: 825–828, 2001; 50 Years of Atomic Time-Keeping: 1955 to 2005 (Special Issue), *Metrologia*, 42(3):S1–S153, June 2005; H. S. Margolis et al., Hertz-level measurement of the optical frequency in a single $^{88}Sr^{+}$ ion, *Science*, 306:1355–1358, 2004; *Proceedings of the First ESA Optical Clock Workshop*, 2005; F. Riehle, *Frequency Standards*, Wiley-VCH, 2004.

Vehicle-highway automation systems

Vehicle-highway automation systems support drivers by automating portions of the driving task, and eventually will be able to take over the entire driving task. This support includes warning of an imminent hazard, assisting the driver in controlling the vehicle's speed and separation from other vehicles or its steering, braking to avoid a crash, and in the longer term taking full responsibility for driving the vehicle.

Vehicle-highway automation systems belong to the broader field of intelligent transportation systems (ITS), which apply information technologies to improve road transportation operations. Intelligent transportation systems use information to integrate the roadway infrastructure, the vehicles that drive on it, and the people and goods that move in those vehicles. The integration of these elements will enable the transportation system to operate more efficiently and safely than it has in the past. Some vehicle-highway automation systems rely primarily on technologies installed in vehicles, while others rely on cooperation between vehicle and roadway infrastructure technologies.

Applications to heavy vehicles. Although the automotive consumer market is the largest ultimate application for vehicle-highway automation systems, the early adopters of these systems will be transit buses and heavy trucks. The owners and operators of these vehicles can gain a faster return on their investment using these technologies because of the higher value, heavier use, and higher operating costs of their vehicles. Urban transit buses operating on fixed routes are particularly appropriate early adopters of systems that involve cooperation with roadside infrastructure devices, which initially can be installed at a limited numbers of locations where they are most needed.

Safety warning systems. The most basic level of automation involves safety warning systems, which alert drivers to hazards in their immediate vicinity, detected by onboard sensors. These alerts are typically auditory but may also be haptic, kinesthetic, or visual. The visual warning is least favored because it is unlikely to attract a fatigued or inattentive driver or it might divert the driver's attention from the driving scene.

Forward collision warning. Forward-ranging sensors, using millimeter-wave radar or laser radar (lidar), detect the vehicle's range and closing rate to obstacles in the forward path of the vehicle, generally within a range of 100–150 m (330–500 ft), and determine whether to alert the driver (**illus.** *a*). These systems are designed to detect other vehicles, but as the sensor technologies improve, they should be able to detect other, less salient obstacles. The major challenge in design of these systems is distinguishing stationary objects in the path of the vehicle from other stationary objects in the roadway environment that are not threats (such as roadside or overhead signs, lamp posts, bridge supports, and guard rails) and would be considered false alarms by the driver. These systems are widely available on heavy trucks but have not yet been introduced to the passenger car market in the United States.

Lane-departure warning. Video-image-processing sensors detect lane-boundary markings and use them as a position reference to predict if a vehicle is drifting out of its lane. If a lane departure is predicted and the driver has not activated the directional signals to indicate an intentional lane change, the system alerts the driver by an auditory or haptic warning. This is typically an in-vehicle representation of the sound or vibration of a rumble strip, which can be conveyed through the entertainment system's speakers on the correct side of the vehicle or by actively vibrating that side of the driver's seat. These systems have been introduced to both the heavy-truck and passenger-car markets in the United States.

Lane-change warning. Video or radar sensors mounted by the side-view mirrors and/or the rear of the vehicle scan the blind spots adjacent to the vehicle and

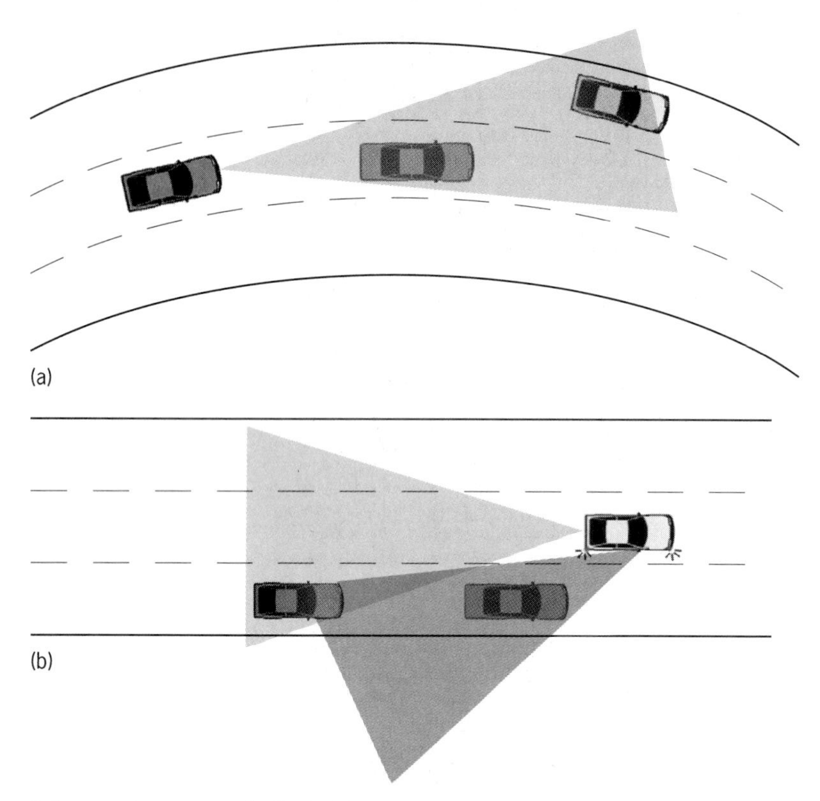

Safety warning systems. (*a*) Field of view of forward collision warning and adaptive cruise control systems, showing how a sensor at the front of a vehicle could detect vehicles ahead of it and how it also would need to distinguish between the vehicle in its lane and the vehicle in the adjacent lane. (*b*) Fields of view of a lane-change warning system, showing a vehicle signaling a lane change to the right, with a sensor mounted near the right side mirror detecting a vehicle in the blind spot and another sensor mounted at the rear of the vehicle detecting a vehicle in the adjacent lane overtaking from behind.

the adjacent lanes behind the vehicle to detect other vehicles that would be threats if the driver changed lanes (illus. *b*). For blind-spot threats, the sensors detect the presence of another vehicle, but for vehicles approaching from behind, they need to be able to detect the range and closing rate in order to predict possible conflicts. The alert typically will be provided when the driver activates the turn signal indicator, so that a warning can be issued before the vehicle has left its original lane and false alarms can be eliminated in situations when the driver does not intend to change lanes. Systems that detect blind-spot threats have been commercially available for heavy trucks for several years, while the more comprehensive lane-change warning systems for trucks and passenger cars are under development.

Intersection collision warning. Intersections are the most complicated part of the roadway environment and therefore the most challenging for collision warning systems. Because the vehicles that could be in conflict at the intersection may be approaching from different directions and not within the line of sight of each other until they are very close to the intersection (and to a collision), intersection collision warnings are more dependent on sensors installed in the roadway infrastructure (such as inductive loops in the pavement, magnetic sensors on the road surface, or video or radar sensors above the roadway).

Intersection-collision-warning systems are being designed to address the conflicts associated with the largest numbers of intersection crashes, which are violations of stop signs and red traffic signals, as well as drivers' misjudgments of the gaps available for making left turns (with respect to cross traffic at stop signs and with respect to opposing traffic at green traffic signals). When impending conflicts are detected, the systems illuminate dynamic signs at the intersection, and with the availability of wireless communications to vehicles, they will be able to provide in-vehicle auditory or haptic alerts as well. Since several challenging technical issues are involved, these systems are currently in the research and development stage.

Control assistance systems. Control assistance systems are designed to improve driving comfort and convenience by relieving drivers of lower-level driving tasks, while leaving them responsible for detecting and responding to hazardous conditions.

Adaptive cruise control (ACC). Adaptive cruise control is an extension of conventional cruise control, providing the additional capability of maintaining a desired separation behind a preceding vehicle. It uses a forward-looking radar (millimeter wave or infrared) to detect the range and closing rate relative to the preceding vehicle, and then uses those measurements to determine how fast the subject (following) vehicle should be driving (illus. *a*). The desired separation is maintained by controlling the throttle and brake within a limited range of accelerations. The driver typically has several choices of separation distance, normally in the range of 1–2 seconds (the ratio of the separation between vehicles to the speed of the subject vehicle). These systems have been available for several years on heavy trucks and passenger cars in the United States.

Lane-keeping assistance. Lane-keeping assistance relies on the same type of lane detection sensors as lane-departure warning, but uses the information about the vehicle's position and orientation, relative to the lane, to steer the vehicle and help center it in the lane. The steering action is taken over by a small motor in the steering system, which provides a moderate torque that can be easily overridden by the driver. An extension of this capability to lane-departure prevention uses differential braking on the opposite sides of the vehicle to provide the steering effect.

Higher-performance systems, using other lane-detection technologies [such as magnetic markers in the pavement, or the Global Positioning System (GPS)] are under development for transit buses to enable them to drive in narrow lanes and approach bus stops with accuracy comparable to the highest-quality rail transit systems, making it easier for passengers to enter and exit across a small gap between the vehicle floor and loading platform.

Collision mitigation and avoidance systems. Collision mitigation and avoidance systems provide a stronger intervention into driving than control-assistance systems, by applying the brakes under emergency conditions when collisions are imminent and possibly unavoidable. They are based on technologies similar to the adaptive cruise control and forward-collision warning, but because they require higher sensor capabilities to provide sufficiently reliable performance (to avoid erroneous hard braking), they are still under development.

Automated highway systems (AHS). The sensor and actuator technologies that are being applied in support of driver-assistance systems also can generally be applied to automating the driving of the vehicles, although for this application the requirements for accuracy, reliability, and fault management are considerably higher. Research and development on automated highway systems have been in progress for many years, and some systems have already been demonstrated and applied on test tracks. Additional development work is needed to improve the reliability and fault tolerance of the systems, but initial applications under the supervision of professional drivers of buses and trucks should be feasible within the next 10 years.

For background information *see* ALARM SYSTEM; AUTOMOBILE; CONTROL SYSTEMS; GUIDANCE SYSTEMS; HIGHWAY ENGINEERING; LIDAR; MICROSENSOR; RADAR; SATELLITE NAVIGATION SYSTEMS; TRAFFIC-CONTROL SYSTEM in the McGraw-Hill Encyclopedia of Science & Technology.

Steven E. Shladover

Bibliography. P. Ioannou (ed.), *Automated Highway Systems*, Plenum Press, New York, 1997; W. D. Jones, Keeping cars from crashing, *IEEE Spectrum*, pp. 40–45, September 2001; M. Parent, L. Vlacic, and F. Harashima (eds.), *Intelligent Vehicle Technologies*, SAE International, Warrendale, PA, 2001; S. E. Shladover, Introducing intelligent transportation systems: Paradigm for 21st century transportation, *Transport. Res. News*, no. 218, pp. 4–9, January/February 2002 (http://trb.org/news/blurb_detail.asp?id=573); S. E. Shladover, What if cars could drive themselves?, *ACCESS*, no. 16, pp. 2–7, University of California Transportation Center, Spring 2000 (http://www.uctc.net/access/access.asp#16).

Fig. 1. Demonstration of an obligate vernalization requirement. At left is a biennial variety cabbage (*Brassica oleracea*) that had been growing for 5 years without cold exposure. The child is holding a summer-annual variety of *B. oleracea* that flowers rapidly without vernalization.

Vernalization

Vernalization is the process by which plant flowering is promoted by prolonged exposure to the cold of a typical winter. Such a system evolved in certain plants to ensure that flowering occurred at the proper time of year. In some plants the requirement for extended exposure to cold is absolute; that is, they will not flower until the proper duration of cold exposure has been perceived (**Fig. 1**). Other plants may exhibit a quantitative response; that is, flowering will eventually occur without vernalization, but flowering is greatly accelerated by cold exposure.

In horticultural terms, plants can be described as annuals (which set seed once and then die) or perennials (which can flower and set seed multiple times over multiple growing seasons). Many annual plants complete their life cycle over a single summer season; such plants are called summer annuals. Winter annuals and biennials typically require two growing seasons to complete their life cycle because of the need for vernalization. Vernalization also promotes flowering in many perennials.

For many winter annuals and biennials, the optimal growing seasons are fall and early spring. These plants are often well adapted to growing in cool weather, and this permits them to occupy a niche that reduces competition from plants that thrive in the summer season (**Fig. 2**). These plants become established in the fall and flower in the spring. Such plants need a mechanism to ensure that flowering does not occur in the fall (flowering before the onset of winter would be futile) and enable flowering to proceed rapidly in spring before the competition becomes intense. Measuring the duration of exposure to the cold of winter provides such a mechanism. In order to be effective, such a system needs to be able to discriminate between temperature fluctuations in the fall and the longer period of cold that comprises a typical winter. Thus short exposures to cold are usually not adequate to promote flowering. There is, however, much variation in the length of cold

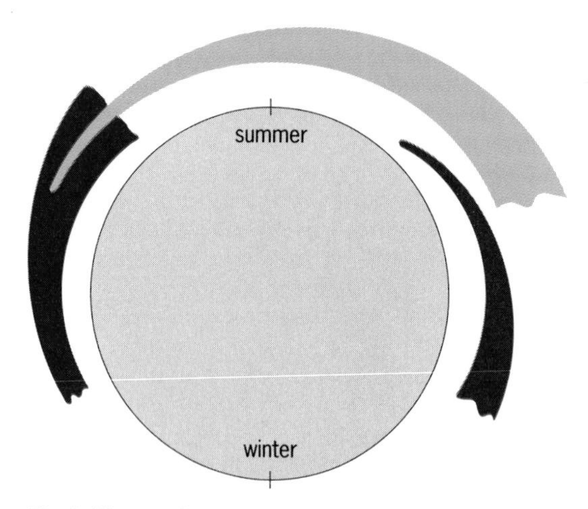

Fig. 2. The growing season of a typical crop plant or deciduous tree (light color) versus the growing season of a typical winter annual (dark color). The growing season of winter annuals is split by winter and somewhat restricted to times when other plants are not growing.

exposure required to promote flowering as well as the optimum cold temperature needed by different species or even different strains of the same species. This variation is not surprising because the vernalization requirement has developed to match the length of winter and the average winter temperatures of the region to which a particular species or strain has become adapted. Genetic flexibility in these parameters is likely to be an important component of successful radiation of species into new climates. It is interesting to note that for many plant species that grow in temperate climates, the release of bud dormancy is also dependent upon exposure to cold of the proper duration. Thus a mechanism to "count" days of cold exposure is used for more than one developmental process in plants.

Physiology of vernalization. As discussed above, the temperature that is most effective in the vernalization-mediated promotion of flowering varies, and the optimum temperature is presumably an adaptation to a particular climate. The flowering response of a particular species to temperature can generally be represented by a bell-shaped curve (**Fig. 3**). Temperatures below the optimum are probably less effective because the plant is too cold to respond. The vernalization process requires metabolic activity and changes gene expression; thus, below the temperature optimum, the ability to accomplish the biochemical changes required to achieve the vernalized state occur more slowly, and at temperatures several degrees below freezing the biochemical changes may not occur at all. Cold temperatures above the optimum may not represent a true winter, and the response to these is attenuated to prevent a plant from flowering at the wrong time of year.

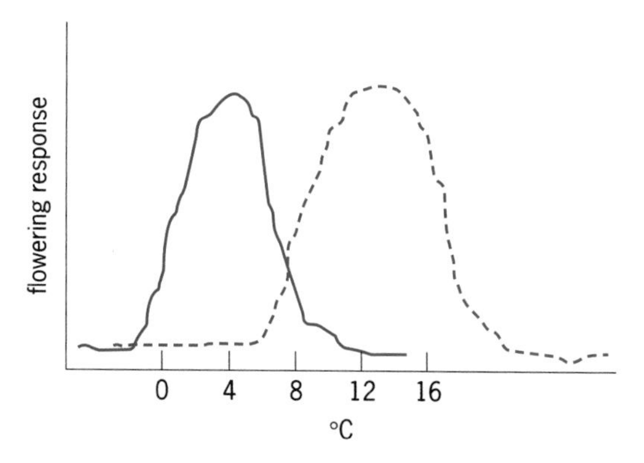

Fig. 3. Typical flowering response versus temperature curves. Shown are idealized data for two species that are adapted to different climates and thus have different temperature optima.

In most vernalization-responding species, the shoot apex is the site of cold perception and response. This is readily demonstrated by grafting. Grafting of vernalized and nonvernalized shoot apices to, respectively, nonvernalized and vernalized stocks shows that a vernalized shoot apex is competent to flower regardless whether or not the plant to which it is grafted is vernalized; and that flowering of a nonvernalized shoot apex is not typically promoted by grafting to a vernalized plant. Thus, in most species, vernalization corresponds to a local change in the ability of a shoot tip to flower.

The vernalized state can be stable in certain species. One of the classic demonstrations of this is from the studies of Anton Lang and Georg Melchers with biennial henbane in the 1960s. For the henbane to flower the plants needed to first be exposed to the proper length of cold and then be exposed to proper daylengths (inductive photoperiods). After vernalization had been completed, if the vernalized henbane plants were grown in noninductive photoperiods (to prevent flowering) for a time, the plants could "remember" the prior cold exposure. This was evident because when the plants grown in noninductive photoperiods were shifted back to inductive photoperiods they flowered. Vernalization therefore establishes a cellular memory that is stable through mitotic cell divisions; this acquisition of cellular memory can be considered an epigenetic switch—that is, it is a stable switch in gene expression (in this case caused by a change in the plant's environment), while the genes themselves remain unchanged.

Mechanism of vernalization. The molecular mechanism of vernalization is being characterized in several plant species. The foundation for such studies is genetic analysis. One type of genetic study is to use natural variation to identify the genes that create a requirement for vernalization (that is, to identify the genes that prevent flowering unless a plant has been exposed to winter). Such studies can be performed in species where there are both summer-annual and winter-annual varieties by simply crossing the two types and analyzing the subsequent generations for the segregation pattern of the summer and winter habit. The first example of this was the demonstration in the early 1900s by C. Correns, one of the rediscoverers of Mendel's principles of genetics, that there was a single gene difference between summer-annual and biennial varieties of henbane. In this case a dominant version (allele) of this gene created a block to flowering that required vernalization to overcome it.

This approach has also been performed in *Arabidopsis thaliana* and has resulted in the identification of two genes, *FRIGIDA (FRI)* and *FLOWERING LOCUS C (FLC)*, for which dominant alleles cooperate to create a vernalization requirement. The cloning of *FLC* revealed the first molecular aspects of vernalization. *FLC* encodes a protein that represses the expression of genes that are necessary for flowering to occur. In vernalization-requiring types, the dominant allele of *FRI* causes *FLC* to be expressed at high levels in the first growing season. Exposure to a long period of cold shuts off *FLC* expression (despite *FRI* expression), enabling plants to flower in the spring. Thus, in the case of *Arabidopsis*, vernalization promotes flowering by causing the repression of a repressor. Once *FLC* is shut off by cold it remains off for the remainder of the plant's life cycle—that is, through a series of mitotic cell divisions and regardless of subsequent temperature shifts. However, *FLC* is expressed again in the next generation to ensure that this next generation does not flower in the fall.

Another genetic approach is to induce mutants by treating plants with mutagenic agents and to screen the resulting mutants for alterations in a process of interest. Screening for mutants in which flowering is not promoted despite vernalizing conditions identifies genes that are required for the vernalization process. These screens have identified genes involved in chromatin remodeling as being necessary for vernalization. Chromatin remodeling involves changing the spectrum of covalent modifications of the histone proteins in the chromatin of a particular gene. The spectrum of histone modifications at a given locus is often referred to as the "histone code" because the specific combination of modifications can create an "on" or "off" state of gene expression. In the fall season, *FLC* chromatin is in an "on" conformation that is associated with an enrichment of specific histone modifications such as acetylation of lysine residues 9 and 14 and trimethylation of lysine 4 of histone 3. After passage through winter and completion of the vernalization process, the levels of these "on" modifications are reduced, and *FLC* chromatin becomes enriched in methylation of lysine 9 and 27 of histone 3. Methylation of lysine 9 and 27 is characteristic of chromatin in the "off" state, and these modifications are thought to lead to a mitotically stable repressed state by recruiting repressor complexes that create a stable autoregulatory loop of gene repression.

One question that is just beginning to be addressed is whether the mechanism of vernalization that operates in *Arabidopsis* is present in other types of plants. Recent studies indicate that, as expected, a similar system based on an *FLC*-like repressor that is shut off by exposure to cold is present in relatives of *Arabidopsis* (the crucifer family) such as cabbage and radish. In cereals (such as wheat and barley), vernalization also appears to involve the cold-mediated shutoff of a gene encoding a repressor of flowering called *VRN2*. However, the *VRN2* repressor is a different type of protein than *FLC*. This indicates that the vernalization response may have evolved independently in crucifers and cereals. Indeed, the major groups of flowering plants may have evolved in warm climates in which a vernalization response was not needed. In this scenario, the vernalization response would have evolved independently as different groups of plants radiated from warm climates into regions where they had to deal with flowering regulation through a winter season.

Agricultural applications. Cabbage, beets, and carrots are examples of crops that are grown for their leaves or underground storage organs, and such crops are ruined by flowering. Breeders have therefore introduced into the genomes of these crops a requirement for a long period of cold exposure to attempt to ensure that they do not flower in the first growing season. In winter cereals, the vernalization requirement permits the crop to be sown in the fall and ensures that flowering and seed set will only commence in the favorable condtions of spring. In certain ornamental crops, such as lilies, vernalization is used commercially to promote rapid and synchronous flowering so that plants in bloom can be marketed at particular times of the year. As the process of vernalization becomes better understood at a molecular level, there may be increased opportunities to use this information to control more precisely when plants flower to optimize yields.

For background information *see* BREEDING(PLANT); GENE ACTION; PHOTOPERIODISM; PLANT GROWTH; PLANT PROPAGATION; VERNALIZATION in the McGraw-Hill Encyclopedia of Science and Technology.

Richard Amasino

Bibliography. R. M. Amasino, Vernalization, competence, and the epigenetic memory of winter, *Plant Cell*, 16:2553–2559, 2004; P. Chouard, Vernalization and its relations to dormancy, *Annu. Rev. Plant Physiol.*, 11:191–238, 1960; S. D. Michaels and R. M. Amasino, Memories of winter: Vernalization and the competence to flower, *Plant Cell Environ.*, 23:1145–1154, 2000.

Very small flying machines

No other air vehicle design space has presented as great a mix of challenges as have miniature flight platforms. Creators of these small crewless or "unmanned" aerial vehicles (UAVs) must address the same physical design constraints that have already been mastered by living airborne organisms, including low-Reynolds-number aerodynamics, high energy density, and extreme miniaturization.

The name "micro air vehicle" (MAV) given to this class of air vehicles is unfortunate because none are truly "micro." Indeed, the original official vehicle definition (about 1995), by the Defense Advanced Research Projects Agency (DARPA), requiring a maximum dimension of 15 cm (6 in.), confirmed the name to be quite inappropriate.

Interest in tiny flying machines had its origins with the notion that a small insectlike flying platform could be devised for covert operations. The Central Intelligence Agency (CIA) dabbled briefly in this area

Fig. 1. Mockup of micro air vehicle (MAV) concept developed at the Massachusetts Institute of Technology's Lincoln Laboratory. (© *R. C. Michelson*)

in an attempt to create a remotely controlled pneumatic "dragon fly." During the early 1990s, the Massachusetts Institute of Technology's Lincoln Laboratory conceived of a small airplane (**Fig. 1**) that could carry a tiny camera system that they had under development. Though it was quite doubtful that this notional platform model could ever fly, it did serve as a catalyst to begin serious discussion within the U.S. government of the roles and missions for a micro air vehicle.

Although researchers continue to work on different MAV configurations, various physical constraints make the design of an air vehicle the size of an insect very difficult. For example, the same forces of nature that discourage insect flight during thunderstorms are also at play when considering MAV flight.

Need for MAVs. In spite of these issues, a strong case for MAVs does exist. The mission space for which size really does matter is "indoors and in confined spaces," where the environment is controlled or at least protected. At present, no assets exist that can rapidly and covertly penetrate buildings, tunnels, caves, bunkers, and other enclosures. MAVs offer the potential to enter enclosures by nonobvious means (upper-story openings) and navigate their interiors more effectively (for example, circumventing obstacles such as stairs and ground objects), and more rapidly than ground robots. They present a new paradigm in reconnaissance where close-in interaction is encouraged rather than a stand-off capability. Key to this behavior is small size, slow flight, and the ability to navigate without signals from the Global Positioning System.

Energy requirements. Beyond the hazards of wind and rain, technology barriers present hurdles as the scale approaches that of insects. Most significant is the inability to store the large amounts of onboard energy required for even moderate endurance—and unlike the insects, harvesting energy from the environment is not yet an efficient option.

Small birds and insects are specialized for the task of energy harvesting: the search for food. Hummingbirds, the smallest of all avians, feeding on dilute nectar, can ingest nearly three times their body mass in nectar per day to sustain life and mobility. Their small bodies cannot carry large amounts of food, so to improve efficiency they choose high-energy foods that provide immediate energy access (sugars) as do many insects. Tiny MAVs suffer from the same need for readily available energy. The energy density of the best battery technologies currently available still cannot match that which is locked chemically in various compounds such as sugars. For example, more energy can currently be extracted from a drop of gasoline than a battery the size of the drop. Some have advocated energy harvesting through the use of solar panels on MAVs. Unfortunately, the efficiency of current solar cells (roughly 5% for common cells, ranging up to 28% for some of the best triple-junction gallium arsenide space-qualified cells) in sizes that could be carried by a MAV is insufficient for sustained flight. The extra weight of such cells negates their use as an endurance extender, and their low voltage output is incompatible with many of the electronic actuators proposed (such as piezoelectric and electro polymers). Finally, night operation or flight through shadows is precluded.

Technological advances. These hurdles have driven designers to focus on larger MAVs in the range of 15–20 cm (6–8 in.). At this scale, current technology lends itself to both chemical and electrical propulsion. However, advances in critical enabling technologies have kept the concept of smaller MAVs on the near horizon. Micro-electro-mechanical systems (MEMS) have enabled the production of extremely small, yet highly functional, navigation hardware. Sensors have steadily decreased in size since the 1990s, with assistance from Japanese consumer camera manufacturers. These sensors include room-temperature infrared sensors, charge-coupled-device (CCD) camera arrays, and microprocessor-driven inertial guidance systems consisting of MEMS accelerometers or piezoelectric rate gyroscopes. The introduction of inexpensive solid-state sensors makes MAVs more of a reality. However, propulsion and energy storage technologies remain critical path items.

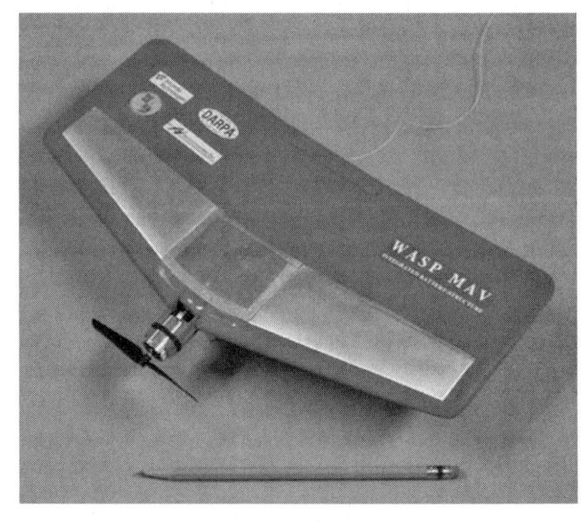

Fig. 2. The Wasp, an electric MAV. (*AeroVironment Inc.*)

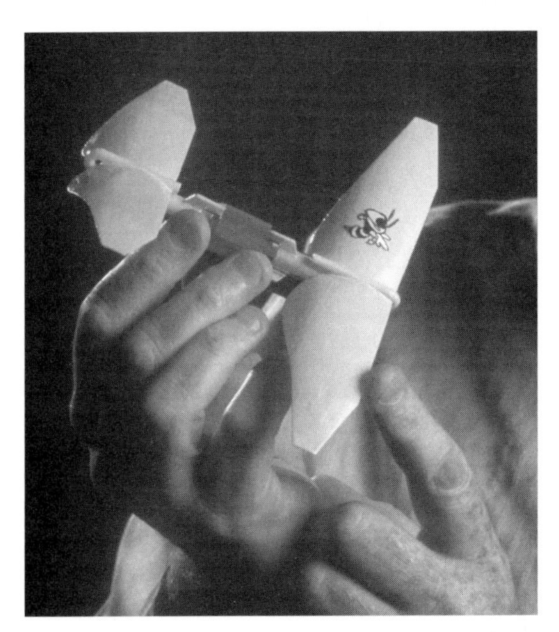

Fig. 3. Georgia Tech Research Institute's Entomopter MAV. (*Georgia Tech, Gary Meeks*)

AeroVironment Inc. of Simi Valley, California, has been heavily involved in MAV and mini research, and makes the experimental Black Widow and Microbat MAVs. This research has focused on miniaturization and efficiency, resulting in some of the smallest flying fixed-wing MAVs so far. Research is being conducted to supply new power sources and a higher level of miniaturization. In particular, AeroVironment's Hornet MAV is powered by a fuel cell, while its Wasp (**Fig. 2**) uses a conformal wing-shaped battery.

Designs. MAVs have taken on many forms. The predominant trend is to use fixed-wing propeller-driven platforms that are in many ways similar to small radio-controlled aircraft used by hobbyists. The fixed-wing MAV systems that have a wingspan of less than 30 cm (1 ft) take slightly different morphological approaches to increase ruggedness and promote ease of employment. MAVs such as the Black Widow, Wasp, German Mikado, and IAI Mosquito 1.5 are designed as "flying wings" with no need of a fuselage, in order to improve ruggedness and reduce drag. Experimental designs by various universities often exhibit closely coupled tail control structures, circular wing planforms, and fuselages. In all cases, these MAVs lack landing gear due to the need to reduce weight and the impracticality of coordinated landings at high speed. MAVs are designed to sustain survivable "crash landings." Some, such as the Black Widow, have parts designed to separate upon landing in an effort to distribute the energy of the "crash." *See* FLYING WINGS AND BLENDED WING BODIES.

Rotary-wing designs. The rotary-wing MAV and mini-UAV platforms are less numerous and often look nothing like a conventional helicopter. With limitations on range, payload, and other parameters, the helicopter and ducted-fan designs tend to be less popular. A counter-rotating rotor device has been demonstrated that uses ultrasonic motors and a mass-offset control scheme (obviating the need for collective or cyclic pitch of the blades). Various versions of a conventional helicopter design named Pixel, constructed of composite materials, have been built and demonstrated. Novel counter-rotating single-blade stopped rotors (wherein the rotor blade can be stopped in flight at an angle of 90° to the fuselage, and then acts as a wing) are under investigation by the U.S. Naval Research Laboratory on an electrically powered MAV known as Samara.

Flapping-wing designs. In an attempt to leverage what already works at small flight scales, many researchers have turned to birds and insects as models, with the hummingbird and the hawk moth being favorite analogs. In most cases, a biomimetic approach has been taken wherein avian or insect analogs are copied. Every living organism that flies under power uses flapping wings, so many flapping-wing machines have been constructed. *See* BIOLOGICALLY INSPIRED ROBOTICS; INSECT FLIGHT.

Designs that do not capitalize on resonance waste energy. All creatures capable of sustained powered flight do so with resonant systems. Flight is expensive from an energy standpoint. Brute-force biological wing flapping can result in flight but at a great cost in fuel. Fuel is heavy, so a practical limit on the fuel that can be carried also places a limit on endurance. The act of flapping a wing involves accelerating a mass in one direction and then decelerating the same mass in preparation for a reversal in direction. Brute-force flapping burns energy to achieve this reversal. Biological systems store kinetic wing energy within their structures as potential energy to be released upon wing reversal. This is the basis for a resonant system that requires only periodic energy input (at the fundamental resonant flapping frequency) rather than a continuous brute-force energy expenditure to accelerate wing mass only to fight its inertia moments later.

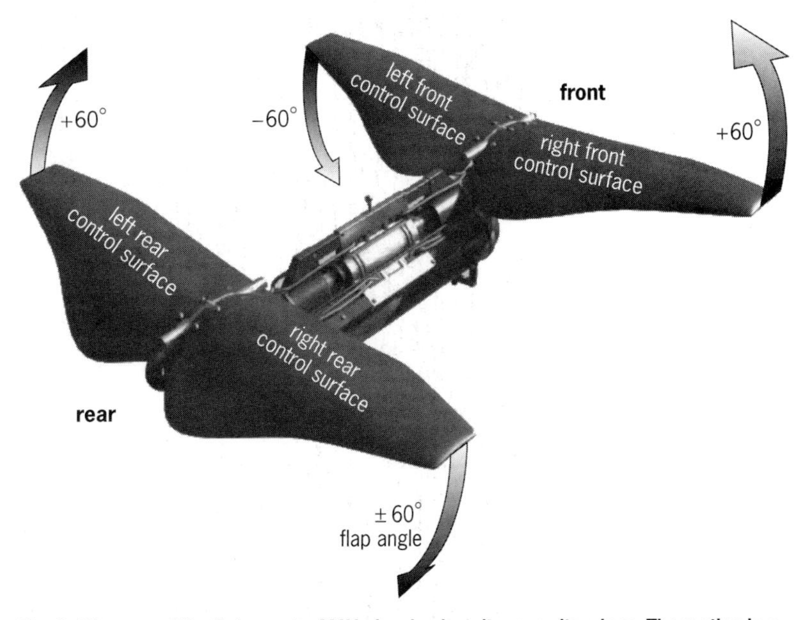

Fig. 4. Diagram of the Entomopter MAV, showing how it moves its wings. The motion is a 25–35-Hz, 180° out-of-phase flapping rotation about the fuselage axis. Lift is controlled independently on each half wing. (© *R. C. Michelson*)

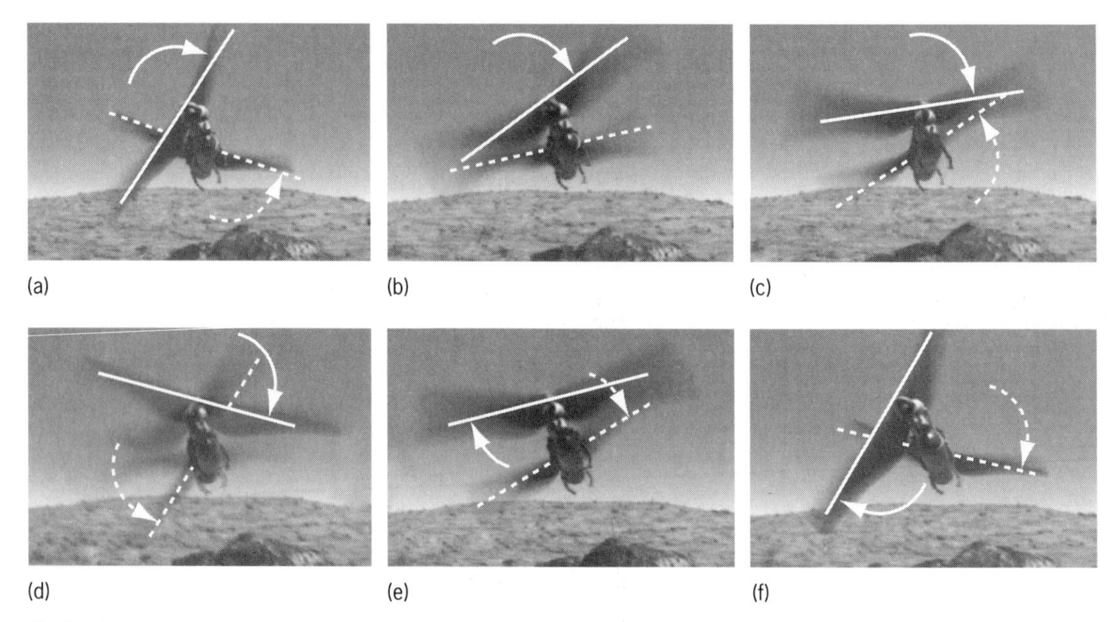

Fig. 5. Entomopter MAV wings in motion during flight. Solid line indicates position of front wing; broken line indicates position of rear wing. (*a*) 0 s. (*b*) 0.005 s. (*c*) 0.010 s. (*d*) 0.015 s. (*e*) 0.030 s, after reversal of motion. (*f*) 0.040 s. (© *R. C. Michelson*)

Flapping-wing MAVs are less conventional than other types in terms of their power source and wing design. Ornithopters, which employ flapping wings, allow relatively high-speed ingress, while still exhibiting abilities such as vertical takeoff and landing (VTOL) as well as hover. The Georgia Tech Research Institute's Entomopter MAV is a leading design in the category of biologically inspired insectlike MAVs (**Fig. 3**).

Other biologically inspired MAVs are under investigation by the U.S. Naval Research Laboratory (Pectenopter, Delphinopter) and the University of California at Berkeley (Robofly). With a reciprocating chemical muscle power source and a set of 180° out-of-phase resonantly flapping wings and (**Figs. 4, 5**), the Entomopter is designed to bridge the gap between fixed and rotary systems and to provide a new platform for urban applications (especially indoors). The reciprocating chemical muscle runs on chemical fuel sources to create a reciprocating motion used to flap wings. It does not involve combustion but relies on rapid chemical decomposition of liquid fuels in the presence of a catalyst to create gas and pressure. These are converted to mechanical energy for wing flapping, heat for thermoelectric power, and ultrasonic emissions for sonar ranging (used for obstacle avoidance); attitude and navigation control are effected by intelligent venting of the gas through the flapping wings.

Prospects. In the future, enabling technologies should allow MAVs to become more practical and prevalent. The lead technologies allowing this are likely to be in the area of propulsion and increased onboard energy density. These advances should allow MAVs (and UAVs in general) to transition from merely crewless aircraft to aerial robots capable of interacting with their environment, and to become fully autonomous thinking machines with a limited "will" and the ability to generate their own mission plans.

For background information *see* AERODYNAMICS; DRONE; FLIGHT; HELICOPTER; MICRO-ELECTRO-MECHANICAL SYSTEMS; RESONANCE (ACOUSTICS AND MECHANICS); ROBOTICS in the McGraw-Hill Encyclopedia of Science & Technology. Robert C. Michelson

Bibliography. R. C. Michelson, The Entomopter, in J. Ayers, J. L. Davis, and A. Rudolph (eds.), *Neurotechnology for Biomimetic Robots*, pp. 481–509, MIT Press, 2002.

Whale-fall worms

When a whale dies, its body may wash ashore, but often it sinks to the ocean bottom. In the deep ocean, which is a dark, cold, and nutrient-poor environment, the arrival of such a massive amount of food, known as a whale fall, represents a bonanza for those creatures able to exploit it. Although there are many opportunistic organisms that will join in the scavenging of the carcass, some animals require whale falls to survive. These whale-fall specialists face a range of problems. Whale falls are rare in terms of the vast expanses of the deep ocean, and their locations are largely a matter of chance. Once a whale fall is colonized, it represents a vast amount of nutrients that may last for years. Thus, whale-fall specialists must have a lifestyle that allows them not only to disperse to find other whales but also to stay and exploit the current whale while it lasts. Another problem is that a considerable amount of the nutrient value of the whale is locked away in the bones (up to 60% of the bone mass may be lipids), but penetrating the skeleton to access this rich food source presents major difficulties.

In 2004, an extraordinary new kind of whale-fall specialist, the worm *Osedax* (derived from the Latin *os* for bone and *edax* to devour), was described that can exploit these hidden lipid reserves and has a reproductive strategy that allows it to successfully colonize other whale falls.

Succession of whale-fall communities. The succession of organisms that exploit whale falls, particularly off the California coast, has been the subject of study for many years. Initially, whale falls tend to have a large amount of flesh and blubber. However, mobile opportunistic scavengers, such as hagfish, sleeper sharks, and invertebrates (for example, crabs), quickly converge on the new food source and strip the flesh from the bones. This initial stage may last only a few months, leaving the whale skeleton. During this period, the surrounding sediment becomes organically enriched from oils and other material leaching out of the carcass.

In the following months (and possibly years), the bones and sediment become infested with numerous polychaete worms, crustaceans, mollusks, and a wide range of other invertebrates. This is known as the enrichment-opportunist stage, and the organisms of the community feed directly on organic material in the bones and sediment. However, the whale fall still holds large amounts of lipids that are not accessible to these enrichment-opportunists, and a third stage of community succession, known as the sulfophilic (sulfur-loving) stage, ensues.

This third period is thought to last up to 50 years or more and is dominated by the breakdown of the large lipid reservoirs in the bones by anaerobic bacteria. A diverse community of organisms then lives on the bones as sulfide is emitted from the breakdown of bone lipids. This ecosystem is largely supported by another group of bacteria, which are chemoautotrophic (derive energy from the oxidation of chemical compounds) and chemically degrade the sulfide. Some animals graze on the bacterial mats that form, whereas others (such as mussels and clams) have symbiotic associations with the bacteria.

This scenario of succession of communities seems to be the norm for whale falls studied to date. However, the discovery of *Osedax*, a group of worms that can penetrate into the whale bone and directly (with the help of bacteria) exploit the lipids within, adds another element to the story.

Discovery of Osedax. In January 2002, researchers from the Monterey Bay Aquarium and Research Institute (MBARI) were exploring the bottom of the Monterey canyon of California using the remote operated vehicle (ROV) *Tiburon*, when they accidentally came across a fairly recent whale fall. The research team immediately noticed that many of the bone surfaces of the juvenile Pacific gray whale were covered with bright red plumes connected to wormlike trunks up to 7 cm (2.8 in.) long (**Fig. 1**). Each worm had four plumes connected to its trunk, which was encased in a transparent jellylike tube. The plumes were actually only the front part of the bizarre worms, which had most of their bodies embedded in the bone.

Fig. 1. Rib of a Pacific gray whale being collected by the ROV *Tiburon*. Arrows indicate patches of *Osedax*. (*Reprinted with permission. © 2003 Monterey Bay Aquarium and Research Institute*)

Osedax morphology. Initial examinations revealed that the worms collected were all females. The main portion of the worm body, which was embedded in the bone, consists of a large ovisac containing thousands of developing eggs. From this ovisac, which is covered in a sheath of bright green tissue, branch off numerous roots of green tissue that extend into the bone marrow (**Fig. 2**). Eggs from the ovisac are carried via an oviduct along the trunk, and the oviduct extends out into the water along with the four plumes (**Fig. 3**). Clearly, large numbers of eggs are spawned into the water by the worms. It was established that the female worms have no mouth or gut, and the plumes are red because they contain blood. Major blood vessels run along the trunk and

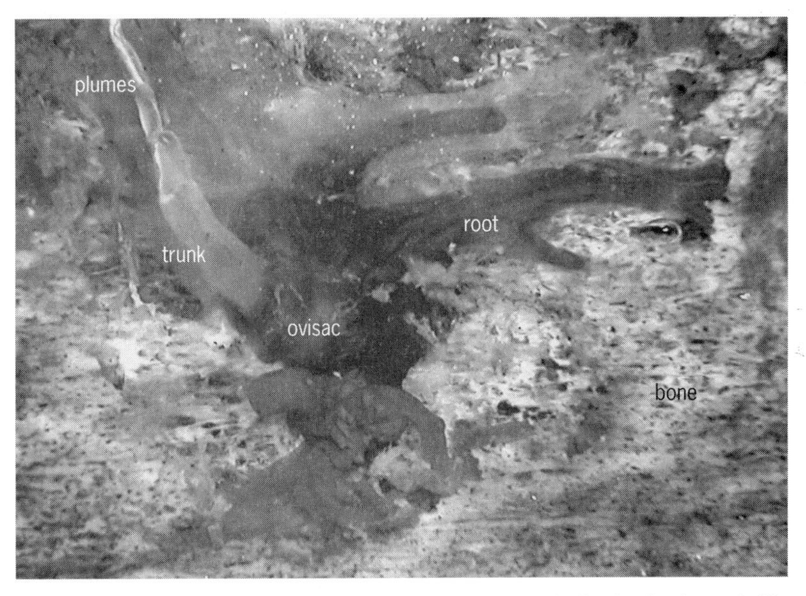

Fig. 2. Whale bone dissected to reveal the main body of a female *Osedax frankpressi* with ovisac and roots.

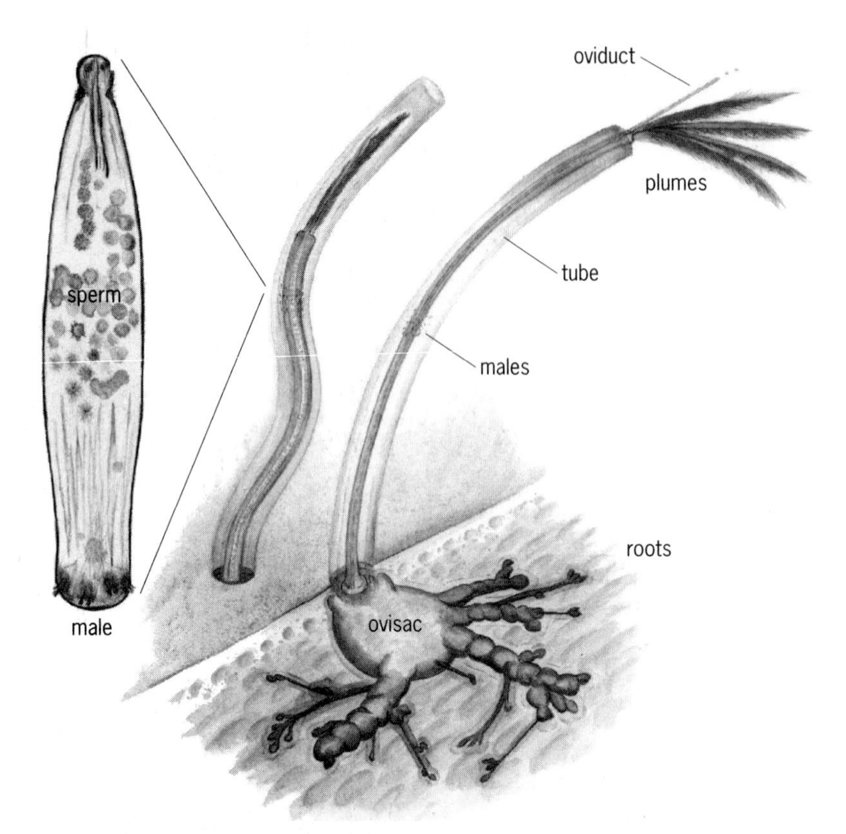

Fig. 3. Painting of *Osedax rubiplumus* showing two females on whale bone. One is cut away to show the ovisac and roots ramifying through the bone. The tubes of each female contain numerous dwarf males lying near the oviduct that carries the eggs into the sea. Note that the plumes and oviduct of the female on the right are extended out of the jellylike tube, as normally seen. The female on the left has been disturbed and has retracted into its tube in response. (*Illustration by Howard Hamon*)

through the ovisac and branch out into the roots, providing the worms with oxygen.

Osedax nutrition and relatives. Microscopic and molecular analyses of the green sheath and roots revealed a layer of cells, known as bacteriocytes, housing large rod-shaped bacteria of the order Oceanospirillales, which are characterized by their ability to break down complex organic compounds. Chemical analyses revealed that these aerobic bacteria not only are responsible for the nutrition of *Osedax* but also derive their own nutrition from the whale bone. The worms provide oxygen for the bacteria to respire and either digest the excess bacteria or somehow derive nutrition from them. This heterotrophic symbiosis differs from the typical symbioses found between bacteria and other deep-sea annelids and mollusks, in which the bacteria are chemoautotrophic (using sulfide or methane).

Reliance on whale bones, heterotrophic symbiosis, and the unique morphology of the ovisac and root system of these worms make this particular symbiotic association unique in the animal kingdom. The symbiosis provided a guide to the possible relatives of *Osedax*. The anatomy of the worms combined with deoxyribonucleic acid (DNA) evidence strongly supports the idea that their closest relatives are polychaete annelid tubeworms, known as vestimentiferans, that live in hydrothermal vents or cold seeps. Just as *Osedax*, these worms, the most well-known of which is *Riftia*, lack a mouth and gut and are filled with symbiotic bacteria.

Osedax males. The fact that all of the worms collected in Monterey canyon were female made researchers question the whereabouts of the males. Close examination of the tubes of the female *Osedax* showed that they often contained numerous tiny males (Fig. 3) that were filled with developing sperm and often contained yolk droplets (**Fig. 4**). The males retained morphological traits typical of worm larvae, and the presence of yolk also strongly suggested that they were in fact pedomorphic, or dwarf, males. The little males were lying on top of or close to the oviduct carrying the eggs into the surrounding sea. The site of egg fertilization has yet to be established.

The tubes of larger females were found to contain more than 100 males each, with an average male-to-female ratio of 17:1. Either females accumulate males over time, or larger females attract more males (because the number of males is correlated with female size). Sex determination in *Osedax* may depend on the environment, as larvae settling from the water on exposed bones mature as females, and those landing on females become males. Such environmental sex determination is known in a few other worm groups.

Age of Osedax. Anatomical analysis and DNA sequence data reveal that the gray whale carcass was in fact occupied by two different *Osedax* species,

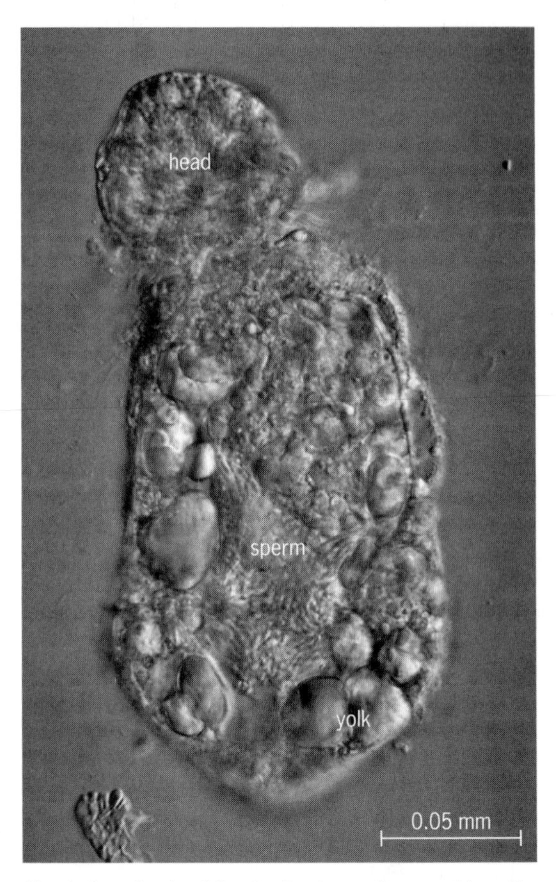

Fig. 4. Dwarf male of *Osedax frankpressi* removed from the tube of a female. The male is only 0.2 mm long and still contains yolk, as well as sperm.

O. rubiplumus and O. frankpressi. DNA sequence analysis revealed that the two species differ substantially (by more than 17%) for a mitochondrial gene known as cytochrome oxidase subunit I (COI). Assuming that nucleotide bases mutate at a constant rate, a molecular clock can be applied to estimate how long it took for the two species to evolve a 17% difference in COI sequence. Based on estimates from calibrated molecular clocks for other worms, it was estimated that the COI of the two species is diverging at a rate of around 0.4 % per million years. This suggests that their most recent common ancestor would have existed approximately 42 million years ago, in the late Eocene—when it may have exploited the bones of archeocete whales, such as *Basilosaurus*. Thus, *Osedax* appears to have been exploiting the bones of marine mammals for millions of years. The members have solved the problem of accessing the lipids locked away in whale bone and have developed a reproductive strategy that allows them to produce large numbers of offspring, some of which will be lucky enough to land on another whale fall.

For background information *see* BACTERIA; DEEP-SEA FAUNA; ECOSYSTEM; HYDROTHERMAL VENT; MARINE MICROBIOLOGY; VESTIMENTIFERA in the McGraw-Hill Encyclopedia of Science & Technology.

G. W. Rouse

Bibliography. S. K. Goffredi et al., *Evolutionary Innovation: A Bone-eating Marine Symbiosis*, 2005; G. W. Rouse, S. K. Goffredi, and R. C. Vrijenhoek, *Osedax*: Bone-eating marine worms with dwarf males, *Science*, 305:668–671, 2004; G. W. Rouse and F. Pleijel, *Polychaetes*, Oxford University Press, 2001; C. R. Smith and A. R. Baco, Ecology of whale falls at the deep sea floor, *Oceanogr. Mar. Biol. Annu. Rev.*, 41:311–354, 2003; C. L. Van Dover et al., Marine biology—Evolution and biogeography of deep-sea vent and seep invertebrates, *Science*, 295:1253–1257, 2002.

World Wide Web search engines

A search engine permits users to search the contents of a repository. Most search engines today provide the ability to search content from the World Wide Web, not the Internet. While search engines exist for all sorts of content (such as images, music, video, news, and advertising), this article will focus on searching Web pages.

Crawling the Web. When a search engine gets a query from a user, it does not literally "search" the Web interactively. Instead, it finds the results in a precompiled index of Web pages that the engine has visited in the past.

In order to collect pages from the Web, a search engine uses a Web crawler (also known as a spider or robot). A Web crawler starts from a small list of known Web sites (such as http://www.yahoo.com), selects the address of one Web page from the list, and attempts to retrieve and store it in a local repository (see **illustration**). If successful, the Web page is then examined for additional Web addresses within hyperlinks. A hyperlink, or more simply, a link, is a reference to the Web address of another document and is usually represented in a Web browser by highlighted words or images that can be clicked. Any page addresses that have not yet been retrieved are added to the list for subsequent crawling.

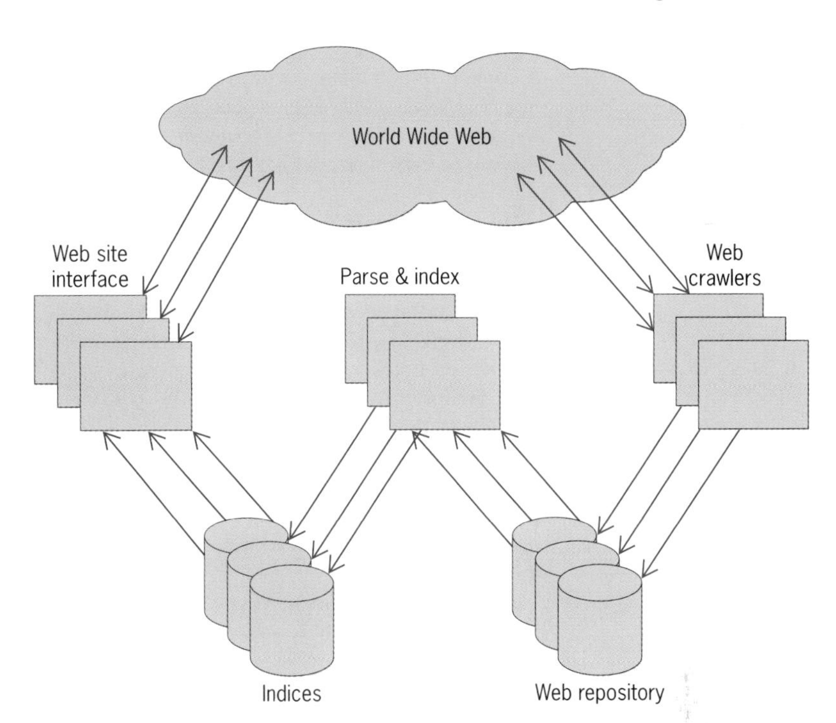

Structure of a Web search engine.

Given the billions of pages accessible on the Web today, most search engines do not want to crawl all the pages. Thus, one significant concern is how to choose which pages to crawl or to recrawl to get the latest version. The typical approach is to crawl pages deemed to be popular.

Text parsing. In addition to finding links for additional crawling, a search engine must parse each page to extract the keywords used in its index. For example, valid "words" might include a sequence of alphanumeric characters, delimited by everything else such as white space, punctuation, and so on. This simple definition quickly leads to important exceptions, such as how to represent words with punctuation (for example, contractions as in "don't"), abbreviations (for example, I.B.M.), or words with symbols (for example, AT&T). It also lacks a limit on the length of valid words, for example, to 100 characters or less.

Text analysis. A simple search engine might take the set of keywords provided by a user and, like a database, return all pages that include each of the keywords. However, a searcher typically wants to know which documents best match the query's topic, so simply keeping track of the words in each document does not suffice.

As it parses each document, the search engine marks the number of times each word appears, and sometimes even whether the word appears in special contexts (such as in the title or a heading). Thus, documents that contain many instances of a query word or emphasize the query word in the title will be ranked higher than documents that do not. In practice, Web search engines record not only the frequency of a term in a document but also the position of each word, which lets the engine match exact phrases.

Analysis of the text in a document collection and determination of optimal weightings in formulas for ranking has been a focus of the information retrieval community for decades. One well-known scheme, TF*IDF, weights each word proportionately to the number of times it appears in a document (TF: Term Frequency), and inversely to the fraction of documents that contain it (IDF: Inverse Document Frequency). Thus, words that are found in all documents are given low weight since they are not likely to help in retrieval, and those words central to the topic of a document will have higher weight because of repetition.

Link analysis. Document text analysis is insufficient on the Web. This is a result of many factors, including the uncontrolled nature of Web content, the size of the Web, and the size of Web queries. In terms of Web content, documents on a particular topic may be of widely varying quality (ranging from research reports to online diaries), and a greater number of documents may be considered relevant for a particular query as the Web grows in size. In addition, short queries, which typically are given by Web searchers, match more documents than longer, more specific queries.

The Web is not just a collection of documents, because the hyperlinks between documents provide a tremendous source of additional information. For example, the highlighted text used in a hyperlink can be considered a short description of the target document and is often used as if that text appeared in the target. Similarly, using the assumption that the existence of a link is similar to a recommendation (from the source to the target), modern search engines use link analysis techniques to improve rankings. They consider not only a document's relevance to a searcher's query but also the popularity, quality, or authority of the page. Given two pages with similar text analysis scores, the one with higher popularity should be ranked first. Counting the number of pages that link to a given Web page is a simple way of determining the page's popularity. Most link analysis techniques go further, giving more weight to links from pages that themselves have high authority. Discoveries of such link analysis techniques by J. Kleinberg and the founders of Google, S. Brin and L. Page, made it possible for Web search engines to provide significantly better results than had been possible by text analysis alone.

In addition to link analysis, search engines use other measures of popularity. One is the tracking of clicks on search engine results, so pages that are often selected will tend to be pushed higher in future rankings for that query. Most search engines also have the ability to track browser activity through the use of toolbar extensions that they supply. This gives the search engine the ability to capture a more dynamic measure of popularity, which can be incorporated into ranking algorithms.

Search engine spam. Given the size and unorganized nature of the Web, search engines are used hundreds of millions of times per day to find pages and answer questions. For Web pages that are ranked highly for popular queries, this query activity translates into substantial traffic as users click through via the result links. This provides significant incentive to some content providers to do whatever is necessary to rank highly in search engine results. The use of techniques that push rankings higher than they belong, including textual as well as link-based methods, is often called spamming a search engine. Using such techniques is discouraged by the major vendors and can result in a Web site being penalized with a much lower query result ranking. However, the benefits of an increased ranking are sufficiently valuable that an industry of search-engine optimization experts has emerged and includes people willing to break the rules set by the engines.

One particularly visible effort has been the automated generation of "comment" spam by adding to publicly accessible Web sites, links back to a target Web site. Such comments are typically off-topic and exist only to promote the target page in link analysis computations by search engines. Some sites have implemented procedures to verify that the comment is authored by a person, such as registration systems and CAPTCHAs (completely automated public Turing tests to tell computers and humans apart), but the process can be unwieldy.

In early 2005, many of the search engines and Website software developers announced the introduction of a special attribute for links called "nofollow." Links marked "nofollow" (such as those in comments that are created by site visitors) are no longer used in link analysis computations, making automated comment generation less beneficial for link spammers.

Search engine architecture. Popular search engines must be designed to scale for both large amounts of data and large numbers of queries. Instead of a few large computer systems, Web search engines consist of large clusters (thousands) of computers very similar to common desktop computers. These machines have the raw power to crawl the billions of pages on the Web, as well as to calculate and store a massive index of those pages. A single machine has the capacity to calculate and store only a portion of the data, so the data are partitioned into slices across many machines, with each slice replicated multiple times to provide robustness in case of machine failure and to divide the query load. Caching, the storing of results in case they are needed again, is also incorporated at multiple levels so that the search engine needs to process only one instance of a query, even when issued by many searchers.

Research. Research in technologies relevant to Web searching, such as text mining, machine learning, and computational linguistics, continues to flourish in academia and via new products and features in commercial services. For example, search queries are often ambiguous. Does the query "apple" refer to the fruit or the company? Tools are now available for searchers to personalize their search results (for example, by biasing the results toward topics covered in previous searches) or to customize the ranking algorithm by selecting topics of interest. Another approach to the problem of ambiguity is to cluster the results so that it is easy for the searcher to focus on the appropriate cluster. Some engines present these clusters textually, while others incorporate novel graphical interfaces to make topical browsing more effective.

The ability to answer queries posed as natural-language questions is another long-standing goal. In practice, most Web search engines accept simple questions (such as "Who killed Abraham Lincoln?"), and some use specialized databases of people, locations, music, movies, and frequently-asked-questions to give explicit answers, in addition to Web search results.

For background information *see* ALGORITHM; DATA MINING; DATABASE MANAGEMENT SYSTEM; HUMAN-COMPUTER INTERACTION; INFORMATION MANAGEMENT; NATURAL LANGUAGE PROCESSING; WORLD WIDE WEB in the McGraw-Hill Encyclopedia of Science & Technology. Brian D. Davison

Bibliography. R. Baeza-Yates and B. Ribeiro-Neto, *Modern Information Retrieval*, Addison Wesley, 1999; S. Chakrabarti, *Mining the Web: Discovering Knowledge from Hypertext Data*, Morgan Kaufmann, 2003; I. H. Witten, A. Moffat, and T. C. Bell, *Managing Gigabytes: Compressing and Indexing Documents and Images*, 2d ed., Morgan Kaufmann, 1999.

Contributors

Contributors

The affiliation of each Yearbook contributor is given, followed by the title of his or her article. An article title with the notation "coauthored" indicates that two or more authors jointly prepared an article or section.

A

Aga, Dr. Diana. *Chemistry Department, University at Buffalo, New York.* PHARMACEUTICAL RESIDUES IN THE ENVIRONMENT.

Amasino, Dr. Richard. *Department of Biochemistry, University of Wisconsin-Madison.* VERNALIZATION.

Andreae, Prof. Dr. Meinrat O. *Max Planck Institute for Chemistry, Mainz, Germany.* FIRE-INDUCED CONVECTION—coauthored.

Andrews, Prof. David. *Department of Mechanical Engineering, University College London, United Kingdom.* NAVAL AVIATION SHIP DESIGN.

B

Baiden, Dr. Gregory. *Laurentian University, School of Engineering, Sudbury, Ontario, Canada.* TELEROBOTICS (MINING).

Bailey, Dr. David M. *Marine Biology Research Division, Scripps Institution of Oceanography, La Jolla, California.* DEEP-SEA VERTEBRATE SPECIES.

Barbieri, Prof. Roberto. *Dipartimento di Scienze Della Terra, University of Bologna, Italy.* FOSSIL COLD-SEEP ECOSYSTEM.

Baumgartner, Dr. Eric T. *Mechanical & Robotic Technologies Group, Jet Propulsion Laboratory, Pasadena, California.* MARS ROVERS.

Bell, Dr. James F., III. *Department of Astronomy, Cornell University, Ithaca, New York.* MINERALOGY OF MARS.

Ben-Akiva, Prof. Moshe. *Department of Civil and Environmental Engineering, Massachusetts Institute of Technology, Cambridge.* DYNAMIC TRAFFIC MANAGEMENT—coauthored.

Bielawski, Dr. C. W. *Department of Chemistry and Biochemistry, The University of Texas at Austin.* POLYMERS FOR MICROELECTRONICS—coauthored.

Bier, Prof. Vicki M. *Department of Industrial and Systems Engineering, University of Wisconsin-Madison.* RISK ANALYSIS (HOMELAND SECURITY).

Bjorhovde, Dr. Reidar. *The Bjorhovde Group, Tucson, Arizona.* STEEL CONSTRUCTION.

Blake, Prof. John T. *Department of Industrial Engineering, Dalhousie University, Halifax, Nova Scotia, Canada.* QUEUING THEORY (HEALTH CARE).

Blancato, Dr. Jerry. *U.S. Environmental Protection Agency National Exposure Research Laboratory, Research Triangle Park, North Carolina.* COMPUTATIONAL ENVIRONMENTAL TOXICOLOGY—coauthored.

Bosworth, Michael. *Vienna, Virginia.* CREWLESS SURFACE WATERCRAFT.

Bottom, Dr. Jon. *Charles River Associates, Boston, Massachusetts.* DYNAMIC TRAFFIC MANAGEMENT—coauthored.

Bowser, Dr. Michael T. *University of Minnesota, Department of Chemistry, Minneapolis.* CAPILLARY ELECTROPHORESIS-SELEX.

Bransden, Tania G. *Clayton School of Information Technology, Monash University, Melbourne, Australia.* COMPLEXITY THEORY—coauthored.

Brodsky, Dr. Jeffrey L. *Department of Biological Sciences, University of Pittsburgh, Pennsylvania.* ENDOPLASMIC RETICULUM QUALITY CONTROL—coauthored.

Butler, Dr. James. *Spey Fishery Board, United Kingdom.* SALMON FARMING.

C

Chan, Dr. Ching-Yao. *University of California, Berkeley.* AUTOMOTIVE RESTRAINT SYSTEMS.

Clack, Dr. Jennifer A. *Department of Zoology, University of Cambridge, United Kingdom.* TETRAPOD ORIGINS.

Coffman, Thayne R. *Principal Scientist, 21st Century Technologies, Austin, Texas.* GRAPH-BASED INTELLIGENCE ANALYSIS.

Cohen, Sarah. *Medical Research Council, Institute of Psychiatry, King's College London, United Kingdom.* PSYCHOPATHOLOGY AND THE HUMAN GENOME PROJECT—coauthored.

Conboy, Dr. John C. *University of Utah, Department of Chemistry, Salt Lake City.* SUM-FREQUENCY VIBRATIONAL SPECTROSCOPY.

Cooke, Dr. Marcus S. *Senior Lecturer, Department of Cancer Studies and Genetics, University of Leicester, United Kingdom.* Reactive oxygen species—coauthored.

Cronin, Prof. Thomas W. *Department of Biological Sciences, University of Maryland, Baltimore.* Fluorescent signaling in mantis shrimps.

D

Dagle, Dr. Jeff. *Pacific Northwest National Laboratory, Richland, Washington.* Electric power system security.

Dahdouh-Guebas, Dr. Farid. *Biocomplexity Research Team, General Botany and Nature Management, Mangrove Management Group, Vrije Universiteit Brussel, Belgium.* Mangrove forests and tsunami protection.

Davidson, Sarah Nell. *Department of Plant Biology, Cornell University, Ithaca, New York.* Plant cell walls—coauthored.

Davis, Dr. Donald W. *Department of Geology, University of Toronto, Ontario, Canada.* Geochronology.

Davison, Dr. Brian D. *Department of Computer Science and Engineering, Lehigh University, Bethlehem, Pennsylvania.* World wide web search engines.

Dell'Aversana, Dr. Pasquale. *Microgravity Advanced Research and Support Center (MARS), Napoli, Italy.* Non-coalescence of droplets.

Devreese, Prof. Jozef T. *Universiteit Antwerpen, Departiment Natuurkunde, Belgium.* Bose-Einstein condensation—coauthored.

DiMarzio, Prof. Charles A. *Department of Electrical and Computer Engineering, Northeastern University, Boston, Massachusetts.* Acoustooptic imaging.

E

Eastment, Dr. Jonathan. *Radio Communications Research Unit, Rutherford Appleton Laboratory, United Kingdom.* Radar meteorology.

Edwards, Dr. Charles. *Jet Propulsion Laboratory, Pasadena, California.* Space probe communications.

Ekdale, Dr. Allan A. *Department of Geology and Geophysics, University of Utah, Salt Lake City.* Paleozoic bivalve borings.

Elenitoba-Johnson, Dr. Kojo S. J. *Department of Pathology, Associated Regional and University Pathologists, Institute for Clinical and Experimental Pathology, University of Utah Health Sciences Center, Salt Lake City.* Proteomics—coauthored.

Evans, Dr. Mark D. *Lecturer, Department of Cancer Studies and Molecular Medicine, University of Leicester, United Kingdom.* Reactive oxygen species—coauthored.

F

Faith, Daniel P. *Australian Museum, Sydney.* Phylogenetic diversity and conservation—coauthored.

Fang, Dr. Liping. *Ryerson University, Department of Mechanical and Industrial Engineering, Toronto, Ontario, Canada.* Conflict analysis and resolution—coauthored.

Favela, Dr. Jesus. *Department of Computer Science, Centro de Investigacion Cientifica y de Educacion Superior de Endenada (CICESE), Mexico.* Context-aware mobile communication systems—coauthored.

Fielding, Dr. Christopher J. *Cardiovascular Research Institute, University of California, San Francisco.* Liver x receptor (lxr).

Finn, Prof. M. G. *Department of Chemistry, The Scripps Research Institute, La Jolla, California.* Click chemistry.

Forgacs, Prof. Gabor. *Department of Physics & Astronomy, University of Missouri-Columbia.* Tissue and organ printing—coauthored.

Fraiser, Dr. Margaret L. *Department of Earth Sciences, University of Southern California, Los Angeles.* Permian-triassic mass extinction.

Frangopol, Dr. Dan M. *Professor of Civil Engineering, Department of Civil, Environmental & Architectural Engineering, University of Colorado, Boulder.* Life cycle analysis of civil structures—coauthored.

Friesen, Lyle F. *Department of Plant Science, Faculty of Agricultural & Food Sciences, University of Manitoba—Winnipeg, Canada.* Herbicide resistance—coauthored.

G

Garcia-Macias, Dr. J. Antonio. *Department of Computer Science, Centro de Investigacion Cientifica y de Educacion Superior de Endenada (CICESE), Mexico.* Context-aware mobile communication systems—coauthored.

Gnanou, Dr. Yves. *Laboratoire de Chimie des Polymères Organiques, Université Bordeaux, Pessac, France.* Controlled/living radical polymerization—coauthored.

Goeckeler, Jennifer L. *Department of Biological Sciences, University of Pittsburgh, Pennsylvania.* Endoplasmic reticulum quality control—coauthored.

Gouma, Dr. Pelagia-Irene. *Materials Science and Engineering Department, State University of New York at Stony Brook.* Electronic olfaction and taste systems.

Green, Prof. David G. *Clayton School of Information Technology, Monash University, Melbourne, Australia.* Complexity theory—coauthored.

Gross, Briana L. *Department of Biology, Indiana University, Bloomington.* Transgressive segregation (plant breeding)—coauthored.

Gunsch, Prof. Claudia K. *Department of Civil and Environmental Engineering, Duke University, Durham, North Carolina.* Microbial dechlorination.

H

Handelsman, Prof. Jo. *Howard Hughes Medical Institute, Department of Plant Pathology, University of Wisconsin-Madison.* Metagenomics.

Harley, Dr. John P. *Department of Biological Sciences, Eastern Kentucky University, Richmond.* Human papillomaviruses.

Harrison, Prof. Terry. *Department of Anthropology, New York University, New York.* Homo floresiensis.

Haskins, Dr. Mark. *Professor of Pathology and Medical Genetics, School of Veterinary Medicine, University of Pennsylvania, Philadelphia.* Gene therapy (veterinary medicine).

Hedlin, Michael A. H. *Laboratory for Atmospheric Acoustics, University of California, San Diego.* Infrasonic monitoring.

Heer, Daniel. *Lucent Technologies, Westford, Massachusetts.* Internet communications—coauthored.

Henke, Harold. *Chartula Consulting and Press, Niwot, Colorado.* Printing on demand.

Hepburn, Capt. Richard. *United States Navy (retired), P.E., Manassas, Virginia.* Dry-docking and heavylift ships.

Hipel, Dr. Keith W. *Department of Systems Design Engineering, University of Waterloo, Ontario, Canada.* Conflict analysis and resolution—coauthored.

I

Ikeya, Dr. Motoji. *Professor Emeritus, Osaka University, Japan.* Earthquake sensory perception in vertebrates.

Illman, Dr. Barbara L. *USDA Forest Service, Forest Products Laboratory, University of Wisconsin-Madison.* Invasive forest species.

J

Johnson, Dr. Kathleen R. *Department of Earth Sciences, Oxford University, United Kingdom.* Caves and climate change.

Jurasek, Prof. Milan. *Department of Mechanics, Faculty of Civil Engineering, Czech Technical University in Prague, Czech Republic.* Size effect on strength (civil engineering).

K

Kane, Dr. Nolan C. *Department of Biology, Indiana University, Bloomington.* Transgressive segregation (plant breeding)—coauthored.

Kienberger, Reinhard. *Max-Planck-Institut für Quantenoptik, Ludwig-Maximilans-Univeristät München, Germany.* Attosecond laser pulses—coauthored.

Kilgour, Dr. D. Marc. *Department of Mathematics, Wilfrid Laurier University, Waterloo, Ontario, Canada.* Conflict analysis and resolution—coauthored.

Klein, Dr. Hugh. *National Physical Laboratory, Teddington, Middlesex, United Kingdom.* Trapped-ion optical clocks.

Klein, Melinda A. *U.S. Plant, Soil, and Nutrition Laboratory and Department of Plant Biology, Cornell University, Ithaca, New York.* Phytoremediation—coauthored.

Kobylarz, Dr. Thaddeus J. A. *Retired, Bell Laboratories, New Jersey.* Compound wireless services.

Kochian, Dr. Leon V. *U.S. Plant, Soil, and Nutrition Laboratory and Department of Plant Biology, Cornell University, Ithaca, New York.* Phytoremediation—coauthored.

Koltsov, Dr. Denis. *Department of Engineering, Lancaster University, United Kingdom.* Nanometer magnets.

Kotsialos, Dr. Apostolos. *University of Durham, United Kingdom.* Dynamic traffic management—coauthored.

Krausz, Prof. Ferenc. *Max-Planck-Institut für Quantenoptik, Ludwig-Maximilans-Univeristät München, Germany.* Attosecond laser pulses—coauthored.

Kumar, Dr. Nitkin. *Intermolecular, Santa Clara, California.* Spider silk—coauthored.

L

Lay, Prof. Thorne. *Earth Sciences Department, University of California, Santa Cruz.* Post-perovskite.

Lee, Dr. Tatia M. C. *Neuropsychology Laboratory, The University of Hong Kong.* Brain imaging of deception.

Lee, Prof. Steven B. *Associate Professor, Director of Forensic Science, Justice Studies Department, San Jose State University, California.* Forensic DNA testing.

Liew, Prof. Jat-Yuen Richard. *Department of Civil Engineering, National University of Singapore.* Fire safety (building design)—coauthored.

Lim, Dr. Megan S. *Department of Pathology, Associated Regional and University Pathologists, Institute for Clinical and Experimental Pathology, University of Utah Health Sciences Center, Salt Lake City.* Proteomics—coauthored.

Lin, Prof. Chentao. *Department of Molecular, Cell, and Developmental Biology, University of California, Los Angeles.* Cryptochrome.

Liu, Dr. Min. *Research Associate, Department of Civil, Environmental & Architectural Engineering, University of Colorado, Boulder.* Life cycle analysis of civil structures—coauthored.

Lysak, Dr. Martin A. *Jodrell Laboratory, Royal Botanic Gardens, Kew, United Kingdom.* Chromosome painting.

M

Maayan, Dr. Lawrence. *Yale University, Child Study Center, New Haven, Connecticut.* Antidepressant use in minors—coauthored.

MacDonald, Prof. Glenn M. *Department of Geography, University of California, Los Angeles.* Arctic treeline change.

Madani, Vahid. *Pacific Gas and Electric, Oakland, California.* Blackout prevention—coauthored.

Madou, Prof. Marc. *Department of Mechanical and Aeropace Engineering, University of California, Irvine.* Carbon MEMS—coauthored.

Malinowski, John. *Baldor Electric Company, Fort Smith, Arkansas.* Energy-efficient motors.

Martin, Dr. Andrés. *Yale University, Child Study Center, New Haven, Connecticut.* ANTIDEPRESSANT USE IN MINORS—coauthored.

Martínez-Garcia, Dra. Ana I. *Department of Computer Science, Centro de Investigacion Cientifica y de Educacion Superior de Endenada (CICESE), Mexico.* CONTEXT-AWARE MOBILE COMMUNICATION SYSTEMS—coauthored.

Maurer, Dr. Edwin P. *Civil Engineering Department, Santa Clara University, California.* DOWNSCALING CLIMATE MODELS.

McCarton, Matt. *Stevensburg, Virginia.* AMPHIBIOUS ASSAULT SHIPS.

McGuffin, Dr. Peter. *Medical Research Council, Institute of Psychiatry, King's College London, United Kingdom.* PSYCHOPATHOLOGY AND THE HUMAN GENOME PROJECT—coauthored.

McKinley, Prof. Gareth. *Department of Mechanical Engineering, Massachusetts Institute of Technology, Cambridge.* SPIDER SILK—coauthored.

Michelson, Robert C. *Georgia Tech Research Institute, Aerospace, Transportation & Advanced Systems Laboratory, Smyrna.* VERY SMALL FLYING MACHINES.

Milner, Dr. Angela C. *Department of Palaeontology, The Natural History Museum, London, United Kingdom.* DINO-BIRDS.

Mirkin, Prof. Chad A. *Northwestern University, Chemistry Department, Evanston, Illinois.* DIP-PEN NANOLITHOGRAPHY.

Mironov, Dr. Vladimir. *Medical University of South Carolina, Charleston.* TISSUE AND ORGAN PRINTING—coauthored.

Morin, Dr. Alain. *Instructor, Behavioral Sciences, Mount Royal College, Psychology (General and Experimental), Calgary, Alberta, Canada.* SELF-RECOGNITION.

Murch, Dr. Randall S. *Virginia Polytechnic Institute and State University, National Capital Region, Alexandria.* MICROBIAL FORENSICS.

N

Nolet, Prof. Guust. *Department of Geosciences, Princeton University, New Jersey.* PLUME IMAGERY.

Novosol, Damir. *KEMA, T and D Consulting, Raleigh, North Carolina.* BLACKOUT PREVENTION—coauthored.

O

Okamoto, Dr. Yoshio. *School of Engineering, Nagoya University, Japan.* POLYMER STEREOCHEMISTRY AND PROPERTIES.

Ortner, Dr. Donald J. *Kensington, Maryland.* INFECTIOUS DISEASE AND HUMAN EVOLUTION.

P

Page, Prof. John. *University of Manitoba, Department of Physics and Astronomy, Winnipeg, Manitoba, Canada.* PHONONIC CRYSTALS—coauthored.

Papageorgiou, Dr. Markos. *Dynamic Systems and Simulation Laboratory, Technical University of Crete, Greece.* DYNAMIC TRAFFIC MANAGEMENT—coauthored.

Papoyan, Ashot. *U.S. Plant, Soil, and Nutrition Laboratory and Department of Plant Biology, Cornell University, Ithaca, New York.* PHYTOREMEDIATION—coauthored.

Paranthaman, Dr. M. Parans. *Oak Ridge National Laboratory, Oak Ridge, Tennessee.* SUPERCONDUCTOR WIRES.

Park, Prof. Jeffrey. *Department of Geology & Geophysics, Yale University, New Haven, Connecticut.* SUMATRA-ANDAMAN EARTHQUAKE.

Pekarovic, Dr. Jan. *Research Associate, Western Michigan University, Department of Paper Engineering, Chemical Engineering and Imaging, Kalamazoo.* 3D INKJET PRINTING—coauthored.

Pekarovicova, Dr. Alexandra. *Associate Professor, Western Michigan University, Department of Paper Engineering, Chemical Engineering and Imaging, Kalamazoo.* 3D INKJET PRINTING—coauthored.

Phaneuf, Prof. Ronald A. *Department of Physics, University of Nevada, Reno.* INTERACTION OF PHOTONS WITH IONIZED MATTER.

Pop, Dr. Mihai. *Bioinformatics Scientist, The Institute for Genomic Research, Rockville, Maryland.* DNA SEQUENCE ASSEMBLY ALGORITHMS.

Prieto-Diaz, Dr. Ruben. *Department of Computer Science, James Madison University, Harrisonburg, Virginia.* ONTOLOGY (INFORMATION TECHNOLOGY).

Prudhomme, Prof. Thomas I. *University of Illinois at Urbana-Champaign.* ENVIRONMENTAL ENGINEERING INFORMATICS.

Q

Qin, Prof. Ning. *Department of Mechanical Engineering, The University of Sheffield, England.* NONCONVENTIONAL AIRCRAFT DESIGNS.

Quinn, Mark A. *Willis Inspace, Bethesda, Maryland.* COMMUNICATIONS SATELLITE FAILURES.

Quinn, Prof. Roger D. *Department of Mechanical and Aerospace Engineering, Case Western Reserve University, Cleveland, Ohio.* BIOLOGICALLY INSPIRED ROBOTICS—coauthored.

R

Rauchwerk, Dr. Michael D. *Lucent Technologies, Holmdel, New Jersey.* INTERNET COMMUNICATIONS—coauthored.

Ray, Trina L. *Cassini Science Planning, Jet Propulsion Laboratory, California Institute of Technology, Pasadena.* CASSINI-HUYGENS MISSION—coauthored.

Reber, Dr. Rolf. *Associate Professor of Cognitive Psychology, University of Bergen, Norway.* HEURISTICS IN JUDGMENT AND DECISION MAKING.

Rhodes, Dr. Donna H. *Engineering Systems Division, Massachusetts Institute of Technology, Cambridge.* PROCESS IMPROVEMENT METHODOLOGIES.

Rieseberg, Dr. Loren H. *Department of Biology, Indiana University, Bloomington.* Transgressive segregation (plant breeding)—coauthored.

Ritzmann, Prof. Roy E. *Department of Biology, Case Western Reserve University, Cleveland, Ohio.* Biologically inspired robotics—coauthored.

Rose, Dr. Jocelyn K. C. *Department of Plant Biology, Cornell University, Ithaca, New York.* Plant cell walls—coauthored.

Rouse, Dr. Greg W. *South Australian Museum, Adelaide.* Whale-fall worms.

Rouse, Prof. William B. *School of Industrial and Systems Engineering,Georgia Institute of Technology, Atlanta.* Enterprise transformation.

Rowell, Dr. Roger M. *Forest Products Laboratory, University of Wisconsin-Madison.* Forest water contamination.

S

Schlachter, Dr. Alfred S. *Lawrence Berkeley National Laboratory, California.* Quantum chaos in a three-body system.

Schmitke, Barry. *Cameco Corporation, Saskatoon, Saskatchewan, Canada.* Jet boring (mining).

Sheiko, Dr. Sergei. *Department of Chemistry, University of North Carolina, Chapel Hill.* Proximal probe polymer analysis.

Shladover, Dr. Steven E. *Research Engineer, Path Program, University of California, Berkeley.* Vehicle-highway automation systems.

Shu, Prof. Lily H. *Department of Mechanical and Industrial Engineering, University of Toronto, Ontario, Canada.* Biomimetic design for remanufacture.

Smith, Dr. Timothy D. *NASA Glenn Research Center, Cleveland, Ohio.* Hydrogen-powered flight—coauthored.

Snowdon, Dr. Charles T. *Department of Psychology, University of Wisconsin-Madison.* Primate communications.

Soligo, Dr. Christophe. *Department of Palaeontology, The Natural History Museum, London, United Kingdom.* Primate origins.

Sollazzo, Dr. Claudio. *Huygens Mission Operations Manager, European Space Agency.* Cassini-huygens mission—coauthored.

Spilker, Dr. Linda. *Cassini Deputy Project Scientist, Jet Propulsion Laboratory, California Institute of Technology, Pasadena.* Cassini-huygens mission—coauthored.

Stansberry, Dr. John. *Steward Observatory, University of Arizona, Tucson.* Sedna.

Stanton, Prof. Anthony. *Carnegie Mellon University, Tepper School of Business, Pittsburgh, Pennsylvania.* Digital cinema.

Sukhovich, Alexey. *University of Manitoba, Department of Physics and Astronomy, Winnipeg, Manitoba, Canada.* Phononic crystals—coauthored.

Sun, Dr. Wen-Yih. *Department of Earth and Atmospheric Sciences, Purdue University, West Lafayette, Indiana.* Regional climate modeling.

T

Tampi, Dr. Rajesh R. *Assistant Professor of Psychiatry, Yale University School of Medicine, Co-Service Manager Geriatric Services, Yale New Haven Psychiatric Hospital, New Haven, Connecticut.* Alzheimer's disease.

Taton, Dr. Daniel. *Laboratoire de Chimie des Polymères Organiques, ENSCPB, Université Bordeaux, Pessac, France.* Controlled/living radical polymerization—coauthored.

Tautz, Dr. Diethard. *Institut für Genetik, Universität Köln, Germany.* DNA barcoding.

Tear, Dr. Timothy H. *Director of Conservation Science, Eastern New York Chapter, The Nature Conservancy.* Nitrogen deposition and forest degradation.

Tempere, Dr. Jacques. *Universiteit Antwerpen, Department Natuurkunde, Belgium.* Bose-einstein condensation—coauthored.

Travis, Dominic A. *Veterinary Epidemiologist, Davee Center for Veterinary Epidemiology, Lincoln Park Zoo, Chicago, Illinois.* Animal disease surveillance and monitoring.

Trentmann, Dr. Jörg. *Institute for Atmospheric Physics, Johannes Gutenberg University, Mainz, Germany.* Fire-induced convection—coauthored.

U

Urbanowitz, Breeana Rae. *Department of Plant Biology, Cornell University, Ithaca, New York.* Plant cell walls—coauthored.

V

Vallero, Dr. Daniel. *U.S. Environmental Protection Agency, National Exposure Research Laboratory, Research Triangle Park, North Carolina.* Computational environmental toxicology—coauthored.

Van Acker, Dr. Rene C. *Department of Plant Science, Faculty of Agricultural & Food Sciences, University of Manitoba—Winnipeg, Canada.* Herbicide resistance—coauthored.

Von Puttkamer, Dr. Jesco. *NASA Headquarters, Office of Space Flight, Washington, DC.* Space flight.

W

Walker, James F. *NASA Glenn Research Center, Cleveland, Ohio.* Hydrogen-powered flight—coauthored.

Wang, Dr. Chunlei. *Department of Mechanical and Aeropace Engineering, University of California, Irvine.* Carbon mems—coauthored.

Wang, Dr. Yibing. *Dynamic Systems and Simulation Laboratory, Technical University of Crete, Greece.* Dynamic traffic management—coauthored.

Wang, Dr. Z. Jane. *Department of Theoretical and Applied Mechanics, Cornell University, Ithaca, New York.* Insect flight.

Watts, Sean M. *Department of Ecology, Evolution and Marine Biology, University of California, Santa Barbara.* Belowground herbivory.

Weitz, Dr. Hedda J. *Department of Plant and Soil Science, University of Aberdeen, Scotland, United Kingdom.* Bioluminescent fungi.

Wilkinson, Dr. Matthew T. *Department of Zoology, University of Cambridge, United Kingdom.* Pterosaur.

Willenborg, Chris. *Department of Plant Science, Faculty of Agricultural & Food Sciences, University of Manitoba—Winnipeg, Canada.* Herbicide resistance—coauthored.

Williams, Dr. Kristen. *Ecological Geographer, CSIRO Sustainable Ecosystems, Queensland Bioscience Precinct, St. Lucia, Brisbane, Australia.* Phylogenetic diversity and conservation—coauthored.

Willson, Prof. C. Grant. *Department of Chemistry and Biochemistry, The University of Texas at Austin.* Polymers for microelectronics—coauthored.

Wilson, Christopher. *DaimlerChrysler, Research and Technology Center, Palo Alto, California.* Intelligent vehicles and infrastructure.

Winguth, Dr. Arne. *Department of Atmospheric and Oceanic Sciences, University of Wisconsin-Madison.* Global biogeochemical cycle.

Y

Yang, Prof. Louie H. *Section of Evolution and Ecology, University of California, Davis.* Cicada.

Yu, H. X. *Department of Civil Engineering, National University of Singapore.* Fire safety (building design)—coauthored.

Z

Zerbe, Dr. John I. *Forest Products Laboratory, Madison, Wisconsin.* Alcohol production from wood.

Index

Index

Asterisks indicate page references to article titles.

A

B

D

F

G

H

I

M

O

P

S

T

U

V

W

X

Y

Z